CONVERSION FACTORS

Length

1 in = 2.54 cm

1 m = 39.4 in = 3.28 ft

1 mi = 5280 ft = 1609 m

1 km = 0.621 mi

1 angstrom (Å) = 10^{-10} m

1 light year = 9.46×10^{15} m

Volume

1 liter = 1000 cm^3

1 gallon = 3.79 liters

Speed

1 mi/h = 1.61 km/h = 0.447 m/s

Mass

1 atomic mass unit (u) = 1.660×10^{-27} kg

(Earth exerts a 2.205 lb force on an object with 1 kg mass)

Force

1 lb = 4.45 N

Work and Energy

1 ft·lb = 1.356 N·m = 1.356 J

1 cal = 4.180 J

1 eV = 1.60×10^{-19} J

1 kWh = 3.60×10^6 J

Power

1 W = 1 J/s = 0.738 ft·lb/s

1 hp (U.S.) = 746 W = 550 ft·lb/s

1 hp (metric) = 750 W

Pressure

1 atm = 1.01×10^5 N/m^2 = 14.7 lb/in^2

= 760 torr

1 Pa = 1 N/m^2

PHYSICAL CONSTANTS

Gravitational constant on Earth g	9.81 N/kg
Universal gravitational constant G	6.67×10^{-11} N·m^2/kg^2
Mass of Earth	5.97×10^{24} kg
Average radius of Earth	6.38×10^6 m
Density of dry air (STP)	1.3 kg/m^3
Density of water (4°C)	1000 kg/m^3
Avogadro's number N_A	6.02×10^{23} particles (g atom)
Boltzmann's constant k	1.38×10^{-23} J/K
Gas constant R	8.3 J/mol·K
Speed of sound in air (0°)	340 m/s
Coulomb's law constant k	$9.0 \times {}^9 10$ N·m^2/C^2
Speed of light c	3.00×10^8 m/s
Fundamental electric charge e	1.60×10^{-19} C
Electron mass m_e	9.11×10^{-31} kg = 5.4858×10^{-4} u
Proton mass m_p	1.67×10^{-27} kg = 1.00727 u
Neutron mass m_n	1.67×10^{-27} kg = 1.00866 u
Planck's constant h	6.63×10^{-34} J·s

POWER OF TEN PREFIXES

Prefix	Abbreviation	Value
Exa	E	10^{18}
Peta	P	10^{15}
Tera	T	10^{12}
Giga	G	10^{9}
Mega	M	10^{6}
Kilo	k	10^{3}
Hecto	h	10^{2}
Deka	da	10^{1}
Deci	d	10^{-1}
Centi	c	10^{-2}
Milli	m	10^{-3}
Micro	μ	10^{-6}
Nano	n	10^{-9}
Pico	p	10^{-12}
Femto	f	10^{-15}
Atto	a	10^{-18}

SOME USEFUL FACTS

Area of circle (radius r) πr^2

Area of sphere (radius r) $4\pi r^2$

Volume of sphere $\dfrac{4}{3}\pi r^3$

Trig definitions:
$\sin\theta = $ (opposite side)/(hypotenuse)
$\cos\theta = $ (adjacent side)/(hypotenuse)
$\tan\theta = $ (opposite side)/(adjacent side)

Quadratic equation:

$0 = ax^2 + bx + c,$
where $x = \dfrac{-b \pm \sqrt{b^2 - 4ac}}{2a}$

COLLEGE
PHYSICS

COLLEGE
PHYSICS

VOLUME 1

EUGENIA ETKINA
Rutgers University

MICHAEL GENTILE
Rutgers University

ALAN VAN HEUVELEN
Rutgers University

PEARSON

Boston   Columbus   Indianapolis   New York   San Francisco   Upper Saddle River
Amsterdam   Cape Town   Dubai   London   Madrid   Milan   Munich   Paris   Montréal   Toronto
Delhi   Mexico City   São Paulo   Sydney   Hong Kong   Seoul   Singapore   Taipei   Tokyo

Publisher: Jim Smith
Executive Editor: Becky Ruden
Project Managers: Katie Conley and Beth Collins
Managing Development Editor: Cathy Murphy
Associate Content Producer: Kelly Reed
Assistant Editor: Kyle Doctor
Team Lead, Program Management, Physical Sciences: Corinne Benson
Full-Service Production and Composition: PreMediaGlobal
Copy Editor: Joanna Dinsmore
Illustrator: Rolin Graphics
Photo Researcher: Eric Shrader
Image Lead: Maya Melenchuk
Manufacturing Buyer: Jeff Sargent
Marketing Manager: Will Moore
Text Designer: tani hasegawa
Cover Designer: Tandem Creative, Inc.
Cover Photo Credit: © Markus Altmann/Corbis

Credits and acknowledgments borrowed from other sources and reproduced, with permission, in this textbook appear on p. C-1.

Library of Congress Cataloging-in-Publication data
Etkina, Eugenia.
   College physics / Eugenia Etkina, Michael Gentile, Alan Van Heuvelen.
        pages cm
   ISBN-13: 978-0-321-71535-7
   ISBN-10: 0-321-71535-7
   1. Physics—Textbooks.   I. Gentile, Michael J.
   II. Van Heuvelen, Alan.   III. Title.
   QC21.3.E85 2012
   530—dc23

                                                    2012035388

1 2 3 4 5 6 7 8 9 10—DOW—16 15 14 13

www.pearsonhighered.com

ISBN 10: 0-321-88593-7; **ISBN 13:** 978-0-321-88593-7 (Volume 1)
ISBN 10: 0-321-90181-9; **ISBN 13:** 978-0-321-90181-1 (Instructor's resource copy)
ISBN 10: 0-321-87970-8; **ISBN 13:** 978-0-321-87970-7 (Books a la carte edition)

Brief Contents

About the Authors

Eugenia Etkina holds a PhD in physics education from Moscow State Pedagogical University and has more than 30 years experience teaching physics. She currently teaches at Rutgers University, where she received the highest teaching award in 2010 and the New Jersey Distinguished Faculty award in 2012. Professor Etkina designed and now coordinates one of the largest programs in physics teacher preparation in the United States, conducts professional development for high school and university physics instructors, and participates in reforms to the undergraduate physics courses. In 1993 she developed a system in which students learn physics using processes that mirror scientific practice. That system serves as the basis for this textbook. Since 2000, Professors Etkina and Van Heuvelen have conducted over 60 workshops for physics instructors and co-authored *The Physics Active Learning Guide* (a companion edition to *College Physics* is now available). Professor Etkina is a dedicated teacher and an active researcher who has published over 40 peer-refereed articles.

Michael Gentile is an Instructor of Physics at Rutgers University. He has a masters degree in physics from Rutgers University, where he studied under Eugenia Etkina and Alan Van Heuvelen, and has also completed postgraduate work in education, high energy physics, and cosmology. He has been inspiring undergraduates to learn and enjoy physics for more than 15 years. Since 2006 Professor Gentile has taught and coordinated a large-enrollment introductory physics course at Rutgers where the approach used in this book is fully implemented. He also assists in the mentoring of future physics teachers by using his course as a nurturing environment for their first teaching experiences. Since 2007 his physics course for the New Jersey Governor's School of Engineering and Technology has been highly popular and has brought the wonders of modern physics to more than 100 gifted high school students each summer.

Alan Van Heuvelen holds a PhD in physics from the University of Colorado. He has been a pioneer in physics education research for several decades. He taught physics for 28 years at New Mexico State University where he developed active learning materials including the *Active Learning Problem Sheets* (the *ALPS Kits*) and the *ActivPhysics* multimedia product. Materials such as these have improved student achievement on standardized qualitative and problem-solving tests. In 1993 he joined Ohio State University to help develop a physics education research group. He moved to Rutgers University in 2000 and retired in 2008. For his contributions to national physics education reform, he won the 1999 AAPT Millikan Medal and was selected a fellow of the American Physical Society. Over the span of his career he has led over 100 workshops on physics education reform. In the last ten years, he has worked with Professor Etkina in the development of the Investigative Science Learning Environment (*ISLE*), which integrates the results of physics education research into a learning system that places considerable emphasis on helping students develop science process abilities while learning physics.

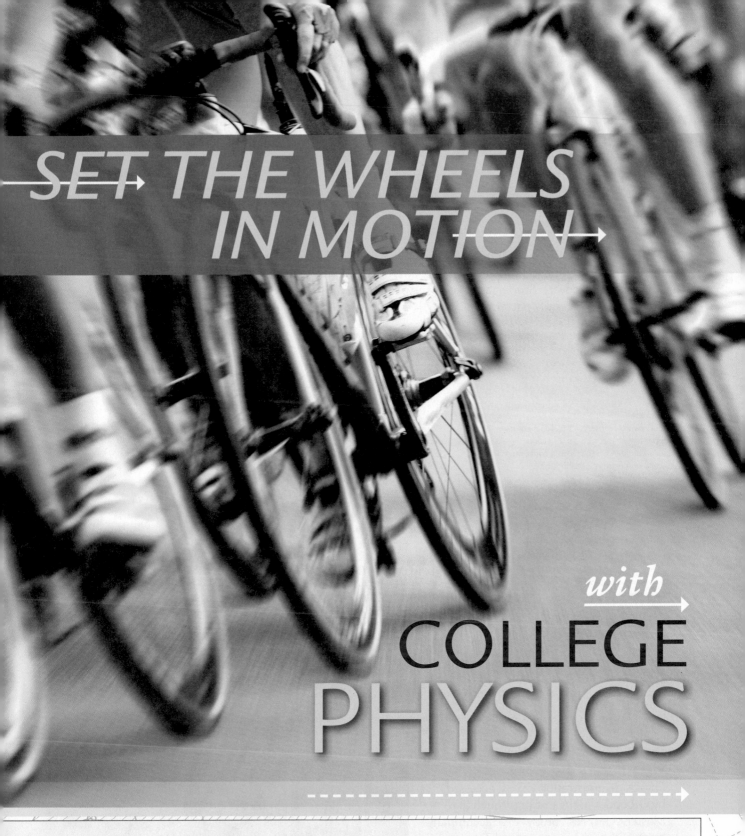

SET THE WHEELS IN MOTION

with COLLEGE PHYSICS

"This is an excellent way to teach physics. The approach is so logical that students will feel they are a) discovering physics themselves, and b) reaching the best conclusions... The style is approachable, consistent, systematic, engaging. I think [this textbook] teaches more than physics—it also gets at the core of the scientific process and that will be just as valuable for the students as any of the physics content."

—Andy Richter, *Valparaiso University*

BUILD A DEEP UNDERSTANDING OF PHYSICS AND THE SCIENTIFIC PROCESS

An active learning approach encourages students to construct an understanding of physics concepts and laws in the same ways that scientists acquire knowledge. Students learn physics by doing physics.

OBSERVATIONAL EXPERIMENT TABLES

Observational Experiment Tables engage students through active discovery. Students make observations, analyze data, and identify patterns.

Scan this QR code with your smartphone to view the video that accompanies this table.

VIDEO 2.3

OBSERVATIONAL EXPERIMENT TABLE

2.3 Two observers watch the same coffee mug.

Observational experiment	Analysis done by each observer
Experiment 1. Observer 1 is slouched down in the passenger seat of a car and cannot see outside the car. Suddenly he observes a coffee mug sliding toward him from the dashboard.	Observer 1 creates a motion diagram and a force diagram for the mug as he observes it. On the motion diagram, increasingly longer $\vec{v}$ arrows indicate that the mug's speed changes from zero to nonzero as seen by observer 1 even though no external object is exerting a force on it in that direction. $\Delta\vec{v}$ $\vec{v}$ $\vec{v}$ $\vec{N}_{D\ on\ M}$ $\vec{F}_{E\ on\ M}$
Experiment 2. Observer 2 stands on the ground beside the car. She observes that the car starts moving forward at increasing speed and that the mug remains stationary with respect to her.	Observer 2 creates a motion diagram and force diagram for the mug as she observes it. There are no $\vec{v}$ or $\Delta\vec{v}$ arrows on the diagram and the mug is at rest relative to her. $\Delta\vec{v}=0$ $\vec{v}=0$ $\vec{N}_{D\ on\ M}$ $\vec{F}_{E\ on\ M}$

Pattern

Observer 1: The forces exerted on the mug by Earth and by the dashboard surface add to zero. But the velocity of the mug increases as it slides off the dashboard. This is inconsistent with the rule relating the sum of the forces and the change in velocity.

Observer 2: The forces exerted on the mug by Earth and by the dashboard surface add to zero. Thus the velocity of the mug should not change, and it does not. This is consistent with the rule relating the sum of the forces and the change in velocity.

VIDEOS

Physics demonstration videos, accessed by QR codes in the text or through the MasteringPhysics® Study Area, accompany most of the Observational and Testing Experiment Tables. Students can observe the exact experiment described in the table.

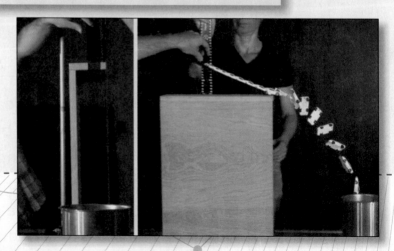

TESTING EXPERIMENT TABLES

Each testing experiment evaluates a hypothesis arising from the observational experiment, and includes the experimental setup, one or more predictions, and the outcome of the experiment. A conclusion summarizes the result of the experimental process.

Scan this QR code with your smartphone to view the video shown below.

TESTING EXPERIMENT TABLE

5.2 Testing the idea that $\Sigma m\vec{v}$ in an isolated system remains constant (all velocities are with respect to the track).

 VIDEO 5.2

Testing experiment	Prediction	Outcome
Cart A (0.40 kg) has a piece of modeling clay attached to its front and is moving right at 1.0 m/s. Cart B (0.20 kg) is moving left at 1.0 m/s. The carts collide and stick together. Predict the velocity of the carts after the collision.	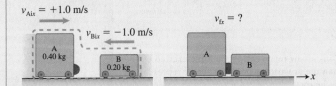 $v_{Aix} = +1.0$ m/s, $v_{Bix} = -1.0$ m/s, $v_{fx} = ?$	After the collision, the carts move together toward the right at close to the predicted speed.

The system consists of the two carts. The direction of velocity is noted with a plus or minus sign of the velocity component:

$$(0.40 \text{ kg})(+1.0 \text{ m/s}) + (0.20 \text{ kg})(-1.0 \text{ m/s})$$
$$= (0.40 \text{ kg} + 0.20 \text{ kg})v_{fx}$$

or

$$v_{fx} = (+0.20 \text{ kg} \cdot \text{m/s})/(0.60 \text{ kg}) = +0.33 \text{ m/s}$$

After the collision, the two carts should move right at a speed of about 0.33 m/s.

Conclusion

Our prediction matched the outcome. This result gives us increased confidence that this new quantity $m\vec{v}$ might be the quantity whose sum is constant in an isolated system.

DEVELOP ADVANCED PROBLEM-SOLVING SKILLS

Students learn to represent physical phenomena in multiple ways using words, figures, and equations, including qualitative diagrams and innovative bar charts that create a foundation for quantitative reasoning and problem solving.

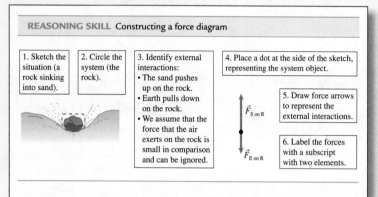

REASONING SKILL Constructing a force diagram

1. Sketch the situation (a rock sinking into sand).

2. Circle the system (the rock).

3. Identify external interactions:
   • The sand pushes up on the rock.
   • Earth pulls down on the rock.
   • We assume that the force that the air exerts on the rock is small in comparison and can be ignored.

4. Place a dot at the side of the sketch, representing the system object.

$\vec{F}_{\text{S on R}}$

$\vec{F}_{\text{E on R}}$

5. Draw force arrows to represent the external interactions.

6. Label the forces with a subscript with two elements.

Notice that the upward-pointing arrow representing the force exerted by the sand on the rock is longer than the downward-pointing arrow representing the force exerted by Earth on the rock. The difference in lengths reflects the difference in the magnitudes of the forces. Later in the chapter we will learn why they have different lengths. For now, we just need to include arrows for all external forces exerted on the system object (the rock).

REASONING SKILL BOXES

These boxes reinforce a particular skill, such as drawing a motion diagram, force diagram, or work-energy bar chart.

BAR CHARTS

Innovative bar charts help to create a foundation for quantitative reasoning and problem solving.

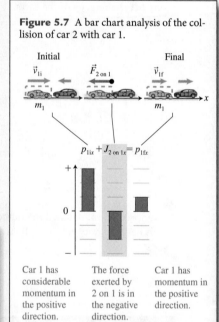

Figure 5.7 A bar chart analysis of the collision of car 2 with car 1.

Initial $\vec{v}_{1i}$ $\vec{F}_{\text{2 on 1}}$ Final $\vec{v}_{1f}$

m_1 m_1 x

$p_{1ix} + J_{\text{2 on 1}x} = p_{1fx}$

Car 1 has considerable momentum in the positive direction.

The force exerted by 2 on 1 is in the negative direction.

Car 1 has momentum in the positive direction.

PROBLEM-SOLVING STRATEGY

The Problem-Solving Strategy boxes walk students step-by-step through the process of solving a worked example, applying concepts covered in the text.

PROBLEM-SOLVING STRATEGY Applying Static Equilibrium Conditions

EXAMPLE 7.6 Use the biceps muscle to lift
Imagine that you hold a 6.0-kg lead ball in your hand with your arm bent. The ball is 0.35 m from the elbow joint. The biceps muscle attaches to the forearm 0.050 m from the elbow joint and exerts a force on the forearm that allows it to support the ball. The center of mass of the 12-N forearm is 0.16 m from the elbow joint. Estimate the magnitude of (a) the force that the biceps muscle exerts on the forearm and (b) the force that the upper arm exerts on the forearm at the elbow.

Sketch and translate

- Construct a labeled sketch of the situation. Include coordinate axes and choose an axis of rotation.
- Choose a system for analysis.

We choose the axis of rotation to be where the upper arm bone (the humerus) presses on the forearm at the elbow joint. This will eliminate from the torque equilibrium equation the unknown force that the upper arm exerts on the forearm.

We choose the system of interest to be the forearm and hand.

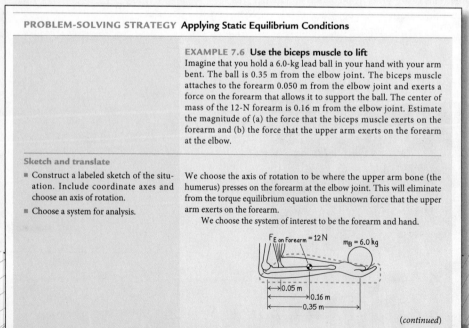

$F_{\text{E on Forearm}} = 12\ N$ $m_B = 6.0\ kg$

0.05 m
0.16 m
0.35 m

(continued)

INSPIRE HIGHER-LEVEL REASONING

Innovative, widely praised examples, exercises, and problems engage students, assess learning, and promote higher-level reasoning.

* **Equation Jeopardy 1** The equation below describes a rotational dynamics situation. Draw a sketch of a situation that is consistent with the equation and construct a word problem for which the equation might be a solution. There are many possibilities.

$$-(2.2\,\text{N})(0.12\,\text{m}) = \left[(1.0\,\text{kg})(0.12\,\text{m})^2\right]\alpha$$

* **Equation Jeopardy 2** The equation below describes a rotational dynamics situation. Draw a sketch of a situation that is consistent with the equation and construct a word problem for which the equation might be a solution. There are many possibilities.

$$-(2.0\,\text{N})(0.12\,\text{m}) + (6.0\,\text{N})(0.06\,\text{m}) = \left[(1.0\,\text{kg})(0.12\,\text{m})^2\right]\alpha$$

JEOPARDY-STYLE END-OF-CHAPTER PROBLEMS
Unique, Jeopardy-style end-of-chapter problems ask students to work backwards from an equation to craft a problem statement. Chapters also include "what if" problems, estimating problems, and qualitative/quantitative multi-part problems.

REVIEW QUESTIONS
Questions at the end of each section of the chapter encourage critical thinking and synthesis rather than recall.

Review Question 1.5 Why is the following statement true? "Displacement is equal to the area between a velocity-versus-time graph line and the time axis with a positive or negative sign."

ESTIMATION PROBLEMS
Estimation problems ask students to make reasonable assumptions and estimates in problem solving as a scientist would do.

EXAMPLE 5.6 Bone fracture estimation[1]
A bicyclist is watching for traffic from the left while turning toward the right. A street sign hit by an earlier car accident is bent over the side of the road. The cyclist's head hits the pole holding the sign. Is there a significant chance that his skull will fracture?

Sketch and translate The process is sketched at the right. The initial state is at the instant that the head initially contacts the pole; the final state is when the head

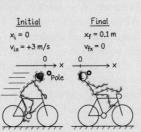

Initial	Final
$x_i = 0$	$x_f = 0.1\,\text{m}$
$v_{ix} = +3\,\text{m/s}$	$v_{fx} = 0$

and body have stopped. The person is the system. We have been given little information, so we'll have to make some reasonable estimates of various quantities in order to make a decision about a possible skull fracture.

Simplify and diagram The bar chart illustrates the momentum change of the system and the impulse exerted by the pole that caused the change. The person was initially moving in the horizontal x-direction with respect to Earth, and not moving after the collision. The pole exerted an impulse in the negative x-direction on the cyclist. We'll need to estimate the following quantities: the mass and speed of the cyclist in this situation, the stopping time interval, and the area of contact. Let's assume that this is a 70-kg cyclist moving at about 3 m/s. The person's body keeps moving forward for a short distance after the bone makes contact with the pole. The skin indents some during the collision. Because of these two factors, we assume

[1]This is a true story—it happened to one of the book's authors, Alan Van Heuvelen.

Real-world applications relate physics concepts and laws to everyday experiences and apply them to problems in diverse fields such as biology, medicine, and astronomy.

8.7 Rotational motion: Putting it all together

We can use our knowledge of rotational motion to analyze a variety of phenomena that are part of our world. In this section, we consider two examples—the effect of the tides on the period of Earth's rotation (the time interval for 1 day) and the motion of bowling (also called pitching) in the sport of cricket.

Tides and Earth's day

The level of the ocean rises and falls by an average of 1 m twice each day, a phenomenon known as the tides. Many scientists, including Galileo, tried to explain this phenomenon and suspected that the Moon was a part of the answer. Isaac Newton was the first to explain how the motion of the Moon actually creates tides. He noted that at any moment, different parts of Earth's surface are at different distances from the Moon and that the distance from a given location on Earth to the Moon varied as Earth rotated. As illustrated in **Figure 8.18**, point A is closer to the Moon than the center of Earth or point B are, and therefore the gravitational force exerted by the Moon on point A is greater than the gravitational force exerted on point B. Due to the difference

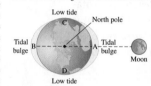

Figure 8.18 The ocean bulges on both sides of Earth along a line toward the Moon.

PUTTING IT ALL TOGETHER
These sections help students synthesize chapter content within real-world applications such as avoiding "the bends" in scuba diving (Chapter 10), making automobiles more efficient (Chapter 13), and building liquid crystal displays (Chapter 24).

Reading Passage Problems

BIO **Muscles work in pairs** Skeletal muscles produce movements by pulling on tendons, which in turn pull on bones. Usually, a muscle is attached to two bones via a tendon on each end of the muscle. When the muscle contracts, it moves one bone toward the other. The other bone remains in nearly the original position. The point where a muscle tendon is attached to the stationary bone is called the *origin*. The point where the other muscle tendon is attached to the movable bone is called the *insertion*. The origin is like the part of a door spring that is attached to the doorframe. The insertion is similar to the part of the spring that is attached to the movable door.

During movement, bones act as levers and joints act as axes of rotation for these levers. Most movements require several skeletal muscles working in groups, because a muscle can only exert a pull and not a push. In addition, most skeletal muscles are arranged in opposing pairs at joints. Muscles that bring two limbs together are called flexor muscles (such as the biceps muscle in the upper arm in **Figure 7.26**). Those that cause the limb to extend outward are called extensor muscles (such as the triceps muscle in the upper arm). The flexor muscle is used when you hold a heavy object in your hand; the extensor muscle can be used, for example, to extend your arm when you throw a ball.

MCAT-STYLE READING PASSAGE PROBLEMS
Help students prepare for the MCAT exam. Because so many students who take this course are planning to study medicine, each chapter includes MCAT-style reading passages and related multiple-choice questions to help prepare students for this important test.

BIOLOGICAL AND MEDICAL EXAMPLES
Examples throughout the text provide relevance for life science majors and include topics such as understanding the effect of radon on the lungs (Chapter 5), controlling body temperature (Chapter 12), and measuring the speed of blood flow (Chapter 20).

Figure 7.26 Muscles often come in flexor-extensor pairs.

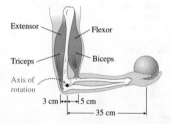

Reading Passage Problems

BIO **Head injuries in sports** A research group at Dartmouth College has developed a Head Impact Telemetry (HIT) System that can be used to collect data about head accelerations during impacts on the playing field. The researchers observed 249,613 impacts from 423 football players at nine colleges and high schools and collected collision data from participants in other sports. The accelerations during most head impacts ($>89\%$) in helmeted sports caused head accelerations less than a magnitude of 400 m/s². However, a total of 11 concussions were diagnosed in players whose impacts caused accelerations between 600 and 1800 m/s², with most of the 11 over 1000 m/s².

REINFORCE SCIENTIFIC THINKING →

Active Learning Guide for College Physics
by Eugenia Etkina, Michael Gentile, and Alan Van Heuvelen
© 2014 • Paper • 400 pages
978-0-321-86445-1 • 0-321-86445-X

Discovery-based activities supplement the knowledge-building
approach of the textbook. This workbook is organized in
parallel with the textbook's chapters.

Blue labels, located in the text's margins, link the discovery-based
activities in the *Active Learning Guide* to concepts covered in
College Physics.

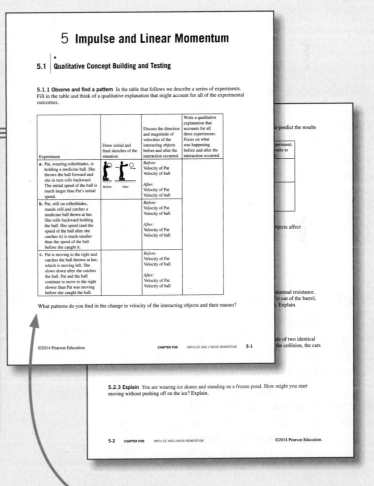

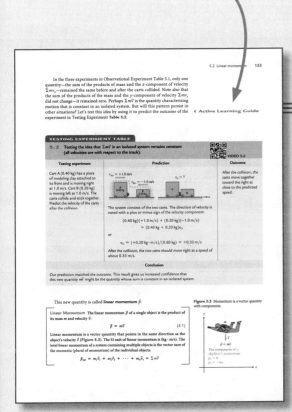

These activities provide an opportunity for
further observation, testing, sketching, and
analysis.

MAKE A DIFFERENCE WITH MasteringPhysics®

The Mastering platform is the most effective and widely used online homework, tutorial, and assessment system for physics. It delivers self-paced tutorials that focus on instructors' course objectives, provides individualized coaching, and responds to each student's progress. The Mastering system helps instructors maximize class time with easy-to-assign, customizable, and automatically graded assessments that motivate students to learn outside of class and arrive prepared for lecture and lab.

www.masteringphysics.com

Prelecture Concept Questions

Assignable Prelecture Concept Questions encourage students to read the textbook prior to lecture so they're more engaged in class.

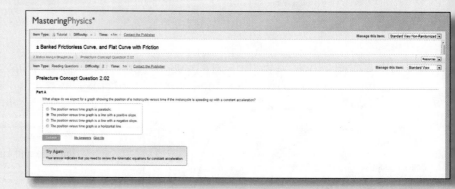

Gradebook Diagnostics

The Gradebook Diagnostics screen provides your favorite weekly diagnostics. With a single click, charts summarize the most difficult problems, vulnerable students, and grade distribution.

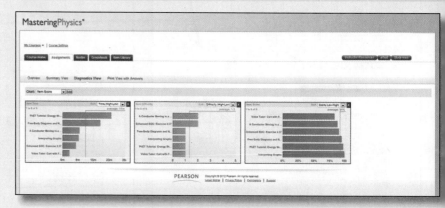

Tutorials with Hints and Feedback

Easily assign tutorials that provide individualized coaching.

- Mastering's hallmark Hints and Feedback offer "scaffolded" instruction similar to what students would experience in an office hour.
- Hints (declarative and Socratic) can provide problem-solving strategies or break the main problem into simpler exercises.
- Wrong-answer-specific feedback gives students exactly the help they need by addressing their particular mistake without giving away the answer.

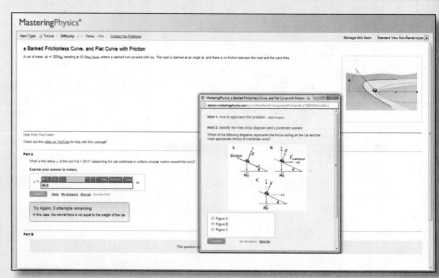

Preface

To the student

College Physics is more than just a book. It's a learning *companion*. As a companion, the book won't just tell you about physics; it will act as a guide to help you build physics ideas using methods similar to those that practicing scientists use to construct knowledge. The ideas that you build will be *yours*, not just a copy of someone else's ideas. As a result, the ideas of physics will be much easier for you to use when you need them: to succeed in your physics course, to obtain a good score on exams such as the MCAT, and to apply to everyday life.

Although few, if any, textbooks can honestly claim to be a pleasure to read, *College Physics* is designed to make the process interesting and engaging. The physics you learn in this book will help you understand many real-world phenomena, from why giant cruise ships are able to float to how telescopes work.

A great deal of research has been done over the past few decades on how students learn. We, as teachers and researchers, have been active participants in investigating the challenges students face in learning physics. We've developed unique strategies that have proven effective in helping students think like physicists. These strategies are grounded in *active learning*, deliberate, purposeful action on your part to learn something new. It's not passively memorizing so that you can repeat it later. When you learn actively you engage with the material. You relate it to what you already know. You think about it in as many different ways as you can. You ask yourself questions such as "Why does this make sense?" and "Under what circumstances does this not apply?"

This book (your learning companion) includes many tools to support the active learning process: each problem-solving strategies tool, worked example, observational experiment, testing experiment, review question, and end-of-chapter question and problem is designed to help you build your understanding of physics. To get the most out of these tools and the course, stay actively engaged in the process of developing ideas and applying them. When things get challenging, don't give up.

At this point you should turn to the chapter Introducing Physics and begin reading. That's where you'll learn the details of the approach that the book uses, what physics is, and how to be successful in the physics course you are taking.

To the instructor

In writing *College Physics,* our main goal was to produce an effective learning companion for students that incorporates results from the last few decades of physics education research. This research has shown that there is a dramatic difference between how physicists construct new ideas and how students traditionally learn physics. Students often leave their physics course thinking of physics as a disconnected set of facts that has little to do with the real world, rather than as a framework for understanding it.

To address this problem we have based this book on a framework known as ISLE (Investigative Science Learning Environment) developed by authors Etkina and Van Heuvelen. In ISLE, the construction of new ideas begins with observational experiments. Students are explicitly presented with simple experiments from which they discern patterns using available tools (diagrams, graphs, bar charts, etc.). To explain the patterns, students devise explanations (hypotheses) for their observations. They then use these explanations in testing experiments to make predictions about the outcomes of these new experiments. If the prediction does not match the outcome of the experiment, the explanation needs to be reevaluated. Explanations that survive this testing process are the physics ideas in which we then have more confidence.

The goal of this approach is to help students understand physics as a process by which knowledge of the natural world is constructed, rather than as a body of given laws and facts. This approach also helps students reason using the tools that physicists and physics educators have developed for the analysis of phenomena—for example, motion and force diagrams, kinematics and thermodynamics graphs, energy and momentum bar charts, and many other visual representations. Using these tools helps students bridge the gap between words and mathematical equations. Along the way, they develop independent and critical thinking skills that will allow them to build their own understanding of physics principles.

All aspects of *College Physics* are grounded in ISLE and physics education research. As a result, all of the features of the text have been designed to encourage students to investigate, test ideas, and apply scientific reasoning.

Key learning principles

To achieve these goals we adhere to five key learning principles:

1. **Concept first, name second:** The names we use for physics concepts have everyday-life meanings that may differ from the meanings they have when used in physics. For example, in physics *flux* refers to the amount to which a directed quantity (such as the magnetic field) points through a surface, but in everyday-life *flux* refers to continuous change. Confusion over the meaning of terms can get in the way of learning. We address this difficulty by developing the concept first and only then assigning a name to it.

2. **Careful language**: The vernacular physicists use is rooted in history and tradition. While physicists have an internal "dictionary" that lets them understand the meaning of specific terms, students do not. We are extremely careful to use language that promotes understanding. For example: physicists would say that "heat flows from a hot object to a cool object." Heat isn't a substance that objects possess; heat is the *flow of energy*. In this book we only use the word *heat* to refer to the process of energy transfer.

3. **Bridging words and mathematics**: Words and mathematics are very abstract representations of physical phenomena. We help students translate between these abstractions by using concrete representations such as force diagrams and energy bar charts as intermediate steps.

4. **Making sense of mathematics**: We explicitly teach students how to evaluate the results of their quantitative reasoning so they can have confidence in that reasoning. We do this by building qualitative understanding first and then explicitly teaching students how to use that understanding to check for quantitative consistency. We also guide students to use limiting cases to evaluate their results.

5. **Moving away from plug-and-chug problem solving approaches**: In this book you will find many non-traditional examples and end-of-chapter problems that require students to use higher-level reasoning skills and not just plug numbers into equations that have little meaning for them. Jeopardy problems (where a solution is given and students must invent a problem that leads to it), "tell-all" problems (where students must determine everything possible), and estimation problems (where students do not have quantities given to them) are all designed to encourage higher reasoning and problem solving skills.

These key principles are described in greater detail in the Introduction to the *Instructor's Guide* that accompanies *College Physics*—please read that introduction. It elaborates on the implementation of the methodology that we use in this book and provides guidance on how to integrate the approach into your course.

While our philosophy informs *College Physics*, you need not fully subscribe to it to use this textbook. We've organized the book to fit the structure of most algebra-based physics courses: We begin with kinematics and Newton's laws, then move on to conserved quantities, statics, gases, fluids, thermodynamics, electricity and magnetism, vibrations and waves, optics, and finally modern physics. The structure of each chapter will work with any method of instruction. You can assign all of the innovative experimental tables and end-of-chapter problems, or only a few. The text provides thorough treatment of fundamental principles, supplementing this coverage with experimental evidence, new representations, an effective approach to problem solving, and interesting and motivating examples.

Real-world applications

To effectively teach physics, especially to the non-physics student, a textbook must actively engage the student's interest. To that effect, *College Physics* includes a wealth of real-world applications. Each chapter begins with a brief vignette designed to intrigue the reader. For example, Chapter 11 opens with a description of plaque build-up in arteries that can lead to stroke. Chapters also open with a set of motivating questions that are answered as students read subsequent sections. In each chapter, worked examples and exercises cover such topics as what keeps a car on the road when spinning around a circular track (Chapter 4), why air bags are so effective (Chapter 5), and why your ears pop when you change altitude (Chapter 10). A Putting it all together section applies concepts from the chapter to complex phenomena such as collisions (Chapter 6), lightning (Chapter 15), and the Doppler Effect (Chapter 20). Many applications are grounded in biology or medicine. Approximately eight percent of end-of-chapter problems are on biomedical topics. A complete list of applications appears on pages xxiv–xxvi.

Chapter features

Chapter-opening features engage the student in the chapter topic.

Each chapter opens with a bridge from the concepts and skills that students will have learned in previous chapters. This bridge takes the form of **"Be sure you know how to"** statements with cross-references to the relevant material in previous chapters.

Each chapter also includes a set of **Motivating questions** to capture student interest. These questions are answered within the chapter content. Two examples are, "Why do people snore?" and "How does a refrigerator stay cold inside?"

A brief **vignette** opens each chapter with a real-world story related to one of the motivating questions.

In-chapter features encourage the active construction of knowledge about physics and support students as they read and review the material.

Experimental tables help students explore science as a process of inquiry (e.g., making observations, analyzing data, identifying patterns, testing hypotheses, etc.) and develop reasoning skills they can use to solve physics problems.

- **Observational experiment tables** engage students in an active discovery process as they learn about key physics ideas. By analyzing and finding patterns for the experiments, students learn the process of science.

- **Testing experiment tables** allow students to test hypotheses by predicting an outcome, conducting the experiment, and forming a conclusion that compares the prediction to the outcome and summarizes the results.

Many of these tables are accompanied by videos of the experiments. Students can view them through a QR code on the table using their smartphone, or online in the MasteringPhysics study area.

- **Section review questions** encourage critical thinking and synthesis rather than recall. Answers to review questions are given at the end of each chapter.

Three types of worked examples guide students through the problem-solving process.

Examples are complete problems that utilize the four-step problem-solving strategy.

- **Sketch and Translate:** This step teaches students to translate the problem statement into the language of physics. Students read the problem, sketch the situation and include known values, and identify the unknown(s).

- **Simplify and Diagram:** Students simplify the physics problem with an appropriate physical representation, a force diagram or other representation that reflects the situation in the problem and helps them construct a mathematical equation to solve it.

- **Represent Mathematically:** In this step, students apply the relevant mathematical equations. For example, they use the force diagram to apply Newton's second law in component form.

- **Solve and Evaluate:** In the last step, students rearrange equations and insert known values to solve for the unknown(s).

Conceptual exercises focus on developing students' conceptual understanding. These exercises utilize two of the problem-solving steps: *Sketch and Translate,* and *Simplify and Diagram.*

Quantitative exercises develop students' ability to solve unknowns quantitatively. These exercises utilize the problem-solving steps *Represent Mathematically* and *Solve and Evaluate.*

Each worked example ends with a **Try It Yourself** question, an additional exercise that builds on the worked example and asks students to solve a similar problem without the scaffolding.

The text also includes additional support for problem solving.

Problem-solving strategy boxes work through an example, explaining as well as applying the four-step strategy.

Reasoning skill boxes summarize the use of a particular skill, such as drawing a motion diagram, a force diagram, or a work-energy bar chart.

Tips within the text encourage the use of particular strategies or caution the reader about common misconceptions.

Key equations and **definitions boxes** highlight important laws or principles that govern the physics concepts developed in the chapters.

The Putting it all together section (found in most chapters) focuses on two to four real-world applications of the physics learned in the chapter.

Putting it all together contains conceptual explanations and capstone worked examples that will often draw on more than one principle. The goal of this section is to help students synthesize what they have learned and broaden their understanding of the phenomena they are exploring.

End-of-chapter features

The **chapter summary** reviews key concepts presented in the chapter. The summary utilizes the multiple representation approach, displaying the concepts in words, figures, and equations.

Each chapter includes the authors' widely-lauded and highly creative problem sets, including their famous Jeopardy-style problems, "tell-all" problems, and estimating problems. The authors have written every end-of-chapter item themselves. Each question and problem is thoroughly grounded in their deep understanding of how students learn physics.

Questions: The question section includes both multiple choice and conceptual short answer questions, to help build students' fluency in the words, symbols, pictures, and graphs used in physics. Most of the questions are qualitative.

Problems: Chapters include an average of 60 section-specific problems. Approximately eight percent of the problems are drawn from biology or medicine; other problems relate to astronomy, geology, and everyday life.

General Problems: These challenging problems often involve multiple parts and require students to apply conceptual knowledge learned in previous chapters. As much as possible, the problems have a real-world context to enhance the connection between physics and students' daily lives.

MCAT reading passage and related multiple choice questions and problems: Because so many students who take this course are planning to study medicine, each chapter includes MCAT-style reading passages with related multiple-choice questions to help prepare students for their MCAT exam.

Instructor supplements

The *Instructor's Guide*, written by Eugenia Etkina, Alan Van Heuvelen, and highly respected physics education researcher David Brookes, walks you through the innovative approaches they take to teaching physics. Each chapter of the *Instructor's Guide* contains a roadmap to assigning chapter content, *Active Learning Guide* assignments, homework, and videos of the demonstration experiments. In addition, the authors call out common pitfalls to mastering physics concepts and describe techniques that will help your students identify and overcome their misconceptions. Tips include how to manage the complex vocabulary of physics, when to use classroom-response tools, and how to organize lab, lecture, and small group learning time. Drawing from their extensive experience as teachers and researchers, the authors give you the support you need to make *College Physics* work for you.

The cross-platform **Instructor Resource DVD** (ISBN 0-321-88897-9) provides invaluable and easy-to-use resources for your class, organized by textbook chapter. The contents include a comprehensive library of more than 220 applets from **ActivPhysics OnLine™**, as well as all figures, photos, tables, and summaries from the textbook in JPEG and Power-Point formats. A set of editable **Lecture Outlines** and **Classroom Response System "Clicker" Questions** on PowerPoint will engage your students in class.

MasteringPhysics® (www.masteringphysics.com) is a powerful, yet simple, online homework, tutorial, and assessment system designed to improve student learning and results. Students benefit from wrong-answer specific feedback, hints, and a huge variety of educationally effective content while unrivalled gradebook diagnostics allow an instructor to pinpoint the weaknesses and misconceptions of their class.

NSF-sponsored published research (and subsequent studies) show that MasteringPhysics has dramatic educational results. MasteringPhysics allows instructors to build wide-ranging homework assignments of just the right difficulty and length and provides them with efficient tools to analyze in unprecedented detail both class trends and the work of any student.

In addition to the textbook's end-of-chapter problems, MasteringPhysics for *College Physics* also includes tutorials, prelecture concept questions, and Test Bank questions for each chapter. MasteringPhysics also now has the following learning functionalities:

- **Prebuilt Assignments:** These offer instructors a mix of end-of-chapter problems and tutorials for each chapter.
- **Learning Outcomes:** In addition to being able to create their own learning outcomes to associate with questions in an assignment, professors can now select content that is tagged to a large number of publisher-provided learning outcomes. They can also print or export student results based on learning outcomes for their own use or to incorporate into reports for their administration.

- **Quizzing and Testing Enhancements:** These include options to hide item titles, add password protection, limit access to completed assignments, and randomize question order in an assignment.
- **Math Remediation:** Found within selected tutorials, special links provide just-in-time math help and allow students to brush up on the most important mathematical concepts needed to successfully complete assignments. This new feature links students directly to math review and practice, helping students make the connection between math and physics.

The **Test Bank** contains more than 2,000 high-quality problems, with a range of multiple-choice, true/false, short-answer, and regular homework-type questions. Test files are provided in both TestGen® (an easy-to-use, fully networkable program for creating and editing quizzes and exams) and Word format, and can be downloaded from www.pearsonhighered.com/educator.

Student supplements

The *Active Learning Guide* workbook by Eugenia Etkina, Michael Gentile, and Alan Van Heuvelen consists of carefully-crafted activities that provide an opportunity for further observation, sketching, analysis, and testing. Marginal "Active Learning Guide" icons throughout *College Physics* indicate content for which a workbook activity is available. Whether the activities are assigned or not, students can always use this workbook to reinforce the concepts they have read about in the text, to practice applying the concepts to real-world scenarios, or to work with sketches, diagrams, and graphs that help them visualize the physics.

Physics demonstration videos, accessed with a smartphone through QR codes in the text or online in the MasteringPhysics study area, accompany most of the Observational and Testing Experiment Tables. Students can observe the exact experiment described in the table.

MasteringPhysics® (www.masteringphysics.com) is a powerful, yet simple, online homework, tutorial, and assessment system designed to improve student learning and results. Students benefit from wrong-answer specific feedback, hints, and a huge variety of educationally effective content while unrivalled gradebook diagnostics allow an instructor to pinpoint the weaknesses and misconceptions of their class. The individualized, 24/7 Socratic tutoring is recommended by 9 out of 10 students to their peers as the most effective and time-efficient way to study.

Pearson eText is available through MasteringPhysics, either automatically when MasteringPhysics is packaged with new books, or available as a purchased upgrade online. Allowing students access to the text wherever they have access to the Internet, Pearson eText comprises the full text, including figures that can be enlarged for better viewing. Within eText, students are also able to pop up definitions and terms to help with vocabulary and the reading of the material. Students can also take notes in eText using the annotation feature at the top of each page.

Pearson Tutor Services (www.pearsontutorservices.com) Each student's subscription to MasteringPhysics also contains complimentary access to Pearson Tutor services, powered by Smarthinking, Inc. By logging in with their MasteringPhysics ID and password, they will be connected to highly qualified e-instructors™ who provide additional, interactive online tutoring on the major concepts of physics. Some restrictions apply; offer subject to change.

Acknowledgments

We wish to thank the many people who helped us create this textbook and its supporting materials. First and foremost, we want to thank our team at Pearson Higher Education, especially Cathy Murphy, who provided constructive, objective, and untiring feedback on every aspect of the book; Jim Smith, who has been a perpetual cheerleader for the project; Katie Conley and Beth Collins, who shepherded the book through production, and Becky Ruden and Kelly Reed, who oversaw the media component of the program. Kyle Doctor attended to many important details, including obtaining reviews of the text and production of the *Active Learning Guide*. Special thanks to Margot Otway, Brad Patterson, Gabriele Rennie, and Michael Gillespie, who helped shape the book's structure and features in its early stages. We also want to thank Adam Black for believing in the future of the project. We are indebted to Frank Chmely and Brett Kraabel, who checked and rechecked every fact and calculation in the text. Sen-Ben Liao and Brett Kraabel prepared detailed solutions for every end-of-chapter problem for the *Instructor's Solutions Manual*. We also want to thank all of the reviewers who put their time and energy to providing thoughtful, constructive, and supportive feedback on every chapter.

Our infinite thanks go to Suzanne Brahmia who came up with the Investigative Science Learning Environment acronym "ISLE," and was and is an effective user of the ISLE learning strategy. Her ideas about relating physics and mathematics are reflected in many sections of the book. We thank Kruti Singh for her help with the class testing of the book and all of Eugenia's students for providing feedback and ideas. We also want to thank David Brookes, co-author of the *Instructor's Guide*, whose research affected the treatment of language in the textbook; Gorazd Planinšič, who provided feedback on many chapters; Paul

Bunson, for using first drafts of the book's chapters in his courses; and Dedra Demaree, who has supported ISLE for many years. Dedra also prepared the extensive set of multiple-choice Clicker questions and the PowerPoint lecture outline available on our Instructor Resource DVD.

We have been very lucky to belong to the physics teaching community. Ideas of many people in the field contributed to our understanding of how people learn physics and what approaches work best. These people include Arnold Arons, Fred Reif, Jill Larkin, Lillian McDermott, David Hestenes, Joe Redish, Jim Minstrell, David Maloney, Fred Goldberg, David Hammer, Andy Elby, Tom Okuma, Curt Hieggelke, and Paul D'Alessandris. We thank all of them and many others.

Personal notes from the authors

We wish to thank Valentin Etkin (Eugenia's father), an experimental physicist whose ideas gave rise to the ISLE philosophy many years ago, Inna Vishnyatskaya (Eugenia's mother), who never lost faith in the success of our book, and Dima and Sasha Gershenson (Eugenia's sons), who provided encouragement to Eugenia and Alan over the years. While teaching Alan how to play violin, Alan's uncle Harold Van Heuvelen provided an instructional system very different from that of traditional physics teaching. In Harold's system, many individual abilities (skills) were developed with instant feedback and combined over time to address the process of playing a complex piece of music. We tried to integrate this system into our ISLE physics learning system.

—Eugenia Etkina and Alan Van Heuvelen

First, thanks to my co-authors Alan and Eugenia for bringing me onto this project and giving me the opportunity to fulfill a life goal of writing a book, and for being cherished friends and colleagues these many years. Thanks also to my students and teaching assistants in Physics for Sciences at Rutgers University these last three years for using the book as their primary text and giving invaluable feedback. Thanks eternally to my parents for unquestioning support in all my endeavors, and gifting me with the attitude that all things can be achieved.

For my beloved partner Christine, as with this project and all other challenges in life: Team Effort. Best Kind.

—Mike Gentile

Reviewers and classroom testers

Ricardo Alarcon
Arizona State University

Eric Anderson
University of Maryland, Baltimore County

James Andrews
Youngstown State University

David Balogh
Fresno City College

Linda Barton
Rochester Institute of Technology

Ian Beatty
University of North Carolina at Greensboro

Robert Beichner
North Carolina State University, Raleigh

Aniket Bhattacharya
University of Central Florida

Luca Bombelli
University of Mississippi

Scott Bonham
Western Kentucky University

Gerald Brezina
San Antonio College

Paul Bunson
Lane Community College

Hauke Busch
Georgia College & State University

Rebecca Butler
Pittsburg State University

Paul Camp
Spelman College

Amy Campbell
Georgia Gwinnett College

Juan Catala
Miami Dade College North

Colston Chandler
University of New Mexico

Soumitra Chattopadhyay
Georgia Highlands College

Betsy Chesnutt
Itawamba Community College

Chris Coffin
Oregon State University

Lawrence Coleman
University of California, Davis

Michael Crivello
San Diego Mesa College

Elain Dahl
Vincennes University

Charles De Leone
California State University, San Marcos

Carlos Delgado
Community College of Southern Nevada

Christos Deligkaris
Drury University

Dedra Demaree
Oregon State University

Karim Diff
Santa Fe Community College

Kathy Dimiduk
Cornell University

Diana Driscoll
Case Western Reserve University

Raymond Duplessis
Delgado Community College

Taner Edis
Truman State University

Bruce Emerson
Central Oregon Community College

Xiaojuan Fan
Marshall University

Nail Fazleev
University of Texas at Arlington

Gerald Feldman
George Washington University

Jane Flood
Muhlenberg College

Tom Foster
Southern Illinois University, Edwardsville

Richard Gelderman
Western Kentucky University

Anne Gillis
Butler Community College

Martin Goldman
University of Colorado, Boulder

Greg Gowens
University of West Georgia

Michael Graf
Boston College

Alan Grafe
University of Michigan, Flint

Recine Gregg
Fordham University

Elena Gregg
Oral Roberts University

John Gruber
San Jose State University

Arnold Guerra
Orange Coast College

Edwin Hach III
Rochester Institute of Technology

Steve Hagen
University of Florida, Gainesville

Thomas Hemmick
State University of New York, Stony Brook

Scott Hildreth
Chabot College

Zvonimir Hlousek
California State University, Long Beach

Mark Hollabaugh
Normandale Community College

Klaus Honscheid
Ohio State University

Kevin Hope
University of Montevallo

Joey Huston
Michigan State University

Richard Ignace
East Tennessee State University

Doug Ingram
Texas Christian University

George Irwin
Lamar University

Darrin Johnson
University of Minnesota

Adam Johnston
Weber State University

Mikhail Kagan
Pennsylvania State University

David Kaplan
Southern Illinois University, Edwardsville

James Kawamoto
Mission College

Julia Kennefick
University of Arkansas

Casey King
Horry-Georgetown Technical College

Patrick Koehn
Eastern Michigan University

Victor Kriss
Lewis-Clark State College

Peter Lanagan
College of Southern Nevada, Henderson

Albert Lee
California State University, Los Angeles

Todd Leif
Cloud County Community College

Eugene Levin
Iowa State University

Jenni Light
Lewis-Clark State College

Curtis Link
Montana Tech of The University of Montana

Donald Lofland
West Valley College

Susannah Lomant
Georgia Perimeter College

Rafael Lopez-Mobilia
University of Texas at San Antonio

Kingshuk Majumdar
Berea College

Gary Malek
Johnson County Community College

Eric Mandell
Bowling Green State University

Lyle Marschand
Northern Illinois University

Donald Mathewson
Kwantlen Polytechnic University

Mark Matlin
Bryn Mawr College

Timothy McKay
University of Michigan

David Meltzer
Arizona State University

William Miles
East Central Community College

Rabindra Mohapatra
University of Maryland, College Park

Enrique Moreno
Northeastern University

Joe Musser
Stephen F. Austin State University

Charles Nickles
University of Massachusetts, Dartmouth

Gregor Novak
United States Air Force Academy

John Ostendorf
Vincennes University

Philip Patterson
Southern Polytechnic State University

Jeff Phillips
Loyola Marymount University

Francis Pichanick
University of Massachusetts, Amherst

Dmitri Popov
State University of New York, Brockport

Matthew Powell
West Virginia University

Roberto Ramos
Indiana Wesleyan University

Greg Recine
Fordham University

Edward Redish
University of Maryland, College Park

Lawrence Rees
Brigham Young University

Lou Reinisch
Jacksonville State University

Andrew Richter
Valparaiso University

Melodi Rodrigue
University of Nevada, Reno

Charles Rogers
Southwestern Oklahoma State University

David Rosengrant
Kennesaw State University

Alvin Rosenthal
Western Michigan University

Lawrence Rowan
University of North Carolina at Chapel Hill

Roy Rubins
University of Texas at Arlington

Otto Sankey
Arizona State University

Rolf Schimmrigk
Indiana University, South Bend

Brian Schuft
North Carolina Agricultural and Technical State University

Sara Schultz
Montana State University, Moorhead

Bruce Schumm
University of California, Santa Cruz

David Schuster
Western Michigan State University

Bart Sheinberg
Houston Community College, Northwest College

Carmen Shepard
Southwestern Illinois College

Douglas Sherman
San Jose State University

Chandralekha Singh
University of Pittsburgh

David Snoke
University of Pittsburgh

David Sokoloff
University of Oregon

Mark Stone
Northwest Vista College

Bernhard Stumpf
University of Idaho

Tatsu Takeuchi
Virginia Polytechnic Institute and State University

Julie Talbot
University of West Georgia

Colin Terry
Ventura College

Beth Ann Thacker
Texas Tech University

John Thompson
University of Maine, Orono

Som Tyagi
Drexel University

David Ulrich
Portland Community College

Eswara P. Venugopal
University of Detroit, Mercy

James Vesenka
University of New England

William Waggoner
San Antonio College

Jing Wang
Eastern Kentucky University

Tiffany Watkins
Boise State University

Laura Weinkauf
Jacksonville State University

William Weisberger
State University of New York, Stony Brook

John Wernegreen
Eastern Kentucky University

Daniel Whitmire
University of Louisiana at Lafayette

Brian Woodahl
Indiana University-Purdue University Indianapolis

Gary Wysin
Kansas State University, Manhattan

Real-World Applications

Contents

Introducing Physics

Why do we need to use models to explain the world around us?

How is the word "law" used differently in physics than in the legal system?

How do we solve physics problems?

I.1   What is physics?

Physics is a fundamental experimental science encompassing subjects such as mechanical motion, waves, light, electricity, magnetism, atoms, and nuclei. Knowing physics allows you to understand many aspects of the world, from why bending over to lift a heavy load can injure your back to why Earth's climate is changing. Physics explains the very small—atoms and subatomic particles—and the very large—planets, galaxies, and celestial bodies such as white dwarfs, pulsars, and black holes.

In each chapter, we will apply our knowledge of physics to other fields of science and technology such as biology, medicine, geology, astronomy, architecture, engineering, agriculture, and anthropology. For instance, in this book you will learn about techniques used by archeologists to determine the age of

Physics in the field. Archaeologists applied principles from physics to determine that this skeleton of *Australopithecus afarensis*, nicknamed "Lucy," lived about 3.2 million years ago.

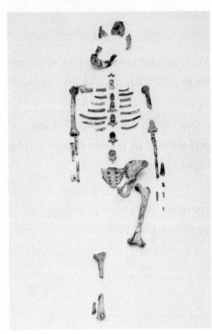

bones, about electron microscopes and airport metal detectors, about ways in which thermal energy is gained and lost in homes, about the development of stresses and tensions in body muscles, and why high blood pressure indicates problems with the circulatory system.

In this book we will concentrate not only on developing an understanding of the important basic laws of physics but also on the processes that physicists employ to discover and use these laws. The processes (among many) include:

- Collecting and analyzing experimental data.
- Making explanations and experimentally testing them.
- Creating different representations (pictures, graphs, bar charts, etc.) of physical processes.
- Finding mathematical relations between different variables.
- Testing those relations in new experiments.

The search for rules

Physicists search for general rules or **laws** that bring understanding to the chaotic behavior of our surroundings. In physics the word *law* means a mathematical relation between variables inferred from the data or through some reasoning process. The laws, once discovered, often seem obvious, yet their discovery usually requires years of experimentation and theorizing. Despite being called "laws," these laws are temporary in the sense that new information often leads to their modification, revision, and, in some cases, abandonment.

For example, in 200 B.C. Apollonius of Perga watched the Sun and the stars moving in arcs across the sky and adopted the concept that Earth occupied the center of a revolving universe. Three hundred years later, Ptolemy provided a theory to explain the complicated motion of the planets in that Earth-centered universe. Ptolemy's theory, which predicted with surprising accuracy the changing positions of the planets, was accepted for the next 1400 years. However, as the quality of observations improved, discrepancies between the predictions of Ptolemy's theory and the real positions of the planets became bigger and bigger. A new theory was needed. Copernicus, who studied astronomy at the time that Columbus sailed to America, developed a theory of motion for the heavenly bodies in which the Sun resided at the center of the universe while Earth and the other planets moved in orbits around it. More than 100 years later the theory was revised by Johannes Kepler and later supported by careful experiments by Galileo Galilei. Finally, 50 years after Galileo's death, Isaac Newton formulated three simple laws of motion and the universal law of gravitation, which together provided a successful explanation for the orbital motion of Earth and the other planets. These laws also allowed us to predict the positions of new planets, which at the time were not yet known. For nearly 300 years Newton's ideas went unaltered until Albert Einstein made several profound improvements to our understanding of motion and gravitation at the beginning of the 20th century.

Newton's inspiration provided not only the basic resolution of the 1800-year-old problem of the motion of the planets but also a general framework for analyzing the mechanical properties of nature. Newton's simple laws give us the understanding needed to guide rockets to the moon, to build skyscrapers, and to lift heavy objects safely without injury.

It is difficult to appreciate the great struggles our predecessors endured as they developed an understanding that now seems routine. Today, similar struggles occur in most branches of science, though the questions being investigated have changed. How does the brain work? What causes Earth's magnetism? What is the nature of the pulsating sources of X-ray radiation in our galaxy? Is the recently discovered accelerated expansion of the universe really caused by a mysterious "dark energy," or is our interpretation of the observations of distant supernovae that revealed the acceleration incomplete?

Does this understanding make the world a better place?

The pursuit of basic understanding often seems greatly removed from the activities of daily living. If J. J. Thomson's peers had asked him in 1897 if there was any practical application for his discovery of the electron, he probably could not have provided a satisfactory answer. Yet a little over a century later, the electron plays an integral part in our everyday technology. Moving electrons in electric circuits produce light for reading and warmth for cooking. Knowledge of the electron has made it possible for us to transmit the information that we see as images on our smart phones and hear as sound from our MP3 players. Could the discovery of dark energy mentioned above lead to a similar technological revolution sometime in the future? It is certainly possible, but the details of that revolution would be very difficult to envision today, just as Thomson could not envision the impact his discovery of the electron would have throughout the 20th and 21st centuries.

The processes for devising and using new rules

Physics is an experimental science. To answer questions, physicists do not just think and dream in their offices but constantly engage in experimental investigations. Physicists use special measuring devices to observe phenomena (natural and planned), describe their observations (carefully record them using words, numbers, graphs, etc.), find repeating features called patterns (for example, the distance traveled by a falling object is directly proportional to the square of the time in flight), and then try to explain these patterns. By doing this, physicists answer the questions of "why" or "how" the phenomenon happened and then deduce the rules that explain the phenomenon.

However, a deduced rule is not automatically accepted as true. Every rule needs to undergo careful testing. When physicists test a rule, they use the rule to predict the outcomes of new experiments. As long as there is no experiment whose outcome is inconsistent with predictions made using the rule, the rule is not disproved. The rule is consistent with all experimental evidence gathered so far. However, a new experiment could be devised tomorrow whose outcome is not consistent with the prediction made using the rule. The point is that there is no way to "prove" a rule once and for all. At best, the rule just hasn't been disproven yet.

A simple example will help you understand some processes that physicists follow when they study the world. Imagine that you walk into the house of your acquaintance Bob and see 10 tennis rackets of different quality and sizes. This is an **observational experiment**. During an observational experiment a scientist collects data that seem important. Sometimes it is an accidental or unplanned experiment. The scientist has no prior expectation of the outcome. In this case the number of tennis rackets and their quality and sizes represent

Newton's laws. Thanks to Newton, we can explain the motion of the Moon. We can also build skyscrapers.

the data. Having so many tennis rackets seems unusual to you, so you try to explain the data you collected (or, in other words, to explain why Bob has so many rackets) by devising several hypotheses. A **hypothesis** is an explanation of some sort that usually is based on some mechanism that is behind what is going on. One hypothesis is that Bob has lots of children and they all play tennis. A second hypothesis is that he makes his living by fixing tennis rackets. A third hypothesis is that he is a thief and he steals tennis rackets.

How do you decide which hypothesis is correct? You reason: if Bob has many children, and I walk around the house checking the sizes of clothes that I find, then I will find clothes of different sizes. Checking the clothing sizes is a new experiment, called a **testing experiment**. A testing experiment is different from an observational experiment. In a testing experiment, a specific hypothesis is being "put on trial." This hypothesis is used to construct a clear expectation of the outcome of the experiment. This clear expectation (based on the hypothesis being tested) is called a **prediction**. So, you conduct the testing experiment and walk around the house checking the closets. You do find clothes of different sizes. This is the outcome of your testing experiment. Does it mean for absolute certain that Bob has the rackets because all of his children play tennis? He could still be a racket repairman or a thief. Therefore, if the outcome of the testing experiment matches the prediction based on your hypothesis, you cannot say that you proved the hypothesis. All you can say is that you failed to disprove it. However, if you walk around the house and do not find any children's clothes, you can say with more confidence that the number of rackets in the house is not due to Bob having lots of children who play tennis. Still, this conclusion would only be valid if you made an **assumption:** Bob's children live in the house and wear clothes of different sizes. Generally, in order to reject a hypothesis you need to check the additional assumptions you made and determine if they are reasonable.

Imagine you have rejected the first hypothesis (you didn't find any children's clothes). Next you wish to test the hypothesis that Bob is a thief. This is your reasoning: If Bob is a thief (the hypothesis), and I walk around the house checking every drawer (the testing experiment), I should not find any receipts for the tennis rackets (the prediction). You perform the experiment and you find no receipts. Does it mean that Bob is a thief? He might just be a disorganized father of many children or a busy repairman. However, if you find all of the receipts, you can say with more confidence that he is not a thief (but he could still be a repairman). Thus it is possible to disprove (rule out) a hypothesis, but it is not possible to prove it once and for all. The process that you went through to create and test your hypotheses is depicted in **Figure 1.1**. At the end of your investigation you might be left with a hypothesis that you failed to disprove. As a physicist you would now have some confidence in this hypothesis and start using it to solve other problems.

> **TIP** Notice the difference between a hypothesis and a prediction. A *hypothesis* is an idea that explains why or how something that you observe happens. A *prediction* is a statement of what should happen in a particular experiment if the hypothesis being tested were true. The prediction is based on the hypothesis and cannot be made without a specific experiment in mind.

Using this book you will learn physics by following a process similar to that described above. Throughout the book are many observational experiments and descriptions of the patterns that emerge from the data. After you read about the observational experiments and patterns, think about possible

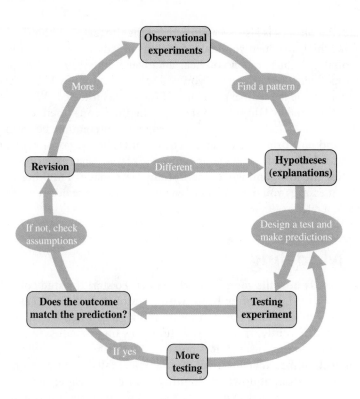

Figure I.1 Science is a cyclical process for creating and testing knowledge.

explanations for these patterns. The book will then describe possible experiments to test the proposed explanations and also the predicted outcomes based on the hypotheses being tested. Then it will describe the outcomes of the actual experiments. Sometimes the outcomes of the actual experiments will match the predicted outcomes, and sometimes they will not. Based on the experimental results and the analysis of the assumptions that were made, the book will help you make a judgment about the hypothesis being tested.

What language do physicists use?

Physicists use words and the language of mathematics to express ideas about the world. But they also represent these ideas and the world itself in other ways—sketches, diagrams, and even cut-out paper models (James Watson made a paper model of DNA when trying to determine its structure). In physics, however, the ultimate goal is to understand the mechanisms behind physical phenomena and to devise mathematical rules that allow for quantitative predictions of new phenomena. Thus, a big part of physics is identifying measurable properties of the phenomena (**physical quantities**, such as mass, speed, force), collecting quantitative data, and finding the patterns in that data.

How will learning physics change your interactions with the world?

Even if you do not plan on becoming a professional physicist, learning physics can change the way you think about the world. For example, why do you feel cold when you wear wet clothes? Why is it safe to sit in a car during a lightning storm? Why, when people age, do they have trouble reading small-sized fonts? Why are parts of the world experiencing more extreme climate events? Knowing physics will also help you understand what underlies many important technologies. How does an MRI work? How can a GPS know your present position and guide you to a distant location? How do power plants generate electric energy?

Studying physics is also a way to acquire the processes of knowledge construction, which will help you make decisions based on evidence rather than on personal opinions. When you hear an advertisement for a shampoo that makes your hair 97.5% stronger, you will ask: How do they know this? Did this number come from an experiment? If it did, was it tested? What assumptions did they make? Did they control for the food consumed, exercise, air quality, etc.? Understanding physics will help you differentiate between actual evidence and unsubstantiated claims. For instance, before you accept a claim, you might ask about the data supporting the claim, what experiments were used to test the idea, and what assumptions were made. Thinking critically about the messages you hear will change the way you make decisions as a consumer and a citizen.

Modeling. Physicists often model complex structures as point-like objects.

(a)

(b)

(c)

I.2   Modeling

Physicists study how the complex world works. To start the study of some aspect of that world, they often begin with a simplified version. Take a common phenomenon: a falling leaf. It twists and zigzags when falling. Different leaves move differently. If we wish to study how objects fall, a leaf is a complicated object to use. Instead, it is much easier to start by observing a small, round object falling. When a small, round object falls, all of its points move the same way—straight down. As another example, consider how you move your body when you walk. Your back foot on the pavement lifts and swings forward, only to stop for a short time when it again lands on the pavement, now ahead of you. Your arms swing back and forth. The trunk of your body moves forward steadily. Your head also moves forward but bobs up and down slightly, especially if you run. It would be very difficult to start our study of motion by analyzing all these complicated parts and movements. Thus, physicists create in their minds simplified representations (called **models**) of physical phenomena and then think of the phenomena in terms of those models. Physicists begin with very simple models and then add complexity as needed to investigate more detailed aspects of the phenomena.

A simplified object

To simplify real objects, physicists often neglect both the dimensions of objects (their sizes) and their structures (the different parts) and instead regard them as single **point-like objects**.

Is modeling a real object as a point-like object a good idea? Imagine a 100-meter race. The winner is the first person to get a body part across the finish line. It might be a runner's toe or it might be the head. The judge needs to observe the movement of all body parts (or a photo of the parts) across a very small distance near the finish line to decide who wins. Here, that very small distance near the finish line is small compared to the size of the human body. This is a situation where modeling the runners as point-like objects is not reasonable. However, if you are interested in how long it takes a person to run 100 meters, then the movement of different body parts is not as important, since 100 meters is much larger than the size of a runner. In this case, the runners can be modeled as point-like objects. Even though we are talking about the same situation (a 100-meter race), the aspect of the situation that interests us determines how we choose to model the runners.

Consider an airplane landing on a runway (**Figure I.2a**). We want to determine how long it takes for it to stop. Since all of its parts move together, the part we study does not matter. In that case it is reasonable to model the airplane as a point-like object. However, if we want to build a series of gates for planes to unload passengers (Figure I.2b), then we need to consider the

Figure I.2 An airplane can be considered a point-like object (a) when landing, (b) but not when parking.

(a)   **(b)**

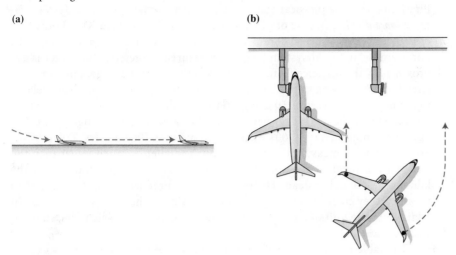

motion of the different parts of the airplane. For example, there must be enough room for an airplane to turn while maneuvering into and out of the gate. In this case the airplane cannot be modeled as a point-like object.

> **Point-like object** A point-like object is a simplified representation of a real object. As a rule of thumb, you can model a real object as a point-like object when one of the following two conditions are met: (a) when all of its parts move in the same way, or (b) when the object is much smaller than the other relevant lengths in the situation. The same object can be modeled as a point-like object in some situations but not in others.

Modeling

The process that we followed to decide when a real object could be considered a point-like object is an example of what is called **modeling**. The modeling of objects is a first step that physicists use when they study natural phenomena. In addition to simplifying the objects that they study, scientists simplify the interactions between objects and also the processes that are occurring in the real world. Then they add complexity as their understanding grows. Galileo Galilei is believed to be the first scientist to consciously model a phenomenon. In his studies of falling objects in the early 17th century, he chose to simplify the real phenomenon by ignoring the interactions of the falling objects with the air.

> **Modeling** A model is a simplified representation of an object, a system (a group of objects), an interaction, or a process. A scientist creating the model decides which features to include and which to neglect.

I.3 Physical quantities

To describe physical phenomena quantitatively, physicists construct **physical quantities**: features or characteristics of phenomena that can be measured experimentally. Measurement means comparing the characteristic to an assigned **unit** (a chosen standard).

Units of measure

Physicists describe physical quantities using the **SI system**, or *Le Système international d'unités*, whose origin goes back to the 1790s when King Louis XVI of France created a special commission to invent a new metric system of units. For example, in the SI system length is measured in meters. One meter is approximately the distance from your nose to the tip of the fingers of your outstretched arm. A long step is about one meter. Other units of length are related to the meter by powers of 10 using prefixes (milli, kilo, nano, ...). These prefixes relate smaller or bigger versions of the same unit to the basic unit. For example, 1 millimeter is 0.001 meter; 1 kilometer is 1000 meters. The prefixes are used when a measured quantity is much smaller or much larger than the basic unit. If the distance is much larger than 1 m, you might want to use the kilometer (10^3 m) instead. The most common prefixes and the powers of 10 to which they correspond are given on the inside of the book's back cover. In addition to the unit of length, the SI system has six other basic units, summarized in **Table 1.1**.

The table provides a "feel" for some of the units but does not say exactly how each unit is defined. More careful definitions are important in order that measurements made by scientists in different parts of the world are consistent. However, to understand the precise definitions of these units, one needs to know more physics. We will learn how each unit is precisely defined when we investigate the concepts on which the definition is based.

Measuring instruments

Physicists use a measuring instrument to compare the quantity of interest with a standardized unit. Each measuring instrument is calibrated so that it reads in multiples of that unit. Some examples of measuring instruments are a thermometer to measure temperature (calibrated in degrees Celsius or degrees Fahrenheit), a watch to measure time intervals (calibrated in seconds), and a

Table 1.1 Basic SI physical quantities and their units.

Physical quantity	Unit name and symbol	Physical description
Time	Second, s	One second is the time it takes for the heart to beat once.
Length	Meter, m	One meter is the length of one stride.
Mass	Kilogram, kg	One kilogram is the mass of 1 liter of water.
Electric current	Ampere, A	One ampere is the electric current through a 100-watt lightbulb in an American household
Temperature	Kelvin, K	One Kelvin degree is the same as 1 degree on the Celsius scale or about 2 degrees on the Fahrenheit scale.
Amount of matter	Mole, mol	One mole of oxygen is about 32 g.
Intensity of light	Candela, cd	One candela is the intensity of light produced by a relatively large candle at a distance of 1 m.

meter stick to measure the height of an object (calibrated in millimeters). We can now summarize these ideas about physical quantities and their units.

> **Physical quantity** A physical quantity is a feature or characteristic of a physical phenomenon that can be measured in some unit. A measuring instrument is used to make a quantitative comparison of this characteristic with a unit of measure. Examples of physical quantities are your height, your body temperature, the speed of your car, and the temperature of air or water.

Significant digits

When we measure a physical quantity, the instrument we use and the circumstances under which we measure it determine how precisely we know the value of that quantity. Imagine that you wear a pedometer (a device that measures the number of steps that you take) and wish to determine the number of steps on average that you take per minute. You walk for 26 min (as indicated by your analog wristwatch) and see that the pedometer shows 2254 steps. You divide 2254 by 26 using your calculator, and it says 86.692307692307692. If you accept this number, it means that you know the number of steps per minute within plus or minus 0.000000000000001 steps/min. If you accept the number 86.69, it means that you know the number of steps to within 0.01 steps/min. If you accept the number 90, it means that you know the number of steps within 10 steps/min. Which answer should you use?

To answer this question, let's first focus on the measurements. Although your watch indicated that you walked for 26 min, you could have walked for as few as 25 min or for as many as 27 min. The number 26 does not give us enough information to know the time more precisely than that. The time measurement 26 min has two **significant digits**, or two numbers that carry meaning contributing to the precision of the result. The pedometer measurement 2254 has four significant digits. Should the result of dividing the number of steps by the amount of time you walked have two or four significant digits? If we accept four, it means that the number of steps per minute is known more precisely than the time measurement in minutes. This does not make sense. The number of significant digits in the final answer should be the same as the number of significant digits of the quantity used in the calculation that has the *smallest number of significant digits*. Thus, in our example, the average number of steps per minute should be 86, plus or minus 1 step/min: 86 ± 1.

Let's summarize the rule for determining significant digits. The precision of the value of a physical quantity is determined by one of two cases. If the quantity is measured by a single instrument, its precision depends on the instrument used to measure it. If the quantity is calculated from other measured quantities, then its precision depends on the least precise instrument out of all the instruments used to measure a quantity used in the calculation.

Another issue with significant digits arises when a quantity is reported with no decimal points. For example, how many significant digits does 6500 have—two or four? This is where scientific notation helps. **Scientific notation** means writing numbers in terms of their power of 10. For example, we can write 6500 as 6.5×10^3. This means that the 6500 actually has two significant digits: 6 and 5. If we write 6500 as 6.50×10^3, it means 6500 has three significant digits: 6, 5, and 0. The number 6.50 is more precise than the number 6.5, because it means that you are confident in the number to the hundredths place. Scientific notation provides a compact way of writing large and small numbers and also allows us to indicate unambiguously the number of significant digits a quantity has.

Measuring and estimating. In everyday life, rough estimates are often sufficient.

I.4   Making rough estimates

Sometimes we are interested in making a rough estimate of a physical quantity. The ability to make rough estimates is useful in a variety of situations, such as the following. (1) You need to decide whether a goal is worth pursuing—for example, can you make a living as a piano tuner in a town that already has a certain number of tuners? (2) You need to know roughly the amount of material needed for some activity—for instance, the food needed for a party or the number of bags of fertilizer needed for your lawn. (3) You want to estimate a number before it is measured—for example, how rapid a time-measuring device should be to detect laser light reflected from a distant object. (4) You wish to check whether a measurement you have made is reasonable—for instance, the measurement of the time for light to travel to a mountain and back or the mass of oxygen consumed by a hummingbird. (5) You wish to determine an unknown quantity—for example, an estimation of the number of cats in the United States or the compression force on the disks in your back when lifting a box of books in different ways.

The procedure for making rough estimates usually means selecting some basic physical quantities whose values are known or can be estimated and then combining the numbers using a mathematical procedure that leads to the desired answer. For example, suppose we want to estimate the number of pounds of food that an average person eats during a lifetime. First, assume that the average person consumes about 2000 calories/day to maintain a healthy metabolism. This food consists of carbohydrates, proteins, and fat. Using the labels of food packaging we find that 1 gram of carbohydrate or 1 gram of protein gives us about 4 calories, and 1 g of fat gives us about 9 calories. Thus, we assume that each gram of food consumed gives us on average 5 calories. Thus, each day, according to our assumptions, the person consumes about

$$(\, 2000 \text{ calories/day} \,)(\, 1 \text{ g/5 calories} \,) \; = \; 400 \text{ g/day}$$

There are 365 days in a year, and we assume that the average life expectancy of a person is 70 years. Thus, the total food consumed during a lifetime is

$$(\, 2000 \text{ calories/day} \,)(\, 1 \text{ g/5 calories} \,)(\, 365 \text{ days/year} \,)(\, 70 \text{ years/lifetime} \,)$$
$$= \; 10{,}080{,}000 \text{ g/lifetime}$$

From the conversion table on the inside front cover, we see that 2.2 lb is 1000 g. Thus, our estimate of the number of pounds of food consumed in a lifetime is

$$(\, 10{,}080{,}000 \text{ g/lifetime} \,)(\, 2.2 \text{ lb/1000 g} \,) \; = \; 22{,}176 \text{ lb/lifetime}$$

Our estimated result has five significant digits. Is this appropriate? To answer this, we need to look at the number of significant digits in each quantity used in the estimate. The calories/day quantity is probably uncertain by about 500 calories. That means the 2000 calories/day has just one significant digit. The ratio $(\, 1 \text{ g/5 calories} \,) = 0.20$ could probably be $(\, 1 \text{ g/6 calories} \,) = 0.17$, or about 0.03 different from our estimate. So that quantity has one or two significant digits. The life expectancy (70 years) could be off by about 10 years. Again, that's just one significant digit. Since the least certain quantity used in the calculation (the calories/day) has one significant digit, the final result should also be reported with just one significant digit. That would be 20,000 lb/lifetime.

I.5   Vector and scalar physical quantities

There are two general types of physical quantities—those that contain information about magnitude as well as direction and those that contain magnitude information only. Physical quantities that do not contain information about direction are called **scalar quantities** and are written using *italic* symbols (m, T, etc). Mass is a scalar quantity, as is temperature. To manipulate scalar quantities, you use standard arithmetic and algebra rules—addition, subtraction, multiplication, division, etc. You add, subtract, multiply, and divide scalars as though they were ordinary numbers.

Physical quantities that contain information about magnitude and direction are called **vector quantities** and are represented by italic symbols with an arrow on top ($\vec{F}$, $\vec{v}$, etc.). The little arrow on top of the symbol always points to the right. The actual direction of the vector quantity is shown in a diagram. For example, force is a physical quantity with both magnitude and direction (direction is very important if you are trying to hammer a nail into the wall). When you push a door, your push can be represented with a force arrow on a diagram; the stronger you push, the longer that arrow must be. The direction of the push is represented by the direction of that arrow (**Figure I.3**). The arrow's direction indicates the direction of the vector, and the arrow's relative length indicates the vector's magnitude. The methods for manipulating vector quantities (adding and subtracting them as well as multiplying a vector quantity by a scalar quantity and multiplying two vector quantities) are introduced as needed in the following chapters. Such manipulations are also summarized in the appendix Working with Vectors.

Figure I.3 The force that your hand exerts on a door is a vector quantity represented by an arrow.

$\vec{F}_{\text{Hand on Door}}$

Scalar quantities. Temperature is a scalar quantity; it has magnitude, but not direction.

I.6 How to use this book to learn physics

A textbook is only one part of a learning system, but knowing how to use it most effectively will make it easier to learn and to succeed in the course. This textbook will help you construct understanding of some of the most important ideas in physics, learn to use physics knowledge to analyze physical phenomena, and develop the general process skills that scientists use in the practice of science.

Learning new material

Read the book as soon as possible after new material is discussed in class while the material is still fresh in your mind. First, scan the relevant sections and, if necessary, the whole chapter. Does it appear that the material involves completely new ideas, or is it just the application of what you have already learned? If the material does involve new ideas, how do the new ideas fit into what you have already learned? Then, read the relevant new section(s) slowly. Keep relating what you read to your current understanding. Pay attention to the Tips—they will help prevent confusion and future difficulties.

The most important strategy that will help you learn better is called **interrogation.** Interrogation means continually asking yourself the same question when reading the text. This question is so important that we put it in the box below:

> ## Why is this true?

Make sure that you ask yourself this question as often as possible so that eventually it becomes a habit. Out of all the strategies that are recommended for reading comprehension, this is the one that is directly connected to better learning outcomes. For example, consider the first sentence of the next paragraph: "Solving physics problems is much more than plugging numbers into an equation." Ask yourself, "Why is this true?" Possibly, because one needs to understand what physics concepts are relevant, or what simplifying assumptions are important. There can be other reasons. By just stopping and interrogating yourself as often as possible about what is written in the book you will be able to understand and remember this information better.

Problem solving

Solving physics problems is much more than plugging numbers into an equation. To use the book for problem-solving practice, focus on the problem-solving steps used in the worked examples.

Step 1: *Sketch and Translate* First, read the text of the problem several times slowly to make sure you understand what it says. Next, try to visualize the situation or process described in the text of the problem. Try to imagine what is happening. Draw a sketch of the process and label it with any information you have about the situation. This often involves an initial situation and a final situation. Often, the information in the problem statement is provided in words and you will need to *translate* it into physical quantities. Having the problem information in a visual sketch also frees some of your mind so that you can use its resources for other parts of the problem solving.

Step 2: *Simplify and Diagram* Decide how you can simplify the process. How will you model the object of interest (the object you are investigating)? What interactions can you neglect? To diagram means to represent the problem process using some sort of diagram, bar chart, graph, or picture that includes physics quantities. Diagrams bridge the gap between the verbal and sketch representations of the process and the mathematical representation of it.

Step 3: *Represent Mathematically* Construct a mathematical description of the process. You will use the sketch from Step 1 and the diagram(s) from Step 2 to help construct this mathematical description and evaluate it to see if it is reasonable. By representing the situation in these multiple ways and learning to translate from one way to the other, you will start giving meaning to the abstract symbols used in the mathematical description of the process.

Step 4: *Solve and Evaluate* Finally, solve the mathematical equations and evaluate the results. Do the numbers and signs make sense? Are the units correct? Another method involves evaluating whether the answer holds in extreme cases—you will learn more about this technique as you progress through the book.

Try to solve the example problems that are provided in the chapters by using this four-step strategy without looking at the solution. After finishing, compare your solution to the one described in the book. Then do the *Try It Yourself* part of the example problem and compare your answer to the book's answer. If you are still having trouble, try to use the same strategy to actively solve other example problems in the text or from the Active Learning Guide (if you are using that companion book). Uncover the solutions to these worked examples only after you have tried to complete the problem on your own. Then try the same process on the homework problems assigned by your instructor.

Notice that quantitative exercises and conceptual exercises in the book have fewer steps: *Represent Mathematically* and *Solve and Evaluate* for the quantitative exercises, and *Sketch and Translate* and *Simplify and Diagram* for the conceptual exercises. Sketching the process and representing it in different ways is an important step in solving any problem.

Summary

We are confident that this book will act as a useful companion in your study of physics and that you will take from the course not just the knowledge of physics but also an understanding of the process of science that will help you in all your scientific endeavors. Learning physics through the approach used in this book builds a deeper understanding of physics concepts and an improved ability to solve difficult problems compared to traditional learning methods. In addition, you will learn to reason scientifically and be able to transfer those reasoning skills to many other aspects of your life.

1 Kinematics: Motion in One Dimension

What is a safe following distance between you and the car in front of you?

Can you be moving and not moving at the same time?

Why do physicists say that an upward thrown object is falling?

Be sure you know how to:

- Define what a point-like object is (Introducing Physics).
- Use significant digits in calculations (Introducing Physics).

When you drive, you are supposed to follow the 3-second tailgating rule. When the car in front of you passes some fixed sign at the side of the road, your car should be far enough behind so that it takes you 3 seconds to reach the same sign. You then have a good chance of avoiding a collision if the car in front stops abruptly. If you are 3 seconds behind the car in front of you when you see its brake lights, you should be able to step on the brake and avoid a collision. If you are closer than 3 seconds away, a collision is likely. In this chapter we will learn the physics behind the 3-second rule.

Scientists often ask questions about things that most people accept as being "just the way it is." For example, in the northern hemisphere, we have more hours of daylight in June than in December. In the southern hemisphere, it's just the opposite. Most people simply accept this fact. However, scientists want explanations for such simple phenomena. In this chapter, we learn to describe a phenomenon that we encounter every day but rarely question—motion.

1.1 What is motion?

When describing motion, we need to focus on two important aspects: the object whose motion we are describing (**the object of interest**) and the person who is doing the describing (**the observer**). Consider Observational Experiment **Table 1.1**, which analyzes how the description of an object's motion depends on the observer.

OBSERVATIONAL EXPERIMENT TABLE

1.1   Different observers describe an object's motion.

Observational experiment	Analysis
Experiment 1. Jan observes a ball in her hands as she walks across the room. Tim, sitting at a desk, also observes the ball. Jan reaches the other side of the room without taking her eyes from the ball; her head did not turn. Tim's head has turned in order to follow the ball.	The two observers (Jan and Tim) see the same object of interest (the ball) differently. With respect to Jan, the ball's position does not change. With respect to Tim, its position does change.

(continued)

Observational experiment	Analysis
Experiment 2. Ted and Sue are passengers on the same train. Ted does not have to turn his head to keep his eyes on Sue. Joan, standing on the station platform, turns her head to follow Sue.	Ted and Joan see the same object of interest (Sue) differently. With respect to Ted, Sue's position does not change. With respect to Joan, Sue's position changes.

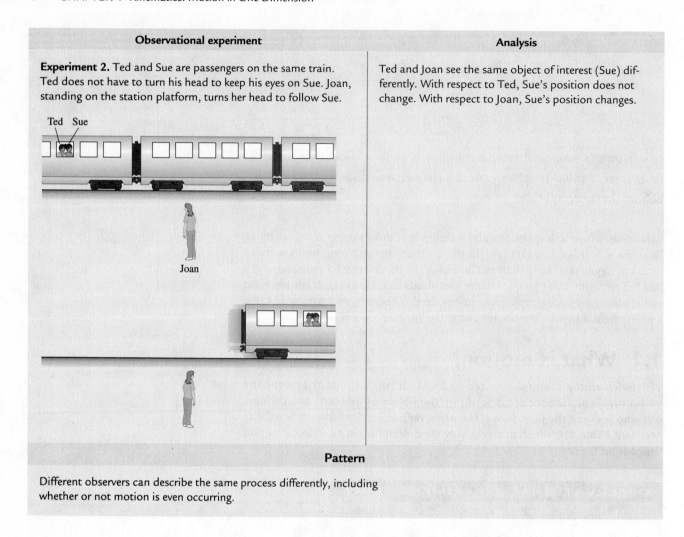

Pattern
Different observers can describe the same process differently, including whether or not motion is even occurring.

In Table 1.1, we saw that different observers can describe the same process differently. One person sees the object of interest moving while another does not. They are both correct from their own perspectives. In order to describe the motion of something, we need to identify the observer.

> **Motion** is a change in an object's position relative to a given observer during a certain change in time. Without identifying the observer, it is impossible to say whether the object of interest moved. Physicists say *motion is relative,* meaning that the motion of any object of interest depends on the point of view of the observer.

Are you moving as you read this book? Your friend walking past you first sees you in front of her, then she sees you next to her, and finally she sees you behind her. Though you are sitting in a chair, you definitely are moving with respect to your friend. You are also moving with respect to the Sun or with respect to a bird flying outside.

What makes the idea of relative motion confusing at first is that people intuitively use Earth as the object of reference—the object with respect to which they describe motion. If an object does not move with respect to Earth, many people would say that the object is not moving. That is why it took scientists thousands of years to understand the reason for days and nights on Earth. An observer on Earth uses Earth as the object of reference and sees the Sun moving in an arc across the sky (**Figure 1.1a**). An observer on a distant spaceship sees Earth rotating on its axis so that different parts of its surface face the Sun at different times (Figure 1.1b).

Reference frames

Specifying the observer before describing the motion of an object of interest is an extremely important part of constructing what physicists call a **reference frame.** A reference frame includes an object of reference, a coordinate system with a scale for measuring distances, and a clock to measure time. If the object of reference is large and cannot be considered a point-like object, it is important to specify where on the object of reference the origin of the coordinate system is placed. For example, if you want to describe the motion of a bicyclist and choose your object of reference to be Earth, you place the origin of the coordinate system at the surface, not at Earth's center.

> **Reference frame** A reference frame includes three essential components:
> (a) An *object of reference* with a specific *point of reference* on it.
> (b) A *coordinate system*, which includes one or more coordinate axes, such as, *x, y, z,* and an origin located at the point of reference. The coordinate system also includes a unit of measurement (a scale) for specifying distances along the axes.
> (c) A *clock,* which includes an origin in time called $t = 0$ and a unit of measurement for specifying times and time intervals.

Modeling motion

When we model objects, we make simplified assumptions in order to analyze complicated situations. Just as we simplified an object to model it as a point-like object, we can also simplify a process involving motion. What is the simplest way an object can move?

  Imagine that you haven't ridden a bike in a while. You would probably start by riding in a straight line before you attempt a turn. This kind of motion is called **linear motion** or **one-dimensional motion**.

> **Linear motion** is a model of motion that assumes that an object, considered as a point-like object, moves along a straight line.

  For example, we want to model a car's motion along a straight stretch of highway. We can assume the car is a point-like object (it is small compared to the length of the highway) and the motion is linear motion (the highway is long and straight).

Review Question 1.1 Physicists say, "Motion is relative." Why is this true?

1.2 A conceptual description of motion

To describe linear motion more precisely, we start by devising a visual representation. Consider Observational Experiment **Table 1.2**, in which a bowling ball rolls on a smooth floor.

Figure 1.1 Motion is relative. Two observers explain the motion of the Sun relative to Earth differently.

(a)

An observer on Earth sees the Sun move in an arc across the sky.

(b)

An observer in a spaceship sees the person on Earth as rotating under a stationary Sun.

OBSERVATIONAL EXPERIMENT TABLE

1.2   Using dots to represent motion.

VIDEO 1.2

Observational experiment	Analysis
Experiment 1. You push a bowling ball (the object of interest) once and let it roll on a smooth linoleum floor. You place beanbags each second beside the bowling ball. The beanbags are evenly spaced.	We can represent the locations of the bags each second for the slow-moving bowling ball as dots on a diagram.
Experiment 2. You repeat Experiment 1, but you push the ball harder before you let it roll. The beanbags are farther apart but are still evenly spaced.	The dots in this diagram represent the evenly spaced bags, which are separated by a greater distance than the bags in Experiment 1.
Experiment 3. You push the bowling ball and let it roll on a carpeted floor instead of a linoleum floor. The distance between the beanbags decreases as the ball rolls.	The dots in this diagram represent the decreasing distance between the bags as the ball rolls on the carpet.
Experiment 4. You roll the ball on the linoleum floor and gently and continually push on it with a board. The beanbag separation spreads farther apart as the pushed ball rolls.	The dots in this diagram represent the increasing distance between bags as the ball is continually pushed across the linoleum floor.

Pattern

- The spacing of the dots allows us to visualize motion.
- When the object travels without speeding up or slowing down, the dots are evenly spaced.
- When the object slows down, the dots get closer together.
- When the object moves faster and faster, the dots get farther apart.

Motion diagrams

In the experiments in Table 1.2, the beanbags were an approximate record of where the ball was located as time passed and help us visualize the motion of the ball. We can represent motion in even more detail by adding **velocity arrows** to each dot that indicate which way the object is moving and how fast it is moving as it passes a particular position (see **Figure 1.2**). These new diagrams are called **motion diagrams.** The longer the arrow, the faster the motion. The small arrow above the letter v indicates that this characteristic of motion has a direction as well as a magnitude—called a **vector quantity.** In Figure 1.2a, the dots are evenly spaced, and the velocity arrows all have the same length and point in the same direction. This means that the ball was moving equally fast in the same direction at each point. Similar diagrams with velocity arrows for the other three experiments in Table 1.2 are shown in Figures 1.2b–d.

Velocity change arrows

In Experiment 4 the bowling ball was moving increasingly fast while being pushed. The velocity arrows in the motion diagram thus got increasingly longer. We can represent this change with a **velocity change arrow** $\Delta \vec{v}$. The Δ (delta) means a change in whatever quantity follows the Δ, a change in $\vec{v}$ in this case. The $\Delta \vec{v}$ doesn't tell us the exact increase or decrease in the velocity; it only indicates a qualitative difference between the velocities at two adjacent points in the diagram.

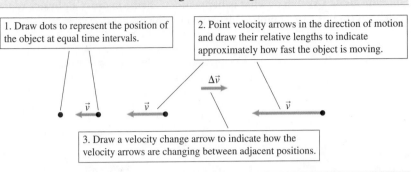

REASONING SKILL Constructing a motion diagram.

1. Draw dots to represent the position of the object at equal time intervals.

2. Point velocity arrows in the direction of motion and draw their relative lengths to indicate approximately how fast the object is moving.

3. Draw a velocity change arrow to indicate how the velocity arrows are changing between adjacent positions.

Figure 1.2 Motion diagrams represent the types of motion shown in Table 1.2.

(a)

The dots represent the positions of the ball at regular times.

The velocity arrows represent how fast the ball is moving and its direction.

(b)

This ball is moving at a constant but faster speed than the ball in (a).

(c)

The ball is slowing down.

(d)

The ball is speeding up.

Note that we have redrawn the diagram shown in Figure 1.2d in **Figure 1.3a**. For illustration purposes only, we number the $\vec{v}$ arrows consecutively for each position: $\vec{v}_1, \vec{v}_2, \vec{v}_3$, etc. To draw the velocity change arrow as the ball moves from position 2 to position 3 in Figure 1.3a, we place the second arrow $\vec{v}_3$ directly above the first arrow $\vec{v}_2$, as shown. The $\vec{v}_3$ arrow is longer than the $\vec{v}_2$ arrow. This tells us that the object was moving faster at position 3 than at position 2. To visualize the change in velocity, we need to think about how arrow $\vec{v}_2$ can be turned into $\vec{v}_3$. We can do it by placing the tail of a velocity change arrow $\Delta\vec{v}_{23}$ at the head of $\vec{v}_2$ so that the head of $\Delta\vec{v}_{23}$ makes the combination $\vec{v}_2 + \Delta\vec{v}_{23}$ the same length as $\vec{v}_3$ (Figure 1.3b). Since they are the same length and in the same direction, the two vectors $\vec{v}_2 + \Delta\vec{v}_{23}$ and $\vec{v}_3$ are equal:

$$\vec{v}_2 + \Delta\vec{v}_{23} = \vec{v}_3$$

Note that if we move $\vec{v}_2$ to the other side of the equation, then

$$\Delta\vec{v}_{23} = \vec{v}_3 - \vec{v}_2$$

Thus, $\Delta\vec{v}_{23}$ is the difference of the third velocity arrow and the second velocity arrow—the change in velocity between position 2 and position 3. (To learn more about vector addition, read the appendix Graphical Addition and Subtraction of Vectors.)

Making a complete motion diagram

We now place the $\Delta\vec{v}$ arrows above and between the dots in our diagrams where the velocity change occurred (see **Figure 1.4a**). The dots in these more detailed motion diagrams indicate the object's position at equal time intervals; velocity arrows and velocity change arrows are also included. A $\Delta\vec{v}$ arrow points in the same direction as the $\vec{v}$ arrows when the object is speeding up; the $\Delta\vec{v}$ arrow points in the opposite direction of the $\vec{v}$ arrows when the object is slowing down. When velocity changes by the same amount during each consecutive time interval, the $\Delta\vec{v}$ arrows for each interval are the same length. In such cases we need only one $\Delta\vec{v}$ arrow for the entire motion diagram (see Figure 1.4b).

The Reasoning Skill box summarizes the procedure for constructing a motion diagram. Notice that in the experiment represented in this diagram, the object is moving from right to left and slowing down.

Figure 1.3 Determining the magnitude and the direction of the velocity change arrow in a motion diagram.

(a)

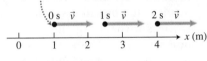

The ball is speeding up.

$\vec{v}_1$ $\vec{v}_2$ $\vec{v}_3$ $\vec{v}_4$

$\vec{v}_3$

How can we represent the change in velocity from 2 to 3?

$\vec{v}_2$

(b)

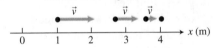

$\vec{v}_3$ $\vec{v}_2 + \Delta\vec{v}_{23} = \vec{v}_3$

$\vec{v}_2$ $\Delta\vec{v}_{23}$ $\Delta\vec{v}_{23} = \vec{v}_3 - \vec{v}_2$

Add the $\Delta\vec{v}_{23}$ change arrow to $\vec{v}_2$ to get $\vec{v}_3$.

Figure 1.4 Two complete motion diagrams, including position dots, $\vec{v}$ arrows, and $\Delta\vec{v}$ arrows.

(a)

A motion diagram showing the velocity change for each consecutive position change

$\Delta\vec{v}_{12}$ $\Delta\vec{v}_{23}$ $\Delta\vec{v}_{34}$

$\vec{v}_1$ $\vec{v}_2$ $\vec{v}_3$ $\vec{v}_4$

1   2   3   4

(b)

When the velocity change is constant from time interval to time interval, we need only one $\Delta\vec{v}$ for the diagram.

$\Delta\vec{v}$

$\vec{v}$ $\vec{v}$ $\vec{v}$ $\vec{v}$

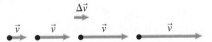

TIP When drawing a motion diagram, always specify the position of the observer. In the Reasoning Skill box, the observer is on the ground.

Read Conceptual Exercise 1.1 several times and visualize the situation. If possible, draw a sketch of what is happening. Then construct a physics representation (in this case, a motion diagram) for the process.

CONCEPTUAL EXERCISE 1.1
Driving in the city

A car at rest at a traffic light starts moving faster and faster when the light turns green. The car reaches the speed limit in 4 seconds, continues at the speed limit for 3 seconds, then slows down and stops in 2 seconds while approaching the second stoplight. There, the car is at rest for 1 second until the light turns green. Meanwhile, a cyclist approaching the first green light keeps moving without slowing down or speeding up. She reaches the second stoplight just as it turns green. Draw a motion diagram for the car and another for the bicycle as seen by an observer on the ground. If you place one diagram below the other, it will be easier to compare them.

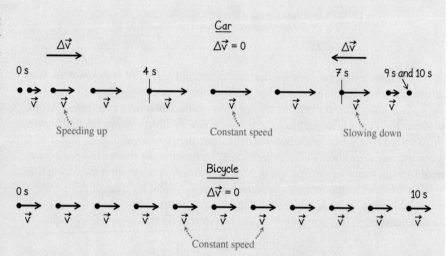

Sketch and translate Visualize the motion for the car and for the bicycle as seen by the observer on the ground. The car and the bicycle will be our objects of interest.

The motion of the car has four distinct parts:

1. starting at rest and moving faster and faster for 4 seconds;

2. moving at a constant rate for 3 seconds;

3. slowing down to a stop for 2 seconds; and

4. sitting at rest for 1 second.

The bicycle moves at a constant rate with respect to the ground for the entire time.

Simplify and diagram We can model the car and the bicycle as point-like objects (dots). In each motion diagram, there will be 11 dots, one for each second of time (including one for time zero). The last two dots for the car will be on top of each other since the car was at rest from time = 9 s to time = 10 s. The dots for the bicycle are evenly spaced.

Try it yourself: Two bowling balls are rolling along a linoleum floor. One of them is moving twice as fast as the other. At time zero, they are next to each other on the floor. Construct motion diagrams for each ball's motion during a time of 4 seconds, as seen by an observer on the ground. Indicate on the diagrams the locations at which the balls were next to each other at the same time. Indicate possible mistakes that a student can make answering the question above.

Answer: See the figure below. The balls are side by side only at time zero—the first dot for each ball. It looks like they are side by-side when at the 2-m position, but the slow ball is at the 2-m position at 2 s and the faster ball is there at 1 s. Similar reasoning applies for the 4-m positions—the balls reach that point at different times.

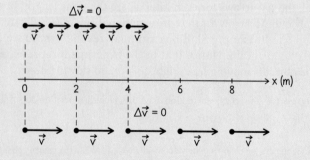

Review Question 1.2 What information about a moving object can we extract from a motion diagram?

1.3 Quantities for describing motion

A motion diagram helps represent motion qualitatively. To analyze situations more precisely, for example, to determine how far a car will travel after the brakes are applied, we need to describe motion quantitatively. In this section, we devise some of the quantities we need to describe linear motion.

Time and time interval

People use the word "time" to talk about the reading on a clock and how long a process takes. Physicists distinguish between these two meanings with different terms: time (a clock reading) and time interval (a difference in clock readings).

> **Time and time interval** *Time* (clock reading) t is the reading on a clock or some other time-measuring instrument. *Time interval* $(t_2 - t_1)$ or Δt is the difference of two times. In the SI system (metric units), the unit of time and of time interval is the second. Other units are minutes, hours, days, and years. Time and time interval are both scalar quantities.

Position, displacement, distance, and path length

Along with a precise definition for time and time interval, we need to precisely define four quantities that describe the location and motion of an object: position, displacement, distance, and path length.

> **Position, displacement, distance, and path length** The *position* of an object is its location with respect to a particular coordinate system (usually indicated by x or y). The *displacement* of an object, usually indicated by $\vec{d}$, is a vector that starts from an object's initial position and ends at its final position. The magnitude (length) of the displacement vector is called *distance d*. The *path length l* is how far the object moved as it traveled from its initial position to its final position. Imagine laying a string along the path the object took. The length of string is the path length.

Figure 1.5a shows a car's initial position x_i at initial time t_i. The car first backs up (moving in the negative direction) toward the origin of the coordinate system at $x = 0$. The car stops and then moves in the positive x-direction to its final position x_f. Notice that the *initial position* and the *origin* of a coordinate system are not necessarily the same points! The displacement $\vec{d}$ for the whole trip is a vector that points from the starting position at x_i to the final position at x_f (Figure 1.5b). The distance for the trip is the magnitude of the displacement (always a positive value). The path length l is the distance from x_i to 0 plus the distance from 0 to x_f (Figure 1.5c). Note that the path length does not equal the distance.

Scalar component of displacement for motion along one axis

To describe linear motion quantitatively we first specify a reference frame. For simplicity we can point one coordinate axis either parallel or antiparallel (opposite in direction) to the object's direction of motion. For linear motion,

Figure 1.5 Position, displacement, distance, and path length for a short car trip.

(a) Positions x_i and x_f

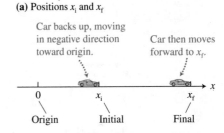

Car backs up, moving in negative direction toward origin.

Car then moves forward to x_f.

Origin   Initial   Final

(b) Displacement $\vec{d}$ and distance d

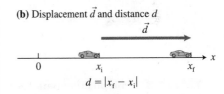

$$d = |x_f - x_i|$$

(c) Path length l

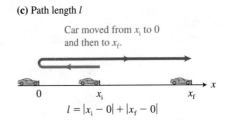

Car moved from x_i to 0 and then to x_f.

$$l = |x_i - 0| + |x_f - 0|$$

> **TIP** Sometimes we use the subscripts 1, 2, and 3 for times and the corresponding positions to communicate a sequence of different and distinguishable stages in any process, and sometimes we use i (initial) and f (final) to communicate the sequence.

Figure 1.6 Indicating an object's position at a particular time t, for example, $x(t)$.

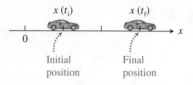

Figure 1.7 The x-component of displacement is (a) positive; (b) negative.

(a)

Positive displacement when the person moves in the positive direction

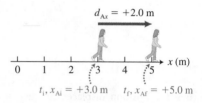

$d_{Ax} = +2.0$ m

$t_i, x_{Ai} = +3.0$ m $t_f, x_{Af} = +5.0$ m

(b)

Negative displacement when the person moves in the negative direction

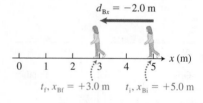

$d_{Bx} = -2.0$ m

$t_f, x_{Bf} = +3.0$ m $t_i, x_{Bi} = +5.0$ m

Table 1.3 Time-position data for linear motion.

Clock reading (time)	Position
$t_0 = 0.0$ s	$x_0 = 1.00$ m
$t_1 = 1.0$ s	$x_1 = 2.42$ m
$t_2 = 2.0$ s	$x_2 = 4.13$ m
$t_3 = 3.0$ s	$x_3 = 5.52$ m
$t_4 = 4.0$ s	$x_4 = 7.26$ m
$t_5 = 5.0$ s	$x_5 = 8.41$ m
$t_6 = 6.0$ s	$x_6 = 10.00$ m

Active Learning Guide›

we need only one coordinate axis to describe the object's changing position. In the example of the car trip, at the initial time t_i the car is at position $x(t_i)$, and at the final time t_f the car is at $x(t_f)$ (**Figure 1.6**). The notation $x(t)$ means the position x is a function of clock reading t (spoken "x of t"), not x multiplied by t. When we need to note a specific value of position x at a specific clock reading t_1, instead of writing, $x(t_1)$ we will write x_1. The same applies to x_i, x_f, etc. The vector that points from the initial position x_i to the final position x_f is the **displacement vector**.

The quantity that we determine through the operation $x_f - x_i$ is called the **x-scalar component of the displacement** vector and is abbreviated d_x (usually, we drop the term "scalar" and just call this the **x-component of the displacement**). **Figure 1.7a** shows that the initial position of person A is $x_{Ai} = +3.0$ m and the final position is $x_{Af} = +5.0$ m; thus the x-component of the person's displacement is

$$d_{Ax} = x_{Af} - x_{Ai} = (+5.0 \text{ m}) - (+3.0 \text{ m}) = +2.0 \text{ m}$$

The displacement is positive since the person moved in the positive x-direction. In Figure 1.7b, person B moved in the negative direction from initial position of $+5.0$ m to the final position of $+3.0$ m; thus the x-component of displacement of the person is negative:

$$d_{Bx} = x_{Bf} - x_{Bi} = (+3.0 \text{ m}) - (+5.0 \text{ m}) = -2.0 \text{ m}$$

Distance is always positive, as it equals the absolute value of the displacement $|x_f - x_i|$. In the example above, the displacements for A and B are different, but the distances are both $+2.0$ m (always positive).

Significant digits

Note that in Figure 1.7 the positions were written as $+3.0$ m, $+5.0$ m, etc. Could we have written them instead as $+3$ m and $+5$ m, or as $+3.00$ m and $+5.00$ m? The thickness of a human body from the back to the front is about 0.2 m (20 cm). Thus, we should be able to measure the person's location at one instant of time to within about 0.1 m but not to 0.01 m (1 cm). Thus, the locations can reasonably be given as $+3.0$ m, which implies an accuracy of ± 0.1 m. (For more on significant digits, see Chapter I, Introducing Physics.)

Review Question 1.3 Sammy went hiking between two camps that were separated by about 10 kilometers (km). He hiked approximately 16 km to get from one camp to the other. Translate 10 km and 16 km into the language of physical quantities.

1.4 Representing motion with data tables and graphs

So far, we have learned how to represent linear motion with motion diagrams. In this section, we learn to represent linear motion with data tables and graphs.

Imagine your friend (the object of interest) walking across the front of your classroom. To record her position, you drop a beanbag on the floor at her position each second (**Figure 1.8a**). The floor is the object of reference. The origin of the coordinate system is 1.00 m from the first beanbag, and the position axis points in the direction of motion. **Table 1.3** shows each bag's positions and the corresponding clock reading. Do you see a pattern in the table's data? One way to determine if there is a pattern is to plot the data on a graph (Figure 1.8b). This graph is called a **kinematics position-versus-time**

graph. In physics, the word **kinematics** means description of motion. Kinematics graphs contain more precise information about an object's motion than motion diagrams can.

Time t is usually considered to be the independent variable, as time progresses even if there is no motion, so the horizontal axis will be the t axis. Position x is the dependent variable (position changes with time), so the vertical axis will be the x-axis.

A row in Table 1.3 turns into two points, one on each axis. Each point on the horizontal axis represents a time (clock reading). Each point on the vertical axis represents the position of a beanbag. When we draw lines through these points and perpendicular to the axes, they intersect at a single location—a dot on the graph that simultaneously represents a time and the corresponding position of the object. This dot *is not* a location in real space but rather a representation of the position of the beanbag at a specific time.

Is there a trend in the locations of the dots on the graph? We see that the position increases as the time increases. This makes sense. We can draw a smooth best-fit curve that passes as close as possible to the data points—a **trendline** (Figure 1.8c). It looks like a straight line in this particular case—the position is linearly dependent on time.

Correspondence between a motion diagram and position-versus-time graph

To understand how graphs relate to motion diagrams, consider the motion represented by the data in Table 1.3 and in Figure 1.8c. **Figure 1.9** shows a modified motion diagram for the data in Table 1.3 (the dot times are shown and the $\Delta \vec{v}$ arrows have been removed for simplicity) and the corresponding position-versus-time graph. The position of each dot on the motion diagram corresponds to a point on the position axis. The graph line combines the information about the position of an object and the clock reading when this position occurred. Note, for example, that the $t = 4.0$ s dot on the motion diagram at position $x = 7.26$ m is at 7.26 m on the position axis. The corresponding dot on the graph is at the intersection of the vertical line passing through 4.0 s and the horizontal line passing though 7.26 m.

> **TIP** The quantity that appears on the vertical axis of the graph can represent the position of an object whose actual position is changing along a horizontal axis (or along a vertical axis or along an inclined axis). The position on the vertical axis does not mean the object is moving in the vertical direction.

Figure 1.8 Constructing a kinematics position-versus-time graph.

(a)

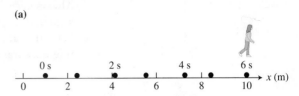

(b)

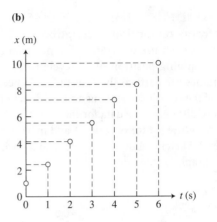

(c)

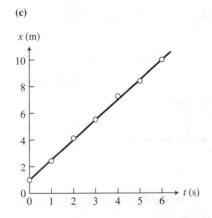

Figure 1.9 Correspondence between a motion diagram and the position-versus-time graph.

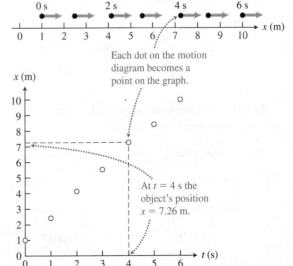

The role of a reference frame

Always keep in mind that representations of motion (motion diagrams, tables, kinematics graphs, equations, etc.) depend on the reference frame chosen. Let's look at the representations of the motion of a cyclist using two different reference frames.

CONCEPTUAL EXERCISE 1.2 Effect of reference frame on motion description

Two observers each use different reference frames to record the changing position of a bicycle rider. Both reference frames use Earth as the object of reference, but the origins of the coordinate systems and the directions of the x-axes are different. The data for the cyclist's trip are presented in **Table 1.4** for observer 1 and in **Table 1.5** for observer 2. Sketch a motion diagram and a position-versus-time graph for the motion when using each reference frame.

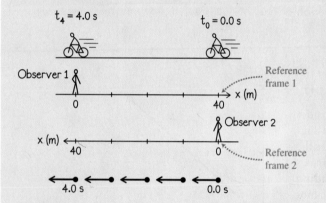

Sketch and translate According to Table 1.4, the observer in reference frame 1 sees the cyclist (the object of interest) at time $t_0 = 0.0$ s at position $x_0 = 40.0$ m and at $t_4 = 4.0$ s at position $x_4 = 0.0$ m. Thus, the cyclist is moving in the negative direction relative to the coordinate axis in reference frame 1. Meanwhile, according to Table 1.5, the observer in reference frame 2 sees the cyclist at time $t_0 = 0.0$ s at position $x_0 = 0.0$ m and at time $t_4 = 4.0$ s at position $x_4 = 40.0$ m. Thus, the cyclist is moving in the positive direction relative to reference frame 2.

Simplify and diagram Since the size of the cyclist is small compared to the distance he is traveling, we can represent him as a point-like object. The motion diagram for the cyclist is the same for both observers, as they are using the same object of reference. Using the data in the tables, we plot kinematics position-versus-time graphs for each observer, below. Although the graph for observer 1 looks very different from the group for observer 2, they represent the same motion. The graphs look different because the reference frames are different.

Table 1.4 Time–position data for cyclist when using reference frame 1.

Clock reading (time)	Position
$t_0 = 0.0$ s	$x_0 = 40.0$ m
$t_1 = 1.0$ s	$x_1 = 30.0$ m
$t_2 = 2.0$ s	$x_2 = 20.0$ m
$t_3 = 3.0$ s	$x_3 = 10.0$ m
$t_4 = 4.0$ s	$x_4 = 0.0$ m

Table 1.5 Time–position data for cyclist when using reference frame 2.

Clock reading (time)	Position
$t_0 = 0.0$ s	$x_0 = 0.0$ m
$t_1 = 1.0$ s	$x_1 = 10.0$ m
$t_2 = 2.0$ s	$x_2 = 20.0$ m
$t_3 = 3.0$ s	$x_3 = 30.0$ m
$t_4 = 4.0$ s	$x_4 = 40.0$ m

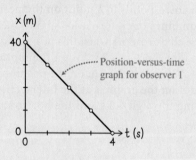

Position-versus-time graph for observer 1

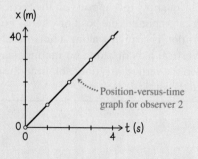

Position-versus-time graph for observer 2

Try it yourself: A third observer recorded the values (in **Table 1.6**) for the time and position of the same cyclist. Describe the reference frame of this observer.

Answer: The point of reference could be another cyclist moving in the opposite positive direction from the direction in which the first cyclist is traveling, with each covering the same distance relative to the ground during the same time interval.

Table 1.6 Data collected by the third observer.

Clock reading (time)	Position
$t_0 = 0.0\,\text{s}$	$x_0 = 0.0\,\text{m}$
$t_1 = 1.0\,\text{s}$	$x_1 = -20.0\,\text{m}$
$t_2 = 2.0\,\text{s}$	$x_2 = -40.0\,\text{m}$
$t_3 = 3.0\,\text{s}$	$x_3 = -60.0\,\text{m}$
$t_4 = 4.0\,\text{s}$	$x_4 = -80.0\,\text{m}$

Review Question 1.4 A position-versus-time graph representing a moving object is shown in **Figure 1.10**. What are the positions of the object at clock readings 2.0 s and 5.0 s?

Figure 1.10 A position-versus-time graph representing a moving object.

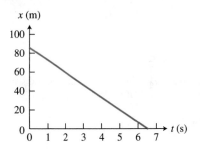

1.5 Constant velocity linear motion

In the last section we devised a graphical representation of motion. Here we will connect graphs to mathematical representations using two motorized toy cars racing toward a finish line. At time 0 they are next to each other, but car B is moving faster than car A and reaches the finish line first (**Figure 1.11**). The data that we collect are shown in Observational Experiment **Table 1.7**. Earth is the object of reference. The origin of the coordinate system (the point of reference) is 1.0 m to the left of the position of the cars at $t_0 = 0$. The positive x-direction points right in the direction of the cars' motions. Now let's use the data to find a pattern.

OBSERVATIONAL EXPERIMENT TABLE

1.7   Graphing the motion of cars.

Observational experiment

Data for car A		Data for car B	
$t_0 = 0.0\,\text{s}$	$x_0 = 1.0\,\text{m}$	$t_0 = 0.0\,\text{s}$	$x_0 = 1.0\,\text{m}$
$t_1 = 1.0\,\text{s}$	$x_1 = 1.4\,\text{m}$	$t_1 = 1.0\,\text{s}$	$x_1 = 1.9\,\text{m}$
$t_2 = 2.0\,\text{s}$	$x_2 = 1.9\,\text{m}$	$t_2 = 2.0\,\text{s}$	$x_2 = 3.0\,\text{m}$
$t_3 = 3.0\,\text{s}$	$x_3 = 2.5\,\text{m}$	$t_3 = 3.0\,\text{s}$	$x_3 = 3.9\,\text{m}$
$t_4 = 4.0\,\text{s}$	$x_4 = 2.9\,\text{m}$	$t_4 = 4.0\,\text{s}$	$x_4 = 5.0\,\text{m}$
$t_5 = 5.0\,\text{s}$	$x_5 = 3.5\,\text{m}$	$t_5 = 5.0\,\text{s}$	$x_5 = 6.0\,\text{m}$

Analysis

We graph the data with the goal of finding a pattern. The trendlines for both cars are straight lines. The line for car B has a bigger angle with the time axis than the line for car A.

Pattern

It looks like a straight line is the simplest reasonable choice for the best-fit curve in both cases (the data points do not have to be exactly on the line).

Figure 1.11 Positions of cars A and B at 0 s and 5.0 s.

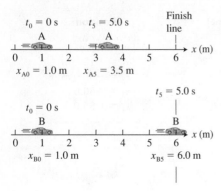

Figure 1.12 The sign of the slope indicates the direction of motion.

(a)

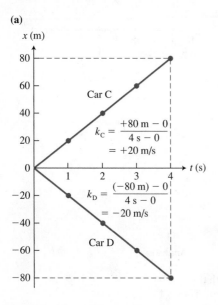

(b)

The position-versus-time graph for car D has negative slope.

The position-versus-time graph for car C has positive slope.

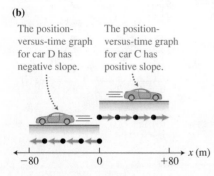

In Table 1.7, the slope of the line representing the motion of car B is greater than the slope of the line representing the motion of car A. What is the physical meaning of this slope? In mathematics the value of a dependent variable is usually written as y and depends on the value of an independent variable, usually written as x. A function $y(x) = f(x)$ is an operation that one needs to do to x as an input to have y as the output. For a straight line, the function $y(x)$ is $y(x) = kx + b$, where k is the slope and b is the y intercept—the value of the y when $x = 0$.

In the case of the cars, the independent variable is time t and the dependent variable is position x. The equation of a straight line becomes $x(t) = kt + b$, where b is the x-intercept of the line, and k is the slope of the line. The x-intercept is the x-position when $t = 0$, also called the initial position of the car x_0. Both cars started at the same location: $x_{A0} = x_{B0} = 1.0$ m.

To find the **slope** k of a straight line, we can choose *any* two points on the line and divide the change in the vertical quantity (Δx in this case) by the change in the horizontal quantity (Δt in this case):

$$k = \frac{x_2 - x_1}{t_2 - t_1} = \frac{\Delta x}{\Delta t}.$$

For example, for car A the slope of the line is

$$k_A = \frac{3.5\text{ m} - 1.0\text{ m}}{5.0\text{ s} - 0.0\text{ s}} = +0.5\text{ m/s}.$$

The slope of the line for car B is

$$k_B = \frac{6.0\text{ m} - 1.0\text{ m}}{5.0\text{ s} - 0.0\text{ s}} = +1.0\text{ m/s}.$$

Now we have all the information we need to write mathematical equations that describe the motion of each of the two cars:

Car A: $x_A = (+0.5\text{ m/s})t + (1.0\text{ m})$

Car B: $x_B = (+1.0\text{ m/s})t + (1.0\text{ m})$

Notice that the units of the slope are meters per second. The slope indicates how the object's position changes with respect to time. The slope of the line contains more information than just how fast the car is going. It also tells us the direction of motion relative to the coordinate axis.

Consider the motions represented graphically in **Figure 1.12a**. The slope of the position-versus-time graph for car C is $+20$ m/s, but the slope of the position-versus-time graph for car D is -20 m/s. What is the significance of the minus sign? Car C is moving in the positive direction, but car D is moving in the negative direction (check the motion diagrams in Figure 1.12b). The magnitudes of the slopes of their position-versus-time graphs are the same, but the signs are different. Thus, in addition to the information about how fast the car is traveling (its **speed**), the slope tells in what direction it is traveling. Together, speed and direction are called **velocity**, and this is what the slope of a position-versus-time graph represents. You are already familiar with the term "velocity arrow" used on motion diagrams. Now you have a formal definition for velocity as a physical quantity.

Velocity and speed for constant velocity linear motion For constant velocity linear motion, the component of velocity v_x along the axis of motion can be found as the slope of the position-versus-time graph or the ratio of the component of the displacement of an object $x_2 - x_1$ during *any* time interval $t_2 - t_1$:

$$v_x = \frac{x_2 - x_1}{t_2 - t_1} = \frac{\Delta x}{\Delta t} \tag{1.1}$$

Examples of units of velocity are m/s, km/h, and mi/h (which is often written as mph). *Speed* is the magnitude of the velocity and is always a positive number.

Note that velocity is a vector quantity. In vector form, motion at constant velocity is $\vec{v} = \vec{d}/\Delta t$. Here we divide a vector by a scalar. As scalars are just numbers that do not have directions, when we multiply or divide a vector by a scalar, all we need to do is to change the magnitude accordingly without changing direction (unless the scalar is negative, in which case it changes the direction of the vector by 180 degrees). Therefore, in our case the velocity vector has the same direction as the displacement vector. This means that the direction for the velocity vector shows the direction of motion (same as the direction of the displacement vector) and the magnitude shows the speed. But since it is difficult to operate mathematically with vectors, we will work with components.

> **TIP** Eq. (1.1) allows you to use *any* change in position divided by the time interval during which that change occurred to obtain the same number—as long as the position-versus-time graph is a straight line (the object is moving at constant velocity). Later in the chapter, you will learn how to modify this equation for cases in which the velocity is not constant.

Equation of motion for constant velocity linear motion

We can rearrange Eq. (1.1) into a form that allows us to determine the position of an object at time t_2 knowing only its position at time t_1 and the x-component of its velocity: $x_2 = x_1 + v_x(t_2 - t_1)$. If we apply this equation for time zero ($t_0 = 0$) when the initial position is x_0, then the position x at any later position time t can be written as follows.

Position equation for constant velocity linear motion

$$x = x_0 + v_x t \qquad (1.2)$$

where x is the function $x(t)$, position x_0 is the position of the object at time $t_0 = 0$ with respect to a particular reference frame, and the (constant) x-component of the velocity of the object v_x is the slope of the position-versus-time graph.

Below you see a new type of task—a quantitative exercise. Quantitative exercises include two steps of the problem-solving process: Represent mathematically and Solve and evaluate. Their purpose is to help you practice using new equations right away.

QUANTITATIVE EXERCISE 1.3 A cyclist
In Conceptual Exercise 1.2, you constructed graphs for the motion of a cyclist using two different reference frames. Now construct mathematical representations (equations) for the cyclist's motion for each of the two graphs. Do the equations indicate the same position for the cyclist at time $t = 6.0$ s?

Represent mathematically The cyclist moves at constant velocity; thus the general mathematical description of his motion is $x = x_0 + v_x t$, where

$$v_x = \frac{x_2 - x_1}{t_2 - t_1}.$$

Solve and evaluate Using the graph for reference frame 1 in Conceptual Exercise 1.2, we see that the cyclist's initial position is $x_0 = +40$ m. The velocity along the x-axis (the slope of the graph line) is

$$v_x = \frac{0\,\text{m} - 40\,\text{m}}{4\,\text{s} - 0\,\text{s}} = -10\,\text{m/s}$$

The minus sign indicates that the velocity points in the negative x-direction (toward the left) relative to that axis. The motion of the bike with respect to reference frame 1 is described by the equation

$$x = x_0 + v_x t = 40\,\text{m} + (-10\,\text{m/s})t$$

Using the graph for reference frame 2, we see that the cyclist's initial position is $x_0 = 0$ m. The x-component of the velocity along the axis of motion is

$$v_x = \frac{40\,\text{m} - 0\,\text{m}}{4\,\text{s} - 0\,\text{s}} = +10\,\text{m/s}$$

(continued)

The positive sign indicates that the velocity points in the positive x-direction (toward the left). The motion of the bike relative to reference frame 2 is described by the equation

$$x = x_0 + v_x t = 0\,\text{m} + (10\,\text{m/s})t$$

The position of the cyclist at time $t_1 = 6\,\text{s}$ with respect to reference frame 1 is

$$x = 40\,\text{m} + (-10\,\text{m/s})(6\,\text{s}) = -20\,\text{m}.$$

With respect to reference frame 2:

$$x = 0\,\text{m} + (+10\,\text{m/s})(6\,\text{s}) = +60\,\text{m}.$$

How can the position of the cyclist be both $-20\,\text{m}$ and $+60\,\text{m}$? Remember that the description of motion of an object depends on the reference frame. If you put a dot on coordinate axis 1 at the $-20\,\text{m}$ position and a dot on coordinate axis 2 at the $+60\,\text{m}$ position, you find that the dots are in fact at the same location, even though that location corresponds to a different position in each reference frame. Both descriptions of the motion are correct and consistent; but each one is meaningful only with respect to the corresponding reference frame.

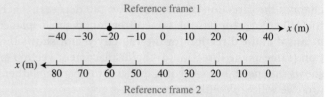

Try it yourself: Use the data for the motion of the cyclist as seen by the third observer in the Try It Yourself part of Conceptual Exercise 1.2 to write the equation of motion. Why is the magnitude of the cyclist's velocity different than the 10 m/s in the example above?

Answer: $x = (0\,\text{m}) + (-20\,\text{m/s})t$. The observer is moving with respect to Earth at the same speed in the direction opposite to the cyclist.

Below you see another new type of task—a worked example. The worked examples include all four steps of the problem-solving strategy we use in this book. (See Introducing Physics for descriptions of these steps.)

EXAMPLE 1.4 You chase your sister

Your young sister is running at 2.0 m/s toward a mud puddle that is 6.0 m in front of her. You are 10.0 m behind her running at 5.0 m/s to catch her before she enters the mud. Will she need a bath?

Sketch and translate We start by drawing a sketch of what is happening. Your sister and you are two objects of interest. Next, we choose a reference frame with Earth as the object of reference. The origin of the coordinate system will be your initial position and the positive direction will be toward the right, in the direction that you both run, as shown below.

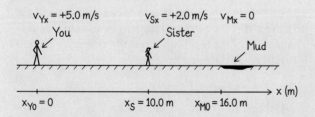

We can now mathematically describe the positions and velocities of you and your sister at the beginning of the process. The initial clock reading is zero at the moment that you are at the origin. Note that you were both running before time zero; this just happened to be the time when we started analyzing the process. We want to know the time when you and your sister are at the same position. This will be the position where you catch up to her.

Simplify and diagram We assume that you and your sister are point-like objects. To sketch graphs of the motions, find the sister's position at 1 second by multiplying her speed by 1 second and adding it to her initial position. Do it for 2 seconds and for 3 seconds as well. Plot these values on a graph for the corresponding clock readings (1 s, 2 s, 3 s, etc.) and draw a line that extends through these points. Repeat this for yourself.

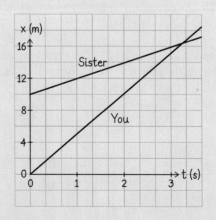

Represent mathematically Use Eq. (1.2) to construct mathematical representations of motion. The form of the equation is the same for both ($x = x_0 + v_x t$); however, the values for the initial positions and the components of the velocities along the axis are different.

Sister: $x_S = (10.0 \text{ m}) + (2.0 \text{ m/s})t$

You: $x_Y = (0.0 \text{ m}) + (5.0 \text{ m/s})t$

From the graphs, we see that the distance between you and your sister is shrinking with time. Do the equations tell the same story? For example, at time $t = 2.0$ s, your sister is at position

$$x_S(2 \text{ s}) = (10.0 \text{ m}) + (2.0 \text{ m/s})(2.0 \text{ s}) = 14.0 \text{ m}$$

and you are at

$$x_Y(2 \text{ s}) = (0.0 \text{ m}) + (5.0 \text{ m/s})(2.0 \text{ s}) = 10.0 \text{ m}.$$

You are catching up to your sister.

Solve and evaluate The time t at which the two of you are at the same position can be found by setting $x_S(t) = x_Y(t)$:

$$(10.0 \text{ m}) + (2.0 \text{ m/s})t = (0.0 \text{ m}) + (5.0 \text{ m/s})t$$

Rearrange the above to determine the time t when you are both at the same position:

$$(2.0 \text{ m/s})t - (5.0 \text{ m/s})t = (0.0 \text{ m}) - (10.0 \text{ m})$$
$$(-3.0 \text{ m/s})t = -(10.0 \text{ m})$$
$$t = 3.33333333 \text{ s}$$

The 3.33333333 s number produced by our calculator has many more significant digits than the givens. Should we round it to have the same number of significant digits as the given quantities? The rule of thumb is that if it is the final result, you need to round this number to 3.3, as the answer cannot be more precise than the given information. However, we do not round the result

of an intermediate calculation. We use the result as is to calculate the next quantity needed to get the final answer and then round the final result.

Sister: $x_S(t) = (10.0 \text{ m}) + (2.0 \text{ m/s})(3.33333333 \text{ s})$
$$= 16.7 \text{ m}$$

You: $x_Y(t) = (0.0 \text{ m}) + (5.0 \text{ m/s})(3.33333333 \text{ s})$
$$= 16.7 \text{ m}$$

Note that if you used the rounded number 3.3 s, you would get 16.5 m for your sister and 16.6 m for you. These would be slightly less than the result calculated above. However, for our purposes it does not matter, as the goal of this example was to decide if you could catch your sister before she reaches the puddle. Since you caught her at a position of about 16.7 m, with the uncertainty of about 0.1 m, this position is slightly greater than the 16.0 m distance to the puddle. Therefore, your sister reaches the puddle before you. This answer seems consistent with the graphical representation of the motion shown above.

Try it yourself: Describe the problem situation using a reference frame with the sister (not Earth) as the object and point of reference and the positive direction pointing toward the puddle.

Answer: With respect to this reference frame, the sister is at position 0 and at rest; you are initially at -10.0 m and moving toward your sister with velocity $+3.0$ m/s; and the mud puddle is initially at $+6.0$ m and moving toward your sister with velocity -2.0 m/s.

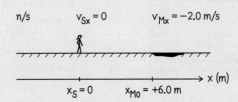

TIP In the reference frame we chose in Example 1.4, the positions of you and your sister are always positive, as are the components of your velocities. Also, your initial position will be zero. Thus the calculations are the easiest. Often the description of the motion of object(s) will be simplest in one particular reference frame.

Graphing velocity

So far, we have learned to make position-versus-time graphs. We could also construct a graph of an object's velocity as a function of time. Consider Example 1.4, in which you chase your sister. Again we will use Earth as the object of reference. You are moving at a constant velocity whose x-component is $v_x = +5$ m/s. For your sister, the x-component of velocity is $+2.0$ m/s. We

Figure 1.13 Velocity-versus-time graphs.

(a) v-versus-t graph lines with Earth as object of reference

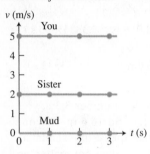

(b) v-versus-t graph lines with Sister as object of reference

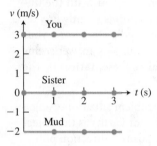

Figure 1.14 Using the v_x-versus-t graph to determine displacement $x - x_0$.

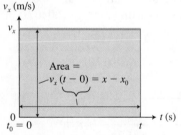

An object's displacement $x - x_0$ between $t_0 = 0$ and time t is the area between the v_x-versus-t curve and the t axis.

place clock readings on the horizontal axis and the x-component of your and your sister's velocities on the vertical axis; then we plot points for these velocities at each time (see **Figure 1.13a**). The best-fit line for each person is a horizontal straight line, which makes sense since neither of their velocities are changing. For you, the equation of the best-fit line is $v_{Yx}(t) = +5.0$ m/s, and for your sister, it is $v_{Sx}(t) = +2.0$ m/s, where $v_x(t)$ represents the x-component of velocity as a function of time.

If instead we choose your sister as the object of reference, her velocity with respect to herself is zero, so the best-fit curve is again a horizontal line but at a value of 0.0 m/s instead of +2.0 m/s; your velocity is +3.0 m/s (see Figure 1.13b); and the mud's velocity is −2.0 m/s. The minus sign indicates that from your sister's point of view, the mud is moving in the negative direction toward her at speed +2.0 m/s.

> **TIP** Notice that a horizontal line on a position-versus-time graph means that the object is at rest (the position is constant with time). The same horizontal line on the velocity-versus-time graph means that the object is moving at constant velocity (its velocity does not change with time).

Finding displacement from a velocity graph

We have just learned to construct a velocity-versus-time graph. Can we get anything more out of such graphs besides being able to represent velocity graphically? As you know, for constant velocity linear motion, the position of an object changes with time according to $x = x_0 + v_x t$. Rearranging this equation a bit, we get $x - x_0 = v_x t$. The left side is the displacement of the object from time zero to time t. Now look at the right side: v_x is the vertical height of the velocity-versus-time graph line and t is the horizontal width from time zero to time t (see **Figure 1.14**). We can interpret the right side as the shaded area between the velocity-versus-time graph line and the time axis. In equation $x - x_0 = v_x t$, this area (the right side) equals the displacement of the object from time zero to time t on the left side of the equation. Here the displacement is a positive number because we chose an example with positive velocity.

Let's extend this reasoning to more general cases: an object initially at position x_1 at time t_1 and later at position x_2 at time t_2 and moving in either the positive or negative direction.

> **Displacement is the area between a velocity-versus-time graph line and the time axis** For motion with constant velocity, the magnitude of the displacement $x_2 - x_1$ (the distance traveled) of an object during a time interval from t_1 to t_2 is the area between a velocity-versus-time graph line and the time axis between those two clock readings. The displacement is the area with a plus sign when the velocity is positive and the area with a negative sign negative when velocity is negative.

QUANTITATIVE EXERCISE 1.5 Displacement of you and your sister

Use the velocity-versus-time graphs shown in Figure 1.13a for you and your sister (see Example 1.4) to find your displacements with respect to Earth for the time interval from 0 to 3.0 s. The velocity components relative to Earth are +2.0 m/s for your sister and +5.0 m/s for you.

Represent mathematically For constant velocity motion, the object's displacement $x - x_0$ is the area of a rectangle whose vertical side equals the object's velocity v_x and the horizontal side equals the time interval $t - t_0 = t - 0$ during which the motion occurred.

Solve and evaluate The displacement of your sister heading toward the mud puddle during the 3.0-s time interval is the product of her velocity and the time interval:

$$d_S = (x - x_0)_S = (+2.0 \text{ m/s})(3.0 \text{ s}) = +6.0 \text{ m}$$

She was originally at position 10.0 m, so she is now at position $(10.0 + 6.0)$ m $= +16.0$ m. Your displacement during that same 3.0-s time interval is the product of your velocity and the time interval:

$$d_Y = (x - x_0)_Y = (+5.0 \text{ m/s})(3.0 \text{ s}) = +15.0 \text{ m}$$

You were originally at 0.0 m, so you are now at +15.0 m—one meter behind your sister.

Try it yourself: Determine the magnitudes of the displacements of you and your sister from time zero to time 2.0 s and your positions at that time. Your initial position is zero and your sister's is 10 m.

Answer: The sister's values are $d_S = 4.0$ m and $x_S = 14.0$ m, and your values are $d_Y = 10.0$ m and $x_Y = 10.0$ m. She is 4.0 m ahead of you at that time.

Review Question 1.5 Why is the following statement true? "Displacement is equal to the area between a velocity-versus-time graph line and the time axis with a positive or negative sign."

1.6 Motion at constant acceleration

In the last section, the function $v_x(t)$ was a horizontal line on the velocity-versus-time graph because the velocity was constant. How would the graph look if the velocity were changing? One example of such a graph is shown in **Figure 1.15**. A point on the curve indicates the velocity of the object shown on the vertical axis at a particular time shown on the horizontal axis. In this case, the velocity is continually changing and is positive.

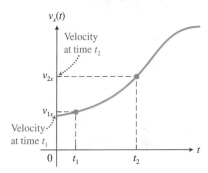

Figure 1.15 Velocity-versus-time graph for motion with changing velocity.

Instantaneous velocity and average velocity

The velocity of an object at a particular time is called the **instantaneous velocity**. Figure 1.15 shows a velocity-versus-time graph for motion with continually changing instantaneous velocity. When an object's velocity is changing, we cannot use Eq. (1.1) to determine its instantaneous velocity, because the ratio

$$v_x = \frac{x_2 - x_1}{t_2 - t_1} = \frac{\Delta x}{\Delta t}$$

is not the same for different time intervals the way it was when the object was moving at constant velocity. However, we can still use the this equation to determine the **average velocity**, which is the ratio of the change in position and the time interval during which this change occurred. For motion at constant velocity, the instantaneous and average velocity are equal; for motion with changing velocity, they are not.

When an object moves with changing velocity, its velocity can change quickly or slowly. To characterize the rate at which the velocity of an object is changing, we need a new physical quantity.

Acceleration

To analyze motion with changing velocity, we start by looking for the simplest type of linear motion with changing velocity. This occurs when the velocity of the object increases or decreases by the same amount during the same time interval (a *constant rate* of change). Imagine that a cyclist is speeding up so that his velocity is increasing at a constant rate with respect to an observer on the ground.

Figure 1.16 Velocity-versus-time graph when the velocity is changing at a constant rate.

(a)

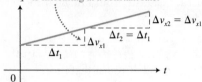

The velocity in the positive direction v_x is increasing at a constant rate.

(b)

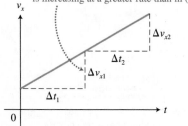

The velocity in the positive direction is increasing at a greater rate than in (a).

(c)

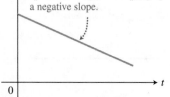

The velocity is in the positive direction ($v_x > 0$ and is above the t axis). The acceleration is negative ($a_x < 0$) since v_x is decreasing and its $v_x(t)$ graph has a negative slope.

(d)

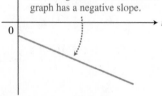

The velocity is in the negative direction (below the t axis). The acceleration is negative ($a_x < 0$) since v_x is increasing in the negative direction and its $v_x(t)$ graph has a negative slope.

TIP If it is difficult for you to think about velocity and acceleration in abstract terms, try calculating the acceleration for simple integer velocities.

Graphically, this process is represented as the velocity-versus-time graph shown in **Figure 1.16a**. The graph is a straight line that is not horizontal. Now imagine that a car next to the cyclist is also speeding up, but at a greater rate than the biker. Its velocity-versus-time graph is shown in Figure 1.16b. The larger slope indicates that the car's velocity increases at a faster rate. The physical quantity that characterizes the change in velocity during a particular time interval is called **acceleration** $\vec{a}$. When the object is moving along a straight line and the slope of the velocity-versus-time graph is constant, the acceleration of an object is equal to the slope:

$$a_x = \frac{v_{2x} - v_{1x}}{t_2 - t_1} = \frac{\Delta v_x}{\Delta t}$$

The acceleration can be either positive or negative. In the previous examples, if the cyclist or the car had been slowing down, their velocity-versus-time graph would instead have a negative slope, which corresponds to a decreasing speed and a negative acceleration, as v_{2x} is smaller than v_{1x} (Figure 1.16c). However, an object can have a negative acceleration and speed up! This happens when the object is moving in the negative direction and has a negative component of velocity, but its speed in the negative direction is increasing in magnitude (see Figure 1.16d).

Because velocity is a vector quantity and the acceleration shows how quickly the velocity changes as time progresses, acceleration is also a vector quantity. We can define acceleration in a more general way. The **average acceleration** of an object during a time interval is the following:

$$\vec{a} = \frac{\vec{v}_2 - \vec{v}_1}{t_2 - t_1} = \frac{\Delta \vec{v}}{\Delta t}$$

To determine the acceleration, we need to determine the velocity change vector $\Delta \vec{v} = \vec{v}_2 - \vec{v}_1$. This equation involves the subtraction of vectors. You can think about the same equation as addition by rearranging it to be $\vec{v}_1 + \Delta \vec{v} = \vec{v}_2$. Note that $\Delta \vec{v}$ is the vector that we add to $\vec{v}_1$ to get $\vec{v}_2$ (**Figure 1.17**). We did this when making motion diagrams in Section 1.3, only then we were not concerned with the exact lengths of the vectors. The acceleration vector $\vec{a} = \Delta \vec{v}/\Delta t$ is in the same direction as the velocity change vector $\Delta \vec{v}$, as the time interval Δt is a scalar quantity.

Acceleration An object's average acceleration during a time interval Δt is the change in its velocity $\Delta \vec{v}$ divided by that time interval:

$$\vec{a} = \frac{\vec{v}_2 - \vec{v}_1}{t_2 - t_1} = \frac{\Delta \vec{v}}{\Delta t} \qquad (1.3)$$

If Δt is very small, then the acceleration given by this equation is the **instantaneous acceleration** of the object. For one-dimensional motion, the component of the average acceleration along a particular axis (for example, for the x-axis) is

$$a_x = \frac{v_{2x} - v_{1x}}{t_2 - t_1} = \frac{\Delta v_x}{\Delta t} \qquad (1.4)$$

The unit of acceleration is $(\text{m/s})/\text{s} = \text{m/s}^2$.

Note that if an object has an acceleration of $+6 \text{ m/s}^2$, it means that its velocity changes by $+6 \text{ m/s}$ in 1 s, or by $+12 \text{ m/s}$ in 2 s $[(+12 \text{ m/s})/(2 \text{ s}) = +6 \text{ m/s}^2]$.

It is possible for an object to have a zero velocity and nonzero acceleration—for example, at the moment when an object starts moving from rest. An object can also have a nonzero velocity and zero acceleration—for example, an object moving at constant velocity. Note that the acceleration of an object depends on the observer. For example, a car is accelerating for an observer on the ground but is not accelerating for the driver of that car.

Determining the velocity change from the acceleration

If, at $t_0 = 0$ for linear motion, the x-component of the velocity of some object is v_{0x} and its acceleration a_x is constant, then its velocity v_x at a later time t can be determined by substituting these quantities into Eq. (1.4):

$$a_x = \frac{v_x - v_{0x}}{t - 0}$$

Rearranging, we get an expression for the changing velocity of the object as a function of time:

$$v_x = v_{0x} + a_x t \qquad (1.5)$$

For one-dimensional motion, the directions of the vector components a_x, v_x, and v_{0x} are indicated by their signs relative to the axis of motion—positive if in the positive x- direction and negative if in the negative x-direction.

Figure 1.17 How to determine the change in velocity $\Delta \vec{v}$

We add $\Delta \vec{v}$ to $\vec{v}_1$ to get $\vec{v}_2$.

EXAMPLE 1.6 Bicycle ride

Suppose that you are sitting on a bench watching a cyclist riding a bicycle on a flat, straight road. In a 2.0-s time interval, the velocity of the bicycle changes from −4.0 m/s to −7.0 m/s. Describe the motion of the bicycle as fully as possible.

Sketch and translate We can sketch the process as shown below. The bicycle (the object of interest) is moving in the negative direction with respect to the chosen reference frame. The components of the bicycle's velocity along the axis of motion are negative: $v_{0x} = -4.0$ m/s at time $t_0 = 0.0$ s and $v_x = -7.0$ m/s at $t = 2.0$ s. The speed of the bicycle (the magnitude of its velocity) increases. It is moving faster in the negative direction. We can determine the acceleration of the bicycle and describe the changes in its velocity.

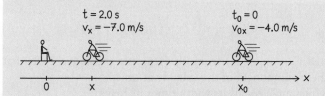

Simplify and diagram Our sketch of the motion diagram for the bicycle is shown below. Note that $\Delta \vec{v}$ points in the negative x-direction.

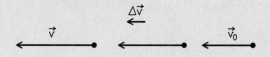

Represent mathematically We apply Eq. (1.4) to determine the acceleration:

$$a_x = \frac{v_x - v_{0x}}{t - t_0}$$

Solve and evaluate Substituting the given velocities and times, we get

$$a_x = \frac{(-7.0\,\text{m/s}) - (-4.0\,\text{m/s})}{2.0\,\text{s} - 0.0\,\text{s}} = -1.5\,\text{m/s}^2$$

The bicycle's x-component of velocity at time zero was −4.0 m/s. Its velocity was changing by −1.5 m/s each second. So 1 s later, its velocity was

$$\begin{aligned} v_{1s\,x} &= v_{0x} + \Delta v_x \\ &= (-4.0\,\text{m/s}) + (-1.5\,\text{m/s}) \\ &= -5.5\,\text{m/s}. \end{aligned}$$

During the second 1-s time interval, the velocity changed by another −1.5 m/s and was now $(-5.5\,\text{m/s}) + (-1.5\,\text{m/s}) = (-7.0\,\text{m/s})$. In this example, the bicycle was speeding up by 1.5 m/s each second in the negative direction.

Try it yourself: A car's acceleration is −3.0 m/s². At time 0 its velocity is +14 m/s. What happens to the velocity of the car? What is its velocity after 3 s?

Answer: The car's velocity in the positive x-direction is decreasing. After 3 s $v_x = 5.0$ m/s.

TIP It is possible for an object to have a positive acceleration and slow down and to have a negative acceleration and speed up. When an object is speeding up, the acceleration vector is in the same direction as the velocity vector, and the velocity and acceleration components along the same axis have the same sign. When an object slows down, the acceleration is in the opposite direction relative to the velocity; their components have opposite signs.

Displacement of an object moving at constant acceleration

Active Learning Guide›

For motion at constant velocity, we know that the area between the velocity-versus-time graph line and the time axis between two clock readings equal the magnitude of the object's displacement. Is this still the case when the velocity is changing?

Consider the displacement of an object during the short shaded time interval Δt shown in **Figure 1.18a**. The velocity is almost constant during that time interval. Note that $v_x = \Delta x / \Delta t$ or $v_x \cdot \Delta t = \Delta x$. Thus, the small displacement Δx during that time interval Δt is the small shaded area $v_x \cdot \Delta t$ between that curve and the time axis (the height times the width of the narrow rectangle). We can repeat the same procedure for many successive short time intervals (Figure 1.18b), building up the area between the curve and the time axis as a sum of areas of small rectangles of different heights. The total area, shown in Figure 1.18c, is the total displacement of the object during the time interval between the initial time t_0 and the final time t. A negative area (the rectangle is below the time axis) corresponds to a negative displacement.

Figure 1.18 The magnitude of the object's displacement is the area under a velocity-versus-time graph.

(a)

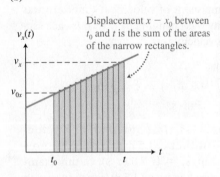

The displacement Δx during a short time interval Δt is the area of the shaded rectangle.

$\Delta x = v_x \times \Delta t$
(height)(width)

Displacement from a v-versus-t graph The magnitude of the displacement $x - x_0$ (distance) of an object during a time interval $t - t_0$ is the area between the velocity-versus-time curve and the time axis between those time readings. The displacement is negative for areas below the time axis and positive for areas above.

Equation of motion—position as a function of time

(b)

We can use the preceding idea to find an equation for the position x of an object at different times t. Consider **Figure 1.19a**. We can find the area between this curve and the time axis by breaking the trapezoidal area into a triangle on top and a rectangular below (Figure 1.19b). The rectangle represents the displacement for motion at constant velocity. The triangle represents the additional displacement caused by acceleration. The area of a triangle is $\frac{1}{2} \times$ base $\times$ height. The base of the triangle is $t - 0$ and the height is $v_x - v_{0x}$. So the area of the triangle is

Displacement $x - x_0$ between t_0 and t is the sum of the areas of the narrow rectangles.

$$A_{\text{triangle}} = \frac{1}{2}(t - 0)(v_x - v_{0x}) = \frac{1}{2}(t)(a_x t) = \frac{1}{2}a_x t^2,$$

where we substituted $v_x - v_{0x} = a_x t$ from Eq. (1.5) into the above. Note that v_x is the value of the x-component of the velocity at time t.

The area of the rectangle equals its width times its height:

(c)

$$A_{\text{rectangle}} = v_{x0}(t - 0)$$

Displacement $x - x_0$ between t_0 and t is the area between the v_x-versus-t graph line and the t axis.

The total area between the curve and the time axis (the displacement $x - x_0$ of the object) is

$$x - x_0 = A_{\text{rectangle}} + A_{\text{triangle}} = v_{0x}t + \frac{1}{2}a_x t^2$$

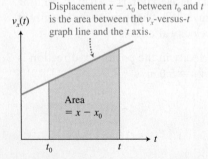

Area
$= x - x_0$

$$\Rightarrow x = x_0 + v_{0x}t + \frac{1}{2}a_x t^2$$

(The symbol $\Rightarrow$ means that this equation follows from the previous equation.)

Does the above result make sense? Consider a limiting case, for example, when the object is traveling at a constant velocity (when $a_x = 0$). In this case the equation should reduce to the result from our investigation of linear motion with constant velocity ($x = x_0 + v_{x0}t$). It does. We can also check the units of each term in this equation for consistency (when terms in an equation are added or subtracted, each of those terms must have the same units). Each term has units of meters, so the units also check.

Position of an object during linear motion with constant acceleration For any initial position x_0 at clock reading $t_0 = 0$, we can determine the position x of an object at any later time t, provided we also know the initial velocity v_{0x} of the object and its constant acceleration a_x:

$$x = x_0 + v_{0x}t + \frac{1}{2}a_x t^2 \qquad (1.6)$$

Since t^2 appears in Eq. (1.6), the position-versus-time graph for this motion will not be a straight line but will be a parabola (a parabola is a graph line for a quadratic function) (**Figure 1.20a**). Unlike the position-versus-time graph for constant velocity motion where $\Delta x/\Delta t$ is the same for any time interval, the position-versus-time graph line for accelerated motion is not a straight line; it does not have a constant slope. At different times, the change in position Δx has a different value for the same time interval Δt (Figure 1.20b). The line tangent to the position-versus-time graph line at a particular time has a slope $\Delta x/\Delta t$ that equals the velocity v_x of the object at that time. The slopes of the tangent lines at different times for the graph line in Figure 1.20b differ—they are greater when the position changes more during the same time interval.

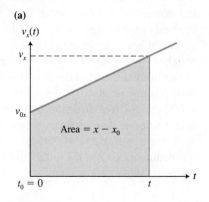

Figure 1.19 The magnitude of the total displacement $x - x_0$ equals the sum of the two areas.

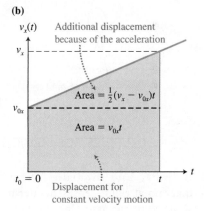

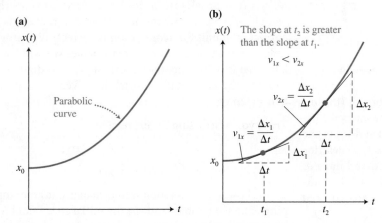

Figure 1.20 (a) A position-versus-time graph for constant acceleration motion has a parabolic shape. (b) The velocity v_x is the slope of the position-versus-time graph.

The next example involves vertical motion. Thus, we will use a vertical y-coordinate axis and apply the equations of motion.

EXAMPLE 1.7 Acceleration estimate for stopping a falling person

A woman jumps off a large boulder. She is moving at a speed of about 5 m/s when she reaches the ground. *Estimate* her acceleration during landing. Indicate any other quantities you use in the estimate.

Sketch and translate To solve the problem, we assume that the woman's knees bend while landing so that her body travels 0.4 m closer to the ground compared to the standing upright position. That is, 0.4 m is the stopping distance for the main part of her body. Below we have sketched the initial and final situations during the landing.

(continued)

We chose the central part of her body as the object of interest. We use a vertical reference with a y-axis pointing down and the final location of the central part of her body as the origin of the coordinate axis. The initial values of her motion at the time $t_0 = 0$ when she first touches the ground are $y_0 = -0.4$ m (the central part of her body is 0.4 m in the negative direction above her final position) and $v_{0y} = +5$ m/s (she is moving downward, the positive direction relative to the y-axis). The final values at some unknown time t at the instant she stops are $y = 0$ and $v_y = 0$. Note: It is important to have a coordinate axis and the initial and final values with appropriate signs.

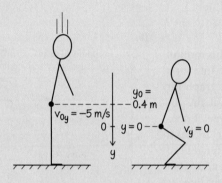

Simplify and diagram The motion diagram at right\ represents her motion while stopping, assuming constant acceleration. We cannot model the woman as a point-like object in this situation, so we will focus on the motion of her midsection.

Represent mathematically The challenge in representing this situation mathematically is that there are two unknowns: the magnitude of her acceleration a_y (the unknown we wish to determine) and the time interval t between when she first contacts the ground and when she comes to rest. However, both Eqs. (1.5) and (1.6) describe linear motion with constant acceleration and have a_y and t in them. Since we have two equations and two unknowns, we can handle this challenge by solving Eq. (1.5) for the time t

$$t = \frac{v_y - v_{0y}}{a_y}$$

and substitute it into rearranged Eq. (1.6):

$$y = y_0 + v_{0y}t + \frac{1}{2}a_y t^2$$

The result is:

$$y - y_0 = v_{0y}\left(\frac{v_y - v_{0y}}{a_y}\right) + \frac{a_y(v_y - v_{0y})^2}{2a_y{}^2}$$

Using algebra, we can simplify the above equation:

$$\Rightarrow 2a_y(y - y_0) = 2v_{0y}(v_y - v_{0y}) + (v_y - v_{0y})^2$$
$$\Rightarrow 2a_y(y - y_0) = (2v_{0y}v_y - 2v_{0y}^2) $$
$$+ (v_y^2 - 2v_y v_{0y} + v_{0y}^2)$$
$$\Rightarrow 2a_y(y - y_0) = v_y^2 - v_{0y}^2$$
$$\Rightarrow a_y = \frac{v_y^2 - v_{0y}^2}{2(y - y_0)}$$

Solve and evaluate Now we can use the above equation to find her acceleration:

$$a_y = \frac{v_y^2 - v_{0y}^2}{2(y - y_0)}$$
$$= \frac{(0)^2 - (5 \text{ m/s})^2}{2[0 - (-0.4 \text{ m})]} = -31 \text{ m/s}^2$$
$$\approx -30 \text{ m/s}^2$$

Is the answer reasonable? The sign is negative. This means that acceleration points upward, as does the velocity change arrow in the motion diagram. This is correct. The unit for acceleration is correct. We cannot judge yet if the magnitude is reasonable. We will learn later that it is. The answer has one significant digit, as it should—the same as the information given in the problem statement.

Try it yourself: Using the expression

$$a_y = \frac{v_y^2 - v_{0y}^2}{2(y - y_0)},$$

decide how the acceleration would change if (a) the stopping distance doubles and (b) the initial speed doubles. Note that the final velocity is $v_y = 0$.

Answer: (a) a_y would be half the magnitude; (b) a_y would be four times the magnitude.

In the Represent Mathematically step above, we developed a new equation for the acceleration by combining Eqs. (1.5) and (1.6). This useful equation is rewritten and described briefly below.

Alternate equation for linear motion with constant acceleration:

$$2a_x(x - x_0) = v_x^2 - v_{0x}^2 \qquad (1.7)$$

This equation is useful for situations in which you do not know the time interval during which the changes in position and velocity occurred.

Review Question 1.6 (a) Give an example in which an object with negative acceleration is speeding up. (b) Give an example in which an object with positive acceleration is slowing down.

1.7 Skills for analyzing situations involving motion

To help analyze physical processes involving motion, we will represent processes in multiple ways: the words in the problem statement, a sketch, one or more diagrams, possibly a graph, and a mathematical description. Different representations have to agree with each other; in other words, they need to be consistent.

Motion at constant velocity

EXAMPLE 1.8 Two walking friends

You stand on a sidewalk and observe two friends walking at constant velocity. At time zero Jim is 4.0 m east of you and walking away from you at speed 2.0 m/s. At time zero, Sarah is 10.0 m east of you and walking toward you at speed 1.5 m/s. Represent their motions with an initial sketch, with motion diagrams, and mathematically.

Sketch and translate We choose Earth as the object of reference with your position as the reference point. The positive direction will point to the east. We have two objects of interest here: Jim and Sarah. Jim's initial position is $x_0 = +4.0$ m and his constant velocity is $v_x = +2.0$ m/s. Sarah's initial position is $X_0 = +10.0$ m and her constant velocity is $V_x = -1.5$ m/s (the velocity is negative since she is moving westward). In our sketch we are using capital letters to represent Sarah and lowercase letters to represent Jim.

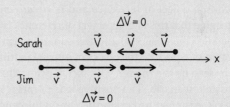

Simplify and diagram We can model both friends as point-like objects since the distances they move are somewhat greater than their own sizes. The motion diagrams below represent their motions.

Represent mathematically Now construct equations to represent Jim's and Sarah's motion:

Jim: $\qquad x = +4.0$ m $+ (2.0$ m/s$)t$

Sarah: $\qquad X = +10.0$ m $+ (-1.5$ m/s$)t$

Solve and evaluate We were not asked to solve for any quantity. We will do it in the Try it yourself exercise.

Try it yourself: Determine the time when Jim and Sarah are at the same position, and where that position is.

Answer: They are at the same position when $t = 1.7$ s and when $x = X = 7.4$ m.

Equation Jeopardy problems

Learning to read the mathematical language of physics with understanding is an important skill. To help develop this skill, this text includes Jeopardy-style problems. In this type of problem, you have to work backwards: you are given one or more equations and are asked to use them to construct a consistent sketch of a process. You then convert the sketch into a diagram of a process that is consistent with the equations and sketch. Finally, you invent a word problem that the equations could be used to solve. Note that there are often many possible word problems for a particular mathematical description.

CONCEPTUAL EXERCISE 1.9 Equation Jeopardy

The following equation describes an object's motion:

$$x = (5.0\ \text{m}) + (-3.0\ \text{m/s})\,t$$

Construct a sketch, a motion diagram, kinematics graphs, and a verbal description of a situation that is consistent with this equation. There are many possible situations that the equation describes equally well.

Sketch and translate This equation looks like a specific example of our general equation for the linear motion of an object with constant velocity: $x = x_0 + v_x t$. The minus sign in front of the 3.0 m/s indicates that the object is moving in the negative x-direction. At time zero, the object is located at position $x_0 = +5.0$ m with respect to some chosen object of reference and *is already moving*. Let's imagine that this chosen object of reference is a running person (the observer) and the equation represents the motion of a person (the object of interest) sitting on a bench as seen by the runner. The sketch below illustrates this possible scenario. The positive axis points from the observer (the runner) toward the person on the bench, and at time zero the person sitting on it is 5.0 m in front of the runner and coming closer to the runner as time elapses.

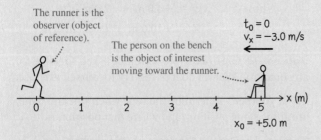

Simplify and diagram Model the object of interest as a point-like object. A motion diagram for the situation is shown below. The equal spacing of the dots and the equal lengths of velocity arrows both indicate that the object of interest is moving at constant velocity with respect to the observer.

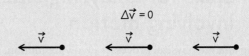

Motion diagram for bench relative to runner

Position-versus-time and velocity-versus-time kinematics graphs of the process are shown in below. The position-versus-time graph has a constant -3.0 m/s slope and a $+5.0$ m intercept with the vertical (position) axis. The velocity-versus-time graph has a constant value (-3.0 m/s) and a zero slope (the velocity is not changing). The following verbal description describes this particular process: A jogger sees a person on a bench in the park 5.0 m in front of him. The bench is approaching at a speed of 3.0 m/s as seen in the jogger's reference frame. The direction pointing from the jogger to the bench is positive.

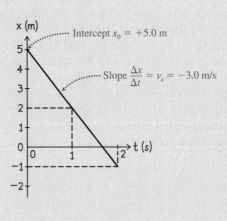

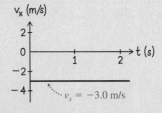

Try it yourself: Suppose we switch the roles of observer and object of reference in the last example. Now the person on the bench is the object of reference and observes the runner. We choose to describe the process by the same equation as in the example:

$$x = (5.0\,\text{m}) + (-3.0\,\text{m/s})\,t$$

Construct an initial sketch and a motion diagram that are consistent with the equation and with the new observer and new object of reference.

Answer: An initial sketch for this process and a consistent motion diagram are shown to the right.

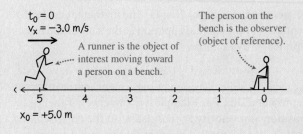

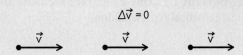

Motion diagram for runner relative to person on bench

Motion at constant nonzero acceleration

Now let's apply some representation techniques to linear motion with constant (nonzero) acceleration.

EXAMPLE 1.10   Equation Jeopardy

A process is represented mathematically by the following equation:

$$x = (-60\,\text{m}) + (10\,\text{m/s})\,t + (1.0\,\text{m/s}^2)\,t^2$$

Use the equation to construct an initial sketch, a motion diagram, and words to describe a process that is consistent with this equation.

Sketch and translate The above equation appears to be an application of Eq. (1.6), which we constructed to describe linear motion with constant acceleration, if we assume that the $1.0\,\text{m/s}^2$ in front of t^2 is the result of dividing $2.0\,\text{m/s}^2$ by 2:

$$x = x_0 + v_{0x}t + \frac{1}{2}a_x t^2$$

$$x = (-60\,\text{m}) + (10\,\text{m/s})\,t + \frac{1}{2}(2.0\,\text{m/s}^2)\,t^2$$

It looks like the initial position of the object of interest is $x_0 = -60\,\text{m}$, its initial velocity is $v_{0x} = +10\,\text{m/s}$, and its acceleration is $a_x = +2.0\,\text{m/s}^2$. Let's imagine that this equation describes the motion of a car passing a van in which you, the observer, are riding on a straight highway. The car is 60 m behind you and moving 10 m/s faster than your van. The car speeds up at a rate of $2.0\,\text{m/s}^2$ with respect to the van. The object of reference is you in the van; the positive direction is the direction in which the car and van are moving. A sketch of the situation appears below.

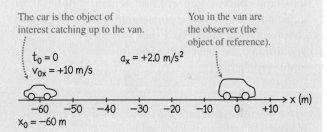

Simplify and diagram The car can be considered a point-like object—much smaller than the dimensions of the path it travels. The car's velocity and acceleration are both positive. Thus, the car's velocity in the positive x-direction is increasing as it moves toward the van (toward the origin). Below is a motion diagram for the car's motion as seen from the van. The successive dots in the diagram are spaced increasingly farther apart as the velocity increases; the velocity arrows are drawn increasingly longer. The velocity arrow (and the acceleration) point in the positive x-direction, that is, in same direction as the velocity arrows.

(continued)

Represent mathematically The mathematical representation of the situation appears at the start of the Equation Jeopardy example.

Solve and evaluate To evaluate what we have done, we can check the consistency (agreement) of the different representations. For example, we can check if the initial position and velocity are consistent in the equation, the sketch, and the motion diagram. In this case, they are.

Try it yourself: Describe a different scenario for the same mathematical representation.

Answer: This mathematical representation could describe the motion of a cyclist moving on a straight path as seen by a person standing on a sidewalk 60 m in front of the cyclist. The positive direction is in the direction the cyclist is traveling. When the person starts observing the cyclist, she is moving at an initial velocity of $v_{0x} = +10$ m/s and speeding up with acceleration $a_x = +2.0$ m/s².

PROBLEM-SOLVING STRATEGY Kinematics

Our four-step problem-solving procedure uses a multiple representation strategy that has proven successful in solving physics problems. In this chapter and many others we will walk you through the strategy with an example problem. In the left-hand column, you will find general guidelines for solving the problems in that chapter. On the right-hand side, we walk you through the process of solving the example problem.

EXAMPLE 1.11 Car arriving at a red light
The velocity-versus-time graph shown here represents a car's motion. What time interval is needed for the car to stop, and how far does it travel while stopping?

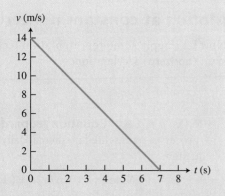

Sketch and translate

- Sketch the situation described in the problem. Choose the object of interest.
- Include an object of reference and a coordinate system. Indicate the origin and the positive direction.
- Label the sketch with relevant information.

From the graph, we see that the car's velocity at time zero is $v_{0x} = +14$ m/s. The car is the object of interest. The object of reference is the ground. The car's initial position is unknown—we'll choose to place it at location $x_0 = 0$ at $t_0 = 0$. The plus sign means the car is moving in the positive x-direction. From the graph, we see that the car's velocity in the positive x-direction decreases by 2.0 m/s for each second; thus the slope of the graph $\Delta v_x / \Delta t$ is $(-2.0$ m/s$)/$s. We create an initial sketch.

$t_0 = 0$
$x_0 = 0$
$v_{0x} = +14$ m/s

Simplify and diagram

- Decide how you will model the moving object (for example, as a point-like object).
- Can you model the motion as constant velocity or constant acceleration?
- Draw motion diagrams and kinematics graphs if needed.

We model the car as a point-like object moving along a straight line at constant acceleration. The velocity arrows get increasingly smaller since the magnitude of the velocity is decreasing. We draw a motion diagram.

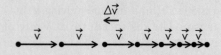

Represent mathematically

■ Use the sketch(es), motion diagram(s), and kinematics graph(s) to construct a mathematical representation (equations) of the process. Be sure to consider the sign of each quantity.

Rearrange Eq. (1.5) to determine the time at which the velocity decreases to zero:

$$t = \frac{v_x - v_{0x}}{a_x}$$

We can then determine the position of the car at that time using the position equation of motion:

$$x = 0 + (14\,\text{m/s})t + \frac{1}{2}(-2.0\,\text{m/s}^2)t^2.$$

Solve and evaluate

■ Solve the equations to find the answer to the question you are investigating.

■ Evaluate the results to see if they are reasonable. Check the units and decide if the calculated quantities have reasonable values (sign, magnitude). Check limiting cases: Examine whether the final equation leads to a reasonable result if one of the quantities is zero or infinity. This strategy applies when you derive a new equation while solving a problem.

Substituting the known information in the first equation above:

$$t = \frac{(0\,\text{m/s}) - (14\,\text{m/s})}{-2.0\,\text{m/s}^2} = 7.0\,\text{s}$$

The car stops after a time interval $(t - t_0) = 7.0\,\text{s}$.

The car's position when it stops is

$$x = 0 + (14\,\text{m/s})(7.0\,\text{s}) + \frac{1}{2}(-2.0\,\text{m/s}^2)(7.0\,\text{s})^2 = 49\,\text{m}$$

The units are correct and the magnitudes are reasonable.

In the limiting case of zero acceleration, the car should never stop. Our equation for the time it takes to stop gives

$$t = \frac{v_x - v_{0x}}{a_x} = \frac{v_x - v_{0x}}{0}$$

The result of dividing of a nonzero quantity by zero is infinity. Thus our equation predicts that it takes an infinite time for the car to stop. The limiting case checks out.

Try it yourself: A cyclist is moving in the negative x-direction at a speed of 6.0 m/s. He sees a red light and stops in 3.0 s. What is his acceleration?

Answer: $a_x = +2.0\,\text{m/s}^2$. The acceleration is positive even though the cyclist's speed decreased.

Review Question 1.7 A car's motion with respect to the ground is described by the following function:

$$x = (-48\,\text{m}) + (12\,\text{m/s})t + (-2.0\,\text{m/s}^2)t^2$$

Mike says that its original position is $(-48\,\text{m})$ and its acceleration is $(-2.0\,\text{m/s}^2)$. Do you agree? If yes, explain why; if not, explain how to correct his answer.

1.8 Free fall

In this chapter we have learned about two simple models of motion—linear motion with constant velocity and linear motion with constant acceleration. Now we will look at a special case of linear motion—the motion of falling objects.

Let's start with the following observational experiment. Tear out a sheet of paper from your notebook and hold the paper in one hand. Hold a textbook parallel to the floor in the other hand and then drop both side by side.

Figure 1.21 The position of a falling ball every 0.100 s.

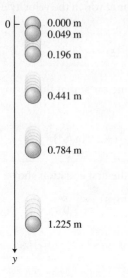

Active Learning Guide➤

Figure 1.22 A velocity-versus-time graph for a falling ball.

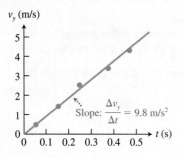

The book lands first. Next, crumple the paper into a tight ball. Now drop the book and the crumpled paper side by side. This time they land at about the same time. Does the motion depend on how heavy the objects are or on their shapes? Galileo Galilei (1564–1642) was the first to realize that it was easier to answer this question if he considered first the motion of falling objects in the absence of air. This hypothetical motion was *a model of the real process* and became known as **free fall.** Galileo hypothesized, based on a series of experiments, that free fall occurs exactly the same way for all objects regardless of their mass and shape.

Based on observations of falling objects, Galileo thought that the speed of freely falling objects was increasing as they moved closer to the surface of Earth. He hypothesized that the speed increases in the simplest way—linearly with time of flight or, in other words, the acceleration of free-falling objects was constant. Galileo did not have a video camera or a watch with a second hand to test his hypothesis. But we do! Imagine that we videotape a small metal ball that is dropped beside a ruler (**Figure 1.21**). Using the small ball allows us to create a situation very close to the model as the air has very little effect on the ball's motion. If the hypothesis is correct, the speed of the ball should increase linearly with time. After recording the fall, we step through the video frame by frame and record the ball's position every 0.100 s (**Table 1.8**). Earth is our object of reference; the origin of the coordinate axis is at the initial location of the ball. The positive direction points down.

To determine the average velocity during each time interval, we calculate the displacement of the ball between consecutive times and then divide by the time interval. For example, the average speed between 0.100 s and 0.200 s is $(0.196\,\text{m} - 0.049\,\text{m})/(0.200\,\text{s} - 0.100\,\text{s}) = 1.47\,\text{m/s}$. These calculated velocities in the last row are associated with the clock readings t^* at the middle of each time interval (the third row). Finally, we use these velocities and t^* times to make a velocity-versus-time graph (**Figure 1.22**). The best-fit curve for this data is a straight line. Therefore, we model the motion of the metal ball as motion with constant acceleration. The slope of the line equals $9.8\,\text{m/s}^2$.

We can represent the motion of a falling ball mathematically using the equations of motion for constant acceleration [Eqs. (1.5) and (1.6)] with $a_y = 9.8\,\text{m/s}^2$:

$$v_y = v_{0y} + a_y t = v_{0y} + (9.8\,\text{m/s}^2)\,t \tag{1.8}$$

$$y = y_0 + v_{0y} t + \frac{1}{2}(9.8\,\text{m/s}^2)\,t^2 \tag{1.9}$$

where y_0 and v_{0y} are the position and instantaneous velocity, respectively, of the object at the clock reading $t_0 = 0$. These equations apply if the positive y-direction is down. When using an upward pointing y-axis, we place a minus sign in front of the $9.8\,\text{m/s}^2$. The magnitude of the object's acceleration while falling without air resistance is given a special symbol, g, where $g = 9.8\,\text{m/s}^2$.

Table 1.8 Position and time data for a small falling ball.

t (s)	0.000	0.100	0.200	0.300	0.400	0.500
y (m)	0.000	0.049	0.196	0.441	0.784	1.225
t* (s)		0.050	0.150	0.250	0.350	0.450
v_av (m/s)		0.49	1.47	2.45	3.43	4.41

If we videotape a small object thrown upward and then use the data to construct a velocity-versus-time graph, we find that its acceleration is still the same at all clock readings. For an upward-pointing axis, the object's acceleration is -9.8 m/s^2 on the way up, -9.8 m/s^2 on the way down, and even -9.8 m/s^2 at the instant when the object is momentarily at rest at the highest point of its motion. At all times during the object's flight, its velocity is changing at a rate of -9.8 m/s each second. A motion diagram and the graphs representing the position-versus-time, velocity-versus-time, and acceleration-versus-time are shown in **Figure 1.23**. The positive direction is up. Notice that when the position-versus-time graph is at its maximum (object is at maximum height), the velocity is instantaneously zero (the slope of the position-versus-time graph is zero). The acceleration is never zero, even at the moment when the velocity of the object is zero.

It might be tempting to think that at the instant an object is not moving, its acceleration must be zero. This is only true for an object that is at rest and remains at rest. In the case of an object thrown upward, if its acceleration at the top of the flight is zero, it would never descend (it would remain at rest at its highest point).

> **TIP** Physicists say that an object is in a state of free fall even when it is thrown upward, because its acceleration is the same on the way up as on the way down.

Review Question 1.8 Free-fall acceleration can be both positive or negative. Why is this true?

1.9 Tailgating: Putting it all together

Drivers count on their ability to apply the brakes in time if the car in front of them suddenly slows. However, if you are following too closely behind another car, you may not be able to stop in time. Let's look at the motion of two vehicles in what appears to be a safe driving situation.

Figure 1.23 Free-fall motion for an upward thrown ball: (a) a motion diagram, (b) position-versus-time graph, (c) velocity-versus-time graph, and (d) acceleration-versus-time graph.

(a)

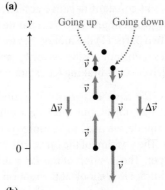

(b)

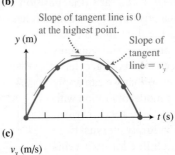

(c)

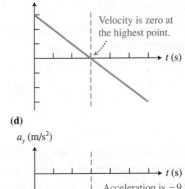

(d)

EXAMPLE 1.12 An accident involving tailgating
A car follows about two car lengths (10.0 m) behind a van. At first, both vehicles are traveling at a conservative speed of 25 m/s (56 mi/h). The driver of the van suddenly slams on the brakes to avoid an accident, slowing down at 9.0 m/s^2. The car driver's reaction time is 0.80 s and the car's maximum acceleration while slowing down is also 9.0 m/s^2. Will the car be able to stop before hitting the van?

Sketch and translate At right we represent this situation for each vehicle (we have two objects of interest). We'll use capital letters to indicate quantities referring to the van and lowercase letters for quantities referring to the car. We use the coordinate system shown with the

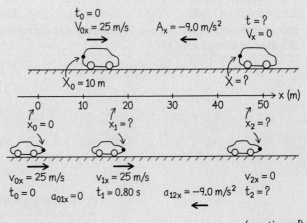

(continued)

origin of the coordinates at the initial position of the car's front bumper. The positive direction is in the direction of motion.

The process starts when the van starts braking. It moves at constant negative acceleration throughout the entire problem process. We separate the motion of the car into two parts: (1) the motion before the driver applies the brakes (constant positive velocity) and (2) its motion after the driver starts braking (constant negative acceleration).

Simplify and diagram We model each vehicle as a point-like object, but since we are trying to determine if they collide, we need to be more specific about their positions. The position of the car will be the position of its front bumper. The position of the van will be the position of its rear bumper. We look at the motion of each vehicle separately. If the car's final position is greater than the van's final position, then a collision has occurred at some point during their motion. Assume that the vehicles have constant acceleration so that we can apply our model of motion with constant acceleration. A velocity-versus-time graph line for each vehicle is shown at right.

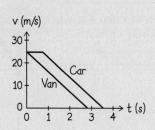

Represent mathematically Equation (1.7) can be used to determine the distance the van travels while stopping:

$$2A_x(X - X_0) = V_x^2 - V_{0x}^2$$

$$\Rightarrow X = \frac{V_x^2 - V_{0x}^2}{2A_x} + X_0$$

The car part 1 Since the car is initially traveling at constant velocity, we use Eq. (1.2). The subscript 0 indicates the moment the driver sees the van start slowing down. The subscript 1 indicates the moment the car driver starts braking.

$$x_1 = x_0 + v_{0x}t_1$$

The car part 2 After applying the brakes, the car has an acceleration of $-9.0\,\text{m/s}^2$. The subscript 2 indicates the moment the car stops moving. We represent this part of the motion using Eq. (1.7):

$$2a_x(x_2 - x_1) = v_{2x}^2 - v_{1x}^2$$

$$\Rightarrow x_2 = \frac{v_{2x}^2 - v_{1x}^2}{2a_x} + x_1$$

$$\Rightarrow x_2 = \frac{v_{2x}^2 - v_{1x}^2}{2a_x} + (x_0 + v_{0x}t_1)$$

The last step came from inserting the result from Part 1 for x_1.

Solve and evaluate The van's initial velocity is $V_0 = +25\,\text{m/s}$, its final velocity is $V_x = 0$, and its acceleration is $A = -9.0\,\text{m/s}^2$. Its initial position is two car lengths in front of the front of the car, so $X_0 = 2 \times 5.0\,\text{m} = 10\,\text{m}$. The final position of the van is

$$X = \frac{V_x^2 - V_{0x}^2}{2A_x} + X_0$$

$$= \frac{0^2 - (25\,\text{m/s})^2}{2(-9.0\,\text{m/s}^2)} + 10\,\text{m}$$

$$= 45\,\text{m}$$

The car's initial position is $x_0 = 0$, its initial velocity is $v_{0x} = 25\,\text{m/s}$, its final velocity is $v_{2x} = 0$, and its acceleration when braking is $a_x = -9.0\,\text{m/s}^2$. The car's final position is

$$x_2 = \frac{v_{2x}^2 - v_{1x}^2}{2a_x} + (x_0 + v_{0x}t_1)$$

$$= \frac{0^2 - (25\,\text{m/s})^2}{2(-9.0\,\text{m/s}^2)} + [0\,\text{m} + (25\,\text{m/s})(0.8\,\text{s})]$$

$$= 55\,\text{m}$$

The car would stop about 10 m beyond where the van would stop. There will be a collision between the two vehicles.

This analysis illustrates why tailgating is such a big problem. The car traveled at a 25-m/s constant velocity during the relatively short 0.80-s reaction time. During the same 0.80 s, the van's velocity decreased by $(0.80\,\text{s})(-9.0\,\text{m/s}^2) = 7.2\,\text{m/s}$ from 25 m/s to about 18 m/s. So the van was moving somewhat slower than the car when the car finally started to brake. Since they were both slowing down at about the same rate, the tailgating vehicle's velocity was always greater than that of the vehicle in front until they hit.

Try it yourself: Two cars, one behind the other, are traveling at 30 m/s (13 mi/h). The front car hits the brakes and slows down at the rate of $10\,\text{m/s}^2$. The driver of the second car has a 1.0-s reaction time. The front car's speed has decreased to 20 m/s during that 1.0 s. The rear car traveling at 30 m/s starts braking, slowing down at the same rate of $10\,\text{m/s}^2$. How far behind the front car should the rear car be so it does not hit the front car?

Answer: The rear car should be at least 30 m behind the front car.

Review Question 1.9 Explain, using physics terms, why tailgating accidents occur.

Summary

Words	Pictorial and physical representations	Mathematical representation
A **reference frame** consists of an object of reference, a point of reference on that object, a coordinate system whose origin is at the point of reference, and a clock. (Section 1.1)	Object of reference Clock Object of interest 0 — Point of reference (origin) Coordinate axis → x	
■ **Time or clock reading** t (a scalar quantity) is the reading on a clock or another time measuring instrument. (Section 1.1) ■ **Time interval** Δt (a scalar quantity) is the difference of two times. (Section 1.3)	t t_1 t_2	Time: t Time interval: $$\Delta t = t_2 - t_1$$
■ **Position** x (a scalar quantity) is the location of an object relative to the chosen origin. (Section 1.3) ■ **Displacement** $\vec{d}$ is a vector drawn from the initial position of an object to its final position. The x-component of the displacement d_x is the change in position of the object along the x-axis. (Section 1.3) ■ **Distance** d (a scalar quantity) is the magnitude of the displacement and is always positive. (Section 1.3) ■ **Path length** l is the length of a string laid along the path the object took. (Section 1.3)	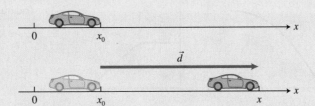 0 x_0 x $\vec{d}$ 0 x_0 x x	$d_x = x - x_0 > 0$ if $\vec{d}$ points in the positive direction of x-axis. $d_x = x - x_0 < 0$ if $\vec{d}$ points in the negative direction. $d = |x - x_0|$
■ **Velocity** $\vec{v}$ (a vector quantity) is the displacement of an object during a time interval divided by that time interval. The velocity is *instantaneous* if the time interval is very small and *average* if the time interval is longer. (Section 1.5) ■ **Speed** v (a scalar quantity) is the magnitude of the velocity. (Section 1.5)	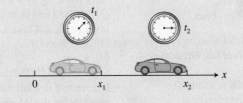 t_1 t_2 0 x_1 x_2 → x	For linear constant velocity, motion $\vec{v} = \vec{d}/\Delta t$ $v_x = \dfrac{\Delta x}{\Delta t} = \dfrac{x_2 - x_1}{t_2 - t_1}$ Eq. (1.1) $v = \left| \dfrac{\Delta x}{\Delta t} \right|$

(continued)

Words	Pictorial and physical representations	Mathematical representation
■ **Acceleration** $\vec{a}$ (a vector quantity) is the change in an object's velocity $\Delta\vec{v}$ during a time interval Δt divided by the time interval. The acceleration is *instantaneous* if the time interval is very small and *average* if the time interval is longer. (Section 1.6)	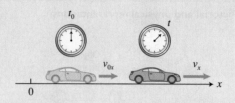	$a_x = \dfrac{v_{2x} - v_{1x}}{t_2 - t_1} = \dfrac{\Delta v_x}{\Delta t}$ Eq. (1.4) (rearranged) $a_x = \dfrac{\Delta v_x}{\Delta t} = \dfrac{v_x - v_{0x}}{t - t_0}$ Eq. (1.4)
Motion with constant velocity or constant acceleration can be represented with a sketch, a motion diagram, kinematics graphs, and mathematically. (Sections 1.5–1.6)	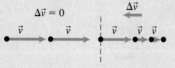 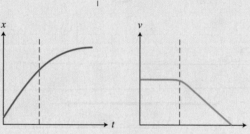	$v_x = v_{0x} + a_x t$ Eq. (1.5) $x = x_0 + v_{0x}t + \dfrac{1}{2}a_x t^2$ Eq. (1.6) $2a_x(x - x_0) = v_x^2 - v_{0x}^2$ Eq. (1.7) (rearranged) $x - x_0 = \dfrac{v_x^2 - v_{0x}^2}{2a_x}$ Eq. (1.7)

 For instructor-assigned homework, go to MasteringPhysics.

Questions

Multiple Choice Questions

1. Match the general elements of physics knowledge (left) with the appropriate examples (right).

   Model of a process   Free fall
   Model of an object   Acceleration
   Physical quantity   Rolling ball
   Physical phenomenon   Point-like object

   (a) Model of a process—Acceleration; Model of an object—Point-like object; Physical quantity—Free fall; Physical phenomenon—Rolling ball.
   (b) Model of a process—Rolling ball; Model of an object—Point-like object; Physical quantity—Acceleration; Physical phenomenon—Free fall.
   (c) Model of a process—Free fall; Model of an object—Point-like object; Physical quantity—Acceleration; Physical phenomenon—Rolling ball.

2. Which group of quantities below consists only of scalar quantities?
   (a) Average speed, displacement, time interval
   (b) Average speed, path length, clock reading
   (c) Temperature, acceleration, position

3. Which of the following are examples of time interval?
   (1) I woke up at 7 am. (2) The lesson lasted 45 minutes. (3) Svetlana was born on November 26. (4) An astronaut orbited Earth in 4 hours.
   (a) 1, 2, 3, and 4   (b) 2 and 4   (c) 2
   (d) 4   (e) 3

4. A student said, "The displacement between my dorm and the lecture hall is 1 kilometer." Is he using the correct physical quantity for the information provided? What should he have called the 1 kilometer?
   (a) Distance   (b) Path length   (c) Position
   (d) Both a and b are correct.

5. An object moves so that its position depends on time as $x = +12 - 4t + t^2$. Which statement below is not true?
   (a) The object is accelerating.
   (b) The speed of the object is always decreasing.
   (c) The object first moves in the negative direction and then in the positive direction.
   (d) The acceleration of the object is $+2 \text{ m/s}^2$.
   (e) The object stops for an instant at 2.0 s.

6. Choose a correct approximate velocity-versus-time graph for the following hypothetical motion: a car moves at constant velocity, and then slows to a stop and without a pause moves in the opposite direction with the same acceleration (**Figure Q1.6**).

Figure Q1.6

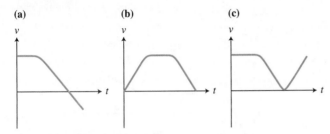

7. In which case are average and instantaneous velocities the same? Explain.
   (a) When the object moves at constant velocity
   (b) When the object moves at constant acceleration
   (c) When the object does not move
   (d) a and c
   (e) a, b, and c

8. You drop a small ball, and then a second small ball. When you drop the second ball, the distance between them is 3 cm. What statement below is correct? Explain.
   (a) The distance between the balls stays the same.
   (b) The distance between the balls decreases.
   (c) The distance between the balls increases.

9. Your car is traveling west at 12 m/s. A stoplight (the origin of the coordinate axis) to the west of you turns yellow when you are 20 m from the edge of the intersection (see **Figure Q1.9**). You apply the brakes and your car's speed decreases. Your car stops before it reaches the stoplight. What are the signs for the components of kinematics quantities?

Figure Q1.9

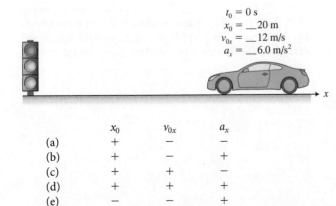

$$t_0 = 0 \text{ s}$$
$$x_0 = \_\_20 \text{ m}$$
$$v_{0x} = \_\_12 \text{ m/s}$$
$$a_x = \_\_6.0 \text{ m/s}^2$$

	x_0	v_{0x}	a_x
(a)	+	−	−
(b)	+	−	+
(c)	+	+	−
(d)	+	+	+
(e)	−	−	+

10. Which velocity-versus-time graph in **Figure Q1.10** best describes the motion of the car in the previous problem (see Figure Q1.9) as it approaches the stoplight?

Figure Q1.10

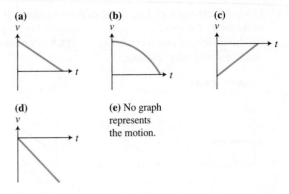

(e) No graph represents the motion.

11. Suppose that (c) in Figure Q1.10 represents the velocity-versus-time graph for a moving object. Which of the following gives the correct signs for the velocity and acceleration components [positive (+) or negative (−)] of this motion?

	v_x	a_x
(a)	+	+
(b)	+	−
(c)	−	+
(d)	−	−

12. A sandbag hangs from a rope attached to a rising hot air balloon. The rope connecting the bag to the balloon is cut. How will two observers see the motion of the sandbag? Observer 1 is in the hot air balloon and observer 2 is on the ground.
    (a) Both 1 and 2 will see it go down.
    (b) 1 will see it go down and 2 will see it go up.
    (c) 1 will see it go down and 2 will see it go up and then down.

13. An apple falls from a tree. It hits the ground at a speed of about 5.0 m/s. What is the approximate height of the tree?
    (a) 2.5 m   (b) 1.2 m   (c) 10.0 m   (d) 2.4 m

14. You have two small metal balls. You drop the first ball and throw the other one in the downward direction. Choose the statements that are not correct.
    (a) The second ball will spend less time in flight.
    (b) The first ball will have a slower final speed when it reaches the ground.
    (c) The second ball will have larger acceleration.
    (d) Both balls will have the same acceleration.

15. You throw a small ball upward. Then you throw it again, this time at twice the initial speed. Choose the correct statement.
    (a) The second time, the ball travels twice as far up as the first time.
    (b) The second time, the ball has twice the magnitude of acceleration while in flight that it did the first time.
    (c) The second time, the ball spends twice as much time in flight.
    (d) All of the choices are correct.

16. You throw a small ball upward and notice the time it takes to come back. If you then throw the same ball so that it takes twice as much time to come back, what is true about the motion of the ball the second time?
    (a) Its initial speed was twice the speed in the first experiment.
    (b) It traveled an upward distance that is twice the distance of the original toss.
    (c) It had twice as much acceleration on the way up as it did the first time.
    (d) The ball stopped at the highest point and had zero acceleration at that point.

Conceptual Questions

17. Lance Armstrong is cycling along an 800-m straight stretch of the track. His speed is 13 m/s. Choose all of the graphical representations of motion from **Figure Q1.17** that correctly describe Armstrong's motion.

Figure Q1.17

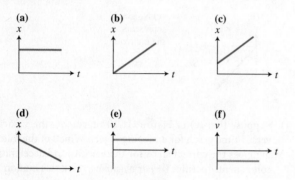

18. In what reasonable ways can you represent or describe the motion of a car traveling from one stoplight to the next? Construct each representation for the moving car.
19. What is the difference between speed and velocity? Between path length and distance? Between distance and displacement? Give an example of each.
20. What physical quantities do we use to describe motion? What does each quantity characterize? What are their SI units?
21. Devise stories describing each of the motions shown in each of the graphs in **Figure Q1.21**. Specify the object of reference.

Figure Q1.21

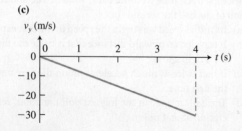

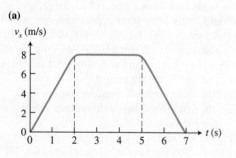

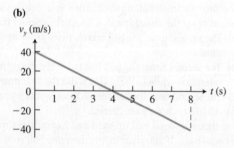

22. For each of the position-versus-time graphs in **Figure Q1.22**, draw velocity-versus-time graphs and acceleration-versus-time graphs.

Figure Q1.22

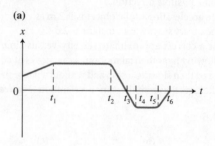

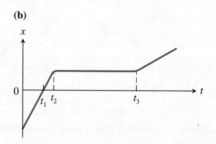

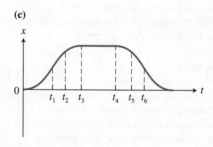

23. Can an object have a nonzero velocity and zero acceleration? If so, give an example.
24. Can an object at one instant of time have zero velocity and nonzero acceleration? If so, give an example.
25. Your little sister has a battery-powered toy truck. When the truck is moving, how can you determine whether it has constant velocity, constant speed, constant acceleration, or changing acceleration? Explain in detail.
26. You throw a ball upward. Your friend says that at the top of its flight the ball has zero velocity and zero acceleration. Do you agree or disagree? If you agree, explain why. If you disagree, how would you convince your friend of your opinion?

Problems

Below, BIO indicates a problem with a biological or medical focus. Problems labeled EST ask you to estimate the answer to a quantitative problem rather than derive a specific answer. Asterisks indicate the level of difficulty of the problem. Problems marked with / require you to make a drawing or graph as part of your solution. Problems with no * are considered to be the least difficult. A single * marks moderately difficult problems. Two ** indicate more difficult problems.

1.2 A conceptual description of motion

1. A car starts at rest from a stoplight and speeds up. It then moves at constant speed for a while. Then it slows down until reaching the next stoplight. Represent the motion with a motion diagram as seen by the observer on the ground.

2. * / You are an observer on the ground. (a) Draw two motion diagrams representing the motions of two runners moving at the same constant speeds in opposite directions toward you. Runner 1, coming from the east, reaches you in 5 s, and runner 2 reaches you in 3 s. (b) Draw a motion of diagram for the second runner as seen by the first runner.

3. * A car is moving at constant speed on a highway. A second car catches up and passes the first car 5 s after it starts to speed up. Represent the situation with a motion diagram. Specify the observer with respect to whom you drew the diagram.

4. * / A hat falls off a man's head and lands in the snow. Draw a motion diagram representing the motion of the hat as seen by the man.

1.3 and 1.4 Quantities for describing motion and Representing motion with data tables and graphs

5. * You drive 100 km east, do some sightseeing, and then turn around and drive 50 km west, where you stop for lunch. (a) Represent your trip with a displacement vector. Choose an object of reference and coordinate axis so that the scalar component of this vector is (b) positive; (c) negative; (d) zero.

6. * Choose an object of reference and a set of coordinate axes associated with it. Show how two people can start and end their trips at different locations but still have the same displacement vectors in this reference frame.

7. The scalar x-component of a displacement vector for a trip is -70 km. Represent the trip using a coordinate axis and an object of reference. Then change the axis so that the displacement component becomes $+70$ km.

8. * You recorded your position with respect to the front door of your house as you walked to the mailbox. Examine the data presented in **Table 1.9** and answer the following questions: (a) What instruments did you use to collect data? (b) What are the uncertainties in your data? (c) Represent your motion using a position-versus-time graph. (d) Tell the story of your motion in words. (e) Show on the graph the displacement, distance, and path length.

Table 1.9

t (s)	1	2	3	4	5	6	7	8	9
x (steps)	2	4	9	13	18	20	16	11	9

1.5 Constant velocity linear motion

9. * You need to determine the time interval (in seconds) needed for light to pass an atomic nucleus. What information do you need? How will you use it? What simplifying assumptions about the objects and processes do you need to make? What approximately is that time interval?

10. A speedometer reads 65 mi/h. (a) Use as many different units as possible to represent the speed of the car. (b) If the speedometer reads 100 km/h, what is the car's speed in mi/h?

11. Convert the following record speeds so that they are in mi/h, km/h, and m/s. (a) Australian dragonfly—36 mi/h; (b) the diving peregrine falcon—349 km/h; and (c) the Lockheed SR-71 jet aircraft—980 m/s (about three times the speed of sound).

12. EST **Hair growth speed** Estimate the rate that your hair grows in meters per second. Indicate any assumptions you made.

13. * EST / A kidnapped banker looking through a slit in a van window counts her heartbeats and observes that two highway exits pass in 80 heartbeats. She knows that the distance between the exits is 1.6 km (1 mile). (a) Estimate the van's speed. (b) Choose and describe a reference frame and draw a position-versus-time graph for the van.

14. EST Make a simplified map of the path from where you live to your physics classroom. (a) Label your path and your displacement. (b) Estimate the time interval that you need to reach the classroom from where you live and your average speed.

15. * **Equation Jeopardy** Two observers observe two different moving objects. However, they describe their motions mathematically with the same equation: $x(t) = 10 \text{ km} - (4 \text{ km/h})t$. (a) Write two short stories about these two motions. Specify where each observer is and what she is doing. What is happening to the moving object at $t = 0$? (b) Use significant digits to determine the interval within which the initial position is known.

16. * Your friend's pedometer shows that he took 17,000 steps in 2.50 h during a hike. Determine everything you can about the hike. What assumptions did you make? How certain are you in your answer? How would the answer change if the time were given as 2.5 h instead of 2.50 h?

17. During a hike, two friends were caught in a thunderstorm. Four seconds after seeing lightning from a distant cloud, they heard thunder. How far away was the cloud (in kilometers)? Write your answer as an interval using significant digits as your guide. Sound travels in air at about 340 m/s.

18. Light travels at a speed of 3.0×10^8 m/s in a vacuum. The approximate distance between Earth and the Sun is 150×10^6 km. How long does it take light to travel from the Sun to Earth? What are the margins within which you know the answer?

19. Proxima Centauri is 4.22 ± 0.01 light-years from Earth. Determine the length of 1 light-year and convert the distance to the star into meters. What is the uncertainty in the answer?

20. * Spaceships traveling to other planets in the solar system move at an average speed of 1.1×10^4 m/s. It took Voyager about 12 years to reach the orbit of Uranus. What can you learn about the solar system using these data? What assumption did you make? How did this assumption affect the results?

21. ** **Figure P1.21** shows a velocity-versus-time graph for the bicycle trips of two friends with respect to the parking lot where they started. (a) Determine their displacements in 20 s. (b) If Xena's position at time zero is 0 and Gabriele's position is 60 m, what time interval is needed for Xena to catch Gabriele? (c) Use the information from (b) to write function $x(t)$ for Gabriele with respect to Xena.

Figure P1.21

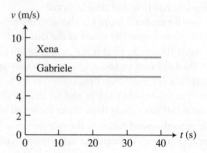

22. * / **Table 1.10** shows position and time data for your walk along a straight path. (a) Tell everything you can about the walk. Specify the object of reference. (b) Draw a motion diagram, draw a graph $x(t)$, and write a function $x(t)$ that is consistent with the data and the chosen reference frame.

Table 1.10

Time (s)	Position (m)
0	80
10	40
20	0
30	−40
40	−80
50	−120

23. * / **Table 1.11** shows position and time data for your friend's bicycle ride along a straight bike path. (a) Tell everything you can about his ride. Specify the observer. (b) Draw a motion diagram, draw a graph $x(t)$, and write a function $x(t)$ that is consistent with the ride.

Table 1.11

Time (s)	Position (m)
0	−200
10	−120
20	−40
30	40
40	120
50	200

24. * You are walking to your physics class at speed 1.0 m/s with respect to the ground. Your friend leaves 2.0 min after you and is walking at speed 1.3 m/s in the same direction. How fast is she walking with respect to you? How far does your friend travel before she catches up with you? Indicate the uncertainty in your answers. Describe any assumptions that you made.

25. * Gabriele enters an east–west straight bike path at the 3.0-km mark and rides west at a constant speed of 8.0 m/s. At the same time, Xena rides east from the 1.0-km mark at a constant speed of 6.0 m/s. (a) Write functions $x(t)$ that describe their positions as a function of time with respect to Earth. (b) Where do they meet each other? In how many different ways can you solve this problem? (c) Write a function $x(t)$ that describes Xena's motion with respect to Gabriele.

26. * Jim is driving his car at 32 m/s (72 mi/h) along a highway where the speed limit is 25 m/s (55 mi/h). A highway patrol car observes him pass and quickly reaches a speed of 36 m/s.

At that point, Jim is 300 m ahead of the patrol car. How far does the patrol car travel before catching Jim?

27. * You hike two thirds of the way to the top of a hill at a speed of 3.0 mi/h and run the final third at a speed of 6.0 mi/h. What was your average speed?

28. * Olympic champion swimmer Michael Phelps swam at an average speed of 2.01 m/s during the first half of the time needed to complete a race. What was his average swimming speed during the second half of the race if he tied the record, which was at an average speed of 2.05 m/s?

29. * A car makes a 100-km trip. It travels the first 50 km at an average speed of 50 km/h. How fast must it travel the second 50 km so that its average speed is 100 km/h?

30. * Jane and Bob see each other when 100 m apart. They are moving toward each other, Jane at speed 4.0 m/s and Bob at speed 3.0 m/s with respect to the ground. What can you determine about this situation using these data?

31. * The graph in **Figure P1.31** represents four different motions. (a) Write a function $x(t)$ for each motion. (b) Use the information in the graph to determine as many quantities related to the motion of these objects as possible. (c) Act out these motions with two friends. (Hint: think of what each object was doing at $t = 0$.)

Figure P1.31

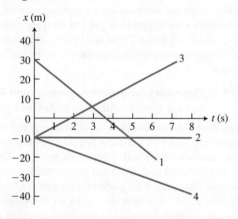

1.6 Motion at constant acceleration

32. A car starts from rest and reaches the speed of 10 m/s in 30 s. What can you determine about the motion of the car using this information?

33. A truck is traveling east at +16 m/s. (a) The driver sees that the road is empty and accelerates at $+1.0 \ m/s^2$ for 5.0 s. What can you determine about the truck's motion using these data? (b) The driver then sees a red light ahead and decelerates at $-2.0 \ m/s^2$ for 3.0 s. What can you determine about the truck's motion using these data? (c) Determine the values of the quantities you listed in (a) and (b).

34. **Bumper car collision** On a bumper car ride, friends smash their cars into each other (head-on) and each has a speed *change* of 3.2 m/s. If the magnitudes of acceleration of each car during the collision averaged 28 m/s^2, determine the time interval needed to stop and the stopping distance for each car while colliding. Specify your reference frame.

35. A bus leaves an intersection accelerating at $+2.0 \text{ m/s}^2$. Where is the bus after 5.0 s? What assumption did you make? If this assumption is not valid, would the bus be closer or farther away from the intersection compared to your original answer? Explain.

36. A jogger is running at $+4.0$ m/s when a bus passes her. The bus is accelerating from $+16.0$ m/s to $+20.0$ m/s in 8.0 s. The jogger speeds up with the same acceleration. What can you determine about the jogger's motion using these data?

37. * / The motion of a person as seen by another person is described by the equation $v = -3.0 \text{ m/s} + (0.5 \text{ m/s}^2)t$. (a) Represent this motion with a motion diagram and position-, velocity-, and acceleration-versus-time graphs. (b) Say everything you can about this motion and describe what happens to the person when his speed becomes zero.

38. **Tour de France** While cycling at speed of 10 m/s, Lance Armstrong starts going downhill with an acceleration of magnitude 1.2 m/s^2. The descent takes 10.0 s. What can you determine about Lance's motion using these data? What assumptions did you make?

39. An automobile engineer found that the impact of a truck colliding at 16 km/h with a concrete pillar caused the bumper to indent only 6.4 cm. The truck stopped. Determine the acceleration of the truck during the collision.

40. **BIO Squid propulsion** *Lolliguncula brevis* squid use a form of jet propulsion to swim—they eject water out of jets that can point in different directions, allowing them to change direction quickly. When swimming at a speed of 0.15 m/s or greater, they can accelerate at 1.2 m/s^2. (a) Determine the time interval needed for a squid to increase its speed from 0.15 m/s to 0.45 m/s. (b) What other questions can you answer using the data?

41. **Dragster record on the desert** In 1977, Kitty O'Neil drove a hydrogen peroxide–powered rocket dragster for a record time interval (3.22 s) and final speed (663 km/h) on a 402-m-long Mojave Desert track. Determine her average acceleration during the race and the acceleration while stopping (it took about 20 s to stop). What assumptions did you make?

42. * Imagine that a sprinter accelerates from rest to a maximum speed of 10.8 m/s in 1.8 s. In what time interval will he finish the 100-m race if he keeps his speed constant at 10.8 m/s for the last part of the race? What assumptions did you make?

43. ** / Two runners are running next to each other when one decides to accelerate at a constant rate of a. The second runner notices the acceleration after a short time interval Δt when the distance between the runners is Δx. The second runner accelerates at the same acceleration. Represent their motions with a motion diagram and position-versus-time graph (both graph lines on the same set of axes). Use any of the representations to predict what will happen to the distance between the runners—will it stay Δx, increase, or decrease? Assume that the runners continue to have the same acceleration for the duration of the problem.

44. **Meteorite hits car** In 1992, a 14-kg meteorite struck a car in Peekskill, NY, leaving a 20-cm-deep dent in the trunk. (a) If the meteorite was moving at 500 m/s before striking the car, what was the magnitude of its acceleration while stopping? Indicate any assumptions you made. (b) What other questions can you answer using the data in the problem?

45. **BIO Froghopper jump** A spittlebug called the froghopper (*Philaenus spumarius*) is believed to be the best jumper in the animal world. It pushes off with muscular rear legs for 0.0010 s, reaching a speed of 4.0 m/s. Determine its acceleration during this launch and the distance that the froghopper moves while its legs are pushing.

46. **Tennis serve** The fastest server in women's tennis is Venus Williams, who recorded a serve of 130 mi/h (209 km/h) in 2007. If her racket pushed on the ball for a distance of 0.10 m, what was the average acceleration of the ball during her serve? What was the time interval for the racket-ball contact?

47. * **Shot from a cannon** In 1998, David "Cannonball" Smith set the distance record for being shot from a cannon (56.64 m). During a launch in the cannon's barrel, his speed increased from zero to 80 km/h in 0.40 s. While he was being stopped by the catching net, his speed decreased from 80 km/h to zero with an average acceleration of 180 m/s^2. What can you determine about Smith's flight using this information?

48. **Col. John Stapp's final sled run** Col. John Stapp led the U.S. Air Force Aero Medical Laboratory's research into the effects of higher accelerations. On Stapp's final sled run, the sled reached a speed of 284.4 m/s (632 mi/h) and then stopped with the aid of water brakes in 1.4 s. Stapp was barely conscious and lost his vision for several days but recovered. Determine his acceleration while stopping and the distance he traveled while stopping.

49. * Sprinter Usain Bolt reached a maximum speed of 11.2 m/s in 2.0 s while running the 100-m dash. (a) What was his acceleration? (b) What distance did he travel during this first 2.0 s of the race? (c) What assumptions did you make? (d) What time interval was needed to complete the race, assuming that he ran the last part of the race at his maximum speed? (e) What is the total time for the race? How certain are you of the number you calculated?

50. * Imagine that Usain Bolt can reach his maximum speed in 1.7 s. What should be his maximum speed in order to tie the 19.5-s record for the 200-m dash?

51. A bus is moving at a speed of 36 km/h. How far from a bus stop should the bus start to slow down so that the passengers feel comfortable (a comfortable acceleration is 1.2 m/s^2)?

52. * **EST** You want to estimate how fast your car accelerates. What information can you collect to answer this question? What assumptions do you need to make to do the calculation using the information?

53. * In your car, you covered 2.0 m during the first 1.0 s, 4.0 m during the second 1.0 s, 6.0 m during the third 1.0 s, and so forth. Was this motion at constant acceleration? Explain.

54. (a) Determine the acceleration of a car in which the velocity changes from -10 m/s to -20 m/s in 4.0 s. (b) Determine the car's acceleration if its velocity changes from -20 m/s to -18 m/s in 2.0 s. (c) Explain why the sign of the acceleration is different in (a) and (b).

1.7 Skills for analyzing situations involving motion

55. * Use the velocity-versus-time graph lines in **Figure P1.55** to determine the change in the position of each car from 0 s to 60 s. Represent the motion of each car mathematically as a function $x(t)$. Their initial positions are A (200 m) and B (−200 m).

Figure P1.55

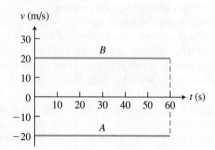

56. * ✐ An object moves so that its position changes in the following way: $x = 10 \text{ m} - (4 \text{ m/s})t$. (a) Describe all of the known quantities for this motion. (b) Invent a story for the motion. (c) Draw a position-versus-time graph, and use the graph to determine when the object reaches the origin of the reference frame. (e) Act out the motion.

57. ** ✐ An object moves so that its position changes in the following way: $x(t) = -100 \text{ m} + (30 \text{ m/s})t + (3.0 \text{ m/s}^2)t^2$. (a) What kind of motion is this (constant velocity, constant acceleration, or changing acceleration)? (b) Describe all of the known quantities for this motion. (c) Invent a story for the motion. (d) Draw a velocity-versus-time graph, and use it to determine when the object stops. (e) Use equations to determine when and where it stops. Did you get the same answer using graphs and equations?

58. ** ✐ The position of an object changes according to the functions listed below. For each case, determine the known quantities concerning the motion, devise a story describing the motion consistent with the functions, and draw position-versus-time, velocity-versus-time, and acceleration-versus-time graphs: (a) $x(t) = 15.0 \text{ m} - (-3.0 \text{ m/s}^2)t^2$; (b) $x(t) = 30.0 \text{ m} - (1.0 \text{ m/s})t$; and (c) $x = -10 \text{ m}$.

59. * ✐ The positions of objects A and B with respect to Earth depend on time as follows: $x(t)_A = 10.0 \text{ m} - (4.0 \text{ m/s})t$; $x(t)_B = -12 \text{ m} + (6 \text{ m/s})t$. Represent their motions on a motion diagram and graphically (position-versus-time and velocity-versus-time graphs). Use the graphical representations to find where and when they will meet. Confirm the result with mathematics.

60. * Two cars on a straight road at time zero are beside each other. The first car, traveling at speed 30 m/s, is passing the second car, which is traveling at 24 m/s. Seeing a cow on the road ahead, the driver of each car starts to slow down at 6.0 m/s². Represent the motions of the cars mathematically and on a velocity-versus-time graph from the point of view of a pedestrian. Where is each car when it stops?

61. * The changing velocity of a car is represented in the velocity-versus-time graph shown in **Figure P1.61**. (a) Describe everything you

Figure P1.61

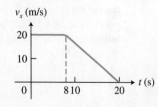

can about the motion of the car using the graph. (b) What is the displacement of the car between times 10 s and 20 s? (c) What was the average speed of the car?

62. * The changing velocity of a car is represented in the velocity-versus-time graph shown in **Figure P1.62**. (a) Describe everything you can about the motion of the car using the graph. (b) What is the displacement of the car between times 0 s and 45 s? What is the path traveled? (c) What is the average speed of the car during all 70 s? What is the average velocity?

Figure P1.62

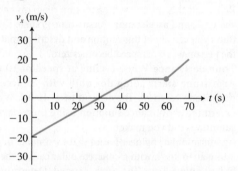

63. A diagram representing the motion of two cars is shown in **Figure P1.63**. The number near each dot indicates the clock reading in seconds when the car passes that location. (a) Indicate times when the cars have the same speed. (b) Indicate times when they have the same position.

Figure P1.63

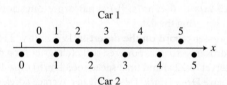

64. ** Solve the equations below for the unknown quantities and then describe a possible process that is consistent with the equations. There are many possibilities. The object is moving on an inclined surface. This is a two-part process.

Part I: $x_1 = 0 + (0)t_1 + (2.5 \text{ m/s}^2)t_1^2$

Part II: $x_2 = x_1 + (20 \text{ m/s})(0.40 \text{ s}) + (1/2)a_{x2}(0.40 \text{ s})^2$

1.8 and 1.9 Free fall and Tailgating: Putting it all together

65. You accidentally drop an eraser out the window of an apartment 15 m above the ground. (a) How long will it take for the eraser to reach the ground? (b) What speed will it have just before it reaches the ground? (c) If you multiply the time interval answer from (a) and the speed answer from (b), why is the result much more than 15 m?

66. What is the average speed of the eraser in the previous problem?

67. You throw a tennis ball straight upward. The initial speed is about 12 m/s. Say everything you can about the motion of the ball. Is 12 m/s a realistic speed for an object that you can throw with your hands?

68. While skydiving, your parachute opens and you slow from 50.0 m/s to 8.0 m/s in 0.80 s. Determine the distance you fall while the parachute is opening. Some people faint if they experience acceleration greater than 5 g (5 times 9.8 m/s²). Will you feel faint? Explain and discuss simplifying assumptions inherent in your explanation.

69. After landing from your skydiving experience, you are so excited that you throw your helmet upward. The helmet rises 5.0 m above your hands. What was the initial speed of the helmet when it left your hands? How long was it moving from the time it left your hands until it returned?

70. You are standing on the rim of a canyon. You drop a rock and in 7.0 s hear the sound of it hitting the bottom. How deep is the canyon? What assumptions did you make? Examine how each assumption affects the answer. Does it lead to a larger or smaller depth than the calculated depth? (The speed of sound in air is about 340 m/s.)

71. You are doing an experiment to determine your reaction time. Your friend holds a ruler. You place your fingers near the sides of the lower part of the ruler without touching it. The friend drops the ruler without warning you. You catch the ruler after it falls 12.0 cm. What was your reaction time?

72. **EST Cliff divers** Divers in Acapulco fall 36 m from a cliff into the water. Estimate their speed when they enter the water and the time interval needed to reach the water. What assumption did you make? Does this assumption make the calculated speed larger or smaller than actual speed?

73. * Galileo dropped a light rock and a heavy rock from the Leaning Tower of Pisa, which is about 55 m high. Suppose that Galileo dropped one rock 0.50 s before the second rock. With what initial velocity should he drop the second rock so that it reaches the ground at the same time as the first rock?

74. A person holding a lunch bag is moving upward in a hot air balloon at a constant speed of 7.0 m/s. When the balloon is 24 m above the ground, she accidentally releases the bag. What is the speed of the bag just before it reaches the ground?

75. A parachutist falling vertically at a constant speed of 10 m/s drops a penknife when 20 m above the ground. What is the speed of the knife just before it reaches the ground?

76. * You are traveling in your car at 20 m/s a distance of 20 m behind a car traveling at the same speed. The driver of the other car slams on the brakes to stop for a pedestrian who is crossing the street. Will you hit the car? Your reaction time is 0.60 s. The maximum acceleration of each car is 9.0 m/s^2.

77. * You are driving a car behind another car. Both cars are moving at speed 80 km/h. What minimum distance behind the car in front should you drive so that you do not crash into the car's rear end if the driver of that car slams on the brakes? Indicate any assumptions you made.

78. A driver with a 0.80-s reaction time applies the brakes, causing the car to have 7.0-m/s^2 acceleration opposite the direction of motion. If the car is initially traveling at 21 m/s, how far does the car travel during the reaction time? How far does the car travel after the brakes are applied and while skidding to a stop?

79. ** Some people in a hotel are dropping water balloons from their open window onto the ground below. The balloons take 0.15 s to pass your 1.6-m-tall window. Where should security look for the raucous hotel guests? Indicate any assumptions that you made in your solution.

80. ** **BIO EST Avoiding injury from hockey puck** Hockey players wear protective helmets with facemasks. Why? Because the bone in the upper part of the cheek (the zygomatic bone) can fracture if the acceleration of a hockey puck due to its interaction with the bone exceeds 900 g for a time lasting 6.0 ms or longer. Suppose a player was not wearing a facemask. Is it likely that the acceleration of a hockey puck when hitting the bone would exceed these numbers? Use some reasonable numbers of your choice and estimate the puck's acceleration if hitting an unprotected zygomatic bone.

81. ** **EST** A bottle rocket burns for 1.6 s. After it stops burning, it continues moving up to a maximum height of 80 m above the place where it stopped burning. Estimate the acceleration of the rocket during launch. Indicate any assumptions made during your solution. Examine their effect.

82. * **Data from state driver's manual** The state driver's manual lists the reaction distances, braking distances, and total stopping distances for automobiles traveling at different initial speeds (**Table 1.12**). Use the data determine the driver's reaction time interval and the acceleration of the automobile while braking. The numbers assume dry surfaces for passenger vehicles.

Table 1.12 Data from driver's manual.

Speed (mi/h)	Reaction distance (m)	Braking distance (m)	Total stopping distance (m)
20	7	7	14
40	13	32	45
60	20	91	111

83. ** **EST** Estimate the time interval needed to pass a semi-trailer truck on a highway. If you are on a two-lane highway, how far away from you must an approaching car be in order for you to safely pass the truck without colliding with the oncoming traffic? Indicate any assumptions used in your estimate.

84. * Car A is heading east at 30 m/s and Car B is heading west at 20 m/s. Suddenly, as they approach each other, they see a one-way bridge ahead. They are 100 m apart when they each apply the brakes. Car A's speed decreases at 7.0 m/s each second and Car B decreases at 9.0 m/s each second. Do the cars collide?

Reading Passage Problems

BIO Head injuries in sports A research group at Dartmouth College has developed a Head Impact Telemetry (HIT) System that can be used to collect data about head accelerations during impacts on the playing field. The researchers observed 249,613 impacts from 423 football players at nine colleges and high schools and collected collision data from participants in other sports. The accelerations during most head impacts (>89%) in helmeted sports caused head accelerations less than a magnitude of 400 m/s^2. However, a total of 11 concussions were diagnosed in players whose impacts caused accelerations between 600 and 1800 m/s^2, with most of the 11 over 1000 m/s^2.

85. Suppose that the magnitude of the head velocity change was 10 m/s. Which time interval below for the collision would be closest to producing a possible concussion with an acceleration of 1000 m/s^2?
    (a) 1 s (b) 0.1 s (c) 10^{-2} s
    (d) 10^{-3} s (e) 10^{-4} s

86. Using numbers from the previous problem, which answer below is closest to the average speed of the head while stopping?
    (a) 50 m/s (b) 10 m/s (c) 5 m/s
    (d) 0.5 m/s (e) 0.1 m/s

87. Suppose the average speed while stopping was 4 m/s (not necessarily the correct value) and the collision lasted 0.01 s. Which answer below is closest to the head's stopping distance (the distance it moves while stopping)?

    (a) 0.04 m   (b) 0.4 m   (c) 4 m
    (d) 0.02 m   (e) 0.004 m

88. Use Eq. (1.7) and the numbers from Problem 85 to determine which stopping distance below is closest to that which would lead to a 1000 m/s^2 head acceleration.

    (a) 0.005 m   (b) 0.5 m   (c) 0.1 m
    (d) 0.01 m   (e) 0.05 m

89. Choose from the list below the changes in the head impacts that would *reduce* the acceleration during the impact.

    1. A shorter impact time interval
    2. A longer impact time interval
    3. A shorter stopping distance
    4. A longer stopping distance
    5. A smaller initial speed
    6. A larger initial speed

    (a) 1, 4, 6   (b) 1, 3, 5   (c) 1, 4, 5
    (d) 2, 4, 5   (e) 2, 4, 6

Sending rockets to observe X-ray sources Before 1962, few astronomers believed that the universe contained celestial bodies that were hot enough to emit X-rays—about 10,000 times hotter than the surface of the Sun.

Because the atmosphere absorbs the X-rays produced by such sources, they can only be detected beyond Earth's atmosphere, 200 km or more above Earth's surface. Before satellites were available in the 1970s, scientists searched for X-ray sources by launching rockets (the first in 1962 from White Sands Missile Range in New Mexico) that contained detectors that could sample the skies for the short time interval that the rocket remained above the atmosphere—less than 10 min. Such a Terrier-Sandhawk rocket was flown on May 11, 1970 from the Kauai Test Range in Hawaii. Modern satellites can collect data continuously. Satellite observations and analysis have now identified several types of celestial bodies that emit X-rays, including X-ray pulsars in the constellations of Cygnus and Hercules, supernovae remnants, and quasars.

90. Detectors on rockets moving above Earth's atmosphere can detect X-ray sources, but similar detectors on Earth cannot because

    (a) light from the Sun overwhelms the X-ray signals in the detectors.
    (b) air in the atmosphere absorbs the X-rays before they reach Earth-based detectors.
    (c) the rocket can see the X-ray sources more easily because it is nearer them.
    (d) Earth is much heaver than a rocket, and hence the X-rays affect it less.

91. During fuel burn, the vertically launched Terrier-Sandhawk rocket had an acceleration of 300 m/s^2 (30 times free-fall acceleration—called 30 *g*). The fuel burned for 8 s. About how fast was the rocket moving at the end of the burn?

    (a) 2400 m/s (b) 40 m/s   (c) 240 *g*   (d) 4 *g*

92. Which answer below is closest to the height of the Terrier-Sandhawk rocket at the end of fuel burn?

    (a) 20,000 m (b) 10,000 m (c) 1000 m (d) 300 m

93. Which number below is closest to the time interval after blast-off that the Terrier-Sandhawk rocket reached its maximum height?

    (a) 19,000 s (b) 2400 s   (c) 250 s   (d) 10 s

94. Which number below is closest to the maximum height reached by the Terrier-Sandhawk rocket?

    (a) 300,000 m (b) 200,000 m (c) 12,000 m (d) 9600 m

Newtonian Mechanics

2

Why do seat belts and air bags save lives?

If you stand on a bathroom scale in a moving elevator, does its reading change?

Can a parachutist survive a fall if the parachute does not open?

Seat belts and air bags save about 250,000 lives worldwide every year by preventing a seated driver or passenger from flying forward into the hard-surfaced steering wheel or dashboard after a vehicle stops abruptly. Air bags combined with seat belts significantly reduce the risk of injury (belts alone by about 40% and belts with air bags by about 54%). How do seat belts and air bags provide this protection?

Be sure you know how to:

- Draw a motion diagram for a moving object (Section 1.2).
- Determine the direction of acceleration using a motion diagram (Section 1.6).
- Add vectors graphically and by components for one-dimensional motion (Section 1.2 and Mathematics Review appendix).

In the last chapter, we learned to *describe* motion—for example, to determine a car's acceleration when stopped abruptly during a collision. However, we did not discuss the causes of the acceleration. In this chapter, we will learn *why* an object has a particular acceleration. This knowledge will help us *explain* the motion of many objects: cars, car passengers, elevators, skydivers, and even rockets.

2.1 Describing and representing interactions

What causes objects to accelerate or maintain a constant velocity? Consider a simple experiment—standing on Rollerblades® on a horizontal floor. No matter how hard you swing your arms or legs you cannot start moving by yourself; you need to either push off the floor or have someone push or pull you. Physicists say that the floor or the other person *interacts* with you, thus changing your motion. Objects can interact directly, when they touch each other, or at a distance—a magnet attracts or repels another magnet without touching it.

Choosing a system in a sketch of a process

We learned (in Chapter 1) that the first step in analyzing any process is sketching it. **Figure 2.1** shows a sketch of a car skidding to avoid a collision with a van. In this and later chapters we choose in the sketch one particular object for detailed analysis. We call this object the **system**. All other objects that are not part of the system can interact with it (touch it, pull it, and push it) and are in the system's **environment**. Interactions between the system object and objects in the environment are called **external interactions**.

Figure 2.1 A sketch of a car skidding to avoid a collision with a van.

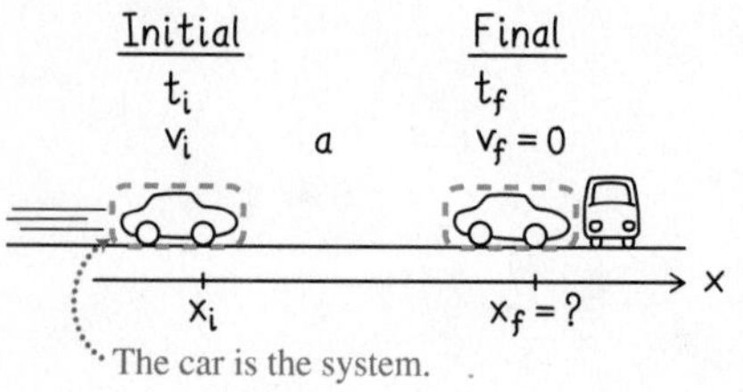

> **System** A system is the object that we choose to analyze. Everything outside that system is called its environment and consists of objects that might interact with the system (touch, push, or pull it) and affect its motion through external interactions.

On the sketch we make a light boundary (a closed dashed line) around the system object to emphasize the system choice (the skidding car in Figure 2.1). In this chapter systems consist of only one object. In later chapters systems can have more than one object. Sometimes, a single system object has parts—like the wheels on the car and its axles. The parts interact with each other. Since both parts are in the system, these are called **internal interactions**. In this chapter we will model an object like a car as a point-like object and ignore such internal interactions.

Representing interactions

External interactions affect the motion or lack of motion of a system object. Consider holding a bowling ball in one hand and volleyball in the other. Each ball is considered as a system object (**Figure 2.2a**). What objects interact with each ball? Your hand pushes up hard to keep the bowling ball steady and much less to keep the volleyball steady (Figure 2.2b). We use an arrow to represent the upward push exerted by each hand on one of the balls. Notice that the arrow is longer for the interaction of the hand with the bowling ball than for the hand with the volleyball.

The arrow represents a "force" that is exerted by the hand on the ball. A **force** is a physical quantity that characterizes how hard and in what direction an external object pushes or pulls on the system object. The symbol for force is $\vec{F}$ with a subscript that identifies the external object that exerts the force and the system object on which the force is exerted. For example, the hand pushing on the bowling ball is represented as $\vec{F}_{\text{H on B}}$. The force that the hand exerts on the volleyball is $\vec{F}_{\text{H on V}}$. The arrow above the symbol indicates that force is a vector quantity with a magnitude *and* direction. The SI unit of a force is called a newton (N). When you hold a 100-g ball, you exert an upward force that is a little less than 1.0 N. We will devise a formal definition for force later in this chapter.

Do any other objects exert forces on the balls? Intuitively, we know that something must be pulling down to balance the upward force your hands exert on the balls. The word **gravity** represents the interaction of planet Earth with the ball. Earth pulls downward on an object toward Earth's center. Because of this interaction we need to include a second arrow representing the force that Earth exerts on the ball $\vec{F}_{\text{E on B}}$ (see Figure 2.2c). Intuitively, we know that the two arrows for each ball should be of the same length, since the ball is not moving anywhere.

Do other objects besides the hand and Earth interact with the ball? Air surrounds everything close to Earth. Does it push down or up on the balls? Let's hypothesize that the air does push down. If air pushes down, then our hands have to push up harder to balance the combined effect of the downward push of the air and the downward pull of Earth on the balls. Let's test the hypothesis that air pushes down on the balls.

Testing a hypothesis

To test a hypothesis in science means to first accept it as a true statement (even if we disagree with it); then design an experiment whose outcome we can predict using this hypothesis (a testing experiment); then compare the outcome of the experiment and the prediction; and, finally, make a preliminary judgment about the hypothesis. If the outcome matches the prediction, we can say that the hypothesis has not been disproved by the experiment. When this happens, our confidence in the hypothesis increases. If the outcome and prediction are inconsistent, we need to reconsider the hypothesis and possibly reject it.

To test the hypothesis that the air exerts a downward force on objects, we attach a ball to a spring and let it hang; the spring stretches (**Figure 2.3a**). Next we place the ball and spring inside a large jar that is connected to a vacuum pump and pump the air out of the inside of the jar. We predict that *if* the air

Figure 2.2 Representing external interactions (forces exerted on a system).

(a)

A bowling ball   A volleyball
(System 1)   (System 2)

(b)

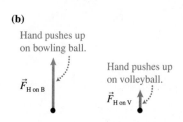

Hand pushes up on bowling ball.

$\vec{F}_{\text{H on B}}$

Hand pushes up on volleyball.

$\vec{F}_{\text{H on V}}$

(c)

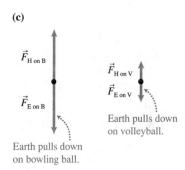

$\vec{F}_{\text{H on B}}$

$\vec{F}_{\text{E on B}}$

Earth pulls down on bowling ball.

$\vec{F}_{\text{H on V}}$

$\vec{F}_{\text{E on V}}$

Earth pulls down on volleyball.

Figure 2.3 A testing experiment to determine the effect of air on the ball.

(a)   **(b)**

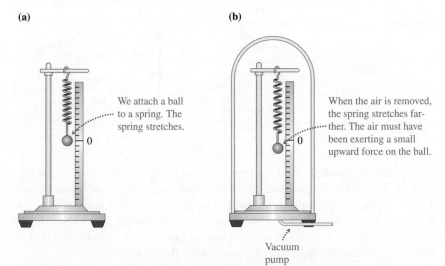

We attach a ball to a spring. The spring stretches.

When the air is removed, the spring stretches farther. The air must have been exerting a small upward force on the ball.

Vacuum pump

inside the jar pushes down on the ball (the hypothesis), *then* when we pump the air out of the jar, it should be easier to support the ball—the spring should stretch less (the prediction that follows from the hypothesis).

When we do the experiment, the outcome does not match the prediction— the spring actually stretches *slightly more* when the air is pumped out of the jar (Figure 2.3b). Evidently the air does not push down on the ball; instead, it helps support the ball by exerting an upward force on the ball. This outcome is surprising. When you study fluids, you will learn the mechanism by which air pushes up on objects.

Reflect

Let's reflect on what we have done here. We formulated an initial hypothesis— air *pushes down* on objects. Then we designed an experiment whose outcome we could predict using the hypothesis—the ball on a spring in a vacuum jar. We used the hypothesis to make a prediction of the outcome of the testing experiment—the spring should stretch less in a vacuum. We then performed the experiment and found that something completely different happened. We revised our hypothesis—air *pushes up slightly* on objects. Note that air's upward push on the ball is very small. For many situations, the effect of air on objects can be ignored.

Drawing force diagrams

A force diagram (sometimes called a free-body diagram) represents the forces that objects in a system's environment exert on it (see Figure 2.2c). We represent the system object by a dot to show that we model it as a point-like object. Arrows represent the forces. Unlike a motion diagram, a force diagram does not show us how a process changes with time; it shows us only the forces at a single instant. For processes in which no motion occurs, this makes no difference. But when motion does occur, we need to know if the force diagram is changing as the object moves.

Consider a rock dropped from above and sinking into sand, making a small crater. We construct a force diagram for shortly after the rock touches the sand but before it completely stops moving.

REASONING SKILL Constructing a force diagram

1. Sketch the situation (a rock sinking into sand).	2. Circle the system (the rock).	3. Identify external interactions: • The sand pushes up on the rock. • Earth pulls down on the rock. • We assume that the force that the air exerts on the rock is small in comparison and can be ignored.	4. Place a dot at the side of the sketch, representing the system object.

5. Draw force arrows to represent the external interactions.

6. Label the forces with a subscript with two elements.

Notice that the upward-pointing arrow representing the force exerted by the sand on the rock is longer than the downward-pointing arrow representing the force exerted by Earth on the rock. The difference in lengths reflects the difference in the magnitudes of the forces. Later in the chapter we will learn why they have different lengths. For now, we just need to include arrows for all external forces exerted on the system object (the rock).

> **TIP** Remember that on the force diagram, you only draw forces exerted *on* the system object. Do not draw forces that the system object exerts on other objects! For example, the rock exerts a force on the sand, but we do not include this force in the force diagram since the sand is not part of the system.

CONCEPTUAL EXERCISE 2.1 Force diagram for a book

Book A sits on a table with book B on top of it. Construct a force diagram for book A.

Sketch and translate We sketch the situation below. We choose book A as the system object. Notice that the dashed line around book A passes between the table and book A, and between book B and book A. It's important to be precise in the way you draw this line so that the separation between the system and the environment is clear. In this example, Earth, the table, and book B are external environmental objects that exert forces on book A.

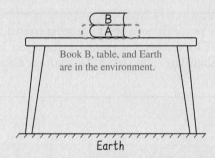

Book B, table, and Earth are in the environment.

Earth

Simplify and diagram Draw a force diagram for book A, which is represented by a dot. Two objects

in the environment touch book A. The table pushes up on the bottom surface of the book, exerting a force $\vec{F}_{\text{T on A}}$, and book B pushes down on the top surface of book A, exerting force $\vec{F}_{\text{B on A}}$. In addition, Earth exerts a downward force on book A $\vec{F}_{\text{E on A}}$.

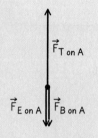

Try it yourself: Construct a force diagram for book A assuming that another book C is placed on top of book B.

Answer: The same three objects interact with book A. Earth exerts the same downward force on book A ($\vec{F}_{\text{E on A}}$). C does not directly touch A and exerts no force on A. However, C does push down on B, so B exerts a greater force on A ($\vec{F}_{\text{B on A}}$). Because the downward force of B on A is greater, the table exerts a greater upward force on book A ($\vec{F}_{\text{T on A}}$).

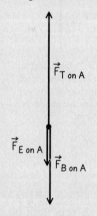

Normal force

In the previous example, the force that the table exerts on book A and the force that book B exerts on book A are both perpendicular to their touching surfaces with A. Such perpendicular touching forces are called **normal forces**. Normal does not mean vertical, although in this example they were vertical forces. In the future, these forces will be labeled using the letter N instead of F. Normal forces are *contact forces* (due to touching objects), as opposed to the force that Earth exerts on the book.

Review Question 2.1 You slide toward the right at decreasing speed on a horizontal wooden floor. Choose yourself as the system and list the external objects that interact with and exert forces on you.

2.2 Adding and measuring forces

Most often, more than one environmental object exerts a force on a system object. How can we add them to find the total or net force exerted on the system object? In this chapter we restrict our attention to forces that are exerted and point along one axis. Consider the process of lifting a suitcase.

Figure 2.4 The sum of the forces (the net force) exerted on the suitcase.

(a)

Sketch the situation and choose a system.

(b)

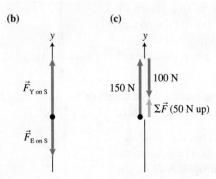

$\vec{F}_{Y \text{ on } S}$

$\vec{F}_{E \text{ on } S}$

Draw a force diagram for the system object showing the external forces exerted on the system.

(c)

y

150 N

100 N

$\Sigma\vec{F}$ (50 N up)

Place the arrows head to tail. The sum of the forces (the net force) goes from the tail of the first arrow to the head of the last arrow.

Adding forces graphically

You lift a suitcase straight up (**Figure 2.4a**). Earth pulls down on the suitcase, exerting a force of magnitude $F_{E \text{ on } S} = 100$ N, and you exert an upward force of magnitude $F_{Y \text{ on } S} = 150$ N (Figure 2.4b). What is the magnitude and direction of the total force exerted on the suitcase? The net effect of the two forces exerted along a vertical axis is the same as a 50-N force pointed straight up. Why? Remember that force is a vector. To add two vectors to find the sum $\Sigma\vec{F} = \vec{F}_{E \text{ on } S} + \vec{F}_{Y \text{ on } S}$, we place them head to tail (see Figure 2.4c) and draw the vector that goes from the tail of the first vector to the head of the second vector. This new vector is the sum vector, often called the resultant vector. In the case of forces it is called the **net force**. In our example, the net force points upward and its length is 50 N. If we assign the upward y-direction to be positive, then the sum of the forces has a y-component $\Sigma F_y = +50$ N. If the positive direction is down, the sum of the forces has a y-component $\Sigma F_y = -50$ N.

If several external objects in the environment exert forces on the system object, we still use vector addition to find the *sum* Σ of the forces exerted on the object:

$$\Sigma\vec{F}_{\text{on O}} = \vec{F}_{1 \text{ on O}} + \vec{F}_{2 \text{ on O}} + \cdots + \vec{F}_{n \text{ on O}} \qquad (2.1)$$

> **TIP** The sum of the force vectors is not a new force being exerted. Rather, it is the combined effect of all the forces being exerted on the object. Because of this, the resultant vector should never be included in the force diagram for that object.

CONCEPTUAL EXERCISE 2.2 Measuring forces
A very light plastic bag hangs from a light spring. The spring is not stretched. We place one golf ball into the bag and observe that the spring stretches to a new length. We add a second ball and observe that the spring stretches twice as far. We add a third ball and observe that the spring stretches three times as far. How can we use this experiment to develop a method to measure the magnitude of a force?

Sketch and translate First, draw sketches to represent the four situations as shown below. On each sketch,

carefully show the change in the length of the spring. Choose the bag with the golf balls as the system and analyze the forces exerted in each case.

Simplify and diagram Assume that Earth exerts the same force on each ball $\vec{F}_{E \text{ on } 1B}$ independent of the presence of other balls. Thus, the total force exerted by Earth on the three-ball system is three times greater than the force exerted on the one-ball system. Draw a force diagram for each case. Assume that in each case the spring exerts a force on the system $\vec{F}_{S \text{ on } \#B}$ that is equal in magnitude and opposite in direction to the force that Earth exerts on the system $\vec{F}_{E \text{ on } \#B}$ so that the sum of the forces exerted on the system with a number # of golf balls is zero.

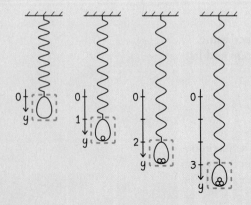

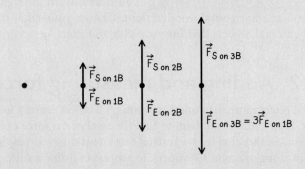

Try it yourself: Represent the relation between the force that the spring exerts on the bag with a number # of golf balls ($F_{S \text{ on } \#B}$) and spring stretch (y) with an F-versus-y graph. Draw a trendline.

Answer: Based on the graph's trendline, we see that the spring elongates until the force it exerts on the system object balances the force that Earth exerts on it.

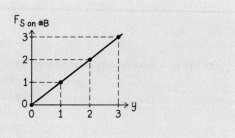

Measuring force magnitudes

Figure 2.5 A spring scale.

Force is a vector physical quantity with a magnitude and a direction. In the next conceptual exercise, we develop a method for determining the magnitude of a force.

Conceptual Exercise 2.2 provides us with one method to measure an unknown force that an object exerts on a system. We calibrate a spring in terms of some standard force, such as Earth's pull on one or more golf balls. Then if some unknown force is exerted on a system object, we can use the spring to exert a balancing force on that object. The unknown force is equal in magnitude to the force exerted by the spring and opposite in direction. In this case, we would be measuring force in units equal to Earth's pull on a golf ball. We could use any spring to balance a known standard force (1 N or approximately the force that Earth exerts on a 100-g object) and then calibrate this spring in newtons by placing marks at equal stretch distances as we pull on its end with increasing force. We thus build a spring scale—a simple instrument to measure forces (**Figure 2.5**).

TIP In physics, force is a physical quantity that characterizes an interaction between two objects—its direction and magnitude. For a force to exist there must be two objects that interact, just like a hug requires the interaction of two people. Force does not reside in an object. However, in everyday language the idea that force resides in an object remains very strong; people say, "The truck's force caused a lot of damage to the telephone pole." We will be careful in this book to always identify the two interacting objects when speaking about any force. Remember, if you are thinking about a force that is exerted on a moving object and cannot find another object that interacts with it, then you are thinking of something else, not force.

Review Question 2.2 A book bag hanging from a spring scale is partially supported by a platform scale. The platform scale reads about 36 units of force and the spring scale reads about 28 units of force. What is the magnitude of the force that Earth exerts on the bag? Explain.

2.3 Conceptual relationship between force and motion

Active Learning Guide›

When we drew a force diagram for a ball held by a person, we intuitively drew the forces exerted on the ball as being equal in magnitude. What if the person throws the ball upward? Or slowly lifts it upward? Or if she catches a ball falling from above? Would she still need to exert a force on the ball of magnitude equal to that Earth exerts on the ball? In other words, is there a relationship between the forces that are exerted on an object and the way the object moves?

Consider the three simple experiments in Observational Experiment **Table 2.1** involving a bowling ball (the system) rolling on a smooth surface. Is there is a pattern between its motion diagram and the force diagram?

OBSERVATIONAL EXPERIMENT TABLE

2.1 How are motion and forces related?

VIDEO 2.1

Observational experiment	Analysis	
	Motion diagram	Force diagrams for first and third positions
Experiment 1. A bowling ball B rolls on a very hard, smooth surface S without slowing down. 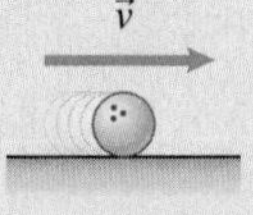 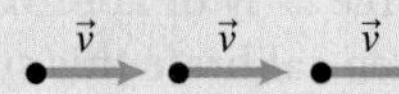 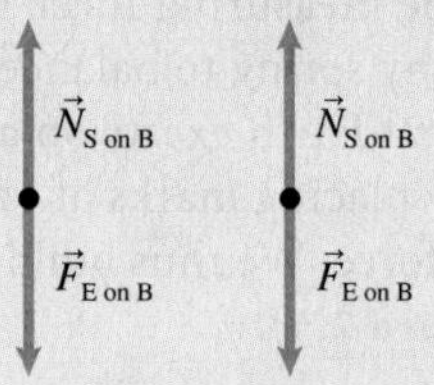	$\Delta \vec{v} = 0$	$\vec{N}_{S\,on\,B}$ $\vec{F}_{E\,on\,B}$
Experiment 2. A ruler R lightly pushes the rolling bowling ball opposite the ball's direction of motion. The ball continues to move in the same direction, but slows down. 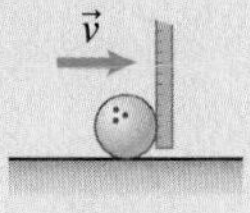 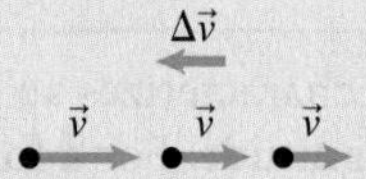 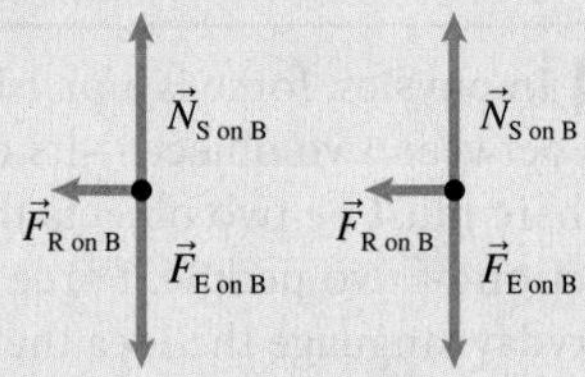	$\Delta \vec{v}$	$\vec{N}_{S\,on\,B}$ $\vec{F}_{R\,on\,B}$ $\vec{F}_{E\,on\,B}$
Experiment 3. A ruler R lightly pushes the rolling bowling ball in the direction of its motion. 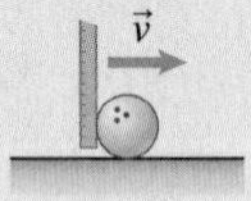 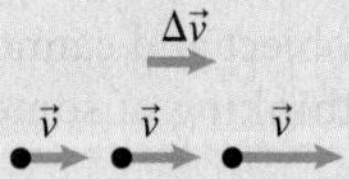 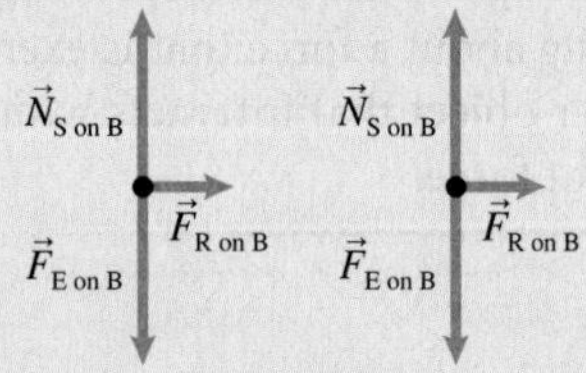	$\Delta \vec{v}$	$\vec{N}_{S\,on\,B}$ $\vec{F}_{E\,on\,B}$ $\vec{F}_{R\,on\,B}$

Pattern

- In all the experiments, the vertical forces add to zero and cancel each other. We consider only forces exerted on the ball in the horizontal direction.
- In the first experiment, the sum of the forces exerted on the ball is zero; the ball's velocity remains constant.
- In the second and third experiments, when the ruler pushes the ball, the velocity change arrow ($\Delta \vec{v}$ arrow) points in the same direction as the sum of the forces.

Summary: The $\Delta \vec{v}$ arrow in all experiments is in the same direction as the sum of the forces. Notice that there is no pattern relating the *direction* of the velocity $\vec{v}$ to the direction of the sum of the forces. In Experiment 2, the velocity and the sum of the forces are in opposite directions, but in Experiment 3, they are in the same direction.

In each of the experiments in Table 2.1, the $\Delta\vec{v}$ arrow for a system object and the sum of the forces $\Sigma\vec{F}$ that external objects exert on that object are in the same direction. That is one idea. A second idea is that we often observe that the $\vec{v}$ arrow for a system object (the direction the object is moving) is in the same direction as the sum of the forces exerted on it. For example, a grocery cart moves in the direction the shopper pushes it and a soccer ball moves in the direction the player kicks it. We should test both ideas.

Testing possible relationships between force and motion

We have two possible ideas that relate motion and force:

1. An object's velocity $\vec{v}$ always points in the direction of the sum of the forces $\Sigma\vec{F}$ that other objects exert on it.
2. An object's velocity *change* $\Delta\vec{v}$ always points in the direction of the sum of the forces $\Sigma\vec{F}$ that other objects exert on it.

To test these two relationships, we use each to predict the outcome of the experiments in Testing Experiment **Table 2.2**. Then we perform the experiments and compare the outcomes with the predictions. From this comparison, we decide if we can reject one or both of the relationships.

TESTING EXPERIMENT TABLE

2.2 **Testing ideas of how velocity and the sum of the forces $\Sigma\vec{F}$ are related.**

VIDEO 2.2

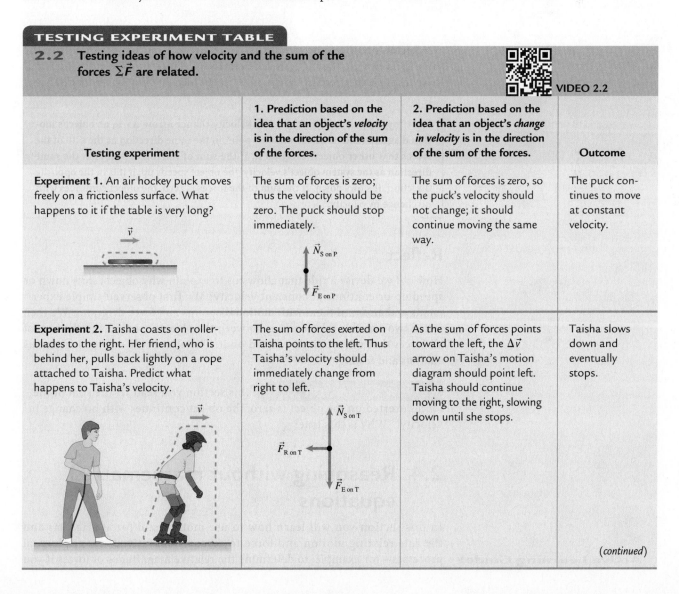

Testing experiment	1. Prediction based on the idea that an object's *velocity* is in the direction of the sum of the forces.	2. Prediction based on the idea that an object's *change in velocity* is in the direction of the sum of the forces.	Outcome
Experiment 1. An air hockey puck moves freely on a frictionless surface. What happens to it if the table is very long? $\vec{v}$	The sum of forces is zero; thus the velocity should be zero. The puck should stop immediately. $\vec{N}_{\text{S on P}}$ $\vec{F}_{\text{E on P}}$	The sum of forces is zero, so the puck's velocity should not change; it should continue moving the same way.	The puck continues to move at constant velocity.
Experiment 2. Taisha coasts on rollerblades to the right. Her friend, who is behind her, pulls back lightly on a rope attached to Taisha. Predict what happens to Taisha's velocity. $\vec{v}$	The sum of forces exerted on Taisha points to the left. Thus Taisha's velocity should immediately change from right to left. $\vec{N}_{\text{S on T}}$ $\vec{F}_{\text{R on T}}$ $\vec{F}_{\text{E on T}}$	As the sum of forces points toward the left, the $\Delta\vec{v}$ arrow on Taisha's motion diagram should point left. Taisha should continue moving to the right, slowing down until she stops.	Taisha slows down and eventually stops.

(continued)

Testing experiment	Prediction 1	Prediction 2	Outcome
Experiment 3. You throw a ball upward. What happens to the ball after it leaves your hand? 	The only force exerted on the ball after it leaves your hand points down. Thus the ball should immediately begin moving downward after you release it. 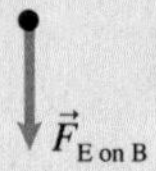 $\vec{F}_{\text{E on B}}$	The only force exerted on the ball after it leaves your hand points down. The $\Delta\vec{v}$ arrow on the motion diagram should point down, and the ball should slow down until it stops and then start moving back down at increasing speed.	The ball moves up at decreasing speed and then reverses direction and starts moving downward.

Conclusion

- All outcomes contradict the predictions based on idea 1—we can reject it.
- All outcomes are consistent with the predictions based on idea 2. This does not necessarily mean it is true, but it does mean our confidence in the idea increases.

Recall that the direction of the $\Delta\vec{v}$ arrow in a motion diagram is in the same direction as the object's acceleration $\vec{a}$. Thus, based on this idea and these testing experiments, we can now accept idea 2 with greater confidence.

> **Relating forces and motion** The velocity change arrow $\Delta\vec{v}$ in an object's motion diagram (and its acceleration $\vec{a}$) point in the same direction as the sum of the forces that other objects exert on it. If the sum of the forces points in the same direction as the system object's velocity, the object speeds up; if it is in the opposite direction, it slows down. If the sum of the forces is zero, the object continues with no change in velocity.

Reflect

How did we devise a rule that allows us to explain why objects slow down or speed up or continue at constant velocity? We first observed simple experiments and analyzed them with motion diagrams and force diagrams. We then tested two possible relationships between the objects' motion and the sum of all forces that other objects exerted on it. The above rule emerged from this analysis and testing.

Review Question 2.3 In this section you read "If the sum of the forces exerted on the object is zero, the object continues with no change in velocity." Why is this true?

2.4 Reasoning without mathematical equations

In this section you will learn how to use motion and force diagrams and the rule relating motion and force to reason qualitatively about physical processes—for example, to determine the relative magnitudes of forces if you

Active Learning Guide›

have information about motion or to estimate velocity changes if you have information about forces. The key here is to make sure that the two representations (the force diagrams and the motion diagrams) are consistent.

In the next conceptual exercise, we use information about the forces that external objects exert on a woman to answer a question about her motion. In the Try it Yourself question we reverse the process—we use known information about the motion to answer a question about an unknown force. Try to answer the questions yourself before looking at the solutions.

CONCEPTUAL EXERCISE 2.3

Diagram Jeopardy

The force diagram shown here describes the forces that external objects exert on a woman (in this scenario, the force diagram does not change with time). Describe three different types of motion that are consistent with the force diagram.

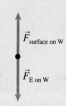
$\vec{F}_{\text{surface on W}}$

$\vec{F}_{\text{E on W}}$

Sketch and translate Two equal-magnitude oppositely directed forces are being exerted on the woman ($\Sigma \vec{F} = 0$). Thus, a motion diagram for the woman must have a zero velocity change ($\Delta \vec{v} = 0$).

Simplify and diagram Three possible motions consistent with this idea are shown at the right.

1. She stands at rest on a horizontal surface.

2. She glides at constant velocity on rollerblades on a smooth horizontal surface.

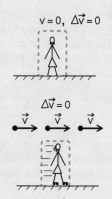
$v = 0, \ \Delta\vec{v} = 0$

$\Delta\vec{v} = 0$
$\vec{v} \quad \vec{v} \quad \vec{v}$

3. She stands on the floor of an elevator that moves up or down at constant velocity.

Note that in all three of the above, the velocity change arrow is zero. This is consistent with the sum of the forces being zero.

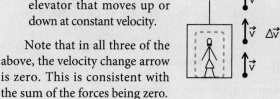

Try it yourself: Suppose that the elevator described above was moving up at decreasing speed instead of at constant speed. How then would the force diagram be different?

Answer: A velocity change $\Delta \vec{v}$ arrow for her motion would now point down opposite the direction of her velocity. Thus, the sum of the forces $\Sigma \vec{F}$ that other objects exert on her must also point down. This means that the magnitude of the upward force $\vec{F}_{\text{S on W}}$ that the elevator floor (surface) exerts on her must now be less than the magnitude of the downward force $\vec{F}_{\text{E on W}}$ that Earth exerts on her ($F_{\text{S on W}} < F_{\text{E on W}}$).

Review Question 2.4 An elevator in a tall office building moves downward at constant speed. How does the magnitude of the upward force exerted by the cable on the elevator $\vec{F}_{\text{C on El}}$ compare to the magnitude of the downward force exerted by Earth on the elevator $\vec{F}_{\text{E on El}}$? Explain your reasoning.

‹ Active Learning Guide

2.5 Inertial reference frames and Newton's first law

Our description of the motion of an object depends on the observer's reference frame. However, in this chapter we have tacitly assumed that all observers were standing on Earth's surface. For example, in **Section 2.3**, we analyzed several experiments and concluded that if the forces exerted on one object by other objects add to zero, then the chosen object moves at constant velocity. Are there any observers who will see a chosen object moving with changing velocity even though the sum of the forces exerted on the object appears to be zero?

Inertial reference frames

In Observational Experiment **Table 2.3**, we consider two different observers analyzing the same situation.

OBSERVATIONAL EXPERIMENT TABLE

2.3   Two observers watch the same coffee mug.

VIDEO 2.3

Observational experiment	Analysis done by each observer
Experiment 1. Observer 1 is slouched down in the passenger seat of a car and cannot see outside the car. Suddenly he observes a coffee mug sliding toward him from the dashboard.	Observer 1 creates a motion diagram and a force diagram for the mug as he observes it. On the motion diagram, increasingly longer $\vec{v}$ arrows indicate that the mug's speed changes from zero to nonzero as seen by observer 1 even though no external object is exerting a force on it in that direction. 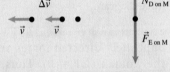
Experiment 2. Observer 2 stands on the ground beside the car. She observes that the car starts moving forward at increasing speed and that the mug remains stationary with respect to her.	Observer 2 creates a motion diagram and force diagram for the mug as she observes it. There are no $\vec{v}$ or $\Delta\vec{v}$ arrows on the diagram and the mug is at rest relative to her. 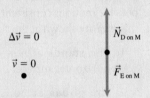

Pattern

Observer 1: The forces exerted on the mug by Earth and by the dashboard surface add to zero. But the velocity of the mug increases as it slides off the dashboard. This is inconsistent with the rule relating the sum of the forces and the change in velocity.
Observer 2: The forces exerted on the mug by Earth and by the dashboard surface add to zero. Thus the velocity of the mug should not change, and it does not. This is consistent with the rule relating the sum of the forces and the change in velocity.

Observer 2 in Table 2.3 can account for what is happening using the rule relating the sum of the forces and changing velocity, but observer 1 cannot. For observer 1, the mug's velocity changes for no apparent reason.

Similarly, a passenger on a train (observer 1) might suddenly see her laptop computer start to slide forward off her lap. A person on the platform (observer 2) can explain this event using the rule we developed. The train's velocity started decreasing as it approached the station, but the computer continued forward at constant velocity.

It appears that the applicability of the rule depends on the reference frame of the observer. Observers (like observer 2) who *can* explain the behavior of the mug and the computer by using the rule relating the sum of the forces and changing velocity are said to be observers in **inertial reference**

frames. Those (like observer 1) who *cannot* explain the behavior of the mug and the computer using the rule are said to be observers in **noninertial reference frames**.

> **Inertial reference frame** An inertial reference frame is one in which an observer sees that the velocity of the system object does not change if no other objects exert forces on it or if the sum of all forces exerted on the system object is zero. For observers in noninertial reference frames, the velocity of the system object can change even though the sum of forces exerted on it is zero.

A passenger in a car or train that is speeding up or slowing down with respect to Earth is an observer in a noninertial reference frame. When you are in a car that abruptly stops, your body jerks forward—yet nothing is pushing you forward. When you are in an airplane taking off, you feel pushed back into your seat, even though nothing is pushing you in that direction. In these examples, you are an observer in a noninertial reference frame. Observers in inertial reference frames can explain the changes in velocity of objects by considering the forces exerted on them by other objects. Observers in noninertial reference frames cannot. From now on, we will always analyze phenomena from the point of view of observers in inertial reference frames. This idea is summarized by Isaac Newton's first law.

> **Newton's first law of motion** For an observer in an inertial reference frame when no other objects exert forces on a system object or when the forces exerted on the system object add to zero, the object continues moving at constant velocity (including remaining at rest). **Inertia** is the phenomenon in which a system object continues to move at constant velocity when the sum of the forces exerted on it by other objects is zero.

Physicists have analyzed the motion of thousands of objects from the point of view of observers in inertial reference frames and found no contradictions to the rule. Newton's first law of motion limits the reference frames with respect to which the other laws that you will learn in this chapter are valid—these other laws work only for the observers in inertial reference frames.

Review Question 2.5 What is the main difference between inertial and noninertial reference frames? Give an example.

Isaac Newton. Isaac Newton (1643–1727) invented differential and integral calculus, formulated the law of universal gravitation, developed a new theory of light, and put together the ideas for his three laws of motion. His work on mechanics was presented in the book entitled *Philosophiae Naturalis Principia Mathematica (Mathematical Principles of Natural Philosophy)*.

2.6 Newton's second law

Our conceptual analyses in Sections 2.3 and 2.4 indicated that an object does not change velocity and does not accelerate when the sum of all forces exerted on it is zero. We also learned that an object's velocity change and acceleration are in the same direction as the sum of the forces that other objects exert on it. In this section we will learn how to predict the magnitude of an object's acceleration if we know the forces exerted on it. The experiments in Observational Experiment **Table 2.4** will help us construct this quantitative relationship.

OBSERVATIONAL EXPERIMENT TABLE

2.4   Forces and resulting acceleration.

VIDEO 2.4

Observational experiment	Analysis

Experiment 1. A cart starts at rest on a low-friction horizontal track. A force probe continuously exerts one unit of force in the positive direction. The same experiment is repeated five times. Each time, the force probe exerts one more unit of force on the cart (up to five units, shown in the last diagram). A computer records the value of the force, and a motion detector on the track records the cart's speed and acceleration.

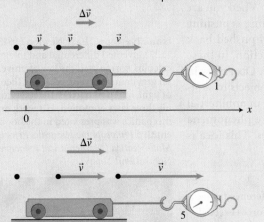

Using this information, we create velocity-versus-time and acceleration-versus-time graphs for two of the five different magnitudes of force. Note that the greater the force, the greater the acceleration.

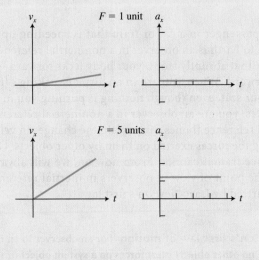

Experiment 2. We repeat the same five experiments, only this time the cart is moving in the positive direction, and the probe pulls back on the cart in the negative direction so that the cart slows down.

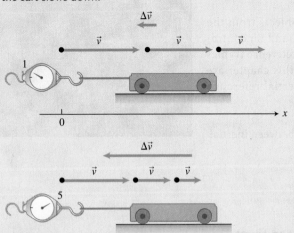

We create velocity-versus-time and acceleration-versus-time graphs for the cart when forces of two different magnitudes oppose the cart's motion.

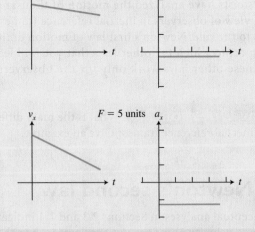

Pattern

- When the sum of the forces exerted on the cart is constant, its acceleration is constant—the cart's speed increases at a constant rate.
- Plotting the acceleration-versus-force using the five positive and five negative values of the force, we obtain the graph at the right.
- The acceleration is directly proportional to the force exerted by the force probe (in this case it is the sum of all forces) and points in the direction of the force.

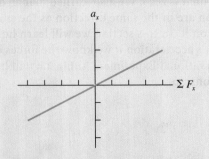

The outcome of these experiments expressed mathematically is as follows:

$$\vec{a} \propto \Sigma\vec{F} \tag{2.1}$$

where $\Sigma\vec{F}$ is the sum of all the forces that other objects exert on the object (not an additional force), and $\vec{a}$ is the object's acceleration. The symbol $\propto$ means "is proportional to." In other words, if the sum of the forces doubles, then the acceleration doubles. When the sum of the forces is zero, the acceleration is zero. When the sum of the forces exerted on an object is constant, the object's resulting acceleration (not velocity) is constant.

Mass, another physical quantity

Do other physical quantities affect acceleration? Note that the strong man shown in **Figure 2.6** can only cause a very small acceleration while pulling the bus but could easily cause a large acceleration when pulling a small wagon. The amount of matter being pulled must affect the acceleration.

Let's perform another experiment to find the quantitative effect of the amount of *matter* being pulled. We use the force probe to pull one cart, then two carts stacked on top of each other, and then three and four carts on top of each other. In each case, the force probe exerts the same force on the carts, regardless of how many carts are being pulled. The experiment is summarized in Observation Experiment **Table 2.5**.

Figure 2.6 A strong man pulls a bus.

‹Active Learning Guide

OBSERVATIONAL EXPERIMENT TABLE

2.5 Amount of matter and acceleration.

Observational experiment	Analysis

We pull the indicated number of stacked carts using an identical pulling force and measure the acceleration with a motion detector.

Number n of carts	Acceleration (m/s^2)
1	1.00
2	0.49
3	0.34
4	0.25

Motion detector   Carts   Force probe

We can graph the acceleration-versus-number of carts for constant pulling force.

From the graph we see that increasing the number of carts decreases the acceleration. To check whether this relationship is inversely proportional, plot a versus $\frac{1}{n}$.

Pattern

Since the graph a versus $\dfrac{1}{n}$ is a straight line, we conclude that $a \propto \dfrac{1}{n}$.

From the pattern observed in Table 2.5, we conclude that the greater the amount of matter being pulled, the smaller the object's acceleration when the same force is exerted on it. This property of an object, which affects its acceleration, is called **mass**.

To measure the mass of an object quantitatively, you first define a standard unit of mass. The choice for the unit of mass is arbitrary, but after the unit has been chosen, the masses of all other objects can be determined from this unit. The SI standard of mass, the kilogram (kg), is a cylinder made of a platinum-iridium alloy stored in a museum of measurements near Paris. Copies of this cylinder are available in most countries. A quart of milk has a mass of about 1 kg. Suppose, for example, that you exert a constant pulling force on a 1.0-kg object (and that all other forces exerted on this object are balanced), and you measure its acceleration. You then exert the same pulling force on another object of unknown mass. Your measurement indicates that it has half the acceleration of the standard 1.0-kg object. Thus, its mass is twice the standard mass (2.0 kg). This method is not practical for everyday use. Later we will learn another method to measure the mass of an object, a method that is simple enough to use in everyday life.

Our experiments indicate that when the same force is exerted on two objects, the one with the greater mass will have a smaller acceleration. Mathematically:

$$a \propto \frac{1}{m} \tag{2.2}$$

> **Mass** m characterizes the amount of matter in an object. When the same unbalanced force is exerted on two objects, the object with greater mass has a smaller acceleration. The unit of mass is called the kilogram (kg). Mass is a scalar quantity, and masses add as scalars.

Newton's second law

We have found that the acceleration $\vec{a}$ of a system is proportional to the vector sum of the forces $\Sigma \vec{F}$ exerted on it by other objects [Eq. (2.1)] and inversely proportional to the mass m of the system [Eq. (2.2)]. We can combine these two proportionalities into a single equation.

$$\vec{a}_{\text{System}} \propto \frac{\Sigma \vec{F}_{\text{on System}}}{m_{\text{System}}} \tag{2.3}$$

Rearrange the above to get $m_{\text{System}} \vec{a}_{\text{System}} \propto \Sigma \vec{F}_{\text{on System}}$. We can turn this into an equation if we choose the unit of force to be $\text{kg} \cdot \text{m/s}^2$. Because force is such a ubiquitous quantity, physicists have given the force unit a special name called a newton (N). A force of 1 newton (1 N) causes an object with a mass of 1 kg to accelerate at 1 m/s^2.

$$1 \text{ N} = 1 \text{ kg} \cdot \text{m/s}^2. \tag{2.4}$$

Eq. (2.3), rewritten with the equality sign

$$\vec{a}_{\text{System}} = \frac{\Sigma \vec{F}_{\text{on System}}}{m_{\text{System}}}$$

is called Newton's second law. As noted earlier in the chapter, the symbol Σ (the Greek letter sigma) means that you must add what follows the Σ.

> **Newton's second law** The acceleration $\vec{a}_{\text{System}}$ of a system object is proportional to the vector sum of all forces being exerted on the object and inversely proportional to the mass m of the object:
>
> $$\vec{a}_{\text{System}} = \frac{\Sigma \vec{F}_{\text{on System}}}{m_{\text{System}}} = \frac{\vec{F}_{1 \text{ on System}} + \vec{F}_{2 \text{ on System}} + \cdots}{m_{\text{System}}} \tag{2.5}$$
>
> The vector sum of all the forces being exerted on the system by other objects is
>
> $$\Sigma \vec{F}_{\text{on System}} = \vec{F}_{1 \text{ on System}} + \vec{F}_{2 \text{ on System}} + \cdots$$
>
> The acceleration of the system object points in the same direction as the vector sum of the forces.

> **TIP** Notice that the "vector sum of the forces" mentioned in the defini-
> tion above does not mean the sum of their magnitudes. Vectors are not
> added as numbers; their directions affect the magnitude of the vector sum.

Does this new equation make sense? For example, does it work in extreme
cases? First, imagine an object with an infinitely large mass. According to
the law, it will have zero acceleration for any process in which the sum of the
forces exerted on it is finite:

$$\vec{a}_{\text{System}} = \frac{\Sigma \vec{F}_{\text{on System}}}{\infty} = 0$$

This seems reasonable, as an infinitely massive object would not change
motion due to finite forces exerted on it. On the other hand, an object with
a zero mass will have an infinitely large acceleration when a finite magnitude
force is exerted on it:

$$\vec{a}_{\text{System}} = \frac{\Sigma \vec{F}_{\text{on System}}}{0} = \infty$$

Both extreme cases make sense. Newton's second law is a so-called *cause-
effect relationship*. The right side of the equation (the sum of the forces being
exerted) is the cause of the effect (the acceleration) on the left side.

$$\underset{\text{Effect}}{\nearrow \vec{a}_{\text{System}}} = \frac{\Sigma \vec{F}_{\text{on System}}}{m_{\text{System}} \underset{\text{Cause}}{\searrow}}$$

On the other hand, $\vec{a} = \Delta \vec{v}/\Delta t$ is called an *operational definition* of acceleration. It tells us how to determine the quantity acceleration but does not
tell us *why* it has a particular value. For example, suppose that an elevator's
speed changes from 2 m/s to 5 m/s in 3 s as it moves vertically along a straight
line in the positive *x*-direction. The elevator's acceleration (using the definition of acceleration) is

$$a_x = \frac{5\,\text{m/s} - 2\,\text{m/s}}{3\,\text{s}} = +1\,\text{m/s}^2$$

This operational definition does not tell you the reason for the acceleration. If you know that the mass of the elevator is 500 kg and that Earth exerts
a 5000-N downward force on the accelerating elevator and the cable exerts a
5500-N upward force on it, then using the cause-effect relationship of Newton's
second law:

$$\frac{5500\,\text{N} + (-5000\,\text{N})}{500\,\text{kg}} = +1\,\text{m/s}^2$$

Thus, you obtain the same number using two different methods—one
from kinematics (the part of physics that *describes* motion) and the other from
dynamics (the part of physics that *explains* motion).

Force components used for forces along one axis

When the vector sum of the forces exerted on a system object points along
one direction (for example, the *x*-direction), you can use the component form
of the Newton's second law equation for the *x*-direction instead of the vector
equation [Eq. (2.5)]:

$$a_{\text{System }x} = \frac{\Sigma F_{\text{on System }x}}{m_{\text{System}}} \tag{2.6}$$

A similar equation applies for the y-direction, if the forces are all along the y-axis. To use Eq. (2.6) (or the y-version of the equation), you first need to identify the positive direction of the axis. Then find the components of all the forces being exerted on the system. Forces that point in the positive direction have a positive component, and forces that point in the negative direction have a negative component.

In this chapter, we will analyze situations in which (a) the forces that external objects exert on the system are all along the y-axis or (b) the forces pointing along the y-axis balance and don't contribute to the acceleration of the system along the x-axis, allowing us to analyze the situation along the x-axis only. Consider first a situation where all forces are along the y-axis.

EXAMPLE 2.4 Lifting a suitcase

Earth exerts a downward 100-N force on a 10-kg suitcase. Suppose you exert an upward 120-N force on the suitcase. If the suitcase starts at rest, how fast is it traveling after lifting for 0.50 s?

Sketch and translate First, we make a sketch of the initial and final states of the process, choosing the suitcase as the system object. The sketch helps us visualize the process and also brings together all the known information, letting our brains focus on other aspects of solving the problem. One common aspect of problems like this is the use of a two-step strategy. Here, we use Newton's second law to determine the acceleration of the suitcase and then use kinematics to determine the suitcase's speed after lifting 0.50 s.

Simplify and diagram Next, we construct a force diagram for the suitcase while being lifted. The y-components of the forces exerted on the suitcase are your upward pull on the suitcase $F_{\text{Y on S}y} = +F_{\text{Y on S}} = +120$ N and Earth's downward pull on the suitcase $F_{\text{E on S}y} = -F_{\text{E on S}} = -100$ N. Because the upward force is larger, the suitcase will have an upward acceleration $\vec{a}$.

Represent mathematically Since all the forces are along the y-axis, we apply the y-component form of Newton's second law to determine the suitcase's acceleration (notice how the subscripts in the equation below change from step to step):

$$a_{\text{S}y} = \frac{\Sigma F_{\text{on S}y}}{m_{\text{S}}} = \frac{F_{\text{Y on S}y} + F_{\text{E on S}y}}{m_{\text{S}}}$$

$$= \frac{(+F_{\text{Y on S}}) + (-F_{\text{E on S}})}{m_{\text{S}}}$$

$$= \frac{F_{\text{Y on S}} - F_{\text{E on S}}}{m_{\text{S}}}$$

After using Newton's second law to determine the acceleration of the suitcase, we then use kinematics to determine the suitcase's speed after traveling upward for 0.50 s:

$$v_y = v_{0y} + a_y t$$

The initial velocity is $v_{0y} = 0$.

Solve and evaluate Now substitute the known information in the Newton's second law y-component equation above to find the acceleration of the suitcase:

$$a_{\text{S}y} = \frac{F_{\text{Y on S}} - F_{\text{E on S}}}{m_{\text{S}}} = \frac{120 \text{ N} - 100 \text{ N}}{10 \text{ kg}} = +2.0 \text{ m/s}^2$$

Insert this and other known information into the kinematics equation to find the vertical velocity of the suitcase after lifting for 0.50 s:

$$v_y = v_{0y} + a_y t = 0 + (+2.0 \text{ m/s}^2)(0.50 \text{ s}) = +1.0 \text{ m/s}$$

The unit for time is correct and the magnitude is reasonable.

Try it yourself: How far up did you pull the suitcase during this 0.50 s?

Answer: The average speed while lifting it was $(0 + 1.0 \text{ m/s})/2 = 0.50 \text{ m/s}$. Thus you lifted the suitcase $y - y_0 = (0.50 \text{ m/s})(0.50 \text{ s}) = 0.25$ m.

Now let's consider a case in which an object moves on a horizontal surface, Earth exerts a downward gravitational force on it, and the surface exerts an upward normal force of the same magnitude.

EXAMPLE 2.5   Pulling a lawn mower

You pull horizontally on the handle of a lawn mower that is moving across a horizontal grassy surface. The lawn mower's mass is 32 kg. You exert a force of magnitude 96 N on the mower. The grassy surface exerts an 83-N resistive friction-like force on the mower. Earth exerts a downward force on the mower of magnitude 314 N, and the grassy surface exerts an upward normal force of magnitude 314 N. What is the acceleration of the mower?

Sketch and translate We choose the mower as the system and sketch the situation, as shown below.

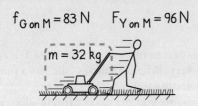

$$f_{\text{G on M}} = 83\ \text{N} \qquad F_{\text{Y on M}} = 96\ \text{N}$$

$$m = 32\ \text{kg}$$

Simplify and diagram We model the mower as a point-like object and draw a force diagram for it. As the forces do not change during motion, the diagram can represent the motion at any instant. Three external objects interact with the system—the grassy surface, Earth, and you. You exert a horizontal force on the mower $\vec{F}_{\text{Y on M}}$; Earth exerts a downward gravitational force on the mower $\vec{F}_{\text{E on M}}$. The grassy surface exerts a force on the mower that we will represent with two force arrows—a normal force $\vec{N}_{\text{G on M}}$ perpendicular to the surface (in this case it points upward) and a horizontal friction force $\vec{f}_{\text{G on M}}$ opposite the direction of motion.

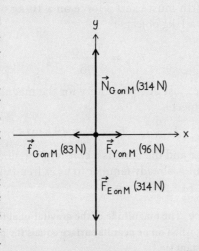

The downward force $\vec{F}_{\text{E on M}}$ and the upward normal force $\vec{N}_{\text{G on M}}$ have the same magnitudes and point in opposite directions. This means that these two forces cancel and do not contribute to the horizontal acceleration of the mower. We can ignore them.

Represent mathematically Since the acceleration of the system is along the x-axis and the forces perpendicular to the grassy surface cancel, we can use the x-component form of Newton's second law to determine the acceleration of the mower:

$$a_{\text{M}x} = \frac{\Sigma F_x}{m_{\text{M}}} = \frac{F_{\text{Y on M}x} + f_{\text{G on M}x}}{m_{\text{M}}}$$

The x-component of the force you exert on the mower is positive since that force points in the positive x-direction ($F_{\text{Y on M}x} = +F_{\text{Y on M}}$). The x-component of the friction force exerted by the grass on the mower is negative since that force points in the negative x-direction ($f_{\text{G on M}x} = -f_{\text{G on M}}$). Therefore, the acceleration of the mower is

$$a_{\text{M}x} = \frac{F_{\text{Y on M}} + (-f_{\text{G on M}})}{m_{\text{M}}} = \frac{F_{\text{Y on M}} - f_{\text{G on M}}}{m_{\text{M}}}$$

Solve and evaluate We can now determine the mower's acceleration by substituting the known information into the preceding equation:

$$a_{\text{M}x} = \frac{F_{\text{Y on M}} - f_{\text{G on M}}}{m_{\text{M}}} = \frac{96\ \text{N} - 83\ \text{N}}{32\ \text{kg}}$$

$$= +0.40625\ \text{N/kg} = +0.41\ \text{m/s}^2$$

The known information had two significant digits, so we rounded the answer to two significant digits. The units for acceleration are correct, and the magnitude is reasonable.

Try it yourself: Imagine that you are pulling the mower as described above. After the mower has accelerated to a velocity with which you are comfortable, what is the magnitude of the force you should exert on the mower so that it now moves at constant velocity?

Answer: 83 N.

Review Question 2.6 Jim says that $m\vec{a}$ is a special force exerted on an object and it should be represented on the force diagram. Do you agree or disagree with Jim? Explain your answer.

2.7 Gravitational force law

In the last example, we were given the mass of the lawn mower (32 kg) and the magnitude of the force that Earth exerts on the mower (314 N). Is it possible to determine the magnitude of this force by just knowing the mass of the mower? In fact, it is.

Imagine that we evacuate all the air from a 3.0-m-long Plexiglas tube, place a motion sensor at the top, and drop objects of various sizes, shapes, and compositions through the tube. The measurements taken by the motion sensor reveal that all objects fall straight down with the same acceleration, 9.8 m/s^2. Earth exerts the only force on the falling object $\vec{F}_{\text{E on O}}$ during the entire flight. If we choose the positive y-axis pointing down and apply the y-component form of Newton's second law, we get

$$a_{\text{O }y} = \frac{1}{m_{\text{O}}} F_{\text{E on O }y} = \frac{1}{m_{\text{O}}}(+F_{\text{E on O}}) = +\frac{F_{\text{E on O}}}{m_{\text{O}}}$$

Every object dropped in our experiment had the same free-fall acceleration, $g = 9.8 \text{ m/s}^2$, even those with very different masses (such as a ping-pong ball and a lead ball). Thus, the gravitational force that Earth exerts on each object must be proportional to its mass so that the mass cancels when we calculate the acceleration. Earth must exert a force on a 10-kg object that is 10 times greater than that on a 1-kg object:

$$a_{\text{O }y} = \frac{F_{\text{E on O}}}{m_{\text{O}}} = g$$

This reasoning leaves just one possibility for the magnitude of the force that Earth exerts on an object:

$$F_{\text{E on O}} = m_{\text{O}}g = m_{\text{O}}(9.8 \text{ m/s}^2)$$

The ratio of the force and the mass is a constant for all objects—the so-called gravitational constant g, already familiar to us as free-fall acceleration.

Gravitational force The magnitude of the gravitational force that Earth exerts on any object $F_{\text{E on O}}$ when on or near its surface equals the product of the object's mass m and the constant g:

$$F_{\text{E on O}} = m_{\text{O}}g \tag{2.7}$$

where $g = 9.8 \text{ m/s}^2 = 9.8 \text{ N/kg}$ on or near Earth's surface. This force points toward the center of Earth.

The value of the free-fall acceleration g in the above gravitational force law [Eq. (2.7)] does not mean that the object is actually falling. The g is just used to determine the magnitude of the gravitational force exerted on the object by Earth whether the object is falling or sitting at rest on a table or moving down a water slide. To avoid confusion, we will use $g = 9.8 \text{ N/kg}$ rather than 9.8 m/s^2 when calculating the gravitational force.

We learn in the chapter on circular motion (Chapter 4) that the gravitational constant g at a particular point depends on the mass of Earth and on how far this point is from the center of Earth. On Mars or the Moon, the gravitational constant depends on the mass of Mars or the Moon, respectively. The gravitational constant is 1.6 N/kg on the Moon and 3.7 N/kg on Mars. You could throw a ball upward higher on the Moon since g is smaller there, resulting in a smaller force exerted downward on the ball.

Active Learning Guide›

Newton's second law says that the acceleration of an object is inversely proportional to its mass. However, the acceleration with which all objects fall in the absence of air is the same. How can this be?

2.8 Skills for applying Newton's second law for one-dimensional processes

In this section we develop a strategy that can be used whenever a process involves force and motion. We will introduce the strategy by applying it to the 2007 sky dive of diving champion Michael Holmes. After more than 1000 successful jumps, Holmes jumped from an airplane 3700 m above Lake Taupo in New Zealand. His main parachute failed to open, and his backup chute became tangled in its cords. The partially opened backup parachute slowed his descent to about 36 m/s (80 mi/h) as he reached a 2-m-high thicket of wild shrubbery. Holmes plunged through the shrubbery, which significantly decreased his speed before he reached the ground. Holmes survived with a collapsed right lung and a broken left ankle. The general steps of a problem-solving strategy for force-motion problems are described on the left side of Example 2.6 and applied to Holmes's landing in the shrubbery on the right side.

The language of physics
The *weight* of an object on a planet is the force that the planet exerts on the object. We will not use the term "weight of an object" because it implies that weight is a property of the object rather than an interaction between two objects.

‹ Active Learning Guide

PROBLEM-SOLVING STRATEGY Applying Newton's Laws For One-Dimensional Processes

EXAMPLE 2.6 Holmes's sky dive
Michael Holmes (70 kg) was moving downward at 36 m/s (80 mi/h) and was stopped by 2.0-m-high shrubbery and the ground. Estimate the average force exerted by the shrubbery and ground on his body while stopping his fall.

Sketch and translate
- Make a sketch of the process.
- Choose the system object.
- Choose a coordinate system.
- Label the sketch with everything you know about the situation.

We sketch the process, choosing Holmes as the system object H. We want to know the average force that the shrubbery and ground S-G exert on him from when he first touches the shrubbery to the instant when he stops. We choose the y-axis pointing up and the origin at the ground where Holmes comes to rest. We use kinematics to find his acceleration while stopping and Newton's second law to find the average force that the shrubbery and ground exerted on him while stopping him.

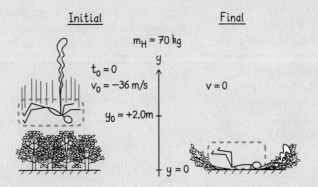

Initial   Final

$m_H = 70$ kg

$t_0 = 0$
$v_0 = -36$ m/s $v = 0$

$y_0 = +2.0$ m

$y = 0$

(continued)

Simplify and diagram

- Make appropriate simplifying assumptions about the process. For example, can you neglect the size of the system object or neglect frictional forces? Can you assume that forces or acceleration are constant?
- Then represent the process with a motion diagram and/or a force diagram(s). Make sure the diagrams are consistent with each other.

We model Holmes as a point-like object and assume that the forces being exerted on him are constant so that they lead to a constant acceleration. A motion diagram for his motion while stopping is shown along with the corresponding force diagram. To draw the force diagram we first identify the objects interacting with Holmes as he slows down: the shrubbery and the ground (combined as one interaction) and Earth. The shrubbery and ground exert an upward normal force $\vec{N}_{\text{S-G on H}}$ on Holmes. Earth exerts a downward gravitational force $\vec{F}_{\text{E on H}}$. The force diagram is the same for all points of the motion diagram because the acceleration is constant. On the force diagram the arrow for $\vec{N}_{\text{S-G on H}}$ must be longer to match the motion diagram, which shows the velocity change arrow pointing up.

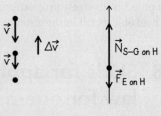

Represent mathematically

- Convert these qualitative representations into quantitative mathematical descriptions of the situation using kinematics equations and Newton's second law for motion along the axis. After you make the decision about the positive and negative directions, you can determine the signs for the force components in the equations. Add the force components (with either positive or negative signs) to find the sum of the forces.

The y-component of Holmes's average acceleration is

$$a_y = \frac{v_y^2 - v_{0y}^2}{2(y - y_0)}$$

The y-component of Newton's second law with the positive y-direction up is

$$a_y = \frac{\Sigma F_{\text{on H}\,y}}{m_{\text{H}}}$$

The y-component of the force exerted by the shrubbery-ground on Holmes is $N_{\text{S-G on H}\,y} = +N_{\text{S-G on H}}$ and the y-component of the force exerted by Earth is $F_{\text{E on H}\,y} = -F_{\text{E on H}} = -m_{\text{H}}g$. Therefore,

$$a_y = \frac{N_{\text{S-G on H}\,y} + F_{\text{E on H}\,y}}{m_{\text{H}}} = \frac{(+N_{\text{S-G on H}}) + (-F_{\text{E on H}})}{m_{\text{H}}} = \frac{+N_{\text{S-G on H}} - m_{\text{H}}g}{m_{\text{H}}}$$

$$\Rightarrow N_{\text{S-G on H}} = m_{\text{H}}a_y + m_{\text{H}}g$$

Solve and evaluate

- Substitute the known values into the mathematical expressions and solve for the unknowns.
- Finally, evaluate your work to see if it is reasonable (check units, limiting cases, and whether the answer has a reasonable magnitude). Check whether all representations—mathematical, pictorial, and graphical—are consistent with each other.

Holmes's average acceleration was

$$a_y = \frac{0^2 - (-36\ \text{m/s})^2}{2(0 - 2.0\ \text{m})} = +324\ \text{m/s}^2$$

Holmes's initial velocity is negative, since he is moving in the negative direction. His initial position is $+2.0$ m at the top of the shrubbery, and his final position is zero at the ground. His velocity in the negative direction is decreasing, which means the velocity change and the acceleration point in the opposite direction (positive). The average magnitude of the force exerted by the shrubbery and ground on Holmes is

$$N_{\text{S-G on H}} = m_{\text{H}}a_y + m_{\text{H}}g = (70\ \text{kg})(324\ \text{m/s}^2) + (70\ \text{kg})(9.8\ \text{N/kg})$$

$$= 22{,}680\ \text{kg}\cdot\text{m/s}^2 + 686\ \text{N} = 23{,}366\ \text{N} = 23{,}000\ \text{N}$$

The force has a magnitude greater than the force exerted by Earth—thus the results are consistent with the force diagram and motion diagram. The magnitude is huge and the units are correct. A limiting case for zero acceleration gives us a correct prediction—the force exerted on Holmes by the shrubbery and ground equals the force exerted by Earth.

Try it yourself: Use the strategy discussed above to estimate the average force that the ground would have exerted on Holmes if he had stopped in a conservative 0.20 m with no help from the shrubbery.

Answer: 230,000 N (over 50,000 lb).

The force when Holmes landed in the shrubbery had a magnitude of 23,000 N, greater than 5000 lb! The force was exerted over a significant area of his body, thanks to the shrubbery. Most of all, the shrubbery increased the stopping distance. If Holmes had landed directly on dirt or something harder that would compress less than the shrubbery, his acceleration would have been much greater, as would the force exerted on him.

> **TIP** In the last equation we used an N in italics to indicate the magnitude of the normal force exerted by the shrubbery and ground on Holmes. On the right side, we used an N in roman type to indicate the unit of force. Be careful not to confuse these two similar looking notations.

An elevator ride standing on a bathroom scale

In Example 2.7, we consider a much less dangerous process, one you could try the next time you ride an elevator. When you stand on a bathroom scale, the scale reading indicates how hard you are pushing on the scale. The normal force that it exerts on you balances the downward force that Earth exerts on you (called your *weight*, in everyday language), resulting in your zero acceleration. What will the scale read if you stand on it in a moving elevator?

EXAMPLE 2.7 Elevator ride

You stand on a bathroom scale in an elevator as it makes a trip from the first floor to the tenth floor of a hotel. Your mass is 50 kg. When you stand on the scale in the stationary elevator, it reads 490 N (110 lb). What will the scale read (a) early in the trip while the elevator's upward acceleration is 1.0 m/s^2, (b) while the elevator moves up at a constant speed of 4.0 m/s, and (c) when the elevator slows to a stop with a downward acceleration of 1.0 m/s^2 magnitude?

Sketch and translate We sketch the situation as shown at right, choosing you as the system object. The coordinate axis points upward with its origin at the first floor of the elevator shaft. Your mass is $m_Y = 50 \text{ kg}$, the magnitude of the force that Earth exerts on you is $F_{\text{E on Y}} = m_Y g = 490 \text{ N}$, and your acceleration is (a) $a_y = +1.0 \text{ m/s}^2$ (the upward velocity is increasing); (b) $a_y = 0$ (v is a constant 4.0 m/s upward); and (c) $a_y = -1.0 \text{ m/s}^2$ (the upward velocity is decreasing, so the acceleration points in the opposite, negative direction).

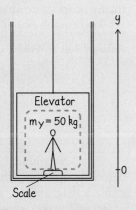

Simplify and diagram We model you as a point-like object and represent you as a dot in both the motion and force diagrams, shown for each part of the trip in Figures a, b, and c. On the diagrams, E represents Earth, Y

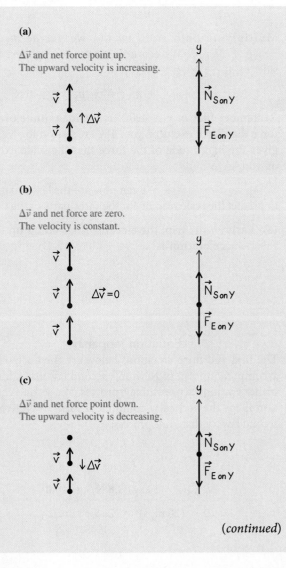

(a)

$\Delta \vec{v}$ and net force point up.
The upward velocity is increasing.

(b)

$\Delta \vec{v}$ and net force are zero.
The velocity is constant.

$\Delta \vec{v} = 0$

(c)

$\Delta \vec{v}$ and net force point down.
The upward velocity is decreasing.

(continued)

is you, and S is the scale. The magnitude of the downward force that Earth exerts does not change (it equals $m_Y\vec{g}$, and neither m_Y nor $\vec{g}$ change). Notice that the force diagrams and motion diagrams are consistent with each other for each part of the trip. The length of the normal force arrows representing the force that the scale exerts on you changes from one case to the next so that the sum of the forces point in the same direction as your velocity change arrow.

Represent mathematically The motion and the forces are entirely along the vertical y-axis. Thus, we use the vertical y-component form of Newton's second law [Eq. (2.6)] to analyze the process. There are two forces exerted on you (the system object) so there will be two vertical y-component forces on the right side of the equation: the y-component of the force that Earth exerts on you, $F_{E\,on\,Y\,y} = -m_Y g$, and the y-component of the normal force that the scale exerts on you, $N_{S\,on\,Y\,y} = +N_{S\,on\,Y}$:

$$a_{Y\,y} = \frac{\Sigma F_y}{m_Y} = \frac{F_{E\,on\,Y\,y} + N_{S\,on\,Y\,y}}{m_Y} = \frac{-m_Y g + N_{S\,on\,Y}}{m_Y}.$$

Multiplying both sides by m_Y, we get $a_{Y\,y}m_Y = -m_Y g + N_{S\,on\,Y}$. We can now move $-m_Y g$ to the left side: $m_Y a_{Y\,y} + m_Y g = N_{S\,on\,Y}$, or

$$N_{S\,on\,Y} = m_Y a_{Y\,y} + m_Y g = m_Y a_{Y\,y} + 490\,\text{N}$$

Remember that $m_Y g = 490\,\text{N}$ is the magnitude of the force that Earth exerts on you. The expression for $N_{S\,on\,Y}$ gives the magnitude of the force that the scale exerts on you.

Solve and evaluate We can now use the last equation to predict the scale reading for the three parts of the trip.

(a) Early in the trip, the elevator is speeding up and its acceleration is $a_{Y\,y} = +1.0\,\text{m/s}^2$. During that

time interval, the force exerted by the scale on you should be

$$\begin{aligned} N_{S\,on\,Y} &= m_Y\,a_{Y\,y} + 490\,\text{N} \\ &= (50\,\text{kg})(+1.0\,\text{m/s}^2) + 490\,\text{N} = 540\,\text{N} \end{aligned}$$

(b) In the middle of the trip, when the elevator moves at constant velocity, your acceleration is zero and the scale should read:

$$\begin{aligned} N_{S\,on\,Y} &= m_Y\,a_{Y\,y} + 490\,\text{N} \\ &= (50\,\text{kg})(0\,\text{m/s}^2) + 490\,\text{N} = 490\,\text{N} \end{aligned}$$

(c) When the elevator is slowing down near the end of the trip, its acceleration points downward and is $a_y = -1.0\,\text{m/s}^2$. Then the force exerted by the scale on you should be

$$\begin{aligned} N_{S\,on\,Y} &= m_Y\,a_{Y\,y} + 490\,\text{N} \\ &= (50\,\text{kg})(-1.0\,\text{m/s}^2) + 490\,\text{N} = 440\,\text{N} \end{aligned}$$

When the elevator is at rest or moving at constant speed, the scale reading equals the magnitude of the force that Earth exerts on you. When the elevator accelerates upward, the scale reads more. When it accelerates downward, even if you are moving upward, the scale reads less. What is also important is that the motion and force diagrams in Figures a, b, and c are consistent with each other and the force diagrams are consistent with the predicted scale readings—an important consistency check of the motion diagrams, force diagrams, and math.

Try it yourself: What will the scale read when the elevator starts from rest on the tenth floor and moves downward with increasing speed and a downward acceleration of $-1.0\,\text{m/s}^2$, then moves down at constant velocity, and finally slows its downward trip with an acceleration of $+1.0\,\text{m/s}^2$ until it stops at the first floor?

Answer: 440 N, 490 N, and 540 N.

EXAMPLE 2.8 Equation Jeopardy

The first and third equations below are the horizontal x-component form of Newton's second law and a kinematics equation representing a process. The second equation is for the vertical y-component form of Newton's second law for that same process:

$$a_x = \frac{-f_{S\,on\,O}}{(50\,\text{kg})}$$

$$N_{S\,on\,O} - (50\,\text{kg})(9.8\,\text{N/kg}) = 0$$

$$0 - (20\,\text{m/s})^2 = 2a_x(+25\,\text{m})$$

First, determine the values of all unknown quantities in the equations. Then work backward and construct a force diagram and a motion diagram for a system object and invent a process that is consistent with the equations (there are many possibilities). Reverse the problem-solving strategy you used in Example 2.6.

Solve The second equation can be solved for the normal force:

$$N_{S\,on\,O} = (50\,\text{kg})(9.8\,\text{N/kg}) = 490\,\text{N}$$

We solve the third equation for the acceleration a_x:

$$a_x = \frac{-(20 \text{ m/s})^2}{2(25 \text{ m})} = -8.0 \text{ m/s}^2$$

Now substitute this result in the first equation to find the magnitude of the force $f_{\text{S on O}}$:

$$-f_{\text{S on O}} = (50 \text{ kg})(-8.0 \text{ m/s}^2) = -400 \text{ N}$$

Represent mathematically The third equation in the problem statement looks like the application of the following kinematics equation to the problem process:

$$v_x^2 - v_{0x}^2 = 2a_x(x - x_0)$$

By comparing the above to the third provided equation, we see that the final velocity was $v_x = 0$, the initial velocity was $v_{0x} = +20$ m/s, and the displacement of the object was $(x - x_0) = +25$ m. The first equation indicates that there is only one force exerted on the system object in the horizontal x-direction. It has a magnitude of 400 N and points in the negative direction (the same direction as the object's acceleration), which is opposite the direction of the initial velocity. The second equation indicates that the system object has a 50-kg mass. It could be some kind of friction force that causes a 50-kg object to slow down and stop while it is traveling 25 m horizontally.

Simplify and diagram We can now construct a motion diagram and a force diagram for the object (see **a** and **b**).

(a)   **(b)**

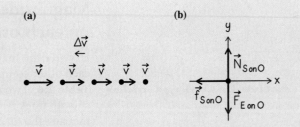

Sketch and translate This could have been a 50-kg snowboarder moving at 20 m/s after traveling down a steep hill and then skidding to a stop in 25 m on a horizontal surface (see below). You might have been asked to determine the average friction force that the snow exerted on the snowboarder while he was slowing down.

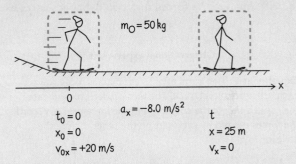

Try it yourself: Suppose the friction force remained the same but the snowboarder's initial speed was 10 m/s instead of 20 m/s. In what other ways would the trip be affected?

Answer: The snowboarder would stop in $(25 \text{ m})/4 = 6$ m.

Review Question 2.8 Three friends argue about the type of information a bathroom scale reports: Eugenia says that it reads the weight of a person, Alan says that it reads the sum of the forces exerted on the person by Earth and the scale, and Mike says that the scale reads the force that the person exerts on the scale. Who do you think is correct? Why?

2.9 Forces come in pairs: Newton's third law

So far, we have analyzed a system's acceleration due to the forces exerted on it by external objects. What effect does the system have on these other external objects? To help answer this question, we observe the interaction of two objects and analyze what happens to each of them. Suppose you wear rollerblades and push abruptly on a wheeled cart that is loaded with a heavy box. If you and the cart are on a hard smooth floor, the cart starts moving away (it accelerates), and you also start to move and accelerate in the opposite direction. Evidently, you exerted a force on the cart and the cart exerted a force on you. Since the accelerations were in opposite directions, the forces must point in opposite directions. Let's consider more quantitatively the effect of such mutual interactions between two objects.

Magnitudes of the forces that two objects exert on each other

Active Learning Guide>

How do the magnitudes of the forces that two interacting objects exert on each other compare? Consider the experiments in Observational Experiment Table 2.6. Two dynamics carts with very low-friction wheels roll freely on a smooth track before colliding. We mount force probes on each cart in order to measure the forces that each cart exerts on the other while colliding. A motion sensor on each end of the track records the initial velocity of each cart before the collision.

OBSERVATIONAL EXPERIMENT TABLE

2.6   Analyze the forces that two dynamics carts exert on each other.

VIDEO 2.6

Observational experiment	Analysis
Experiment 1. Two carts of different masses move toward each other on a level track. The motion detector indicates their speed before the collision and the force probes record the forces exerted by each cart on the other. Before the collision: $m_1 = 2\text{ kg}, v_{1x} = +2\text{ m/s}$ $m_2 = 1\text{ kg}, v_{2x} = -2\text{ m/s}$	As both carts changed velocities due to the collision, they must have exerted forces on each other. The computer recordings from the force probes show that the forces that the carts exert on each other vary with time and at each time have the same magnitude and point in the opposite direction. Cart 1 exerts a force on cart 2 toward the right, and cart 2 exerts a force on cart 1 toward the left. 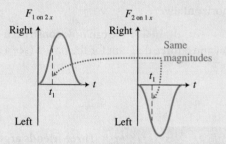
Experiment 2. Cart masses and velocities before collision: $m_1 = 2\text{ kg}, v_{1x} = 0\text{ m/s (at rest)}$ $m_2 = 1\text{ kg}, v_{2x} = -1\text{ m/s}$	Although the forces that the carts exert on each other are smaller than in the first experiment, the magnitudes of the forces at each time are still the same. 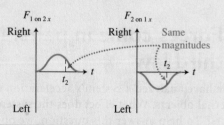
Experiment 3. Cart masses and velocities before collision: $m_1 = 2\text{ kg}, v_{1x} = +2\text{ m/s}$ $m_2 = 1\text{ kg}, v_{2x} = -1\text{ m/s}$	The same analysis applies. 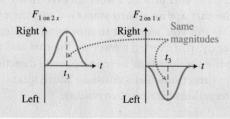

Pattern

In each experiment, independent of the masses and velocities of the carts before the collisions, at every instant during the collision the force that cart 1 exerted on cart 2 $\vec{F}_{1 \text{ on } 2}$ had the same magnitude as but pointed in the opposite direction from the force that cart 2 exerted on cart 1 $\vec{F}_{2 \text{ on } 1}$.

The cart collisions in Table 2.6 indicate that the force that one cart exerts on the other is equal in magnitude and opposite in direction to the force that the other cart exerts on the first.

$$\vec{F}_{\text{object 1 on object 2}} = -\vec{F}_{\text{object 2 on object 1}}$$

Will the pattern that we found allow us to correctly predict the results of some new experiment?

Testing the idea

Imagine that you have two spring scales. Attach one scale to a hook on the wall and pull on its other end with a second scale (**Figure 2.7**). If the preceding equation is correct, then the scale you pull should have the same reading as the scale fixed to the wall. When you do the experiment, you find that the scales do indeed have the same readings. If you reverse the scales and repeat the experiment, you find that they *always* have the same readings.

The outcome of the experiment matched the prediction we made based on the preceding equation. The objects exert equal-magnitude, oppositely directed forces on each other. To be convinced of the validity of this outcome, we need many more testing experiments. So far, physicists have found no experiments involving the dynamics of everyday processes that violate this equation. This rule is called Newton's third law.

Figure 2.7 Spring scales exert equal-magnitude forces on each other.

Newton's third law When two objects interact, object 1 exerts a force on object 2. Object 2 in turn exerts an equal-magnitude, oppositely directed force on object 1:

$$\vec{F}_{\text{object 1 on object 2}} = -\vec{F}_{\text{object 2 on object 1}} \qquad (2.8)$$

Note that these forces are exerted on different objects and *cannot* be added to find the sum of the forces exerted on one object.

It seems counterintuitive that two interacting objects always exert forces of the same magnitude on each other. Imagine playing ping-pong. A paddle hits the ball and the ball flies rapidly toward the other side of the table. However, the paddle seems to move forward with little change in motion. How is it possible that the light ball exerted a force of the same magnitude on the paddle as the paddle exerted on the ball?

To resolve this apparent contradiction, think about the masses of the interacting objects and their corresponding accelerations. If the forces are the same, the object with larger mass has smaller magnitude acceleration than the object with smaller mass:

$$a_{\text{paddle}} = \frac{F_{\text{ball on paddle}}}{m_{\text{paddle}}} \quad \text{and} \quad a_{\text{ball}} = \frac{F_{\text{paddle on ball}}}{m_{\text{ball}}}$$

Because the mass of the ball is so small, the same force leads to a large change in velocity. The paddle's mass, on the other hand, is much larger. Thus, the same magnitude force leads to an almost zero velocity change. We observe the velocity change and incorrectly associate that alone with the force exerted on the object.

> **TIP** Remember that the forces in Newton's third law are exerted on two different objects. This means that the two forces will never show up on the same force diagram, and they should not be added together to find the sum of the forces. You have to choose the system object and consider only the forces exerted on *it*!

CONCEPTUAL EXERCISE 2.9 A book on the table

A book sits on a tabletop. Identify the forces exerted on the book by other objects. Then, for each of these forces, identify the force that the book exerts on another object. Explain why the book is not accelerating.

Sketch and translate Draw a sketch of the situation and choose the book as the system.

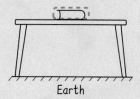

Earth

Simplify and diagram Assume that the tabletop is perfectly horizontal and model the book as a point-like object. A force diagram for the book is shown at right. Earth exerts a downward gravitational force on the book $\vec{F}_{\text{E on B}}$, and the table exerts an upward normal (contact) force on the book $\vec{N}_{\text{T on B}}$. Newton's second law explains why the book is not accelerating; the forces exerted on it by other objects are balanced and add to zero.

The subscripts on each force identify the two objects involved in the interaction. The Newton's third law pair force will have its subscripts reversed. For example, Earth exerts a downward gravitational force on the book ($\vec{F}_{\text{E on B}}$). According to Newton's third law, the book must exert an equal-magnitude upward gravitational force on Earth ($\vec{F}_{\text{B on E}} = -\vec{F}_{\text{E on B}}$), as shown at right. The table exerts an upward contact force on the book ($\vec{N}_{\text{T on B}}$), so the book must exert an equal-magnitude downward contact force on the table ($\vec{N}_{\text{B on T}} = -\vec{N}_{\text{T on B}}$).

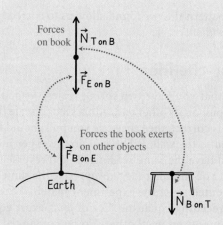

Forces on book

$\vec{N}_{\text{T on B}}$

$\vec{F}_{\text{E on B}}$

Forces the book exerts on other objects

$\vec{F}_{\text{B on E}}$

Earth

$\vec{N}_{\text{B on T}}$

Try it yourself: A horse pulls on a sled that is stuck in snow and not moving. Your friend Chris says this happens because the horse exerts on the sled the same magnitude force that the sled exerts on the horse. Since the sum of the forces is zero, there is no acceleration. What is wrong with Chris' reasoning?

Answer: Chris added the forces exerted on two different objects and did not consider all forces exerted on the sled. If you choose the sled as the system object, then the horse pulls forward on the sled, and the snow exerts a backward, resistive force. If these two horizontal forces happen to be of the same magnitude, they add to zero, and the sled does not accelerate horizontally. If, on the other hand, we choose the horse as the system, the ground exerts a forward force on the horse's hooves (since the horse is exerting a force backward on the ground), and the sled pulls back on the horse. If those forces have the same magnitude, the net horizontal force is again zero, and the horse does not accelerate.

EXAMPLE 2.10 Froghopper jump

The froghopper (*Philaenus spumarius*), an insect about 6 mm in length, is considered by some scientists to be the best jumper in the animal world. The froghopper can jump 0.7 m vertically, over 100 times higher than its length.

The froghopper's speed can change from zero to 4 m/s in about 1 ms = 1×10^{-3} s—an acceleration of 4000 m/s². The froghopper achieves this huge acceleration using its leg muscles, which occupy 11% of its 12-mg body mass. What is the average force that the froghopper exerts on the surface during the short time interval while pushing off during its jump?

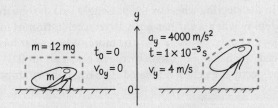

Sketch and translate We construct an initial-final sketch of the process, choosing the froghopper as the system object. We use a vertical y-axis with the positive direction pointing up and the origin of the coordinate system at the surface. We know the froghopper's acceleration while pushing off.

Simplify and diagram We can make a motion diagram for the pushing off process and a force diagram. The surface S exerts an upward normal force $\vec{N}_{S\,on\,F}$ on the froghopper, and Earth exerts a downward gravitational force $\vec{F}_{E\,on\,F}$. Note that because the acceleration is upward, the sum of the y-components of the

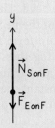

forces must also be upward; thus, the upward normal force exerted by the surface is greater in magnitude than the downward gravitational force exerted by Earth.

The force diagram shows the normal force exerted by the surface on the froghopper $\vec{N}_{S\,on\,F}$. We are asked to determine the force that the froghopper exerts on the surface $\vec{N}_{F\,on\,S}$. According to Newton's third law, the magnitude of $\vec{N}_{S\,on\,F}$ will be the same as the magnitude of $\vec{N}_{F\,on\,S}$.

Represent mathematically The process occurs along the vertical direction, so we apply the vertical y-component form of Newton's second law:

$$a_y = \frac{\Sigma F_y}{m_F} = \frac{F_{E\,on\,F\,y} + N_{S\,on\,F\,y}}{m_F} = \frac{-m_F g + N_{S\,on\,F}}{m_F}$$

Rearranging the above, we get an expression for the force of the surface on the froghopper while pushing off:

$$N_{S\,on\,F} = m_F a_y + m_F g$$

Solve and evaluate We can now substitute the known information in the above, converting the froghopper mass into kg:

$$12\,mg = (12\,mg)\left(\frac{1\,g}{10^3\,mg}\right)\left(\frac{1\,kg}{10^3\,g}\right) = 12 \times 10^{-6}\,kg$$

$$\begin{aligned} N_{S\,on\,F} &= (12 \times 10^{-6}\,kg)(4000\,m/s^2) \\ &\quad + (12 \times 10^{-6}\,kg)(9.8\,m/s^2) \\ &= 0.048\,N + 0.00012\,N = 0.048\,N \approx 0.05\,N \end{aligned}$$

According to Newton's third law, the magnitude of the force exerted on the surface by the froghopper is also 0.05 N.

The force that the surface exerts on the froghopper is 400 times greater than the force that Earth exerts on the froghopper. Also, notice that the force magnitudes and both the motion and force diagrams are consistent with each other—a nice check on our work.

Try it yourself: Suppose an 80-kg basketball player could push off a gym floor exerting a force (like the froghopper's) that is 400 times greater than the force Earth exerts on him and that the push off lasted 0.10 s (unrealistically short). How fast would he be moving when he left contact with the floor?

Answer: About 400 m/s (900 mi/h). He would be moving faster if he took longer to push off while exerting the same force.

Review Question 2.9 Identify force pairs for the following interactions and compare the force magnitudes: A rollerblader and the floor; you pushing a refrigerator across the kitchen floor; and a tow truck pulling a car.

Figure 2.8 An air bag stops a crash test dummy during a collision.

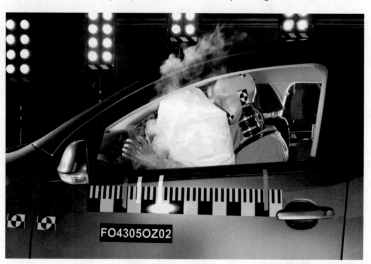

2.10 Seat belts and air bags: Putting it all together

At the beginning of the chapter we posed a question about airbags. How do they save lives? We now have all the physics needed to investigate this question. Consider **Figure 2.8**. An air bag is like a balloon with heavy-walled material that is packed in a small box in the steering wheel or the passenger side dashboard. Air bags are designed to deploy when a car has a negative acceleration of magnitude $10g[10(9.8 \text{ m/s}^2) = 98 \text{ m/s}^2 \approx 100 \text{ m/s}^2]$ or more. When a car has such a rapid decrease in speed, the bag inflates with nitrogen gas in about 0.04 s and forms a cushion for the occupant's chest and head. The bag has two important effects:

1. It spreads the force that stops the person over a larger area of the body.

2. It increases the stopping distance, and consequently the stopping time interval, thus reducing the average force stopping the occupant.

Why is spreading the stopping force over the air bag an advantage? If a person uses only seat belts, his head is not belted to the seat and tends to continue moving forward during a collision, even though his chest and waist are restrained. To stop the head without an air bag, the neck must exert considerable force on the head. This can cause a dangerous stretching of the spinal cord and muscles of the neck, a phenomenon known as "whiplash." The air bag exerts a more uniform force across the upper body and head and helps make all parts stop together.

How does the air bag increase the stopping distance? Suppose a test car is traveling at a constant speed of 13.4 m/s (30 mi/h) until it collides head-on into a concrete wall. The front of the car crumples about 0.65 m. A crash test dummy is rigidly attached to the car's seat and is further protected by the rapidly inflating airbag. The dummy also travels about 0.65 m before coming to rest. Without an air bag or a seat belt, the dummy would continue to move forward at the initial velocity of the car. The dummy would then crash into the steering wheel or windshield of the stopped car and stop in a distance much less than 0.65 m—like flying into a rigid wall. The smaller the stopping distance, the greater the acceleration, and therefore the greater the force that is exerted on the dummy. Let's estimate the average force exerted by the air bag on the body during a collision.

Active Learning Guide›

EXAMPLE 2.11 Force exerted by air bag on driver during collision

A 60-kg crash test dummy moving at 13.4 m/s (30 mi/h) stops during a collision in a distance of 0.65 m. Estimate the average force that the air bag and seat belt exert on the dummy.

Sketch and translate We sketch and label the situation as shown below. choosing the crash test dummy as the system object. The positive x-direction will be in the direction of motion, and the origin will be at the position of the dummy at the start of the collision.

$t_0 = 0$ $a_x = ?$ t
$x_0 = 0$ $x = 0.65$ m
$v_{0x} = 13.4$ m/s $v_x = 0$

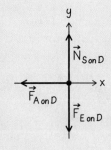

Simplify and diagram We model the dummy D as a point-like object and assume that the primary force exerted on the dummy while stopping is due to the air bag and seat belt's $\vec{F}_{A \, on \, D}$, shown in the force diagram. We can ignore the downward gravitational force that Earth exerts on the dummy $\vec{F}_{E \, on \, D}$ and the upward normal force that the car seat exerts on the dummy $\vec{N}_{S \, on \, D}$ since they balance and do not contribute to the acceleration.

Represent mathematically To determine the dummy's acceleration, we use kinematics:

$$a_x = \frac{v_x^2 - v_{0x}^2}{2(x - x_0)}$$

Once we have the dummy's acceleration, we apply the x-component form of Newton's second law to find the force exerted by the air bag and seat belts on the dummy:

$$a_x = \frac{F_{A \, on \, D \, x}}{m_D} = \frac{-F_{A \, on \, D}}{m_D} = -\frac{F_{A \, on \, D}}{m_D}$$

$$\Rightarrow F_{A \, on \, D} = -m_D a_x$$

Solve and evaluate We know that $v_{0x} = +13.4$ m/s and $v_x = 0$ (the dummy has stopped). The initial position of the dummy is $x_0 = 0$ and the final position is $x = 0.65$ m. The acceleration of the dummy while in contact with the air bag and seat belt is:

$$a_x = \frac{0^2 - (13.4 \, \text{m/s})^2}{2(0.65 \, \text{m} - 0 \, \text{m})} = -138 \, \text{m/s}^2$$

Thus, the *magnitude* of the average force exerted by the air bag and seat belt on the dummy is

$$F_{A \, on \, D} = -(60 \, \text{kg})(-138 \, \text{m/s}^2) = 8300 \, \text{N}$$

This force [$8300 \, \text{N}(1 \, \text{lb}/4.45 \, \text{N}) = 1900 \, \text{lb}$] is almost 1 ton. Is this estimate reasonable? The magnitude is large, but experiments with crash test dummies in the real world are consistent with a force this large in magnitude, a very survivable collision.

Try it yourself: Find the acceleration of the dummy and the magnitude of the average force needed to stop the dummy if it is not belted, has no air bag, and stops in 0.1 m when hitting a hard surface.

Answer: $-900 \, \text{m/s}^2$ and 54,000 N.

Review Question 2.10 Explain how an air bag and seat belt reduce the force exerted on the driver of a car during a collision.

Summary

Words	Pictorial and physical representations	Mathematical representation
System and environment A system object is circled in a sketch of a process. Environmental objects that are not part of the system are external and might interact with the system and affect its motion. (Section 2.1)	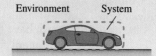 Environment System	
The force that one object exerts on another characterizes an interaction between the two objects (a pull or a push) denoted by a symbol $\vec{F}$ with a subscript with two elements indicating the two interacting objects. (Section 2.1)	 $\vec{F}_{\text{R on W}}$	$\vec{F}_{\text{E on O}}$ Interacting objects
A force diagram represents the forces that external objects exert on the system object. The arrows in the diagram point in the directions of the forces, and their lengths indicate the relative magnitudes of the forces. The unit of force is the newton (N); $1\,\text{N} = (1\,\text{kg})(1\,\text{kg}\cdot\text{m/s}^2)$. (Sections 2.1 and 2.6)	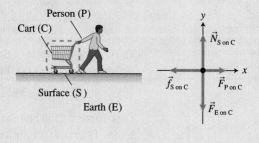 Person (P) Cart (C) Surface (S) Earth (E)	
Relating forces and motion The $\Delta\vec{v}$ arrow in an object's motion diagram is in the same direction as the sum of the forces that other objects exert on it. (Section 2.3)	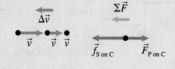	
In an **inertial reference frame,** the velocity of a system object does not change if no other objects exert forces on it or if the sum of all the forces exerted on the system object is zero. (Section 2.5)		
Newton's first law of motion If no other objects exert forces on a system object or if the forces on the system object add to zero, then the object continues moving at constant velocity (as seen by observers in inertial reference frames). (Section 2.5)	$\Delta\vec{v} = 0$ $\vec{N}_{\text{S on C}}$ $\vec{f}_{\text{S on C}}$ $\vec{F}_{\text{P on C}}$ $\vec{F}_{\text{E on C}}$	If $\sum F_x = 0$, then v_x = constant (same for y-direction)
Mass m characterizes the amount of "material" in an object and its resistance to a change in motion. (Section 2.6)		

Newton's second law The acceleration a_O of a system object is proportional to the sum of the forces that other objects exert on the system object and inversely proportional to its mass m. (Section 2.6)

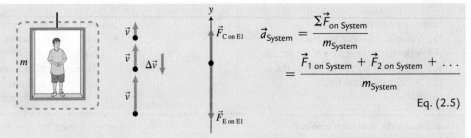

$$\vec{a}_{System} = \frac{\Sigma \vec{F}_{on\ System}}{m_{System}}$$

$$= \frac{\vec{F}_{1\ on\ System} + \vec{F}_{2\ on\ System} + \cdots}{m_{System}}$$

Eq. (2.5)

In the **component form of Newton's second law,** the force components along each axis are included with $+$ or $-$ signs times their magnitudes. (Section 2.6)

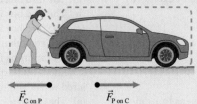

$$ma_x = (+F_{P\ on\ C}) + (-f_{S\ on\ C})$$
$$0 = (+N_{S\ on\ C}) + (-F_{E\ on\ C})$$

$$a_{System\ x} = \frac{\Sigma F_{on\ System\ x}}{m_{System}}$$

$$= \frac{F_{1\ on\ System\ x} + F_{2\ on\ System\ x} + \cdots}{m_{System}}$$

$$a_{System\ y} = \frac{\Sigma F_{on\ System\ y}}{m_{System}}$$

$$= \frac{F_{1\ on\ System\ y} + F_{2\ on\ System\ y} + \cdots}{m_{System}}$$

Eq. (2.6)

Newton's third law Two objects exert equal-magnitude and opposite direction forces of the same type on each other. (Section 2.9)

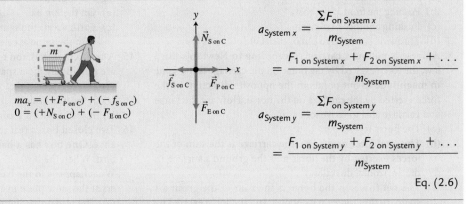

$$\vec{F}_{1\ on\ 2} = -\vec{F}_{2\ on\ 1} \qquad \text{Eq. (2.8)}$$

The gravitational force $\vec{F}_{E\ on\ O}$ that Earth exerts on an object of mass m when on or near Earth's surface depends on the gravitational constant g of Earth. If on or near another planet or the Moon, the gravitational constant near those objects is different. (Section 2.7)

Magnitude

$$F_{E\ on\ O} = m_O g \qquad \text{Eq. (2.7)}$$

where

$$g = 9.8\ \text{m/s}^2 = 9.8\ \text{N/kg}$$

when the object is on or near Earth's surface.

 For instructor-assigned homework, go to **MasteringPhysics.**

Questions

Multiple Choice Questions

1. An upward-moving elevator slows to a stop as it approaches the top floor. Which answer below best describes the relative magnitudes of the upward force that the cable exerts on the elevator $\vec{F}_{C\ on\ EI}$ and the downward gravitational force that Earth exerts on the elevator $\vec{F}_{E\ on\ EI}$?
   (a) $F_{C\ on\ EI} > F_{E\ on\ EI}$ (b) $F_{C\ on\ EI} = F_{E\ on\ EI}$
   (c) $F_{C\ on\ EI} < F_{E\ on\ EI}$ (d) None of these

2. You apply the brakes of your car abruptly and your book starts sliding off the front seat. Three observers explain this differently. Observer A says that the book continued moving and the car accelerated from underneath it. Observer B says that the car pushed forward on the book. Observer C says that she must be in a noninertial reference frame because the book started moving without any extra objects interacting with it. Which of the observers is correct?
   (a) A   (b) B   (c) C
   (d) A and C   (e) All of the observers

3. Which of the statements below explains why a child lurches forward in a stroller when you abruptly stop the stroller?
   (a) The child does not lurch forward but instead continues her motion.
   (b) Your pull on the stroller causes the child to move in the opposite direction.
   (c) Newton's third law

4. Which observers can explain the phenomenon of whiplash, which occurs when a car stops abruptly using Newton's laws?
   (a) The driver of the car   (b) A passenger in the car
   (c) An observer on the sidewalk beside the car and road

5. Which vector quantities describing a moving object are always in the same direction?
   (a) Velocity and acceleration
   (b) Velocity and the sum of the forces
   (c) Acceleration and the sum of the forces
   (d) Acceleration and force
   (e) Both b and c are correct.

6. You are standing in a boat. Which of the following strategies will make the boat start moving?
   (a) Pushing its mast
   (b) Pushing the front of the boat
   (c) Pushing another passenger
   (d) Throwing some cargo out of the boat

7. A horse is pulling a carriage. According to Newton's third law, the force exerted by the horse on the carriage is the same in magnitude as but points in the opposite direction of the force exerted by the carriage on the horse. How are the horse and carriage able to move forward?
   (a) The horse is stronger.
   (b) The total force exerted on the carriage is the sum of the forces exerted by the horse and the ground's surface in the horizontal direction.
   (c) The net force on the horse is the sum of the ground's static friction force on its hooves and the force exerted by the carriage.
   (d) Both b and c are correct.

8. A book sits on a tabletop. Which of Newton's laws explains its equilibrium?
   (a) First   (b) Second   (c) Third
   (d) Both the first and the second

9. A spaceship moves in outer space. What happens to its motion if there are no external forces exerted on it? If there is a constant force exerted on it in the direction of its motion? If something exerts a force opposite its motion?
   (a) It keeps moving; it speeds up with constant acceleration; it slows down with constant acceleration.
   (b) It slows down; it moves with constant velocity; it slows down.
   (c) It slows down; it moves with constant velocity; it stops instantly.

10. A 0.10-kg apple falls on Earth, whose mass is about 6×10^{24} kg. Which is true of the gravitational force that Earth exerts on the apple?
   (a) It is bigger than the force that the apple exerts on Earth by almost 25 orders of magnitude.
   (b) It is the same magnitude.
   (c) We do not know the magnitude of the force the apple exerts on Earth.

11. A man stands on a scale and holds a heavy object in his hands. What happens to the scale reading if the man quickly lifts the object upward and then stops lifting it?
   (a) The reading increases, returns briefly to the reading when standing stationary, then decreases.
   (b) The reading decreases, returns briefly to the reading when standing stationary, then increases.
   (c) Nothing, since the mass of the person with the object remains the same. Thus the reading does not change.

12. You stand on a bathroom scale in a moving elevator. What happens to the scale reading if the cable holding the elevator suddenly breaks?
   (a) The reading will increase.
   (b) The reading will not change.
   (c) The reading will decrease a little.
   (d) The reading will drop to 0 instantly.

13. A person pushes a 10-kg crate exerting a 200-N force on it, but the crate's acceleration is only 5 m/s². Explain.
   (a) The crate pushes back on the person, thus the total force is reduced.
   (b) There are other forces exerted on the crate so that the total force is reduced.
   (c) Not enough information is given to answer the question.

14. Two small balls of the same material, one of mass m and the other of mass $2m$, are dropped simultaneously from the Leaning Tower of Pisa. On which ball does Earth exert a bigger force?
   (a) On the $2m$ ball   (b) On the m ball
   (c) Earth exerts the same force on both balls because they fall with the same acceleration.

15. A box full of lead and a box of the same size full of feathers are floating inside a spaceship that has left the solar system. Choose equipment that you can use to compare their masses.
   (a) A balance scale   (b) A digital scale
   (c) A watch with a second hand and a meter stick

16. Two closed boxes rest on the platforms of an equal arm balance. One box has a ball in it and the other box contains a bird. When the bird sits still in the box, the scale is balanced. What happens to the balance if the bird flies in the box (hovers at the same place inside the box)?
   (a) The bird box is lighter.
   (b The bird box is heavier.
   (c) The bird box is the same.
   (d) Not enough information is provided.

17. A person jumps from a wall and lands stiff-legged. Which statement best explains why the person is less likely to be injured when landing on soft sand than on concrete?
   (a) The concrete can exert a greater force than the sand.
   (b) The person sinks into the sand, increasing the stopping distance.
   (c) The upward acceleration of the person in the sand is less than on concrete, thus the force that the sand exerts on the person is less.
   (d) b and c   (e) a, b, and c

18. What do objects that are already in motion on a smooth surface need in order to maintain the same motion?
   (a) A constant push exerted by another object
   (b) A steadily increasing push   (c) Nothing

Conceptual Questions

19. **Figure Q2.19** is a velocity-versus-time graph for the vertical motion of an object. Choose a correct combination (a, b, or c) of

Figure Q2.19

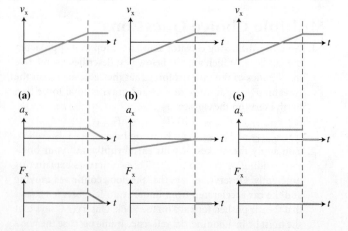

an acceleration-versus-time graph and a force-versus-time graph for the object.

20. Explain the purpose of crumple zones, that is, the front of a car that collapses during a collision.

21. Explain why when landing on a firm surface after a fall you should not land with stiff legs.

22. A small car bumps into a large truck. Compare the forces that the truck exerts on the car and the car exerts on the truck if before the collision (a) the truck was stationary and the car was moving; (b) the car and the truck were moving in opposite directions; (c) the car and the truck were moving in the same direction.

23. You are pulling a sled. Compare the forces that you exert on the sled and the sled exerts on you if you (a) move at constant velocity; (b) speed up; (c) slow down.

24. ✔ You stand on a bathroom scale in a moving elevator. The elevator is moving up at increasing speed. The acceleration is constant. Draw three consecutive force diagrams for you.

Problems

Below, BIO indicates a problem with a biological or medical focus. Problems labeled EST ask you to estimate the answer to a quantitative problem rather than derive a specific answer. Problems marked with ✔ require you to make a drawing or graph as part of your solution. Asterisks indicate the level of difficulty of the problem. Problems with no * are considered to be the least difficult. A single * marks moderately difficult problems. Two ** indicate more difficult problems.

2.1 and 2.2 Describing and representing interactions and Adding and measuring forces

1. In **Figure P2.1** you see unlabeled force diagrams for balls in different situations. Match the diagrams with the following descriptions. (1) A ball is moving upward after it leaves your hand. (2) You hold a ball in your hand. (3) A ball is falling down. (4) You are throwing a ball (still in your hand) straight up. (5) You are lifting a ball at a constant pace. Explain your choices. Label the forces on the diagrams.

Figure P2.1

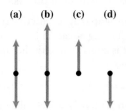

(a)   (b)   (c)   (d)

2. ✔ Draw a force diagram (a) for a bag hanging at rest from a spring; (b) for the same bag sitting on a table; and (c) for the same bag that you start to lift so it moves up faster and faster.

3. ✔ For each of the following situations, draw the forces exerted on the moving object and identify the other object causing each force. (a) You pull a wagon along a level floor using a rope oriented horizontally. (b) A bus moving on a horizontal road slows down in order to stop. (c) You lift your overnight bag into the overhead compartment on an airplane.

4. ✔ You hang a book bag on a spring scale and place the bag on a platform scale so that the platform scale reads 25.7 N and the spring scale reads 17.6 N. (a) Draw a force diagram to represent the situation. (b) What is the magnitude of the force that Earth exerts on the bag?

2.3 and 2.4 Conceptual relationship between force and motion and Reasoning without mathematical equations

5. ✔ A block of dry ice slides at constant velocity along a smooth, horizontal surface (no friction). (a) Construct a motion diagram. (b) Draw position- and velocity-versus-time graphs. (c) Construct a force diagram for the block for three instances represented by dots on the motion diagram. Are the diagrams consistent with each other?

6. * ✔ You throw a ball upward. (a) Draw a motion diagram and two force diagrams for the ball on its way up and another motion diagram and two force diagrams for the ball on its way down. (b) Represent the motion of the ball with a position-versus-time graph and velocity-versus-time graph.

7. ✔ A string pulls horizontally on a cart so that it moves at increasing speed along a smooth, frictionless, horizontal surface. When the cart is moving medium-fast, the pulling is stopped abruptly. (a) Describe in words what happens to the cart's motion when the pulling stops. (b) Illustrate your description with motion diagrams, force diagrams, and position-versus-time and velocity-versus-time graphs. Indicate on the graphs when the pulling stopped. What assumptions did you make?

8. * Solving the previous problem, your friend says that after the string stops pulling, the cart starts slowing down. (a) Give a reason for his opinion. (b) Do you agree with him? Explain your opinion. (c) Explain how you can design an experiment to test his idea.

9. * ✔ A string pulls horizontally on a cart so that it moves at increasing speed along a smooth, frictionless, horizontal surface. When the cart is moving medium-fast, the magnitude of the pulling force is reduced to half its former magnitude. (a) Describe what happens to the cart's motion after the reduction in the string pulling. (b) Illustrate your description with motion diagrams, force diagrams, and position-versus-time and velocity-versus-time graphs.

10. * Solving the previous problem, your friend says that if the string pulls half as hard, the cart should move half as fast compared to the speed it moved when the string was pulling twice as hard. (a) Explain why your friend would think this way. (b) Do you agree with his opinion? (c) Explain how you would convince him that he is incorrect.

11. ✔ Three motion diagrams for a moving elevator are shown in **Figure P2.11**. Construct two force diagrams for the elevator

for *each* motion diagram. Be sure that the lengths of the force arrows are the appropriate relative lengths and that there is consistency between the force diagrams and the motion diagrams. What assumptions did you make?

Figure P2.11

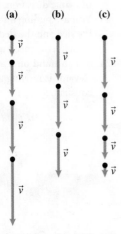

(a)   (b)   (c)

12. * An elevator is pulled upward so it moves with increasing upward speed—the force exerted by the cable on the elevator is constant and greater than the downward gravitational force exerted by Earth. When the elevator is moving up medium-fast, the force exerted by the cable on the elevator changes abruptly to just balance the downward gravitational force of Earth—the sum of the forces that the cable and Earth exert on the elevator is now zero. Now what happens to the elevator? Explain. Represent your answer with position-versus-time and velocity-versus-time graphs. What assumptions did you make?

13. * Solving the previous problem, your friend says that the elevator will stop if the two forces have equal magnitudes. (a) Why would he think this way? (b) Do you agree with his opinion? (c) If you disagree, how would you convince him that he is incorrect?

14. A block of dry ice slides at a constant velocity on a smooth, horizontal, frictionless surface. A second block of dry ice slides twice as fast on the same surface (at a higher constant velocity). Compare the resultant forces exerted on each block. Explain your reasoning.

15. An elevator moves downward at constant velocity. Construct a motion diagram and three consecutive force diagrams for the elevator (ignore the resistive force exerted by the air on the elevator) as it is moving. Make the relative lengths of the force arrows consistent with the motion diagram.

16. * Figures P2.11a, b, and c show three motion diagrams for an elevator moving downward. (a) For each diagram, say everything you can about the elevator's motion. (b) Draw a force diagram for each motion diagram. (c) Could you draw a different motion diagram for each force diagram? Explain how it is possible.

2.5 Inertial reference frames and Newton's first law

17. * Your friend has a pie on the roof of his van. You are standing on the ground and observe the van stopping abruptly for a red light. The pie does not slip off the roof. (a) Construct a motion diagram and a force diagram for the pie as the van approaches the red light, from your frame of reference and from the driver's frame of reference. (b) Repeat part (a) for the case when the light turns green. Be sure to specify the observer and identify the other object causing each force. (c) Are the motion diagrams consistent with the force diagrams for each case?

18. * A train traveling from New York to Philadelphia is passing a station. A ball is sitting on the floor of the train not moving with respect to the train. (a) Draw a force diagram and a motion diagram for the ball as seen by the observers on the train and on the platform. (b) The ball now starts accelerating forward relative to the floor. Draw force and motion diagrams for the ball as seen by the observers on the train and on the platform. Which of the observers can use Newton's first law to explain the ball's acceleration? Explain.

19. * Explain the phenomenon of whiplash from two points of view: that of an observer on the ground and an observer in the car.

2.6 Newton's second law

20. An astronaut exerts a 100-N force pushing a beam into place on the International Space Station. The beam accelerates at 0.10 m/s^2. Determine the mass of the beam. What is the percent uncertainty in your answer?

21. Four people participate in a rope competition. Two of them pull the rope right, exerting forces of magnitude 330 N and 380 N. The other two pull left, exerting forces of magnitude 300 N and 400 N. What is the sum of the forces exerted on a short section in the middle of the rope?

22. * **Shot put throw** During a practice shot put throw, the 7.0-kg shot left world champion C. J. Hunter's hand at speed 13 m/s. While making the throw, his hand pushed the shot a distance of 1.7 m. Describe all the physical quantities you can determine using this information. Describe the assumptions you need to make to determine them.

23. * You know the sum of the forces $\Sigma \vec{F}$ exerted on an object of mass m during Δt seconds. The object is at rest at the beginning of the time interval. List three physical quantities that you can determine about that object's motion using this information. Then explain how you will determine them.

24. * You record the displacement of an object as a constant force is exerted on it. (a) If the time interval during which the force is exerted doubles, how does the object's displacement change? Indicate all the assumptions that you made. (b) Explain how your answer changes if one of the assumptions is not valid.

25. * **Equation Jeopardy 1** Invent a problem for which the following equation can be a solution:

$$200 \text{ N} - 40 \text{ N} = (40 \text{ kg})a_x$$

2.7 and 2.8 Gravitational force law and Skills for applying Newton's second law for one-dimensional processes

26. * **Equation Jeopardy 2** Describe in words a problem for which the following equation is a solution and draw a force diagram that is consistent with the equation (specify the direction of the axis):

$$+29.4 \text{ N} - F_{\text{R on O}} = (3.0 \text{ kg})(3.0 \text{ m/s}^2)$$

27. * **Equation Jeopardy 3** Describe in words a problem for which the following equation is a solution and draw a force diagram that is consistent with the equation (specify the direction of the axis):

$$100 \text{ N} - f_{\text{S on O}} = (30 \text{ kg})(-1.0 \text{ m/s}^2)$$

28. * **Equation Jeopardy 4** Describe in words a problem for which the following equation is a solution and draw a force diagram that is consistent with the equation (specify the direction of the axis):

$$-196 \text{ N} + F_{\text{P on O}} = (20 \text{ kg})(-2.0 \text{ m/s}^2)$$

29. * **Spider-Man** Spider-Man holds the bottom of an elevator with one hand. With his other hand, he holds a spider cord attached to a 50-kg box of explosives at the bottom of the cord. Determine the force that the cord exerts on the box if (a) the elevator is at rest; (b) the elevator accelerates up at 2.0 m/s^2; (c) the upward-moving elevator's speed decreases at a rate of 2.0 m/s^2; and (d) the elevator falls freely.

30. * A farmer pushes his 500-kg wagon along a horizontal level icy road, exerting a 125-N horizontal force on the wagon. (a) Determine the acceleration of the wagon. How certain are you about your answer? What assumptions did you make? Would the number be higher or lower if you did not make those assumptions? (b) If the wagon started at rest, how fast was it moving after being pushed for 5.0 s?

31. * **Stuntwoman** The downward acceleration of a 60-kg stuntwoman near the end of a fall from a very high building is 7.0 m/s^2. What resistive force does the air exert on her body at that point?

32. EST Estimate the average force that a baseball pitcher's hand exerts on a 0.145-kg baseball as he throws a 40 m/s (90 mi/h) pitch. Indicate all of the assumptions you made.

33. * **Super Hornet jet takeoff** A 2.1×10^4-kg F-18 Super Hornet jet airplane (see **Figure P2.33**) goes from zero to 265 km/h in 90 m during takeoff from the flight deck of the USS Nimitz aircraft carrier. What physical quantities can you determine using this information? Make a list and determine the values of three of them.

Figure P2.33

34. * **Lunar Lander** The Lunar Lander of mass 2.0×10^4 kg made the last 150 m of its trip to the Moon's surface in 120 s, descending at approximately constant speed. The *Handbook of Lunar Pilots* indicates that the gravitational constant on the Moon is 1.633 N/kg. Using these quantities, what can you learn about the Lunar Lander's motion?

35. * A Navy Seal of mass 80 kg parachuted into an enemy harbor. At one point while he was falling, the resistive force of air exerted on him was 520 N. What can you determine about his motion?

36. * **Astronaut** Karen Nyberg, a 60-kg astronaut, sits on a bathroom scale in a rocket that is taking off vertically with an acceleration of 3 g. What does the scale read?

37. * A 0.10-kg apple falls off a tree branch that is 2.0 m above the grass. The apple sinks 0.060 m into the grass while stopping. Determine the force that the grass exerts on the apple while stopping it. Indicate any assumptions you made.

38. ** An 80-kg fireman slides 5.0 m down a fire pole. He holds the pole, which exerts a 500-N steady resistive force on the fireman. At the bottom he slows to a stop in 0.40 m by bending his knees. What can you determine using this information? Determine it.

2.9 Forces come in pairs: Newton's third law

39. Earth exerts a 1.0-N gravitational force on an apple as it falls toward the ground. (a) What force does the apple exert on Earth? (b) Compare the accelerations of the apple and Earth due to these forces. The mass of the apple is about 100 g and the mass of Earth is about 6×10^{24} kg.

40. * / You push a bowling ball down the lane toward the pins. Draw force diagrams for the ball (a) just before you let it go; (b) when the ball is rolling (for two clock readings); (c) as the ball is hitting a bowling pin. (d) For each force exerted on the ball in parts (a)–(c), draw the Newton's third law force beside the force diagram, and indicate the object on which these third law forces are exerted.

41. * EST / (a) A 50-kg skater initially at rest throws a 4-kg medicine ball horizontally. Describe what happens to the skater and to the ball. (b) Estimate the acceleration of the ball during the throw and of the skater using a reasonable value for the force that a skater can exert on the medicine ball. (c) The skater moving to the right catches the ball moving to the left. After the catch, both objects move to the right. Draw force diagrams for the skater and for the ball while the ball is being caught.

42. ** EST Basketball player LeBron James can jump vertically over 0.9 m. Estimate the force that he exerts on the surface of the basketball court as he jumps. (a) Compare this force with the force that the surface exerts on James. Describe all assumptions used in your estimate and state how each assumption affects the result. (b) Repeat the problem looking at the time interval when he is landing back on the floor.

43. * EST The Scottish Tug of War Association contests involve eight-person teams pulling on a rope in opposite directions. Estimate the force that the rope exerts on each team. Indicate any assumptions you made and include a force diagram for a short section of the rope.

44. * / A bowling ball hits a pin. (a) Draw a force diagram and a motion diagram for the ball during the collision and separate diagrams for the pin. (b) Explain why your friend who has not taken physics would insist that the bowling ball hits the pin harder than the pin hits the ball.

2.10 Seat belts and air bags: Putting it all together

45. * **Car safety** The National Transportation Safety Bureau indicates that a person in a car crash has a reasonable chance of survival if his or her acceleration is less than 300 m/s^2. (a) What magnitude force would cause this acceleration in such a collision? (b) What stopping distance is needed if the initial speed before the collision is 20 m/s (72 km/h or 45 mi/h)? (c) Indicate any assumptions you made.

46. * A 70-kg person in a moving car stops during a car collision in a distance of 0.60 m. The stopping force that the air bag exerts on the person is 8000 N. Name at least three physical quantities describing the person's motion that you can determine using this information, and then determine them.

General Problems

47. * BIO EST **Left ventricle pumping** The lower left chamber of the heart (the left ventricle) pumps blood into the aorta. According to biophysical studies, a left ventricular contraction lasts about 0.20 s and pumps 88 g of blood. This blood starts at rest and after 0.20 s is moving through the aorta at about 2 m/s. (a) Estimate the force exerted on the blood by the left ventricle. (b) What is the percent uncertainty in your answer? (c) What assumptions did you make? Did the assumptions increase or decrease the calculated value of the force compared to the actual value?

48. **** EST Acorn hits deck** You are sitting on a deck of your house surrounded by oak trees. You hear the sound of an acorn hitting the deck. You wonder if an acorn will do much damage if instead of the deck it hits your head. Make appropriate estimations and assumptions and provide a reasonable answer.

49. **** EST Olympic dive** During a practice dive, Olympic diver Fu Mingxia reached a maximum height of 5.0 m above the water. She came to rest 0.40 s after hitting the water. Estimate the average force that the water exerted on her while stopping her.

50. **** / EST** The brakes on a bus fail as it approaches a turn. The bus is traveling at the speed limit before it moves about 23 m across grass before hitting a wall. A bicycle on the bike rack on the front of the bus is crushed, but there is little damage to the bus. (a) Draw force diagrams for the bus and the wall during the collision. (b) Estimate the average force that the bicycle and bus exert on the wall while stopping. Indicate any assumptions made in your estimate.

51. **** EST** You are doing squats on a bathroom scale. You decide to push off the scale and jump up. Estimate the reading as you push off and as you land. Indicate any assumptions you made.

52. **** EST** Estimate the horizontal speed of the runner shown in **Figure P2.52** at the instant she leaves contact with the starting blocks. Indicate any assumptions you made.

Figure P2.52

53. **** EST** Estimate the maximum acceleration of Earth if all people got together and jumped up simultaneously.

54. **** EST** Estimate how much Earth would move during the jump described in Problem 53.

Reading Passage Problems

Col. John Stapp crash tests From 1946 through 1958, Col. John Stapp headed the U.S. Air Force Aero Medical Laboratory's studies of the human body's ability to tolerate high accelerations during plane crashes. Conventional wisdom at the time indicated that a plane's negative acceleration should not exceed 180 m/s^2 (18 times gravitational acceleration, or $18\ g$). Stapp and his colleagues built a 700-kg "Gee Whiz" rocket sled, track, and stopping pistons to measure human tolerance to high acceleration. Starting in June 1949, Stapp and other live subjects rode the sled. In one of Stapp's rides, the sled started at rest and 360 m later was traveling at speed 67 m/s when its braking system was applied, stopping the sled in 6.0 m. He had demonstrated that $18\ g$ was not a limit for human deceleration.

55. In an early practice run while the rocket sled was stopping, a passenger dummy broke its restraining device and the window of the rocket sled and stopped after skidding down the track. What physics principle best explains this outcome?
    (a) Newton's first law   (b) Newton's second law
    (c) Newton's third law   (d) The first and second laws
    (e) All three laws

56. Which answer below is closest to Stapp's 67 m/s speed in miles per hour?
    (a) 30 mi/h   (b) 40 mi/h   (c) 100 mi/h
    (d) 120 mi/h   (e) 150 mi/h

57. Which answer below is closest to the magnitude of the acceleration of Stapp and his sled as their speed increased from zero to 67 m/s?
    (a) 5 m/s^2 (b) 6 m/s^2 (c) 10 m/s^2
    (d) 12 m/s^2 (e) 14 m/s^2

58. Which answer below is closest to the magnitude of the acceleration of Stapp and his sled as their speed decreased from 67 m/s to zero?
    (a) $12\ g$ (b) $19\ g$ (c) $26\ g$
    (d) $38\ g$ (e) $48\ g$

59. Which answer below is closest to the average force exerted by the restraining system on 80-kg Stapp while his speed decreased from 67 m/s to zero in a distance of 6.0 m?
    (a) 10,000 N   (b) 20,000 N   (c) 30,000 N
    (d) 40,000 N   (e) 50,000 N

60. Which answer below is closest to the time interval for Stapp and his sled to stop as their speed decreased from 67 m/s to zero?
    (a) 0.09 s   (b) 0.18 s   (c) 0.34 s
    (d) 5.4 s   (e) 10.8 s

Using proportions A proportion is defined as an equality between two ratios; for instance, $a/b = c/d$. Proportions can be used to determine the expected change in one quantity when another quantity changes. Suppose, for example, that the speed of a car doubles. By what factor does the stopping distance of the car change? Proportions can also be used to answer everyday questions, such as whether a large container or a small container of a product is a better buy on a cost-per-unit-mass basis.

Suppose that a small pizza costs a certain amount. How much should a larger pizza of the same thickness cost? If the cost depends on the amount of ingredients used, then the cost should increase in proportion to the pizza's area and not in proportion to its diameter:

$$Cost = k(Area) = k(\pi r^2) \qquad (2.9)$$

where r is the radius of the pizza and k is a constant that depends on the price of the ingredients per unit area. If the area of the pizza doubles, the cost should double, but k remains unchanged.

Let us rearrange Eq. (2.9) so the two variable quantities (cost and radius) are on the right side of the equation and the constants are on the left:

$$k\pi = \frac{Cost}{r^2}$$

This equation should apply to any size pizza. If r increases, the cost should increase so that the ratio $Cost/r^2$ remains constant. Thus, we can write a proportion for pizzas of different sizes:

$$k\pi = \frac{Cost}{r^2} = \frac{Cost'}{r'^2}$$

For example, if a 3.5-in.-radius pizza costs $4.00, then a 5.0-in. radius pizza should cost

$$Cost' = \frac{r'^2}{r^2}Cost = \frac{(5.0 \text{ in})^2}{(3.5 \text{ in})^2}(\$4.00) = \$8.20$$

This process can be used for most equations relating two quantities that change while all other quantities remain constant.

61. The downward distance d that an object falls in a time interval t if starting at rest is $d = \frac{1}{2}at^2$. On the Moon, a rock falls 10.0 m in 3.50 s. How far will the object fall in 5.00 s, assuming the same acceleration?
    (a) 14.3 m   (b) 20.4 m   (c) 4.90 m
    (d) 7.00 m   (e) 10.0 m

62. The downward distance d that an object falls in a time interval t if starting at rest is $d = \frac{1}{2}at^2$. On the Moon, a rock falls 10.0 m in 3.50 s. What time interval t is needed for it to fall 15.0 m, assuming the same acceleration?
    (a) 2.33 s   (b) 2.86 s   (c) 3.50 s
    (d) 4.29 s   (e) 5.25 s

63. A car's braking distance d (the distance it travels if rolling to a stop after the brakes are applied) depends on its initial speed v_0, the maximum friction force $\vec{f}_{\text{R on C}}$ exerted by the road on the car, and the car's mass m according to the equation

$$\frac{2f_{s\,\text{max}}}{m}d = v_0^2.$$

Suppose the braking distance for a particular car and road surface is 26 m when the initial speed is 18 m/s. What is the braking distance when traveling at 27 m/s?

    (a) 59 m   (b) 39 m   (c) 26 m
    (d) 17 m   (e) 12 m

64. You decide to open a pizza parlor. The ingredients require that you charge $4.50 for a 7.0-in.-diameter pizza. How large should you make a pizza whose price is $10.00, assuming the cost is based entirely on the cost of ingredients?
    (a) 1.4 in.   (b) 3.1 in.   (c) 7.0 in.
    (d) 10 in.   (e) 16 in.

65. A circular wool quilt of 1.2 m diameter costs $200. What should the price of a 1.6-m-diameter quilt be if it is to have the same cost per unit area?
    (a) $110   (b) $150   (c) $270   (d) $360

3

Applying Newton's Laws

How does knowing physics help human cannonballs safely perform their tricks?

Would an adult or a small child win a race down a water slide?

How does friction help us walk?

Be sure you know how to:

- Use trigonometric functions (Mathematics appendix).
- Use motion diagrams and mathematical equations to describe motion (Sections 1.2, 1.5, and 1.6).
- Identify a system, construct a force diagram for it, and use the force diagram to apply Newton's second law (Sections 2.1, 2.6, and 2.8).

On March 10, 2011, David Smith, Jr., set the record for the distance traveled by a human cannonball— 59 meters, improving the record set by his father by more than 2 meters. A human cannonball is a performance trick in which a person is ejected from a cannon by a compressed spring or compressed air and lands either on a horizontal net or on an inflated bag. To make a human cannonball shot successful, a designer must apply Newton's laws and kinematics equations to the complex process. Most importantly, the designer must be able to predict where the cannonball will land to make sure there is a supporting cushion there. We'll learn how to do it in this chapter.

When we first studied force and motion dynamics processes (in Chapter 2), we learned that a system object's acceleration depends on the sum of the forces that other objects exert on it and on the mass of the object. We considered processes in which the forces exerted on an object by other objects were mainly along one axis—the axis along which motion occurred. These processes are common, but most everyday life processes involve forces that are *not* all directed only along the axis of motion. In this chapter we will learn to apply Newton's laws to those more complex processes.

3.1 Force components

To apply Newton's second law in situations in which the force vectors do not all point along the coordinate axes, we need to learn how to break forces into their components. We start by using a vector quantity with which we are already very familiar—displacement $\vec{d}$, a vector that starts from an object's initial position and ends at its final position.

Graphical vector addition and components of a vector

Suppose you want to take a trip and can reach your final destination in two different ways (see **Figure 3.1**). Route 1 is a direct path that is represented by a displacement vector $\vec{C}$. Its tail represents your initial location and its head represents your final destination. Route 2 goes along two roads represented by displacement vectors $\vec{A}$ and $\vec{B}$ that are perpendicular to each other. You first travel along $\vec{A}$, then along $\vec{B}$, and end at the same final destination.

Route 2 is an example of how to add displacement vectors graphically. You place the tail of $\vec{A}$ at your initial position; then place the tail of $\vec{B}$ at the head of $\vec{A}$. In this example, the head of $\vec{B}$ will be at your final position. Because the initial and final positions are the same for either route, we say that $\vec{C} = \vec{A} + \vec{B}$. Any displacement vector (for example, $\vec{C}$) can be replaced by two perpendicular displacement vectors (for example, $\vec{A}$ and $\vec{B}$) if these two vectors graphically add to equal $\vec{C}$, as illustrated in Figure 3.1.

Because forces are vectors, we can add them graphically as well. As we did with the travel routes above, we can replace a force $\vec{F}$ by two perpendicular forces $\vec{F}_x$ and $\vec{F}_y$ that graphically add to equal the original force. Suppose, for example, we place a small box on a very smooth surface. The box is stationary. Then we attach strings and spring scales to the box and pull in opposite directions, exerting a 5-N force on each string at angles of 37° relative to the plus or minus x-axis (**Figure 3.2a**). These two forces balance, and the box remains stationary.

Now, let's replace the string and spring scale pulling at 37° relative to the positive x-axis with two strings and two spring scales, one pulling along the x-axis and the other along the y-axis (Figure 3.2b). We find that if the x-axis scale pulls exerting a 4-N force and the y-axis scale pulls exerting a 3-N force, the box again remains stationary. Thus, these two forces have the same effect on the box as the single 5-N force exerted on the box at an angle of 37° above the positive x-direction.

Figure 3.1 Two routes to get to the same destination.

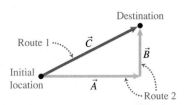

Figure 3.2 Graphically adding force components. We can always replace any force $\vec{F}$ with two perpendicular forces $\vec{F}_x$ and $\vec{F}_y$, as long as the perpendicular forces graphically add to $\vec{F}$.

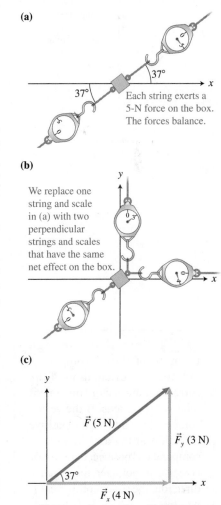

(a)

Each string exerts a 5-N force on the box. The forces balance.

(b)

We replace one string and scale in (a) with two perpendicular strings and scales that have the same net effect on the box.

(c)

$\vec{F}$ (5 N)

$\vec{F}_y$ (3 N)

$\vec{F}_x$ (4 N)

The perpendicular 4-N and 3-N force component vectors graphically add to form the 5-N force.

This experimental result is consistent with graphical vector addition, where we find that a 4-N force in the positive x-direction plus a 3-N force in the positive y-direction add to produce a 5-N force at 37° relative to the positive x-axis (Figure 3.2c). The vectors form a "3-4-5" right triangle.

In summary, we can always replace any force $\vec{F}$ with two perpendicular forces $\vec{F}_x$ and $\vec{F}_y$, as long as the perpendicular forces graphically add to $\vec{F}$ (Figure 3.2c). In this case, the perpendicular forces are along the perpendicular x- and y-axes and are called the x- and y-**vector components** of the original force $\vec{F}$. Since the vector components are also vectors, they are written with a vector symbol above them. Note that $\vec{F}_x + \vec{F}_y = \vec{F}$, just as displacement $\vec{C}$ was identical to the displacement $\vec{A} + \vec{B}$ in Figure 3.1.

Scalar components of a vector

When working with x- and y-axes that are perpendicular to each other, we do not need to use the vector components of $\vec{F}$ but can instead provide the same information about the vector by specifying what are called the **scalar components** F_x and F_y of the force $\vec{F}$. The advantage of the scalar components is that they are numbers with signs, which can be added and subtracted more easily than vector quantities. **Figure 3.3** shows how to find the scalar components. The force vector $\vec{F}$ is 5 N and points 37° above the negative x-axis. The x-vector component of force $\vec{F}$ is 4 N long and points in the negative x-direction. Thus, the x-scalar component of $\vec{F}$ is $F_x = -4$ N. Similarly, the y-vector component of $\vec{F}$ is 3 N in the positive y-direction—it pulls upward in the positive y-direction. Thus, we can specify the y-scalar component of $\vec{F}$ as $F_y = +3$ N. Thus, the x- and y-components $F_x = -4$ N and $F_y = +3$ N tell us everything we need to know about the force $\vec{F}$. Note that scalars are *not* written with a vector symbol above them.

Finding the scalar components of a force from its magnitude and direction

If we know the magnitude of a force F and the angle θ that the force makes above or below the positive or negative x-axis, we can determine its scalar components. The Skill box summarizes how to calculate the scalar components of the force.

Figure 3.3 The vector and scalar components of a vector.

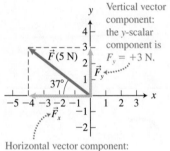

TIP Note that the sign (+ or −) of a scalar component indicates the orientation of the corresponding vector component relative to the axis. If the vector component points in the positive direction of the axis, the scalar component is positive. If the vector component points in the negative direction, the scalar component is negative.

REASONING SKILL Determining the scalar components of a vector.

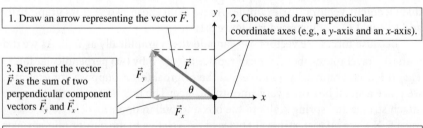

1. Draw an arrow representing the vector $\vec{F}$.

2. Choose and draw perpendicular coordinate axes (e.g., a y-axis and an x-axis).

3. Represent the vector $\vec{F}$ as the sum of two perpendicular component vectors $\vec{F}_y$ and $\vec{F}_x$.

4. Finally, use the right triangle and trigonometry to calculate the values of the scalar components:
$$F_x = \pm F \cos \theta \qquad (3.1x)$$
$$F_y = \pm F \sin \theta \qquad (3.1y)$$
where F is the magnitude of the force vector and θ is the angle (90° or less) that $\vec{F}$ makes with respect to the $\pm x$-axis. F_x is positive if in the positive x-direction and negative if in the negative x-direction. F_y is positive if in the positive y-direction and negative if in the negative y-direction.

QUANTITATIVE EXERCISE 3.1 Components of forces exerted on a knot

Three ropes pull on the knot shown below. We place the force diagram on a grid so that you can see the components of the three forces that the ropes exert on the knot. Use the diagram to visually determine the x- and y-scalar components of the force that rope 2 exerts on the knot. Then use the mathematical method described in the Skill box to calculate the components of this force. The results of using the two methods should be consistent with each other.

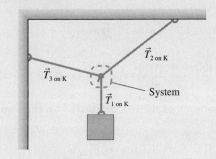

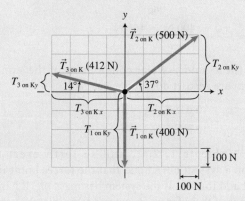

Represent mathematically We label forces that ropes exert on the knot using a letter T instead of F. This is a common way of noting the forces that ropes or strings exert on an object—tension forces. From the force diagram, we see that rope 2 pulls on the knot at a 37° angle above the positive x-axis, exerting a force of magnitude $T_{2\,on\,K} = 500$ N. The x-scalar component of this force is +400 N (4 grid units in the positive x-direction) and the y-scalar component of this force is +300 N (3 grid units in the positive y-direction). In symbols, $T_{2\,on\,K\,x} = +400$ N and $T_{2\,on\,K\,y} = +300$ N.

Solve and evaluate Now use the mathematical method to calculate the components of the force that rope 2 exerts on the knot:

$$T_{2\,on\,K\,x} = +(500\text{ N})\cos 37° = +400\text{ N}$$
$$T_{2\,on\,K\,y} = +(500\text{ N})\sin 37° = +300\text{ N}$$

The diagrammatical and the mathematical methods are consistent.

Try it yourself: Use the same two methods to determine the components of the forces exerted by ropes 1 and 3 on the knot.

Answer: $T_{3\,on\,K\,x} = -400$ N and $T_{3\,on\,K\,y} = +100$ N, and $T_{1\,on\,K\,x} = 0$ and $T_{1\,on\,K\,y} = -400$ N.

TIP Be careful when applying Eqs. (3.1x) and (3.1y). The angles that appear in those equations must be measured with respect to the positive or negative x-axis.

Review Question 3.1 When does a vector have a positive scalar component? When does a vector have a negative scalar component?

3.2 Newton's second law in component form

Now that we have learned how to work with vectors and their components, we can start to apply Newton's second law to more complex situations in which one or more of the forces exerted on the system do not point along one of the coordinate axes. The figure in Quantitative Exercise 3.1 is an example of such a situation. We know that the knot is not accelerating. Thus, the sum of the forces exerted on the knot is zero:

$$\Sigma \vec{F}_{\text{on K}} = \vec{T}_{1\,\text{on K}} + \vec{T}_{2\,\text{on K}} + \vec{T}_{3\,\text{on K}} = 0$$

It is difficult to use this vector equation to analyze the situation further—for example, to determine one of the forces that is unknown if you know the other two forces. However, we can do such tasks if this situation is represented in scalar component form.

Notice in the force diagram in Quantitative Exercise 3.1 that the y-scalar components of the forces that the three ropes exert on the knot add to zero. Ropes 2 and 3 exert forces that have positive y-scalar components $+300\,\text{N} + 100\,\text{N} = +400\,\text{N}$, and rope 1 exerts a force with a negative y-scalar component $-400\,\text{N}$. Consequently:

$$a_{\text{K}y} = \frac{T_{1\,\text{on K}\,y} + T_{2\,\text{on K}\,y} + T_{3\,\text{on K}\,y}}{m_{\text{K}}}$$

$$= \frac{-400\,\text{N} + 300\,\text{N} + 100\,\text{N}}{m_{\text{K}}} = 0$$

Similarly, the x-scalar components of the forces exerted on the knot also add to zero:

$$a_{\text{K}x} = \frac{T_{1\,\text{on K}\,x} + T_{2\,\text{on K}\,x} + T_{3\,\text{on K}\,x}}{m_{\text{K}}}$$

$$= \frac{+0 + 400\,\text{N} - 400\,\text{N}}{m_{\text{K}}} = 0$$

Thus, we infer that when the x- and y-scalar components of the sum of the forces exerted on the system are zero, it does not accelerate:

$$a_x = 0 \text{ if } \Sigma F_{\text{on K}\,x} = 0$$

$$a_y = 0 \text{ if } \Sigma F_{\text{on K}\,y} = 0$$

Suppose that in general objects 1, 2, 3, and so forth exert forces $\vec{F}_{1\,\text{on S}}, \vec{F}_{2\,\text{on S}}, \vec{F}_{3\,\text{on S}}, \ldots$ on a system object and that the forces point in arbitrary directions and do not add to zero. If the system has mass m_{S}, it has an acceleration $\vec{a}_{\text{S}}$ as a consequence of these forces (Newton's second law):

$$\vec{a}_{\text{S}} = \frac{\vec{F}_{1\,\text{on S}} + \vec{F}_{2\,\text{on S}} + \vec{F}_{3\,\text{on S}} + \cdots}{m_{\text{S}}}$$

Because this equation involves vectors, we can't work with it directly. However, we can split it into its x- and y-scalar component forms to use Newton's second law to determine the system's acceleration.

Newton's second law rewritten in scalar component form becomes:

$$a_{\text{S}x} = \frac{F_{1\,\text{on S}\,x} + F_{2\,\text{on S}\,x} + F_{3\,\text{on S}\,x} + \cdots}{m_{\text{S}}} \tag{3.2 x}$$

$$a_{\text{S}y} = \frac{F_{1\,\text{on S}\,y} + F_{2\,\text{on S}\,y} + F_{3\,\text{on S}\,y} + \cdots}{m_{\text{S}}} \tag{3.2 y}$$

In practice, we usually choose the axes so that the object accelerates along only one of these axes and has zero acceleration along the other axis. The process of analyzing a situation using Newton's second law in component form is illustrated in the Reasoning Skill box on the next page. (Note that for convenience we can drop the word *scalar* and use the simpler term *component form*).

REASONING SKILL Using Newton's second law in component form.

1. Draw a force diagram (the force arrows represent the forces that other objects exert on the system).

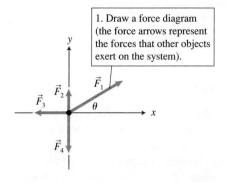

2. Visualize the x-components of the forces and apply the x-component form of Newton's second law to the force diagram:

$$a_x = \frac{\Sigma F_x}{m} = \frac{F_{1x} + F_{2x} + F_{3x} + F_{4x}}{m} = \frac{+F_1 \cos\theta + 0 + (-F_3) + 0}{m}$$

Notice that: $F_{1x} = +F_1 \cos\theta$
$F_{2x} = +F_2 \cos 90° = 0$
$F_{3x} = -F_3 \cos 0° = -F_3$
$F_{4x} = +F_4 \cos 90° = 0$

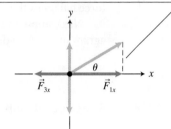

3. Visualize the y-components of the forces and apply the y-component form of Newton's second law to the force diagram:

$$a_y = \frac{\Sigma F_y}{m} = \frac{F_{1y} + F_{2y} + F_{3y} + F_{4y}}{m} = \frac{+F_1 \sin\theta + F_2 + 0 + (-F_4)}{m}$$

Notice that: $F_{1y} = +F_1 \sin\theta$
$F_{2y} = +F_2 \sin 90° = +F_2$
$F_{3y} = -F_3 \sin 0° = 0$
$F_{4y} = -F_4 \sin 90° = -F_4$

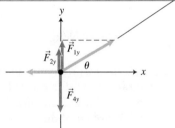

Figure 3.4 Calculating the normal force. (a) Alice lowers a book down the wall; (b) a force diagram for the situation.

(a)

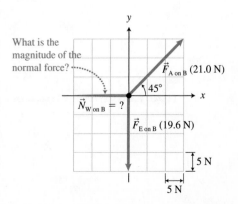

Using components to solve a problem

Now, let's use this force component method to analyze the following process. Alice pushes on a 2.0-kg book, exerting a 21.0-N force as it slides down a slippery vertical wall (**Figure 3.4a**). She pushes 45° above the horizontal. We use the above reasoning skill to determine the magnitude of the force that the wall exerts on the book and the acceleration of the book down the wall's surface.

Choose the book as the system, model it as a point-like object, and draw a force diagram for it (Figure 3.4b). The force diagram includes three forces: the 21.0-N force that Alice exerts on the book $\vec{F}_{\text{A on B}}$, the downward gravitational force that Earth exerts on the book $\vec{F}_{\text{E on B}}$ of magnitude $mg = (2.0 \text{ kg})(9.8 \text{ N/kg}) = 19.6 \text{ N}$, and the normal force $\vec{N}_{\text{W on B}}$ that the wall exerts on the book (its magnitude is not yet known—we did not include in the force diagram the end of the arrow representing that force). Notice that the normal force that the wall exerts on the book is perpendicular to the wall's surface and points horizontally away from the wall (the normal force is not vertical).

The book does not leave contact with the wall; therefore, the x-component of its acceleration is zero, as is the x-component of the sum of the forces exerted on the book. We apply the x-component form of Newton's second law to the force diagram to determine the magnitude of the normal force so that the book stays on the surface:

$$a_{\text{B}x} = \frac{1}{m_\text{B}}\Sigma F_x = \frac{F_{\text{A on B}x} + N_{\text{W on B}x} + F_{\text{E on B}x}}{m_\text{B}} = 0$$

(b)

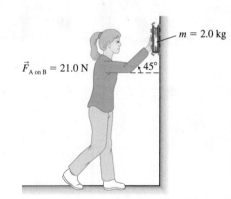

What is the magnitude of the normal force?

Now, insert expressions for the x-scalar components into the above equation:

$$+(21.0\,\text{N})\cos 45° + (-N_{\text{W on B}}\cos 0°) + (19.6\,\text{N})\cos 90° = 0$$

Since $\cos 90° = 0$ and $\cos 0° = 1.0$, we find that

$$(21.0\,\text{N})\cos 45° - N_{\text{W on B}} + 0 = 0$$

or

$$N_{\text{W on B}} = (21.0\,\text{N})\cos 45° = 14.8\,\text{N}$$

Notice in the force diagram with the grid that the x-component of the pushing force that Alice exerts on the book has magnitude 14.8 N—the wall normal force balances the x-component of that pushing force.

Next, apply the y-component form of Newton's second law to the force diagram in order to determine the y-component of the acceleration of the book:

$$a_{\text{B}\,y} = \frac{\Sigma F_y}{m_\text{B}} = \frac{F_{\text{A on B}\,y} + N_{\text{W on B}\,y} + F_{\text{E on B}\,y}}{m_\text{B}}$$

Substitute into this equation expressions for the y-components of the forces:

$$a_{\text{B}\,y} = \frac{+(21.0\,\text{N})\sin 45° + N_{\text{W on B}}\sin 0° + [-(19.6\,\text{N})\sin 90°]}{2.0\,\text{kg}}$$

Noting that $N_{\text{W on B}} = 14.8\,\text{N}$, $\sin 0° = 0$, and $\sin 90° = 1.0$, we get

$$a_{\text{B}\,y} = \frac{+(21.0\,\text{N})\sin 45° + (14.8\,\text{N})(0) + (-19.6\,\text{N})(1)}{2.0\,\text{kg}}$$

$$= \frac{14.8\,\text{N} + 0 - 19.6\,\text{N}}{2.0\,\text{kg}} = -2.4\,\text{N/kg} = -2.4\,\text{m/s}^2$$

For this problem with the force diagram on a grid, you can check your work visually by looking at the force diagram. Note, for example, that the pushing force has a +14.8-N vertical component and the gravitational force is $-19.6 \approx -20$ N. Thus, the net vertical force is about -5 N and the vertical acceleration should be about $-5\,\text{N}/(2.0\,\text{kg}) \approx 2.5\,\text{m/s}^2$, as calculated.

> **TIP** Perhaps the part of this procedure that causes the most difficulty is the construction of the force diagram. Be sure to include in the diagram only forces exerted *on* the system by external objects (outside the system). Do *not* include forces that the system exerts on external objects (objects that are not included in the system).

Review Question 3.2 Apply Newton's second law in component form for the force diagram and process sketched in **Figure 3.5** (both x- and y-axes).

3.3 Problem-solving strategies for analyzing dynamics processes

Our analysis of processes in this and the previous chapter often involves the application of both Newton's second law and kinematics equations, thus relating the external forces exerted on an object and its changing motion. We call these **dynamics processes**. Below we describe a general method for analyzing dynamics processes and illustrate its use for a simple example of pulling a sled.

Figure 3.5

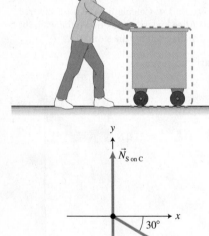

Active Learning Guide›

PROBLEM-SOLVING STRATEGY Analyzing dynamics processes

Active Learning Guide>

EXAMPLE 3.2 Pulling a sled

You pull a sled across a hard snowy surface. The sled and the two children sitting on it have a total mass of 60.0 kg. The rope you pull exerts a 100-N force on the sled and is oriented 37° above the horizontal. If the sled starts at rest, how fast is it moving after being pulled 10.0 m?

Sketch and translate

- Make a sketch of the process.

- Choose a system.

- Choose coordinate axes with one axis in the direction of acceleration and the other axis perpendicular to that direction.

- Indicate in the sketch everything you know about the process relative to these axes. Identify the unknown quantity of interest.

We first sketch the process.

Choose the sled and children as the system.

The process starts with the sled at rest ($v_{0\,x} = 0$) and ends when it has traveled $x_1 - x_0 = 10.0$ m and has an unknown final speed $v_{1\,x}$.

Simplify and diagram

- Simplify the process. For example, can you model the system as a point-like object? Can you ignore friction?

- Represent the process diagrammatically with a motion diagram and/or a force diagram.

- Check for consistency of the diagrams—for example, is the sum of the forces in the direction of the acceleration?

Consider the system as a point-like object. Since we have no information about friction, assume its effects on the sled are minor.

A motion diagram for the sled is shown below. The sled moves at increasing speed toward the right.

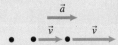

Draw a force diagram for the system. Earth exerts a downward gravitational force on the system $\vec{F}_{\text{E on SL}}$, the rope pulls on the sled $\vec{T}_{\text{R on SL}}$ at a 37° angle above the horizontal, and the snow exerts a normal force on the sled perpendicular to the surface (in this case upward) $\vec{N}_{\text{S on SL}}$.

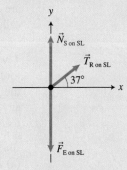

Are the diagrams consistent? The sum of the forces is toward the right in the direction of the acceleration.

(*continued*)

Represent mathematically

- Convert these qualitative representations into quantitative mathematical descriptions of the process using Newton's second law and kinematics equations.

The horizontal x-component form of Newton's second law is

$$ma_x = \Sigma F_x$$

$$m_{SL}a_x = T_{R \text{ on } SL\, x} + N_{S \text{ on } SL\, x} + F_{E \text{ on } SL\, x}$$

Substitute expressions for the x-components of these three forces:

$$m_{SL}a_x = T_{R \text{ on } SL} \cos 37° + N_{S \text{ on } SL} \cos 90° + F_{E \text{ on } SL} \cos 90°$$

Noting that $\cos 90° = 0$ and dividing both sides by the mass of the sled with the children, we find that

$$a_x = \frac{T_{R \text{ on } SL} \cos 37° + 0 + 0}{m_{SL}}$$

We could at this point use the y-component equation to find the magnitude of the normal force that the surface exerts on the sled, but we don't need to do that to answer the question of interest.

The above equation can be used to determine the sled's acceleration. We are then left with a kinematics problem to determine the speed $v_{1\,x}$ of the sled after pulling it for $x_1 - x_0 = 10.0$ m. We can use kinematics Eq. (1.7):

$$v_{1x}^2 = v_{0x}^2 + 2a_x(x_1 - x_0)$$

Solve and evaluate

- Substitute the given values into the mathematical expressions and solve for the unknowns.
- Decide whether the assumptions that you made were reasonable.
- Finally, evaluate your work to see if it is reasonable (check units, limiting cases, and whether the answer has a reasonable magnitude).
- Make sure the answer is consistent with other representations.

The acceleration is

$$a_x = \frac{+T_{R \text{ on } SL} \cos 37°}{m_{SL}} = \frac{+(100\text{ N})(\cos 37°)}{60.0\text{ kg}} = +1.33 \text{ m/s}^2$$

The speed of the sled after being pulled for 10.0 m will be

$$v_{1x}^2 = v_{0x}^2 + 2a_x(x_1 - x_0) = 0^2 + 2(1.33\text{ m/s}^2)(10.0\text{ m} - 0)$$

or

$$v_{1\,x} = 5.2 \text{ m/s}$$

This is fast but not unreasonable. In real life, friction between the sled and the snow would probably cause the speed to be slower. The units are correct. If we examine a limiting case, in which the rope pulls vertically, the horizontal acceleration is zero as it should be.

Try it yourself: (a) Determine the y-component of the gravitational force that Earth exerts on the sled-children system. (b) Then determine the y-component of the force that the rope exerts on the system. (c) Finally, use the y-component form of Newton's second law to determine the y-component of the normal force that the snow exerts on the sled.

Answers:

(a) $F_{E \text{ on } SL\, y} = -m_{SL}g \sin 90° = -590$ N;

(b) $T_{R \text{ on } SL\, y} = +T_{R \text{ on } SL} \sin 37° = +60$ N;

(c) $N_{S \text{ on } SL\, y} = +530$ N.

Looking at the force diagram, we see that these numbers make sense. Why is the magnitude of the normal force less than $m_{SL}g$?

EXAMPLE 3.3 Acceleration of a train

You carry a pendulum and a protractor with you onto a train. As the train starts moving, the pendulum string swings back to an angle of 8.0° with respect to vertical and remains at that angle. Determine the acceleration of the train at that moment.

Sketch and translate We sketch the situation as shown below. Choose the pendulum bob attached to the bottom of the string as the system. The train station is our object of reference. The pendulum bob is accelerating horizontally; thus, the sum of the forces exerted on it should point horizontally. We choose the x-axis to point in the horizontal direction and the y-axis to point perpendicular in the vertical direction.

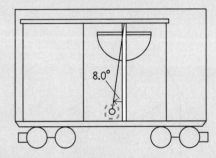

Simplify and diagram Assume that the pendulum bob is a point-like object and that its acceleration equals the acceleration of the train. Next, construct a force diagram for the pendulum bob. The string exerts a force $\vec{T}_{\text{S on B}}$ that is oriented at an 8.0° angle relative to the vertical or 82.0° relative to the horizontal. Earth exerts a downward gravitational force on the bob $\vec{F}_{\text{E on B}}$.

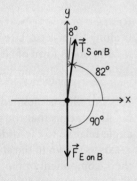

Represent mathematically Use the force diagram to help apply Newton's second law in the x- and y-component forms:

x-component form of Newton's second law:

$$m_{\text{B}}a_{\text{B }x} = T_{\text{S on B }x} + F_{\text{E on B }x}$$
$$m_{\text{B}}a_{\text{B }x} = +T_{\text{S on B}}\cos 82° + F_{\text{E on B}}\cos 90°$$

y-component form of Newton's second law:

$$m_{\text{B}}a_{\text{B }y} = T_{\text{S on B }y} + F_{\text{E on B }y}$$
$$m_{\text{B}}a_{\text{B }y} = +T_{\text{S on B}}\sin 82° + (-F_{\text{E on B}}\sin 90°)$$

Solve and evaluate Note that the bob's velocity is not changing in the vertical direction; thus, $a_{\text{B }y} = 0$. The bob's velocity is changing in the horizontal direction, so the x-component of acceleration is not zero. Also, recall that the gravitational force that Earth exerts on the bob has magnitude $F_{\text{E on B}} = m_{\text{B}}g$. Finally, note that $\cos 90° = 0$ and $\sin 90° = 1.0$. With these substitutions and a bit of algebra, the above x- and y-component equations become

$$m_{\text{B}}\,a_{\text{B }x} = T_{\text{S on B}}\cos 82°$$
$$0 = T_{\text{S on B}}\sin 82° - m_{\text{B}}g$$

We want to determine $a_{\text{B }x}$ but do not know $T_{\text{S on B}}$ and m_{B}. In this particular case, a bit of mathematical creativity helps. Move the $m_{\text{B}}g$ in the second equation to the left side, and divide the left side of the first equation by the left side of the second equation; then divide the right side of the first equation by the right side of the second equation (we can do it as we are dividing each side of the first equation by terms that are equal and nonzero). We get

$$\frac{m_{\text{B}}a_{\text{B }x}}{m_{\text{B}}g} = \frac{T_{\text{S on B}}\cos 82°}{T_{\text{S on B}}\sin 82°}$$

The masses of the bob and the force exerted by the string cancel, thus:

$$a_{\text{B }x} = \frac{\cos 82°}{\sin 82°}g = \cot 82°(9.8\ \text{m/s}^2) = 1.4\ \text{m/s}^2$$

The units are correct, and the magnitude is reasonable. As a limiting case analysis, we find from the above equation that if the acceleration of the train is zero, then the string hangs straight down at a 90° angle relative to the horizontal

$$\frac{\cos 90°}{\sin 90°} = 0 \text{ as } \cos 90° = 0 \quad \text{and} \quad \sin 90° = 1.$$

This is what we expect if the train is parked at the train station.

Try it yourself: Suppose the train is moving forward at a constant velocity of magnitude 18 m/s. At what angle will the pendulum bob string hang?

Answer: The bob is not accelerating ($a_{\text{B }x} = 0$) and the left side of the final equation in the solution is zero. The bob should hang straight down, as in a stationary train.

Processes involving inclines

Processes involving inclines are common in everyday life. For example, a skier skis down a slope or a wagon is pulled uphill. How do we use the component form of Newton's second law and kinematics to help analyze such processes?

EXAMPLE 3.4 Who wins a water slide race?

A child of mass m and an adult of mass $4m$ simultaneously start to slide from the top of a water slide that has a downward slope of 30° relative to the horizontal. Which person reaches the bottom first?

Sketch and translate We can choose either the child or the adult as the system. Our sketch shows the person at one moment during the downhill slide and includes the known information. The person who has the greater acceleration along the slide will win the race, so we will construct an expression for the person's acceleration and see how that acceleration depends on the person's mass. The object of reference is Earth. We choose the axes later when constructing the force diagram.

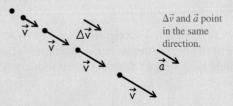

Simplify and diagram The person is modeled as a point-like object. Since a water slide is very smooth and slippery, we assume that we can disregard the friction force exerted by the slide on the person.

Next, we build a motion diagram for the person, as shown below. The velocity arrows are parallel to the slide, and the person's velocity is increasing so the velocity change arrow $\Delta\vec{v}$ (and the acceleration $\vec{a}$) points down the slide.

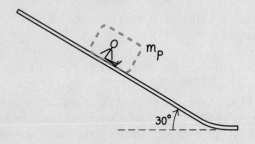

$\Delta\vec{v}$ and $\vec{a}$ point in the same direction.

A force diagram shows that Earth exerts a downward gravitational force $\vec{F}_{\text{E on P}}$ on the person and the slide exerts a normal force $\vec{N}_{\text{S on P}}$ on the person perpendicular to the surface. We choose the x-axis parallel

to the inclined surface and the y-axis perpendicular to the inclined surface. With this choice, the acceleration points in the positive x-direction. If we had chosen the standard horizontal and vertical directions for our coordinate axes, then the acceleration would have nonzero x- and y-components, making the situation more difficult to analyze.

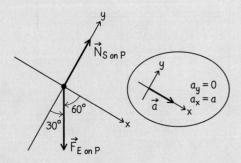

Represent mathematically The motion of the person is along the x-axis. Use the force diagram to apply the x-component form of Newton's second law:

$$m_{\text{P}}a_x = F_{\text{E on P }x} + N_{\text{S on P }x}$$

The slide is inclined at 30° relative to the horizontal. This means that the gravitational force makes a 60° angle relative to the slide (and therefore relative to the x-axis as shown in the above diagram). We can calculate the x-component of the gravitational force in two ways: $F_{\text{E on P }x} = m_{\text{P}}g\cos 60°$ or $F_{\text{E on P }x} = m_{\text{P}}g\sin 30°$. Both methods give the same result. The above component equation becomes

$$m_{\text{P}}a_x = +(m_{\text{P}}g)\cos 60° + N_{\text{S on P}}\cos 90°$$

Solve and evaluate Substituting $\cos 90° = 0$, we get

$$m_{\text{P}}a_x = +(m_{\text{P}}g)\cos 60° + 0$$

Dividing by the m_{P} on each side of the equation, we get $a_x = g\cos 60°$. The person's acceleration is independent of her/his mass! Thus, the acceleration of the child and that of the adult are the same. If they start at the same time at the same position, both initially at rest, and have the same acceleration, they arrive at the bottom at the same time. The race is a tie (see the photo).

Try it yourself: Suppose the slide above is 20 m long. How fast will the person be moving when they reach the bottom of the slide?

Answer: 14 m/s.

The result of Example 3.4 may surprise you, but it is reasonable. The acceleration depends on both the gravitational force and the mass. The gravitational force that Earth exerts on the adult (only the *x*-component of this force determines the acceleration in this case) is four times that exerted by Earth on the child, but because the adult's mass is four times greater than the child's, the accelerations of the adult and the child are equal. This result is similar to what we learned in Chapters 1 and 2—objects of all different masses have the same free-fall acceleration.

Figure 3.6 Complementary angles.

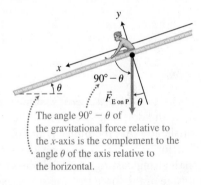

The angle $90° - \theta$ of the gravitational force relative to the *x*-axis is the complement to the angle θ of the axis relative to the horizontal.

TIP Notice in the last example that the inclined surface and the *x*-axis were at a 30° angle relative to the horizontal. The gravitational force exerted by Earth on the system then makes a 60° angle relative to the *x*-axis (60° is the complement of 30°). In general, the angle θ of the inclined surface relative to the horizontal is the complement $90° - \theta$ of the angle that the gravitational force makes with respect to the inclined surface (see **Figure 3.6**).

Two objects linked together

We can apply Newton's second law to a process in which two objects are connected together by a cable or rope, such as a van pulling a rope connected to a wagon that is pulling a rope connected to a second wagon. Perhaps the first "scientific" quantitative application of this type involved two blocks of different mass at the ends of a string that passed over a pulley (**Figure 3.7**). George Atwood invented this apparatus (now called the Atwood machine) in 1784. At that time, there were no motion sensors or precision stopwatches—nothing that would allow the accurate measurement of the motion of a rapidly accelerating object (9.8 m/s² would have been considered a rapid acceleration in those days). Atwood's machine allowed the determination of *g* despite these difficulties. We'll consider next a modified and simplified version of the Atwood machine, a machine that can serve the same purpose.

Figure 3.7 An Atwood machine.

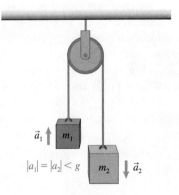

$|a_1| = |a_2| < g$

EXAMPLE 3.5 Slowing the fall

In a modified Atwood machine, one block rests on a horizontal surface and the other hangs off the edge of the surface. A string attached to each block passes over a light, smooth pulley. Determine the force that the string exerts on the hanging block of mass m_2 and the magnitude of the acceleration of each block.

Sketch and translate We first sketch the modified Atwood machine shown below. For analysis we will use two systems—first block 1 and then block 2. The two blocks have masses m_1 and m_2. Block 2 will fall and pull the string attached to block 1. Both blocks will move with increasing speed—block 1 on the horizontal surface and block 2 downward. If the string does not stretch, then at every moment they move with the same speed. The sliding block 1 has the same magnitude acceleration as block 2, but pointing toward the right. Our goal is to find the force exerted by the string on block 2 and the magnitudes of the acceleration of the blocks.

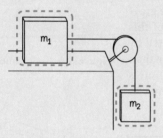

Simplify and diagram Assume that the string is very light and does not stretch, that the pulley's mass is very small, and that the pulley is smooth and rotates free of friction. The pulley changes the direction of the low-mass string passing over it but does not change the magnitude of the force that the string exerts on the objects attached at either end. Assume also that there is no friction between the horizontal surface and block 1.

At right we show our force diagrams for each block and our choice of coordinate systems. For block 1, Earth exerts a downward gravitational force $\vec{F}_{\text{E on 1}}$; the table surface exerts an equal-magnitude upward normal force on block 1 $\vec{N}_{\text{T on 1}}$; and the string exerts an unknown horizontal force toward the right on the block $\vec{T}_{\text{S on 1}}$. For block 2, Earth exerts a downward gravitational force $\vec{F}_{\text{E on 2}}$ and the string exerts an unknown upward force $\vec{T}_{\text{S on 2}}$. As noted earlier, the string exerts the same magnitude force on block 1 as it does on block 2: $T_{\text{S on 1}} = T_{\text{S on 2}} = T$.

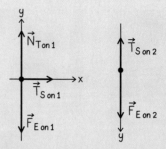

Represent mathematically Use the force diagrams to help apply Newton's second law in component form for each block. For block 1, the forces in the vertical direction balance, since the vertical component of its acceleration is zero. We see by inspection of the force diagram that the gravitational force and the normal force have zero x-components and so we do not include them in the x-component equation. This leaves us with only one force in the horizontal direction—the force exerted by the string. The x-component form of Newton's second law becomes

$$m_1 a_{1\,x} = T_{\text{S on 1}\,x} = +T \qquad (3.3)$$

For block 2, we only need to look at the vertical direction, as no forces are exerted in the horizontal direction. Notice that we've chosen the vertical y-axis pointing downward so that the y-component of block 2's acceleration will equal the x-component of block 1's acceleration. The y-component form of Newton's second law for block 2 is

$$m_2 a_{2\,y} = T_{\text{S on 2}\,y} + F_{\text{E on 2}\,y}$$

Noting that $T_{\text{S on 2}\,y} = -T$ and $F_{\text{E on 2}\,y} = m_2 g$, we get

$$m_2 a_{2\,y} = -T + m_2 g \qquad (3.4)$$

Because the two blocks are connected together by the string, they move with the same speed and have accelerations of the same magnitude, 1 to the right and 2 down.

Solve and evaluate After substituting a for the two accelerations in Eqs. (3.3) and (3.4), the two equations now have the same two unknown quantities T and a.

$$m_1 a = T$$

$$m_2 a = -T + m_2 g$$

We substitute the expression for T from the first equation into the second to get

$$m_2 a = -m_1 a + m_2 g$$

After moving the terms containing the acceleration to the left side, factoring a out, and dividing both sides by the sum of the masses, we have

$$a = \frac{m_2 g}{m_1 + m_2}$$

Note that the acceleration is less than g! Also note that if the mass $m_1 \gg m_2$ (m_1 much greater than m_2), then the acceleration is almost zero and can be measured easily. We can now determine the force exerted by the string on the hanging block by inserting the above expression for the acceleration into $m_1 a = T$:

$$T = m_1 a = m_1 \frac{m_2 g}{m_1 + m_2} = \frac{m_1}{m_1 + m_2} m_2 g$$

Are the results reasonable? Consider the equation for acceleration. The only force pulling the hanging block down is the gravitational force $m_2 g$ that Earth exerts on the hanging block. But because the two blocks are connected, this force has to cause the sum of their masses to accelerate $[m_2 g = (m_1 + m_2)a]$, making the acceleration less than g.

Try it yourself: Determine an expression for acceleration for a regular Atwood machine for which two objects of mass m_1 and m_2 move in the vertical direction ($m_1 > m_2$).

Answer: $a = \dfrac{(m_1 - m_2)g}{m_1 + m_2}$.

We can see now how to use a modified or a regular Atwood machine to determine the acceleration g. If you measure the distance that one of the objects travels (Δy) during a particular time interval (Δt) after being released from rest, you can use the expression

$$\Delta y = a \frac{\Delta t^2}{2}$$

to determine the acceleration of the objects. Using this value for acceleration and the values for their masses, you can use the expression

$$a = \frac{m_2 g}{m_1 + m_2}$$

for a modified machine or

$$a = \frac{(m_1 - m_2)g}{m_1 + m_2}$$

for a regular one to determine the value of g.

Review Question 3.3 For problems involving objects moving upward or downward along inclined surfaces, we choose the x-axis parallel to the surface and the y-axis perpendicular to the surface. Why not use horizontal and vertical axes?

3.4 Friction

Up to this point, we have assumed that objects move across absolutely smooth surfaces with no friction. In reality, the vast majority of situations involve some degree of friction. In this section we examine the phenomenon of friction conceptually and construct mathematical models that allow us to take friction into account quantitatively.

Static friction

Consider a simple experiment as shown in Observational Experiment **Table 3.1**. A spring scale exerts an increasing force on the block. Observe carefully what happens to the block.

‹Active Learning Guide

OBSERVATIONAL EXPERIMENT TABLE

3.1 Pulling a block with a spring scale.

VIDEO 3.1

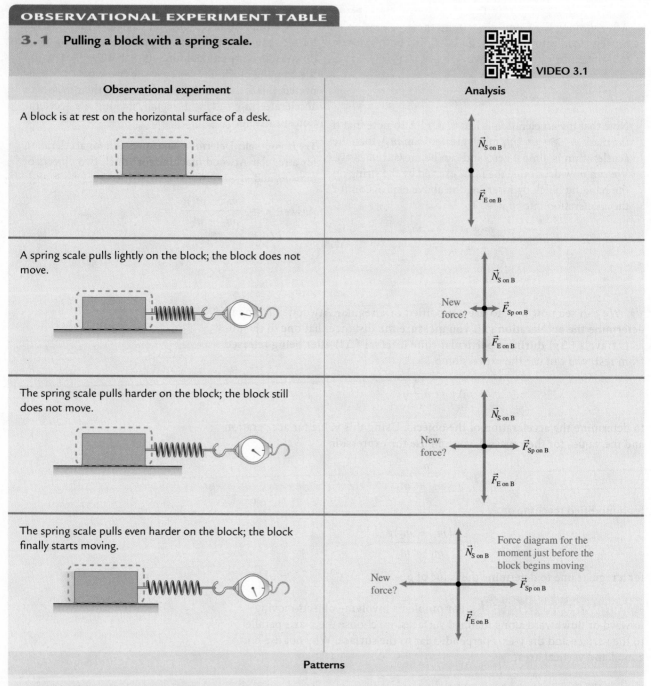

Observational experiment	Analysis
A block is at rest on the horizontal surface of a desk.	$\vec{N}_{\text{S on B}}$ $\vec{F}_{\text{E on B}}$
A spring scale pulls lightly on the block; the block does not move.	New force? $\vec{N}_{\text{S on B}}$ $\vec{F}_{\text{Sp on B}}$ $\vec{F}_{\text{E on B}}$
The spring scale pulls harder on the block; the block still does not move.	New force? $\vec{N}_{\text{S on B}}$ $\vec{F}_{\text{Sp on B}}$ $\vec{F}_{\text{E on B}}$
The spring scale pulls even harder on the block; the block finally starts moving.	New force? $\vec{N}_{\text{S on B}}$ Force diagram for the moment just before the block begins moving $\vec{F}_{\text{Sp on B}}$ $\vec{F}_{\text{E on B}}$

Patterns

- In each of these experiments, the surface exerted a normal force on the block that balanced the downward gravitational force exerted by Earth on the block.
- As the spring scale exerted an increasing force on the block to the right, the block remained stationary (zero acceleration). The surface must have exerted an additional force—an increasing force on the block toward the left.
- Eventually, the spring scale exerted a strong enough force on the block that the block started sliding. Thus, the resistive force must have a maximum value.

The patterns inferred from the experiments in Table 3.1 demonstrated **static friction force**. This force is parallel to the surfaces of two objects that are not moving in relation to each other and opposes the tendency of one object to move across the other. The static friction force changes magnitude to prevent motion—up to a maximum value. When the external force exceeds this static friction force, the block starts moving. This maximum resistive force that the surface can exert on the block is called the **maximum static friction force**.

Figure 3.8 The surface exerts a static friction force that helps you walk or run.

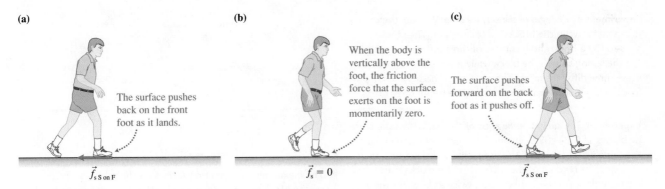

(a)

The surface pushes back on the front foot as it lands.

$\vec{f}_{s\,\text{S on F}}$

(b)

When the body is vertically above the foot, the friction force that the surface exerts on the foot is momentarily zero.

$\vec{f}_s = 0$

(c)

The surface pushes forward on the back foot as it pushes off.

$\vec{f}_{s\,\text{S on F}}$

We often talk about the importance of reducing friction in car engines, in bicycle chains, etc. However, sometimes friction is a necessary phenomenon. For instance, walking on a flat horizontal sidewalk would not be possible if there were no static friction (**Figure 3.8a**). When one foot swings forward and contacts the sidewalk, static friction prevents it from continuing forward and slipping (the way it might on ice). When the front shoe lands, the friction force pushes back on it and prevents it from slipping forward.

When your body is vertically above the foot, the friction force that the sidewalk exerts on your foot is momentarily zero (Figure 3.8b). As your body gets ahead of the back foot, the foot has a tendency to slip backward. However, the static friction force that the sidewalk exerts on your foot now points opposite that slipping direction, that is, forward, in the direction of motion of the body (Figure 3.8c). As long as this foot does not slip, the surface continues exerting a forward static friction force on you, helping your body move forward.

> **TIP** A surface really exerts only one force on an object pressing against it. It is convenient to break this force into two vector components: a component perpendicular to the surface, the normal force $\vec{N}_{\text{S on O}}$, and a component parallel to the surface, the static friction force $\vec{f}_{s\,\text{S on O}}$ (**Figure 3.9**). We can apply Newton's second law in component form to treat these components as separate forces.

What determines the magnitude of the maximum static friction force? To investigate this question, we use an experimental setup (see Observational Experiment **Table 3.2**) similar to that used in Observational Experiment Table 3.1, except that we will vary the characteristics of the block and the surfaces.

Figure 3.9 The force that the surface exerts on an object.

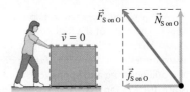

$\vec{v} = 0$ $\vec{F}_{\text{S on O}}$ $\vec{N}_{\text{S on O}}$ $\vec{f}_{s\,\text{S on O}}$

The force of the surface on the block is considered as two forces: the normal force $\vec{N}$ and the friction force $\vec{f}$.

OBSERVATIONAL EXPERIMENT TABLE

3.2 What affects the maximum friction force?

 VIDEO 3.2

Observational experiment	Analysis
Experiment 1. *Changes in the smoothness of the surfaces* We'll use the spring to pull a smooth, plastic block that is resting on three different surfaces: (1) a glass tabletop, (2) a wood tabletop, and (3) a rubber exercise mat. The reading of the spring scale just before the blocks start moving is largest for the rubber mat, next largest for the wood tabletop, and smallest for the glass tabletop.	$\vec{f}_{s\,\text{R on B}} > \vec{f}_{s\,\text{W on B}} > \vec{f}_{s\,\text{G on B}}$

(continued)

Observational experiment	Analysis
Experiment 2. *Changes in the surface area* We vary the contact area between the block and the surface. The block is shaped like a brick and has faces of three different areas. We use the spring to pull the block while it is resting on each of these three different faces. The reading of the scale just before the block starts to move is the same for all three areas.	$$\vec{f}_{s\,R\,on\,BA1} = \vec{f}_{s\,R\,on\,BA2} = \vec{f}_{s\,R\,on\,BA3}$$
Experiment 3. *Changes in the mass of the block* We take plastic blocks of 1.0 kg, 2.0 kg, and 3.0 kg and place them all on the same wood tabletop. We use a spring scale to pull each of them. The reading of the scale when the blocks start moving is smallest for the 1.0-kg block, twice as large for the 2.0-kg block, and three times as large for the 3.0-kg block.	$$\vec{f}_{s\,W\,on\,3\,kg\,B} = 3\vec{f}_{s\,W\,on\,1\,kg\,B}$$ $$\vec{f}_{s\,W\,on\,2\,kg\,B} = 2\vec{f}_{s\,W\,on\,1\,kg\,B}$$ The maximum static friction force that the tabletop exerts on the block is directly proportional to the mass of the block.

Patterns

The maximum static friction force that the surface can exert on the block depends on the roughness of the contacting surfaces and the mass of the block, but does not depend on the area of contact between the surfaces.

The patterns in Table 3.2 make testable hypotheses. To test a hypothesis, we need to design an experiment in which we can vary one of the properties and make a prediction of the outcome based on the hypothesis being tested. We will test one hypothesis in Testing Experiment **Table 3.3**, that the maximum static friction force is directly proportional to the mass of the object.

TESTING EXPERIMENT TABLE

3.3   Does the maximum static friction force depend on mass?

VIDEO 3.3

Testing experiment	Prediction	Outcome
We use a spring attached to a spring scale to pull a 1-kg block. The mass of the block does not change, but we push down on the block with a spring that exerts a series of downward forces on it. For each of these downward forces, we use the pulling string and spring scale to determine the maximum static friction force the surface exerts on the block. 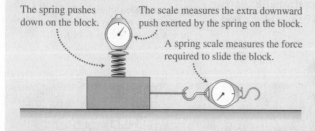 The spring pushes down on the block. The scale measures the extra downward push exerted by the spring on the block. A spring scale measures the force required to slide the block.	If the friction force is proportional to the mass of the block, the friction force should remain constant during the experiment.	The friction force changes—the harder we press on the block, the higher the maximum static friction force.

Conclusion

The outcome of the experiment did not match the prediction; the hypothesis requires a revision.

The unexpected outcome of the testing experiment requires further investigation. In **Table 3.4** we present the detailed data using the apparatus in Table 3.3. Remember that we are keeping the roughness of the surfaces and surface areas the same during all of the experiments.

Table 3.4 Maximum static friction force when a block presses harder against a surface.

Mass of the block	Extra downward force exerted on the 1-kg block	Normal force exerted by the surface on the block	Maximum static friction force	Ratio of maximum static friction force to normal force
1.0 kg	0.0 N	9.8 N	3.0 N	0.31
1.0 kg	5.0 N	14.8 N	4.5 N	0.30
1.0 kg	10.0 N	19.8 N	6.1 N	0.31
1.0 kg	20.0 N	29.8 N	9.1 N	0.31

Table 3.5 The coefficients of kinetic and static friction for two different surfaces.

Contacting surfaces	Coefficient of static friction	Coefficient of kinetic friction
Rubber on concrete (dry)	1	0.6–0.85
Steel on steel	0.74–0.78	0.42–0.57
Aluminum on steel	0.61	0.47
Glass on glass	0.9–1	0.4
Wood on wood	0.25–0.5	0.20
Waxed skis on wet snow	0.14	0.1
Teflon on Teflon	0.04	0.04
Greased metals	0.1	0.06
Surfaces in a healthy human joint	0.01	0.003

The data in Table 3.4 disprove the hypothesis that the maximum static friction force depends on the mass of the block. We have also found a new pattern. The ratio of the maximum static friction force to the normal force

$$\frac{f_{s\,S\,on\,O\,max}}{N_{S\,on\,O}} \approx 0.31$$

and is about the same (within experimental uncertainty) for all the measurements (the last column in the table). It appears that the maximum static friction force is directly proportional to the magnitude of the normal force: $f_{s\,S\,on\,O\,max} \propto N_{S\,on\,O}$. This finding makes sense if you look back at Figure 3.9: the normal force and the friction force are two perpendicular components of the same force—the force that a surface exerts on an object! If the normal force exerted by the surface on an object increases, the maximum static friction force the surface exerts on the object increases proportionally. It also makes sense if you think of pulling a sled over snow. Pulling is easier than pushing because when you pull, you lift the sled a little off the surface, reducing the force that it exerts on the snow and thus reducing the normal force the snow exerts on the sled. As a result, the friction force exerted on the sled decreases and it is easier to pull.

If we repeat the Table 3.4 experiments using a different type of block on a different surface, we get similar results. The ratio of the maximum friction force to the normal force is the same for all of the different values of the normal forces. However, the proportionality constant is different for different surfaces; the proportionality depends on the types of contacting surfaces. The proportionality constant is greater for two rough surfaces contacting each other and less for smoother surfaces.

This ratio

$$\mu_s = \frac{f_{s\,max}}{N}$$

is called the **coefficient of static friction** μ_s for a particular pair of surfaces. The coefficient of static friction is a measure of the relative difficulty of sliding two surfaces across each other. The easier it is to slide one surface on the other, the smaller the value of μ_s. You experience different values of μ_s when you try to walk on ice and on rough snow. Where are you more likely to slide?

The coefficient of static friction μ_s has no units because it is the ratio of two forces. Although μ_s usually has values between 0 and 1, the value can be greater than 1. Its value is about 0.8 for rubber car tires on a dry highway surface and is very small for bones in healthy body joints separated by cartilage and synovial fluid. Some values for different surfaces are listed in **Table 3.5**.

Static friction force When two objects are in contact and we try to pull one across the other, they exert a static friction force on each other. This force is parallel to the contacting surfaces of the two objects and opposes the tendency of one object to move across the other. The static friction force changes magnitude to prevent motion—up to a maximum value. This maximum static friction force depends on the roughness of the two surfaces (on the coefficient of static friction μ_s between the surfaces) and on the magnitude of the normal force N exerted by one surface on the other. The magnitude of the static friction force is always less than or equal to the product of these two quantities:

$$0 \leq f_s \leq \mu_s N \tag{3.5}$$

Keep in mind the assumptions used when constructing this model of friction. We used relatively light objects resting on relatively firm surfaces. The objects never caused the surfaces to deform significantly (for example, it was not a car tire sinking into mud). Equation (3.5) is only reasonable in situations in which these conditions hold.

Kinetic friction

If we repeat the previous friction experiments with a block that is already in motion, we find a similar relationship between the resistive friction force exerted by the surface on the block and the normal force exerted by the surface on the block. There are, however, two differences: (1) under the same conditions, the magnitude of the kinetic friction force is always lower than the magnitude of the maximum static friction force; (2) the resistive force exerted by the surface on the moving object does not vary but has a constant value. As with the static friction force $\vec{f}_s$, the magnitude of this **kinetic friction force** f_k depends on the roughness of the contacting surfaces (indicated by a **coefficient of kinetic friction** μ_k) and on the magnitude N of the normal force exerted by one of the surfaces on the other, but not on the surface area of contact. The word *kinetic* indicates that the surfaces in contact are moving relative to each other.

Kinetic friction force When an object slides along a surface, the surfaces exert kinetic friction forces on each other. These forces are exerted parallel to the contacting surfaces and oppose the motion of one surface relative to the other surface. The kinetic friction force depends on the surfaces themselves (on the coefficient of kinetic friction μ_k) and on the magnitude of the normal force N exerted by one surface on the other:

$$f_k = \mu_k N \tag{3.6}$$

As with any mathematical model, this expression for kinetic friction has its limitations. First, it is applicable only for sliding objects, not rolling objects. Second, it has the same assumption about the rigidity of the surfaces as the model for static friction. Third, predictions based on Eq. (3.6) fail for objects moving at high speed. Although this equation does not have general applicability, it is simple and useful for rigid surfaces and everyday speeds.

What causes friction?

Let's use the example of Velcro fastening material to help understand the phenomenon of friction. The hooks and loops on the two surfaces of Velcro connect with each other, making it almost impossible to slide one surface across the other. All objects are, in a less dramatic way, like Velcro. Even the slickest surfaces have tiny bumps that can hook onto the tiny bumps on another surface (**Figure 3.10**). Logically, smoother surfaces should have reduced friction. For example, a book with a glossy cover slides farther across a table than a book with a rough, unpolished cover. The friction force exerted by the table on the glossy book is less that on the other book.

However, if the surfaces are too smooth (for example, two polished metal blocks), the friction increases again. Why? All substances are made of particles that are attracted to each other. This attraction between particles keeps a solid object in whatever shape it has. But this attraction between particles works only at very short distances. If the two surfaces are very smooth so that the microscopic particles are close enough to attract each other strongly (almost close enough to form chemical bonds with each other), it becomes more difficult to slide one surface relative to the other (for example, two pieces of plastic wrap or two polished metal blocks).

Determining friction experimentally

Let's try to determine the coefficient of static friction between a running shoe and a kitchen floor tile using two independent methods. If we only use one method, it will be difficult to decide if the result is reasonable.

Experiment 1

In the first experiment, we secure a tile to a horizontal tabletop and place a shoe on top of it. The shoe is pulled horizontally with a spring scale, which exerts an increasingly greater force on it until the shoe begins to slide (**Figure 3.11a**). A force diagram for the instant just before it slides is shown in Figure 3.11b. Since the shoe is not accelerating in the vertical direction, the upward normal force exerted on it by the tile equals the downward gravitational force exerted on it by Earth: $N_{\text{T on S}} = m_S g$.

For the horizontal x-direction, only the tension force exerted by the spring scale on the shoe and the static friction force exerted by the tile on the shoe are included in the x-scalar component of Newton's second law. The other two forces exerted on the shoe do not have x-components. Thus, the x-scalar component form of Newton's second law is

$$ma_x = T_{\text{Scale on S}\,x} + f_{s\,\text{T on S}\,x}$$

Just before the shoe starts to slide, its acceleration is zero, and the scale reads the maximum force of static friction that the tile exerts on the shoe:

$$0 = T_{\text{Scale on S max}} - f_{s\,\text{T on S max}}$$

Since $T_{\text{max Scale on S}} = f_{s\,\text{max T on S}} = \mu_s N_{\text{T on S}}$, we can determine the coefficient of static friction:

$$\mu_s = \frac{f_{s\,\text{T on S max}}}{N_{\text{T on S}}} = \frac{T_{\text{Scale on S max}}}{m_S g}$$

All we need to do is measure the mass of the shoe and record the reading on the spring scale at the moment the shoe starts to slide.

The measured shoe mass is 0.37 kg, and the maximum scale reading is 2.6 N just before the shoe starts to slide. Thus, the coefficient of static friction is

$$\mu_s = \frac{f_{s\,\text{T on S max}}}{N_{\text{T on S}}} = \frac{T_{\text{Scale on S max}}}{m_S g} = \frac{2.6\ \text{N}}{(0.37\ \text{kg})(9.8\ \text{N/kg})} = 0.72$$

Figure 3.10 A microscopic view of contacting surfaces.

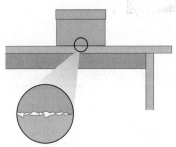

Rough edges on the contacting surfaces cause friction.

‹Active Learning Guide

Figure 3.11 Experiment 1. Determining μ_s for the shoe-tile surface by pulling the shoe across a horizontal tile.

(a)

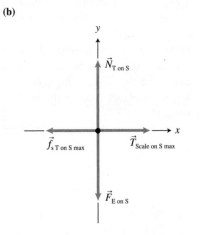

(b)

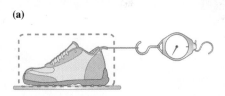

We reported just two significant digits since the measured quantities have just two significant digits. Thus, based on this experiment, the coefficient of static friction is $\mu_s = 0.72 \pm 0.01$.

Experiment 2

For the second experiment, we place the shoe on the tile and tilt the tile until the shoe starts to slide (**Figure 3.12a**). The static friction force that the tile exerts on the shoe increases as the tilt angle increases. Just before the shoe slides, the static friction has its maximum possible value.

We can now use the force diagram in Figure 3.12b to help apply Newton's second law in component form for the shoe just before it starts sliding. Notice that the x-axis of the coordinate system is parallel to the tilted tile and the y-axis is perpendicular to the tile. The magnitude of the gravitational force that Earth exerts on the shoe is $F_{E\,on\,S} = m_S g$, and the shoe's acceleration is zero just before it starts to slide.

y-component equation: $m_S \cdot 0 = N_{T\,on\,S}\sin 90° - m_S g\cos\theta + f_{s\,T\,on\,S\,max}\sin 0°$

x-component equation: $m_S \cdot 0 = N_{T\,on\,S}\cos 90° - m_S g\sin\theta + f_{s\,T\,on\,S\,max}\cos 0°$

Computing the values of the known sines and cosines and inserting the expression for the maximum static friction force $f_{s\,max\,T\,on\,S} = \mu_s N_{T\,on\,S}$, we get

$$0 = N_{T\,on\,S} - m_S g\cos\theta$$
$$0 = -m_S g\sin\theta + \mu_s N_{T\,on\,S}$$

We have two equations with two unknowns, $N_{T\,on\,S}$ and μ_s. Since our interest is in the coefficient of static friction, we solve the first equation for the normal force ($N_{T\,on\,S} = m_S g\cos\theta$) and substitute this into the second equation:

$$0 = -m_S g\sin\theta + \mu_s m_S g\cos\theta$$

Cancel the common $m_S g$ from each term and rearrange the above equation to get an expression for μ_s:

$$\mu_s = \frac{\sin\theta}{\cos\theta} = \tan\theta$$

This is an amazing result—the coefficient of static friction between the shoe and the tile can be determined just from the angle of the tile's tilt when the shoe starts sliding.

When we do the experiment, the shoe starts sliding when the tile is at an angle of about $\theta = 36°$. Thus, the coefficient of static friction determined from this experiment is:

$$\mu_s = \frac{\sin\theta}{\cos\theta} = \tan\theta = \tan 36° = 0.73$$

Again, because of the number of significant digits, this is equivalent to $\mu_s = 0.73 \pm 0.01$. This is consistent with the result from the first experiment, since the ranges of possible values overlap. The coefficient of static friction between the shoe and the tile is between 0.72 and 0.73.

Now, let's consider a real-world situation that involves kinetic friction.

Using skid marks for evidence

When a car stops under normal conditions, it rolls to a stop; the tires do not skid along the road surface. However, if the driver slams on the brakes to stop suddenly, the tires can lock, causing the car to skid. Police officers use the length of skid marks to estimate the speed of the vehicle at the time the driver applied the brakes. Police stations have charts listing the kinetic friction coefficients of various brands of car tires on different types of road surfaces.

Figure 3.12 Experiment 2. Tilting the tile to determine μ_s for the shoe-tile surface.

(a)   **(b)**

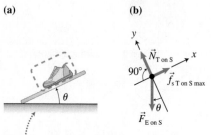

There is a maximum tilt angle at which the static friction has its maximum value just before the shoe slides.

TIP Notice that the magnitude of the normal force that a surface exerts on an object does not necessarily equal the magnitude of the gravitational force that Earth exerts on the object—especially when the object is on an inclined surface!

EXAMPLE 3.6 Was the car speeding?

A car involved in a minor accident left 18.0-m skid marks on a horizontal road. After inspecting the car and the road surface, the police officer decided that the coefficient of kinetic friction was 0.80. The speed limit was 15.6 m/s (35 mi/h) on that street. Was the car speeding?

Sketch and translate We first sketch the process (see the figure below). We choose the car as the system. Earth is the object of reference. The coordinate system consists of a horizontal x-axis pointing in the direction of the velocity and a vertical y-axis pointing upward.

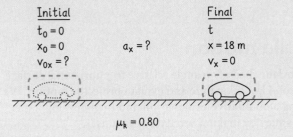

Initial
$t_0 = 0$
$x_0 = 0$
$v_{0x} = ?$
$a_x = ?$

Final
t
$x = 18$ m
$v_x = 0$

$\mu_k = 0.80$

Simplify and diagram Assume that the car can be modeled as a point-like object, that its acceleration while stopping was constant, and that the resistive force exerted by the air on the car is small compared with the other forces exerted on it. We can sketch a motion diagram for the car. Our force diagram below shows three forces exerted on the car. Earth exerts a downward gravitational force on the car $\vec{F}_{E \text{ on } C}$, the road exerts an upward normal force on the car $\vec{N}_{R \text{ on } C}$ (perpendicular to the road surface), and the road also exerts a backward kinetic friction force on the car $\vec{f}_{k \text{ R on } C}$ (parallel to the road's surface and opposite the car's velocity). This friction force causes the car's speed to decrease.

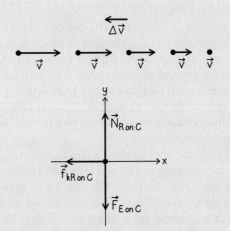

Represent mathematically Use the force diagram as shown above to help apply Newton's second law in component form. Use the expression for the kinetic friction force ($f_{k \text{ R on } C} = \mu_k N_{R \text{ on } C}$), the expression for the gravitational force ($F_{E \text{ on } C} = m_C g$), and one of the kinematics

equations. The car remains in contact with the road surface, so the y-component of its acceleration a_y is zero.

y-component equation:

$$0 = N_{R \text{ on } C y} + F_{E \text{ on } C y} + f_{k R \text{ on } C y}$$

$$0 = N_{R \text{ on } C} \sin 90° - m_C g \sin 90° + f_{k R \text{ on } C} \sin 0°$$

Note that $\sin 90° = 1.0$ and $\sin 0° = 0$. Thus,

$$N_{R \text{ on } C} = m_C g$$

The magnitude of the kinetic friction force is then

$$f_{k R \text{ on } C} = \mu_k N_{R \text{ on } C} = \mu_k m_C g$$

x-component equation:

$$m_C a_x = N_{R \text{ on } C x} + F_{E \text{ on } C x} + f_{k R \text{ on } C x}$$

$$m_C a_x = N_{R \text{ on } C} \cos 90° + m_C g \cos 90° - f_{k R \text{ on } C} \cos 0°$$

Substitute $\cos 90° = 0$ and $\cos 0° = 1.0$ into the above to get

$$m_C a_x = -f_{k R \text{ on } C}$$

Combine the two equations for $f_{k R \text{ on } C}$ to get

$$m_C a_x = -\mu_k m_C g$$

or

$$a_x = -\mu_k g$$

Now use kinematics to determine the car's velocity v_{0x} before the skid started:

$$v_x^2 - v_{0x}^2 = 2(x - x_0) a_x$$

Solve and evaluate The car's acceleration while stopping was

$$a_x = -\mu_k g = -(0.80)(9.8 \text{ m/s}^2) = -7.84 \text{ m/s}^2$$

We use the kinematics equation $0^2 - v_{0x}^2 = 2(x - x_0) a_x$ to determine the initial speed of the car before the skid started (recall that the final speed $v_x = 0$ and that the stopping distance was 18 m):

$$0^2 - v_{0x}^2 = 2(x - x_0) a_x = 2(18 \text{ m} - 0)(-7.84 \text{ m/s}^2)$$
$$= -282 \text{ m}^2/\text{s}^2 = -(16.8 \text{ m/s})^2$$

$$\Rightarrow v_{0x} = 16.8 \text{ m/s}$$
$$= 16.8 \text{ m/s}(3600 \text{ s/h})(1 \text{ mi}/1609 \text{ m})$$
$$= 38 \text{ mi/h}$$

This is slightly over the 15.6 m/s (35 mi/h) speed limit, but probably not enough for a speeding conviction. Note also that the answer had the correct units.

Try it yourself: Your car is moving at a speed 16 m/s on a flat, icy road when you see a stopped vehicle ahead. Determine the distance needed to stop if the effective coefficient of friction between your car tires and the road is 0.40.

Answer: 33 m.

Other types of friction

There are other types of friction besides static and kinetic friction, such as rolling friction. Rolling friction is caused by the surfaces of rolling objects indenting slightly as they turn. This friction is decreased in tires that have been inflated to a higher pressure. In a later chapter (Chapter 11) we learn about another type of friction, the friction that air or water exerts on a solid object moving through the air or water—a so-called drag force.

Review Question 3.4 What is the force of friction that the floor exerts on your refrigerator? Assume that the mass of the refrigerator is 100 kg, the coefficient of kinetic friction is 0.30, and the coefficient of static friction is 0.35. What assumptions did you make in answering this question?

3.5 Projectile motion

Projectiles are objects launched at an angle relative to a horizontal surface. We can use Newton's second law to analyze and explain projectile motion. We will begin by constructing a qualitative explanation and then develop a quantitative description. Let's start by making some observations.

Qualitative analysis of projectile motion

You can easily create your own projectile using a basketball. First, throw the ball straight upward. It moves up and then down with respect to you and with respect to the floor and eventually returns to your hands (**Figure 3.13a**). Next, walk in a straight line at constant speed and throw the ball straight up again. Have a friend videotape you from the side. The ball goes up and returns to your hands as before, but this time you've moved along your walking path. Your friend sees a projectile—a ball traveling in an arc. Finally, put on rollerblades and repeat the experiment several times, each time moving with greater speed. As long as you don't change your speed or direction while the ball is in flight, it lands back in your hands. The video shows that the ball moves in an arc with respect to the ground, and at every frame it is directly above your hands (Figure 3.13b). Why does this happen?

We can analyze projectile motion by independently considering the ball's vertical and horizontal motion. Earth exerts a gravitational force on the ball, so its upward speed decreases until it stops at the highest point, and then its downward speed increases until it returns to your hands (the vertical component of the ball's acceleration is constant due to the gravitational force exerted on it by Earth). With respect to you, skating at constant velocity on rollerblades, the ball simply moves straight up and down like the ball in Figure 3.13a, even when you are moving horizontally.

In addition to this vertical motion, the ball also moves horizontally with respect to the floor. An observer standing at the side is in an inertial reference frame. No object exerts a horizontal force on the ball. Thus, according to Newton's first law, the ball's horizontal velocity does not change once it is released and is the same as your horizontal component of velocity. In every experiment the ball continues moving horizontally as if it were not thrown upward, and it moves up and down as if it does not move horizontally. It seems that the horizontal and vertical motions of the ball are independent of each other. Let's test this explanation in Testing Experiment **Table 3.6**.

Figure 3.13 A projectile launched by a moving person.

(a)

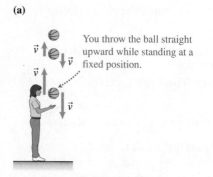

You throw the ball straight upward while standing at a fixed position.

(b)

While moving, you throw the ball straight upward relative to yourself.

Notice that the ball is always above your hands.

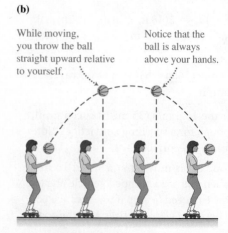

TESTING EXPERIMENT TABLE

3.6 **Testing the independence of horizontal and vertical motions.**

 VIDEO 3.6

Testing experiment	Prediction	Outcome
One ball is shot horizontally when a compressed spring is released. Simultaneously, a second ball is dropped. Which ball hits the surface first? 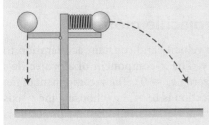	Both balls start with zero initial vertical speed; thus their vertical motions are identical. Since we think that the vertical motion is independent of the horizontal motion, we predict that they will land at the same time.	When we try the experiment, the balls do land at the same time.

Conclusion
The outcome supports the idea of independent horizontal and vertical motions. We've failed to disprove that idea.

The prediction may seem counterintuitive. However, the result matched the prediction. Since the ball on the right travels a longer path than the one on the left, why doesn't it land later? The vertical motions of both projectiles are identical; thus they land at the same time. The one on the right moves forward at constant velocity while it is falling, but this horizontal motion does not affect the vertical fall.

CONCEPTUAL EXERCISE 3.7 **Throwing a ball**
You throw a tennis ball as a projectile. Draw an arrow or arrows representing its instantaneous velocity and acceleration and the force or forces exerted on the ball by other objects when at the three positions shown below.

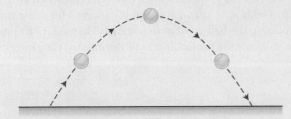

Sketch and translate The ball is the system of interest.

Simplify and diagram To draw velocity arrows, we consider the velocity at each point as consisting of a constant horizontal vector component and a vertical vector component whose magnitude decreases as the ball moves up and increases as it moves down. The vertical component is zero at the highest point in the trajectory. If we do this very carefully, we find that the velocity arrows are tangent to the ball's path at each position. If we ignore air resistance, only

one object exerts a force on the ball—Earth. Thus in our sketch, the force arrows point down at all three positions as shown below. The acceleration arrows point in the same direction as the sum of the forces—downward. Notice that at the top of the path, the velocity of the object is horizontal; however, the acceleration still points downward. The direction of velocity at each point and the direction of the

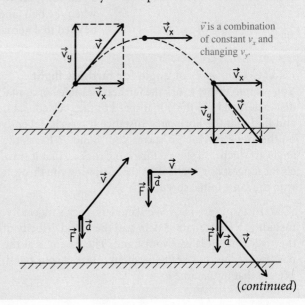

$\vec{v}$ is a combination of constant v_x and changing v_y.

(continued)

sum of the forces exerted on the object at each point do not have to point in the same direction.

Try it yourself: How does the magnitude of the net force and the acceleration compare at the first and third positions shown in diagram?

Answer: The magnitude of the net force is the same at both positions. The magnitude of the acceleration is the same at both positions.

The idea that the motions in two directions are independent of each other will help us to develop a quantitative way to describe projectile motion.

Quantitative analysis of projectile motion

We can use the equations of motion for velocity and constant acceleration to analyze projectile motion quantitatively. The x-component of a projectile's acceleration in the horizontal direction is zero ($a_x = 0$). The y-component of the projectile's acceleration in the vertical direction is $a_y = -g$ (choosing the y-axis to point up). Consider the projectile shown in **Figure 3.14**. When the projectile is launched at speed v_0 at an angle θ relative to the horizontal, its initial x- and y-velocity components are $v_{0x} = v_0 \cos \theta$ and $v_{0y} = v_0 \sin \theta$. The x-component of velocity remains constant during the flight, and the y-component changes in the same way that the velocity of an object thrown straight upward changes. If the motions in the x- and y-directions are independent of each other, then the equations that describe the motion of a projectile become the following:

Figure 3.14 The initial velocity component vectors and their magnitudes.

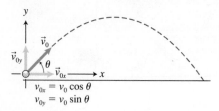

$v_{0x} = v_0 \cos \theta$
$v_{0y} = v_0 \sin \theta$

Projectile motion in the x-direction
$$(a_x = 0)$$
$$v_x = v_{0x} = v_0 \cos \theta \quad (3.7x)$$
$$x = x_0 + v_{0x}t$$
$$= x_0 + (v_0 \cos \theta)t \quad (3.8x)$$

Projectile motion in the y-direction
$$(a_y = -g)$$
$$v_y = v_{0y} + a_y t = v_0 \sin \theta + (-g)t \quad (3.7y)$$
$$y = y_0 + v_{0y}t + \frac{1}{2}a_x t^2$$
$$= y_0 + (v_0 \sin \theta)t - \frac{1}{2}g t^2 \quad (3.8y)$$

Equation (3.8y) can be used to determine the time interval for the projectile's flight, and Eq. (3.8x) can be used to determine how far the projectile travels in the horizontal direction during that time interval. For example, if the projectile leaves ground level and returns to ground level (for example, when you hit a golf ball), then Eq. (3.8y) with $y_0 = 0$ and $y = 0$ can be solved for t, the time when the ball lands (assuming the ball is hit at time zero). Then Eq. (3.8x) can be used to determine the distance the projectile travels during that time interval.

EXAMPLE 3.8 Best angle for farthest flight

You want to throw a rock the farthest possible horizontal distance. You keep the initial speed of the rock constant and find that the horizontal distance it travels depends on the angle at which it leaves your hand. What is the angle at which you should throw the rock so that it travels the longest horizontal distance, assuming you throw it with the same initial speed?

Sketch and translate We sketch the rock's trajectory, including a coordinate system and the initial velocity of the rock when it leaves your hand. The origin is at the position of the rock at the moment it leaves your hand. We call that initial time $t_0 = 0$.

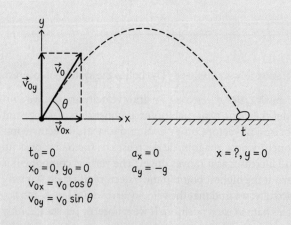

$t_0 = 0$
$x_0 = 0, y_0 = 0$
$v_{0x} = v_0 \cos \theta$
$v_{0y} = v_0 \sin \theta$

$a_x = 0$
$a_y = -g$

$x = ?, y = 0$

Simplify and diagram Assume that the rock leaves at the same elevation as it lands. That's not actually what happens, since the rock leaves your hand several feet above the ground, but it is a reasonable assumption if the rock travels a long distance compared to its initial height above the ground. Assume also that the rock is a point-like object and that the drag force exerted by the air on the rock does not affect its motion.

Represent mathematically First, determine the time interval that the rock is in flight using Eq. (3.8y). Note that because of the coordinate system we chose, $y_0 = y = 0$ and $t_0 = 0$:

$$0 = 0 + (v_0 \sin \theta)t - \frac{1}{2}gt^2$$

Dividing both sides of the equation by t and then solving for t, we obtain

$$t = \frac{2v_0 \sin \theta}{g}$$

This is the time of flight—we see that it depends only on the initial velocity of the rock and the launch angle. To determine the distance that the rock travels in the horizontal direction, insert this expression for t into Eq. (3.8x), and note that $x_0 = 0$:

$$x = (v_0 \cos \theta)t = v_0 \cos \theta \left(\frac{2v_0 \sin \theta}{g}\right) = \frac{v_0^2\, 2 \sin \theta \cos \theta}{g}$$

Solve and evaluate Recall from trigonometry that $2 \sin \theta \cos \theta = \sin(2\theta)$. This lets us rewrite the above equation as

$$x = \frac{v_0^2 \sin 2\theta}{g}$$

Our goal is to determine what launch angle will result in the rock traveling the farthest distance. Remember that the sine of an angle has a maximum value of 1 when the angle is 90°. Thus, $\sin 2\theta = 1$ when $2\theta = 90°$ or when $\theta = 45°$. The rock will travel the maximum horizontal distance when you launch it at an angle of 45° above the horizontal.

This answer seems reasonable. If you throw the rock at a smaller angle, the vertical component of the velocity is small and it spends less time in flight. If you launch it at an angle closer to 90°, it spends more time in flight, but its horizontal velocity component is small so that it does not travel very far horizontally. The 45° angle is a nice compromise between long time of flight and large horizontal velocity.

Try it yourself: Will the rock travel farther if it is launched at a 60° angle or a 30° angle?

Answer: It will travel the same distance in each case (see below.)

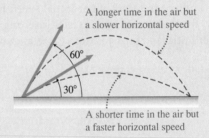

A longer time in the air but a slower horizontal speed

60°

30°

A shorter time in the air but a faster horizontal speed

Human cannonballs

Calculating projectile motion is not always as simple as Eqs. (3.7) and (3.8) would suggest. Human cannonball launches must use complicated formulas to account for the effect of air resistance. Results obtained from Eqs. (3.7) and (3.8) are correct only in the absence of air resistance or if air resistance is small. Even so, we will use the simplified equations in the next example.

> **TIP** Notice that the sign in front of g in the equations depends on the chosen direction of the y-axis. Try to choose the direction of the axis so that it simplifies the situation as much as possible.

EXAMPLE 3.9 Shot from a cannon
Stephanie Smith Havens (sister of David Smith, Jr.) is to be shot from an 8-m-long cannon into a net 40 m from the end of the cannon barrel and at the same elevation (our assumption). The barrel of the cannon is oriented 45 degrees above the horizontal. Estimate the speed with which she needs to leave the cannon to make it to the net.

Sketch and translate We sketch the process as shown on the next page. Smith Havens leaves the barrel traveling at an unknown speed v_0 at a 45° angle above the horizontal. We choose the origin of the coordinate system to be at the end of the barrel. Time zero will be when she leaves the cannon barrel. Thus, $y_0 = y = 0$, where y is her final elevation. The initial x-component of her velocity is $v_{0x} = v_0 \cos 45°$, and the initial y-component

(continued)

of her velocity is $v_{0y} = v_0 \sin 45°$. The x-component of her velocity is $a_x = 0$, and the y-component of her acceleration is $a_y = -g$.

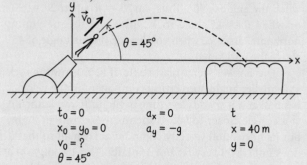

$$t_0 = 0 \qquad a_x = 0 \qquad t$$
$$x_0 = y_0 = 0 \qquad a_y = -g \qquad x = 40\,\text{m}$$
$$v_0 = ? \qquad\qquad\qquad y = 0$$
$$\theta = 45°$$

Simplify and diagram We model Smith Havens as a point-like object and assume that air resistance does not significantly affect her motion. The only force exerted on her while she is in flight is the gravitational force that Earth exerts on her.

$\vec{F}_{\text{E on H}}$ (mg)

Represent mathematically We get an expression for her time of flight in terms of her initial speed by using Eq. (3.8y) for the vertical motion after inserting the values for her initial and final positions (both are zero):

$$0 = 0 + (v_0 \sin 45°)t + \frac{1}{2}(-g)t^2$$

or

$$t = \frac{2v_0 \sin 45°}{g}$$

We can substitute this expression into Eq. (3.8x) and solve for v_0:

$$x = 0 + (v_0 \cos 45°)t$$

$$= 0 + v_0(\cos 45°)\frac{2v_0(\sin 45°)}{g}$$

$$= v_0^2 \frac{2(\cos 45°)(\sin 45°)}{g}$$

Solve and evaluate Multiply each side of the previous equation by g and divide both sides by $2(\cos 45°)(\sin 45°)$ to get an expression for the initial speed that Stephanie needs in order to reach the net:

$$v_0^2 = \frac{xg}{2(\cos 45°)(\sin 45°)}$$

$$= \frac{(40\,\text{m})(9.8\,\text{m/s}^2)}{2(\cos 45°)(\sin 45°)}$$

$$= 392\,\text{m}^2/\text{s}^2$$

Taking the positive square root of both sides of this equation, we find that $v_0 \approx 20$ m/s or 44 mi/h. If we had included the effects of air resistance, her initial speed would have had to be greater than our estimate—so the fact that we underestimated the initial speed makes sense. Also notice that the units of our result are correct.

Try it yourself: Suppose Smith Havens was shot from a horizontal barrel that was 19.6 m higher than the catching net and that the net was placed 40 m horizontally from the end of the barrel. (a) What time interval is needed for her trip to the net? (b) At what speed does she need to leave the barrel in order to reach the net?

Answer: (a) 2.0 s, the time interval needed to fall 19.6 m; (b) 20 m/s, the horizontal speed needed to travel 40 m horizontally in 2.0 s.

Review Question 3.5 Why do we need to resolve the initial velocity of a projectile into components when we analyze situations involving projectiles?

3.6 Using Newton's laws to explain everyday motion: Putting it all together

Each second of our lives is affected by phenomena that can be described and explained using Newton's laws—the cycles of day and night, the four seasonal variations, walking, driving cars, riding bicycles, and throwing and catching balls. We have already explored walking as a function of Newton's laws. In this section, we will analyze another everyday phenomenon, starting and stopping a car.

Static friction helps a car start and stop

When a car is moving, the wheels turn relative to the road surface. The part of the tire immediately behind the part in contact with the surface is lifting up off the road, and the part of the tire immediately in front of the part in contact is moving down to make new contact with the road. But the part of the tire in contact with the road is at rest with respect to the road—for just a short time interval (**Figure 3.15a**)—just as your foot is stationary relative to the ground as your body moves forward while walking or running.

Increasing or decreasing the car's speed involves static friction between the tire's region of contact and the pavement. How can this be? If you want to move faster, the tire turns faster and pushes back harder on the pavement (Figure 3.15b). The pavement in turn pulls forward more on the tire (Newton's third law) and helps accelerate the car forward. If you want to slow down, the tire turns slower and pulls forward on the pavement (Figure 3.15c). The pavement in turn pushes back on the tire (Newton's third law) and helps the car accelerate backwards. Thus, the static friction force exerted by the road on the car can cause the car to speed up or slow down depending on the direction of that force.

Because the coefficient of static friction between the tire and the road is greater than the coefficient of kinetic friction, stopping is more efficient if your tires do not skid but instead roll to a stop. This is why vehicles are equipped with antilock brakes: to keep the tires rolling on the road instead of locking up and skidding.

Static friction not only helps a car speed up and slow down but also plays an important role in maintaining a car's constant speed. Without this additional force, the car would eventually slow down, because oppositely directed forces, such as air resistance and rolling resistance, point opposite the car's forward motion. A car's engine makes the axle and car tires rotate. The tires push back on the road and the road in turn pushes forward on the car in the opposite direction. The forward static friction force that the road exerts on the car keeps the car moving at constant speed. The presence of air resistance and rolling resistance explain why when you are driving you must continually give the engine gas; otherwise, you would not be able to maintain constant speed and your car would slow down.

> **TIP** Some people think that the car's engine exerts a force on the car that starts the car's motion and helps it maintain its constant speed despite the air resistance. However, the forces that the engine exerts on other parts of the car are internal forces. Only external forces exerted by objects in the environment can affect the car's acceleration. The engine does rotate the wheels, and the wheels push forward or back on the ground. However, it is the ground (an external object) that pushes backward or forward on the wheels causing the car to slow down or speed up. The force that is responsible for this backward or forward push is the static friction force that the road exerts on the car tires.

Figure 3.15 Static friction helps a car accelerate.

(a) The tire is moving to the right at constant speed.

Tire lifting off road   Tire not moving   Tire making new contact

(b) The tire is moving to the right and turning faster.

$\vec{f}_{\text{s Road on Tire}}$

If the tire turns faster, it pushes back harder on the road. The road in turn pushes forward on the tire, causing the car to accelerate to the right.

(c) The tire is moving to the right and turning slower.

$\vec{f}_{\text{s Road on Tire}}$

If the tire turns slower, it pulls forward on the road and the road pushes back on the tire, exerting a force that slows the car, causing it to accelerate backward.

CONCEPTUAL EXERCISE 3.10 Car going down a hill

Your car moves down a 6.0° incline at a constant velocity of magnitude 16 m/s. Describe everything you can about the friction force that the road exerts on your car.

Sketch and translate Sketch the situation as shown. Since the car is moving at constant velocity, the sum of the forces that other objects exert on the car must be zero. We use this information to help construct a force

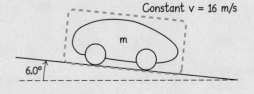

Constant v = 16 m/s

m

6.0°

diagram for the car and then determine the direction and relative magnitude of the friction force that the road exerts on the car.

Simplify and diagram Model the car as a point-like object and assume that air resistance does not significantly affect the motion of the car. Make a force diagram for the car (as shown on the next page). The x-axis points in the direction of motion, and the y-axis points perpendicular to the road surface. The road exerts a normal force on the car $\vec{N}_{\text{R on C}}$ perpendicular to the road (it has only a y-component) and Earth exerts a downward gravitational force on the car $\vec{F}_{\text{E on C}}$ (it has a negative y-component and a small positive x-component). For the car to move at constant velocity, the road must exert a static friction force on the car $\vec{f}_{\text{s R on C}}$ that points in the negative

(continued)

x-direction to balance the x-component of the gravitational force ($mg \cos 84°$) that Earth exerts on the car. If we had included air resistance, the static friction force would not have had to be as large in order to balance the x-component of the gravitational force.

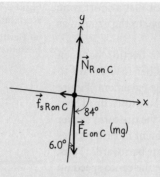

Try it yourself: A car moves at a high, constant velocity on a horizontal, level road. Does the road exert a static friction force on the car, and if so, in what direction?

Answer: Air resistance becomes more important at high speeds and can exert a several hundred newton force on the car that opposes its motion. Thus, the road now has to exert a static friction force on the car tires that pushes the car in the forward direction—the tires push back on the road, and the road in turn pushes forward on the tires.

EXAMPLE 3.11 Equation Jeopardy

The equations below are the horizontal x- and vertical y-component forms of Newton's second law applied to a physical process. Solve for the unknown quantities. Then work backward and construct a force diagram for the system of interest and invent a process and question for which the equations might provide an answer (there are many possibilities). Remember that the italicized N is the symbol for normal force, and the roman N is a symbol for the newton.

x-equation:

$$(200\text{ N})\cos 30° + 0 - 0.40 N_{S \text{ on } O} + 0 = (50\text{ kg})a_x$$

y-equation:

$$(200\text{ N})\sin 30° + N_{S \text{ on } O} + 0 - (50\text{ kg})(9.8\text{ N/kg})$$
$$= (50\text{ kg})0$$

Solve Inserting the cosine and sine values, we get x-equation:

$$(200\text{ N})0.87 + 0 - 0.40 N_{S \text{ on } O} + 0 = (50\text{ kg})a_x$$

y-equation:

$$(200\text{ N})0.50 + N_{S \text{ on } O} + 0 - (50\text{ kg})(9.8\text{ N/kg}) = 0$$

We can solve the y-equation for the magnitude of the normal force: $N_{S \text{ on } O} = 390$ N. Inserting this value into the x-equation produces the following:

$$(200\text{ N})0.87 + 0 - 0.40(390\text{ N}) + 0 = (50\text{ kg})a_x$$

This can now be solved for a_x:

$$a_x = (174\text{ N} - 156\text{ N})/(50\text{ kg})$$
$$= +0.36\text{ N/kg} = +0.36\text{ m/s}^2$$

Simplify and diagram The equations provide the components for each of the four forces exerted on the system object (since there are four terms on the left side of each equation). Consider the x- and y-scalar components of each force:

1. A 200-N force oriented 30° above the positive x-axis—maybe a rope is exerting a force on an object.

2. A 390-N normal force points along the y-axis, perpendicular to a surface.

3. It looks like a 0.40(390 N) = 156 N friction force points in the negative x-direction with a coefficient of friction equal to 0.40.

4. A 490-N gravitational force points in the negative y-direction perpendicular to the surface.

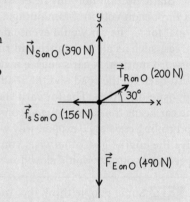

We can now use these forces to construct a force diagram.

Sketch and translate The situation could involve a sled or wagon or crate that is being pulled along a horizontal surface as shown. Note that the rope is at an angle with respect to the horizontal, so the force that it exerts has a +100-N y-component. It combines with the normal force's +390-N y-component to balance the gravitational force's −490-N y-component.

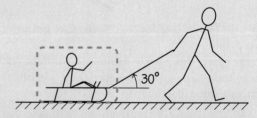

One possible problem statement Determine the acceleration of a 50-kg sled being pulled across a horizontal surface by a rope oriented 30° above the horizontal and pulling with a force of 200 N. The coefficient of kinetic friction between the sled and the surface is 0.40.

Review Question 3.6 You read in this section that it is the road and not the engine that is most directly responsible for a car speeding up. Why is this true?

Summary

Words	Pictorial and physical representations	Mathematical representation
Components of a vector quantity A vector quantity such as force $\vec{F}$ can be broken into its scalar components F_x and F_y, which indicate the effect of the force in two perpendicular directions. (Section 3.1)	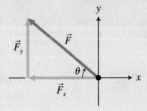	$F_x = \pm F\cos\theta$ Eq. (3.1x) $F_y = \pm F\sin\theta$ Eq. (3.1y) θ is the angle relative to the $\pm x$-axis. A scalar component is positive if it falls along a positive axis and negative if it falls along a negative axis.
Newton's second law in component form The acceleration of an object in the x-direction is the sum of the x-components of the forces exerted on it divided by its mass. The acceleration of the object in the y-direction is the sum of the y-components of the forces divided by its mass. (Section 3.2)	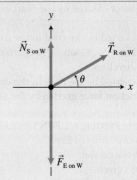	$a_x = \dfrac{\Sigma F_x}{m}$ Eq. (3.2x) $a_y = \dfrac{\Sigma F_y}{m}$ Eq. (3.2y)
Static friction force is the force exerted by a surface on another surface (parallel to both surfaces) when they are not moving relative to each other. The force magnitude adjusts up to a maximum static friction force, depending on the force exerted on the object in an effort to start its motion. (Section 3.4)	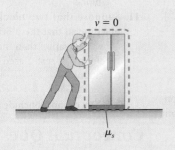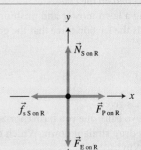	$f_{s\,max} \leq \mu_s N$ Eq. (3.5)
Kinetic friction force is the force exerted by a surface on another surface (parallel to both surfaces) when the surfaces are moving relative to each other. (Section 3.4)	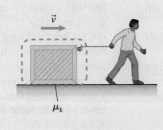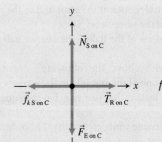	$f_k = \mu_k N$ Eq. (3.6)
Projectile motion Projectiles are objects launched either horizontally or at an angle with the horizontal. Horizontal and vertical motions of projectiles are independent of each other. If we ignore the resistive force that air exerts on the projectile, then its horizontal acceleration is zero ($a_x = 0$) and its vertical acceleration is the free-fall acceleration ($a_y = -g$). (Section 3.5)	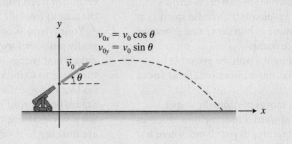	x-direction equations: $x = x_0 + v_{0x}t$ $v_x = v_0\cos\theta = \text{constant}$ y-direction equations: (if positive direction up) $y = y_0 + v_{0y}t + \dfrac{1}{2}(-g)t^2$ Eq. (3.8x) $v_y = v_0\sin\theta + (-g)t$ Eq. (3.8y)

111

For instructor-assigned homework, go to MasteringPhysics.

Questions

Multiple Choice Questions

1. A car accelerates along a road. Identify the best combination of forces exerted on the car in the horizontal direction that explains the acceleration.
   (a) The force exerted by the engine in the direction of motion and the kinetic friction force exerted by the road in the opposite direction
   (b) The static friction force exerted by the road in the direction of motion and the kinetic friction force exerted by the road in the opposite direction
   (c) The static friction force exerted by the road in the direction of motion and the friction force exerted by the air in the opposite direction

2. A person pushes a 10-kg crate exerting a 200-N force on it, but the crate's acceleration is only 5 m/s². Explain.
   (a) The crate pushes back on the person.
   (b) Not enough information is given.
   (c) There are other forces exerted on the crate.

3. Compare the ease of pulling a lawn mower and pushing it. In particular, in which case is the friction force that the grass exerts on the mower greater?
   (a) They are the same.
   (b) Pulling is easier.
   (c) Pushing is easier.
   (d) Not enough information is given.

4. You simultaneously release two balls: one you throw horizontally, and the other one you drop straight down. Which one will reach the ground first? Why?
   (a) The ball dropped straight down lands first, since it travels a shorter distance.
   (b) Neither. Their vertical motion is the same and so they will reach the ground at the same time.
   (c) It depends on the mass of the balls—the heavier ball falls faster.

5. You shoot an arrow with a bow. The following is Newton's second law applied for one of the instants in the arrow's trip: $m\vec{a} = m\vec{g}$. Choose the instant that best matches the equation description:
   (a) The arrow is accelerating while contacting the bowstring.
   (b) The arrow is flying up.
   (c) The arrow is slowing down while sinking into the target.
   (d) This equation does not describe any part of the arrow's trip.

6. In what reference frame does a projectile launched at speed v_0 at angle θ above the horizontal move only in the vertical direction?
   (a) There is no such reference frame.
   (b) In a reference frame that moves with the projectile
   (c) In a reference frame that moves horizontally at speed $v_0 \cos\theta$
   (d) In a reference frame that moves horizontally at speed v_0

7. You throw a ball vertically upward. From the time it leaves your hand until just before it returns to your hand, where is it located when the magnitude of its acceleration is greatest? Do not neglect air resistance.
   (a) Just after it leaves your hand on the way up
   (b) Just before arriving at your hand on the way down

   (c) At the top
   (d) Acceleration is the same during the entire flight.

8. While running, how should you throw a ball with respect to you so that you can catch it yourself?
   (a) Slightly forward   (b) Slightly backward
   (c) Straight up   (d) It is impossible.

9. You hold a block on a horizontal, frictionless surface. It is connected by a string that passes over a pulley to a vertically hanging block (a modified Atwood machine). What is the magnitude of the acceleration of the hanging object after you release the block on the horizontal surface?
   (a) Less than g (b) More than g (c) Equal to g

10. In the process described in the previous question, what is the magnitude of the force exerted by the string on the block on the horizontal plane after it is released?
    (a) Equal to the force that Earth exerts on the hanging block
    (b) Less than the force exerted by Earth on the hanging block
    (c) More than the force exerted by Earth on the hanging block

11. Suppose that two blocks are positioned on an Atwood machine so that the block on the right of mass m_1 hangs at a lower elevation than the block on the left of mass m_2. Both blocks are at rest. Based on this observation, what can you conclude?
    (a) $m_1 > m_2$ (b) $m_1 = m_2$ (c) $m_1 < m_2$
    (d) You cannot conclude anything with the given information.

Conceptual Questions

12. ✎ A box with a heavy television set in it is placed against a box with a toaster oven in it. Both sit on the floor. Draw force diagrams for the TV box and the toaster oven box if you push the TV box to the right, exerting a force $\vec{F}_{\text{Y on TV}}$. Repeat for the situation in which you are exerting a force of exactly the same magnitude and direction, but pushing the toaster oven box, which in turn pushes the TV box.

13. Your friend says that two blocks in an Atwood machine can never have the same acceleration, no matter what we assume. What principles of physics is your friend using as the basis for his opinion?

14. How can an Atwood machine be used to determine the acceleration of freely falling objects if none of the objects used is in the state of free fall?

15. ✎ Your friend is on rollerblades holding a pendulum. You gently push her forward and let go. You observe that the pendulum first swings in the opposite direction (backward) and then returns to the vertical orientation as she coasts forward. (a) Draw a force diagram to explain the behavior of the pendulum bob as your friend is being pushed. (b) How can you use it to determine the acceleration of your friend while you are pushing her? (c) Why does the pendulum return to its vertical orientation after you stop pushing your friend?

16. Explain why a car starts skidding when a driver abruptly applies the brakes.

17. Explain why old tires need to be replaced.

18. Describe two experiments that policemen can perform to determine the coefficient of kinetic friction between car tires and the road. Why would they need to do two independent experiments for the same set of tires?

19. Explain how friction helps you to walk.

20. Explain why you might fall forward when you stumble.

21. Explain why you might fall backward when you slip.

22. Explain why the tires of your car can "spin out" when you are caught in the mud.

23. You throw two identical balls simultaneously at the same initial speed: one downward and the other horizontally. Describe and compare their motions in as much detail as you can.

24. Your friend says that the vertical force exerted on a projectile when at the top of its flight is zero. Why would he say this? Do you agree or disagree? If you disagree, how would you convince your friend that your opinion is correct?

25. Your friend says that a projectile launched at an angle relative to the horizontal moves forward because it retains the force of the launcher. Why would she say this? Do you agree or disagree? If you disagree, how would you convince your friend that your opinion is correct?

26. ✓ An object of mass m_1 placed on an inclined plane (angle θ relative to the horizontal) is connected by a string that passes over a pulley to a hanging object of mass m_2 ($m_2 \gg m_1$). Draw a force diagram for each object. Do not ignore friction.

27. ✓ An object of mass m_1 placed on an inclined plane (angle θ relative to the horizontal) is connected by a string that passes over a pulley to a hanging object of mass m_2 ($m_1 \gg m_2$). Draw a force diagram for each object. Do not assume that friction can be ignored.

Problems

Below, BIO indicates a problem with a biological or medical focus. Problems labeled EST ask you to estimate the answer to a quantitative problem rather than derive a specific answer. Problems marked with ✓ require you to make a drawing or graph as part of your solution. Asterisks indicate the level of difficulty of the problem. Problems with no * are considered to be the least difficult. A single * marks moderately difficult problems. Two ** indicate more difficult problems.

3.1 Force components

1. Determine the x- and y-components of each force vector shown in **Figure P3.1**.

Figure P3.1

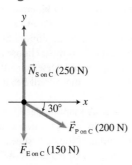

2. Determine the x- and y-components of each force vector shown in **Figure P3.2**.

Figure P3.2

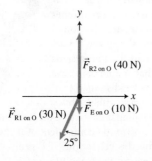

3. Determine the x- and y-components of each displacement shown in **Figure P3.3**.

Figure P3.3

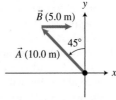

4. *✓ The x- and y-components of several unknown forces are listed below (F_x, F_y). For each force, draw on an x, y coordinate system the components of the force vectors. Determine the magnitude and direction of each force: (a) ($+100$ N, -100 N), (b) (-300 N, -400 N), and (c) (-400 N, $+300$ N).

5. *✓ The x- and y-scalar components of several unknown forces are listed below (F_x, F_y). For each force, draw an x, y coordinate system and the vector components of the force vectors. Determine the magnitude and direction of each force: (a) (-200 N, $+100$ N), (b) ($+300$ N, $+400$ N), and (c) ($+400$ N, -300 N).

3.2 Newton's second law in component form

6. * Three ropes pull on a knot shown in **Figure P3.6a**. The knot is not accelerating. A partially completed force diagram for the knot is shown in Figure P3.6b. Use qualitative reasoning (no math) to determine the magnitudes of the forces that ropes 2 and 3 exert on the knot. Explain in words how you arrived at your answers.

7. * Solve the previous problem quantitatively using Newton's second law.

Figure P3.6a

(a)

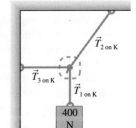

(b)

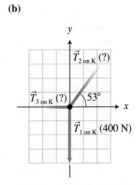

8. / For each of the following situations, draw the forces exerted on the moving object and identify the other object causing each force. (a) You pull a wagon along a level floor using a rope oriented 45° above the horizontal. (b) A bus moving on a horizontal road slows in order to stop. (c) You slide down an inclined water slide. (d) You lift your overnight bag into the overhead compartment on an airplane. (e) A rope connects two boxes on a horizontal floor, and you pull horizontally on a second rope attached to the right side of the right box (consider each box separately).

9. * Write Newton's second law in component form for each of the situations described in Problem 8.

10. / For the situations described here, construct a force diagram for the block, sled, and skydiver. (a) A cinder block sits on the ground. (b) A rope pulls at an angle of 30° relative to the horizontal on a sled moving on a horizontal surface. The sled moves at increasing speed toward the right (the surface is not smooth.) (c) A rope pulls on a sled parallel to an inclined slope (inclined at an arbitrary angle). The sled moves at increasing speed up the slope. (d) A skydiver falls downward at constant terminal velocity (air resistance is present).

11. * Write Newton's second law in component form for each of the situations described in Problem 10.

12. * Apply Newton's second law in component form for the force diagram shown in Figure P3.1.

13. * Apply Newton's second law in component form for the force diagram shown in Figure P3.2.

14. * / **Equation Jeopardy 1** The equations below are the horizontal x- and vertical y- component forms of Newton's second law applied to a physical process. Solve for the unknowns. Then work backward and construct a force diagram for the object of interest and invent a problem for which the equations might be an answer (there are many possibilities).

$$(5.0\,\text{kg})a_x = (50\,\text{N})\cos 30° + N\cos 90°$$
$$+ (5.0\,\text{kg})(9.8\,\text{N/kg})\cos 90°$$

$$(5.0\,\text{kg})0 = (-50\,\text{N})\sin 30° + N\sin 90°$$
$$- (5.0\,\text{kg})(9.8\,\text{N/kg})\sin 90°$$

15. * / **Equation Jeopardy 2** The equations below are the horizontal x- and vertical y-component forms of Newton's second law applied to a physical process for an object on an incline. Solve for the unknowns. Then work backward and construct a force diagram for the object and invent a problem for which the equations might be an answer (there are many possibilities).

$$(5.0\,\text{kg})a_x = (30\,\text{N})\cos 30° + N\cos 90°$$
$$- (5.0\,\text{kg})(9.8\,\text{N/kg})\cos 60°$$

$$(5.0\,\text{kg})0 = (30\,\text{N})\sin 30° + N\sin 90°$$
$$- (5.0\,\text{kg})(9.8\,\text{N/kg})\sin 60°$$

16. ** / **Equation Jeopardy 3** The equations below are the horizontal x- and vertical y-component forms of Newton's second law and a kinematics equation applied to a physical process. Solve for the unknowns. Then work backward and construct a force diagram for the object of interest and invent a problem for which the equations might be an answer (there are many possibilities). Provide all the information you know about your process.

$$(5.0\,\text{kg})a_x = (50\,\text{N})\cos 30° + N\cos 90°$$
$$+ (5.0\,\text{kg})(9.8\,\text{N/kg})\cos 90°$$

$$(5.0\,\text{kg})0 = (-50\,\text{N})\sin 30° + N\sin 90°$$
$$- (5.0\,\text{kg})(9.8\,\text{N/kg})\sin 90°$$

$$x - 0 = (2.0\,\text{m/s})(4.0\,\text{s}) + \frac{1}{2}a_x(4.0\,\text{s})^2$$

17. * You exert a force of 100 N on a rope that pulls a sled across a very smooth surface. The rope is oriented 37° above the horizontal. The sled and its occupant have a total mass of 40 kg. The sled starts at rest and moves for 10 m. List all the quantities you can determine using these givens and determine three of the quantities on the list.

18. * You exert a force of a known magnitude F on a grocery cart of total mass m. The force you exert on the cart points at an angle θ below the horizontal. If the cart starts at rest, determine an expression for the speed of the cart after it travels a distance d. Ignore friction.

19. * **Olympic 100-m dash start** At the start of his race, 86-kg Olympic 100-m champion Usain Bolt from Jamaica pushes against the starting block, exerting an average force of 1700 N. The force that the block exerts on his foot points 20° above the horizontal. Determine his horizontal speed after the force is exerted for 0.32 s. Indicate any assumptions you made.

20. * **Accelerometer** A string with one 10-g washer on the end is attached to the rearview mirror of a car. When the car leaves an intersection, the string makes an angle of 5° with the vertical. What is the acceleration of the car? [*Hint:* Choose the washer as the system object for your force diagram. Use the vertical component equation of Newton's second law to find the magnitude of the force that the string exerts on the washer. Then continue with the horizontal component equation.]

21. * **Your own accelerometer** A train has an acceleration of magnitude $1.4\,\text{m/s}^2$ while stopping. A pendulum with a 0.50-kg bob is attached to a ceiling of one of the cars. Determine everything you can about the pendulum during the deceleration of the train.

3.3 Problem-solving strategies for analyzing dynamics problems

22. * **Skier** A 52-kg skier starts at rest and slides 30 m down a hill inclined at 12° relative to the horizontal. List five quantities that describe the motion of the skier, and solve for three of them (at least one should be a kinematics quantity).

23. * **Ski rope tow** You agree to build a backyard rope tow to pull your siblings up a 20-m slope that is tilted at 15° relative to the horizontal. You must choose a motor that can pull your 40-kg sister up the hill. Determine the force that the rope should exert on your sister to pull her up the hill at constant velocity.

24. * **Soapbox racecar** A soapbox derby racecar starts at rest at the top of a 301-m-long track tilted at an average 4.8° relative to the horizontal. If the car's speed were not reduced by any structural effects or by friction, how long would it take to complete the race? What is the speed of the car at the end of the race?

25. * **BIO** **Whiplash experience** A car sitting at rest is hit from the rear by a semi-trailer truck moving at 13 m/s. The car lurches forward with an acceleration of about 300 m/s². **Figure P3.25** shows an arrow that represents the force that the neck muscle exerts on the head so that it accelerates forward with the body instead of flipping backward. If the head has a mass of 4.5 kg, what is the horizontal component of the force $\vec{F}$ required to cause this head acceleration? If $\vec{F}$ is directed 37° below the horizontal, what is the magnitude of $\vec{F}$?

Figure P3.25

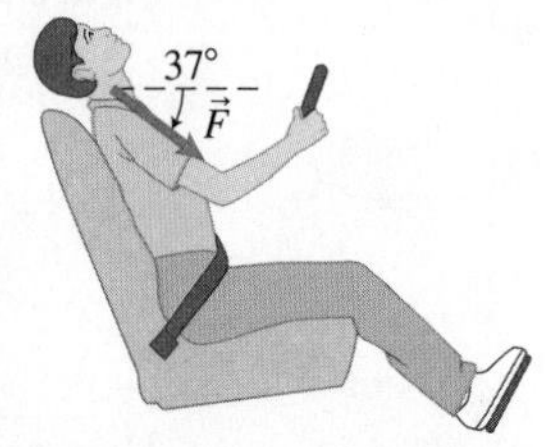

26. * **Iditarod race practice** The dogs of four-time Iditarod Trail Sled Dog Race champion Jeff King pull two 100-kg sleds that are connected by a rope. The sleds move on an icy surface. The dogs exert a 240-N force on the rope attached to the front sled. Find the acceleration of the sleds and the force the rope between the sleds exerts on each sled. The front rope pulls horizontally.

27. * You pull a rope oriented at a 37° angle above the horizontal. The other end of the rope is attached to the front of the first of two wagons that have the same 30-kg mass. The rope exerts a force of magnitude T_1 on the first wagon. The wagons are connected by a second horizontal rope that exerts a force of magnitude T_2 on the second wagon. Determine the magnitudes of T_1 and T_2 if the acceleration of the wagons is $2.0 \, \text{m/s}^2$.

28. ** Rope 1 pulls horizontally, exerting a force of 45 N on an 18-kg wagon attached by a second horizontal rope to a second 12-kg wagon. Make a list of physical quantities you can determine using this information, and solve for three of them, including one kinematics quantity.

29. * Three sleds of masses m_1, m_2, m_3 are on a smooth horizontal surface (ice) and connected by ropes, so that if you pull the rope connected to sled 1, all the sleds start moving. Imagine that you exert a force of a known magnitude on the rope attached to the first sled. What will happen to all of the sleds? Provide information about their accelerations and all the forces exerted on them. What assumptions did you make?

30. ** Repeat Problem 29, only this time with the sleds on a hill.

31. ** Your daredevil friends attach a rope to a 140-kg sled that rests on a frictionless icy surface. The rope extends horizontally to a smooth dead tree trunk lying at the edge of a cliff. Another person attaches a 30-kg rock at the end of the rope after it passes over the tree trunk and then releases the rock—the rope is initially taut. Determine the acceleration of the sled, the force that the rope exerts on the sled and on the rock, and the time interval during which the person can jump off the sled before it reaches the cliff 10 m ahead. There is no friction between the rope and the tree trunk.

32. * Assume the scenario described in Problem 31, but in this case a hanging rock of unknown mass accelerates downward at $2.7 \, \text{m/s}^2$ and pulls the sled with it. Determine the mass of the hanging rock and the force that the rope exerts on the sled. There is no friction between the rope and the tree trunk.

33. ** The 20-kg block shown in **Figure P3.33** accelerates down and to the left, and the 10-kg block accelerates up. Find the magnitude of this acceleration and the force that the cable exerts on a block. There is no friction between the block and the inclined plane, and the pulley is frictionless and light.

Figure P3.33

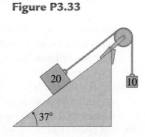

34. ** A person holds a 200-g block that is connected to a 250-g block by a string going over a light pulley with no friction in the bearing (an Atwood machine). After the person releases the 200-g block, it starts moving upward and the heavier block descends. (a) What is the acceleration of each block? (b) What is the force that the string exerts on each block? (c) How long will it take each block to traverse 1.0 m?

35. ** Two blocks of masses m_1 and m_2 are connected to each other on an Atwood machine. A person holds one of the blocks with her hand. When the system is released, the heavier block moves down with an acceleration of $2.3 \, \text{m/s}^2$ and the lighter object moves up with an acceleration of the same magnitude. What is one possible set of masses for the blocks?

3.4 Friction

36. * / **Equation Jeopardy 4** The equations below are the horizontal x- and vertical y-component forms of Newton's second law applied to a physical process. Solve for the unknowns. Then work backward and construct a force diagram for the object of interest and invent a problem for which the equations might provide an answer (there are many possibilities).

$$(5.0 \, \text{kg})a_x = (50 \, \text{N})\cos 30° + N\cos 90° - 0.5 \, N\cos 0° + (5.0 \, \text{kg})(9.8 \, \text{N/kg})\cos 90°$$

$$(5.0 \, \text{kg})0 = (-50 \, \text{N})\sin 30° + N\sin 90° + 0.5 \, N\sin 0° - (5.0 \, \text{kg})(9.8 \, \text{N/kg})\sin 90°$$

37. * / **Equation Jeopardy 5** The equations below are the x- and y-component forms of Newton's second law applied to a physical process for an object on an incline. Solve for the unknowns. Then work backward and construct a force diagram for the object and invent a problem for which the equations might provide an answer (there are many possibilities).

$$(5.0 \, \text{kg})0 = +F\cos 0° + N\cos 90° - 0.50 \, N\cos 0° - (5.0 \, \text{kg})(9.8 \, \text{N/kg})\cos 60°$$

$$(5.0 \, \text{kg})0 = +F\sin 0° + N\sin 90° - 0.50 \, N\sin 0° - (5.0 \, \text{kg})(9.8 \, \text{N/kg})\sin 60°$$

38. A 91.0-kg refrigerator sits on the floor. The coefficient of static friction between the refrigerator and the floor is 0.60. What is the minimum force that one needs to exert on the refrigerator to start the refrigerator sliding?

39. A 60-kg student sitting on a hardwood floor does not slide until pulled by a 240-N horizontal force. Determine the coefficient of static friction between the student and floor.

40. * **Racer runs out of gas** James Stewart, 2002 Motocross/Supercross Rookie of the Year, is leading a race when he runs out of gas near the finish line. He is moving at 16 m/s when he enters a section of the course covered with sand where the effective coefficient of friction is 0.90. Will he be able to coast through this 15-m-long section to the finish line at the end? If yes, what is his speed at the finish line?

41. * **Car stopping distance and friction** A certain car traveling at 60 mi/h (97 km/h) can stop in 48 m on a level road. Determine the coefficient of friction between the tires and the road. Is this kinetic or static friction? Explain.

42. * A 50-kg box rests on the floor. The coefficients of static and kinetic friction between the bottom of the box and the floor are 0.70 and 0.50, respectively. (a) What is the minimum force a person needs to exert on the box to start it sliding? (b) After the box starts sliding, the person continues to push it, exerting the same force. What is the acceleration of the box?

43. * Marsha is pushing down and to the right on a 12-kg box at an angle of 30° below horizontal. The box slides at constant velocity across a carpet whose coefficient of kinetic friction with the box is 0.70. Determine three physical quantities using this information, one of which is a kinematics quantity.

44. * A wagon is accelerating to the right. A book is pressed against the back vertical side of the wagon and does not slide down. Explain how this can be.

45. * In Problem 44, the coefficient of static friction between the book and the vertical back of the wagon is μ_s. Determine an expression for the minimum acceleration of the wagon in terms of μ_s so that the book does not slide down. Does the mass of the book matter? Explain.

46. * A car has a mass of 1520 kg. While traveling at 20 m/s, the driver applies the brakes to stop the car on a wet surface with a 0.40 coefficient of friction. (a) How far does the car travel before stopping? (b) If a different car with a mass 1.5 times greater is on the road traveling at the same speed and the coefficient of friction between the road and the tires is the same, what will its stopping distance be? Explain your results.

47. * A 20-kg wagon accelerates on a horizontal surface at 0.50 m/s^2 when pulled by a rope exerting a 120-N force on the wagon at an angle of 25° above the horizontal. Determine the magnitude of the effective friction force exerted on the wagon and the effective coefficient of friction associated with this force.

48. * You want to use a rope to pull a 10-kg box of books up a plane inclined 30° above the horizontal. The coefficient of kinetic friction is 0.30. What force do you need to exert on the other end of the rope if you want to pull the box (a) at constant speed and (b) with a constant acceleration of 0.50 m/s^2 up the plane? The rope pulls parallel to the incline.

49. * A car with its wheels locked rests on a flatbed of a tow truck. The flatbed's angle with the horizontal is slowly increased. When the angle becomes 40°, the car starts to slide. Determine the coefficient of static friction between the flatbed and the car's tires.

50. * **Olympic skier** Olympic skier Lindsey Vonn skis down a steep slope that descends at an angle of 30° below the horizontal. The coefficient of sliding friction between her skis and the snow is 0.10. Determine Vonn's acceleration, and her speed 6.0 s after starting.

51. * **Another Olympic skier** Bode Miller, 80-kg downhill skier, descends a slope inclined at 20°. Determine his acceleration if the coefficient of friction is 0.10. How would this acceleration compare to that of a 160-kg skier going down the same hill? Justify your answer using sound physics reasoning.

52. * / A crate of mass m sitting on a horizontal floor is attached to a rope that pulls at an angle θ above the horizontal. The coefficient of static friction between the crate and floor is μ_s. (a) Construct a force diagram for the crate when being pulled by the rope but not sliding. (b) Determine an expression for the smallest force that the rope needs to exert on the crate that will cause the crate to start sliding.

53. * EST You absentmindedly leave your book bag on the top of your car. (a) Estimate the safe acceleration of the car needed for the bag to stay on the roof. Describe the assumptions that you made. (b) Estimate the safe speed. Describe the assumptions that you made.

54. A book slides off a desk that is tilted 15° degrees relative to the horizontal. What information about the book or the desk does this number provide?

55. * Block 1 is on a horizontal surface with a 0.29 coefficient of kinetic friction between it and the surface. A string attached to the front of block 1 passes over a light frictionless pulley and down to hanging block 2. Determine the mass of block 2 in terms of block 1 so that the blocks move at constant non-zero speed while sliding.

3.5 Projectile motion

56. * **Equation Jeopardy 6** The equations below describe a projectile's path. Solve for the unknowns and then invent a process that the equations might describe. There are many possibilities.

$$x = 0 + (20 \text{ m/s})(\cos 0°)t$$

$$0 = 8.0 \text{ m} + (20 \text{ m/s})(\sin 0°) - \frac{1}{2}(9.8 \text{ m/s}^2)t^2$$

57. / A bowling ball rolls off a table. Draw a force diagram for the ball when on the table and when in the air at two different positions.

58. / A ball moves in an arc through the air (see **Figure P3.58**). Construct a force diagram for the ball when at positions (a), (b), and (c). Ignore the resistive force exerted by the air on the ball.

Figure P3.58

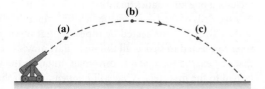

59. / A marble is thrown as a projectile at an angle above the horizontal. Draw its path during the flight. Choose six different positions along the path so that one of them is at the highest point. For each position, indicate the direction of the marble's velocity, acceleration, and all of the forces exerted on it by other objects.

60. A baseball leaves a bat and flies upward and toward center field. After it leaves the bat, are any forces exerted on the ball in the horizontal direction? In the direction of motion? In the vertical direction? If so, identify the other object that causes each force. Do not ignore air resistance.

61. * **Robbie Knievel ride** On May 20, 1999, Robbie Knievel easily cleared a narrow part of the Grand Canyon during a world-record–setting long-distance motorcycle jump—69.5 m. He left the jump ramp at a 10° angle above the horizontal. How fast was he traveling when he left the ramp? Indicate any assumptions you made.

62. * Daring Darless wishes to cross the Grand Canyon of the Snake River by being shot from a cannon. She wishes to be launched at 60° relative to the horizontal so she can spend more time in the air waving to the crowd. With what speed must she be launched to cross the 520-m gap?

63. ** A football punter wants to kick the ball so that it is in the air for 4.0 s and lands 50 m from where it was kicked. At what angle and with what initial speed should the ball be kicked? Assume that the ball leaves 1.0 m above the ground.

64. * A tennis ball is served from the back line of the court such that it leaves the racket 2.4 m above the ground in a horizontal direction at a speed of 22.3 m/s (50 mi/h). (a) Will the ball cross a 0.91-m-high net 11.9 m in front of the server? (b) Will the ball land in the service court, which is within 6.4 m of the net on the other side of the net?

65. * EST An airplane is delivering food to a small island. It flies 100 m above the ground at a speed of 160 m/s. (a) Where should the parcel be released so it lands on the island? Neglect

air resistance. (b) Estimate whether you should release the parcel earlier or later if there is air resistance. Explain.

66. * If you shoot a cannonball from the same cannon first at 30° and then at 60° relative to the horizontal, which orientation of the cannon will make the ball go farther? How do you know? Under what circumstances is your answer valid? Explain.

67. When you actually perform the experiment described in Problem 66, the ball shot at a 60° angle lands closer to the cannon than the ball shot at a 30° angle. Explain why this happens.

68. ** You can shoot an arrow straight up so that it reaches the top of a 25-m-tall building. (a) How far will the arrow travel if you shoot it horizontally while pulling the bow in the same way? The arrow starts 1.45 m above the ground. (b) Where do you need to put a target that is 1.45 m above the ground in order to hit it if you aim 30° above the horizontal? (c) Determine the maximum distance that you can move the target and hit it with the arrow.

69. * Robin Hood wishes to split an arrow already in the bull's-eye of a target 40 m away. If he aims directly at the arrow, by how much will he miss? The arrow leaves the bow horizontally at 40 m/s.

3.6 Using Newton's laws to explain everyday motion: Putting it all together

70. Three force diagrams for a car are shown in **Figure P3.70**. Indicate as many situations as possible for the car in terms of its velocity and acceleration at that instant for each diagram.

Figure P3.70

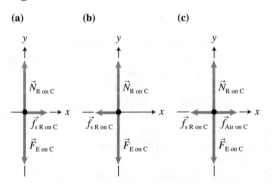

71. * A minivan of mass 1560 kg starts at rest and then accelerates at 2.0 m/s². (a) What is the object exerting the force on the minivan that causes it to accelerate? What type of force is it? (b) Air resistance and other opposing resistive forces are 300 N. Determine the magnitude of the force that causes the minivan to accelerate in the forward direction.

72. ** A daredevil motorcycle rider hires you to plan the details for a stunt in which she will fly her motorcycle over six school buses. Provide as much information as you can to help the rider successfully complete the stunt.

General Problems

73. * EST Estimate the range of the horizontal force that a sidewalk exerts on you during every step while you are walking. Indicate clearly how you made the estimate.

74. * Two blocks of masses m_1 and m_2 hang at the ends of a string that passes over the very light pulley with low friction

bearings shown in **Figure P3.74**. Determine an expression in terms of the masses and any other needed quantities for the magnitude of the acceleration of each block and the force that the string exerts on each block. Apply the equation for two cases: (a) the blocks have the same mass, but one is positioned lower than the other and (b) the blocks have different masses, but the heavier block is positioned higher than the light one. What assumptions did you make?

Figure P3.74

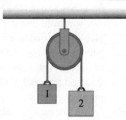

75. ** A 0.20-kg block placed on an inclined plane (angle 30° above the horizontal) is connected by a string going over a pulley to a 0.60-kg hanging block. Determine the acceleration of the system if there is no friction between the block and the surface of the inclined plane.

76. ** A 3.5-kg object placed on an inclined plane (angle 30° above the horizontal) is connected by a string going over a pulley to a 1.0-kg hanging block. (a) Determine the acceleration of the system if there is no friction between the object and the surface of the inclined plane. (b) Determine the magnitude of the force that the string exerts on both objects.

77. ** A 3.5-kg object placed on an inclined plane (angle 30° above the horizontal) is connected by a string going over a pulley to a 1.0-kg object. Determine the acceleration of the system if the coefficient of static friction between object 1 and the surface of the inclined plane is 0.30 and equals the coefficient of kinetic friction.

78. ** An object of mass m_1 placed on an inclined plane (angle θ above the horizontal) is connected by a string going over a pulley to a hanging object of mass m_2. Determine the acceleration of the system if there is no friction between object 1 and the surface of the inclined plane. If the problem has multiple answers, explore all of them.

79. ** An object of mass m_1 placed on an inclined plane (angle θ above the horizontal) is connected by a string going over a pulley to a hanging object of mass m_2. Determine the acceleration of the system if the coefficient of static friction between object 1 and the surface of the inclined plane is μ_s, and the coefficient of kinetic friction is μ_k. If the problem has multiple answers, explore all of them.

80. ** You are driving at a reasonable constant velocity in a van with a windshield tilted 120° relative to the horizontal (see **Figure P3.80**). As you pass under a utility worker fixing a power line, his wallet falls onto the windshield. Determine the acceleration needed by the van so that the wallet stays in place. When choosing your coordinate axes, remember that you want the wallet's acceleration to be horizontal rather than vertical. What assumptions and approximations did you make?

Figure P3.80

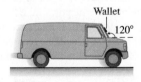

81. ** A ledge on a building is 20 m above the ground. A taut rope attached to a 4.0-kg can of paint sitting on the ledge passes up over a pulley and straight down to a 3.0-kg can of nails on the ground. If the can of paint is accidentally knocked off the ledge, what time interval does a carpenter have to catch the can of paint before it smashes on the floor?

82. ** **EST** **Bicycle ruined** The brakes on a bus fail as it approaches a turn. The bus was traveling at the speed limit before it moved about 24 m across grass and hit a brick wall. A bicycle attached to a rack on the front of the bus was crushed between the bus and the brick wall. There was little damage to the bus. Estimate the average force that the bicycle and bus exert on the wall while stopping. Indicate any assumptions made in your estimate.

83. * You are hired to devise a method to determine the coefficient of friction between the ground and the soles of a shoe and of its competitors. Explain your experimental technique and provide a physics analysis that could be used by others using this method.

84. * The mass of a spacecraft is about 480 kg. An engine designed to increase the speed of the spacecraft while in outer space provides 0.09-N thrust at maximum power. By how much does the engine cause the craft's speed to change in 1 week of running at maximum power? Describe any assumptions you made.

85. * **EST** A 60-kg rollerblader rolls 10 m down a 30° incline. When she reaches the level floor at the bottom, she applies the brakes. Use Newton's second law to estimate the distance she will move before stopping. Justify any assumptions you made.

86. Design, perform, and analyze the results of an experiment to determine the coefficient of static friction and the coefficient of kinetic friction between a penny and the cover of this textbook.

87. * **Tell all** A sled starts at the top of the hill shown in **Figure P3.87**. Add any information that you think is reasonable about the process that ensues when the sled goes down the hill and finally stops. Then tell everything you can about this process.

Figure P3.87

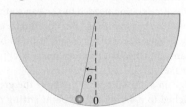

Reading Passage Problems

Professor tests airplane takeoff speed D. A. Wardle, a professor of physics from the University of Auckland, New Zealand, tested the takeoff speed of a commercial airliner. The pilot had insisted that the takeoff speed had to be 232 km/h.[1] To perform the testing experiment, Wardle used a pendulum attached to stiff cardboard (**Figure 3.16**). Prior to takeoff, when the plane was stationary, he marked the position of the pendulum bob on the cardboard to provide a vertical reference line (the dashed line in Figure 3.16). During the takeoff, he recorded the position of the bob at 5-s intervals. The results are shown in the table.

Figure 3.16 Wardle's device.

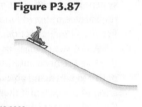

[1]The information is taken from the article by D. A. Wardle "Measurement of aeroplane takeoff speed and cabin pressure" published in *The Physics Teacher*, 37 410–411 (1999).

t (seconds)	θ (degrees)
0	9.9
5	14.8
10	13.8
15	13.0
20	12.0
25	11.4

Using these data, Professor Wardle determined the acceleration at takeoff to be greater than $g/4$. Then he plotted an acceleration-versus-time graph and used it to find the takeoff speed. It turned out to be about 201 km/h. He was very satisfied—the day was windy, and the speed of the breeze was about 15–20 km/h. Thus the takeoff speed predicted by his simple pendulum was 215–220 km/h, very close to what the pilot said.

88. Choose the best force diagram for the pendulum bob as the plane is accelerating down the runway (**Figure P3.88**).

Figure P3.88

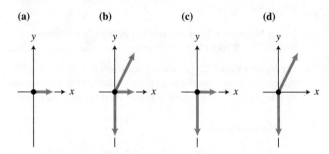

89. The professor used which of the following expression for the pendulum bob acceleration (θ is the angle of the pendulum bob string relative to the vertical)?
    (a) $a = g\sin\theta$ (b) $a = g\cos\theta$
    (c) $a = g\tan\theta$ (d) None of the choices

90. Approximately when did the peak acceleration occur?
    (a) 25 s   (b) 20 s   (c) 10 s   (d) 5 s

91. Approximately when did the peak speed occur?
    (a) 25 s   (b) 20 s   (c) 10 s   (d) 5 s

92. Choose the best velocity-versus-time graph below for the airplane (**Figure P3.92**).

Figure P3.92

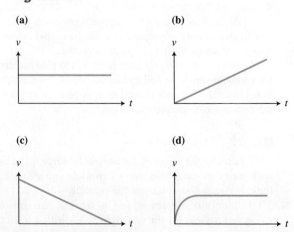

Ski jumping in Vancouver The 2010 Olympic ski jumping contest was held at Whistler Mountain near Vancouver (**Figure 3.17**). During a jump, a skier starts near the top of the in-run, the part down which the skier glides at increasing speed before the jump. The Whistler in-run is 116 m long and for the first part is tilted down at about 35° below the horizontal. There is then a curve that transitions into a takeoff ramp, which is tilted 11° below the horizontal. The skier flies off this ramp at high speed, body tilted forward and skis separated (**Figure 3.18**). This position exposes a large surface area to the air, which creates lift, extends the time of the jump, and allows the jumper to travel farther. In addition, the skier pushes off the exit ramp of the in-run to get a vertical component of velocity when leaving the ramp. The skier lands 125 m or more from the end of the in-run. The landing surface has a complex shape and is tilted down at about 35° below the horizontal. The skier moves surprisingly close (2 to 6 m) above the snowy surface for most of the jump. The coefficient of kinetic friction between the skis and the snow on the in-run is about 0.05 ± 0.02 and skiers' masses are normally small—about 60 kg. We can make some rough estimates about an idealized ski jump with an average in-run inclination of $(35° + 11°)/2 = 23°$.

Figure 3.17 Whistler Mountain ski jump in-run and landing area.

Figure 3.18 A ski jumper exposes a large surface area to the air in order to get more lift.

93. Which answer below is closest to the magnitude of the normal force that the idealized in-run exerts on the 60-kg skier?
    (a) 590 N   (b) 540 N   (c) 250 N   (d) 230 N
94. Which numbers below are closest to the magnitudes of the kinetic friction force and the component of the gravitational force parallel to the idealized inclined in-run?
    (a) 30 N, 540 N   (b) 27 N, 540 N   (c) 12 N, 540 N
    (d) 30 N, 230 N   (e) 27 N, 230 N   (f) 12 N, 230 N
95. Which answers below are closest to the magnitude of the skier's acceleration while moving down the idealized in-run and to the skier's speed when leaving its end?
    (a) 9.8 m/s², 48 m/s (b) 4.3 m/s², 32 m/s
    (c) 4.3 m/s², 28 m/s (d) 3.4 m/s², 32 m/s
    (e) 3.4 m/s², 28 m/s
96. Assume that the skier left the ramp moving horizontally. Treat the skier as a point-like particle and assume the force exerted by air on him is minimal. If he landed 125 m diagonally from the end of the in-run and the landing region beyond the in-run was inclined 35° below the horizontal for its entire length, which answer below is closest to the time interval that he was in the air?
    (a) 1.9 s   (b) 2.4 s   (c) 3.1 s
    (d) 3.8 s   (e) 4.3 s
97. Using the same assumptions as stated in Problem 96, which answer below is closest to the jumper's speed when leaving the in-run?
    (a) 37 m/s   (b) 31 m/s   (c) 27 m/s
    (d) 24 m/s   (e) 21 m/s
98. Which factors below would keep the skier in the air longer and contribute to a longer jump?
    1. The ramp at the end of the in-run is level instead of slightly tilted down.
    2. The skier extends his body forward and positions his skis in a V shape.
    3. The skier has wider and longer than usual skis.
    4. The skier pushes upward off the end of the ramp at the end of the in-run.
    5. The skier crouches in a streamline position when going down the in-run.
    (a) 1   (b) 5   (c) 1, 3, 5
    (d) 2, 3, 4   (e) 1, 2, 4, 5   (f) 1, 2, 3, 4, 5

4 Circular Motion

Why do pilots sometimes black out while pulling out at the bottom of a power dive?

Are astronauts really "weightless" while in orbit?

Why do you tend to slide across the car seat when the car makes a sharp turn?

Be sure you know how to:

- Find the direction of acceleration using a motion diagram (Section 1.6).
- Draw a force diagram (Section 2.1).
- Use a force diagram to help apply Newton's second law in component form (Sections 3.1 and 3.2).

Kruti Patel, a civilian test pilot, wears a special flight suit and practices special breathing techniques to prevent dizziness, disorientation, and possibly passing out as she comes out of a power dive. These symptoms characterize a blackout, which can occur when the head and brain do not received a sufficient amount of blood. Why would pulling out of a power dive cause a blackout, and why does a special suit prevent it from occurring? Our study of circular motion in this chapter will help us understand blackouts and other interesting phenomena.

In previous chapters we studied motion in situations in which the sum of the forces exerted on a system had a constant magnitude and direction. In fact, in real life we encounter relatively few situations where motion is this simple. More often the forces exerted on an object continually change direction and magnitude as time passes. In this chapter we will focus on circular motion. It is the simplest example of motion in which the sum of the forces exerted on a system object by other objects continually changes.

4.1 The qualitative velocity change method for circular motion

Consider the motion of the car shown in **Figure 4.1** as it travels at constant speed around a circular track. The instantaneous velocity of the car is tangent to the circle at every point. Recall from Chapter 1 that an object has acceleration if its velocity changes—in magnitude *or* in direction or in both. Even though the car is moving at constant speed, it is accelerating because the *direction* of its velocity changes from moment to moment.

To estimate the direction of acceleration of any object (including that of the car) while passing a particular point on its path (see **Figure 4.2a**), we use the **velocity change method**. Consider a short time interval $\Delta t = t_f - t_i$ during which the car passes that point. The velocity arrows $\vec{v}_i$ and $\vec{v}_f$ represent the initial and final velocities of the car, a little before and a little after the point where we want to estimate its acceleration (note that the velocity arrows are tangent to the curve in the direction of motion). The velocity change vector is $\Delta \vec{v} = \vec{v}_f - \vec{v}_i$. You can think of $\Delta \vec{v}$ as the vector that you need to add to the initial velocity $\vec{v}_i$ in order to get the final velocity $\vec{v}_f$, that is, $\vec{v}_i + \Delta \vec{v} = \vec{v}_f$. Rearranging this, we get $\Delta \vec{v} = \vec{v}_f - \vec{v}_i$ (the change in velocity).

To estimate $\Delta \vec{v}$, place the $\vec{v}_i$ and $\vec{v}_f$ arrows tail to tail (Figure 4.2b) without changing their magnitudes or directions. $\Delta \vec{v}$ starts at the head of $\vec{v}_i$ and ends at the head of $\vec{v}_f$. The car's acceleration $\vec{a}$ is in the direction of the $\Delta \vec{v}$ arrow (Figure 4.2c) and is the ratio of the velocity change and the time interval needed for that change:

$$\vec{a} = \frac{\Delta \vec{v}}{\Delta t}$$

Figure 4.1 Since the direction of the velocity is changing, the car is accelerating.

The magnitude of the car's velocity is constant but its direction is changing.

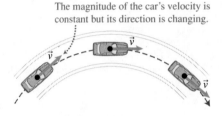

‹ Active Learning Guide

Figure 4.2 Estimating the direction of acceleration during two-dimensional motion.

(a)

What is $\vec{a}$ as the car passes this point?

t_i　　　　t_f

(b)

Place the $\vec{v}_i$ and $\vec{v}_f$ arrows tail-to-tail. Draw a $\Delta \vec{v}$ velocity change arrow from the head of $\vec{v}_i$ to the head of $\vec{v}_f$.

$\vec{v}_f$

$\Delta \vec{v}$

$\vec{v}_i$

$\vec{v}_i + \Delta \vec{v} = \vec{v}_f$
or
$\Delta \vec{v} = \vec{v}_f - \vec{v}_i$

(c)

The acceleration arrow $\vec{a}$ is in the direction of $\Delta \vec{v}$.

$$\vec{a} = \frac{\Delta \vec{v}}{\Delta t} = \frac{\vec{v}_f - \vec{v}_i}{t_f - t_i}$$

> **TIP** When using this diagrammatic method to estimate the acceleration direction during circular motion, make sure that you choose initial and final points at the same distance before and after the point at which you are estimating the acceleration direction. Draw long velocity arrows so that when you put them tail to tail, you can clearly see the direction of the velocity change arrow. Also, be sure that the velocity change arrow points from the head of the initial velocity to the head of the final velocity.

CONCEPTUAL EXERCISE 4.1 Direction of racecar's acceleration

Determine the direction of the racecar's acceleration at points A, B, and C in the figure below as the racecar travels at constant speed on the circular path.

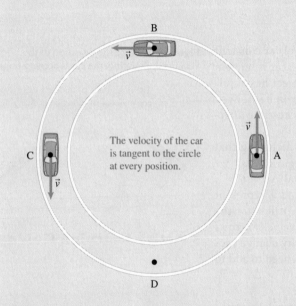

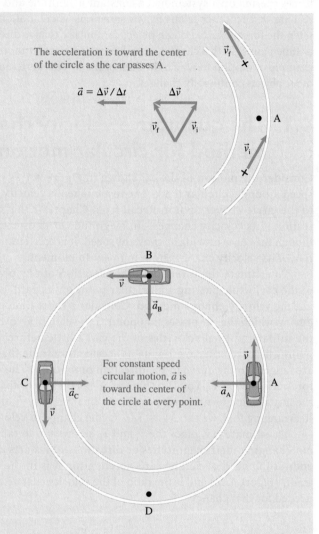

The acceleration is toward the center of the circle as the car passes A.

$$\vec{a} = \Delta\vec{v} / \Delta t \qquad \Delta\vec{v}$$

Sketch and translate A top view of the car's path is shown above. We are interested in the car's acceleration as it passes points A, B, and C.

Simplify and diagram To find the direction of the car's acceleration at each point, we use the velocity change method (shown for point A above, right). When done for all three points, notice that a pattern emerges: the acceleration at different points along the car's path has a different direction, but in every case it points toward the center of the circular path.

Try it yourself: Find the direction of the car's acceleration at point D on the track.

Answer: The car's acceleration points toward the center of the circle here as well.

In Conceptual Exercise 4.1 we found an important pattern: when an object is moving in a circle at *constant speed*, its acceleration at any position points toward the center of the circle. This acceleration is called **radial acceleration**.

> **Review Question 4.1** Why is it true that when an object is moving in a circle at constant speed, its acceleration at any point points toward the center of the circle?

4.2 Qualitative dynamics of circular motion

Newton's second law tells us that an object's acceleration during linear motion is caused by the forces exerted on the object and is in the direction of the vector sum of all forces ($\vec{a} = \Sigma\vec{F}/m$). By applying this idea to constant speed circular motion, we can devise the following hypothesis:

> *Hypothesis* The sum of the forces exerted on an object moving at constant speed along a circular path points toward the center of that circle in the same direction as the object's acceleration.

Notice that the hypothesis mentions only the sum of the forces pointing toward the center (and no forces in the direction of motion). Consider the two experiments described in Testing Experiment **Table 4.1**. Are the net forces exerted on these objects consistent with the above hypothesis?

TESTING EXPERIMENT TABLE

4.1 **Does the net force exerted on an object moving at constant speed in a circle point toward the center of the circle?**

Testing experiment	Predictions based on hypothesis	Outcome
Experiment 1. You swing a pail at the end of a rope in a horizontal circle at a constant speed.	We predict that as the pail passes any point along its path, the sum of the forces exerted on the pail by other objects should point toward the center of the circle—in the direction of the acceleration.	The vertical force components balance; the sum of the forces points along the radial axis toward the center of the circle in agreement with the prediction. No force pushes the pail in the direction of motion.
Side view 		Side view 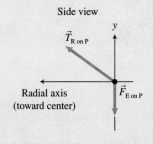
Experiment 2. A metal ball rolls in a circle on a flat, smooth surface against the inside wall of a metal ring.	We predict that the sum of the forces that other objects exert on the ball points toward the center of the circle—in the direction of the acceleration.	The vertical force components balance. The sum of the forces points along the radial axis toward the center of the circle in agreement with the prediction. No force pushes in the direction of motion.
Top view 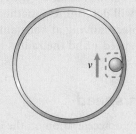		Side view 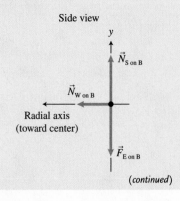

(continued)

Conclusion

In both cases, the sum of the forces exerted on the system object by other objects points toward the center of the circle in the direction of the acceleration—consistent with the hypothesis.

The results of our experiments were consistent with our hypothesis. If you were to repeat the analysis for other points on the path, you would get the same result. This outcome is consistent with Newton's second law—an object's acceleration equals the sum of the forces (the net force) that other objects exert on it divided by the mass of the object:

$$\vec{a} = \frac{\vec{F}_{1\,\text{on object}} + \vec{F}_{2\,\text{on object}} + \cdots}{m_{\text{object}}}$$

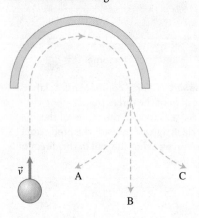

Figure 4.3 Top view of a ball rolling toward a circular ring.

> **TIP** Notice that when the object moves at constant speed along the circular path, the net force has no tangential component.

Review Question 4.2 A ball rolls at a constant speed on a horizontal table toward a semicircular barrier as shown in **Figure 4.3**. Is there a nonzero net force exerted on the ball (1) before it contacts the barrier, (2) while it is in contact with the barrier, and (3) after it no longer contacts the barrier? If so, what is the direction of the net force? When the ball leaves the barrier, in what direction will it move: A, B, or C?

4.3 Radial acceleration and period

So far we have learned how to qualitatively determine the direction of the acceleration of an object moving in a circle and how that acceleration relates to the forces exerted on the object. Let's now look at how to determine the magnitude of the object's acceleration. We will begin by thinking about factors that might affect its acceleration.

Imagine a car following the circular curve of a highway. Our experience indicates that the faster a car moves along a highway curve, the greater the risk that the car will skid off the road. So the car's speed v matters. Also, the tighter the turn, the greater the risk that the car will skid. So the radius r of the curve also matters. In this section we will determine a mathematical expression relating the acceleration of an object moving at constant speed in a circular path to these two quantities (its speed v and the radius r of the circular path).

Dependence of acceleration on speed

Let's begin by investigating the dependence of the acceleration on the object's speed. In Observational Experiment **Table 4.2**, we use the diagrammatic velocity change method to investigate how the acceleration differs for objects moving at speeds v, $2v$, and $3v$ while traveling along the same circular path of radius r.

OBSERVATIONAL EXPERIMENT TABLE

4.2   How does an object's speed affect its radial acceleration during constant speed circular motion?

Observational experiment	Analysis using velocity change method
Experiment 1. An object moves in a circle at constant speed.	The acceleration is toward the center of the circular path.

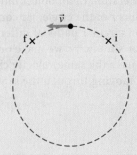
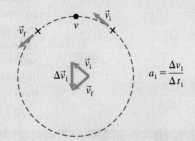

$$a_1 = \frac{\Delta v_1}{\Delta t_1}$$

Experiment 2. An object moves in the same circle at a constant speed that is twice as fast as in Experiment 1.

When the object moves twice as fast between the same two points on the circle, the velocity change doubles. In addition, the velocity change occurs in one-half the time interval since it is moving twice as fast. Hence, the acceleration increases by a factor of 4.

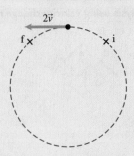
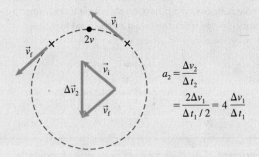

$$a_2 = \frac{\Delta v_2}{\Delta t_2}$$
$$= \frac{2\Delta v_1}{\Delta t_1 / 2} = 4\frac{\Delta v_1}{\Delta t_1}$$

Experiment 3. An object moves in the same circle at a constant speed that is three times as fast as in Experiment 1.

Tripling the speed triples the velocity change and reduces by one-third the time interval needed to travel between the points—the acceleration increases by a factor of 9.

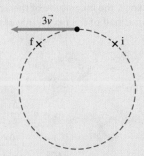
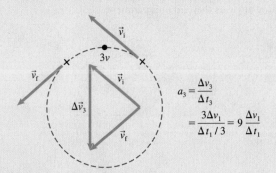

$$a_3 = \frac{\Delta v_3}{\Delta t_3}$$
$$= \frac{3\Delta v_1}{\Delta t_1 / 3} = 9\frac{\Delta v_1}{\Delta t_1}$$

Pattern

We find that doubling the speed of the object results in a fourfold increase of its radial acceleration; tripling the speed leads to a ninefold increase. Therefore, the radial acceleration of the object is proportional to its speed squared.

$$a_r \propto v^2$$

From the pattern in Table 4.2 we conclude that the magnitude of radial acceleration is proportional to the speed squared. We can express this pattern mathematically:

$$a_r \propto v^2 \tag{4.1}$$

Dependence of acceleration on radius

To find how the magnitude of acceleration of an object moving in a circle at constant speed depends on the radius r of the circle, we consider two objects moving with the same speed but on circular paths of different radii (Observational Experiment **Table 4.3**). For simplicity, we make one circle twice the radius of the other. We arrange to have the same velocity change for the two objects by considering them while moving through the same angle θ (rather than through the same distance).

OBSERVATIONAL EXPERIMENT TABLE

4.3   How does the acceleration depend on the radius of the curved path?

Observational experiment	Analysis using velocity change method
Experiment 1. An object moves in a circle of radius r at speed v. Choose two points on the circle to examine the velocity change from the initial to the final location.	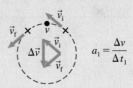 $a_1 = \dfrac{\Delta v}{\Delta t_1}$
Experiment 2. An object moves in a circle of radius $2r$ at speed v. Choose the points for the second experiment so that the velocity change is the same as in Experiment 1. This occurs if the radii drawn to the location of the object at the initial position and to the final position make the same angle as they did in Experiment 1. 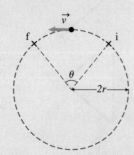	To have the same velocity change as in Experiment 1, the object has to travel twice the distance, as the radius is twice as long. Since the speed of the object is the same as in the first experiment, it takes the object twice as long to travel that distance. Hence, the magnitude of the acceleration in this experiment is half of the magnitude of the acceleration in Experiment 1. 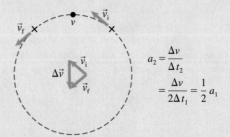 $a_2 = \dfrac{\Delta v}{\Delta t_2}$ $= \dfrac{\Delta v}{2\Delta t_1} = \dfrac{1}{2}\, a_1$

Pattern

When an object moves in a circle at constant speed, its acceleration decreases by half when the radius of the circular path doubles—the bigger the circle, the smaller the acceleration. It appears that the acceleration of the object is proportional to the inverse of the radius of its circular path:

$$a_r \propto \frac{1}{r}$$

If we carried out the same thought experiment with circular paths of other radii, we would get similar results. The magnitude of the radial component of the acceleration of an object moving in a circular path is inversely proportional to the radius of the circle:

$$a_r \propto \frac{1}{r} \qquad\qquad (4.2)$$

We now combine the two proportionalities in Eqs. (4.1) and (4.2):

$$a_r \propto \frac{v^2}{r}$$

If we were to do a detailed mathematical derivation, we would find that the constant of proportionality is 1. Thus, we can turn this into an equation for the magnitude of the radial acceleration.

Radial acceleration For an object moving at constant speed v on a circular path of radius r, the magnitude of the radial acceleration is

$$a_r = \frac{v^2}{r} \qquad\qquad (4.3)$$

The acceleration points toward the center of the circle. The units for v^2/r are the correct units for acceleration:

$$\left(\frac{m^2/s^2}{m}\right) = \frac{m}{s^2}$$

This expression for radial acceleration agrees with our everyday experience. When a car is going around a highway curve at high speed, a large v^2 in the numerator leads to its large acceleration. When it is going around a sharp turn, the small radius r in the denominator also leads to a large acceleration.

Consider a limiting case: when the radius of the circular path is infinite—equivalent to the object moving in a straight line. The velocity of an object moving at constant speed in a straight line is constant, and its acceleration is zero. Notice that the acceleration in Eq. (4.3) is zero if the radius in the denominator is infinite.

‹Active Learning Guide

EXAMPLE 4.2 Blackout

When a fighter pilot pulls out from the bottom of a power dive, his body moves at high speed along a segment of an upward-bending approximately circular path. However, while his body moves up, his blood tends to move straight ahead (tangent to the circle) and begins to fill the easily expandable veins in his legs. This can deprive his brain of blood and cause a blackout if the radial acceleration is 4 g or more and lasts several seconds. Suppose during a dive an airplane moves at a modest speed of $v = 80$ m/s (180 mi/h) through a circular arc of radius $r = 150$ m. Is the pilot likely to black out?

Sketch and translate We sketch the situation above, right. To determine if blackout occurs, we will estimate the radial acceleration of the pilot as he passes along the lowest point of the circular path.

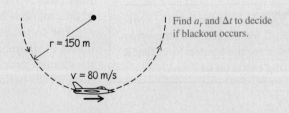

Find a_r and Δt to decide if blackout occurs.

Simplify and diagram Assume that at the bottom of the plane's power dive, the plane is moving in a circle at constant speed and the point of interest is the lowest point on the circle. Assume also that the magnitude of the plane's acceleration is constant for a quarter circle, that is, one-eighth of a circle on each side of the bottom point. The acceleration points up (see the velocity change method shown on the next page).

(continued)

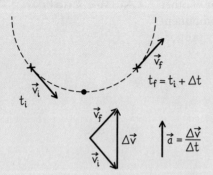

Represent mathematically We calculate the magnitude of the radial acceleration using Eq. (4.3):

$$a_r = \frac{v^2}{r}$$

To estimate the time interval that the pilot will experience this acceleration, we calculate the time interval for the plane to move through the part of the power dive represented by a quarter circle (arc length $l = 2\pi r/4$):

$$\Delta t = \frac{l}{v} = \frac{2\pi r/4}{v}$$

Solve and evaluate Inserting the given quantities, we find:

$$a_r = \frac{(80 \text{ m/s})^2}{150 \text{ m}} = 42 \text{ m/s}^2$$

The acceleration is more than four times greater than g (more than $4 \times 9.8 \text{ m/s}^2$). Thus, the pilot could potentially black out if the acceleration lasts too long. The time interval during which the pilot will experience this acceleration is about

$$\Delta t = \frac{[2\pi(150 \text{ m})]/4}{80 \text{ m/s}} = \frac{240 \text{ m}}{80 \text{ m/s}} \approx 3 \text{ s}$$

This is long enough that blackout is definitely a concern. Special flight suits are made that exert considerable pressure on the legs during such motion. This pressure prevents blood from accumulating in the veins of the legs.

Try it yourself: Imagine that you are a passenger on a roller coaster with a double loop-the-loop—two consecutive loops. You are traveling at speed 24 m/s as you move along the bottom part of a loop of radius 10 m. Determine (a) the magnitude of your radial acceleration while passing the lowest point of that loop and (b) the time interval needed to travel along the bottom quarter of that approximately circular loop. (c) Are you at risk of having a blackout? Explain.

Answer: (a) 58 m/s² or almost six times the 9.8 m/s² free-fall acceleration; (b) about 0.7 s; (c) the acceleration is more than enough to cause blackout but does not last long enough—you are safe from blacking out and will get a good thrill!

Period

When an object repeatedly moves in a circle, we can describe its motion with another useful physical quantity, its **period** T. The period equals the time interval that it takes an object to travel around the entire circular path one time and has the units of time. For example, suppose that a bicyclist racing on a circular track takes 24 s to complete a circle that is 400 m in circumference. The period T of the motion is 24 s.

For constant speed circular motion we can determine the speed of an object by dividing the distance traveled in one period (the circumference of the circular path, $2\pi r$) by the time interval T it took the object to travel that distance (its period), or $v = 2\pi r/T$. Thus,

$$T = \frac{2\pi r}{v} \qquad (4.4)$$

TIP Do not confuse the symbol T for period with the symbol T for the tension force that a string exerts on an object.

We can express the radial acceleration of the object in terms of its period by inserting this special expression for speed in terms of period $v = 2\pi r/T$ into Eq. (4.3):

$$a_r = \frac{v^2}{r} = \left(\frac{2\pi r}{T}\right)^2 \frac{1}{r} = \frac{4\pi^2 r^2}{T^2 r} = \frac{4\pi^2 r}{T^2} \qquad (4.5)$$

Let's use limiting case analysis to see if Eq. (4.5) is reasonable. For example, if the speed of the object is extremely large, its period would be very short ($T = 2\pi r/v$). Thus, according to Eq. (4.5), its radial acceleration would be very large. That makes sense—high speed and large radial acceleration. Similarly, if the speed of the object is small, its period will be very large and its radial acceleration will be small. That also makes sense.

QUANTITATIVE EXERCISE 4.3
Singapore hotel

What is your radial acceleration when you sleep in a hotel in Singapore at Earth's equator? Remember that Earth turns on its axis once every 24 hours and everything on its surface actually undergoes constant speed circular motion with a period of 24 hours. A picture of Earth with you as a point at the equator is shown below.

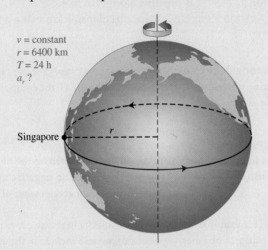

$v = $ constant
$r = 6400$ km
$T = 24$ h
a_r ?

Singapore r

Represent mathematically Since you are in constant speed circular motion, Eq. (4.5) can be used to determine your radial acceleration. Your period T is the time interval needed to travel once in this circle (24 h). Thus the magnitude of your radial acceleration is

$$a_r = \frac{v^2}{r} = \frac{4\pi^2 r}{T^2}$$

Solve and evaluate At the equator, $r = 6400$ km and $T = 24$ h. So, making the appropriate unit conversions,

we get the magnitude of the acceleration:

$$a_r = \frac{4\pi^2 (6.4 \times 10^6 \text{m})}{(24 \text{ h} \times 3600 \text{ s/h})^2} = 0.034 \text{ m/s}^2$$

Is this result reasonable? Compare it to the much greater free-fall acceleration of objects near Earth's surface—9.8 m/s². Your radial acceleration due to Earth's rotation when in Singapore is tiny by comparison. Because it is so small, the radial acceleration due to Earth's rotation around its axis can be ignored under most circumstances.

Try it yourself: Use the figure below to estimate what your radial acceleration would be if you were living in Anchorage, Alaska.

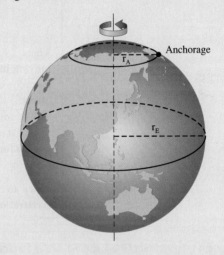

Anchorage

r_A

r_E

Answer: About 0.016 m/s². Note that the period is the same, but the distance of Anchorage from the axis of Earth's rotation (the radial distance used in this estimate) is about one-half of Earth's radius.

Remember that Newton's laws were formulated and are valid only for observers in inertial reference frames, that is, for observers who are not accelerating. But we have just discovered that observers on Earth's surface are accelerating due to Earth's rotation. Does this mean that Newton's laws do not apply? As we found in Quantitative Exercise 4.3, the acceleration due to Earth's rotation is much smaller than the accelerations we experience from other types of motion. Thus, in most situations we can assume that Earth is not rotating and therefore *does* count as an inertial reference frame. This means that Newton's laws do apply with a high degree of accuracy when using Earth's surface as a reference frame.

Review Question 4.3 Use dimensional analysis (inspecting the units) to evaluate whether or not the two expressions for radial acceleration (v^2/r and $4\pi^2 r/T^2$) have the correct units of acceleration.

4.4 Skills for analyzing processes involving circular motion

The strategy for analyzing processes involving constant speed circular motion is similar to that used to analyze linear motion processes. However, when we analyze constant speed circular motion processes, we do *not* use traditional x- and y-axes but instead use a radial r-axis and sometimes a vertical y-axis. The radial r-axis should always point *toward the center of the circle*. The radial acceleration of magnitude $a_r = v^2/r$ is positive along this axis.

The application of Newton's second law for circular motion using a radial axis is summarized below.

Circular motion component form of Newton's second law For the radial direction (the axis pointing toward the center of the circular path), the component form of Newton's second law is

$$a_r = \frac{\Sigma F_r}{m} \quad \text{or} \quad ma_r = \Sigma F_r \tag{4.6}$$

where ΣF_r is the sum of the radial components of all forces exerted on the object moving in the circle (positive toward the center of the circle and negative away from the center) and $a_r = v^2/r$ is the magnitude of the radial acceleration of the object.

For some situations (for example, a car moving around a highway curve or a person standing on the platform of a merry-go-round), we also include in the analysis the force components along a perpendicular vertical y-axis:

$$ma_y = \Sigma F_y = 0 \tag{4.7}$$

When an object moves with uniform circular motion, both the y-component of its acceleration and the y-component of the net force exerted on it are zero.

PROBLEM-SOLVING STRATEGY **Processes involving constant speed circular motion**

The following example explains how to solve circular motion problems. We describe the general strategy on the left side of the table and apply it specifically to the problem in Example 4.4 on the right side.

EXAMPLE 4.4 Driving over a hump in the road
Josh drives his car at a constant 12 m/s speed over a bridge whose shape is bowed in the vertical arc of a circle. Find the direction and the magnitude of the force exerted by the car seat on Josh as he passes the top of the 30-m-radius arc. His mass is 60 kg.

Sketch and translate

- Sketch the situation described in the problem statement. Label it with all relevant information.

- Choose a system object and a specific position to analyze its motion.

The sketch includes all of the relevant information: Josh's speed, his mass, and the radius of the arc along which the car moves. Josh is the system object.

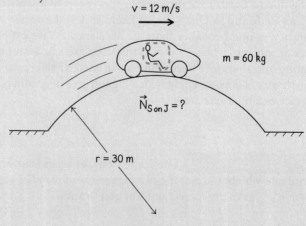

Simplify and diagram

- Decide if the system can be modeled as a point-like object.

- Determine if the constant speed circular motion approach is appropriate.

- Indicate with an arrow the direction of the object's acceleration as it passes the chosen position.

- Draw a force diagram for the system object at the instant it passes that position.

- On the force diagram, draw an axis in the radial direction toward the center of the circle.

Consider Josh as a point-like object and analyze him as he passes the highest part of the vertical circle along which he travels.

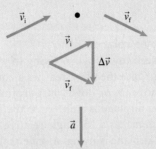

A velocity change diagram as he passes the top of the circular path indicates that he has a downward radial acceleration toward the center of the circle.

   The forces exerted on him are shown in the force diagram. The net force must also point down. Thus, the upward normal force $\vec{N}_{\text{S on J}}$ that the car seat exerts on Josh is smaller in magnitude than the downward gravitational force $\vec{F}_{\text{E on J}}$ that Earth exerts on Josh. A radial r-axis points down toward the center of the circle.

Represent mathematically

- Convert the force diagram into the radial r-component form of Newton's second law.

- For objects moving in a horizontal circle (unlike this example), you may also need to apply a vertical y-component form of Newton's second law.

Apply the radial form of Newton's second law $ma_r = \Sigma F_r$.

   The radial components of the two forces exerted on Josh are $F_{\text{E on J }r} = +F_{\text{E on J}} = +mg$ and $N_{\text{S on J }r} = -N_{\text{S on J}}$. The magnitude of the radial acceleration is $a_r = v^2/r$. Therefore, the radial form of Newton's second law is

$$m\frac{v^2}{r} = +mg + (-N_{\text{S on J}})$$

Solving for $N_{\text{S on J}}$, we have

$$N_{\text{S on J}} = mg - m\frac{v^2}{r}$$

Solve and evaluate

- Solve the equations formulated in the previous two steps.

- Evaluate the results to see if they are reasonable (the magnitude of the answer, its units, limiting cases).

Substituting the known information into the previous equation, we get

$$N_{\text{S on J}} = (60\,\text{kg})(9.8\,\text{m/s}^2) - (60\,\text{kg})\frac{(12\,\text{m/s})^2}{(30\,\text{m})} = 588\,\text{N} - 288\,\text{N} = 300\,\text{N}$$

The seat exerts a smaller upward force on Josh than Earth pulls down on him. You have probably noticed this effect when going over a smooth hump in a roller coaster or while in a car or on a bicycle when crossing a hump in the road—it almost feels like you are leaving the seat or starting to float briefly above it. This feeling is caused by the reduced upward normal force that the seat exerts on you.

Try it yourself: Imagine that Josh drives on a road that has a dip in it. The speed of the car and the radius of the dip are the same as in Example 4.4. Find the direction and the magnitude of the force exerted by the car seat on Josh as he passes the bottom of the 30-m-radius dip.

Answer: The seat exerts an upward force of 880 N. This is almost 50% more than the 590-N force that Earth exerts on Josh—he sinks into the seat.

The next three examples involve processes in which some of the forces have nonzero radial *r*-components *and* some of the forces have nonzero vertical *y*-components. To solve these problems we will use the component form of Newton's second law in both the radial *r*- and vertical *y*-directions.

Active Learning Guide ➤

EXAMPLE 4.5   Toy airplane

You have probably seen toy airplanes flying in a circle at the end of a string. Once the plane reaches a constant speed, it does not move up or down, just around in a horizontal circle. Our airplane is attached to the end of a 46-cm string, which makes a 25° angle relative to the horizontal while the airplane is flying. A scale at the top of the string measures the force that the string exerts on the airplane. Predict the period of the airplane's motion (the time interval for it to complete one circle).

Sketch and translate We sketch the situation, including the known information: the length of the string and the angle of the string relative to the horizontal. The airplane is the system object.

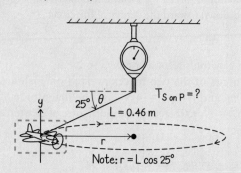

Note: $r = L \cos 25°$

Simplify and diagram Neglect the airplane's interaction with the air and assume that it is a point-like particle moving at constant speed in a horizontal circle. To draw a force diagram, identify objects that interact with the airplane—in this case, Earth and the string. Include a vertical *y*-axis and a radial *r*-axis; note that the radial axis points horizontally toward the center of the circle—not along the string. Because the airplane moves at constant speed, its acceleration points toward the center of the circle—it has a radial component but does not have a vertical component. Thus, the vertical component of the tension force that the string exerts on the plane $\vec{T}_{S \text{ on } P\, y}$ must balance the downward force that Earth exerts on the plane $\vec{F}_{E \text{ on } P}$, which has magnitude $m_{P}g$ (see the force diagram broken into components at right). The radial component of the force that the string exerts on the plane is the only force with a nonzero radial component and causes the plane's radial acceleration.

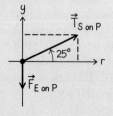

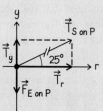

Represent mathematically Now, we use the force diagram to help apply Newton's second law in component form. First, apply the vertical *y*-component form of Newton's second law: $m_{P}a_{y} = \Sigma F_{y}$. The *y*-components of the forces exerted on the plane are $T_{S \text{ on } P\, y} = +T_{S \text{ on } P} \sin 25°$ and $F_{E \text{ on } P\, y} = -m_{P}g$, and the *y*-component of the plane's acceleration is zero. We can now apply the *y*-component form of Newton's second law:

$$0 = +T_{S \text{ on } P} \sin 25° + (-m_{P}g)$$

Next, apply the radial *r*-component version of Newton's second law: $m_{P}a_{r} = \Sigma F_{r}$. The force exerted by Earth on the plane does not have a radial component. The radial component of the force exerted by the string on the plane is $T_{S \text{ on } P\, r} = +T_{S \text{ on } P} \cos 25°$, and the magnitude of the radial acceleration is $a_{r} = 4\pi^{2}r/T^{2}$ [Eq. (4.5)]. Thus:

$$m_{P}\frac{4\pi^{2}r}{T^{2}} = +T_{S \text{ on } P} \cos 25°$$

Note that $T_{S \text{ on } P}$ represents the magnitude of the tension force that the string exerts on the airplane and T represents the period of the plane's motion—the time interval it takes the plane to travel once around its circular path. Note also that the radius of the circular path is not the length of the string but is instead

$$r = L \cos 25° = (0.46 \text{ m})\cos 25° = 0.417 \text{ m}$$

where *L* is the length of the string (46 cm).

Solve and evaluate The goal is to predict the period *T* of the plane's circular motion. Unfortunately, the above radial application of Newton's second law has three unknowns: $T_{S \text{ on } P}$, *T*, and the mass of the plane. However, we can rearrange the vertical *y*-equation to get an expression for the magnitude of the tension force exerted by the string on the plane:

$$T_{S \text{ on } P} = \frac{m_{P}g}{\sin 25°}$$

Insert this expression for $T_{S \text{ on } P}$ into the radial component equation to get an expression that does *not* involve the force that the string exerts on the plane:

$$m_{P}\frac{4\pi^{2}r}{T^{2}} = \frac{m_{P}g}{\sin 25°} \cos 25°$$

Divide both sides by m_{P}, multiply both sides by T^{2}, and do some rearranging to get

$$T^{2} = \frac{4\pi^{2}r \sin 25°}{g \cos 25°}$$

Taking the square root of each side, we find that

$$T = 2\pi\sqrt{\frac{r\sin 25°}{g\cos 25°}}$$

Substituting the known information, we predict the period to be

$$T = 2\pi\sqrt{\frac{(0.417\text{ m})\sin 25°}{(9.8\text{ m/s}^2)\cos 25°}} = 0.88\text{ s}$$

Try it yourself: Imagine that another plane of mass 0.12 kg moves with a speed of 6.2 m/s. A spring scale measures the tension force that the string exerts on the plane as 5.0 N. Predict the radius of the plane's orbit.

Answer: 0.95 m.

Many carnivals and amusement parks have a "rotor ride" in which people stand up against the wall of a spinning circular room. The room (also called the "drum") spins faster until, at a certain speed, the floor drops out! Amazingly, the people remain up against the wall with their feet dangling. How is this possible?

EXAMPLE 4.6 Rotor ride

A 62-kg woman is a passenger in a rotor ride. A drum of radius 2.0 m rotates faster and faster about a vertical axis until it reaches a period of 1.7 s. When the drum reaches this turning rate, the floor drops away but the woman does not slide down the wall of the drum. Imagine that you were one of the engineers who designed this ride. Which characteristics of the ride would ensure that the woman remained stuck to the wall? Justify your answer quantitatively.

Sketch and translate Representing the situation in different ways should help us decide which characteristics of the drum are important. We start with a sketch along with all the relevant information. The woman will be the system. Her mass is $m_W = 62$ kg, the radius of her circular path is $r = 2.0$ m, and the period of her circular motion is $T = 1.7$ s.

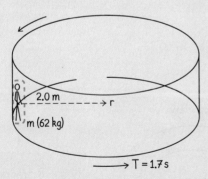

Simplify and diagram Model the woman as a point-like object and consider the situation once the drum has reached its maximum (constant) speed. Because she is moving in a circular path at constant speed, her acceleration points toward the center of the circle and we can use our understanding of constant speed circular motion to analyze the situation.

Next, draw a force diagram for the woman as she passes one point along the circular path. She interacts with

two objects—Earth and the drum. Earth exerts a downward gravitational force $\vec{F}_{E\text{ on }W}$ on her. The drum exerts a force that we can resolve into two vector components: a normal force $\vec{N}_{D\text{ on }W}$ perpendicular to the drum's surface and toward the center of the circle and an upward static friction force $\vec{f}_{s\,D\text{ on }W}$ parallel to its surface. Include a radial r-axis that points toward the center of the drum and a vertical y-axis pointing upward. Examining the force diagram, we see that if the maximum upward static friction force is less than the downward gravitational force, the woman will slip. Thus the engineer's rotor ride design problem can be formulated as follows. How large must the coefficient of static friction be so that the static friction force exerted on the woman by the drum balances the force that Earth exerts on her?

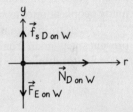

Represent mathematically Use the force diagram to help apply the component form of Newton's second law for the inward radial r-direction and for the vertical y-direction. The normal force is the only force that has a nonzero radial component.

Radial r-equation: $\quad m_W a_r = \Sigma F_r$

$$\Rightarrow m_W\frac{v^2}{r} = N_{D\text{ on }W}$$

Now, consider the y-component form of Newton's second law. The static friction force $\vec{f}_{s\,D\text{ on }W}$ points in the positive y-direction, and the gravitational force $\vec{F}_{E\text{ on }W}$ points in the negative y-direction. Since these forces must balance for the woman not to slip, the y-component of the woman's acceleration should be zero.

Vertical y-equation: $\quad m_W a_y = \Sigma F_y$

$$\Rightarrow 0 = f_{s\,D\text{ on }W} + (-F_{E\text{ on }W})$$

Assume that the static friction force is at its maximum possible value, meaning $f_{s\,D\text{ on }W} = \mu_s N_{D\text{ on }W}$. We wish

(continued)

to find the smallest (minimum) coefficient of static friction $\mu_{s\,min}$ that results in the woman remaining stationary. Substitute this minimum coefficient of static friction and the earlier expression for the normal force exerted by the drum into this expression:

$$f_{s\,max\,D\,on\,W} = \mu_{s\,min}N_{D\,on\,W} = \mu_{s\,min}m_W\frac{v^2}{r}$$

Substituting this expression for $f_{s\,max\,D\,on\,W}$ and $F_{E\,on\,W} = m_Wg$ into the vertical y-component application of Newton's second law, we get

$$0 = \mu_{s\,min}m_W\frac{v^2}{r} - m_Wg$$

Notice that the woman's mass cancels out of this equation. This means that the rotor ride works equally well for any person, independent of the person's mass and for many people at once as well!

To determine the minimum value of μ_s needed to prevent the woman from sliding, we divide both sides by m_Wv^2 and multiply by r and rearrange to get

$$\mu_{s\,min} = \frac{gr}{v^2}.$$

We find the woman's speed by using the information we have about the radius and period of her circular motion:

$$v = \frac{2\pi r}{T}.$$

Plug this expression for the speed into the equation above for the minimum coefficient of friction:

$$\mu_{s\,min} = \frac{gr}{v^2} = \frac{grT^2}{4\pi^2r^2} = \frac{gT^2}{4\pi^2r}$$

Solve and evaluate We can now use the given information to find the minimum coefficient of friction:

$$\mu_{s\,min} = \frac{gT^2}{4\pi^2r} = \frac{9.8\ \text{m/s}^2 \times (1.7\ \text{s})^2}{4 \times 9.87 \times 2.0\ \text{m}} = 0.36$$

The coefficient of friction is a number with no units (also called *dimensionless*), and our answer also has no units. The magnitude 0.36 is reasonable and easy to obtain with everyday materials (wood on wood ranges from 0.25–0.5). To be on the safe side, we probably want to make the surface rougher, with twice the minimum coefficient of static friction. The limiting case analysis of the final expression also supports the result. If the drum is stationary ($T = \infty$), the coefficient of static friction would have to be infinite—the person would have to be glued permanently to the vertical surface. Perhaps most important, the required coefficient of static friction does not depend on the mass of the rider. Earth exerts a greater gravitational force on a more massive person. But the drum also pushes toward the center with a greater normal force, which leads to an increased friction force. The effects balance, and the mass does not matter.

Try it yourself: Merry-go-rounds have special railings for people to hold when riding. What is the purpose of the railings? Answer this question by drawing a force diagram for an adult standing on a rotating merry-go-round.

Answer: The railings exert a force on the person in the radial direction. This force helps provide the necessary radial acceleration.

EXAMPLE 4.7 Texas Motor Speedway
The Texas Motor Speedway is a 2.4-km (1.5-mile)-long oval track. One of its turns is about 200 m in radius and is banked at 24° above the horizontal. How fast would a car have to move so that no friction is needed to prevent it from sliding sideways off the raceway (into the infield or off the track)?

Sketch and translate We start by drawing top view and rear view sketches of the situation. The car is the system. For simplicity, we draw the speedway as a circular track.

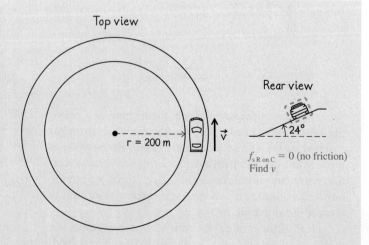

Top view

Rear view

$r = 200\ \text{m}$

$\vec{v}$

24°

$f_{s\,R\,on\,C} = 0$ (no friction)
Find v

Simplify and diagram Model the car as a point-like object that moves along a circular path at constant speed. Earth exerts a downward force $\vec{F}_{E\,on\,C}$. The surface of the road R exerts on the car a normal force $\vec{N}_{R\,on\,C}$ perpendicular to the surface (see the force diagram at right). We assume the sideways friction force exerted by the road on the car is zero (see the problem statement). For constant speed circular motion, the net force exerted on the car should point toward the center of the circle. The horizontal radial component of the normal force $N_{R\,on\,C\,r}$ causes the radial acceleration toward the center. Since the vertical y-component of the car's acceleration is zero, the vertical y-component of the normal force $N_{R\,on\,C\,y}$ balances the downward gravitational force $\vec{F}_{E\,on\,C}$ that Earth exerts on the car.

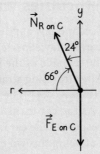

Represent mathematically Use the force diagram to help apply the radial r- and vertical y-component forms of Newton's second law. Let's start with the vertical y-equation.

Vertical y-equation: $m_C a_y = \Sigma F_y = N_{R\,on\,C\,y} + F_{E\,on\,C\,y}$

Since $a_y = 0$, $N_{R\,on\,C\,y} = N_{R\,on\,C}\sin 66°$, and $F_{E\,on\,C\,y} = -m_C g$, we get $0 = +N_{R\,on\,C}\sin 66° - m_C g$, or

$$N_{R\,on\,C}\sin 66° = m_C g \qquad (4.8)$$

Now, apply the radial component equation.

Radial r-equation: $m_C a_r = \Sigma F_r = N_{R\,on\,C\,r} + F_{E\,on\,C\,r}$

Since $a_r = v^2/r$, $N_{R\,on\,C\,r} = N_{R\,on\,C}\cos 66°$, and $F_{E\,on\,C\,r} = 0$, we get:

$$m_C \frac{v^2}{r} = N_{R\,on\,C}\cos 66° + 0 \qquad (4.9)$$

Now combine Eqs. (4.8) and (4.9) to eliminate $N_{R\,on\,C}$ and determine an expression for the speed of the car. To do this, divide the first equation by the second with the $N_{R\,on\,C}$ sides of each equation on the left:

$$\frac{N_{R\,on\,C}\sin 66°}{N_{R\,on\,C}\cos 66°} = \frac{m_C g}{m_C \dfrac{v^2}{r}}$$

Canceling $N_{R\,on\,C}$ on the left side and the m_C on the right side and remembering that

$$\frac{\sin 66°}{\cos 66°} = \tan 66°,$$

we get

$$\tan 66° = \frac{gr}{v^2}$$

Rearrange the above to find v^2:

$$v^2 = \frac{gr}{\tan 66°}$$

Solve and evaluate The road has a radius $r = 200$ m. Thus,

$$v = \sqrt{\frac{gr}{\tan 66°}} = \sqrt{\frac{(9.8\ \text{m/s}^2)(200\ \text{m})}{\tan 66°}}$$

$$= 29.5\ \text{m/s}$$

$$= 30\ \text{m/s} = 66\ \text{mi/h}$$

This is clearly much less than the speed of actual racecars. See the Try It Yourself question.

Try it yourself: Construct a force diagram for a racecar that is traveling somewhat faster around the circular track and indicate what keeps the racecar moving in a circle at this higher speed.

Answer: If the racecar is going faster than 30 m/s, static friction has to push in on the car parallel to the road surface and toward the center of the track, as shown. The friction force has a negative y-component and causes the normal force to be bigger. The combination of the radial components of the increased magnitude normal force and the static friction force provides the much greater net radial force needed to keep the car moving in a circle.

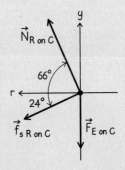

TIP Remember that there is no special force that causes the radial acceleration of an object moving at constant speed along a circular path. This acceleration is caused by all of the forces exerted on the system object by other objects (Earth, the surface of a road, a rope...). Add the radial components of these regular forces. This sum is what causes the radial acceleration of the system object.

Conceptual difficulties with circular motion

Although Newton's second law applies to circular motion just as it applies to linear motion, some everyday experiences seem to contradict that fact. Imagine that you sit on the left side of the back seat of a taxi moving at high speed on a straight road. After traveling straight for a while, the taxi starts to make a high speed *left* turn (**Figure 4.4**). The car seat is slippery and you aren't holding on to anything. As the car turns left, you start sliding across the seat *toward the right* with respect to the car until you hit the door on the other side of the car. This feeling of being thrown outward in a turning car seems inconsistent with the idea that the net force points toward the center of the circle during circular motion. The feeling that there is a force pushing outward on you seems very real, but what object is exerting that force?

Remember that Newton's laws explain motion only when made by an observer in an inertial reference frame. Since the car is accelerating as it goes around the curve, it is not an inertial reference frame. To a roadside observer, you and the car were moving at constant velocity before the car started turning. The forces exerted on you by Earth and the car seat balanced. The roadside observer saw the car begin turning to the left and you continuing to travel with constant velocity because the net force that other objects exerted on you was zero (we assume that the seat is slippery and the friction force that the surface of the seat exerted on you was very small). When the door finally intercepted your forward-moving body, it started exerting a force on you toward the center of the circle. At that moment, you started moving with the car at constant speed but changing velocity along the circular path. Your velocity now changes because of the normal force exerted on you by the door.

Review Question 4.6   Think back to Example 4.6 (the rotor ride). Use your understanding of Newton's laws and constant speed circular motion to explain why the woman in the ride seems to feel a strong force pushing her against the wall of the drum.

4.5 The law of universal gravitation

So far, we have investigated examples of circular motion on or near the surface of Earth. Now let's look at circular motion as it occurs for planets moving around the Sun and satellites and our Moon moving around Earth.

Observations and explanations of planetary motion

By Newton's time, scientists already knew a great deal about the motion of the planets in the solar system. Planets were known to move around the Sun at approximately constant speed in almost circular orbits. (Actually, the orbits have a slightly elliptical shape.) However, no one had a scientific explanation for what was causing the Moon and the planets to travel in their nearly circular orbits.

Newton was among the first to hypothesize that the Moon moved in a circular orbit around Earth because Earth pulled on it, continuously changing the direction of the Moon's velocity. He wondered if the force exerted by Earth on the Moon was the same type of force as the force that Earth exerted on falling objects, such as an apple falling from a tree.

The Dependence of Gravitational Force on Distance Newton realized that directly measuring the gravitational force exerted by Earth on another object was impossible. However, his second law ($\vec{a} = \Sigma \vec{F}/m$) provided an indirect way to determine the gravitational force exerted by Earth: measure the acceleration

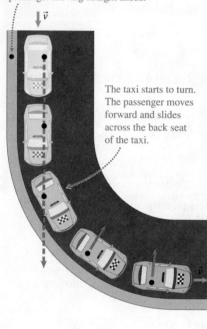

Figure 4.4 An observer's view of a passenger as a taxi makes a high-speed turn.

An observer sees the taxi and passenger moving straight ahead.

$\vec{v}$

The taxi starts to turn. The passenger moves forward and slides across the back seat of the taxi.

of the object and use the second law to determine the gravitational force. He knew that the Moon was separated from Earth by about $r_M = 3.8 \times 10^8$ m, roughly 60 Earth radii ($60R_E$). With this distance and the Moon's orbital period of 27.3 days, Newton was able to use Eq. (4.5) to determine the Moon's radial acceleration:

$$a_{r\,at\,R=60R_E} = \frac{4\pi^2 R}{T^2} = \frac{4\pi^2(3.8 \times 10^8 \text{ m})}{[(27.3 \text{ days})(86,400 \text{ s/day})]^2}$$

$$= \frac{4\pi^2(3.8 \times 10^8 \text{ m})}{(2.36 \times 10^6 \text{ s})^2} = 2.69 \times 10^{-3} \text{ m/s}^2$$

He then did a similar analysis for a second situation. He performed a thought experiment in which he imagined what would happen if the Moon were condensed to a small point-like object (while keeping its mass the same) located near Earth's surface. This would place the "Moon particle" at a distance of one Earth radius ($1\,R_E$) from Earth's center. If the Moon only interacted with Earth, it would have the same free-fall acceleration as any object near Earth's surface (9.8 m/s^2), since this acceleration is independent of mass. Now Newton knew the Moon's acceleration at two different distances from Earth's center.

Newton then used this information to determine how the gravitational force that one object exerts on another depended on the separation of the objects, assuming that the force causing the acceleration of the Moon is the gravitational force exerted on it by Earth and that this force changes with the separation of the objects. He took the ratio of these two accelerations using his second law:

$$\frac{a_{r\,at\,R=60R_E}}{a_{r\,at\,R=1R_E}} = \frac{F_{E\,on\,Moon\,at\,R=60R_E}/m_{Moon}}{F_{E\,on\,Moon\,at\,R=1R_E}/m_{Moon}} = \frac{F_{E\,on\,Moon\,at\,R=60R_E}}{F_{E\,on\,Moon\,at\,R=1R_E}}$$

Notice that the Moon's mass cancels when taking this ratio. Substituting the two accelerations he had observed, Newton made the following comparison between the forces:

$$\frac{F_{E\,on\,Moon\,at\,R=60R_E}}{F_{E\,on\,Moon\,at\,R=1R_E}} = \frac{a_{r\,at\,R=60R_E}}{a_{r\,at\,R=1R_E}} = \frac{2.69 \times 10^{-3} \text{ m/s}^2}{9.8 \text{ m/s}^2} = \frac{1}{3600} = \frac{1}{60^2}$$

$$\Rightarrow F_{E\,on\,Moon\,at\,R=60R_E} = \frac{1}{60^2}F_{E\,on\,Moon\,at\,R=1R_E}$$

The force exerted on the Moon when the Moon is at a distance of 60 times Earth's radius is $1/60^2$ times the force exerted on the Moon when near Earth's surface (about 1 times Earth's radius). It appears that as the distance from Earth's center to the Moon *increases* by a factor of 60, the force decreases by a factor of $1/60^2 = 1/3600$. This suggests that the gravitational force that Earth exerts on the Moon depends on the inverse square of its distance from the center of Earth:

$$F_{E\,on\,Moon\,at\,r} \propto \frac{1}{r^2} \tag{4.10}$$

> **TIP** You might be wondering why if Earth pulls on the Moon, the Moon does not come closer to Earth in the same way that an apple falls from a tree. The difference in these two cases is the speed of the objects. The apple is at rest with respect to Earth before it leaves the tree, and the Moon is moving tangentially. Think of what would happen to the Moon if Earth stopped pulling on it—it would fly away along a straight line.

The Dependence of Gravitational Force on Mass Newton next wanted to determine in what way the gravitational force depended on the masses of the interacting objects. Remember that the free-fall acceleration of an object near

Earth's surface does not depend on the object's mass. All objects have the same free-fall acceleration of 9.8 m/s²:

$$a_O = \frac{F_{E\,on\,O}}{m_O} = g$$

Mathematically, this only works if the magnitude of the gravitational force that Earth exerts on a falling object $F_{E\,on\,O}$ is proportional to the mass of that object: $F_{E\,on\,O} = m_O g$. Newton assumed that the gravitational force that Earth exerts on the Moon has this same feature—the gravitational force must be proportional to the Moon's mass:

$$F_{E\,on\,M} \propto m_{Moon} \tag{4.11}$$

The last question was to decide whether the gravitational force exerted by Earth on the Moon depends only on the Moon's mass or on the mass of Earth as well. Here, Newton's own third law suggested an answer. If Earth exerts a gravitational force on the Moon, then according to Newton's third law the Moon must also exert a gravitational force on Earth that is equal in magnitude (see **Figure 4.5**). Because of this, Newton decided that the gravitational force exerted by Earth on the Moon should also be proportional to the mass of Earth:

$$F_{E\,on\,M} = F_{M\,on\,E} \propto m_{Earth} \tag{4.12}$$

Newton then combined Eqs. (4.10)–(4.12), obtaining a mathematical relationship for the dependence of the gravitational force on all these factors:

$$F_{E\,on\,M} = F_{M\,on\,E} \propto \frac{m_{Earth}m_{Moon}}{r^2} \tag{4.13}$$

Newton further proposed that this was not just a mathematical description of the gravitational interaction between Earth and the Moon, but that it was more general and described the gravitational interaction between any two objects; he called this relation the **law of universal gravitation**. Newton's extension of the applicability of Eq. (4.13) to celestial objects was based on his analysis of Kepler's laws, which you will read about in the next section.

When Newton devised Eq. (4.13), he did not know the masses of Earth and the Moon and could not measure the force that one exerted on the other. Thus, the best he could do was to construct the above proportionality relationship. It was much later that scientists determined the value of the constant of proportionality:

$$G = 6.67 \times 10^{-11}\,\mathrm{N \cdot m^2/kg^2}$$

G, known as **the universal gravitational constant**, allows us to write a complete mathematical equation for the gravitational force that one mass exerts on another:

$$F_{g\,1\,on\,2} = G\frac{m_1 m_2}{r^2} \tag{4.14}$$

Notice that the value of the universal gravitation constant is quite small. If you take two objects, each with a mass of 1.0 kg and separate them by 1.0 m, the force that they exert on each other equals 6.67×10^{-11} N. This means that the gravitational force that objects like us (with masses of about 50–100 kg) exert on each other is extremely weak. However, because the masses of celestial objects are huge (Earth's mass is approximately 6.0×10^{24} kg and the Sun's mass is 2.0×10^{30} kg), the gravitational forces that they exert on each other are very large, despite the small value of the gravitational constant.

Newton's third law and the gravitational force

The mass of Earth is approximately 6.0×10^{24} kg. The mass of a tennis ball is approximately 50 g, or 5.0×10^{-2} kg. How does the gravitational force that

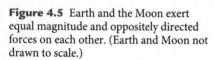

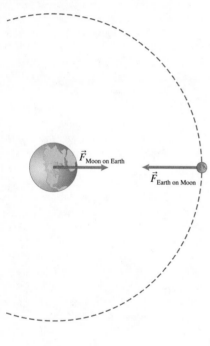

Figure 4.5 Earth and the Moon exert equal magnitude and oppositely directed forces on each other. (Earth and Moon not drawn to scale.)

$\vec{F}_{Moon\,on\,Earth}$

$\vec{F}_{Earth\,on\,Moon}$

Earth exerts downward on the ball compare to the upward force that the ball exerts on Earth? According to Newton's third law, interacting objects exert forces of equal magnitude and opposite direction on each other. The gravitational force exerted by Earth on the ball has the same magnitude as the force that the ball exerts on Earth. Equation (4.14) is consistent with this idea:

$$F_{\text{E on B}} = G\frac{M_E m_B}{r^2}$$

$$= (6.67 \times 10^{-11}\,\text{N}\cdot\text{m}^2/\text{kg}^2)\frac{(6.0 \times 10^{24}\,\text{kg})(5.0 \times 10^{-2}\,\text{kg})}{(6.4 \times 10^6\,\text{m})^2} = 0.49\,\text{N}$$

$$F_{\text{B on E}} = G\frac{m_B M_E}{r^2}$$

$$= (6.67 \times 10^{-11}\,\text{N}\cdot\text{m}^2/\text{kg}^2)\frac{(5.0 \times 10^{-2}\,\text{kg})(6.0 \times 10^{24}\,\text{kg})}{(6.4 \times 10^6\,\text{m})^2} = 0.49\,\text{N}$$

(In the denominator we use the distance between the center of Earth and the center of the ball.)

It might seem counterintuitive that the ball exerts a gravitational force on Earth. After all, Earth doesn't seem to react every time someone drops something. But according to Newton's second law, if the ball exerts a nonzero net force on Earth, then Earth should accelerate. What is the magnitude of this acceleration?

$$a_E = \frac{F_{\text{B on E}}}{M_E} = \frac{0.49\,\text{N}}{6.0 \times 10^{24}\,\text{kg}} = 8.2 \times 10^{-26}\,\text{m/s}^2$$

Earth's acceleration is so tiny that there is no known way to observe it. On the other hand, the acceleration of the ball is easily noticeable since the ball's mass is so much smaller.

$$a_B = \frac{F_{\text{E on B}}}{m_B} = \frac{0.49\,\text{N}}{(5.0 \times 10^{-2}\,\text{kg})} = 9.8\,\text{m/s}^2$$

Notice that the ball has the familiar free-fall acceleration. In fact, the discussion above allows us to understand why free-fall acceleration on Earth equals $9.8\,\text{m/s}^2$ and not some other number. Use Newton's second law and the law of gravitation for an object falling freely near Earth's surface to determine its acceleration:

$$a_{\text{object}} = \frac{F_{\text{Earth on object}}}{m_{\text{object}}} = \frac{(GM_{\text{Earth}}m_{\text{object}})/R^2_{\text{Earth}}}{m_{\text{object}}} = \frac{GM_{\text{Earth}}}{R^2_{\text{Earth}}}$$

Inserting Earth's mass $M_{\text{Earth}} = 5.97 \times 10^{24}\,\text{kg}$ and its radius $R_{\text{Earth}} = 6.37 \times 10^6\,\text{m}$, we have

$$a_{\text{object}} = \frac{GM_{\text{Earth}}}{R^2_{\text{Earth}}}$$

$$= \frac{(6.67 \times 10^{-11}\,\text{N}\cdot\text{m}^2/\text{kg}^2)(5.97 \times 10^{24}\,\text{kg})}{(6.37 \times 10^6\,\text{m})^2} = 9.8\,\text{N/kg} = 9.8\,\text{m/s}^2$$

This is exactly the free-fall acceleration of objects near Earth's surface. The fact that the experimentally measured value of $9.8\,\text{m/s}^2$ agrees with the value calculated using the law of gravitation gives us more confidence in the correctness of the law. This consistency check serves the purpose of a testing experiment. The expression $F_{\text{E on O}} = m_O g$ is actually a special case of the law of gravitation used to determine the gravitational force that Earth exerts on objects when near Earth's surface. Using a similar technique we can determine the free-fall acceleration of an object near the surface of the Moon ($m_{\text{Moon}} = 7.35 \times 10^{22}\,\text{kg}$ and $R_{\text{Moon}} = 1.74 \times 10^6\,\text{m}$) to be $1.6\,\text{N/kg} = 1.6\,\text{m/s}^2$.

Kepler's laws and the law of universal gravitation

About half a century before Newton's work, Johannes Kepler (1571–1630) studied astronomical data concerning the motion of the known planets and crafted three laws, called Kepler's laws of planetary motion. Newton knew about Kepler's work and applied the force equation developed for Earth-Moon interaction to the Sun and the known planets. By using the law of universal gravitation he could explain Kepler's laws. This fact contributed to scientists' confidence in the law of universal gravitation.

Kepler's First Law of Planetary Motion The orbits of all planets are ellipses with the Sun located at one of the ellipse's foci (**Figure 4.6a**).[1]

Kepler's Second Law of Planetary Motion When a planet travels in an orbit, an imaginary line connecting the planet and the Sun continually sweeps out the same area during the same time interval, independent of the planet's distance from the Sun (Figure 4.6b).

Kepler's Third Law of Planetary Motion The square of the period T of the planet's motion (the time interval to complete one orbit) divided by the cube of the semimajor axis of the orbit (which is half the maximum diameter of an elliptical orbit or the radius r of a circular one) equals the same constant for all the known planets:

$$\frac{T^2}{r^3} = K \tag{4.15}$$

Newton was able use his laws of motion and gravitation to derive all three of Kepler's laws of planetary motion. In addition, astronomers later used Newton's law of universal gravitation to predict the locations of the then-unknown planets Neptune and Pluto. They did it by observing subtle inconsistencies between the observed orbits of known planets and the predictions of the law of universal gravitation. At present, physicists and engineers use the law of universal gravitation to decide how to launch satellites into their desired orbits and how to successfully send astronauts to the Moon and beyond. Newton's law of universal gravitation has been tested countless times and is consistent with observations to a very high degree of accuracy.

Figure 4.6 Orbital models used by Kepler to develop his laws of planetary motion.

(a) The orbits of all planets are ellipses with the Sun located at one focus.

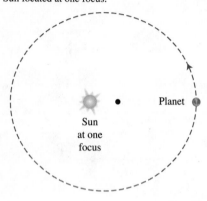

(b) An imaginary line connecting the planet and the Sun continually sweeps out the same area during the same time interval.

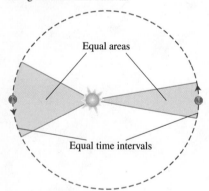

Newton's law of universal gravitation The magnitude of the attractive gravitational force that an object with mass m_1 exerts on an object with mass m_2 separated by a center-to-center distance r is:

$$F_{g\,1\,on\,2} = G\frac{m_1 m_2}{r^2} \tag{4.14}$$

where $G = 6.67 \times 10^{-11}\,\text{N} \cdot \text{m}^2/\text{kg}^2$ is known as the gravitational constant.

However, just because this relationship is called a "law" does not mean that the law is always valid. Every mathematical expression in physics is applicable only for certain circumstances, and the law of universal gravitation is no exception.

Limitations of the law of universal gravitation

The law of universal gravitation in the form of Eq. (4.14) only applies to objects with spherical symmetry or whose separation distance is much bigger than their size. An object that is spherically symmetric has the shape of a perfect sphere and a density that only varies with distance from its center. In these cases, we can model the objects as though all of their mass was located

---

[1] The shape of the planetary orbits is close to circular for most of the planets. In those cases, the foci of the ellipse are very close to the center of the circular orbit that most closely approximates the ellipse.

at a single point at their centers. However, if the objects are not spherically symmetric and are close enough to each other so that they cannot be modeled as point-like particles, then the law of universal gravitation does not apply directly. But using calculus, we can divide each object into a collection of very small point-like objects and use the law of universal gravitation on each pair to add by integration the forces of the small point-like particles on each other.

However, even with calculus, there are details of some motion for which the law cannot account. When astronomers made careful observations of the orbit of Mercury, they noticed that its orbit exhibited some patterns that the law of universal gravitation could not explain. It wasn't until the early 20th century, when Einstein constructed a more advanced theory of gravity, that scientists could predict all of the details of the motion of Mercury.

Review Question 4.5 Give an example that would help a friend better understand how the magnitude of the universal gravitation constant $G = 6.67 \times 10^{-11} \, \text{N} \cdot \text{m}^2/\text{kg}^2$ affects everyday life.

4.6 Satellites and astronauts: Putting it all together

The Moon orbits Earth due to the gravitational force that provides the necessary radial acceleration. It is Earth's natural satellite, a celestial object that orbits a bigger celestial object (usually a planet). Artificial satellites are objects that are placed in orbit by humans. The first artificial satellite was launched in 1957, and since then thousands of satellites have been launched. Earth satellites make worldwide communication possible, allow us to monitor Earth's surface and the weather, help us find our way to unknown destinations with our global positioning systems (GPS), and provide access to hundreds of television stations. The special satellites used in many of these applications must orbit at a specific distance above Earth's surface. How do we determine that distance?

Satellites

You may have noticed that the satellite TV receiving dishes on residential rooftops never move. They always point at the same location in the sky. This means that the satellite from which they are receiving signals must always remain at the same location in the sky. In order to do so, the satellite must be placed at a very specific altitude that allows the satellite to travel once around Earth in exactly 24 hours while always remaining above the equator. Such satellites are called **geostationary.** An array of such satellites can provide communications to all parts of Earth.

EXAMPLE 4.8 Geostationary satellite
You are in charge of launching a geostationary satellite into orbit. At what altitude above the equator must the satellite orbit in order to provide continuous communication to a stationary dish antenna on Earth? The mass of Earth is $5.97 \times 10^{24} \, \text{kg}$.

Sketch and translate A geostationary satellite, such as that shown on the top of the next page, completes one orbit around Earth each 24 hours, making the period of its circular motion the same as that of Earth rotating below it, $T = 24 \, \text{h} = 86,400 \, \text{s}$. We use our understanding of circular motion and the law of universal gravitation to decide the radius r of the satellite's orbit from the center of Earth. Then we can determine the altitude of the satellite's orbit above Earth's surface. The satellite is the system object in this problem.

(continued)

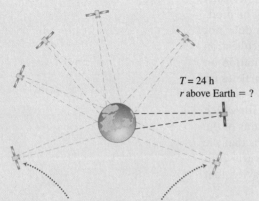

T = 24 h
r above Earth = ?

Each satellite communicates with parts of Earth.

Simplify and diagram Model the satellite as a point-like particle and assume that it moves with constant speed circular motion. Make a force diagram for the satellite.

$\vec{F}_{\text{Earth on Satellite}}$

Represent mathematically The only force exerted on the satellite is the gravitational force exerted by Earth. Use the radial r-component form of Newton's second law and the law of universal gravitation to get an expression for the satellite's radial acceleration:

$$a_{S\,r} = \frac{F_{\text{E on S}}}{m_{S}} = \frac{Gm_{E}m_{S}/r^2}{m_{S}} = \frac{Gm_{E}}{r^2}$$

Since the satellite travels with constant speed circular motion, its acceleration must also be

$$a_{S\,r} = \frac{v^2}{r} = \frac{(2\pi r/T)^2}{r} = \frac{4\pi^2 r}{T^2}$$

Setting the two expressions for acceleration equal to each other, we get

$$\frac{4\pi^2 r}{T^2} = \frac{Gm_{E}}{r^2}$$

Now, solve this for the radius of the orbit:

$$r^3 = \frac{Gm_{E}T^2}{4\pi^2} \quad \text{or} \quad r = \left(\frac{Gm_{E}T^2}{4\pi^2}\right)^{1/3}$$

Solve and evaluate Inserting all the relevant information gives an answer:

$$r = \left(\frac{(6.67 \times 10^{-11}\,\text{N} \cdot \text{m}^2/\text{kg}^2)(5.97 \times 10^{24}\,\text{kg})(8.64 \times 10^4\,\text{s})^2}{4\pi^2}\right)^{1/3}$$

$$= (75.3 \times 10^{21}\,\text{m}^3)^{1/3} = 4.2 \times 10^7\,\text{m}$$

This is the distance of the satellite from the center of Earth. Since the radius of Earth is $6.4 \times 10^6\,\text{m} = 0.64 \times 10^7\,\text{m}$, the distance of the satellite above Earth's surface is

$$4.2 \times 10^7\,\text{m} - 0.64 \times 10^7\,\text{m} = 3.6 \times 10^7\,\text{m} = 22{,}000\,\text{mi}$$

Try it yourself: Imagine that you want to launch a satellite that moves just above Earth's surface. Determine the speed of such a satellite. (This isn't realistic, since the atmosphere would strongly affect the satellite's orbit, but it's a good practice exercise.)

Answer: 7900 m/s.

Are astronauts weightless?

We are all familiar with videos of astronauts floating in the Space Shuttle or in the International Space Station. News reports commonly say that the astronauts are weightless or experience zero gravity. Is this true?

QUANTITATIVE EXERCISE 4.9 Are astronauts weightless in the International Space Station?
The International Space Station orbits about $0.40 \times 10^6\,\text{m}$ (250 miles) above Earth's surface, or $(6.37 + 0.40) \times 10^6\,\text{m} = 6.77 \times 10^6\,\text{m}$ from Earth's center. Compare the force that Earth exerts on an astronaut in the station to the force when he is on Earth's surface.

Represent mathematically The gravitational force that Earth exerts on the astronaut (system object) in the space station is

$$F_{\text{E on A in station}} = G\frac{m_{E}m_{A}}{(r_{\text{station-Earth}})^2}$$

where $r_{\text{station-Earth}} = 6.77 \times 10^6$ m is the distance from Earth's center to the astronaut in the space station. The gravitational force exerted by Earth on the astronaut when on Earth's surface is

$$F_{\text{E on A on Earth's surface}} = G \frac{m_E m_A}{\left(r_{\text{surface-Earth}}\right)^2}$$

where $r_{\text{surface-Earth}} = 6.37 \times 10^6$ m is the distance from the center of Earth to its surface. To compare these two forces, take their ratio:

$$\frac{F_{\text{E on A in station}}}{F_{\text{E on A on Earth's surface}}} = \frac{G \dfrac{m_E m_A}{r_{\text{station-Earth}}^2}}{G \dfrac{m_E m_A}{r_{\text{surface-Earth}}^2}} = \left(\frac{r_{\text{surface-Earth}}}{r_{\text{station-Earth}}}\right)^2$$

Solve and evaluate Inserting the appropriate values gives

$$\frac{F_{\text{E on A in station}}}{F_{\text{E on A on Earth's surface}}} = \left(\frac{6.37 \times 10^6 \text{ m}}{6.77 \times 10^6 \text{ m}}\right)^2 = 0.89$$

The gravitational force that Earth exerts on the astronaut when in the space station is just 11% less than the force exerted on him when on Earth's surface. The astronaut is far from "weightless."

Try it yourself: At what distance from Earth's center should a person be so that her weight is half her weight when on Earth's surface?

Answer: 1.4 R_E.

Quantitative Exercise 4.9 shows that astronauts are not actually weightless while in the International Space Station. So why do they float? Remember that Earth exerts a gravitational force on both the astronaut and the space station. This force causes them both to fall toward Earth at the same rate while they fly forward, thus staying on the same circular path. Both the astronaut and the space station are in free fall. As a result, if the astronaut stood on a scale placed on the floor of the orbiting space station, her weight, according to the scale, would be zero. The same thing happens on Earth when you stand on a scale inside an elevator that is in free fall, falling at the acceleration $\vec{g}$. Since both you and the elevator fall with the same acceleration, you do not press on the scale, and the scale does not press on you; it reads zero.

The confusion occurs because in physics weight is a shorthand way of referring to the gravitational force being exerted on an object, not the reading of a scale. The scale measures the normal force it exerts on any object with which it is in contact. So astronauts are not "weightless"; the word is actually being misused.

Review Question 4.6 A friend says he has heard that the Moon is falling toward Earth. What can you tell your friend to reassure him that all is well?

Summary

Words	Pictorial and physical representations	Mathematical representation
Velocity change method to find the direction of acceleration We can estimate the direction of an object's acceleration as it passes a point along its circular path by drawing an initial velocity vector v_i just before that point and a final velocity vector v_f just after that point. We place the vectors tail to tail and draw a velocity change vector from the tip of the initial to the tip of the final velocity. The acceleration points in the direction of the velocity change vector. (Section 4.1)	Top view 	$$\vec{v}_i + \Delta\vec{v} = \vec{v}_f$$ **or** $$\Delta\vec{v} = \vec{v}_f - \vec{v}_i$$ $$\vec{a} = \frac{\Delta\vec{v}}{\Delta t} = \frac{\vec{v}_f - \vec{v}_i}{t_f - t_i}$$
Radial acceleration for constant speed circular motion An object moving at constant speed along a circular path has acceleration that points toward the center of the circle and has a magnitude a_r that depends on its speed v and the radius r of the circle. (Section 4.3)	Top view 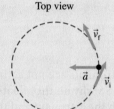	$$a_r = \frac{v^2}{r} \qquad \text{Eq. (4.3)}$$
Period and radial acceleration The radial acceleration can also be expressed using the period T of circular motion, the time interval needed for an object to complete one trip around the circle. (Section 4.3)		$$v = \frac{2\pi r}{T} \qquad \text{Eq. (4.4)}$$ $$a_r = \frac{4\pi^2 r}{T^2} \qquad \text{Eq. (4.5)}$$
Net force for constant speed circular motion The sum of the forces exerted on an object during constant speed circular motion points in the positive radial direction toward the center of the circle. The object's acceleration is the sum of the radial components of all forces exerted on an object divided by its mass—consistent with Newton's second law. In addition, for horizontal circular motion, you sometimes analyze the vertical y-components of forces exerted on an object. (Section 4.4)	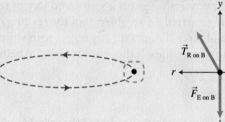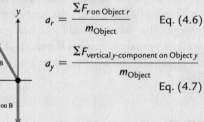	$$a_r = \frac{\Sigma F_{r\text{ on Object }r}}{m_{\text{Object}}} \qquad \text{Eq. (4.6)}$$ $$a_y = \frac{\Sigma F_{\text{vertical }y\text{-component on Object }y}}{m_{\text{Object}}}$$ $$\text{Eq. (4.7)}$$
Law of universal gravitation This force law is used primarily to determine the magnitude of the force that the Sun exerts on planets or that planets exert on satellites or on moons. The force depends on the masses of the objects and on their center-to-center separation r (Section 4.5).	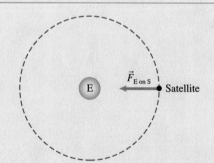	$$F_{g\,1\text{ on }2} = G\frac{m_1 m_2}{r^2} \quad \text{Eq. (4.14)}$$

Questions

Multiple Choice Questions

1. Which of the objects below is accelerating?
   (a) Object moving at constant speed along a straight line
   (b) Object moving at constant speed in a circle
   (c) Object slowing down while moving in a straight line
   (d) Both b and c describe accelerating objects.

2. The circle in **Figure Q4.2** represents the path followed by an object moving at constant speed. At four different positions on the circle, the object's motion is described using velocity and acceleration arrows. Choose the location where the descriptions are correct.
   (a) point A   (b) point B   (c) point C   (d) point D

Figure Q4.2

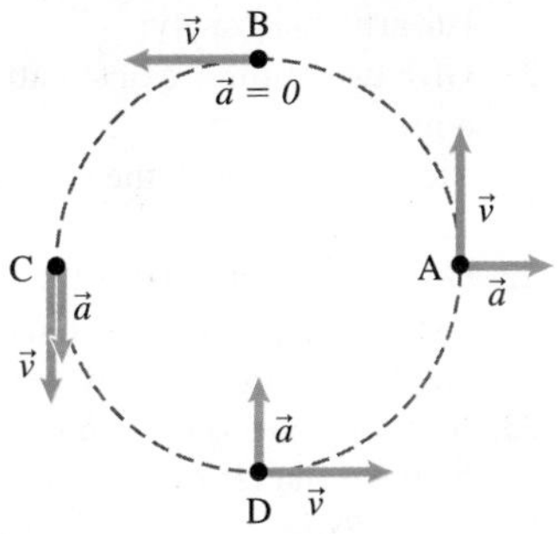

3. One of your classmates drew a force diagram for a pendulum bob at the bottom of its swing. He put a horizontal force arrow in the direction of the velocity. Evaluate his diagram by choosing from the statements below:
   (a) He is incorrect since the force should point partly forward and tilt partly down.
   (b) He is incorrect since there are no forces in the direction of the object's velocity.
   (c) He is correct since any moving object has a force in the direction of its velocity.
   (d) He is correct but he also needs to add an outward (down) force.

4. Why is it difficult for a high-speed car to negotiate an unbanked turn?
   (a) A huge force is pushing the car outward.
   (b) The magnitude of the friction force might not be enough to provide the necessary radial acceleration.
   (c) The sliding friction force is too large.
   (d) The faster the car moves, the harder it is for the driver to turn the steering wheel.

5. How does a person standing on the ground explain why you, sitting on the left side of a slippery back car seat, slide to the right when the car makes a high-speed left turn?
   (a) You tend to move in a straight line and thus slide with respect to the seat that is moving to the left under you.
   (b) There is a net outward force being exerted on you.
   (c) The force of motion propels you forward.
   (d) The car seat pushes you forward.

6. A pilot performs a vertical loop-the-loop at constant speed. The pilot's head is always pointing toward the center of the circle. Where is the blackout more likely to occur?
   (a) At the top of the loop
   (b) At the bottom of the loop
   (c) At both top and bottom
   (d) None of the answers is correct.

7. Why is the following an inaccurate statement about blackout? "As the g forces climb up toward 7 g's, ..."
   (a) There is no such thing as a g force.
   (b) 7 g's is not a big force.
   (c) g forces are constant and do not climb.
   (d) g forces are not responsible for blackout but can cause dizziness.

8. Why do you feel that you are being thrown upward out of your seat when going over an upward arced hump on a roller coaster?
   (a) There is an additional force lifting up on you.
   (b) At the top you continue going straight and the seat moves out from under you.
   (c) You press on the seat less than when the coaster is at rest. Thus the seat presses less on you.
   (d) Both b and c are correct.
   (e) a, b, and c are correct.

9. Compare the magnitude of the normal force of a car seat on you with the magnitude of the force that Earth exerts on you when the car moves across the bottom of a dip in the road.
   (a) The normal force is more than Earth's gravitational force.
   (b) The normal force is less than Earth's gravitational force.
   (c) The two forces are equal.
   (d) It depends on whether the car is moving right or left.

10. If you put a penny on the center of a rotating turntable, it does not slip. However, if you place the penny near the edge, it is likely to slip off. Which answer below explains this observation?
    (a) The penny moves faster at the edge and hence needs a greater force to keep it moving.
    (b) The edge is more slippery since the grooves are farther apart.
    (c) The radial acceleration is greater at the edge and the friction force is not enough to keep the penny in place.
    (d) The outward force responsible for slipping is greater at the edge than at the middle.

11. Where on Earth's surface would you expect to experience the greatest radial acceleration as a result of Earth's rotation?
    (a) On the poles
    (b) On the equator
    (c) On the highest mountain
    (d) Since all points of Earth have the same period of rotation, the acceleration is the same everywhere.

12. What observational data might Newton have used to decide that the gravitational force is inversely proportional to the distance squared between the centers of objects?
    (a) The data on the acceleration of falling apples
    (b) The data describing the Moon's orbit (period and radius) and the motion of falling apples
    (c) The data on comets
    (d) The data on moonrise and moonset times

13. What observations combined with his second and third laws helped Newton decide that the gravitational force of one object on another object is directly proportional to the product of the masses of the interacting objects?
    (a) The data on the acceleration of falling apples
    (b) The data on Moon phases
    (c) The data on comets
    (d) The data on moonrise and moonset times

14. What would happen to the force exerted by the Sun on Earth if the Sun shrank and became half its present size while retaining the same mass?
    (a) The force would be half the present force.
    (b) The force would be one-fourth of the present force.
    (c) The force would double.
    (d) The force would stay the same.

Conceptual Questions

15. Your friend says that an object weighs less on Jupiter than on Earth as Jupiter is far away from the center of Earth. Do you agree or disagree?
16. Your friend says that when an object is moving in a circle, there is a force pushing it out away from the center. Why would he say this? Do you agree or disagree? If you disagree, how would you convince your friend of your opinion?
17. Describe three everyday phenomena that are consistent with your knowledge of the dynamics of circular motion. Specify where the observer is. Then find an observer who will not be able to explain the same phenomena using the knowledge of Newton's laws and circular motion.
18. You place a coin on a rotating turntable. Describe a circular motion experiment to estimate the maximum coefficient of static friction between the coin and the turntable.

19. Astronauts on the space station orbiting Earth are said to be in "zero gravity." Do you agree or disagree? If you disagree, why?
20. In the movies you often see space stations with "artificial gravity." They look like big doughnuts rotating around an axis perpendicular to the plane of the doughnut (see **Figure Q4.21**). People walking on the outer rim inside the turning space station feel the same gravitational effects as if they were on Earth. How does such a station work to simulate artificial gravity?

Figure Q4.21

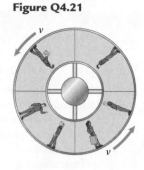

21. Give two examples of situations in which the acceleration of a moving object is zero but the velocity is not zero and two examples in which the velocity is zero but the acceleration is not zero.
22. Give two examples of situations in which an object moves at constant speed and has zero acceleration and two in which the object moves at constant speed and has a nonzero acceleration.
23. Name a planet on which you would weigh less than on Earth. Name a planet on which you would weigh more than on Earth. Explain how you know.

Problems

Below, BIO indicates a problem with a biological or medical focus. Problems labeled EST ask you to estimate the answer to a quantitative problem rather than derive a specific answer. Problems marked with ✏ require you to make a drawing or graph as part of your solution. Asterisks indicate the level of difficulty of the problem. Problems with no * are considered to be the least difficult. A single * marks moderately difficult problems. Two ** indicate more difficult problems.

4.1–4.3 The qualitative velocity change method for circular motion, Qualitative dynamics of circular motion, and Radial acceleration and period

1. ✏ **Mountain biker** While mountain biking, you first move at constant speed along the bottom of a trail's circular dip and then at constant speed across the top of a circular hump. Assume that you and the bike are a system. Determine the direction of the acceleration at each position and construct a force diagram for each position (consistent with the direction of the acceleration). Compare at each position the magnitude of the force of the surface on the bike with the force Earth exerts on the system.

2. * ✏ You swing a rock tied to a string in a vertical circle. (a) Determine the direction of the acceleration of the rock as it passes the lowest point in its swing. Construct a consistent force diagram for the rock as it passes that point. How does the force that the string exerts on the rock compare to the force that Earth exerts on the rock? Explain. (b) Repeat the above analysis as best you can for the rock as it passes the highest point in the swing. (c) If the string is tied around your finger, when do you feel a stronger pull—when the rock is at the bottom of the swing or at the top? Explain.

3. ✏ **Loop-the-loop** You ride a roller coaster with a loop-the-loop. Compare *as best you can* the normal force that the seat exerts on you to the force that Earth exerts on you when you

are passing the bottom of the loop and the top of the loop. Justify your answers by determining the direction of acceleration and constructing a force diagram for each position. Make your answers consistent with Newton's second law.

4. You start an old record player and notice a bug on the surface close to the edge of the record. The record has a diameter of 12 inches and completes 33 revolutions each minute. (a) What are the speed and the acceleration of the bug? (b) What would the bug's speed and acceleration be if it were halfway between the center and the edge of the record?

5. Determine the acceleration of Earth due to its motion around the Sun. What do you need to assume about Earth to make the calculation? How does this acceleration compare to the acceleration of free fall on Earth?

6. The Moon is an average distance of 3.8×10^8 m from Earth. It circles Earth once each 27.3 days. (a) What is its average speed? (b) What is its acceleration? (c) How does this acceleration compare to the acceleration of free fall on Earth?

7. **Aborted plane landing** You are on an airplane that is landing. The plane in front of your plane blows a tire. The pilot of your plane is advised to abort the landing, so he pulls up, moving in a semicircular upward-bending path. The path has a radius of 500 m with a radial acceleration of 17 m/s^2. What is the plane's speed?

8. BIO **Ultracentrifuge** You are working in a biology lab and learning to use a new ultracentrifuge for blood tests. The specifications for the centrifuge say that a red blood cell rotating in the ultracentrifuge moves at 470 m/s and has a radial acceleration of 150,000 g's (that is, 150,000 times 9.8 m/s^2). The radius of the centrifuge is 0.15 m. You wonder if this claim is correct. Support your answer with a calculation.

9. Jupiter rotates once about its axis in 9 h 56 min. Its radius is 7.13×10^4 km. Imagine that you could somehow stand on the surface (although in reality that would not be possible, because Jupiter has no solid surface). Calculate your radial acceleration in meters per second squared and in Earth g's.

10. * Imagine that you are standing on a horizontal rotating platform in an amusement park (like the platform for a merry-go-round). The period of rotation and the radius of the platform are given, and you know your mass. Make a list of the physical quantities you could determine using this information, and describe how you would determine them.

11. * A car moves along a straight line to the right. Two friends standing on the sidewalk are arguing about the motion of a point on the rotating car tire at the instant it reaches the lowest point (touching the road). Jake says that the point is at rest. Morgan says that the point is moving to the left at the car's speed. Justify each friend's opinion. Explain whether it is possible for them to simultaneously be correct.

12. * Three people are standing on a horizontally rotating platform in an amusement park. One person is almost at the edge, the second one is $(3/5)R$ from the center, and the third is $(1/2)R$ from the center. Compare their periods of rotation, their speeds, and their radial accelerations.

13. * / Consider the scenario described in Problem 12. If the platform speeds up, who is more likely to have trouble staying on the platform? Support your answer with a force diagram and describe the assumptions that you made.

14. * **Merry-go-round acceleration** Imagine that you are standing on the rotating platform of a merry-go-round in an amusement park. You have a stopwatch and a measuring tape. Describe how you will determine your radial acceleration when standing at the edge of the platform and when halfway from the edge. What do you expect the ratio of these two accelerations to be?

15. / **Ferris wheel** You are sitting on a rotating Ferris wheel. Draw a force diagram for yourself when you are at the bottom of the circle and when you are at the top.

16. EST * **Estimate** the radial acceleration of the foot of a college football player in the middle of punting a football.

17. EST * **Estimate** the radial acceleration of the toe at the end of the horizontally extended leg of a ballerina doing a pirouette.

4.4 Skills for analyzing processes involving circular motion

18. * Is it safe to drive your 1600-kg car at speed 27 m/s around a level highway curve of radius 150 m if the effective coefficient of static friction between the car and the road is 0.40?

19. * You are fixing a broken rotary lawn mower. The blades on the mower turn 50 times per second. What is the magnitude of the force needed to hold the outer 2 cm of the blade to the inner portion of the blade? The outer part is 21 cm from the center of the blade, and the mass of the outer portion is 7.0 g.

20. * Your car speeds around the 80-m-radius curved exit ramp of a freeway. A 70-kg student holds the armrest of the car door, exerting a 220-N force on it in order to prevent himself from sliding across the vinyl-covered back seat of the car and slamming into his friend. How fast is the car moving in meters per second and miles per hour? What assumptions did you make?

21. How fast do you need to swing a 200-g ball at the end of a string in a horizontal circle of 0.5-m radius so that the string makes a 34° angle relative to the horizontal? What assumptions did you make?

22. ** / Christine's bathroom scale in Maine reads 110 lb when she stands on it. Will the scale read more or less in Singapore if her mass stays the same? To answer the question, (a) draw a force diagram for Christine. (b) With the assistance of this diagram, write an expression using Newton's second law that relates the forces exerted on Christine and her acceleration along the radial direction. (c) Decide whether the reading of the scale is different in Singapore. List the assumptions that you made and describe how your answer might change if the assumptions are not valid.

23. ** A child is on a swing that moves in the horizontal circle of radius 2.0 m depicted in **Figure P4.23**. The mass of the child and the seat together is 30 kg and the two cables exert equal-magnitude forces on the chair. Make a list of the physical quantities you can determine using the sketch and the known information. Determine one kinematics and two dynamics quantities from that list.

Figure P4.23

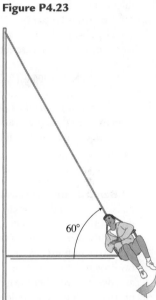

24. * A coin rests on a record 0.15 m from its center. The record turns on a turntable that rotates at variable speed. The coefficient of static friction between the coin and the record is 0.30. What is the maximum coin speed at which it does not slip?

25. **Roller coaster ride** A roller coaster car travels at speed 8.0 m/s over a 12-m-radius vertical circular hump. What is the magnitude of the upward force that the coaster seat exerts on a 48-kg woman passenger?

26. * A person sitting in a chair (combined mass 80 kg) is attached to a 6.0-m-long cable. The person moves in a horizontal circle. The cable angle θ is 62° below the horizontal. What is the person's speed? Note: The radius of the circle is not 6.0 m.

27. * A car moves around a 50-m-radius highway curve. The road, banked at 10° relative to the horizontal, is wet and icy so that the coefficient of friction is approximately zero. At what speed should the car travel so that it makes the turn without slipping?

28. * A 20.0-g ball is attached to a 120-cm-long string and moves in a horizontal circle (see **Figure P4.28**). The string exerts a force on the ball that is equal to 0.200 N. What is the angle θ?

Figure P4.28

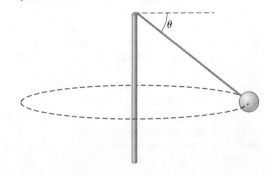

29. A 50-kg ice skater goes around a circle of radius 5.0 m at a constant speed of 3.0 m/s on a level ice rink. What are the magnitude and direction of the horizontal force that the ice exerts on the skates?

30. * A car traveling at 10 m/s passes over a hill on a road that has a circular cross section of radius 30 m. What is the force exerted by the seat of the car on a 60-kg passenger when the car is passing the top of the hill?

31. * A 1000-kg car is moving at 30 m/s around a horizontal level curved road whose radius is 100 m. What is the magnitude of the frictional force required to keep the car from sliding?

32. * / **Equation Jeopardy 1** Describe using words, a sketch, a velocity change diagram, and a force diagram two situations whose mathematical description is presented below.

$$700 \text{ N} - (30 \text{ kg})(9.8 \text{ m/s}^2) = \frac{(30 \text{ kg})v^2}{12 \text{ m}}$$

33. * / **Equation Jeopardy 2** Describe using words, a sketch, a velocity change diagram, and a force diagram two situations whose mathematical description is presented below.

$$\frac{(2.0 \text{ kg})(4.0 \text{ m}^2/\text{s}^2)}{r} = 0.4 \times (2.0 \text{ kg}) \times (9.8 \text{ N/kg})$$

34. ** **Banked curve raceway design** You need to design a banked curve at the new circular Super 100 Raceway. The radius of the track is 800 m and cars typically travel at speed 160 mi/h. What feature of the design is important so that all racecars can move around the track safely in any weather? (a) Provide a quantitative answer. (b) List your assumptions and describe whether the number you provided will increase or decrease if the assumption is not valid.

35. * A circular track is in a horizontal plane, has a radius of r meters, and is banked at an angle θ above the horizontal. (a) Develop an expression for the speed a person should rollerblade on this track so that she needs zero friction to prevent her from sliding sideways off the track. (b) Should another person move faster or slower if her mass is 1.3 times the mass of the first person? Justify your answer.

36. ** Design a quantitative test for Newton's second law as applied to constant speed circular motion. Describe the experiment and provide the analysis needed to make a prediction using the law.

37. * **Spin-dry cycle** Explain how the spin-dry cycle in a washing machine removes water from clothes. Be specific.

4.5–4.6 The law of universal gravitation and Satellites and astronauts: Putting it all together

38. * Your friend says that the force that the Sun exerts on Earth is much larger than the force that Earth exerts on the Sun. (a) Do you agree or disagree with this opinion? (b) If you disagree, how would you convince him of your opinion?

39. Determine the gravitational force that (a) the Sun exerts on the Moon, (b) Earth exerts on the Moon, and (c) the Moon exerts on Earth. List at least two assumptions for each force that you made when you calculated the answers.

40. * (a) What is the ratio of the gravitational force that Earth exerts on the Sun in the winter and the force that it exerts in the summer? (b) What does it tell you about the speed of Earth during different seasons? (c) How many correct answers can you give for part (a)? Hint: Earth's orbit is an ellipse with the Sun located at one of the foci of the ellipse.

41. **Black hole gravitational force** A black hole exerts a 50-N gravitational force on a spaceship. The black hole is 10^{14} m from the ship. What is the magnitude of the force that the black hole exerts on the ship when the ship is one-half that distance from the black hole? [Hint: One-half of 10^{14} m is not 10^7 m.]

42. **EST** * The average radius of Earth's orbit around the Sun is 1.5×10^8 km. The mass of Earth is 5.97×10^{24} kg, and it makes one orbit in approximately 365 days. (a) What is Earth's speed relative to the Sun? (b) Estimate the Sun's mass using Newton's law of universal gravitation and Newton's second law. What assumptions did you need to make?

43. * The Moon travels in a 3.8×10^5 km radius orbit about Earth. Earth's mass is 5.97×10^{24} kg. Determine the period T for one Moon orbit about Earth using Newton's law of universal gravitation and Newton's second law. What assumptions did you make?

44. Determine the ratio of Earth's gravitational force exerted on an 80-kg person when at Earth's surface and when 1000 km above Earth's surface. The radius of Earth is 6370 km.

45. Determine the magnitude of the gravitational force Mars would exert on your body if you were on the surface of Mars.

46. * When you stand on a bathroom scale here on Earth, it reads 540 N. (a) What would your mass be on Mars, Venus, and Saturn? (b) What is the magnitude of the gravitational force each planet would exert on you if you stood on their surface? (c) What assumptions did you make?

47. The free-fall acceleration on the surface of Jupiter, the most massive planet, is 24.79 m/s². Jupiter's radius is 7.0×10^4 km. Use Newtonian ideas to determine Jupiter's mass.

48. A satellite moves in a circular orbit a distance of 1.6×10^5 m above Earth's surface. Determine the speed of the satellite.

49. * Mars has a mass of 6.42×10^{23} kg and a radius of 3.40×10^6 m. Assume a person is standing on a bathroom scale on the surface of Mars. Over what time interval would Mars have to complete one rotation on its axis to make the bathroom scale have a zero reading?

50. * Determine the speed a projectile must reach in order to become an Earth satellite. What assumptions did you make?

51. * Determine the distance above Earth's surface to a satellite that completes two orbits per day. What assumptions did you make?

52. Determine the period of an Earth satellite that moves in a circular orbit just above Earth's surface. What assumptions did you need to make?

53. * A spaceship in outer space has a doughnut shape with 500-m outer radius. The inhabitants stand with their heads toward the center and their feet on an outside rim (see Figure Q4.21). Over what time interval would the spaceship have to complete one rotation on its axis to make a bathroom scale have the same reading for the person in space as when on Earth's surface?

General Problems

54. * **Loop-the-loop** You have to design a loop-the-loop for a new amusement park so that when each car passes the top of the loop inverted (upside-down), each seat exerts a force against a passenger's bottom that has a magnitude equal to 1.5 times the gravitational force that Earth exerts on the passenger. Choose some reasonable physical quantities so these conditions are met. Show that the loop-the-loop will work equally well for passengers of any mass.

55. ** **A Tarzan swing** Tarzan (mass 80 kg) swings at the end of an 8.0-m-long vine (**Figure P4.55**). When directly under the

Figure P4.55

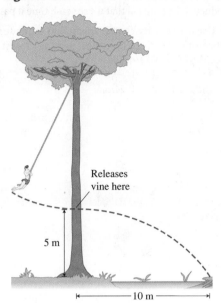

Releases
vine here

5 m

|← 10 m →|

vine's support, he releases the vine and flies across a swamp. When he releases the vine, he is 5.0 m above the swamp and 10.0 m horizontally from the other side. Determine the force the vine exerts on him at the instant before he lets go (the vine is straight down when he lets go).

56. * (a) If the masses of Earth and the Moon were both doubled, by how much would the radius of the Moon's orbit about Earth have to change if its speed did not change? (b) By how much would its speed have to change if its radius did not change? Justify each answer.

57. **EST** * Estimate the radial acceleration of the tread of a car tire. Indicate any assumptions that you made.

58. **EST** ** Estimate the force exerted by the tire on a 10-cm-long section of the tread of a tire as the car travels at speed 80 km/h. Justify any numbers used in your estimate.

59. **EST** ** Estimate the maximum radial force that a football player's leg needs to exert on his foot when swinging the leg to punt the ball. Justify any numbers that you use.

60. **Design 1** Design and solve a circular motion problem for a roller coaster.

61. * **Design 2** Design and solve a circular motion problem for the amusement park ride shown in **Figure P4.61**.

Figure P4.61

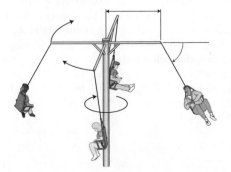

62. ** **Demolition** An old building is being demolished by swinging a heavy metal ball from a crane. Suppose that such a 100-kg ball swings from a 20-m-long wire at speed 16 m/s as the wire passes the vertical orientation. (a) What tension force must the wire be able to withstand in order not to break? (b) Assume the ball stops after sinking 1.5 m into the wall. What was the average force that the ball exerted on the wall? Indicate any assumptions you made for each part of the problem.

63. **Designing a banked roadway** You need to design a banked curve for a highway in which cars make a 90° turn moving at 50 mi/h. Indicate any assumptions you make.

64. * **Evaluation question** You find the following report about blackouts. "The acceleration that causes blackouts in fighter pilots is called the maximum *g*-force. Fighter pilots experience this force when accelerating or decelerating quickly. At high *g*'s, the pilot's blood pressure changes and the flow of oxygen to the brain rapidly decreases. This happens because the pressure outside of the pilot's body is so much greater than the pressure a human is normally accustomed to." Indicate any incorrect physics (including the application of physics to biology) that you find.

65. * Suppose that Earth rotated much faster on its axis—so fast that people were almost weightless when at Earth's surface. How long would the length of a day be on this new Earth?

66. * On Earth, an average person's vertical jump is 0.40 m. What is it on the Moon? Explain.

67. * You read in a science magazine that on the Moon, the speed of a shell leaving the barrel of a modern tank is enough to put the shell in a circular orbit above the surface of the Moon (there is no atmosphere to slow the shell). What should be the speed for this to happen? Is this number reasonable?

Reading Passage Problems

Texas Motor Speedway On Oct. 28, 2000 Gil de Ferran set the single-lap average speed record for the then-named California Speedway—241.428 mi/h. The following year, on April 29, 2001, the Championship Auto Racing Teams (CART) organization canceled the Texas Motor Speedway inaugural Firestone Firehawk 600 race 2 hours before it was to start. During practice, drivers became dizzy when they reached speeds of more than 230 mph on the high-banked track. The Texas Motor Speedway has turns banked at 24° above the horizontal. By comparison, the Indianapolis raceway turns are banked at 9°.

Banking of 24° is unprecedented for Indy-style cars. In qualifying runs for the race, 21 of 25 drivers complained of dizziness and disorientation. They experienced accelerations of 5.5 *g* (5.5 times 9.8 m/s²) that lasted for several seconds. Such accelerations for such long time intervals have caused pilots to black out. Blood drains from their heads to their legs as they move in circles with their heads toward the center of the circle. CART felt that the track was unsafe for their drivers. The Texas Motor Speedway had tested the track with drivers before the race and thought it was safe. The Texas Motor Speedway sued CART.

68. Why did drivers get dizzy and disoriented while driving at the Texas Motor Speedway?
    (a) The cars were traveling at over 200 mi/h.
    (b) The track was tilted at an unusually steep angle.
    (c) On turns the drivers' blood tended to drain from their brains into veins in their lower bodies.
    (d) The *g* force pushed blood into their heads.
    (e) The combination of a and b caused c.

69. What was the time interval needed for Gil de Ferran's car to complete one lap during his record-setting drive ?
    (a) 16.8 s   (b) 18.4 s   (c) 22.4 s
    (d) 25.1 s   (e) 37.3 s

70. If the racecars had no help from friction, which expression below would describe the normal force of the track on the cars while traversing the 24° banked curves?
    (a) $mg\cos 66°$ (b) $mg\sin 66°$ (c) $mg/\cos 66°$
    (d) $mg/\sin 66°$ (e) None of these

71. For the racecars to stay on the road while traveling at high speed, how did a friction force need to be exerted?
    (a) Parallel to the roadway and outward
    (b) Parallel to the roadway and toward the infield
    (c) Horizontally toward the center of the track
    (d) Opposite the direction of motion
    (e) None of the above

72. The average speed reported in the reading passage has six significant digits, implying that the speed is known to within ± 0.001 mi/h. If this is correct, which answer below is closest to the uncertainty in the time needed to travel around the 1.5-mi oval track? (Think about the percent uncertainties.)
    (a) ± 0.0001 s   (b) ± 0.001 s   (c) ± 0.01 s
    (d) ± 0.1 s   (e) ± 1 s

73. What was the approximate radius of the part of the track where the drivers experienced the 5.5 g acceleration?
    (a) 40 m   (b) 80 m   (c) 200 m
    (d) 400 m   (e) 1000 m

Halley's Comet Edmond Halley was the first to realize that the comets observed in 1531, 1607, and 1682 were really one comet (now called Halley's Comet) that moved around the Sun in an elongated elliptical orbit (see **Figure 4.7**). He predicted that the peanut-shaped comet would reappear in 1757. It appeared in March 1759 (attractions to Jupiter and Saturn delayed its trip by 618 days). More recent appearances of Halley's Comet were in 1835, 1910, and 1986. It is expected again in 2061.

Figure 4.7 The elongated orbit of Halley's Comet.

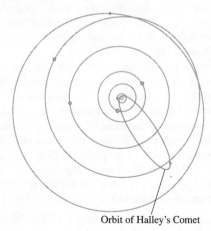

Orbit of Halley's Comet

The nucleus of Halley's Comet is relatively small (15 km long, 8 km wide, and 8 km thick). It has a low 2.2×10^{14} kg mass with an average density of about 600 kg/m³. (The density of water is 1000 kg/m³.) The nucleus rotates once every 52 h. When Halley's Comet is closest to the Sun, temperatures on the comet can rise to about 77 °C and several tons of gas and dust are emitted each second, producing the long tail that we see each time it passes the Sun.

74. **EST** Use the velocity change method to estimate the comet's direction of acceleration when passing closest to the Sun (position I **Figure P4.74**).
    (a) A   (b) B   (c) C   (d) D
    (e) The acceleration is zero.

Figure P4.74

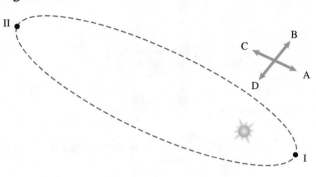

75. What object or objects exert forces on the comet as it passes position I (shown Figure P4.74)?
    (a) The Sun's gravitational force toward the Sun
    (b) The force of motion tangent to the direction the comet is traveling
    (c) An outward force away from the Sun
    (d) a and b
    (e) a, b, and c

76. Suppose that instead of being peanut shaped, Halley's Comet was spherical with a radius of 5.0 km (about its present volume). Which answer below would be closest to your radial acceleration if you were standing on the "equator" of the rotating comet?
    (a) 10^{-5} m/s²   (b) 10^{-3} m/s²   (c) 0.1 m/s²
    (d) 10 m/s²   (e) 1000 m/s²

77. Approximately what gravitational force would the spherical-shaped 5-km radius comet exert on a 100-kg person on the surface of the comet?
    (a) 0.06 N   (b) 0.6 N   (c) 6 N
    (d) 60 N   (e) 600 N

78. The closest distance that the comet passes relative to the Sun is 8.77×10^{10} m (position I in Figure P4.74). Apply Newton's second law and the law of universal gravitation to determine which answer below is closest to the comet's speed when passing position I.
    (a) 1000 m/s   (b) 8000 m/s   (c) 20,000 m/s
    (d) 40,000 m/s   (e) 800,000 m/s

79. The farthest distance that the comet is from the Sun is 5.25×10^{12} m (position II in Figure P4.74). Apply Newton's second law and the law of universal gravitation to determine which answer below is closest to the comet's speed when passing position II.
    (a) 800 m/s   (b) 5000 m/s   (c) 10,000 m/s
    (d) 50,000 m/s   (e) 80,000 m/s

Impulse and Linear Momentum

5

How does jet propulsion work?

How can you measure the speed of a bullet?

Would a meteorite collision significantly change Earth's orbit?

In previous chapters we discovered that the pushing interaction between car tires and the road allows a car to change its velocity. Likewise, a ship's propellers push water backward; in turn, water pushes the ship forward. But how does a rocket, far above Earth's atmosphere, change velocity with no object to push against?

Less than 100 years ago, rocket flight was considered impossible. When U. S. rocket pioneer Robert Goddard published an article

Be sure you know how to:

- Construct a force diagram for an object (Section 2.1).
- Use Newton's second law in component form (Section 3.2).
- Use kinematics to describe an object's motion (Section 1.7).

in 1920 about rocketry and even suggested a rocket flight to the Moon, he was ridiculed by the press. A *New York Times* editorial dismissed his idea, saying, ". . . even a schoolboy knows that rockets cannot fly in space because a vacuum is devoid of anything to push on." We know now that Goddard was correct—but why? What does the rocket push on?

We can use Newton's second law ($\vec{a} = \Sigma \vec{F}/m$) to relate the acceleration of a system object to the forces being exerted on it. However, to use this law effectively we need quantitative information about the forces that objects exert on each other. Unfortunately, if two cars collide, we don't know the force that one car exerts on the other during the collision. When fireworks explode, we don't know the forces that are exerted on the pieces flying apart. In this chapter you will learn a new approach that helps us analyze and predict mechanical phenomena when the forces are not known.

5.1 Mass accounting

We begin our investigation by analyzing the physical quantity of mass. Earlier (in Chapter 2), we found that the acceleration of an object depended on its mass—the greater its mass, the less it accelerated due to an unbalanced external force. We ignored the possibility that an object's mass might change during some process. Is the mass in a system always a constant value?

You have probably observed countless physical processes in which mass seems to change. For example, the mass of a log in a campfire decreases as the log burns; the mass of a seedling increases as the plant grows. What happens to the "lost" mass from the log? Where does the seedling's increased mass come from?

A system perspective helps us understand what happens to the burning log. If we choose only the log as the system, the mass of the system decreases as it burns. However, air is needed for burning. What happens to the mass if we choose the surrounding air and the log as the system?

Suppose that we place steel wool in a closed flask on one side of a balance scale and a metal block of equal mass on the other side (**Figure 5.1a**). In one experiment, we burn the steel wool in the closed flask (the flask also contains air), forming an oxide of iron. We find that the total mass of the closed flask containing burned steel wool (iron oxide) is the same as the mass of the balancing metal block (Figure 5.1b). Next, we burn the steel wool in an open flask and observe that the mass of that flask increases (Figure 5.1c). The steel wool in the open flask burns more completely and absorbs some external oxygen from the air as it burns.

Eighteenth-century French chemist Antoine Lavoisier actually performed such experiments. He realized that the choice of the system was very important. Lavoisier defined an **isolated system** as a group of objects that interact with each other but not with external objects outside the system. The mass of an isolated system is the sum of the masses of all objects in the system. He then used the concept of an isolated system to summarize his (and our) experiments in the following way:

Figure 5.1 The mass is the same in the closed flask (an isolated system) (a) before burning the steel wool and (b) after burning the steel wool. (c) However, the mass increases when the steel wool is burned in the open flask (a non-isolated system).

(a)

The block balances the steel wool and flask.

Closed flask

Steel wool

Balancing block

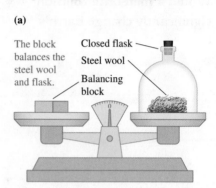

(b)

The steel wool is burned in a closed flask. The block still balances.

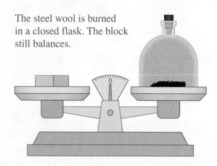

(c)

When the steel wool is burned in the open flask, the mass in the flask increases.

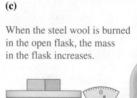

> **Law of constancy of mass** When a system of objects is isolated (a closed container), its mass equals the sum of the masses of its components and does not change—it remains constant in time.

When the system is not isolated (an open container system), the mass might change. However, this change is not random—it is always equal to the amount of mass leaving or entering the system from the environment.

Thus, even when the mass of a system is not constant, we can keep track of the changes if we take into account how much is leaving or entering the system:

$$\begin{pmatrix} initial \text{ mass of} \\ \text{system at earlier} \\ \text{clock reading} \end{pmatrix} + \begin{pmatrix} new \text{ mass entering or} \\ \text{leaving system between} \\ \text{the two clock readings} \end{pmatrix} = \begin{pmatrix} final \text{ mass of} \\ \text{system at later} \\ \text{clock reading} \end{pmatrix}$$

The above equation helps describe the change of mass in any system. The mass is constant if there is no flow of mass in or out of the system, or the mass changes in a predictable way if there is some flow of mass between the system and the environment. Basically, mass cannot appear from nowhere and does not disappear without a trace. Imagine you have a system that has a total mass of $m_i = 3$ kg (a bag of oranges). You add some more oranges to the bag ($\Delta m = 1$ kg). The final mass of the system equals exactly the sum of the initial mass and the added mass: $m_i + \Delta m = m_f$ or $3\,\text{kg} + 1\,\text{kg} = 4\,\text{kg}$ (**Figure 5.2a**). We can represent this process with a bar chart (Figure 5.2b). The bar on the left represents the initial mass of the system, the central bar represents the mass added or taken away, and the bar on the right represents the mass of the system in the final situation. As a result, the height of the left bar plus the height of the central bar equals the height of the right bar. The bar chart allows us to keep track of the changes in mass of a system even if system is not isolated.

Mass is called a **conserved** quantity. A conserved quantity is constant in an isolated system. When the system is not isolated, we can account for the changes in the conserved quantity by what is added to or subtracted from the system.

Just as with every idea in physics, the law of constancy of mass in an isolated system does not apply in all cases. We will discover later in this book (Chapters 28 and 29) that in situations involving atomic particles, mass is not constant even in an isolated system; instead, what is constant is a new quantity that includes mass as a component.

Review Question 5.1 When you burn a log in a fire pit, the mass of wood clearly decreases. How can you define the system so as to have the mass of the objects in that system constant?

5.2 Linear momentum

We now know that mass is an example of a conserved quantity. Is there a quantity related to motion that is conserved? When you kick a stationary ball, there seems to be a transfer of motion from your foot to the ball. When you knock bowling pins down with a bowling ball, a similar transfer occurs. However, motion is not a physical quantity. What physical quantities describing motion are constant in an isolated system? Can we describe the changes in these quantities using a bar chart?

Let's conduct a few experiments to find out. In Observational Experiment **Table 5.1** we observe two carts of different masses that collide on a smooth track. For these experiments, the system will include both carts. A collision is a process that occurs when two (or more) objects come into direct contact with each other. The system is isolated since the forces that the carts exert on each other are internal, and external forces are either balanced (as the vertical forces are) or negligible (the horizontal friction force).

Figure 5.2 (a) The initial mass of the oranges plus the mass of the oranges that were added (or subtracted) equals the final mass of the oranges. (b) The mass change process is represented by a mass bar chart.

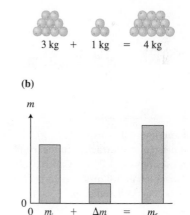

(a)

3 kg   +   1 kg   =   4 kg

(b)

< Active Learning Guide

OBSERVATIONAL EXPERIMENT TABLE

5.1   **Collisions in a system of two carts (all velocities are with respect to the track).**

VIDEO 5.1

Observational experiment	Analysis

Experiment 1. Cart A (0.20 kg) moving right at 1.0 m/s collides with cart B (0.20 kg), which is stationary. Cart A stops and cart B moves right at 1.0 m/s.

The direction of motion is indicated with a plus and a minus sign.

- *Speed:* The sum of the speeds of the system objects is the same before and after the collision: 1.0 m/s + 0 m/s = 0 m/s + 1.0 m/s.
- *Mass · speed:* The sum of the products of mass and speed is the same before and after the collision: 0.20 kg(1.0 m/s) + 0.20 kg(0 m/s) = 0.20 kg(0 m/s) + 0.20 kg(1.0 m/s).
- *Mass · velocity:* The sum of the products of mass and the *x*-component of velocity is the same before and after the collision: 0.20 kg(+1.0 m/s) + 0.20 kg(0) = 0.20 kg(0) + 0.20 kg(+1.0 m/s).

Experiment 2. Cart A (0.40 kg) moving right at 1.0 m/s collides with cart B (0.20 kg), which is stationary. After the collision, both carts move right, cart B at 1.2 m/s, and cart A at 0.4 m/s.

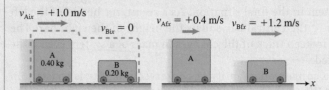

- *Speed:* The sum of the speeds of the system objects is not the same before and after the collision: 1.0 m/s + 0 m/s ≠ 0.4 m/s + 1.2 m/s.
- *Mass · speed:* The sum of the products of mass and speed is the same before and after the collision: 0.40 kg(1.0 m/s) + 0.20 kg(0 m/s) = 0.40 kg(0.4 m/s) + 0.20 kg(1.2 m/s).
- *Mass · velocity:* The sum of the products of mass and the *x*-component of velocity is the same before and after the collision: 0.40 kg(+1.0 m/s) + 0.20 kg(0) = 0.40 kg(+0.4 m/s) + 0.20 kg(+1.2 m/s).

Experiment 3. Cart A (0.20 kg) with a piece of clay attached to the front moves right at 1.0 m/s. Cart B (0.20 kg) moves left at 1.0 m/s. The carts collide, stick together, and stop.

- *Speed:* The sum of the speeds of the system objects is not the same before and after the collision: 1.0 m/s + 1.0 m/s ≠ 0 m/s + 0 m/s.
- *Mass · speed:* The sum of the products of mass and speed is not the same before and after the collision: 0.20 kg(1.0 m/s) + 0.20 kg(1.0 m/s) ≠ 0.20 kg(0 m/s) + 0.20 kg(0 m/s).
- *Mass · velocity:* The sum of the products of mass and the *x*-component of velocity is the same before and after the collision: 0.20 kg(+1.0 m/s) + 0.20 kg(−1.0 m/s) = 0.20 kg(0 m/s) + 0.20 kg(0 m/s).

Patterns

One quantity remains the same before and after the collision in each experiment—the sum of the products of the mass and *x*-velocity component of the system objects.

In the three experiments in Observational Experiment Table 5.1, only one quantity—the sum of the products of mass and the *x*-component of velocity Σmv_x—remained the same before and after the carts collided. Note also that the sum of the products of the mass and the *y*-component of velocity Σmv_y did not change—it remained zero. Perhaps $\Sigma m\vec{v}$ is the quantity characterizing motion that is constant in an isolated system. But will this pattern persist in other situations? Let's test this idea by using it to predict the outcome of the experiment in Testing Experiment **Table 5.2**.

‹ Active Learning Guide

TESTING EXPERIMENT TABLE

5.2 Testing the idea that $\Sigma m\vec{v}$ in an isolated system remains constant (all velocities are with respect to the track).

VIDEO 5.2

Testing experiment	Prediction	Outcome
Cart A (0.40 kg) has a piece of modeling clay attached to its front and is moving right at 1.0 m/s. Cart B (0.20 kg) is moving left at 1.0 m/s. The carts collide and stick together. Predict the velocity of the carts after the collision.	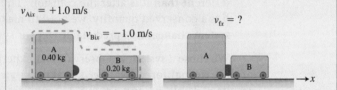	After the collision, the carts move together toward the right at close to the predicted speed.

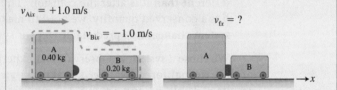

The system consists of the two carts. The direction of velocity is noted with a plus or minus sign of the velocity component:

$$(0.40\,\text{kg})(+1.0\,\text{m/s}) + (0.20\,\text{kg})(-1.0\,\text{m/s})$$
$$= (0.40\,\text{kg} + 0.20\,\text{kg})v_{fx}$$

or

$$v_{fx} = (+0.20\,\text{kg} \cdot \text{m/s})/(0.60\,\text{kg}) = +0.33\,\text{m/s}$$

After the collision, the two carts should move right at a speed of about 0.33 m/s.

Conclusion

Our prediction matched the outcome. This result gives us increased confidence that this new quantity $m\vec{v}$ might be the quantity whose sum is constant in an isolated system.

This new quantity is called **linear momentum** $\vec{p}$.

Linear Momentum The linear momentum $\vec{p}$ of a single object is the product of its mass m and velocity $\vec{v}$:

$$\vec{p} = m\vec{v} \tag{5.1}$$

Linear momentum is a vector quantity that points in the same direction as the object's velocity $\vec{v}$ (**Figure 5.3**). The SI unit of linear momentum is (kg · m/s). The total linear momentum of a system containing multiple objects is the vector sum of the momenta (plural of momentum) of the individual objects.

$$\vec{p}_{net} = m_1\vec{v}_1 + m_2\vec{v}_2 + \cdots + m_n\vec{v}_n = \Sigma m\vec{v}$$

Figure 5.3 Momentum is a vector quantity with components.

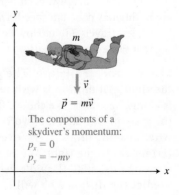

The components of a skydiver's momentum:
$p_x = 0$
$p_y = -mv$

Note the following three important points.

1. Unlike mass, which is a scalar quantity, $\vec{p} = m\vec{v}$ is a vector quantity. Therefore, it is important to consider the direction in which the colliding objects are moving before and after the collision. For example, because cart B in Table 5.2 was moving left along the x-axis, the x-component of its momentum was negative before the collision.

2. Because momentum depends on the velocity of the object, and the velocity depends on the choice of the reference frame, different observers will measure different momenta for the same object. As a passenger, the momentum of a car with respect to you is zero. However, it is not zero for an observer on the ground watching the car move away from him.

3. We chose an isolated system (the two carts) for our investigation. The sum of the products of mass and velocity $\Sigma m\vec{v}$ of all objects in the isolated system remained constant even though the carts collided with each other. However, if we had chosen the system to be just one of the carts, we would see that the linear momentum $\vec{p} = m\vec{v}$ of the cart before the collision is different than it is after the collision. Thus, to establish that momentum $\vec{p}$ is a conserved quantity, we need to make sure that the momentum of a system changes in a predictable way for systems that are not isolated.

We chose a system in Observational Experiment Table 5.1 so that the sum of the external forces was zero, making it an isolated system. Based on the results of Table 5.1 and Table 5.2, it appears that the total momentum of an isolated system is constant.

Momentum constancy of an isolated system The momentum of an isolated system is constant. For an isolated two-object system:

$$m_1\vec{v}_{1i} + m_2\vec{v}_{2i} = m_1\vec{v}_{1f} + m_2\vec{v}_{2f} \tag{5.2}$$

Because momentum is a vector quantity and Eq. (5.2) is a vector equation, we will work with its x- and y-component forms:

$$m_1 v_{1ix} + m_2 v_{2ix} = m_1 v_{1fx} + m_2 v_{2fx} \tag{5.3x}$$

$$m_1 v_{1iy} + m_2 v_{2iy} = m_1 v_{1fy} + m_2 v_{2fy} \tag{5.3y}$$

For a system with more than two objects, we simply include a term on each side of the equation for each object in the system. Let's test the idea that the momentum of an isolated system is constant in another situation.

EXAMPLE 5.1 Two rollerbladers

Jen (50 kg) and David (75 kg), both on rollerblades, push off each other abruptly. Each person coasts backward at approximately constant speed. During a certain time interval, Jen travels 3.0 m. How far does David travel during that same time interval?

Sketch and translate The process is sketched at the right. All motion is with respect to the floor and is along the x-axis. We choose the two rollerbladers as the system. Initially, the two rollerbladers are at rest. After pushing off, Jen (J) moves to the left and David (D) moves to the right. We can use momentum constancy to calculate David's velocity component and predict the distance he will travel during that same time interval.

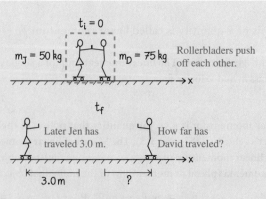

Simplify and diagram We model each person as a point-like object and assume that the friction force exerted on the rollerbladers does not affect their motion. Thus there are no horizontal external forces exerted on the system. In addition, the two vertical forces, an upward normal force $\vec{N}_{\text{F on P}}$ that the floor exerts on each person and an equal-magnitude downward gravitational force $\vec{F}_{\text{E on P}}$ that Earth exerts on each person, cancel, as we see in the force diagrams. Since the net external force exerted on the system is zero, the system is isolated. The forces that the rollerbladers exert on each other are internal forces and should not affect the momentum of the system.

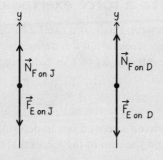

Represent mathematically The initial state (i) of the system is before they start pushing on each other, and the final state (f) is when Jen has traveled 3.0 m.

$$m_{\text{J}}v_{\text{Ji}\,x} + m_{\text{D}}v_{\text{Di}\,x} = m_{\text{J}}v_{\text{Jf}\,x} + m_{\text{D}}v_{\text{Df}\,x}$$

We choose the positive direction toward the right. Because the initial velocity of each person is zero, the above equation becomes

$$0 + 0 = m_{\text{J}}v_{\text{Jf}\,x} + m_{\text{D}}v_{\text{Df}\,x}$$

or

$$m_{\text{D}}v_{\text{Df}\,x} = -m_{\text{J}}v_{\text{Jf}\,x}$$

Solve and evaluate The x-component of Jen's velocity after the push-off is $v_{\text{Jf}\,x} = -(3.0\ \text{m})/\Delta t$, where Δt is the time interval needed for her to travel 3.0 m. We solve the above equation for David's final x-velocity component to determine how far David should travel during that same time interval:

$$v_{\text{Df}\,x} = -\frac{m_{\text{J}}v_{\text{Jf}\,x}}{m_{\text{D}}} = -\frac{m_{\text{J}}}{m_{\text{D}}}v_{\text{Jf}\,x}$$

$$= -\frac{(50\ \text{kg})}{(75\ \text{kg})}\frac{(-3.0\ \text{m})}{\Delta t} = \frac{(2.0\ \text{m})}{\Delta t}$$

Since momentum is constant in this isolated system, we predict that David will travel 2.0 m in the positive direction during Δt. The measured value is very close to the predicted value.

Try it yourself: Estimate the magnitude of your momentum when walking and when jogging. Assume your mass is 60 kg.

Answer: When walking, you travel at a speed of about 1 to 2 m/s. So the magnitude of your momentum will be $p = mv \approx (60\ \text{kg})(1.5\ \text{m/s}) \approx 90\ \text{kg} \cdot \text{m/s}$. When jogging, your speed is about 2 to 5 m/s or a momentum of magnitude $p = mv \approx (60\ \text{kg})(3.5\ \text{m/s}) \approx 200\ \text{kg} \cdot \text{m/s}$.

Notice that in Example 5.1 we were able to determine David's velocity by using the principle of momentum constancy. We did not need any information about the forces involved. This is a very powerful result, since in all likelihood the forces they exerted on each other were not constant. The kinematics equations we have used up to this point have assumed constant acceleration of the system (and thus constant forces). Using the idea of momentum constancy has allowed us to analyze a situation involving nonconstant forces.

So far, we have investigated situations involving isolated systems. In the next section, we will investigate momentum in nonisolated systems.

Review Question 5.2 Two identical carts are traveling toward each other at the same speed. One of the carts has a piece of modeling clay on its front. The carts collide, stick together, and stop. The momentum of each cart is now zero. If the system includes both carts, did the momentum of the system disappear? Explain your answer.

5.3 Impulse and momentum

So far, we have found that the linear momentum of a system is constant if that system is isolated (the net external force exerted on the system is zero). How do we account for the change in momentum of a system when the net external force exerted on it is not zero? We can use Newton's laws to derive an expression relating forces and momentum change.

‹ Active Learning Guide

Impulse due to a force exerted on a single object

When you push a bowling ball, you exert a force on it, causing the ball to accelerate. The average acceleration $\vec{a}$ is defined as the change in velocity $\vec{v}_f - \vec{v}_i$ divided by the time interval $\Delta t = t_f - t_i$ during which that change occurs:

$$\vec{a} = \frac{\vec{v}_f - \vec{v}_i}{t_f - t_i}$$

We can also use Newton's second law to determine an object's acceleration if we know its mass and the sum of the forces that other objects exert on it:

$$\vec{a} = \frac{\Sigma \vec{F}}{m}$$

We now have two expressions for an object's acceleration. Setting these two expressions for acceleration equal to each other, we get

$$\frac{\vec{v}_f - \vec{v}_i}{t_f - t_i} = \frac{\Sigma \vec{F}}{m}$$

Now multiply both sides by $m(t_f - t_i)$ and get the following:

$$m\vec{v}_f - m\vec{v}_i = \vec{p}_f - \vec{p}_i = \Sigma \vec{F}(t_f - t_i) \tag{5.4}$$

The left side of the above equation is the change in momentum of the object. This change depends on the product of the net external force and the time interval during which the forces are exerted on the object (the right side of the equation). Note these two important points:

1. Equation (5.4) is just Newton's second law written in a different form—one that involves the physical quantity momentum.

2. Both force *and* time interval affect momentum—the longer the time interval, the greater the momentum change. A small force exerted for a long time interval can change the momentum of an object by the same amount as a large force exerted for a short time interval.

The product of the external force exerted on an object during a time interval and the time interval gives us a new quantity, the **impulse** of the force. When you kick a football or hit a baseball with a bat, your foot or the bat exerts an impulse on the ball. The forces in these situations are not constant but instead vary in time (see the example in **Figure 5.4**). The shaded area under the varying force curve represents the impulse of the force. We can estimate the impulse by drawing a horizontal line that is approximately the average force exerted during the time interval of the impulse. The area under the rectangular average force-impulse curve equals the product of the height of the rectangle (the average force) and the width of the rectangle (the time interval over which the average force is exerted). The product $F_{av}(t_f - t_i)$ equals the magnitude of the impulse.

Figure 5.4 The impulse of a force is the area under the *F*-versus-*t* graph line.

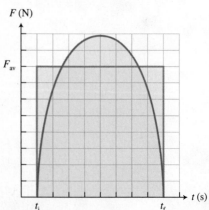

The area of the shaded rectangle is about the same as the area under the curved line and equals the impulse of the force.

Impulse The impulse $\vec{J}$ of a force is the product of the average force $\vec{F}_{av}$ exerted on an object during a time interval $(t_f - t_i)$ and that time interval:

$$\vec{J} = \vec{F}_{av}(t_f - t_i) \tag{5.5}$$

Impulse is a vector quantity that points in the direction of the force. The impulse has a plus or minus sign depending on the orientation of the force relative to a coordinate axis. The SI unit for impulse is $N \cdot s = (kg \cdot m/s^2) \cdot s = kg \cdot m/s$, the same unit as momentum.

It is often difficult to measure directly the impulse of the net average force during a time interval. However, we can determine the net force on the right side of Eq. (5.4) indirectly by measuring or calculating the momentum change on the left side of the equation. For this reason, the combination of impulse and momentum change provides a powerful tool for analyzing interactions between objects. We can now write Eq. (5.4) as the impulse-momentum equation for a single object.

Impulse-momentum equation for a single object If several external objects exert forces on a single-object system during a time interval $(t_f - t_i)$, the sum of their impulses $\Sigma \vec{J}$ causes a change in momentum of the system object:

$$\vec{p}_f - \vec{p}_i = \Sigma \vec{J} = \Sigma \vec{F}_{\text{on System}}(t_f - t_i) \qquad (5.6)$$

The x- and y-scalar component forms of the impulse-momentum equation are

$$p_{fx} - p_{ix} = \Sigma F_{\text{on System }x}(t_f - t_i) \qquad (5.7x)$$

$$p_{fy} - p_{iy} = \Sigma F_{\text{on System }y}(t_f - t_i) \qquad (5.7y)$$

A few points are worth emphasizing. First, notice that Eq. (5.6) is a vector equation, as both the momentum and the impulse are vector quantities. Vector equations are not easy to manipulate mathematically. Therefore, we will use the scalar component forms of Eq. (5.6)—Eqs. (5.7x) and (5.7y).

Second, the time interval in the impulse-momentum equation is very important. When object 2 exerts a force on object 1, the momentum of object 1 changes by an amount equal to

$$\Delta \vec{p}_1 = \vec{p}_{1f} - \vec{p}_{1i} = \vec{F}_{2 \text{ on } 1}(t_f - t_i) = \vec{F}_{2 \text{ on } 1}\Delta t$$

The longer that object 2 exerts the force on object 1, the greater the momentum change of object 1. This explains why a fast-moving object might have less of an effect on a stationary object during a collision than a slow-moving object interacting with the stationary object over a longer time interval. For example, a fast-moving bullet passing through a partially closed wooden door might not open the door (it will just make a hole in the door), whereas your little finger, moving much slower than the bullet, could open the door. Although the bullet moves at high speed and exerts a large force on the door, the time interval during which it interacts with the door is very small (milliseconds). Hence, it exerts a relatively small impulse on the door—too small to significantly change the door's momentum. A photo of a bullet shot through an apple illustrates the effect of a short impulse time (**Figure 5.5**). The impulse exerted by the bullet on the apple was too small to knock the apple off its support.

Third, if the magnitude of the force changes during the time interval considered in the process, we use the average force.

Finally, if the same amount of force is exerted for the same time interval on a large-mass object and on a small-mass object, the objects will have an equal change in momentum (the same impulse was exerted on them). However, the small-mass object would experience a greater change in velocity than the large-mass object.

Figure 5.5 The bullet's time of interaction with the apple is very short, causing a small impulse that does not knock the apple over.

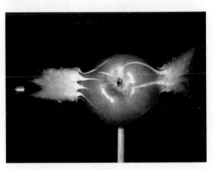

‹ Active Learning Guide

EXAMPLE 5.2 Abrupt stop in a car

A 60-kg person is traveling in a car that is moving at 16 m/s with respect to the ground when the car hits a barrier. The person is not wearing a seat belt, but is stopped by an air bag in a time interval of 0.20 s. Determine the *average* force that the air bag exerts on the person while stopping him.

Sketch and translate First we draw an initial-final sketch of the process. We choose the person as the system since we are investigating a force being exerted on him.

(continued)

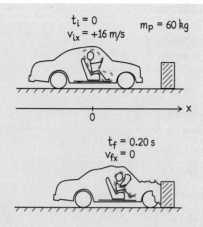

The person's initial x-component of velocity $v_{Pix} = +16$ m/s decreases to the final x-component of velocity $v_{Pfx} = 0$ in a time interval $(t_f - t_i) = 0.20$ s. Thus the average force exerted by the air bag on the person in the x-direction is

$$F_{B\ on\ P\ x} = \frac{(60\ kg)(0 - 16\ m/s)}{(0.20\ s - 0)}$$
$$= -4800\ N$$

The negative sign in -4800 N indicates that the average force points in the negative x-direction. The magnitude of this force is about 1000 lb!

Try it yourself: Suppose a 60-kg crash test dummy is in a car traveling at 16 m/s. The dummy is not wearing a seat belt and the car has no air bags. During a collision, the dummy flies forward and stops when it hits the dashboard. The stopping time interval for the dummy is 0.02 s. What is the average magnitude of the stopping force that the dashboard exerts on the dummy?

Answer: The average force that the hard surface exerts on the dummy would be about 50,000 N, extremely unsafe for a human. Note that the momentum change of the person in Example 5.2 was the same. However, since the change for the dummy occurs during a shorter time interval (0.02 s instead of 0.20 s), the force exerted on the dummy is much greater. This is why air bags save lives.

Simplify and diagram
The force diagram shows the average force $\vec{F}_{B\ on\ P}$ exerted in the negative direction by the bag on the person. The vertical normal force and gravitational forces cancel.

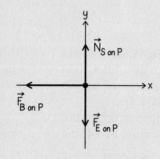

Represent mathematically The x-component form of the impulse-momentum equation is

$$m_P v_{Pi\ x} + F_{B\ on\ P\ x}(t_f - t_i) = m_P v_{Pf\ x}$$

Solve and evaluate Solve for the force exerted by the air bag on the person:

$$F_{B\ on\ P\ x} = \frac{m_P(v_{Pfx} - v_{Pix})}{(t_f - t_i)}$$

Using Newton's laws to understand the constancy of momentum in an isolated system of two or more objects

Let's apply the impulse-momentum equation Eq. (5.4) to the scenario we described in Observational Experiment Table 5.1 in order to explore momentum constancy in a two-object isolated system.

Two carts travel toward each other at different speeds, collide, and rebound backward (**Figure 5.6**). We first analyze each cart as a separate system and then analyze them together as a single system. Assume that the vertical forces exerted on the carts are balanced and that the friction force exerted by the surface on the carts does not significantly affect their motion.

Cart 1: In the initial state, before the collision, cart 1 with mass m_1 travels in the positive direction at velocity $\vec{v}_{1i}$. In the final state, after the collision, cart 1 moves with a different velocity $\vec{v}_{1f}$ in the opposite direction. To determine the effect of the impulse exerted by cart 2 on cart 1, we apply the impulse-momentum equation to cart 1 only:

$$m_1(\vec{v}_{1f} - \vec{v}_{1i}) = \vec{F}_{2\ on\ 1}(t_f - t_i)$$

Figure 5.6 Analyzing the collision of two carts in order to develop the momentum constancy idea.

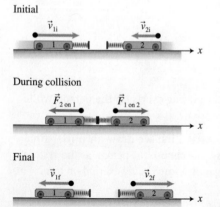

Cart 2: We repeat this analysis with cart 2 as the system. Its velocity and momentum change because of the impulse exerted on it by cart 1:

$$m_2(\vec{v}_{2f} - \vec{v}_{2i}) = \vec{F}_{1 \text{ on } 2}(t_f - t_i)$$

Newton's third law provides a connection between our analyses of the two carts; interacting objects at each instant exert equal-magnitude but oppositely directed forces on each other:

$$\vec{F}_{1 \text{ on } 2} = -\vec{F}_{2 \text{ on } 1}$$

Substituting the expressions for the forces from above and simplifying, we get

$$\frac{m_2(\vec{v}_{2f} - \vec{v}_{2i})}{t_f - t_i} = -\frac{m_1(\vec{v}_{1f} - \vec{v}_{1i})}{t_f - t_i}$$

$$m_2(\vec{v}_{2f} - \vec{v}_{2i}) = -m_1(\vec{v}_{1f} - \vec{v}_{1i})$$

We now move the initial momentum for both objects to the left side and the final momentum for both objects to the right side:

$$\underbrace{m_1\vec{v}_{1i} + m_2\vec{v}_{2i}}_{\text{Initial momentum}} = \underbrace{m_1\vec{v}_{1f} + m_2\vec{v}_{2f}}_{\text{Final momentum}}$$

This is the same equation we arrived at in Section 5.2, where we observed and analyzed collisions to understand the constant momentum of an isolated system. Here we have reached the same conclusions using only our knowledge of Newton's laws, momentum, and impulse.

Review Question 5.3 An apple is falling from a tree. Why does its momentum change? Specify the external force responsible. Find a system in which the momentum is constant during this process.

5.4 The generalized impulse-momentum principle

We can summarize what we have learned about momentum in isolated and non-isolated systems. The change in momentum of a system is equal to the net external impulse exerted on it. If the net impulse is zero, then the momentum of the system is constant. This idea, expressed mathematically as the **generalized impulse-momentum principle,** accounts for situations in which the system includes one or more objects and may or may not be isolated. The generalized impulse-momentum principle means that we can treat momentum as a conserved quantity.

Generalized impulse-momentum principle For a system containing one or more objects, the initial momentum of the system plus the sum of the impulses that external objects exert on the system objects during the time interval $(t_f - t_i)$ equals the final momentum of the system:

$$\underbrace{(m_1\vec{v}_{1i} + m_2\vec{v}_{2i} + \cdots)}_{\substack{\text{Initial momentum of} \\ \text{the system}}} + \underbrace{\Sigma\vec{F}_{\text{on Sys}}(t_f - t_i)}_{\substack{\text{Net impulse exerted on} \\ \text{the system}}} = \underbrace{(m_1\vec{v}_{1f} + m_2\vec{v}_{2f} + \cdots)}_{\substack{\text{Final momentum of} \\ \text{the system}}} \quad (5.8)$$

The *x*- and *y*-component forms of the generalized impulse-momentum principle are

$$(m_1v_{1ix} + m_2v_{2ix} + \cdots) + \Sigma F_{\text{on Sys }x}(t_f - t_i) = (m_1v_{1fx} + m_2v_{2fx} + \cdots) \text{ (5.9x)}$$

$$(m_1v_{1iy} + m_2v_{2iy} + \cdots) + \Sigma F_{\text{on Sys }y}(t_f - t_i) = (m_1v_{1fy} + m_2v_{2fy} + \cdots) \text{ (5.9y)}$$

Note: If the net impulse exerted in a particular direction is zero, then the component of the momentum of the system in that direction is constant.

Equations (5.8) and (5.9) are useful in two ways. First, any time we choose to analyze a situation using the ideas of impulse and momentum, we can start from a single principle, regardless of the situation. Second, the equations remind us that we need to consider all the interactions between the environment and the system that might cause a change in the momentum of the system.

Impulse-momentum bar charts

We can describe an impulse-momentum process mathematically using Eqs. (5.9x and y). These equations help us see that we can represent the changes of a system's momentum using a bar chart similar to the one used to represent the changes of a system's mass. The Reasoning Skill box shows the steps for constructing an **impulse-momentum bar chart** for a simple system of two carts of equal mass traveling toward each other.

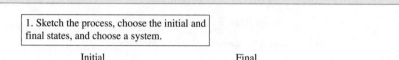

REASONING SKILL   Constructing a qualitative impulse-momentum bar chart.

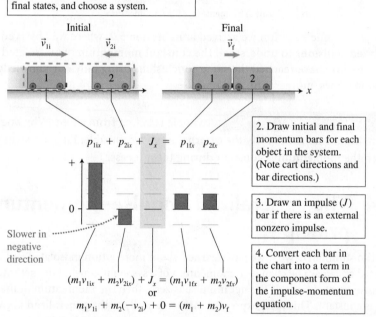

1. Sketch the process, choose the initial and final states, and choose a system.

2. Draw initial and final momentum bars for each object in the system. (Note cart directions and bar directions.)

3. Draw an impulse (J) bar if there is an external nonzero impulse.

4. Convert each bar in the chart into a term in the component form of the impulse-momentum equation.

$$(m_1 v_{1ix} + m_2 v_{2ix}) + J_x = (m_1 v_{1fx} + m_2 v_{2fx})$$
or
$$m_1 v_{1i} + m_2(-v_{2i}) + 0 = (m_1 + m_2)v_f$$

Note that before constructing the bar chart, we represent the process in an initial-final sketch (Step 1 in the Skill box). We then use the sketch to help construct the impulse-momentum bar chart. The lengths of the bars are *qualitative* indicators of the relative magnitudes of the momenta. In the final state in the example shown, the carts are stuck together and are moving in the positive direction. Since they have the same mass and velocity, they each have the same final momentum.

The middle shaded column in the bar chart represents the net external impulse exerted on the system objects during the time interval $(t_f - t_i)$—there is no impulse for the process shown. The shading reminds us that impulse does not reside in the system; it is the influence of the external objects on the momentum of the system. Notice that the sum of the heights of the bars on the left plus the height of the shaded impulse bar should equal the sum of the heights of the bars on the right. This "conservation of bar heights" reflects the conservation of momentum.

We can use the bar chart to apply the generalized impulse-momentum equation (Step 4). Each nonzero bar corresponds to a nonzero term in the equation; the sign of the term depends on the orientation of the bar.

Using impulse-momentum to investigate forces

Can we use the ideas of impulse and momentum to learn something about the forces that two objects exert on each other during a collision? Consider a collision between two cars (**Figure 5.7**).

To analyze the force that each car exerts on the other, we will define the system to include only one of the cars. Let's choose car 1 and construct a bar chart for it. Car 2 exerts an impulse on car 1 during the collision that changes the momentum of car 1. If the initial momentum of car 1 is in the positive direction, then the impulse exerted by car 2 on car 1 points in the negative direction. Because of this, the impulse bar on the bar chart points downward. Note that the total height of the initial momentum bar on the left side of the chart and the height of the impulse bar add up to the total height of the final momentum bar on the right side. Using the bar chart, we can apply the component form of the impulse-momentum equation:

$$m_1 v_{1ix} + J_x = m_1 v_{1fx}$$

The components of the initial and final momentum are positive. As the force is exerted in the negative direction, the x-component of the impulse is negative and equal to $-F_{2\,on\,1}\Delta t$. Thus,

$$+m_1 v_{1i} + (-F_{2\,on\,1}\Delta t) = +m_1 v_{1f}$$

If we know the initial and final momentum of the car and the time interval of interaction, we can use this equation to determine the magnitude of the average force that car 2 exerted on car 1 during the collision.

Figure 5.7 A bar chart analysis of the collision of car 2 with car 1.

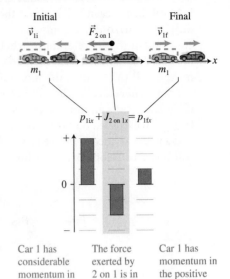

Car 1 has considerable momentum in the positive direction.

The force exerted by 2 on 1 is in the negative direction.

Car 1 has momentum in the positive direction.

‹ Active Learning Guide

> **TIP** When you draw a bar chart, always specify the reference frame (the object of reference and the coordinate system). The direction of the bars on the bar chart (up for positive and down for negative) should match the direction of the momentum or impulse based on the chosen coordinate system.

EXAMPLE 5.3 Happy and sad balls

You have two balls of identical mass and size that behave very differently. When you drop the so-called "sad" ball, it thuds on the floor and does not bounce at all. When you drop the so-called "happy" ball from the same height, it bounces back to almost the same height from which it was dropped. The difference in the bouncing ability of the happy ball is due its internal structure; it is made of different material. You hang each ball from a string of identical length and place a wood board on its end directly below the support for each string. You pull each ball back to an equal height and release the balls one at a time. When each ball hits the board, which has the best chance of knocking the board over: the sad ball or the happy ball?

Sketch and translate Initial and final sketches of the process are shown at the right. The system is just the ball. In the initial state, the ball is just about to hit the board, moving horizontally toward the left (the

balls are moving equally fast). The final state is just after the collision with the board. The happy ball (H) bounces back, whereas the sad ball (S) does not.

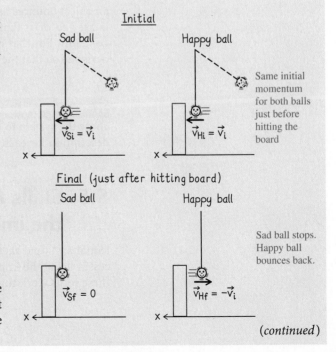

(continued)

Simplify and diagram Assume that the collision time interval Δt for each ball is about the same. We analyze only the horizontal x-component of the process, the component that is relevant to whether or not each of the boards is knocked over. Each board exerts an impulse on the ball that causes the momentum of the ball to change. Therefore, each ball, according to Newton's third law, exerts an impulse on the board that it hits. A larger force exerted on the board means a larger impulse and a better chance to tip the board. A bar chart for each ball-board collision is shown below.

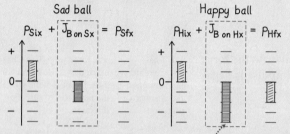

The impulse of the happy ball is twice as large in magnitude as that of the sad ball and causes twice as large a momentum change.

Represent mathematically The x-component form of the impulse-momentum Eq. (5.5x) applied to each ball is as follows:

Sad ball: $mv_i + F_{\text{B on S}x}\Delta t = m \cdot 0$

Happy ball: $mv_i + F_{\text{B on H}x}\Delta t = m(-v_i)$

Note that the x-component of the final velocity of the sad ball is $v_{\text{S}fx} = 0$ (it does not bounce) and that the x-component of the final velocity of the happy ball is $v_{\text{H}fx} = -v_i$ (it bounces).

Solve and evaluate We can now get an expression for the force exerted by each board on each ball:

Sad ball: $F_{\text{B on S}x} = \dfrac{m(0 - v_i)}{\Delta t} = -\dfrac{mv_i}{\Delta t}.$

Happy ball: $F_{\text{B on H}x} = \dfrac{m[(-v_i) - v_i]}{\Delta t} = -\dfrac{2mv_i}{\Delta t}.$

Because we assumed that the time of collision is the same, the board exerts twice the force on the happy ball as on the sad ball, since the board causes the happy ball's momentum to change by an amount twice that of the sad ball. According to Newton's third law, this means that the happy ball will exert twice as large a force on the board as the sad ball. Thus, the happy ball has a greater chance of tipping the board.

Try it yourself: Is it less safe for a football player to bounce backward off a goal post or to hit the goal post and stop?

Answer: Although any collision is dangerous, it is better to hit the goal post and stop. If the football player bounces back off the goal post, his momentum will have changed by a greater amount (like the happy ball in the last example). This means that the goal post exerts a greater force on him, which means there is a greater chance for injury.

The pattern we found in the example above is true for all collisions—when an object bounces back after a collision, we know that a larger magnitude force is exerted on it than if the object had stopped and did not bounce after the collision. For that reason, bulletproof vests for law enforcement agents are designed so that the bullet embeds in the vest rather than bouncing off it.

Review Question 5.4 If in solving the problem in Example 5.3 we chose the system to be the ball and the board, how would the mathematical description for each ball-board collision change?

5.5 Skills for analyzing problems using the impulse-momentum equation

Initial and final sketches and bar charts are useful tools to help analyze processes using the impulse-momentum principle. Let's investigate further how these tools work together. A general strategy for analyzing such processes is described on the left side of the table in Example 5.4 and illustrated on the right side for a specific process.

PROBLEM-SOLVING STRATEGY **Applying the impulse-momentum equation**

Active Learning Guide›	**EXAMPLE 5.4 Bullet hits wood block** A 0.020-kg bullet traveling horizontally at 250 m/s embeds in a 1.0-kg block of wood resting on a table. Determine the speed of the bullet and wood block together immediately after the bullet embeds in the block.

Sketch and translate

- Sketch the initial and final states and include appropriate coordinate axes. Label the sketches with the known information. Decide on the object of reference.

- Choose a system based on the quantity you are interested in; for example, a multi-object isolated system to determine the velocity of an object, or a single-object nonisolated system to determine an impulse or force.

The left side of the sketch below shows the bullet traveling in the positive x-direction with respect to the ground; it then joins the wood. All motion is along the x-axis; the object of reference is Earth. The system includes the bullet and wood; it is an isolated system since the vertical forces balance. The initial state is immediately before the collision; the final state is immediately after.

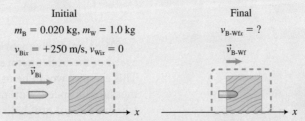

Simplify and diagram

- Determine if there are any external impulses exerted on the system. Drawing a force diagram could help determine the external forces and their directions.

- Draw an impulse-momentum bar chart for the system for the chosen direction(s) to help you understand the situation, formulate a mathematical representation of the process, and evaluate your results.

Assume that the friction force exerted by the tabletop on the bottom of the wood does not change the momentum of the system during the very short collision time interval. The bar chart represents the process. The bar for the bullet is shorter than that for the block—their velocities are the same after the collision, but the mass of the bullet is much smaller. We do not draw a force diagram here, as the system is isolated.

Represent mathematically

- Use the bar chart to apply the generalized impulse-momentum equation along the chosen axis. Each nonzero bar becomes a nonzero term in the equation. The orientation of the bar determines the sign in front of the corresponding term in the equation.

- Remember that momentum and impulse are vector quantities, so include the plus or minus signs of the components based on the chosen coordinate system.

$$m_B v_{Bix} + m_W \cdot 0 + (J_x) = m_B v_{B\text{-}Wfx} + m_W v_{B\text{-}Wfx}$$

$$\text{Since } J_x = 0, \quad v_{B\text{-}Wfx} = \frac{m_B v_{Bix}}{(m_B + m_W)}$$

(continued)

Solve and evaluate

- Insert the known information to determine the unknown quantity.

- Check if your answer is reasonable with respect to sign, unit, and magnitude. Also make sure it applies for limiting cases, such as objects of very small or very large mass.

$$v_{\text{B-Wf}x} = \frac{(0.020\,\text{kg})(250\,\text{m/s})}{(0.020\,\text{kg} + 1.0\,\text{kg})} = +4.9\,\text{m/s}$$

The magnitude of the answer seems reasonable given how fast the bullet was initially traveling. The plus sign indicates the direction, which makes sense, too. The units are also correct (m/s). We can test this using a limiting case: if the mass or speed of the bullet is zero, the block remains stationary after the collision.

Try it yourself: A 0.020-kg bullet is fired horizontally into a 2.00-kg block of wood resting on a table. Immediately after the bullet joins the block, the block and bullet move in the positive *x*-direction at 4.0 m/s. What was the initial speed of the bullet?

Answer: 400 m/s.

We could have worked Example 5.4 backward to determine the initial speed of the bullet before hitting the block (like the Try It Yourself question). This exercise would be useful since the bullet travels so fast that it is difficult to measure its speed. Variations of this method are used, for example, to decide whether or not golf balls conform to the necessary rules. The balls are hit by the same mechanical launching impulse and the moving balls embed in another object. The balls' speeds are determined by measuring the speed of the object they embed in.

Determining the stopping time interval from the stopping distance

When a system object collides with another object and stops—a car collides with a tree or a wall, a person jumps and lands on a solid surface, or a meteorite collides with Earth—the system object travels what is called its **stopping distance**. By estimating the stopping distance of the system object, we can estimate the stopping time interval.

Suppose that a car runs into a large tree and its front end crumples about 0.5 m. This 0.5 m, the distance that the center of the car traveled from the beginning of the impact to the end, is the car's stopping distance. Similarly, the depth of the hole left by a meteorite provides a rough estimate of its stopping distance when it collided with Earth. However, to use the impulse-momentum principle, we need the stopping time interval associated with the collision, not the stopping distance. Here's how we can use a known stopping distance to estimate the stopping time interval.

- Assume that the acceleration of the object while stopping is constant. In that case, the average velocity of the object while stopping is just the sum of the initial and final velocities divided by 2: $v_{\text{average}\,x} = (v_{\text{f}x} + v_{\text{i}x})/2$.

- Thus, the stopping displacement $(x_\text{f} - x_\text{i})$ and the stopping time interval $(t_\text{f} - t_\text{i})$ are related by the kinematics equation

$$x_\text{f} - x_\text{i} = v_{\text{average}\,x}(t_\text{f} - t_\text{i}) = \frac{(v_{\text{f}x} + v_{\text{i}x})}{2}(t_\text{f} - t_\text{i})$$

■ Rearrange this equation to determine the stopping time interval:

$$t_f - t_i = \frac{2(x_f - x_i)}{v_{fx} + v_{ix}} \qquad (5.10)$$

Equation (5.10) provides a method to convert stopping distance $x_f - x_i$ into stopping time interval $t_f - t_i$. Equation (5.10) can be applied to horizontal or vertical stopping.

‹ Active Learning Guide

EXAMPLE 5.5 Stopping the fall of a movie stunt diver

The record for the highest movie stunt fall without a parachute is 71 m (230 ft), held by 80-kg A. J. Bakunas. His fall was stopped by a large air cushion, into which he sank about 4.0 m. His speed was about 36 m/s (80 mi/h) when he reached the top of the air cushion. Estimate the average force that the cushion exerted on his body while stopping him.

Sketch and translate We focus only on the part of the fall when Bakunas is sinking into the cushion. The situation is sketched below. We choose Bakunas as the system and the y-axis pointing up. The initial state is just as he touches the cushion at position $y_i = +4.0$ m, and the final state is when the cushion has stopped him, at position $y_f = 0$. All motion is with respect to Earth. The other information about the process is given in the figure. Be sure to pay attention to the signs of the quantities (especially the initial velocity).

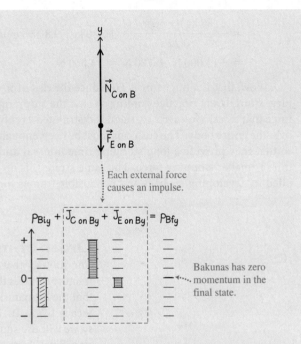

Each external force causes an impulse.

Bakunas has zero momentum in the final state.

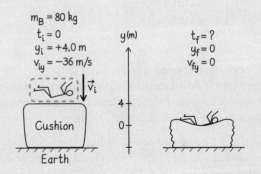

Simplify and diagram We draw a force diagram top right, modeling Bakunas as a point-like object. Since Bakunas's downward speed decreases, the cushion must be exerting an upward force on Bakunas of greater magnitude than the downward force that Earth exerts on him. Thus, the net force exerted on him points upward, in the positive y-direction. Using this information, we can draw a qualitative impulse-momentum bar chart for the process.

Represent mathematically Since all motion and all of the forces are in the vertical direction, we use the bar chart to help construct the vertical y-component form of the impulse-momentum equation [Eq. (5.6y)] to determine the force that the cushion exerts on Bakunas as he sinks into it:

$$m_B v_{iy} + (N_{C\,on\,B\,y} + F_{E\,on\,B\,y})(t_f - t_i) = m_B v_{fy}$$

Using the force diagram, we see that the y-components of the forces are $N_{C\,on\,B\,y} = +N_{C\,on\,B}$ and $F_{E\,on\,B\,y} = -F_{E\,on\,B} = -m_B g$, where $N_{C\,on\,B}$ is the magnitude of the average normal force that the cushion exerts on Bakunas, the force we are trying to estimate. Noting that $v_{fy} = 0$ and substituting the force components into the above equation, we get

$$m_B v_{iy} + [(+N_{C\,on\,B}) + (-m_B g)](t_f - t_i) = m_B \cdot 0$$
$$\Rightarrow m_B v_{iy} + (N_{C\,on\,B} - m_B g)(t_f - t_i) = 0.$$

We can find the time interval that the cushion takes to stop Bakunas using Eq. (5.10) and noting that $v_{fy} = 0$:

$$t_f - t_i = \frac{2(y_f - y_i)}{0 + v_{iy}}$$

(continued)

Solve and evaluate The stopping time interval while Bakunas sinks 4.0 m into the cushion is

$$t_f - t_i = \frac{2(0 - 4.0\,\text{m})}{0 + (-36\,\text{m/s})} = 0.22\,\text{s}$$

Solving for $N_{\text{C on B}}$, we get

$$N_{\text{C on B}} = \frac{-m_B v_{iy}}{(t_f - t_i)} + m_B g$$

$$= \frac{-(80\,\text{kg})(-36\,\text{m/s})}{(0.22\,\text{s})} + (80\,\text{kg})(9.8\,\text{N/kg})$$

$$= +13{,}000\,\text{N} + 780\,\text{N} = 14{,}000\,\text{N}$$

Wow, that is a huge force! To reduce the risk of injury, stunt divers practice landing so that the stopping force that a cushion exerts on them is distributed evenly over the entire body. The cushions must be deep enough so that they provide a long stopping time interval and thus a smaller stopping force. The same strategy is applied to developing air bags and collapsible frames for automobiles to make them safer for passengers during collisions.

Notice four important points. First, we've included only two significant digits since that is how many the data had. Second, it is very easy to make sign mistakes. A good way to avoid these is to draw a sketch that includes a coordinate system and labels showing the values of known physical quantities, including their signs. Third, the impulse due to Earth's gravitational force is small in magnitude compared to the impulse exerted by the air cushion. Lastly, the force exerted by the air cushion would be even greater if the stopping distance and consequently the stopping time interval were shorter.

Try it yourself: Suppose that the cushion in the last example stopped Bakunas in 1.0 m instead of 4.0 m. What would be the stopping time interval and the magnitude of the average force of the cushion on Bakunas?

Answer: The stopping time interval is 0.056 s, and the average stopping force is approximately 50,000 N.

Order-of-magnitude estimate—will bone break?

The strategy that we used in the previous example can be used to analyze skull fracture injuries that might lead to concussions. Laboratory experiments indicate that the human skull can fracture if the compressive force exerted on it per unit area is $1.7 \times 10^8\,\text{N/m}^2$. The surface area of the skull is much smaller than $1\,\text{m}^2$, so we will use square centimeters, a more reasonable unit of area for this discussion. Since $1\,\text{m}^2 = 1 \times 10^4\,\text{cm}^2$, we convert the compressive force per area to

$$(1.7 \times 10^8\,\text{N/m}^2)\left(\frac{1\,\text{m}^2}{1 \times 10^4\,\text{cm}^2}\right) = 1.7 \times 10^4\,\text{N/cm}^2.$$

EXAMPLE 5.6 Bone fracture estimation[1]

A bicyclist is watching for traffic from the left while turning toward the right. A street sign hit by an earlier car accident is bent over the side of the road. The cyclist's head hits the pole holding the sign. Is there a significant chance that his skull will fracture?

Sketch and translate The process is sketched at the right. The initial state is at the instant that the head initially contacts the pole; the final state is when the head

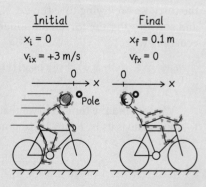

Initial	Final
$x_i = 0$	$x_f = 0.1\,\text{m}$
$v_{ix} = +3\,\text{m/s}$	$v_{fx} = 0$

and body have stopped. The person is the system. We have been given little information, so we'll have to make some reasonable estimates of various quantities in order to make a decision about a possible skull fracture.

Simplify and diagram The bar chart illustrates the momentum change of the system and the impulse exerted by the pole that caused the change. The person was initially moving in the horizontal x-direction with respect to Earth, and not moving after the collision. The pole exerted an impulse in the negative x-direction on the cyclist. We'll need to estimate the following quantities: the mass and speed of the cyclist in this situation, the stopping time interval, and the area of contact. Let's assume that this is a 70-kg cyclist moving at about 3 m/s. The person's body keeps moving forward for a short distance after the bone makes contact with the pole. The skin indents some during the collision. Because of these two factors, we assume

---

[1]This is a true story—it happened to one of the book's authors, Alan Van Heuvelen.

a stopping distance of about 10 cm. Finally, we assume an area of contact of about 4 cm². All of these numbers have large uncertainties and we are not worrying about significant figures, because this is just an estimate.

$$P_{Pix} + \boxed{J_{Pi\ on\ Px}} = P_{Pfx}$$

— The pole's
— negative impulse
— on the person
— causes the
— momentum to
— decrease.

Represent mathematically We now apply the generalized impulse-momentum principle:

$$m_{Person}v_{Person\ i\ x} + (F_{Pole\ on\ Person\ x})(t_f - t_i)$$
$$= m_{Person}v_{Person\ f\ x} = 0$$

$$\Rightarrow F_{Pole\ on\ Person\ x} = -\frac{m_{Person}v_{Person\ i\ x}}{t_f - t_i}$$

We can use the strategy from the last example to estimate the stopping time interval $t_f - t_i$ from the stopping distance $x_f - x_i$:

$$t_f - t_i = \frac{2(x_f - x_i)}{v_{fx} + v_{ix}}$$

where v_{ix} is the initial velocity of the cyclist and $v_{fx} = 0$ is his final velocity.

Solve and evaluate Substituting the estimated initial velocity and the stopping distance into the above, we get an estimate for the stopping time interval:

$$t_f - t_i = \frac{2(x_f - x_i)}{v_{fx} + v_{ix}} = \frac{2(0.1\ m - 0)}{0 + 3\ m/s} = 0.067\ s$$

Since this stopping time interval is an intermediate calculated value, we don't need to worry about its number of significant digits. When we complete our estimate, though, we will keep just one significant digit.

We can now insert our estimated values of quantities in the expression for the force exerted by the pole on the person:

$$F_{Pole\ on\ Person\ x} = -\frac{m_{Person}v_{Person\ i\ x}}{(t_f - t_i)}$$
$$= -\frac{(70\ kg)(3\ m/s)}{(0.067\ s)} = -3000\ N$$

Our estimate of the force per area is

$$\frac{Force}{Area} = \frac{3000\ N}{4\ cm^2} \approx 800\ N/cm^2$$

Is the person likely to fracture his skull? The force per area needed to break a bone is about $1.7 \times 10^4\ N/cm^2 = 17,000\ N/cm^2$. Our estimate could have been off by at least a factor of 10. The force per area is still too little for a fracture.

Try it yourself: What would be the magnitude of the force exerted on the cyclist if he bounced back off the pole instead of stopping, assuming the collision time interval remains the same?

Answer: 6000 N.

Review Question 5.5 As the bullet enters the block in Example 5.4, the block exerts a force on the bullet, causing the bullet's speed to decrease to almost zero. Why did we not include the impulse exerted by the block on the bullet in our analysis of this situation?

5.6 Jet propulsion

Cars change velocity because of an interaction between the tires and the road. Likewise, a ship's propellers push water backward; in turn, water pushes the ship forward. Once the ship or car is moving, the external force exerted by the water or the road has to balance the opposing friction force or the vehicle's velocity will change.

What does a rocket push against in empty space to change its velocity? Rockets carry fuel that they ignite and then eject at high speed out of the exhaust nozzles (see **Figure 5.8**). Could this burning fuel ejected from the rocket provide the push to change its velocity? Choose the system to be the rocket and fuel together. If the rocket and fuel are at rest before the rocket fires its engines, then its momentum is zero. If there are no external impulses, then even after the rocket fires its engines, the momentum of the rocket-fuel system should still be zero. However, the burning fuel is ejected backward at high velocity from the exhaust nozzle and has a backward momentum. The rocket must now have a nonzero forward velocity. We test this idea quantitatively in Testing Experiment **Table 5.3**.

Figure 5.8 A rocket as it expels fuel.

Expelled fuel moves left.   Rocket moves right.

TESTING EXPERIMENT TABLE

5.3   Rocket propulsion.

Testing experiment	Prediction	Outcome
You are traveling through space in a rocket and observe another rocket moving with equal velocity next to you. All of a sudden you notice a burst of burning fuel that is ejected from it. Predict what happens to that rocket's velocity.	Choose the other rocket and its fuel as the system. Your rocket serves as the object of reference and the +x-direction is in the direction of its motion. The other rocket has zero velocity in the initial state with respect to the object of reference. Its final state is just after it expels fuel backward at high speed; the rocket in turn gains an equal magnitude of momentum in the forward direction. We can represent this process with an initial-final sketch and a momentum bar chart for the rocket-fuel system.	The velocity of the other rocket does increase, and we see it move ahead of our rocket.

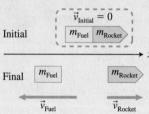

Fuel and rocket system

$$p_{Rix} + p_{Fix} + J_x = p_{Rfx} + p_{Ffx}$$

Assuming that fuel is ejected all at once at constant speed, the velocity of the rocket should be

$$0 = m_{Rocket}v_{Rocket\,x} + m_{Fuel}v_{Fuel\,x}$$

$$\Rightarrow v_{Rocket\,x} = -\frac{m_{Fuel}v_{Fuel\,x}}{m_{Rocket}}$$

Here m_{Rocket} is the mass of the rocket without the fuel. We can also choose the rocket alone as the system. The rocket pushes back on the fuel, expelling it backward at high speed ($v_{Fuel\,x} < 0$); the fuel in turn pushes forward against the rocket, exerting an impulse that causes the rocket's momentum and the velocity (assuming the mass of the rocket itself does not change) to increase ($v_{Rocket\,x} > 0$).

Rocket alone as system

$$p_{Rix} + J_{F\,on\,Rx} = p_{Rfx}$$

Conclusion

The outcome of the experiment is consistent with the prediction, supporting the generalized impulse-momentum principle. We have learned that, independent of the choice of the system, when a rocket expels fuel in one direction, it gains velocity and therefore momentum in the opposite direction. This mechanism of accelerating a rocket or spaceship is called jet propulsion.

The force exerted by the fuel on a rocket during jet propulsion is called **thrust.** Typical large rocket thrusts measure in mega-newtons (10^6 N), and exhaust speeds are more than 10 times the speed of sound. Thrust provides the necessary impulse to change a rocket's momentum. You can observe the principles of jet propulsion using a long, narrow balloon. Blow up the balloon; then open the valve and release it. The balloon will shoot away rapidly in the opposite direction of the air streaming out of the balloon's valve.

In reality, a rocket burns its fuel gradually rather than in one short burst; thus its mass is not a constant number but changes gradually. However, the same methods we used in Testing Experiment Table 5.3, together with some calculus, can be applied to determine the change in the rocket's velocity.

The main idea behind the jet propulsion method is that when an object ejects some of its mass in one direction, it accelerates in the opposite direction. This means that the same method that is used to speed up a rocket can also be used to slow it down. To do this, the fuel needs to be ejected in the same direction that the rocket is traveling.

> **TIP** You can become your own jet propulsion machine by standing on rollerblades or a skateboard and throwing a medicine ball or a heavy book forward or backward.

Review Question 5.6 The following equation is a solution for a problem. State a possible problem.

$$(2.0\,\text{kg})(-8.0\,\text{m/s}) + 0 = (2.0\,\text{kg} + 58\,\text{kg})v_x.$$

5.7 Meteorites, radioactive decay, and two-dimensional collisions: Putting it all together

In this section we apply impulse-momentum ideas to analyze meteorites colliding with Earth, radioactive decay of radon in the lungs, and two-dimensional car collisions. We start by analyzing a real meteorite collision with Earth that occurred about 50,000 years ago.

Canyon Diablo Crater

In this example we use two separate choices of systems to answer different questions about a meteorite collision with Earth.

EXAMPLE 5.7 **Meteorite impact**
Arizona's Meteor Crater (also called Canyon Diablo Crater), shown in **Figure 5.9**, was produced 50,000 years ago by the impact of a 3×10^8-kg meteorite traveling at 1.3×10^4 m/s (29,000 mi/h). The crater is about 200 m deep. Estimate (a) the change in Earth's velocity as a result of the impact and (b) the average force exerted by the meteorite on Earth during the collision.

Sketch and translate A sketch of the process is shown on the next page. To analyze Earth's motion, we choose a coordinate system at rest with respect to Earth. The origin of the coordinate axis is at the point where the meteorite first hits Earth. We keep track of the dot at the bottom of the meteorite. The axis points in the direction of the meteorite's motion.

Figure 5.9 Canyon Diablo Crater, site of a meteorite impact 50,000 years ago.

To answer the first question, we choose Earth and the meteorite as the system and use momentum constancy to determine Earth's change in velocity due to the

(continued)

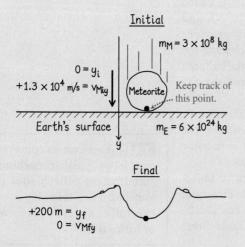

Initial

$m_M = 3 \times 10^8$ kg

$0 = y_i$

$+1.3 \times 10^4$ m/s $= v_{Miy}$

Meteorite

Keep track of this point.

Earth's surface $m_E = 6 \times 10^{24}$ kg

y

Final

$+200$ m $= y_f$
$0 = v_{Mfy}$

collision. To estimate the average force that the meteorite exerted on Earth during the collision (and that Earth exerted on the meteorite), we choose the meteorite alone as the system and use the impulse-momentum equation to answer that question.

Simplify and diagram Assume that the meteorite hits perpendicular to Earth's surface in the positive y-direction. The first impulse-momentum bar chart below represents the process for the Earth-meteorite system to answer the first question. The second bar chart represents the meteorite alone as the system during its collision with Earth to answer the second question.

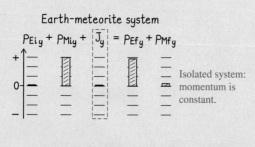

Earth-meteorite system

$PE_{iy} + PM_{iy} + \boxed{J_y} = PE_{fy} + PM_{fy}$

Isolated system: momentum is constant.

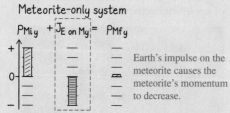

Meteorite-only system

$PM_{iy} + \boxed{J_{E\,on\,My}} = PM_{fy}$

Earth's impulse on the meteorite causes the meteorite's momentum to decrease.

Represent mathematically The y-component of the meteorite's initial velocity is $v_{Miy} = +1.3 \times 10^4$ m/s. Earth's initial velocity is zero (with respect to the object of reference). The y-component of the meteorite's final velocity equals Earth's since the meteorite embeds in Earth. The meteorite's mass is about 3×10^8 kg and Earth's mass is 6×10^{24} kg. We use momentum

constancy to determine the speeds of Earth and the meteorite after they join together:

$$(m_E \cdot 0 + m_M v_{Miy}) + [0(t_f - t_i)] = (m_E v_{Efy} + m_M v_{Mfy})$$
$$\Rightarrow m_M v_{Miy} = m_E v_{Efy} + m_M v_{Mfy} = (m_E + m_M)v_{fy}$$

To estimate the force that the meteorite exerts on Earth during the collision, we use the y-component form of the impulse-momentum equation with the meteorite alone as the system:

$$m_M v_{Miy} + F_{E\,on\,M\,y}(t_f - t_i) = m_M v_{Mfy}$$

The time interval required for the collision [using Eq. (5.9)] is

$$t_f - t_i = \frac{2(y_f - y_i)}{v_{Mfy} + v_{Miy}}$$

Solve and evaluate To answer the first question, we solve for the final velocity of Earth and meteorite together:

$$v_{fy} = \frac{m_M}{m_E + m_M}v_{Miy}$$
$$= \frac{3 \times 10^8 \text{ kg}}{6 \times 10^{24} \text{ kg} + 3 \times 10^8 \text{ kg}}(1.3 \times 10^4 \text{ m/s})$$
$$= 7 \times 10^{-13} \text{ m/s}$$

This is so slow that it would take Earth about 50,000 years to travel just 1 m. Since Earth is so much more massive than the meteorite, the meteorite's impact has extremely little effect on Earth's motion.

For the second question, the time interval for the impact is about

$$t_f - t_i = \frac{2(y_f - y_i)}{v_{Mfy} + v_{Miy}}$$
$$= \frac{2(200 \text{ m})}{(1.3 \times 10^4 \text{ m/s} + 7 \times 10^{-13} \text{ m/s})}$$
$$= 0.031 \text{ s}$$

Like most impulsive collisions, this one was over quickly! Note that we've estimated the displacement of the meteorite to be the depth of the crater. Rearranging the impulse-momentum equation as applied to the collision, we find the average force exerted by Earth on the meteorite:

$$F_{E\,on\,M\,y} = \frac{m_M(v_{Mfy} - v_{Miy})}{(t_f - t_i)}$$
$$= \frac{(3 \times 10^8 \text{ kg})(7 \times 10^{-13} \text{ m/s} - 1.3 \times 10^4 \text{ m/s})}{(0.031 \text{ s})}$$
$$= -1.3 \times 10^{14} \text{ N} \approx -1 \times 10^{14} \text{ N}$$

The force exerted by Earth on the meteorite is negative—it points opposite the direction of the meteorite's initial velocity. According to Newton's third law, the force that the meteorite exerts on Earth is positive and has the same magnitude:

$$F_{M \, on \, E \, y} = +1 \times 10^{14} \, N$$

This sounds like a very large force, but since the mass of Earth is 6×10^{24} kg, this force will cause an acceleration of a little over $10^{-11} \, m/s^2$, a very small number.

Try it yourself: Estimate the change in Earth's velocity and acceleration if it were hit by a meteorite traveling at the same speed as in the last example, stopping in the same distance, but having mass of 6×10^{19} kg instead of 3×10^8 kg.

Answer: About 0.1 m/s and 4 m/s².

TIP Notice how the choice of system in Example 5.7 was motivated by the question being investigated. Always think about your goal when deciding what your system will be.

An object breaks into parts (radioactive decay)

We will learn in the chapter on nuclear physics (Chapter 28) that the nuclei of some atoms are unstable and spontaneously break apart. In a process called alpha decay, the nucleus of the atom breaks into a daughter nucleus that is slightly smaller and lighter than the original parent nucleus and an even smaller alpha particle (symbolized by α; actually a helium nucleus). For example, radon decays into a polonium nucleus (the daughter) and an alpha particle.

Radon is produced by a series of decay reactions starting with heavy elements in the soil, such as uranium. Radon diffuses out of the soil and can enter a home through cracks in its foundation, where it can be inhaled by people living there. Once in the lungs, the radon undergoes alpha decay, releasing fast-moving alpha particles that may cause mutations that could lead to cancer. In the next example, we will analyze alpha decay by radon by using the idea of momentum constancy.

EXAMPLE 5.8 Radioactive decay of radon in lungs

An inhaled radioactive radon nucleus resides more or less at rest in a person's lungs, where it decays to a polonium nucleus and an alpha particle. With what speed does the alpha particle move if the polonium nucleus moves away at 4.0×10^5 m/s relative to the lung tissue? The mass of the polonium nucleus is 54 times greater than the mass of the alpha particle.

Sketch and translate An initial-final sketch of the situation is shown at the right. We choose the system to be the radon nucleus in the initial state, which converts to the polonium nucleus and the alpha particle in the final state. The coordinate system has a positive x-axis pointing in the direction of motion of the alpha particle, with the object of reference being the lung tissue. The initial velocity of the radon nucleus along the x-axis is 0 and the final velocity component of the polonium daughter nucleus is

$v_{Po \, fx} = -4.0 \times 10^5$ m/s. The final velocity component of the alpha particle $v_{\alpha fx}$ is unknown. If the mass of the alpha particle is m, then the mass of the polonium is $54m$.

<u>Initial</u>

$m_{Rn} = 55 \, m$
$v_{Rn \, ix} = 0$

(Rn)

$\longrightarrow x$

<u>Final</u>

$m_{Po} = 54 \, m$ $m_\alpha = m$
$v_{Po \, fx} = -4.0 \times 10^5$ m/s $v_{\alpha \, fx} = ?$

(Po)   (α)

$\overleftarrow{v}_{Po \, f}$ $\overrightarrow{v}_{\alpha \, f}$

(continued)

Simplify and diagram Assume that there are no external forces exerted on the system, meaning that the system is isolated and thus its momentum is constant. The impulse-momentum bar chart below represents the process.

$$P_{\text{Rn ix}} + \boxed{J_x} = P_{\alpha\,\text{fx}} + P_{\text{Po fx}}$$

Isolated system: momentum is constant.

Represent mathematically Use the bar chart to help apply the impulse-momentum equation for the process:

$$m_{\text{Rn}}(0) + (0)(t_f - t_i) = m_{\text{Po}}v_{\text{Pof}x} + m_\alpha v_{\alpha f x}$$

$$\Rightarrow 0 = m_{\text{Po}}v_{\text{Pof}x} + m_\alpha v_{\alpha f x}$$

Solve and evaluate Rearranging, we get an expression for the final velocity of the alpha particle in the x-direction:

$$v_{\alpha f x} = -\frac{m_{\text{Po}}v_{\text{Pof}x}}{m_\alpha}$$

The x-component of the velocity of the alpha particle after radon decay is

$$v_{\alpha f x} = -\frac{m_{\text{Po}}v_{\text{Pof}x}}{m_\alpha} = -\frac{(54 m_\alpha)(-4.0 \times 10^5\,\text{m/s})}{m_\alpha}$$

$$= +2.2 \times 10^7\,\text{m/s}$$

The sign indicates that the alpha particle is traveling in the positive x-direction opposite the direction of the polonium. The magnitude of this velocity is huge—about one-tenth the speed of light! The speeding alpha particle passes through lung tissue and collides with atoms and molecules, dislodging electrons and creating ions. Radon exposure causes approximately 15,000 cases of lung cancer each year.

Try it yourself: Francium nuclei undergo radioactive decay by emitting either an alpha particle or a beta particle (an electron). The alpha particle is about 8000 times more massive than a beta particle. If the particles are emitted with the same speed, in which case is the recoil speed of the nucleus that is left after an alpha or a beta particle is emitted greatest?

Answer: Since the mass of the alpha particle is much greater than the mass of the beta particle, and they are traveling with the same speed, the momentum of the alpha particle is much greater than the momentum of the beta particle. Therefore, the nucleus that is left would have a greater recoil speed during alpha decay.

Collisions in two dimensions

So far, the collisions we have investigated have occurred along one axis. Often, a motor vehicle accident involves two vehicles traveling along perpendicular paths. For these two-dimensional collisions, we can still apply the ideas of impulse and momentum, but we will use one impulse-momentum equation for each coordinate axis.

EXAMPLE 5.9

A 1600-kg pickup truck traveling east at 20 m/s collides with a 1300-kg car traveling north at 16 m/s. The vehicles remain tangled together after the collision. Determine the velocity (magnitude and direction) of the combined wreck immediately after the collision.

Sketch and translate We sketch the initial and final situations of the vehicles. We use a P subscript for the pickup and a C subscript for the car. The initial state is just before the collision; the final state is just after the vehicles collide and are moving together. We choose the two vehicles as the system. The object of reference is Earth; the positive x-axis points east and the positive y-axis points north.

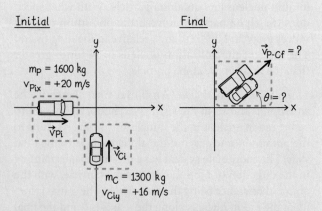

Simplify and diagram Force diagrams represent the side view for each vehicle just before the collision. We assume that the friction force exerted by the road is very small compared to the force that each vehicle exerts on the other. Thus, we ignore the impulse due to surface friction during the short collision time interval of about 0.1 s. We then apply momentum constancy in each direction. Impulse-momentum bar charts for the x-direction and for the y-direction are shown below.

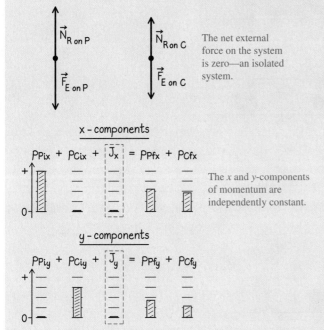

Represent mathematically Now, convert each momentum bar in the x-component bar chart into a term in the x-component form of the impulse-momentum equation [Eq. (5.7x)] and each bar in the y-component bar chart into a term in the y-component form of the impulse-momentum equation [Eq. (5.7y)]. Notice that the x-component of the final velocity vector is $v_{\text{P-Cf}\,x} = v_{\text{P-Cf}} \cos \theta$ and the y-component is $v_{\text{P+Cf}\,x} = v_{\text{P-Cf}} \sin \theta$:

x-component equation:

$$m_{\text{P}}v_{\text{Pi}\,x} + m_{\text{C}}v_{\text{Ci}\,x} = (m_{\text{P}} + m_{\text{C}})v_{\text{P-Cf}} \cos \theta$$

y-component equation:

$$m_{\text{P}}v_{\text{Pi}\,y} + m_{\text{C}}v_{\text{Ci}\,y} = (m_{\text{P}} + m_{\text{C}})v_{\text{P-Cf}} \sin \theta$$

We have two equations and two unknowns ($v_{\text{P-Cf}}$ and θ). We can solve for both unknowns.

Solve and evaluate

x-component equation:

$$(1600 \text{ kg})(20 \text{ m/s}) + (1300 \text{ kg})(0 \text{ m/s})$$
$$= (2900 \text{ kg})v_{\text{P-Cf}} \cos \theta$$

y-component equation:

$$(1600 \text{ kg})(0 \text{ m/s}) + (1300 \text{ kg})(16 \text{ m/s})$$
$$= (2900 \text{ kg})v_{\text{P-Cf}} \sin \theta$$

Divide the left side of the second equation by the left side of the first equation and the right side of the second equation by the right of the first, and cancel the 2900 kg and $v_{\text{P+Cf}}$ on the top and bottom of the right side. We get

$$\frac{(1300 \text{ kg})(16 \text{ m/s})}{(1600 \text{ kg})(20 \text{ m/s})} = \frac{\sin \theta}{\cos \theta} = \tan \theta = 0.65$$

A 33° angle has a 0.65 tangent. Thus, the vehicles move off at 33° above the $+x$-axis (the north of east direction). We can now use this angle with either the x-component equation or the y-component equation above to determine the speed of the two vehicles immediately after the collision. Using the x-component equation, we get

$$v_{\text{P-Cf}} = \frac{(1600 \text{ kg})(20 \text{ m/s})}{(2900 \text{ kg})\cos 33°} = 13 \text{ m/s}$$

From the y-component equation, we have

$$v_{\text{P-Cf}} = \frac{(1300 \text{ kg})(16 \text{ m/s})}{(2900 \text{ kg})\sin 33°} = 13 \text{ m/s}$$

The two equations give the same result for the final speed, a good consistency check. For collisions in which vehicles lock together like this, police investigators commonly use the lengths of the skid marks along with the direction of the vehicles after the collision to determine their initial speeds. This allows them to decide whether either vehicle was exceeding the speed limit before the collision.

Try it yourself: Use a limiting case analysis and the x- and y-component forms of the impulse-momentum equation to predict what would happen during the collision if the pickup had infinite mass. Is the answer reasonable?

Answer: If we place ∞ in

$$\frac{(1300 \text{ kg})(16 \text{ m/s})}{(1600 \text{ kg})(20 \text{ m/s})} = \frac{\sin \theta}{\cos \theta} = \tan \theta$$

in place of the 1600-kg mass of the pickup, the left side of the equation becomes zero. Then $\tan \theta = 0$. The pickup would move straight ahead when hitting the car. In other words, the collision with the car would not change the direction of travel of the pickup. The result seems reasonable if the mass of the pickup was large compared to the mass of the car.

Review Question 5.7 When a meteorite hits Earth, the meteorite's motion apparently disappears completely. How can we claim that momentum is conserved?

Summary

Words	Pictorial and physical representations	Mathematical representation

Isolated system An isolated system is one in which the objects interact only with each other and not with the environment, or the sum of external forces exerted on it is zero. (Sections 5.1–5.2)

Isolated system

Nonisolated system

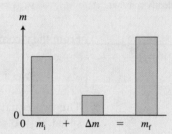

Conservation of mass If the system is isolated, its mass is constant. If the system is not isolated, the change in the system's mass equals the mass delivered to the system or taken away from it. (Section 5.1)

3 kg + 1 kg = 4 kg

$m_i + \Delta m = m_f$

Linear momentum $\vec{p}$ is a vector quantity that is the product of an object's mass m and velocity $\vec{v}$. The total momentum of the system is the sum of the momenta of all objects in the system. (Section 5.2)

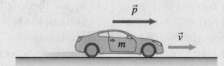

$\vec{p}$ $\vec{v}$

m

$\vec{p} = m\vec{v}$ Eq. (5.1)

$\vec{p}_{system} = \vec{p}_1 + \vec{p}_2 + \cdots$

Impulse $\vec{J}$ Impulse is the product of the average external force $\vec{F}_{av}$ exerted on an object during a time interval Δt and that time interval. (Section 5.3)

$\vec{J} = \vec{F}_{av}(t_f - t_i)$ Eq. (5.5)

Generalized impulse-momentum principle If the system is isolated, its momentum is constant. If the system is not isolated, the change in the system's momentum equals the sum of the impulses exerted on the system during the time interval $\Delta t = (t_f - t_i)$. (Section 5.4)

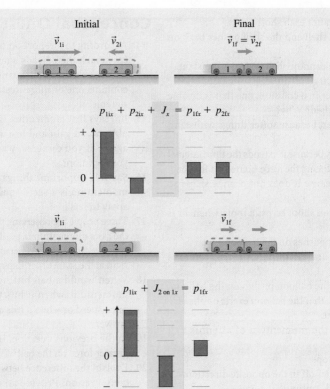

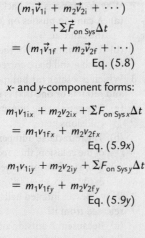

$$(m_1\vec{v}_{1i} + m_2\vec{v}_{2i} + \cdots)$$
$$+ \Sigma \vec{F}_{\text{on Sys}}\Delta t$$
$$= (m_1\vec{v}_{1f} + m_2\vec{v}_{2f} + \cdots)$$
Eq. (5.8)

x- and *y*-component forms:

$$m_1 v_{1ix} + m_2 v_{2ix} + \Sigma F_{\text{on Sys}\,x}\Delta t$$
$$= m_1 v_{1fx} + m_2 v_{2fx}$$
Eq. (5.9x)

$$m_1 v_{1iy} + m_2 v_{2iy} + \Sigma F_{\text{on Sys}\,y}\Delta t$$
$$= m_1 v_{1fy} + m_2 v_{2fy}$$
Eq. (5.9y)

(MP)® **For instructor-assigned homework, go to MasteringPhysics.**

Questions

Multiple Choice Questions

1. The gravitational force that Earth exerts on an object causes an impulse of $+10\,\text{N} \cdot \text{s}$ in one experiment and $+1\,\text{N} \cdot \text{s}$ on the same object in another experiment. How can this be?
   (a) The mass of the obje ct changed.
   (b) The time intervals during which the force was exerted are different.
   (c) The magnitudes of the force were different.

2. A bullet fired at a door makes a hole in the door but does not open it. Your finger does not make a hole in the door but does open it. Why?
   (a) The bullet is too small.
   (b) The force exerted by the bullet is not enough to open the door.
   (c) A finger exerts a smaller force but the time interval is much longer.
   (d) The bullet goes through the door and does not exert a force at all.

3. How would you convince somebody that the momentum of an isolated system is constant?
   (a) It is a law; thus, you do not need to convince anybody.
   (b) Use an example from a textbook to show that the sum of the initial and final velocities of the objects involved in a collision are the same.
   (c) Derive it from Newton's second and third laws.

   (d) Use it to make predictions about a new experiment, and then compare the outcome to the prediction.
   (e) Both (c) and (d) will work.

4. A wagon full of medicine balls is rolling along a street. Suddenly one medicine ball (3 kg) falls off the wagon. What happens to the speed of the wagon?
   (a) The wagon slows down.
   (b) The speed of the wagon does not change.
   (c) The wagon speeds up.
   (d) Additional information about the ball's motion is needed to answer.

5. When can you apply the idea that momentum is constant to solve a problem?
   (a) When the system is isolated
   (b) When the system is not isolated but the time interval when the external forces are exerted is very small
   (c) When the external forces are much smaller than the internal forces

6. Choose an example in which the momentum of a system is not constant.
   (a) A bullet shot from a rifle, with the rifle and the bullet as the system
   (b) A freely falling metal ball, with the ball as the system
   (c) A freely falling metal ball, with the ball and Earth as the system
   (d) It is not possible to give an example since the momentum of a system is always constant.

7. Why do cannons roll back after each shot?
   (a) A cannon pushes on a shell and the shell pushes back on the cannon.
   (b) The momentum of the cannon-shell system is constant.
   (c) Both a and b are correct.
8. Which is a safer car bumper in a collision: one that is flexible and retracts or one that is rigid? Why?
   (a) The retractable bumper, because softer things withstand collisions better
   (b) The retractable bumper, because it extends the time interval of the collision, thus reducing the force exerted on the car
   (c) The rigid bumper, because it does not change shape so easily
9. Why does an inflated balloon shoot across a room when air is released from it?
   (a) Because the outside air pushes on the balloon
   (b) Because the momentum of the balloon-air system is constant
   (c) Because the air inside the balloon pushes on the balloon, exerting the same force that the balloon exerts on the air
   (d) Both b and c are correct.
10. In which situation does the momentum of a tennis ball change more?
    (a) It hits the racket and stops.
    (b) It hits the racket and flies off in the opposite direction.
    (c) It misses the racket and continues moving.
11. A toy car with very low friction wheels and axles rests on a level track. In which situation will its speed increase more?
    (a) It is hit from the rear by a wad of clay that sticks to the car.
    (b) It is hit by a rubber ball with the same mass and velocity of the clay that rebounds in the opposite direction after hitting the car.
12. A meteorite strikes Earth and forms a crater, decreasing the meteorite's momentum to zero. Does this phenomenon contradict the conservation of momentum? Choose as many answers as you think are correct.
    (a) No, because the meteorite system is not isolated
    (b) No, because in the meteorite-Earth system, Earth acquires momentum lost by the meteorite
    (c) No, because the meteorite brings momentum from space
    (d) Yes, because the meteorite is not moving relative to a medium before the collision
13. A 1000-kg car traveling east at 24 m/s collides with a 2000-kg car traveling west at 21 m/s. The cars lock together. What is their velocity immediately after the collision?
    (a) 3 m/s east   (b) 3 m/s west
    (c) 6 m/s east   (d) 6 m/s west
    (e) 15 m/s east

Conceptual Questions

14. According to a report on traumatic brain injury, woodpeckers smack their heads against trees at a force equivalent to 1200 g's without suffering brain damage. This statement contains one or more mistakes. Identify the mistakes in this statement.
15. Jim says that momentum is not a conserved quantity because objects can gain and lose momentum. Do you agree or disagree? If you disagree, what can you do to convince Jim of your opinion?
16. Say five important things about momentum (for example, momentum is a vector quantity). How does each statement apply to real life?
17. Three people are observing the same car. One person claims that the car's momentum is positive, another person claims that it is negative, and the third person says that it is zero. Can they all be right at the same time? Explain.
18. When would a ball hitting a wall have a greater change in momentum: when it hits the wall and bounces back at the same speed or when it hits and sticks to the wall? Explain your answer.
19. In the previous question, in which case does the wall exert a greater force on the ball? Explain.
20. Explain the difference between the concepts of constancy and conservation. Provide an example of a conserved quantity and a nonconserved quantity.
21. Why do you believe that momentum is a conserved quantity?
22. A heavy bar falls straight down onto the bed of a rolling truck. What happens to the momentum of the truck at the instant the bar lands on it? Explain. How many correct answers do you think are possible? Make sure you think of what "falls straight down" means.
23. ✐ Construct impulse-momentum bar charts to represent a falling ball in (a) a system whose momentum is not constant and (b) a system whose momentum is constant. In the initial state, the ball is at rest; in the final state, the ball is moving.
24. ✐ A person moving on rollerblades throws a medicine ball in the direction opposite to her motion. Construct an impulse-momentum bar chart for this process. The person is the system.
25. ✐ A person moving on rollerblades drops a medicine ball straight down relative to himself. Construct an impulse-momentum bar chart for the system consisting of the ball and Earth for this process. The rollerblader is the object of reference, and the final state is just before the ball hits the ground.

Problems

Below, BIO indicates a problem with a biological or medical focus. Problems labeled EST ask you to estimate the answer to a quantitative problem rather than derive a specific answer. Problems marked with ✐ require you to make a drawing or graph as part of your solution. Asterisks indicate the level of difficulty of the problem. Problems with no * are considered to be the least difficult. A single * marks intermediate difficult problems. Two ** indicate more difficult problems.

5.2 Linear momentum

1. You and a friend are playing tennis. (a) What is the magnitude of the momentum of the 0.057-kg tennis ball when it travels at a speed of 30 m/s? (b) At what speed must your 0.32-kg tennis racket move to have the same magnitude momentum as the ball? (c) If you run toward the ball at a speed of 5.0 m/s, and the ball is flying directly at you at a speed of 30 m/s, what

is the magnitude of the total momentum of the system (you and the ball)? Assume your mass is 60 kg. In every case specify the object of reference.

2. You are hitting a tennis ball against a wall. The 0.057-kg tennis ball traveling at 25 m/s strikes the wall and rebounds at the same speed. (a) Determine the ball's original momentum (magnitude and direction). (b) Determine the ball's change in momentum (magnitude and direction). What is your object of reference?

3. A ball of mass m and speed v travels horizontally, hits a wall, and rebounds. Another ball of the same mass and traveling at the same speed hits the wall and sticks to it. Which ball has a greater change in momentum as a result of the collision? Explain your answer.

4. (a) A 145-g baseball travels at 35 m/s toward a baseball player's bat (the bat is the object of reference) and rebounds in the opposite direction at 40 m/s. Determine the ball's momentum change (magnitude and direction). (b) A golfer hits a 0.046-kg golf ball that launches from the grass at a speed of 50 m/s. Determine the ball's change in momentum.

5. * A 1300-kg car is traveling at a speed of 10 m/s with respect to the ground when the driver accelerates to make a green light. The momentum of the car increases by 12,800 kg·m/s. List all the quantities you can determine using this information and determine three of those quantities.

6. * The rules of tennis specify that the 0.057-kg ball must bounce to a height of between 53 and 58 inches when dropped from a height of 100 inches onto a concrete slab. What is the change in the momentum of the ball during the collision with the concrete? You will have to use some free-fall kinematics to help answer this question.

7. A cart of mass m moving right at speed v with respect to the track collides with a cart of mass $0.7m$ moving left. What is the initial speed of the second cart if after the collision the carts stick together and stop?

8. A cart of mass m moving right collides with an identical cart moving right at half the speed. The carts stick together. What is their speed after the collision?

9. EST Estimate your momentum when you are walking at your normal pace.

5.3 Impulse and momentum

10. A 100-g apple is falling from a tree. What is the impulse that Earth exerts on it during the first 0.50 s of its fall? The next 0.50 s?

11. * The same 100-g apple is falling from the tree. What is the impulse that Earth exerts on it during the first 0.50 m of its fall? The next 0.50 m?

12. Why does Earth exert the same impulse during the two time intervals in Problem 10 but different impulses during the same distances traveled in Problem 11?

13. * **Van hits concrete support** In a crash test, a van collides with a concrete support. The stopping time interval for the collision is 0.10 s, and the impulse exerted by the support on the van is 7.5×10^3 N·s. (a) Determine everything you can about the collision using this information. (b) If the van is constructed to collapse more during the collision so that the time interval during which the impulse is exerted is tripled, what is the average force exerted by the concrete support on the van?

14. BIO **Force exerted by heart on blood** About 80 g of blood is pumped from a person's heart into the aorta during each heartbeat. The blood starts at rest with respect to the body and has a speed of about 1.0 m/s in the aorta. If the pumping takes 0.17 s, what is the magnitude of the average force exerted by the heart on the blood?

15. * The train tracks on which a train travels exert a 2.0×10^5 N friction force on the train, causing it to stop in 50 s. (a) Determine the average force needed to stop the train in 25 s. (b) Determine the stopping time interval if the tracks exert a 1.0×10^5-N friction force on the train.

16. ** EST Your friend is catching a falling basketball after it has passed through the basket. Her hands move straight down 0.20 m while catching the ball. Estimate (a) the time interval for the ball to stop as she catches it and (b) the average force that her hands exert on the ball while catching it. Indicate any assumptions or estimates you have to make in order to answer the questions.

17. * BIO **Traumatic brain injury** According to a report on traumatic brain injury, the force that a professional boxer's fist exerts on his opponent's head is equivalent to being hit with a 5.9 kg bowling ball traveling at 8.9 m/s that stops in 0.018 s. Determine the average force that the fist exerts on the head.

18. * A 65-kg astronaut pushes against the inside back wall of a 2000-kg spaceship and moves toward the front. Her speed increases from 0 to 1.6 m/s. (a) If her push lasts 0.30 s, what is the average force that the astronaut exerts on the wall of the spaceship? (b) If the spaceship was initially at rest, with what speed does it recoil? (c) What was the object of reference that you used to answer parts (a) and (b)?

19. * You decide to use your garden hose to wash your garage door. The water shoots out at a rate of 10 kg/s and a speed of 16 m/s with respect to the hose. When the water hits the garage, its speed decreases to zero. Determine the force that the water exerts on the wall. What assumptions did you make?

20. * The air in a windstorm moves at a speed of 30 m/s. When it hits a stop sign, the air stops momentarily. The mass of air hitting the stop sign each second is about 2.0 kg. Make a list of physical quantities you can determine using this information and determine three of them.

21. * An egg rolls off a kitchen counter and breaks as it hits the floor. How large is the impulse that the floor exerts on the egg, and how large is the force exerted on the egg by the floor when stopping it? The counter is 1.0 m high, the mass of the egg is about 50 g, and the time interval during the collision is about 0.010 s.

22. ** **Retractable car bumper** A car bumper exerts an average force on a car as it retracts a certain distance during a collision. Using the impulse-momentum equation, show that the magnitude of the force and the retraction distance are related by the equation $F\Delta x = 0.5mv_0^2$. What assumptions did you make?

23. ** **Proportional reasoning** Use proportional reasoning and the equation from Problem 22 to determine (a) the necessary percent change in the retraction distance so that the average force required to stop a car is reduced by 20% and (b) the percent change in initial to final speed that would produce the same reduction in force.

24. (a) What force is required to stop a 1500-kg car in a distance of 0.20 m if it is initially moving at 2.2 m/s? (b) What if the car is moving at 4.5 m/s?

25. * A boxer delivers a punch to his opponent's head, which has a mass of 7.0 kg. Use the graph in **Figure P5.25** to estimate (a) the impulse of the force exerted by the boxer and (b) the speed of the head after the punch is delivered. What assumptions did you make?

Figure P5.25

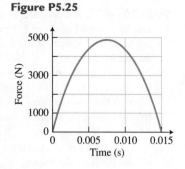

26. * **Air bag force on head** The graph in **Figure P5.26** shows the time variation of the force that an automobile's air bag exerts on a person's head during a collision. The mass of the head is 8.0 kg. Determine (a) the total impulse of the force exerted by the air bag on the person's head and (b) the person's speed just before the collision occurred.

Figure P5.26

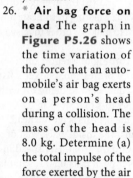

27. * **Equation Jeopardy 1** Invent a problem for which the solution is

$$(27\,\text{kg})(-3.0\,\text{m/s}) + (30\,\text{kg})(+4.0\,\text{m/s}) = (27\,\text{kg} + 30\,\text{kg})v.$$

28. * **Equation Jeopardy 2** Invent a problem for which the solution is

$$(0.020\,\text{kg})(300\,\text{m/s}) - (10\,\text{N})(0.40\,\text{s}) = (0.020\,\text{kg})(100\,\text{m/s}).$$

29. * Write a general impulse-momentum equation that describes the following process: a person skating on rollerblades releases a backpack that falls toward the ground (the process ends before the backpack hits the ground). What is the system, and what are the physical quantities you will use to describe the process?

5.4 Generalized impulse-momentum principle

30. * / Two carts (100 g and 150 g) on an air track are separated by a compressed spring. The spring is released. Represent the process with a momentum bar chart (a) with one cart as the system and (b) with both carts as the system. (c) Write expressions for all of the physical quantities you can from this information. Identify your object of reference.

31. * / A tennis ball of mass m hits a wall at speed v and rebounds at about the same speed. Represent the process with an impulse-momentum bar chart for the ball as the system. Using the bar chart, develop an expression for the change in the ball's momentum. What is the object of reference?

32. * / A tennis ball traveling at a speed of v stops after hitting a net. Represent the process with an impulse-momentum bar chart for the ball as the system. Develop an expression for the ball's change in momentum. What is the object of reference?

33. * / You drop a happy ball and a sad ball of the same mass from height h (see Figure 5.10). One ball hits the ground and rebounds almost to the original height. The other ball does not bounce. Represent each process with a bar chart, starting just before the balls hit the ground to just after the first ball rebounds and when the other ball stops. Choose the ball as the system.

34. * / You experiment again with the balls from Problem 33. You drop them from the same height onto a ruler that is placed on the edge of a table (**Figure P5.34**). One ball knocks the ruler off; the other does not. Represent each process with an impulse-momentum bar chart with (a) the ball as a system and (b) the ball and the ruler as the system. The process starts just before the balls hit the ruler and ends immediately after they hit the ruler. Use the bar charts to help explain the difference in the results of the experiment.

Figure P5.34

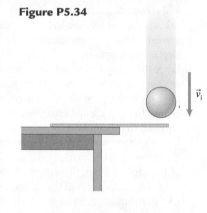

35. ** You demonstrate hitting a board in a karate class. The speed of your hand as it hits the thick board is 14 m/s with respect to the board, and the mass of your hand is about 0.80 kg. How deep does your hand go into the board before stopping if the collision lasts for 2.0×10^{-3} s? What assumptions did you make? What other quantities can you determine using this information?

36. * / You hold a beach ball with your arms extended above your head and then throw it upward. Represent the motion of the ball with an impulse-momentum bar chart for (a) the ball as the system and (b) the ball and Earth as the system.

37. * A basketball player drops a 0.60-kg basketball vertically so that it is traveling at 6.0 m/s when it reaches the floor. The ball rebounds upward at a speed of 4.2 m/s. (a) Determine the magnitude and direction of the ball's change in momentum. (b) Determine the average net force that the floor exerts on the ball if the collision lasts 0.12 s.

38. * **Bar chart Jeopardy** Invent a problem for each of the bar charts shown in **Figure P5.38**.

Figure P5.38

(a)

(b)

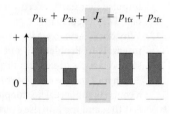

5.5 Skills for solving impulse-momentum problems

39. * A baseball bat contacts a 0.145-kg baseball for 1.3×10^{-3} s. The average force exerted by the bat on the ball is 8900 N. If the ball has an initial velocity of 36 m/s toward the bat and the force of the bat causes the ball's motion to reverse direction, what is the ball's speed as it leaves the bat?

40. * A tennis ball traveling horizontally at a speed of 40.0 m/s hits a wall and rebounds in the opposite direction. The time interval for the collision is about 0.013 s, and the mass of the ball is 0.059 kg. Make a list of quantities you can determine using this information and determine four of them. Assume that the ball rebounds at the same speed.

41. A cannon mounted on the back of a ship fires a 50-kg cannonball in the horizontal direction at a speed of 150 m/s. If the cannon and ship have a combined mass of 40000 kg and are initially at rest, what is the speed of the ship just after shooting the cannon? What assumptions did you make?

42. * A team in Quebec is playing ice baseball. A 72-kg player who is initially at rest catches a 145-g ball traveling at 18 m/s. If the player's skates are frictionless, how much time is required for him to glide 5.0 m after catching the ball?

43. A 10-kg sled carrying a 30-kg child glides on a horizontal, frictionless surface at a speed of 6.0 m/s toward the east. The child jumps off the back of the sled, propelling it forward at 20 m/s. What was the child's velocity in the horizontal direction relative to the ground at the instant she left the sled?

44. A 10,000-kg coal car on the Great Northern Railroad coasts under a coal storage bin at a speed of 2.0 m/s. As it goes under the bin, 1000 kg of coal is dropped into the car. What is the final speed of the loaded car?

45. * **Avoiding chest injury** A person in a car during a sudden stop can experience potentially serious chest injuries if the combined force exerted by the seat belt and shoulder strap exceeds 16,000 N. Describe what it would take to avoid injury by estimating (a) the minimum stopping time interval and (b) the corresponding stopping distance, assuming an initial speed of 16 m/s. Indicate any other assumptions you made.

46. * **Bruising apples** An apple bruises if a force greater than 8.0 N is exerted on it. Would a 0.10-kg apple be likely to bruise if it falls 2.0 m and stops after sinking 0.060 m into the grass? Explain.

47. * **Fast tennis serve** The fastest server in women's tennis is Venus Williams, who recorded a serve of 204 km/h at the French Open in 2007. Suppose that the mass of her racket was 328 g and the mass of the ball was 57 g. If her racket was moving at 200 km/h when it hit the ball, approximately what was the racket's speed after hitting the ball? Indicate any assumptions you made.

48. * You are in an elevator whose cable has just broken. The elevator is falling at 20 m/s when it starts to hit a shock-absorbing device at the bottom of the elevator shaft. If you are to avoid injury, the upward force that the floor of the elevator exerts on your upright body while stopping should be no more than 8000 N. Determine the minimum stopping distance needed to avoid injury (do not forget to include your mass in the calculations). What assumptions did you make? Do these assumptions make the stopping distance smaller or larger than the real-world value?

49. * You jump from the window of a burning hotel and land in a safety net that stops your fall in 1.0 m. Estimate the average force that the net exerts on you if you enter the net at a speed of 24 m/s. What assumptions did you make? If you did not make these assumptions, would the stopping distance be smaller or larger?

50. * **Skid marks** A car skids to a stop. The length of the skid marks is 50 m. What information do you need in order to decide whether the car was speeding before the driver hit the brakes?

51. * BIO **Leg injuries during car collisions** During a car collision, the knee, thighbone, and hip can sustain a force no greater than 4000 N. Forces that exceed this amount could cause dislocations or fractures. Assume that in a collision a knee stops when it hits the car's dashboard. Also assume that the mass of the body parts stopped by the knee is about 20% of the total body mass. (a) What minimum stopping time interval in needed to avoid injury to the knee if the person is initially traveling at 15 m/s (34 mi/h)? (b) What is the minimum stopping distance?

52. * BIO **Bone fracture** The zygomatic bone in the upper part of the cheek can be fractured by a 900-N force lasting 6.0 ms or longer. A hockey puck can easily exert such a force when hitting an unprotected face. (a) What change in velocity of a 0.17-kg hockey puck is needed to provide that impulsive force? What assumptions did you make? (b) A padded facemask doubles the stopping time. By how much does it change the force on the face? Explain.

53. * An impulse of 150 N · s stops your head during a car collision. (a) A crash test dummy's head stops in 0.020 s, when the cheekbone hits the steering wheel. What is the average force that the wheel exerts on the dummy's cheekbone? (b) Would this crash fracture a human cheekbone (see Problem 52)? (c) What is the shortest impact time that a person could sustain without breaking the bone?

54. ✐ A cart is moving on a horizontal track when a heavy bag falls vertically onto it. What happens to the speed of the cart? Represent the process with an impulse-momentum bar chart.

55. * ✐ A cart is moving on a horizontal track. A heavy bag falls off the cart and moves straight down relative to the cart. Describe what happens to the speed of the cart. Represent your answer with the impulse-momentum bar chart. [*Hint:* What reference frame will you use when you draw the bar chart?]

5.6 and 5.7 Jet propulsion and Putting it all together

56. Your friend shoots an 80-g arrow through a 100-g apple balanced on William Tell's head. The arrow has a speed of 50 m/s before passing through the apple and 40 m/s after. Determine the final speed of the apple.

57. * BIO **Potassium decay in body tissue** Certain natural forms of potassium have nuclei that are radioactive. Each radioactive potassium nucleus decays to a slightly less massive daughter nucleus and a high-speed electron called a beta particle. If after the decay the daughter nucleus is moving at speed 200 m/s with respect to the decaying material, how fast is the electron (the beta particle) moving? Indicate any assumptions you made. The mass of the daughter is about 70,000 times greater than the mass of the beta particle.

58. ** **Meteorite impact with Earth** About 65 million years ago a 10-km-diameter 2×10^{15}-kg meteorite traveling at about 10 km/s crashed into what is now the Gulf of Mexico. The impact produced a cloud of debris that darkened Earth and led to the extinction of the dinosaurs. Estimate the speed Earth gained as a result of this impact and the average force that the meteorite exerted on Earth during the collision. Indicate any assumptions made in your calculations.

59. ** Three friends play beach volleyball. The 280 g ball is flying east at speed 8.0 m/s with respect to the ground when one of the players bumps the ball north. The force exerted by the wrist on the ball has an average magnitude of 84 N and lasts for 0.010 s. Determine the ball's velocity (magnitude and direction) following the bump. Does your answer make sense?

60. * **Car collision** A 1180-kg car traveling south at 24 m/s with respect to the ground collides with and attaches to a 2470-kg delivery truck traveling east at 16 m/s. Determine the velocity (magnitude and direction) of the two vehicles when locked together just after the collision.

61. * **Ice skaters collide** While ice skating, you unintentionally crash into a person. Your mass is 60 kg, and you are traveling east at 8.0 m/s with respect to the ice. The mass of the other person is 80 kg, and he is traveling north at 9.0 m/s with respect to the ice. You hang on to each other after the collision. In what direction and at what speed are you traveling just after the collision?

62. **Drifting space mechanic** An astronaut with a mass of 90 kg (including spacesuit and equipment) is drifting away from his spaceship at a speed of 0.20 m/s with respect to the spaceship. The astronaut is equipped only with a 0.50-kg wrench to help him get back to the ship. With what speed and in what direction relative to the spaceship must he throw the wrench for his body to acquire a speed of 0.10 m/s and direct him back toward the spaceship? Explain.

63. * **Astronaut flings oxygen tank** While the astronaut in Problem 62 is trying to get back to the spaceship, his comrade, a 60-kg astronaut, is floating at rest a distance of 10 m from the spaceship when she runs out of oxygen and fuel to power her back to the spaceship. She removes her oxygen tank (3.0 kg) and flings it away from the ship at a speed of 15 m/s relative to the ship. (a) At what speed relative to the ship does she recoil toward the spaceship? (b) How long must she hold her breath before reaching the ship?

64. **Rocket stages** A 5000-kg rocket ejects a 10,000-kg package of fuel. Before ejection, the rocket and the fuel travel together at a speed of 200 m/s with respect to distant stars. If after the ejection, the fuel package travels at 50 m/s opposite the direction of its initial motion, what is the velocity of the rocket?

65. * /️ A rocket has just ejected fuel. With the fuel and the rocket as the system, construct an impulse-momentum bar chart for (a) the rocket's increase in speed and (b) the process of a rocket slowing down due to fuel ejection. (c) Finally, draw bar charts for both situations using the rocket without the fuel as the system.

66. ** You have two carts, a force probe connected to a computer, a motion detector, and an assortment of objects of different masses. Design three experiments to test whether momentum is a conserved quantity. Describe carefully what data you will collect and how you will analyze the data.

General Problems

67. ** **EST** Estimate the recoil speed of Earth if all of the inhabitants of Canada and the United States simultaneously jumped straight upward from Earth's surface (reaching heights from several centimeters to a meter or more). Indicate any assumptions that you made in your estimate.

68. * A cart of mass m traveling in the negative x-direction at speed v collides head-on with a cart that has triple the mass and is moving at 60% of the speed of the first cart. The carts stick together after the collision. In which direction and at what speed will they move?

69. ** Two cars of unequal mass moving at the same speed collide head-on. Explain why a passenger in the smaller mass car is more likely to be injured than one in the larger mass car. Justify your reasoning with the help of physics principles.

70. * **Restraining force during collision** A 1340-kg car traveling east at 13.6 m/s (20 mi/h) has a head-on collision with a 1930-kg car traveling west at 20.5 m/s (30 mi/h). If the collision time is 0.10 s, what is the force needed to restrain a 68-kg person in the smaller car? In the larger car?

71. ** **EST** A carpenter hammers a nail using a 0.40-kg hammerhead. Part of the nail goes into a board. (a) Estimate the speed of the hammerhead before it hits the nail. (b) Estimate the stopping distance of the hammerhead. (c) Estimate the stopping time interval. (d) Estimate the average force that the hammerhead exerts on the nail.

72. ** A 0.020-kg bullet traveling at a speed of 300 m/s embeds in a 1.0-kg wooden block resting on a horizontal surface. The block slides horizontally 4.0 m on a surface before stopping. Determine the coefficient of kinetic friction between the block and surface.

73. ** **EST** Nolan Ryan may be the fastest baseball pitcher of all time. The *Guinness Book of World Records* clocked his fastball at 100.9 mi/h in a 1974 game against the Chicago White Sox. Use the impulse-momentum equation to estimate the force that Ryan exerted on the ball while throwing that pitch. Include any assumptions you made.

74. ** A record rainstorm produced 304.8 mm (approximately 1 ft) of rain in 42 min. Estimate the average force that the rain exerted on the roof of a house that measures 10 m × 16 m. Indicate any assumptions you made.

75. ** **EST** The U.S. Army special units MH-47E helicopter has a mass of 23,000 kg, and its propeller blades sweep out an area of 263 m². It is able to hover at a fixed elevation above one landing point by pushing air downward (the air pushes up on the helicopter blades). Choose a reasonable air mass displaced downward each second and the speed of that air in order for the helicopter to hover. Indicate any assumptions you made.

76. ** A 2045-kg sports utility vehicle hits the rear end of a 1220-kg car at rest at a stop sign. The car and SUV remain locked together and skid 4.6 m before stopping. If the coefficient of friction between the vehicles and the road is 0.70, what was the SUV's initial velocity?

77. ** A car of mass m_1 traveling north at a speed of v_1 collides with a car of mass m_2 traveling east at a speed of v_2. They lock together after the collision. Develop expressions for the direction and the distance the cars will move until they stop if the coefficient of kinetic friction μ_k between the cars' tires and the road is about the same for both cars.

78. ** **Force exerted by wind on Willis Tower** A 10.0-m/s wind blows against one side of the Willis Tower in Chicago. The building is 443 m tall and approximately 80 m wide. Estimate the average force of the air on the side of the building. The density of air is approximately 1.3 kg/m³. Indicate any assumptions that you made.

79. * **Write your own problem.** Write and solve a problem that requires using the law of conservation of momentum in which it is important to know that momentum is a vector quantity.

Reading Passage Problems

BIO Heartbeat detector A prisoner tries to escape from a Nashville, Tennessee prison by hiding in the laundry truck. The prisoner is surprised when the truck is stopped at the gate. A guard enters the truck and handcuffs him. "How did you know I was here?" the prisoner asks. "The heartbeat detector," says the guard.

A heartbeat detector senses the tiny vibrations caused by blood pumped from the heart. With each heartbeat, blood is pumped upward to the aorta, and the body recoils slightly, conserving the momentum of the blood-body system. The body's vibrations are transferred to the inside of the truck. Vibration sensors on the outside of the truck are linked to a geophone, or signal amplifier, attached to a computer. A wave analyzer program in the computer compares vibration signals from the truck to wavelets produced by heartbeat vibrations. The wave analyzer distinguishes a person's heartbeat from other vibrations in the truck or in the surrounding environment, allowing guards to detect the presence of a human in the truck.

80. What does the heartbeat detector sense?
    (a) Electric signals caused by electric dipole charges produced on the heart
    (b) Body vibrations caused by blood pumped from the heart
    (c) Sound caused by breathing
    (d) Slight uncontrollable reflexive motions of an enclosed person
    (e) All of the above

81. A heartbeat detector relies on a geophone placed against the exterior of a truck or car. What can cause the vibrations of the truck or car?
    (a) Wind
    (b) Ground vibrations due to other moving cars or trucks
    (c) Vibrations of passengers in the car or truck
    (d) All of the above

82. What can be used to analyze the motion of a person's body hidden inside a car or truck (choosing the body and blood as a system)?
    (a) The idea that mass of an isolated system is constant
    (b) The idea that momentum of an isolated system is constant
    (c) The impulse-momentum principle
    (d) a and b
    (e) b and c

83. During each heartbeat, about 0.080 kg of blood passes the aorta in about 0.16 s. This blood's velocity changes from about 0.8 m/s upward toward the head to 0.8 m/s down toward the feet. What is the blood's acceleration?
    (a) zero   (b) 5 m/s^2 up   (c) 5 m/s^2 down
    (d) 10 m/s^2 up   (e) 10 m/s^2 down

84. Suppose 0.080 kg of blood moving upward in the aorta at 0.8 m/s reverses direction in 0.16 s when it reaches the aortic arch. If a prisoner is trying to escape from prison by hiding in a laundry truck, and the mass of his body is 70 kg, which is the closest to the speed his body is moving immediately after the blood changes direction passing through the aortic arch?
    (a) 0.0009 m/s   (b) 0.002 m/s   (c) 0.8 m/s
    (d) 0.08 m/s   (e) 0.01 m/s

Space Shuttle launch The mass of the Space Shuttle at launch is about 2.1×10^6 kg. Much of this mass is the fuel used to move the orbiter, which carries the astronauts and various items in the shuttle's payload. The Space Shuttle generally travels from 3.2×10^5 m (200 mi) to 6.2×10^5 m (385 mi) above Earth's surface. The shuttle's two solid fuel boosters (the cylinders on the sides of the shuttle) provide 71.4% of the thrust during liftoff and the first stage of ascent before being released from the shuttle 132 s after launch at 48,000-m above sea level. The boosters continue moving up in free fall to an altitude of approximately 70,000 m and then fall toward the ocean to be recovered 230 km from the launch site. The shuttle's five engines together provide 3.46×10^7 N of thrust during liftoff.

85. Which number below is closest to the acceleration of the shuttle during liftoff? [Hint: Remember the gravitational force that Earth exerts on the shuttle.]
    (a) 3.3 m/s^2 (b) 6.6 m/s^2 (c) 9.8 m/s^2
    (d) 16 m/s^2 (e) 33 m/s^2

86. Which number below is closest to the average vertical acceleration of the shuttle during the first 132 s of its flight?
    (a) 3.3 m/s^2 (b) 5.5 m/s^2 (c) 9.8 m/s^2
    (d) 14 m/s^2 (e) 360 m/s^2

87. The boosters are released from the shuttle 132 s after launch. How do their vertical components of velocity compare to that of the shuttle at the instant of release?
    (a) The boosters' vertical component of velocity is zero.
    (b) The boosters' vertical component of velocity is about −9.8 m/s.
    (c) The vertical component of velocity of the boosters and that of the Shuttle are the same.
    (d) There is too little information to decide.

88. What is the approximate impulse of the jet engine thrust exerted on the shuttle during the first 10 s of flight?
    (a) $980 \text{ N} \cdot \text{s}$ downward   (b) $980 \text{ N} \cdot \text{s}$ upward
    (c) $3.4 \times 10^7 \text{ N} \cdot \text{s}$ upward   (d) $3.4 \times 10^8 \text{ N} \cdot \text{s}$ upward
    (e) $3.4 \times 10^8 \text{ N} \cdot \text{s}$ downward

89. What is the approximate impulse of Earth's gravitational force exerted on the shuttle during the first 10 s of flight?
    (a) $980 \text{ N} \cdot \text{s}$ downward   (b) $980 \text{ N} \cdot \text{s}$ upward
    (c) $2.1 \times 10^7 \text{ N} \cdot \text{s}$ upward   (d) $2.1 \times 10^8 \text{ N} \cdot \text{s}$ upward
    (e) $2.1 \times 10^8 \text{ N} \cdot \text{s}$ downward

90. What is the momentum of the Space Shuttle 10 s after liftoff closest to?
    (a) $2.1 \times 10^6 \text{ kg} \cdot \text{m/s}$ down   (b) $2.1 \times 10^6 \text{ kg} \cdot \text{m/s}$ up
    (c) $2.1 \times 10^7 \text{ kg} \cdot \text{m/s}$ up   (d) $1.3 \times 10^8 \text{ kg} \cdot \text{m/s}$ up
    (e) $1.3 \times 10^8 \text{ kg} \cdot \text{m/s}$ down

91. What answer below is closest to the speed of the shuttle and boosters when they are released? Assume that the free-fall gravitational acceleration at this elevation is about 9.6 m/s^2 down.
    (a) 100 m/s   (b) 300 m/s   (c) 600 m/s
    (d) 1000 m/s   (e) 1400 m/s

6 Work and Energy

Why it is impossible to build a perpetual motion machine?

Why does blood pressure increase when the aorta walls thicken?

If Earth were to become a black hole, how big would it be?

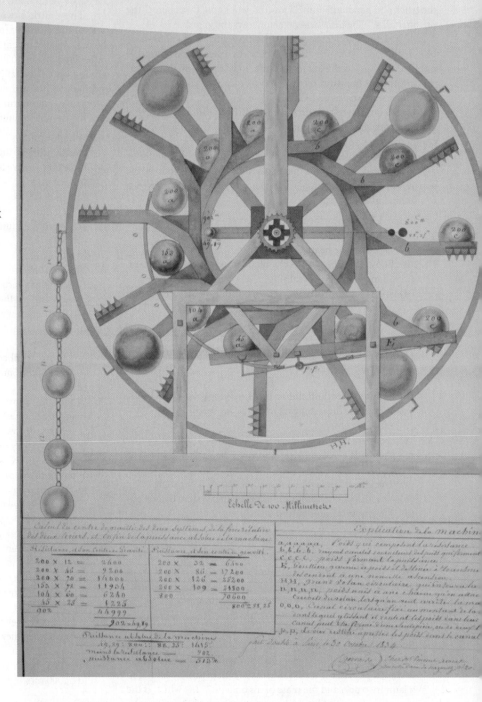

Be sure you know how to:

- Choose a system and the initial and final states of a physical process (Sections 5.2–5.4).
- Use Newton's second law to analyze a physical process (Section 3.3).
- Use kinematics to describe motion (Section 1.7).

For centuries people have dreamed of making a perpetual motion machine. Such a machine would be able to function forever without any sort of power source. Imagine owning a laptop computer that ran continuously year after year with no external power. Although claims of perpetual motion machines go back as far as the 8th century, so far no one has succeeded in inventing one. In 1775 the Royal Academy of Sciences in Paris decreed that the Academy "will no longer accept or deal with proposals concerning perpetual motion." Why has no one been able to build a perpetual motion machine? We will begin investigating this question in this chapter.

We have had good success at analyzing a variety of everyday phenomena using vector quantities (acceleration, force, impulse, and momentum). When working with such vector quantities, we need to apply the component forms of principles such as Newton's second law and the impulse-momentum equation. Can we analyze interesting everyday phenomena using a different type of thinking that depends less on vector quantities? That might make our analysis somewhat easier.

6.1 Work and energy

We will begin our new approach by conducting several experiments and looking for a pattern to help explain what we observe. In each experiment we choose a system of interest, its initial state, its final state, and an external force that is causing the system to change from its initial state to its final state. This change involves a displacement of one of the system objects from one position to another. In the analysis we draw vectors indicating the external force causing the displacement and the resulting displacement of the system object. In Observational Experiment **Table 6.1**, below, we look at the effect of external forces on system objects.

An important pattern in all four observational experiments in Table 6.1 is that an external force $\vec{F}$ was exerted on an object in the system, causing a displacement $\vec{d}$ of the object in the *same direction* as the external force. Physicists

OBSERVATIONAL EXPERIMENT TABLE

6.1 External forces and system changes.

VIDEO 6.1

Observational experiment	Analysis

Experiment 1. You hold a heavy block just above a piece of chalk (the initial state) and then release the block. The chalk does not break. Now you lift the block about 30 cm above the chalk (the final state). If you release the block from the higher elevation final state position, the block falls and smashes the chalk.

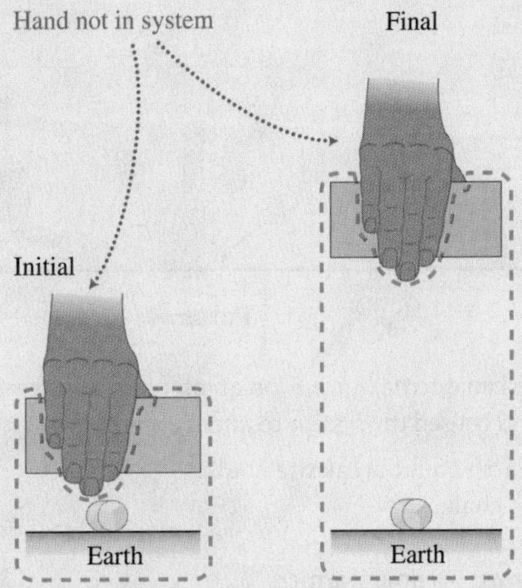

Hand not in system   Final

Initial

Earth   Earth

The force you exerted and the block's displacement while being lifted were in the same direction and caused an increase in the block's elevation and in its ability to break the chalk.

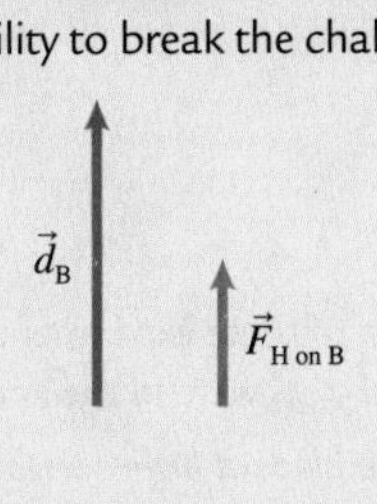

$\vec{d}_B$

$\vec{F}_{H \text{ on } B}$

(continued)

Observational experiments	Analysis

Experiment 2. You push a cart initially at rest (the initial state) until it is moving fast about two-thirds of the way across a smooth track (the final state is where you stop pushing the cart). A piece of chalk is taped to the end of the track. The fast-moving cart (no longer being pushed) collides with the piece of chalk and breaks the chalk.

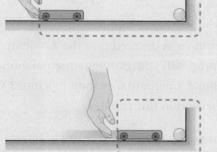

Hand not in system

The force exerted on the cart and the cart's displacement are in the same direction and increased the cart's speed so it could break the chalk.

$\vec{F}_{\text{H on C}}$

$\vec{d}_{\text{C}}$

Experiment 3. A piece of chalk rests in the hanging sling of a slingshot (the initial state). You then pull it back until the slingshot is fully stretched (the final state). When released from the stretched sling, the chalk flies across the room, hits the wall, and smashes.

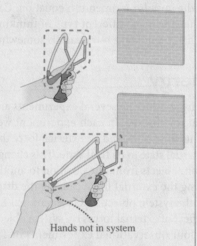

Hands not in system

The force that you exerted on the sling and its displacement are in the same direction and made it possible for the stretched sling to break the chalk.

$\vec{F}_{\text{H on S}}$

$\vec{d}_{\text{S}}$

Experiment 4. A heavy box sits on a shag carpet (the initial state). You pull the box across the carpet to a position several meters from where it started (the final state). When you reach the final position, you touch the bottom of the box and the carpet—they feel slightly warmer.

Initial

Hands not in system

Final

Bottom of box and carpet are warmer.

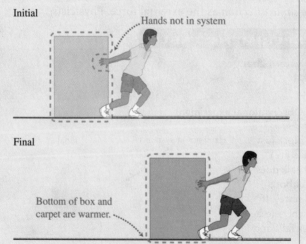

You exerted a force on the box in the direction of its displacement. After several meters of travel across the carpet, the bottom became warmer.

$\vec{F}_{\text{H on B}}$

$\vec{d}_{\text{B}}$

Patterns

In each of these experiments, you exerted an external force $\vec{F}$ on an object in a system. The force $\vec{F}_{\text{You on Object}}$ and the object's displacement $\vec{d}_{\text{Object}}$ were in the *same direction* and caused the system to change so that the system gained the potential to do something new:

1. The *block at higher elevation* above Earth could break the chalk.
2. The *fast-moving cart* could break the chalk.
3. The *stretched slingshot* could break the chalk.
4. The *box and the carpet it was pulled across* became warmer.

say that an object in the environment (in this case, you) did *positive* **work** on the system. In these experiments, the system changed because of the positive work done on it by the external forces. We call those *changes* in the system's **energy**. Four types of energy changed in the systems in the Table 6.1 experiments.

- *Gravitational potential energy*   In Experiment 1, the block in the final state was at a *higher elevation* with respect to Earth than in the initial state. The energy of the object-Earth system associated with the elevation of the object above Earth is called **gravitational potential energy** U_g. The higher above Earth, the greater is the gravitational potential energy.

- *Kinetic energy*   In Experiment 2, the cart was *moving faster* in the final state than when at rest in the initial state. The energy due to an object's motion is called **kinetic energy** K. The faster the object is moving, the greater its kinetic energy.

- *Elastic potential energy*   In Experiment 3, the slingshot in the final state was *stretched more* than in the initial state. The energy associated with an elastic object's degree of stretch is called **elastic potential energy** U_s. The greater the stretch (or compression), the greater is the object's elastic potential energy.

- *Internal energy*   In Experiment 4, the bottom surface of the box *became warmer* as it was pulled across the rough surface. Also, there can be structural changes in the surfaces—part of them can come off. The energy associated with both temperature and structure is called **internal energy** U_{int}. You will learn later that internal energy is the energy of motion and interaction of the microscopic particles making up the objects in the system.

Negative and zero work

Is it possible to devise a process in which an external force causes the energy of a system to decrease or possibly causes no energy change at all? Let's try more experiments to investigate these questions.

Observational Experiment **Table 6.2** shows us that external forces exerted on system objects can have negative or zero effect on a system.

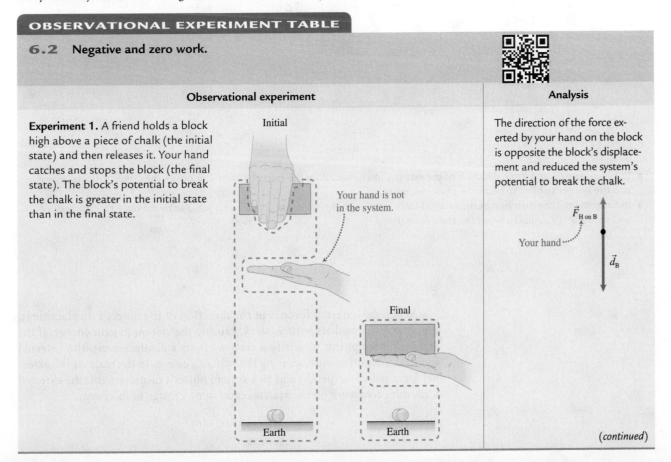

OBSERVATIONAL EXPERIMENT TABLE

6.2   Negative and zero work.

Observational experiment	Analysis
Experiment 1. A friend holds a block high above a piece of chalk (the initial state) and then releases it. Your hand catches and stops the block (the final state). The block's potential to break the chalk is greater in the initial state than in the final state.	The direction of the force exerted by your hand on the block is opposite the block's displacement and reduced the system's potential to break the chalk.

Initial

Your hand is not in the system.

Final

Earth   Earth

$\vec{F}_{\text{H on B}}$

Your hand

$\vec{d}_B$

(continued)

Observational experiments	Analysis
Experiment 2. A cart is moving fast (the initial state) toward a piece of chalk taped on the wall. While it is moving, you push lightly on the moving cart opposite the direction of its motion, causing it to slow down and stop (final state). The cart's potential to break the chalk is greater in the initial state than in the final state. 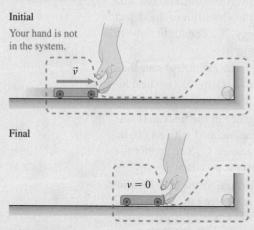	The direction of the force exerted by your hand on the cart is opposite the cart's displacement and caused the moving cart to slow down and stop, thus reducing its potential to break the chalk. 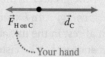
Experiment 3. Your hand holds a block less than 1 cm above a piece of chalk (the initial state)—so close that the chalk would not break if the block is released. Your hand slowly moves the block to the right, keeping the block just above the tabletop until the block is less than 1 cm above a second piece of chalk (the final state), which also would not break if the block were released. 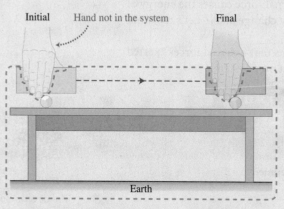	The direction of the force exerted by your hand on the block is perpendicular to the block's displacement and caused no change in the block's potential to break the chalk. 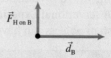

Patterns

- In Experiments 1 and 2 the direction of the external force exerted on the system object is opposite the object's displacement, and the system's ability to break the chalk decreases.
- In Experiment 3 the direction of the external force exerted on the system object is perpendicular to the object's displacement, and the system's ability to break the chalk is unchanged.

When the external force is in the direction of the object's displacement, the external force does *positive work,* causing the system to gain energy. If the external force points opposite a system object's displacement, the external force does *negative work,* causing the system's energy to decrease. If the external force points perpendicular to a system object's displacement, the external force does *zero work* on the system, causing no change to its energy.

Defining work as a physical quantity

We can now create an equation to determine how much work a particular external force does on a system. The equation should be consistent with our observational experiments.

Work The work done by a constant external force $\vec{F}$ exerted on a system object while that system object undergoes a displacement $\vec{d}$ is

$$W = Fd\cos\theta. \tag{6.1}$$

where F is the magnitude of the force in newtons (always positive), d is the magnitude of the displacement in meters (always positive), and θ is the angle between the direction of $\vec{F}$ and the direction of $\vec{d}$. The sign of $\cos\theta$ determines the sign of the work. Work is a scalar physical quantity. The unit of work is the joule (J); $1\,J = 1\,N\cdot m$ (see **Figure 6.1**).

The joule is named in honor of James Joule (1818–1889), one of many physicists who contributed to our understanding of work-energy relationships.

Note that in the four experiments in Table 6.1, the force and displacement were in the same direction: $\theta = 0°$, and $\cos 0° = +1.0$. Positive work was done. In the first two experiments in Table 6.2, the force and displacement were in opposite directions: $\theta = 180°$, and $\cos 180° = -1.0$. Negative work was done. Finally, in Experiment 3 in Table 6.2, the force and displacement were perpendicular to each other: $\theta = 90°$, and $\cos 90° = 0$. Zero work was done.

TIP It is tempting to equate the work done on a system with the force that is exerted on it. However, in physics, there must be a displacement of a system object in order for an external force to do work. Force and work are not the same thing.

Figure 6.1 The definition of work.

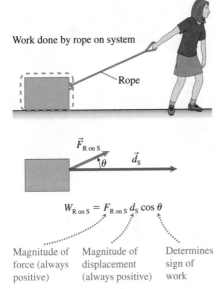

Work done by rope on system

Rope

$\vec{F}_{R\,on\,S}$ θ $\vec{d}_S$

$$W_{R\,on\,S} = F_{R\,on\,S}\,d_S\cos\theta$$

Magnitude of force (always positive) Magnitude of displacement (always positive) Determines sign of work

‹Active Learning Guide

QUANTITATIVE EXERCISE 6.1 Pushing a bicycle uphill

Two friends are cycling up a hill inclined at 8°—steep for bicycle riding. The stronger cyclist helps his friend up the hill by exerting a 50-N pushing force on his friend's bicycle and parallel to the hill while the friend moves a distance of 100 m up the hill. As you can see in the figure below, the force exerted on the weaker cyclist and the displacement are in the same direction. Determine the work done by the stronger cyclist on the weaker cyclist.

$\vec{F}_{S\,on\,W}$ $\vec{d}$

8°

Represent mathematically Choose the system to be the weaker cyclist. The external force that the stronger cyclist (S) exerts on the weaker cyclist (W) $\vec{F}_{S\,on\,W}$ is parallel to the hill, as is the 100-m displacement of the weaker cyclist. The work done by the stronger cyclist on the weaker cyclist is $W = Fd\cos\theta$. The hill is inclined at 8° above the horizontal. Before reading on, decide what angle you would insert in this equation.

Solve and evaluate Note that the angle between $\vec{F}_{S\,on\,W}$ and $\vec{d}$ is 0° and not 8°. In this case, the angle of the hill is not relevant to solving the problem because the force exerted on the cyclist is parallel to the person's displacement. Thus:

$$W_{S\,on\,W} = F_{S\,on\,W}d\cos\theta = (50\,N)(100\,m)\cos 0°$$
$$= +5000\,N\cdot m$$
$$= +5000\,J$$

Try it yourself: You pull a box 20 m up a 10° ramp. The rope is oriented 20° above the surface of the ramp as shown on the next page. The force that the rope exerts on the box is 100 N. What is the work done by the rope on the box?

(continued)

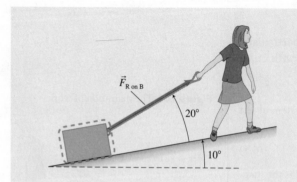

Answer: $W_{\text{R on B}} = (100\,\text{N})(20\,\text{m})\cos 20° = 1900\,\text{J}$. Note that you use the angle between the displacement (parallel to the ramp) and the force the rope exerts on the box (20° above the ramp). You can disregard the angle of the ramp itself.

> **TIP** Remember, the angle that appears in the definition of work is the angle between the external force and the displacement of the system object. It is useful when calculating work to draw tail-to-tail arrows representing the external force doing the work and the system object displacement. Then note the angle between the arrows.

Review Question 6.1 Describe two processes in which an external force is exerted on a system object and no work is done on the system. Explain why no work is done.

6.2 Energy is a conserved quantity

We have found that the work done on a system object by an external force results in a change of one or more types of energy in the system: kinetic energy, gravitational potential energy, elastic potential energy, and internal energy. We can think of the *total energy* U of a system as the sum of all these energies in the system:

$$\text{Total energy} = U = K + U_{\text{g}} + U_{\text{s}} + U_{\text{int}} \qquad (6.2)$$

The energy of a system can be converted from one form to another. For example, the elastic potential energy of a stretched slingshot is converted into the kinetic energy of the chalk when the sling is released. Similarly, the gravitational potential energy of a separated block-Earth system is converted into kinetic energy of the block when the block falls. What happens to the amount of energy when it is converted from one form to another? So far we have one mechanism through which the energy of the system changes—that mechanism is work. Thus it is reasonable to assume that if no work is done on the system, the energy of the system should not change; it should be *constant*. Let's test this hypothesis experimentally (see Testing Experiment **Table 6.3**).

TESTING EXPERIMENT TABLE

6.3   Is the energy of an isolated system constant?

VIDEO 6.3

Testing experiment	Prediction	Outcome
You have a toy car and a frictionless track. The bottom of the track is horizontal to the edge of a table. You can tilt the track at different angles to make the track steeper or shallower. When the car reaches the end of the track, it flies horizontally off the table. Where should you release a car on the track so the car always lands the same distance from the table regardless of the slope of the track? 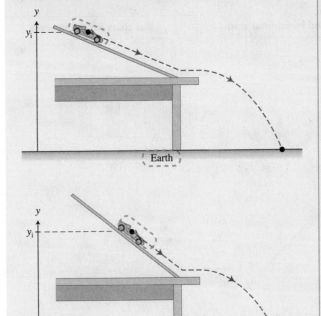	Consider car+Earth as a system. The initial state is just before we release the car and the final state is just as the car leaves the horizontal track. For the car to land on the floor at the same distance from the table's edge, the car needs to have the same horizontal velocity and hence the same kinetic energy when leaving the track. If energy is constant in the isolated system, to get the same kinetic energy at the bottom of the track, the car-Earth system must have the same initial gravitational energy when it starts. Since the gravitational potential energy depends on the separation between the car and Earth, the car should start at the same vertical elevation independent of the tilt angle of the track, $y_i = h$. Our prediction is based on the assumption that the gravitational potential energy of the system is only converted to the kinetic energy of the car and not to internal energy.	When released from the same vertical height with respect to the table, the car lands the same distance from the table.

Conclusion

The outcome of the experiment matches the prediction and thus supports our hypothesis that the energy of an isolated system is constant.

So far we have experimental support for a qualitative hypothesis that the energy of an isolated system is constant and different processes inside the system convert energy from one form to another. We also reason that work is a mechanism through which the energy of a nonisolated system changes. This sounds a lot like our discussion of linear momentum (Chapter 5). In that chapter we described a *conserved quantity* as constant in an isolated system. We also said that when the system is not isolated, we can account for the changes in a conserved quantity by what is added to or subtracted from the system. Based on this reasoning we can hypothesize that energy is a conserved quantity—it is constant in an isolated system and changes as a result of work done on a nonisolated system.

Work-energy bar charts

We can represent work-energy processes with bar charts that are similar to the impulse-momentum bar charts we used to describe momentum. A work-energy bar chart indicates with vertical bars the relative amount of a system's different types of energy in the initial state of a process, the work done on the system by external forces during the process, and the relative amount of different types of energy in the system at the end of the process. The area for the work bar is shaded to emphasize that work does not reside in the system. In **Table 6.4** you see three examples of energy changing from one form to another—represented by words, sketches, and bar charts.

Table 6.4 Three examples of system energy conversions.

Description of process	Sketch of the system and initial-final state	Bar chart for the process
(1) A girl starts at rest at the top of a smooth water slide and is moving fast at the bottom. $$U_g \rightarrow K$$ The system's gravitational potential energy is converted to the girl's kinetic energy as she moves down the slide.		
(2) A fast-moving car skids to a stop on a level road. $$K \rightarrow U_{int}$$ The system's kinetic energy is converted to internal thermal energy due to friction.		
(3) A pop-up toy is compressed and when released pops up to a maximum height of 0.50 m. $$U_s \rightarrow U_g$$ The system's elastic potential energy is converted to gravitational potential energy.		

REASONING SKILL Constructing a qualitative work-energy bar chart.

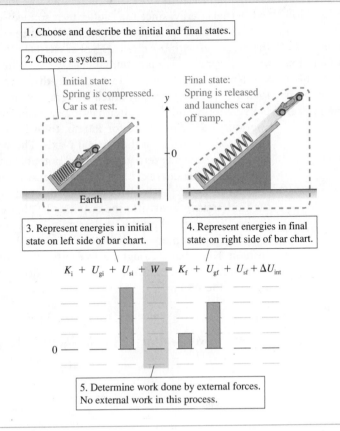

1. Choose and describe the initial and final states.

2. Choose a system.

Initial state:
Spring is compressed.
Car is at rest.

Final state:
Spring is released
and launches car
off ramp.

Earth

3. Represent energies in initial
state on left side of bar chart.

4. Represent energies in final
state on right side of bar chart.

$$K_i + U_{gi} + U_{si} + W = K_f + U_{gf} + U_{sf} + \Delta U_{int}$$

5. Determine work done by external forces.
No external work in this process.

TIP In the Skill box for work-energy bar charts, the system is isolated. No work is done on it. If we do not include Earth in the system, then Earth will do negative work on the cart. However, such a system will not have gravitational potential energy.

The generalized work-energy principle

We can summarize what we have discovered about work and energy in the generalized work-energy principle.

> **Generalized work-energy principle** The sum of the initial energies of a system plus the work done on the system by external forces equals the sum of the final energies of the system:
>
> $$U_i + W = U_f \tag{6.3}$$
>
> or
>
> $$(K_i + U_{gi} + U_{si}) + W = (K_f + U_{gf} + U_{sf} + \Delta U_{int}) \tag{6.3}$$
>
> Note that we have moved $U_{int\,i}$ to the right hand side ($\Delta U_{int} = U_{int\,f} - U_{int\,i}$) since values of internal energy are rarely known, while internal energy changes are.

‹Active Learning Guide

The generalized work-energy principle allows us to define the total energy of a system as the sum of the different types of energy in the system. Total energy is measured in the same units as work and changes when work is done on the system.

The work-energy principle also gives insight into why perpetual motion machines cannot exist. A *perpetual motion machine* is a mechanical device that, once set in motion, continuously and indefinitely does useful things by transferring energy to the environment. This seems impossible because this energy transfer by negative work causes the system's energy to decrease. The machine cannot continue forever.

CONCEPTUAL EXERCISE 6.2 Pole vaulter

A pole vaulter crosses the bar high above the cushion below. Construct a sketch and a work-energy bar chart for two processes relative to the vaulter's jump: (a) the initial state is at the highest point in the jump and the final state is just before the vaulter reaches the cushion below, and (b) the initial state is at the highest point in the jump and the final state is at the instant the jumper has stopped after sinking into the cushion.

Sketch and translate We sketch the processes with different final states (a) just before he hits the cushion and (b) where he is stopped by the cushion. The system includes the vaulter and Earth, but not the cushion. The zero point of the vertical axis is at the vaulter's position after he has sunk into the cushion and stopped.

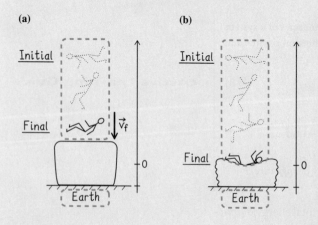

Simplify and diagram We assume that we can ignore air resistance and that the vaulter has zero speed (zero kinetic energy) at the top of the flight. We assume also that the internal energy of the vaulter does not

change significantly (in Part b there is a small increase in internal energy, since the vaulter feels some discomfort when landing on the cushion). We can represent the processes with bar charts. The bar charts both have an initial gravitational potential energy bar. When the vaulter reaches the top of the cushion, the final state for Part a, the system has much less gravitational potential energy and the vaulter has considerable kinetic energy. In the final state for Part b, the vaulter has stopped at the origin of the vertical y-axis. The system in the final state has zero gravitational potential energy and zero kinetic energy. The energy decrease occurred because of the negative work done by the cushion on the vaulter as he sank into the cushion. The force that the cushion exerted on the vaulter pointed up, opposite the displacement of the vaulter sinking into the cushion. Note that although Earth exerts a force on the vaulter, it does not do work on the system, because it is a part of the system.

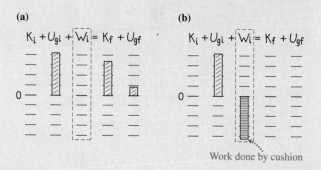

Work done by cushion

Try it yourself: You throw a ball straight upward as shown below. The system is the ball and Earth (but not your hand). Ignore interactions with the air. Draw a work-energy bar chart starting when the ball is at rest in your hand and ending when the ball is at the very top of its flight.

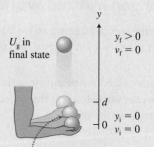

Hand does work on ball only when it is in contact with the ball.

Answer: See the diagram at the right. Note that there is no kinetic energy represented in the bar chart because the ball was not moving in either the initial state or the final state. It does not matter that it was moving in the time interval between those two states.

TIP Note that the amount of gravitational potential energy in a system depends on where the origin is placed on the vertical y-axis. This placement is arbitrary. The important thing is the change in position and the corresponding change in gravitational potential energy.

You might be wondering what objects to include in a system and what objects not to include. Generally, it is preferable to have a larger system so that the changes occurring can be included as energy changes within the system rather than as the work done by external forces. However, often it is best to exclude something like a motor from a system because its energy changes are complex (**Figure 6.2**). For instance, if you are studying a moving elevator, you might choose to exclude from the system the motor that turns the cable while lifting the elevator. However, you can include the motor's effect on the process by determining the magnitude of the force that the cable exerts when pulling up on the elevator.

Figure 6.2 Energy changes in the motor pulling up on the elevator cable are very complicated. It is best to exclude the motor from the system.

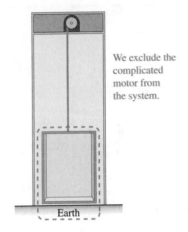

We exclude the complicated motor from the system.

Earth

Review Question 6.2 A system can possess energy but it cannot possess work. Why?

6.3 Quantifying gravitational potential and kinetic energies

In this section we will use what we know about work, forces, and kinematics to devise mathematical expressions for two different types of energy—gravitational potential energy and kinetic energy. We start with gravitational potential energy.

Gravitational potential energy

Imagine that a rope lifts a heavy box upward at a constant negligible velocity (**Figure 6.3a**). The rope is attached to a motor above, which is not shown in the figure. First, we choose only the box as the system and apply Newton's second law to find the magnitude of the force that the rope exerts on the box. Since the box moves up at constant velocity, the upward tension force $\vec{T}_{R\,on\,B}$ exerted by the rope on the box is equal in magnitude to the downward gravitational force $\vec{F}_{E\,on\,B}$ exerted by Earth on the box (see the force diagram in Figure 6.3b). Since the magnitude of the gravitational force is $m_B g$, we find that the magnitude of the tension force for this process is $T_{R\,on\,B} = m_B g$.

To derive an expression for gravitational potential energy, we must change the boundaries of the system to include the box and Earth (if Earth is not included in the system, the system does not have gravitational potential energy). The origin of a vertical y-axis is the ground directly below the box with the positive direction upward. The initial state of the system is the box at position y_i moving upward at a negligible speed $v_i \approx 0$. The final state is the box at position y_f moving upward at the same negligible speed $v_f \approx 0$. According to work-energy Eq. (6.3):

$$U_i + W = U_f$$

The rope does work on the box, lifting the box from vertical position y_i to y_f:

$$W_{R\,on\,B} = T_{R\,on\,B}d\cos\theta = T_{R\,on\,B}(y_f - y_i)\cos 0° = mg(y_f - y_i)$$

‹Active Learning Guide

Figure 6.3 Lifting a box at a negligible constant speed. Notice that the system is box-Earth. (a) Initial and final states; (b) force diagram for the box as the system; (c) energy bar chart for the process.

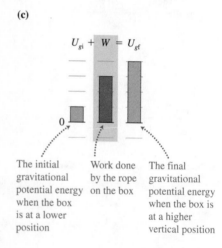

(a)

(b)

The forces have equal magnitudes because the velocity is constant.

(c)

The initial gravitational potential energy when the box is at a lower position

Work done by the rope on the box

The final gravitational potential energy when the box is at a higher vertical position

where we substituted $T_{\text{R on B}} = mg$ and $\cos 0° = 1$. The kinetic energy did not change. Substituting this expression for work into the work-energy equation, we get

$$U_{\text{g i}} + mg(y_{\text{f}} - y_{\text{i}}) = U_{\text{g f}}$$

Figure 6.3c represents this information with an energy bar chart. We now have an expression for the change in the gravitational potential energy of the system: $U_{\text{g f}} - U_{\text{g i}} = mgy_{\text{f}} - mgy_{\text{i}}$. This suggests the following definition for the gravitational potential energy of a system.

> **Gravitational potential energy** The gravitational potential energy of an object-Earth system is
>
> $$U_{\text{g}} = mgy \qquad (6.4)$$
>
> where m is the mass of the object, $g = 9.8 \text{ N/kg}$, and y is the position of the object with respect to the zero of a vertical coordinate system (the origin of the coordinate system is our choice). The units of gravitational potential energy are $\text{kg(N/kg)m} = \text{N} \cdot \text{m} = \text{J (joule)}$, the same unit used to measure work and the same unit for every type of energy.

Kinetic energy

Next, we analyze a simple thought experiment to determine an expression for the kinetic energy of a system that consists of a single object. Imagine that your hand exerts a force $\vec{F}_{\text{H on C}}$ on a cart of mass m while pushing it toward the right a displacement $\vec{d}$ on a horizontal frictionless surface (**Figure 6.4a**). A bar chart for this process is shown in Figure 6.4b. There is no change in gravitational potential energy. The kinetic energy changes from the initial state to the final state because of the work done by the external force exerted by your hand on the cart:

$$K_{\text{i}} + W_{\text{H on C}} = K_{\text{f}}$$

or

$$W_{\text{H on C}} = K_{\text{f}} - K_{\text{i}}$$

We know that the work in this case equals $F_{\text{H on C}}d \cos 0° = F_{\text{H on C}}d$. Thus,

$$F_{\text{H on C}}d = K_{\text{f}} - K_{\text{i}}$$

This does not look like a promising result—the kinetic energy change on the right equals quantities on the left side that do not depend on the mass or speed of the cart. However, we can use dynamics and kinematics to get a result that does depend on these properties of the cart. The horizontal component form of Newton's second law is

$$m_{\text{C}}a_{\text{C}} = F_{\text{H on C}}$$

Figure 6.4 The work done by the hand causes the cart's kinetic energy to increase.

(a)

The hand does positive work on the cart, leading to an increase in its kinetic energy.

(b)

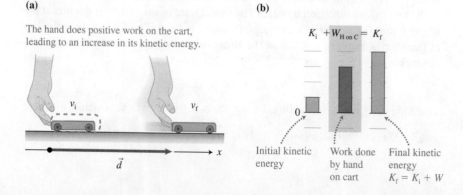

Initial kinetic energy

Work done by hand on cart

Final kinetic energy $K_{\text{f}} = K_{\text{i}} + W$

We can rearrange a kinematics equation ($v_f^2 = v_i^2 + 2ad$) to get an expression for the displacement of the cart in terms of its initial and final speeds and its acceleration:

$$d = \frac{v_f^2 - v_i^2}{2a_C}$$

Now, insert these expressions for force and displacement into the left side of the equation $F_{\text{H on C}}d = K_f - K_i$:

$$F_{\text{H on C}}d = (m_C a_C)\left(\frac{v_f^2 - v_i^2}{2a_C}\right) = m_C\left(\frac{v_f^2}{2} - \frac{v_i^2}{2}\right) = \frac{1}{2}m_C v_f^2 - \frac{1}{2}m_C v_i^2$$

We can now insert this result into the equation $F_{\text{H on C}}d = K_f - K_i$:

$$\frac{1}{2}m_C v_f^2 - \frac{1}{2}m_C v_i^2 = K_f - K_i$$

It appears that $\frac{1}{2}m_C v^2$ is an expression for the kinetic energy of the cart.

> **Kinetic energy** The kinetic energy of an object is
>
> $$K = \frac{1}{2}mv^2 \qquad (6.5)$$
>
> where m is the object's mass and v is its speed relative to the chosen coordinate system.

To check whether the unit of kinetic energy is the joule (J), we use Eq. (6.5) with the units $\frac{\text{kg} \cdot \text{m}^2}{s^2} = \left(\frac{\text{kg} \cdot \text{m}}{s^2}\right)\text{m} = \text{N} \cdot \text{m} = \text{J}$.

EXAMPLE 6.3 An acorn falls

You sit on the deck behind your house. Several 5-g acorns fall from the trees high above, just missing your chair and head. Use the work-energy equation to estimate how fast one of these acorns is moving just before it reaches the level of your head.

Sketch and translate First, we draw a sketch of the process. The system will be the acorn and Earth. The origin of a vertical y-axis will be at your head with the positive y-axis pointing up. The acorn is about 20 m above your head as it begins to fall. We keep track of kinetic energy and gravitational potential energy to find the acorn's speed when it reaches the level of your head. The initial state will be the instant the acorn leaves the tree. The final state is when it reaches the level of your head.

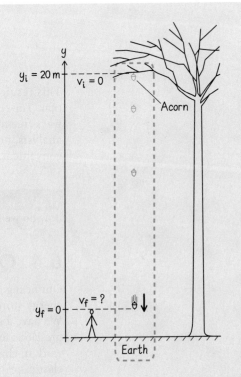

(continued)

Simplify and diagram The acorn is small, and we assume that the air does no significant work on the acorn as it falls. We represent the process with a bar chart.

$$K_i + U_{gi} + W = K_f + U_{gf}$$

The system starts with gravitational potential energy.

The system ends with kinetic energy.

Represent mathematically Use the bar chart to apply the work-energy equation:

$$0 + U_{gi} + 0 = K_f + 0$$

$$mgy_i + 0 = \frac{1}{2}mv_f^2$$

Cancelling the common m on each side and rearranging, we get

$$v_f = \sqrt{2(gy_i)}$$

Solve and evaluate Our estimate of the final speed of the acorn is

$$v_f = \sqrt{2(gy_i)} = \sqrt{2(9.8 \text{ m/s}^2)(20 \text{ m})} = 20 \text{ m/s}$$

That's 45 mi/h—that seems reasonable based on the sound it makes when it hits the deck.

Try it yourself: If you throw an acorn upward at a speed v_i, what is the maximum height above its launching position that the acorn will reach before it starts descending?

Answer: $\dfrac{v_i^2}{2g}$.

Active Learning Guide›

What if in Example 6.3 we had chosen only the acorn as the system of interest? In that case the system would not have any gravitational potential energy. Instead, Earth would be an external object doing positive work on the acorn system. Since the acorn is at rest initially, the system has no initial energy. Earth does work on the system. In the final state, just as the acorn reaches the level of your head, the acorn system has kinetic energy:

$$0 + W = K_f$$

$$Fd\cos\theta = \frac{1}{2}mv_f^2$$

$$(mg)y_i \cos 0° = \frac{1}{2}mv_f^2$$

$$v_f = \sqrt{2(gy_i)}$$

This result is the same as the one we arrived at using the acorn-Earth system. *The choice of system did not affect the result of the analysis.* We are always free to choose the system of interest so that it best suits the goal of our analysis, just as we are free to choose whichever coordinate system is most convenient.

Review Question 6.3 When we use the work-energy equation, how do we incorporate the force that Earth exerts on an object?

6.4 Quantifying elastic potential energy

Our next goal is to construct a mathematical expression for the elastic potential energy stored by an elastic object when it has been stretched or compressed. We have a special problem in deriving this expression. When we derived expressions for gravitational potential and kinetic energies, a constant force did work in changing those energies. However, when you stretch or compress an elastic spring-like object, you have to pull or push harder the more the object is stretched or compressed. The force is not constant. How does the force you exert to stretch a spring-like object change as the object stretches?

Table 6.5 Result of pulling on springs while exerting an increasing force.

Force F exerted by the scale on the spring	Spring 1 stretch distance x	Spring 2 stretch distance x
0.00 N	0.000 m	0.000 m
1.00 N	0.050 m	0.030 m
2.00 N	0.100 m	0.060 m
3.00 N	0.150 m	0.090 m
4.00 N	0.200 m	0.120 m

Hooke's law

To answer this question, we use two springs of the same length: a thinner and less stiff spring 1 and a thicker and stiffer spring 2 (**Figure 6.5a**). The springs are attached at the left end to a rigid object and placed on a smooth surface. We use a scale to pull on the right end of each spring, exerting a force whose magnitude F can be measured by the scale (Figure 6.5b). We record F and the distance x that each spring stretches from its unstretched position (see **Table 6.5**).

Figure 6.6 shows a graph of the Table 6.5 data. We use stretch distance x as an independent variable and the magnitude of the force $\vec{F}_{\text{Scale on Spring}}$ as a dependent variable. The magnitude of the force exerted by the scale on each spring is proportional to the distance that each spring stretches.

$$F_{\text{Scale on Spring}} = kx$$

The coefficient of proportionality k (the slope of the F-versus-x graph) is called the **spring constant.** The slope for the stiffer spring 2 is larger $(33\,\text{N/m})$ than the slope for spring 1 $(20\,\text{N/m})$. In other words, to stretch spring 1 by 1.0 m we would have to exert only a 20-N force, but we would need a 33-N force for spring 2.

Often we are interested not in the force that something exerts on the spring, but in the force that the spring exerts on something else, $\vec{F}_{\text{Spring on Scale}}$. Using Newton's third law we have $\vec{F}_{\text{Spring on Scale}} = -\vec{F}_{\text{Scale on Spring}}$. The x-component of this force is

$$F_{\text{Spring on Scale}\,x} = -kx$$

Note that if an object stretches the spring to the right in the positive direction, the spring pulls back on the object in the opposite negative direction (**Figure 6.7a**, in this case the object is your finger). If the object compresses the spring to the left in the negative direction, the spring pushes back on the object in the opposite positive direction (Figure 6.7b). These observations are the basis for a rule first developed by Robert Hooke (1635–1703), called **Hooke's law.**

Figure 6.5 Measuring spring stretch x caused by a scale pulling with force F on a spring. Notice the difference in thickness and stiffness of the two springs.

(a)

Scale will pull springs (which are now relaxed).

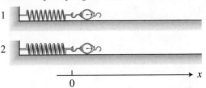

(b)

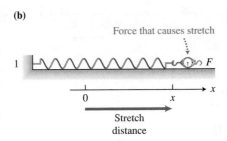

Figure 6.6 A graph of the stretch of springs 1 and 2 when the same force is exerted on each spring. The bigger slope indicates a smaller stretch distance when the same force is exerted on the spring.

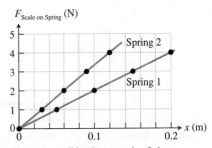

We need to pull harder on spring 2 than on spring 1 to stretch it 0.1 m.

Elastic force (Hooke's law) If any object causes a spring to stretch or compress, the spring exerts an elastic force on that object. If the object stretches the spring along the x-direction, the x-component of the force the spring exerts on the object is

$$F_{\text{S on O}\,x} = -kx \qquad (6.6)$$

The spring constant k is measured in newtons per meter and is a measure of the stiffness of the spring (or any elastic object); x is the distance that the object has been stretched/compressed (not the total length of the object). The elastic force exerted by the spring on the object points in a direction opposite to the direction it was stretched (or compressed)—hence the negative sign in front of kx. The object in turn exerts a force on the spring:

$$F_{\text{O on S}\,x} = +kx$$

Figure 6.7 Hooke's law. (a) The force that the spring exerts on the finger stretching the spring $F_{\text{S on F}}$ points opposite the direction the spring stretches. (b) The force that the spring exerts on the finger compressing the spring $F_{\text{S on F}}$ points opposite the direction of the spring compression.

(a)

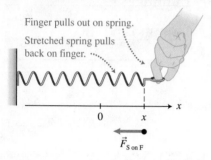

Finger pulls out on spring.

Stretched spring pulls back on finger.

$$\vec{F}_{\text{S on F}}$$

(b)

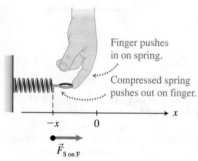

Finger pushes in on spring.

Compressed spring pushes out on finger.

$$\vec{F}_{\text{S on F}}$$

Elastic potential energy

Our goal now is to develop an expression for the elastic potential energy of an elastic stretched or compressed object (such as a stretched spring). Consider the constant slopes of the lines in the graph shown in Figure 6.6. While stretching the spring with your hand from zero stretch ($x = 0$) to some arbitrary stretch distance x, the magnitude of the force your hand exerts on the spring changes in a linear fashion from zero when unstretched to kx when stretched.

To calculate the work done on the spring by such a variable force, we can replace this variable force with the average force $(F_{\text{H on S}})_{\text{average}}$:

$$(F_{\text{H on S}})_{\text{average}} = \frac{0 + kx}{2}$$

The force your hand exerts on the spring is in the same direction as the direction in which the spring stretches. Thus the work done by this force on the spring to stretch it a distance x is

$$W = (F_{\text{H on S}})_{\text{average}}x = \left(\frac{1}{2}kx\right)x = \frac{1}{2}kx^2$$

This work equals the change in the spring's elastic potential energy. Assuming that the elastic potential energy of the unstretched spring is zero, the work we calculated above equals the final elastic potential energy of the stretched spring.

> **Elastic potential energy** The elastic potential energy of a spring-like object with a spring constant k that has been stretched or compressed a distance x from its undisturbed position is
>
> $$U_s = \frac{1}{2}kx^2 \tag{6.7}$$
>
> Just like any other type of energy, the unit of elastic potential energy is the joule (J).

EXAMPLE 6.4 Shooting an arrow

You load an arrow (mass = 0.090 kg) into a bow and pull the bowstring back 0.40 m. The bow has a spring constant $k = 900$ N/m. Determine the arrow's speed as it leaves the bow.

Sketch and translate We sketch the process, as shown below. The system is the bow and arrow. In the initial state, the bowstring is pulled back 0.40 m. In the final state, the string has just relaxed and the arrow has left the string.

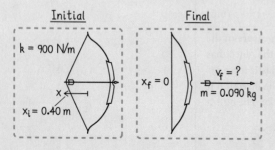

Simplify and diagram Since the arrow moves horizontally, we do not need to keep track of gravitational potential energy. The initial elastic potential energy of the bow is converted into the final kinetic energy of the arrow. We represent the process with the bar chart.

$$K_i + U_{gi} + U_{si} + W = K_f + U_{gf} + U_{sf} + \Delta U_{int}$$

Represent mathematically Use the bar chart to help apply the work-energy equation:

$$U_i + W = U_f$$
$$\Rightarrow K_i + U_{si} + 0 = K_f + U_{sf}$$
$$\Rightarrow 0 + \frac{1}{2}kx^2 = \frac{1}{2}mv^2 + 0$$

Multiply both sides of the above by 2, divide by m, and take the square root to get

$$v = \sqrt{\frac{k}{m}}x$$

Solve and evaluate

$$v = \sqrt{\frac{k}{m}}x = \sqrt{\frac{900 \text{ N/m}}{0.090 \text{ kg}}}(0.40 \text{ m}) = 40 \text{ m/s}.$$

This is reasonable for the speed of an arrow fired from a bow. The units of $\sqrt{\frac{k}{m}}x$ are equivalent to $\frac{\text{m}}{\text{s}}$:

$$\sqrt{\frac{\text{N}}{\text{m} \cdot \text{kg}}}(\text{m}) = \sqrt{\frac{\text{kg} \cdot \text{m}}{\text{s}^2 \cdot \text{m} \cdot \text{kg}}}(\text{m}) = \frac{\text{m}}{\text{s}}$$

Try it yourself: If the same arrow were shot vertically, how high would it go?

Answer: 82 m.

Review Question 6.4 If the magnitude of the force exerted by a spring on an object is kx, why is it that the work done to stretch the spring a distance x is not equal to $kx \cdot x = kx^2$?

6.5 Friction and energy conversion

In nearly every mechanical process, objects exert friction forces on each other. Sometimes the effect of friction is negligible (for example, in an air hockey game), but most often friction is important (for example, a driver applying the brakes to avoid a collision). Our next goal is to investigate how we can incorporate friction into work and energy concepts. Let's analyze a car skidding to avoid an accident.

Can friction force do work?

Imagine that the car's brakes have locked, and the tires are skidding on the road surface (**Figure 6.8a**). Let's first choose the system of interest to be the car. How

‹Active Learning Guide

Figure 6.8 Friction. If we use the car

left side of the above equation add to zero, so we would get $0 = \Delta U_{\text{int}}$, which is not possible.

This same difficulty occurs with another process. Imagine that you pull a rope attached to a box (the system object) so that the box moves at constant very slow velocity on a rough carpet (**Figure 6.9a**). The box is moving so slowly that we will ignore its kinetic energy. A force diagram for the box is shown in Figure 6.9b.

The bar chart (Figure 6.9c) shows that the force exerted by the rope does positive work on the box system; the friction force does negative work. These

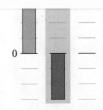

The energy bar chart balances. But after the skid the tire is warmer and some tread has worn off. Thus, ΔU_{int} should be positive.

Figure 6.9 Pulling a box across a rough carpet. The total work done on the box is zero but its internal energy changes. You cannot account for the internal energy change of the box if it alone is the system.

(a)

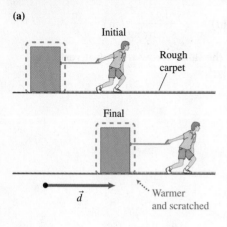

Initial

Rough carpet

Final

$\vec{d}$

Warmer and scratched

(b)

Because the velocity is constant, T and f_k have the same magnitudes.

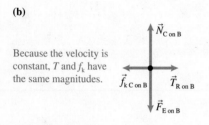

$\vec{N}_{\text{C on B}}$

$\vec{f}_{\text{k C on B}}$ $\vec{T}_{\text{R on B}}$

$\vec{F}_{\text{E on B}}$

(c)

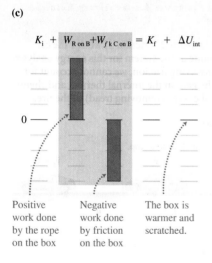

K_i + $W_{\text{R on B}}$+$W_{f \text{k C on B}}$ = K_f + ΔU_{int}

0 —

Positive work done by the rope on the box

Negative work done by friction on the box

The box is warmer and scratched.

horizontal forces have exactly the same magnitude (the box is moving at constant velocity). Thus, the sum of the work done by those forces on the system is zero and the energy of the system should not change.

$$T_{\text{Rope on Box}}d \cos 0° + f_{\text{k Carpet on Box}}d \cos 180° = T_{\text{Rope on Box}}d - f_{\text{k Carpet on Box}}d = 0$$

However, if you touch the bottom of the box at its final position, you find that it is warmer than before you started pulling and the box has scratches on its bottom. Again, the internal energy of the box increased. We get $0 = \Delta U_{\text{int}}$, where ΔU_{int} is greater than zero. Again, the zero work done on the box does *not* equal the positive increase in internal energy. This is a contradiction of the work-energy principle. How can we resolve this?

The effect of friction as a change in internal energy

The key to resolving this problem is to change the system. The new system will include both surfaces that are in contact, for example, the car and the road, or the box and the carpet. The friction force is then an internal force and therefore does no work on the system. But there is a change in the internal energy of the system caused by friction between the two surfaces.

When the rope is pulling the box across the rough carpet at constant velocity, we know that if the rope pulls horizontally on the box, the magnitude of the force that it exerts on the box $T_{\text{R on B}}$ must equal the magnitude of the friction force that the carpet exerts on the box $f_{\text{k C on B}}$. But now the box and carpet are both in the system—so the force exerted by the rope is the only external force. Thus the work done on the box-carpet system is

$$W = T_{\text{R on B}}d \cos 0°$$

Substituting $T_{\text{R on B}} = f_{\text{k C on B}}$ and $\cos 0° = 1.0$ into the above, we get

$$W = +f_{\text{k C on B}}d$$

The only system energy change is its internal energy ΔU_{int}. The work-energy equation for pulling the box across the surface is

$$W = \Delta U_{\text{int}}$$

After inserting the expression for the work done on the box and rearranging, we get

$$\Delta U_{\text{int}} = +f_k d$$

We have constructed an expression for the change in internal energy of a system caused by the friction force that the two contacting surfaces in the system exert on each other when one object moves a distance d across the other.

> **Increase in the system's internal energy due to friction**
>
> $$\Delta U_{\text{int}} = +f_k d \qquad (6.8)$$
>
> where f_k is the magnitude of the average friction force exerted by the surface on the object moving relative to the surface and d is the distance that the object moves across that surface. The increase in internal energy is shared between the moving object and the surface.

Including friction in the work-energy equation as an increase in the system's internal energy produces the same result as calculating the work done by friction force. In this new approach, there is an increase in internal energy $\Delta U_{\text{int}} = +f_k d$ in a system that includes the two surfaces rubbing against each other (this expression goes on the right side of the work-energy equation).

In the work done by friction approach, the negative work done by friction $W_{\text{friction}} = -f_k d$ is included if one of the surfaces is not in the system (this term goes on the left side of the work-energy equation). So mathematically, they have the same effect. When we include the two surfaces in the system, we can see why a skidding tire or the moving box gets warmer and why there might be structural changes. If we include only the car or box in the system and consider work done by friction, the change in the internal energy of the rubbing surfaces is a mystery. In this book we prefer to include both surfaces in the system and consider the increase in internal energy caused by friction.

EXAMPLE 6.5 Skidding to a stop

You are driving your car when another car crosses the road at an intersection in front of you. To avoid a collision, you apply the brakes, leaving 24-m skid marks on the road while stopping. A police officer observes the near collision and gives you a speeding ticket, claiming that you were exceeding the 35 mi/h speed limit. She estimates your car's mass as 1390 kg and the coefficient of kinetic friction μ_k between your tires and this particular road as about 0.70. Do you deserve the speeding ticket?

Sketch and translate We sketch the process. We choose your car and the road surface as the system. We need to decide if your car was traveling faster than 35 mi/h at the instant you applied the brakes.

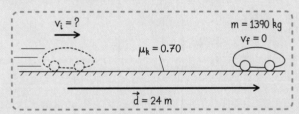

Simplify and diagram Assume that the process occurs on a horizontal level road and neglect interactions with the air. The initial state is just before the brakes are applied. The final state is just after your car has come to rest. We draw the energy bar chart to represent the process. In the initial state, the system has kinetic energy. In the final state, the system has no kinetic energy and has increased internal energy due to friction.

$$K_i + U_{gi} + U_{si} + \boxed{W} = K_f + U_{gf} + U_{sf} + \Delta U_{int}$$

Represent mathematically Convert the bar chart into an equation:

$$K_i + 0 = U_{\text{int f}}$$

$$\Rightarrow \frac{1}{2}m_C v_i^2 = f_{k\,R\,\text{on}\,C}\,d$$

$$\Rightarrow \frac{1}{2}m_C v_i^2 = (\mu_k N_{R\,\text{on}\,C})d$$

The magnitude of the upward normal force $N_{R\,\text{on}\,C}$ that the road exerts on the car equals the magnitude of the downward gravitational force $F_{E\,\text{on}\,C} = m_C g$ that Earth exerts on the car: $N_{R\,\text{on}\,C} = m_C g$. Thus the above becomes

$$\frac{1}{2}m_C v_i^2 = \mu_k m_C g d$$

Solve and evaluate Rearranging the above and canceling the car mass, we get

$$v_i = \sqrt{2\mu_k g d} = \sqrt{2(0.70)(}$$

$$= (18.1 \text{ m/}$$

$$= 41 \text{ mi/h}$$

It looks like you deserve the sistive drag force that air exe be significant for a car travel sistance helps the car's spee actually were traveling fast

Try it yourself: Imagine only you are driving a 20 for the initial speed incr

Answer: The speed doe

Review Question 6.5 Why, when friction cannot be neglected, is it useful to include both surfaces in the system when analyzing processes using the energy approach?

6.6 Skills for analyzing processes using the work-energy principle

In this section, we use a general problem-solving strategy to analyze work-energy processes. The general strategy is described on the left side of the table in Example 6.6 and illustrated on the right side for a specific process.

Active Learning Guide>

PROBLEM-SOLVING STRATEGY   Applying the work-energy principle

EXAMPLE 6.6   An elevator slows to a stop

A 1000-kg elevator is moving downward. While moving down at 4.0 m/s, its speed decreases steadily until it stops in 6.0 m. Determine the magnitude of the tension force that the cable exerts on the elevator ($T_{C\,on\,El}$) while it is stopping.

Sketch and translate

- Sketch the initial and final states of the process, labeling known and unknown information.
- Choose the system of interest.
- Include the object of reference and the coordinate system.

- The elevator and Earth are in the system. We exclude the cable—its effect will be included as the work done by the cable on the elevator.

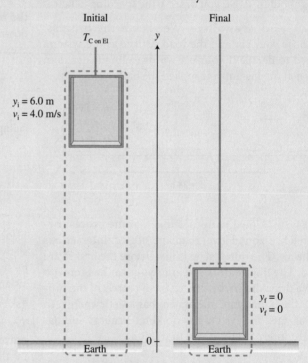

- The observer is on the ground; the coordinate system has a vertical axis pointing up with the zero at the bottom of the shaft.

Simplify and diagram

- What simplifications can you make to the objects, interactions, and processes?
- Decide which energy types are changing.
- Are external objects doing work?

- We assume that the cable exerts a constant force on the elevator and that the elevator can be considered a point-like object since all of its points move the same way (no deformation).
- We will keep track of kinetic energy (since the elevator's speed changes) and gravitational potential energy (since the elevator's vertical position changes).

■ Use the initial-final sketch to help draw a work-energy bar chart. Include work bars (if needed) and initial and final energy bars for the types of energy that are changing. Specify the zero level of gravitational potential energy.

■ The tension force exerted by the cable on the elevator does negative work (the tension force points up, and the displacement of the elevator points down).

■ The zero gravitational potential energy is at the lowest position of the elevator. In its initial state the system has kinetic energy and gravitational potential energy. In its final state the system has no energy.

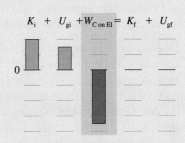

Represent mathematically

■ Convert the bar chart into a mathematical description of the process. Each bar in the chart will appear as a single term in the equation.

$$U_i + W = U_f$$

$$\left(\frac{1}{2}\right)mv_i^2 + mgy_i + T_{C \text{ on El}}(y_i - 0)\cos 180° = 0$$

$$\left(\frac{1}{2}\right)mv_i^2 + mgy_i - T_{C \text{ on El}} y_i = 0$$

Solve and evaluate

■ Solve for the unknown and evaluate the result.

■ Does it have the correct units? Is its magnitude reasonable? Do the limiting cases make sense?

$$T_{C \text{ on El}} = mg + \frac{mv^2}{2y_i}$$

$$= (1000 \text{ kg})(9.8 \text{ N/kg}) + \frac{(1000 \text{ kg})(4.0 \text{ m/s})^2}{2(6.0 \text{ m})}$$

$$= 11{,}000 \text{ N}$$

■ The result has the correct units. The force that the cable exerts is more than the 9800-N force that Earth exerts. This is reasonable since the elevator slows down while moving down; thus the sum of the forces exerted on it should point up.

■ Limiting case: If the elevator slowed down over a much longer distance (y_f = very large number instead of 6.0 m), then the force would be closer to 9800 N.

Try it yourself: Solve the same problem using Newton's second law and kinematics.

Answer: 11,000 N.

Human cannonball

The Try It Yourself exercise demonstrates that we can obtain the same result for Example 6.6 using Newton's second law and kinematics. However, the energy approach is often easier and quicker. Knowing that, let's revisit the human cannonball problem using the ideas of work and energy. (We first analyzed this using Newton's laws and kinematics in Example 3.9.)

EXAMPLE 6.7 The human cannonball again

In order to launch a 60-kg human so that he leaves the cannon moving at a speed of 15 m/s, you need a spring with an appropriate spring constant. This spring will be compressed 3.0 m from its natural length when it is ready to launch the person. The cannon is oriented at an angle of 37° above the horizontal. What spring constant should the spring have so that the cannon functions as desired?

Sketch and translate We draw the sketch first. It shows the system as the person, the cannon (with the spring), and Earth. The initial state is just before the cannon is fired. The final state is when the person is leaving the end of the barrel of the cannon 3.0 m from where he started. All motion is with respect to Earth. The origin of the vertical y-axis is at the initial position of the person.

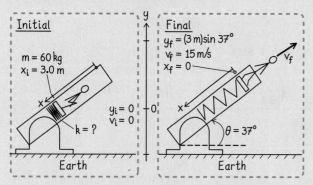

Simplify and diagram Assume that the spring obeys Hooke's law and neglect the relatively small amount of friction between the person and the inside barrel walls of the cannon and between the person and the air. We need to keep track of kinetic energy (the person's speed changes), gravitational potential energy (Earth is in the system and the person's vertical position changes relative to the vertical y-axis), and elastic potential energy changes (the spring compression changes relative to a special x-axis used to keep track of elastic energy). No external forces are being exerted on the system, so the total energy of the system is constant. We draw the bar chart to represent the process.

$$K_i + U_{gi} + U_{si} + \boxed{W} = K_f + U_{gf} + U_{sf}$$

Represent mathematically The work-energy equation for this process is

$$U_i + W = U_f$$
$$\Rightarrow U_{si} + 0 = K_f + U_{gf}$$
$$\Rightarrow \frac{1}{2}kx_i^2 = \frac{1}{2}mv_f^2 + mgy_f$$

Solve and evaluate Dividing both parts of the equation by $(1/2)x_i^2$ we get

$$k = \frac{2\left(\frac{1}{2}mv_f^2 + mgx_i \sin 37°\right)}{x_i^2}$$

$$= \frac{2\left[\frac{1}{2}(60\,\text{kg})(15\,\text{m/s})^2 + (60\,\text{kg})(9.8\,\text{N/kg})(3.0\,\text{m})\sin 37°\right]}{(3.0\,\text{m})^2}$$

$$= 1736\,\text{N/m} \approx 1700\,\text{N/m}$$

The units of the answer are correct for a spring constant. The spring constant is always a positive number and we obtained a positive number. The magnitude of k is quite large, which means this is a stiff spring. This makes sense given that it is launching a person.

Try it yourself: What should the spring constant of a spring be if there is a 100-N friction force exerted on a human cannonball while he is moving up the barrel?

Answer: 1800 N/m.

High blood pressure

We have used the work-energy approach to examine several real-world phenomena. Let's use it to examine how the body pumps blood.

BIO CONCEPTUAL EXERCISE 6.8 Stretching the aorta

Every time your heart beats, the left ventricle pumps about 80 cm³ of blood into the aorta, the largest artery in the human body. This pumping action occurs during a very short time interval, about 0.13 s. The elastic aorta walls stretch to accommodate the extra volume of blood. During the next 0.4 s or so, the walls of the aorta contract, applying pressure on the blood and moving it out of the aorta into the rest of the circulatory system. Represent this process with a qualitative work-energy bar chart.

Sketch and translate We sketch the process, as shown below. Choose the system to be the aorta and the 80 cm³ of blood that is being pumped. The left ventricle is not in the system. Choose the initial state to be just before the left ventricle contracts. Choose the final state to be when the blood is in the stretched aorta before moving out into the rest of the circulatory system.

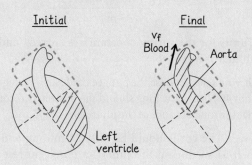

Initial   Final

Simplify and diagram We keep track of the kinetic energy (the blood speed changes) and the elastic potential energy (the aorta wall stretches). We ignore the slight increase in the vertical elevation of the blood. The left ventricle is not

Flexible-walled aorta

$$W = K_f + U_{sf}$$

in the system and does positive work pushing the blood upward in the direction of the blood's displacement. The work-energy bar chart below left represents this process for an aorta with flexible walls.

Try it yourself: Modify the work-energy bar chart for a person with hardened and thickened artery walls.

Answer: If the person has stiff, thick arteries (a condition called atherosclerosis), more energy than normal is required to stretch the aorta walls. This process is represented by the bar chart below. In this case, the blood pressure will be higher than it would be for a healthy cardiovascular system because the heart has to do more work to stretch the walls while pushing blood into the aorta.

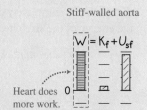

Stiff-walled aorta

$$W = K_f + U_{sf}$$

Heart does more work.

Bungee jumping

On April 1, 1979, four members of the Oxford University Dangerous Sport Club made the first modern bungee jump. They jumped from the 76-m-high (250 ft) Clifton Suspension Bridge in Bristol, England while tied to the bridge with a rubber bungee cord. Let's analyze their jump using work-energy principles.

EXAMPLE 6.9 Bungee jumping

We estimate that the Oxford team used a 40-m-long bungee cord that had stretched another 35 m when the jumper was at the very lowest point in the jump, 1.0 m above the ground. We estimate that the jumper's mass is 70 kg. Imagine that your job is to buy a bungee cord that would provide a safe jump with the above specifications. Specifically, you need to determine the spring constant k of the cord you need to buy.

Sketch and translate We sketch the process, as shown at right. The initial state is just before the jumper jumps, and the final state is when the cord is fully stretched and the jumper is momentarily at rest at the lowest position. The motion is with respect to Earth. We choose a coordinate system with the positive y-direction pointing up and the origin at the jumper's final position. The system is the jumper, the cord, and Earth.

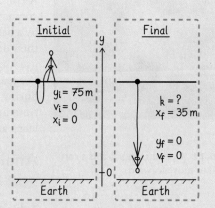

Simplify and diagram We are interested in the gravitational potential energy (the jumper's elevation changes) and the elastic potential energy (the cord stretch changes).

(continued)

In both the initial and final states the system has no kinetic energy. Assume that the bungee cord obeys Hooke's law and that its mass is negligible compared to the mass of the jumper. There are no external forces doing work on the system; thus, the energy of the system is constant. We represent the process with a bar chart, as shown below. It shows us that the initial gravitational potential energy of the system is completely converted into the elastic potential energy of the stretched bungee cord.

$$K_i + U_{gi} + U_{si} + \boxed{W} = K_f + U_{gf} + U_{sf} + \Delta U_{int}$$

Represent mathematically We apply the work-energy equation with one term for each bar in the bar chart:

$$U_{gi} = U_{sf}$$

$$\Rightarrow mgy_i = \frac{1}{2}kx_f^2$$

Solve and evaluate Solving the above for the spring constant gives

$$k = \frac{2mgy_i}{x_f^2}$$

The length of the cord is $L = 40$ m and the cord stretches $x_f = 35$ m. Thus, the distance of the person's initial *position* y_i above the final $y_f = 0$ position is

$$y_i = L + x_f = 40\,\text{m} + 35\,\text{m} = 75\,\text{m}$$

This means that the spring constant has the value

$$k = \frac{2mgy_i}{x_f^2} = \frac{2(70\,\text{kg})(9.8\,\text{N/kg})(75\,\text{m})}{(35\,\text{m})^2} = 84\,\text{N/m}$$

The units for the spring constant are correct, and the magnitude is reasonable.

Try it yourself: Suppose the bungee cord had a spring constant of 40 N/m. How long should the unstretched cord be so that the total distance of the jump remains 75 m?

Answer: The cord should be only 24 m long and will stretch 51 m during the jump—a much more easily stretched cord.

Review Question 6.6 What would change in the solution to the problem in Example 6.9 if we did not include Earth in the system? How would the answer be different?

6.7 Collisions: Putting it all together

A collision is a process that occurs when two (or more) objects are in direct contact with each other for a short time interval, such as when a baseball is hit by a bat (**Figure 6.10**). The ball compresses during the first half of the collision, then decompresses during the second half of the collision. We have already used impulse and momentum principles to analyze collisions (in Chapter 5). We learned that the forces that the two colliding objects exert on each other during the collision are complicated, nonconstant, and exerted for a very brief time interval—roughly 1 ms in the case of the baseball-bat collision. Can we learn anything new by analyzing collisions using the work-energy principle?

Analyzing collisions using momentum and energy principles

Table 6.6 shows three different observational experiments involving collisions. In each case a 1.0-kg object attached to the end of a string (the bob of a pendulum) swings down and hits a 4.0-kg wheeled cart at the lowest point of its swing (**Figure 6.11**). In each experiment, the pendulum bob and cart start at the same initial positions, but the compositions of the pendulum bob and cart are varied. After each collision, the cart moves at a nearly constant speed because of the smoothness of the surface on which it rolls.

Figure 6.10 A baseball is compressed while being hit by a bat.

Figure 6.11 A pendulum bob hits a cart, causing it to move forward.

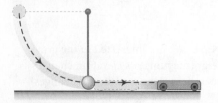

OBSERVATIONAL EXPERIMENT TABLE

6.6 Analyzing energy and momentum during collisions.

 VIDEO 6.6

Observational experiment	Analysis

Experiment 1. A 1.0-kg metal bob swings and hits a 4.0-kg metal cart. Their velocity components just before and just after the collision are shown below:

$$m_1 = 1.0 \text{ kg}; v_{1ix} = 10 \text{ m/s}$$
$$v_{1fx} = -6.0 \text{ m/s}$$
$$m_2 = 4.0 \text{ kg}; v_{2ix} = 0 \text{ m/s}$$
$$v_{2fx} = 4.0 \text{ m/s}$$

Momentum:

Before collision:
$$(1.0 \text{ kg})(+10 \text{ m/s}) + (4.0 \text{ kg}) 0 = +10 \text{ kg} \cdot \text{m/s}$$

After collision:
$$(1.0 \text{ kg})(-6.0 \text{ m/s}) + (4.0 \text{ kg})(+4.0 \text{ m/s}) = +10 \text{ kg} \cdot \text{m/s}$$

Kinetic energy:

Before collision:
$$(1/2)(1.0 \text{ kg})(+10 \text{ m/s})^2 + (1/2)(4.0 \text{ kg}) 0^2 = 50 \text{ J}$$

After collision:
$$(1/2)(1.0 \text{ kg})(-6.0 \text{ m/s})^2 + (1/2)(4.0 \text{ kg})(+4.0 \text{ m/s})^2 = 50 \text{ J}$$

Experiment 2. A 1.0-kg sand-filled balloon swings and hits a 4.0-kg flimsy cardboard cart. After the collision, the damaged cart moves across the table at constant speed. The side of the balloon that hit the cart is flattened in the collision. Their velocity components just before and just after the collision are shown below:

$$m_1 = 1.0 \text{ kg}; v_{1ix} = 10 \text{ m/s}$$
$$v_{1fx} = -4.2 \text{ m/s}$$
$$m_2 = 4.0 \text{ kg}; v_{2ix} = 0 \text{ m/s}$$
$$v_{2fx} = 3.55 \text{ m/s}$$

Momentum:

Before collision:
$$(1.0 \text{ kg})(+10 \text{ m/s}) + (4.0 \text{ kg}) 0 = +10 \text{ kg} \cdot \text{m/s}$$

After collision:
$$(1.0 \text{ kg})(-4.2 \text{ m/s}) + (4.0 \text{ kg})(+3.55 \text{ m/s}) = +10 \text{ kg} \cdot \text{m/s}$$

Kinetic energy:

Before collision:
$$(1/2)(1.0 \text{ kg})(+10 \text{ m/s})^2 + (1/2)(4.0 \text{ kg}) 0^2 = 50 \text{ J}$$

After collision:
$$(1/2)(1.0 \text{ kg})(-4.2 \text{ m/s})^2 + (1/2)(4.0 \text{ kg})(+3.55 \text{ m/s})^2 = 34 \text{ J}$$

Experiment 3. A 1.0-kg sand-filled balloon covered with Velcro swings down and sticks to a 4.0-kg Velcro-covered cardboard cart. The string holding the balloon is cut by a razor blade immediately after the balloon contacts the cart. The damaged cart and flattened balloon move off together across the table. Their velocity components just before and just after the collision are shown below:

$$m_1 = 1.0 \text{ kg}; v_{1ix} = 10 \text{ m/s}$$
$$v_{1fx} = 2.0 \text{ m/s}$$
$$m_2 = 4.0 \text{ kg}; v_{2ix} = 0 \text{ m/s}$$
$$v_{2fx} = 2.0 \text{ m/s}$$

Momentum:

Before collision:
$$(1.0 \text{ kg})(+10 \text{ m/s}) + (4.0 \text{ kg}) 0 = +10 \text{ kg} \cdot \text{m/s}$$

After collision:
$$(1.0 \text{ kg})(+2.0 \text{ m/s}) + (4.0 \text{ kg})(+2.0 \text{ m/s}) = +10 \text{ kg} \cdot \text{m/s}$$

Kinetic energy:

Before collision:
$$(1/2)(1.0 \text{ kg})(+10 \text{ m/s})^2 + (1/2)(4.0 \text{ kg}) 0^2 = 50 \text{ J}$$

After collision:
$$(1/2)(1.0 \text{ kg})(+2.0 \text{ m/s})^2 + (1/2)(4.0 \text{ kg})(+2.0 \text{ m/s})^2 = 10 \text{ J}$$

Patterns

Two important patterns emerge from the data collected from these different collisions.

- The momentum of the system is constant in all three experiments.
- The kinetic energy of the system is constant when no damage is done to the system objects during the collision (Experiment 1 but not in Experiments 2 and 3).

We use momentum and energy principles to analyze the results of the experiments. In all cases, the system is the pendulum bob and the cart. The initial state of the system is the moment just before the collision. The final state is the moment just after the collision ends. The momentum of the system is in the positive x-direction. We keep track of the kinetic energy of both the pendulum bob and the cart. In some of the experiments, one or more objects in the system are deformed; we will discuss internal energy changes after completing our experiment. We do not keep track of gravitational potential energy since the vertical position of the system objects does not change between the initial and final states.

We can understand the first pattern in Table 6.6 using our knowledge of impulse-momentum. The x-component of the net force exerted on the system in all three cases was zero; hence the x-component of momentum should be constant.

What about the second pattern? In Experiment 1, the system objects were very rigid, but in Experiments 2 and 3, they were more fragile and as a result were deformed during the collision. Using the data we can determine the amount of kinetic energy that was converted to internal energy. But it is impossible to predict this amount ahead of time. Unfortunately, this means that in collisions where any deformation of the system objects occurs, we cannot make predictions about the amount of kinetic energy converted to internal energy. However, we now know that even in collisions where the system objects become damaged, the momentum of the system still remains constant. So, even though the work-energy equation is less useful in these types of collisions, the impulse-momentum equation is still very useful.

Types of collisions

The experiments in Observational Experiment Table 6.6 are examples of the three general collision categories: elastic collisions (Experiment 1), inelastic collisions (Experiment 2), and totally inelastic collisions (Experiment 3). **Table 6.7** summarizes these three types of collisions.

Measuring the speed of a fast-moving projectile

We have already encountered a ballistic pendulum (in Chapter 5), a device that measures the speed of fast projectiles, such as golf balls hit in a ball-testing device or bullets fired from a gun. Let's use a ballistic pendulum to learn more about collisions, momentum, work, and energy.

Table 6.7 Types of collisions.

Elastic collisions	Inelastic collisions	Totally inelastic collisions
Both the momentum and kinetic energy of the system are constant. The internal energy of the system does not change. The colliding objects never stick together. Examples: There are no perfectly elastic collisions in nature, although collisions between very rigid objects (such as billiard balls) come close. Collisions between atoms or subatomic particles are almost exactly elastic.	The momentum of the system is constant but the kinetic energy is not. The colliding objects do not stick together. Internal energy increases during the collisions. Examples: A volleyball bouncing off your arms, or you jumping on a trampoline.	These are inelastic collisions in which the colliding objects stick together. Typically, a large fraction of the kinetic energy of the system is converted into internal energy in this type of collision. Examples: You catching a football, or a car collision where the cars stick together.

EXAMPLE 6.10 A ballistic pendulum

A gun is several centimeters from a 1.0-kg wooden block hanging at the end of strings. The gun fires a 10-g bullet that embeds in the block, which swings upward a height of 0.20 m. Determine the speed of the bullet when leaving the gun.

Sketch and translate We sketch the process, below. We use subscripts b for the bullet and B for the block of wood; when they join together, the subscript is bB. The process involves two parts. Part I is the collision of the bullet with the wood block. We know that this is a totally inelastic collision because the bullet combines with the block. Part II is the swinging of the block with the embedded bullet upward to its maximum height.

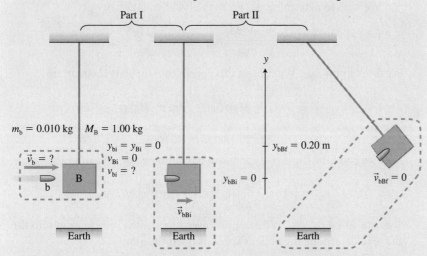

For the Part I collision, sketched on the left side of the figure above, the system is the bullet and the block. The kinetic energy of the system is not constant, but the momentum is. So we use the constant momentum of an isolated system to analyze the collision. The initial state is the instant just before the collision begins, and the final state is the instant just after the bullet joins the block. In the initial state, only the bullet has momentum. In the final state, both the bullet and the wood block have momentum, and their velocities are equal.

The details for the Part II upward swing of the block and bullet to its maximum height are sketched on the right side of the figure above. We choose the bullet, block, and Earth as the system. The initial state is just after the collision is over (the final state of Part I), and the final state is when the block reaches its maximum height.

The two parts are analyzed separately and then combined to determine the bullet's speed before hitting the block.

Simplify and diagram Part I: We draw bar charts to represent each process. The momentum bar chart below, left, represents the x-components of momentum for Part I.

Part II: The string does no work on the block, because it is perpendicular to the block's velocity at every instant. The energy bar chart below, right, shows that the initial kinetic energy of the bullet and block is converted to the gravitational potential energy of the bullet-block-Earth system.

Part I. Momentum

$$p_{bi} + p_{Bi} = p_{bBi}$$

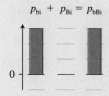

Part II. Energy

$$K_i + U_{gi} + W = K_f + U_{gf} + \Delta U_{int}$$

(*continued*)

Represent mathematically Use the bar charts to help apply momentum constancy to Part I and energy constancy to Part II.

Part I: $$mv_{bi} = (m_b + M_B)v_{bBi}$$

Part II: $$\frac{1}{2}(m_b + M_B)v_{bBi}^2 = (m_b + M_B)gy_{bBf}$$

where m_b is the mass of the bullet, M_B is the mass of the block, v_{bi} is the initial speed of the bullet, v_{bBi} is the speed of the bullet + block immediately after the collision, and y_{bBf} is the y-coordinate of the bullet + block at its highest point.

We wish to determine v_{bi}. From the Part I equation, we get

$$v_{bi} = \frac{m_b + M_B}{m_b}v_{bBi} = \left(1 + \frac{M_B}{m_b}\right)v_{bBi}$$

We don't know v_{bBi}. We can get this speed from the Part II equation:

$$\frac{1}{2}(m_b + M_B)v_{bBi}^2 = (m_b + M_B)gy_{bBf}$$

$$\Rightarrow v_{bBi} = \sqrt{2gy_{bBf}}$$

Now we combine these two equations to eliminate v_{bBi} and solve for v_{bi}:

$$v_{bi} = \left(1 + \frac{M_B}{m_b}\right)v_{bBi} = \left(1 + \frac{M_B}{m_b}\right)\sqrt{2gy_{bBf}}$$

Solve and evaluate We can now insert the known information into the above to determine the bullet's speed as it left the gun:

$$v_{bi} = \left(1 + \frac{1.0\,\text{kg}}{0.010\,\text{kg}}\right)\sqrt{2(9.8\,\text{N/kg})(0.20\,\text{m})} = 200\,\text{m/s}$$

That's close to 450 mi/h—very fast but reasonable for a bullet fired from a gun.

Try it yourself: Determine the initial kinetic energy of the bullet in this example, the final gravitational potential energy of the block-bullet system, and the increase in internal energy of the system.

Answers: $K_i = 200$ J; $U_{gf} = 2$ J; and $\Delta U_{int} = 198$ J.

Review Question 6.7 Imagine that a collision occurs. You measure the masses of the two objects before the collision and measure the velocities of the objects both before and after the collision. Describe how you could use this data to determine which type of collision had occurred.

6.8 Power

Why is it harder for the same person to run up a flight of stairs than to walk if the change in gravitational potential energy of the system person-Earth is the same? The *amount* of internal energy converted into gravitational energy is the same in both cases, but the *rate* of that conversion is not. When you run upstairs you convert the energy at a faster rate. The rate at which the conversion occurs is called the **power**.

> **Power** The power of a process is the amount of some type of energy converted into a different type divided by the time interval Δt in which the process occurred:
>
> $$\text{Power} = P = \left| \frac{\Delta U}{\Delta t} \right| \qquad (6.9)$$
>
> If the process involves external forces doing work, then power can also be defined as the magnitude of the work W done on or by the system divided by the time interval Δt needed for that work to be done:
>
> $$\text{Power} = P = \left| \frac{W}{\Delta t} \right| \qquad (6.10)$$
>
> The SI unit of power is the watt (W). 1 watt is 1 joule/second (1 W = 1 J/s).

A lightbulb with a power of 60 W converts electrical energy into light and internal energy at a rate of 60 J/s. A cyclist in good shape pedaling at moderate speed will convert about 400–500 J of internal chemical energy each second (400–500 W) into kinetic, gravitational potential, and thermal energies.

Power is sometimes expressed in horsepower (hp): 1 hp = 746 W. Horsepower is most often used to describe the power rating of engines or other machines. A 50-hp gasoline engine (typical in cars) converts the internal energy of the fuel into other forms of energy at a rate of $50 \times 746 \text{ W} = 37{,}300 \text{ W}$, or 37,300 J/s.

EXAMPLE 6.11 Lifting weights

Xueli is doing a dead lift. She lifts a 13.6-kg (30-lb) barbell from the floor to just below her waist (a vertical distance of 0.70 m) in 0.80 s. Determine the power during the lift.

Sketch and translate First, we sketch the process, shown below. The system is the barbell and Earth. The initial state is just before Xueli starts lifting. The final state is just after she finishes lifting. A vertical y-axis is used to indicate the change in elevation.

Simplify and diagram We assume that Xueli lifts the barbell so slowly that kinetic energy is zero during the process. She does work on the barbell, causing the system's gravitational potential energy to increase, but no change in kinetic energy. The origin

is at the initial position of the barbell. Since the barbell is moving at a small constant velocity, the external force exerted by Xueli on the system is very nearly constant and equals the gravitational force exerted by Earth on the barbell. We can represent this process with an energy bar chart, as shown.

Represent mathematically The power of this process is

$$P = \left| \frac{W}{\Delta t} \right| = \left| \frac{Fd \cos \theta}{\Delta t} \right| = \left| \frac{mgd \cos \theta}{\Delta t} \right|$$

Solve and evaluate

$$P = \left| \frac{mgd \cos \theta}{\Delta t} \right| = \frac{(13.6 \text{ kg})(9.8 \text{ N/kg})(0.70 \text{ m}) \cos 0°}{0.80 \text{ s}}$$

$$= 120 \text{ W}$$

This is a reasonable power for lifting a barbell. If you use an exercise machine that displays the power output, compare what you can achieve to this number.

Try it yourself: Xueli performs an overhead press—lifting the same barbell from her shoulders to above her head. Estimate the power of this process. The length of her arm is approximately 49 cm. It takes her 1.0 s to lift the bar.

Answer: 65 W.

EXAMPLE 6.12 Power and driving

A 1400-kg car is traveling on a level road at a constant speed of 27 m/s (60 mi/h). The drag force exerted by the air on the car and the rolling friction force exerted by the road on the car tires add to a net force of 680 N pointing opposite the direction of motion of the car. Determine the rate of work done by air and the road on the car. Express the results in watts and in horsepower.

Sketch and translate part First, we sketch the system, choosing the car alone as the system. The initial state is the moment the car passes position x on the x-axis and the final state is a short time interval Δt later when the car has had a displacement Δx parallel to the road.

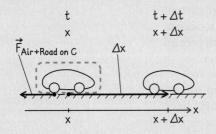

Simplify and diagram We need to determine the magnitude of work per unit time by the combined resistive air and rolling friction forces $(\vec{F}_{\text{Air}+\text{Road on C}})$.

Represent mathematically We determine the work done during displacement Δx and divide by the time interval Δt needed to complete that displacement. The power is the magnitude of that ratio:

$$P = \left| \frac{W}{\Delta t} \right| = \left| \frac{(F_{\text{Air}+\text{Road on C}}) \Delta x \cos 180°}{\Delta t} \right|$$

$$= \left| -(680\,\text{N}) \frac{\Delta x}{\Delta t} \right| = (680\,\text{N})v$$

Solve and Evaluate Substitute the speed into the above to determine the power:

$$P = (680\,\text{N})v = (680\,\text{N})(27\,\text{m/s})$$
$$= 1.8 \times 10^4\,\text{W} = 25\,\text{hp}$$

That's a relatively small power.

Try it yourself: What is the average power needed to cause a 1400-kg car's speed to increase from 20 to 27 m/s in 5 s? Ignore any resistive forces.

Answer: $P = 4.6 \times 10^4$ W or 62 hp.

Review Question 6.8 Jim (mass 80 kg) rollerblades on a smooth linoleum floor a distance of 4.0 m in 5.0 s. Determine the power of this process.

6.9 Improving our model of gravitational potential energy

So far, we have assumed that the gravitational force exerted by Earth on an object is constant $(F_{\text{E on O}} = mg)$. Using this equation we devised the expression for the gravitational potential energy of an object-Earth system as $U_g = mgy$ (with respect to a chosen zero level). This expression is only valid when an object is close to Earth's surface. We know from our study of gravitation (at the end of Chapter 4) that the gravitational force exerted by planetary objects on moons and satellites and by the Sun on the planets changes with the distance between the objects according to Newton's law of gravitation $(F_{1\,\text{on}\,2} = G\,m_1 m_2 / r^2)$ How would the expression for the gravitational potential energy change when we take into account that the gravitational force varies with distance?

Gravitational potential energy for large mass separations

Imagine that a "space elevator" has been built to transport supplies from the surface of Earth to the International Space Station (ISS). The elevator moves at constant velocity, except for the very brief acceleration and deceleration at the beginning and end of the trip. How much work must be done to lift the supplies from the surface to the ISS?

The initial state is the moment just as the supplies leave the surface. The final state is the moment just as they arrive at the ISS (**Figure 6.12a**). We choose Earth and the supplies as the system. The force that the elevator cable exerts on the supplies is an external force that does positive work on the system. We keep track of gravitational potential energy only; it is the only type of energy that changes between the initial and final states. Since the supplies are moving at constant velocity, the force exerted by the elevator cable on the supplies is equal in magnitude to the gravitational force exerted by Earth on the supplies (Figure 6.12b). Earth is the object of reference, and the origin of the coordinate system is at the center of Earth. The process can be described mathematically as follows:

$$U_{gi} + W = U_{gf}$$
$$\Rightarrow W = U_{gf} - U_{gi}$$

At this point it would be tempting to say that the work done by the elevator cable on the system is $W = Fd \cos\theta$, where F is the constant force that the cable exerts on the supplies. However, as the supplies reach higher and higher altitudes, the force exerted by the elevator cable decreases—Earth exerts a weaker and weaker force on the supplies.

Determining the work done by a variable force requires a complex mathematical procedure. The outcome of this procedure is

$$W = \left(-G\frac{m_E m_S}{R_E + h_{ISS}}\right) - \left(-G\frac{m_E m_S}{R_E}\right)$$

where m_E is the mass of Earth, R_E is the radius of Earth, m_S is the mass of the supplies, h_{ISS} is the altitude of the ISS above Earth's surface, and $R_E + h_{ISS}$ is the distance of the ISS from the center of Earth. Before you move on, check whether this complicated equation makes sense (for example, check the units).

The work is written as the difference in two quantities; the latter describes the initial state and the former the final state. If we compare the above result with $W = U_{gf} - U_{gi}$, we see that each term is an expression for the gravitational potential energy of the Earth-object system for a particular object's distance from the center of Earth.

Gravitational potential energy of a system consisting of Earth and any object

$$U_g = -G\frac{m_E m_O}{r_{E\text{-}O}} \qquad (6.11)$$

where m_E is the mass of Earth $(5.97 \times 10^{24}\,\text{kg})$, m_O is the mass of the object, $r_{E\text{-}O}$ is the distance from the center of Earth to the center of the object, and $G = 6.67 \times 10^{-11}\,\text{N}\cdot\text{m}^2/\text{kg}^2$ is Newton's universal gravitational constant.

We can use Eq. (6.11) to find the gravitational potential energy of the system of any two spherical or point-like objects if we know the masses of the objects and the distance between their centers.

Note that the cable did positive work on the system while pulling the supplies away from Earth. When the object (the supplies in this case) is infinitely far away, the gravitational potential energy is zero. The only way to add positive energy to a system and have it become zero is if it started with negative energy, for example, $-5 + 5 = 0$. Thus, for the case of zero energy at infinity, the gravitational potential energy is a negative number when the object is closer to Earth. We can represent the process of pulling an object from the surface of Earth to infinity using a work-energy bar chart (**Figure 6.13**). The initial state is when the object is near Earth and the final state is when it is infinitely far away.

Figure 6.12 A cable lifts supplies to the International Space Station via a space elevator. (a) Determine the work required to lift the supplies. (b) The cable exerts the force in the direction of motion. In this force diagram, the supplies are the system.

(a)

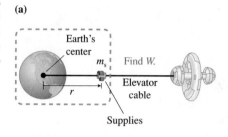

Earth's center

m_s Find W.

Elevator cable

r

Supplies

(b)

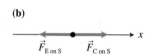

$\vec{F}_{E\,on\,S}$ $\vec{F}_{C\,on\,S}$ x

Figure 6.13 A bar chart representing the work needed to take an object from near Earth to infinitely far away. The system is the supplies and Earth. Note that the final gravitational potential energy of the system is zero.

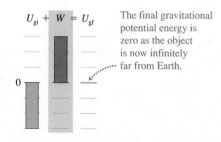

$U_{gi} + W = U_{gf}$

The final gravitational potential energy is zero as the object is now infinitely far from Earth.

0

Now we can determine the amount of work needed to raise 1000 kg of supplies to the International Space Station.

$$W = \left(-G\frac{m_E m_S}{R_E + h_{ISS}}\right) - \left(-G\frac{m_E m_S}{R_E}\right)$$

$$= -Gm_E m_S\left(\frac{1}{R_E + h_{ISS}} - \frac{1}{R_E}\right)$$

$$= -(6.67 \times 10^{-11}\,N\cdot m^2/kg^2)(5.97 \times 10^{24}\,kg)(1000\,kg)$$

$$\times \left(\frac{1}{6.37 \times 10^6\,m + 3.50 \times 10^5\,m} - \frac{1}{6.37 \times 10^6\,m}\right)$$

$$= 3.26 \times 10^9\,J$$

Let's compare this with what would have been calculated had we used our original expression for gravitational potential energy. Choose the zero level at the surface of Earth.

$$W = m_S g y_f - m_S g y_i = m_S g y_f - 0$$

$$= (1000\,kg)(9.8\,N/kg)(3.50 \times 10^5\,m) = 3.43 \times 10^9\,J$$

This differs by only about 5% from the more accurate result. Remember that $U_g = mgy$ is reasonable when the distance above the surface of Earth is a small fraction of the radius of Earth. The altitude of the ISS (350 km) is a small fraction of the radius of Earth (6371 km), so $U_g = mgy$ is still reasonably accurate.

Escape speed

The best Olympic high jumpers can leap over bars that are about 2.5 m (8 ft) above Earth's surface. Let's estimate a jumper's speed when leaving the ground in order to attain that height. We choose the jumper and Earth as the system and the zero level of gravitational potential energy at ground level. The kinetic energy of the jumper as he leaves the ground is converted into the gravitational potential energy of the system.

$$\frac{1}{2}mv^2 = mgy$$

$$\Rightarrow v = \sqrt{2gy} = \sqrt{2(9.8\,N/kg)(2.5\,m)} = 7.0\,m/s$$

How high could he jump if he were on the Moon? The gravitational constant of objects near the Moon's surface is $g_M = \dfrac{Gm_M}{R_M^2} = 1.6\,N/kg$. Using the equation above,

$$\frac{1}{2}mv^2 = mg_M y$$

$$\Rightarrow y = \frac{v^2}{2g_M} = \frac{(7.0\,m/s)^2}{2(1.6\,N/kg)} = 15.3\,m$$

That's about 50 feet!

Is it possible to jump entirely off a celestial body—jumping up and never coming down? What is the minimum speed you would need in order to do this? This minimum speed is called **escape speed**.

EXAMPLE 6.13 Escape speed

What vertical speed must a jumper have in order to leave the surface of a planet and never come back down?

Sketch and translate First, we draw a sketch of the process. The initial state will be the instant after the jumper's feet leave the surface. The final state will be when the jumper has traveled far enough away from the planet to no longer feel the effects of its gravity (at $r = \infty$). Choose the system to be the jumper and the planet.

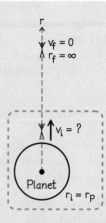

Simplify and diagram We represent the process with the bar chart. In the initial state, the system has both kinetic and gravitational potential energy. In the final state, both the kinetic energy and the gravitational potential energy are zero.

$$K_i + U_{gi} + U_{si} + W = K_f + U_{gf} + U_{sf} + \Delta U_{int}$$

Represent mathematically Using the generalized work-energy equation and the bar chart:

$$U_i + W = U_f$$
$$\Rightarrow K_i + U_{gi} + 0 = K_f + U_{gf}$$
$$\Rightarrow \frac{1}{2}m_j v^2 + \left(-G\frac{m_p m_j}{r_P}\right) + 0 = 0 + 0$$

where m_p is the mass of the planet, m_j is the mass of the jumper, r_P is the radius of the planet, and $G = 6.67 \times 10^{-11}\,\text{N}\cdot\text{m}^2/\text{kg}^2$ is the gravitational constant.

Solve and evaluate Solving for the escape speed of the jumper,

$$v = \sqrt{\frac{2Gm_p}{r_P}} \tag{6.12}$$

We can use the above equation to determine the escape speed for any celestial body. For example, the escape speed for the Moon is

$$v = \sqrt{\frac{2(6.67 \times 10^{-11}\,\text{N}\cdot\text{m}^2/\text{kg}^2)(7.35 \times 10^{22}\,\text{kg})}{1.74 \times 10^6\,\text{m}}}$$
$$= 2370\,\text{m/s}$$

The escape speed for Earth is

$$v = \sqrt{\frac{2Gm_E}{r_E}}$$
$$= \sqrt{\frac{2(6.67 \times 10^{-11}\,\text{N}\cdot\text{m}^2/\text{kg}^2)(5.97 \times 10^{24}\,\text{kg})}{6.37 \times 10^6\,\text{m}}}$$
$$= 11{,}200\,\text{m/s}$$
$$= 11.2\,\text{km/s}$$

Try it yourself: What is the escape speed of a particle near the surface of the Sun? The mass of the Sun is 2.0×10^{30} kg and its radius is 700,000 km.

Answer: 620 km/s.

Black holes

Equation (6.12) for the escape speed suggests something amazing. If the mass of a star or planet were large enough and/or its radius small enough, the escape speed could be made arbitrarily large. What if the star's escape speed were greater than light speed ($c = 3.00 \times 10^8$ m/s)? What would this star look like? Light leaving the star's surface would not be moving fast enough to escape the star. The star would be completely dark.

Let's imagine what would happen if Earth started shrinking so that its material were compressed into a smaller volume. How small would Earth have to be for its escape speed to be greater than light speed? Use Eq. (6.12) with $v = c$ to answer this question.

$$v = c = \sqrt{\frac{2Gm_P}{r_P}} \tag{6.13}$$

or

$$r_P = \frac{2Gm_P}{c^2} = \frac{2(6.67 \times 10^{-11}\,\text{N}\cdot\text{m}^2/\text{kg}^2)(5.97 \times 10^{24}\,\text{kg})}{(3.00 \times 10^8\,\text{m/s})^2} = 8.85\,\text{mm}$$

TIP Notice that the escape speed does not depend on the mass of the escaping object—a tiny speck of dust and a huge boulder would need the same initial speed to leave Earth. Why is that?

Earth would have to be smaller than 9 millimeters! It's difficult to imagine Earth compressed to the size of a marble. Its mass would be the same but it would now be incredibly dense. Equation (6.13) was first constructed by the brilliant astronomer Pierre-Simon Laplace (1749–1827), who used classical mechanics to predict the presence of what he called "dark stars"—now more commonly called black holes.

QUANTITATIVE EXERCISE 6.14 Sun as a black hole

How small would our Sun need to be in order for it to become a black hole?

Represent Mathematically The mass of the Sun is 1.99×10^{30} kg. All we need to do is use Eq. (6.13) to find the radius of this black hole:

$$r_{Sun} = \frac{2Gm_{Sun}}{c^2}$$

Solve and Evaluate

$$r_{Sun} = \frac{2(6.67 \times 10^{-11} \, \text{N} \cdot \text{m}^2/\text{kg}^2)(1.99 \times 10^{30} \, \text{kg})}{(3.00 \times 10^8 \, \text{m/s})^2}$$
$$= 2.95 \times 10^3 \, \text{m} \approx 3 \, \text{km}$$

So, if the Sun collapsed to smaller than 3 km in radius, it would become a black hole. Could this happen? We will return to this question in later chapters.

Try it yourself: Estimate the size to which a human would need to shrink to become invisible in the same sense that a black hole is invisible.

Answer: About 10^{-25} m.

We've been talking about objects whose escape speed is larger than light speed. Physicists once thought that since light has zero mass, the gravitational force exerted on it would always be zero. At the beginning of the 20th century, Albert Einstein's theory of general relativity improved greatly on Newton's law of universal gravitation. According to general relativity, light is affected by gravity. Amazingly, the theory predicts the same size for a black hole that is provided by the Newtonian theory.

Review Question 6.9 In this section you read that the gravitational potential energy of two large bodies (for example, the Sun and Earth) is negative. Why is this true?

Summary

Words	Pictorial and physical representations	Mathematical representation
Work (W) is a way to change the energy of a system. Work is done on a system when an external object exerts a force of magnitude F on an object in the system as it undergoes a displacement of magnitude d. The work depends on the angle θ between the directions of $\vec{F}$ and $\vec{d}$. It is a scalar quantity. (Section 6.1)	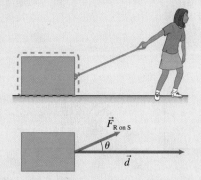	$W = Fd\cos\theta$ Eq. (6.1)
Gravitational potential energy (U_g) is the energy that a system has due to the relative separation of two objects with mass. It is a scalar quantity. U_g depends on the gravitational interaction of the objects. A single object cannot have gravitational potential energy.		$U_g = mgy$ Eq. (6.4) (near Earth's surface, zero level is at the surface) $U_g = -G\dfrac{m_A m_B}{r_{AB}}$ Eq. (6.11) (general expression; zero level is at infinity)
Kinetic energy (K) is the energy of an object of mass m moving at speed v. It is a scalar quantity.		$K = \dfrac{1}{2}mv^2$ Eq. (6.5)
Elastic potential energy (U_s) is the energy of a stretched or compressed elastic object (e.g., coils of a spring or a stretched bow string).		$U_s = \dfrac{1}{2}kx^2$ Eq. (6.7)
Internal energy (U_{int}) is the energy of motion and interaction of the microscopic particles making up the objects in the system. The internal energy of the system changes when the surfaces of the system objects rub against each other. (Section 6.1)		$\Delta U_{int} = f_k d$ Eq. (6.8) (conversion of mechanical energy to internal due to friction) *(continued)*

Words	Pictorial and physical representations	Mathematical representation
Total energy (U) is the sum of all the energies of the system. (Section 6.2)		$U = K + U_g + U_s + U_{int} + \cdots$ Eq. (6.2)
Work-energy principle The energy of a system changes when external forces do work on it. Internal forces do not change the energy of the system. When there are no external forces doing work on the system, the system's energy is constant. (Section 6.2)	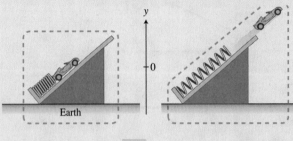 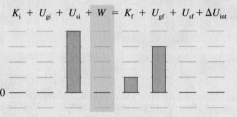 $K_i + U_{gi} + U_{si} + W = K_f + U_{gf} + U_{sf} + \Delta U_{int}$	$U_i + W = U_f$ Eq. (6.3) $(K_i + U_{gi} + U_{si}) + W$ $= (K_f + U_{gf} + U_{sf} + \Delta U_{int})$
Collisions ■ **Elastic:** momentum and kinetic energy of the system are constant—no changes in internal energy. ■ **Inelastic:** momentum is constant but not kinetic energy—internal energy increases and kinetic energy decreases. ■ **Totally inelastic:** an inelastic collision in which the colliding objects stick together. (Section 6.7)		All collisions: $\Sigma \vec{p}_i = \Sigma \vec{p}_f$ Elastic collisions only: $\Sigma K_i = \Sigma K_f$ For other collisions, $\Delta U_{int} > 0$
Power (P) is the rate of energy conversion, or rate of work done on or by a system during a process. (Section 6.8)	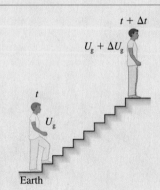	$P = \left\lvert \dfrac{\Delta U}{\Delta t} \right\rvert$ or $P = \left\lvert \dfrac{W}{\Delta t} \right\rvert$ Eq. (6.9 or 6.10)

MP® For instructor-assigned homework, go to
MasteringPhysics.

Questions

Multiple Choice Questions

1. In which of the following is positive work done by a person on a suitcase?
   (a) The person holds a heavy suitcase.
   (b) The person lifts a heavy suitcase.
   (c) The person stands on a moving walkway carrying a heavy suitcase.
   (d) All of the above   (e) None of the above

2. Which answer best represents the system's change in energy for the following process? The system includes Earth, two carts, and a compressed spring between the carts. The spring is released, and in the final state, one cart is moving up a frictionless ramp until it stops. The other cart is moving in the opposite direction on a horizontal frictionless track.
   (a) Kinetic energy to gravitational potential energy
   (b) Elastic potential energy to gravitational potential energy
   (c) Elastic potential energy to kinetic energy and gravitational potential energy

3. Choose a process and system that match the following energy description: The kinetic energy of an object becomes gravitational potential energy of the system.
   (a) A pendulum bob released from a certain height swings to the lowest position; the system is the pendulum bob and the string.
   (b) A pendulum bob released from a certain height swings to the lowest position; the system is the pendulum bob and Earth.
   (c) A pendulum moves from the bottom of its swing to the top; the system is the pendulum bob and Earth.
   (d) A pendulum bob moves from the bottom of its swing to the top; the system is the pendulum bob and the string.

4. Three processes are described below. Choose one process in which there is work done on the system. The spring, Earth, and the cart are part of the system.
   (a) A relaxed spring rests upright on a tabletop. You slowly compress the spring. You then release the spring and it flies up several meters to its highest point.
   (b) A cart at the top of a smooth inclined surface coasts at increasing speed to the bottom (ignore friction).
   (c) A cart at the top of a smooth inclined surface slides at increasing speed to the bottom where it runs into and compresses a spring (ignore friction).

5. Choose which statement describes a process in which an external force does negative work on the system. The person is not part of the system.
   (a) A person slowly lifts a box from the floor to a tabletop.
   (b) A person slowly lowers a box from a tabletop to the floor.
   (c) A person carries a bag of groceries horizontally from one location to another.
   (d) A person holds a heavy suitcase.

6. Which example(s) below involve zero physics work? Choose all that apply.
   (a) A person holds a child.
   (b) A person pushes a car stuck in the snow but the car does not move.
   (c) A rope supports a swinging chandelier.
   (d) A person uses a self-propelled lawn mower on a level lawn.
   (e) A person pulls a sled uphill.

7. Estimate the change in gravitational potential energy when you rise from bed to a standing position.
   (a) No change (0 J)   (b) About 250 J
   (c) About 2500 J   (d) About 25 J

8. What does it mean if object 1 does $+10$ J of work on object 2?
   (a) Object 1 exerts a 10-N force on object 2 in the direction of its 1-m displacement.
   (b) Object 1 exerts a 1-N force on object 2 in the direction of its 10-m displacement.
   (c) Object 1 exerts a 10-N force on object 2 at a 60° angle relative to its 2-m displacement.
   (d) All of the above   (e) None of the above

9. What does it mean if the gravitational potential energy of an Earth-apple system is 10 J?
   (a) Someone did 10 J of work lifting the apple.
   (b) Earth did negative work when someone was lifting the apple.
   (c) A 100-g apple is 10 m above the ground.
   (d) Parts (a) and (c) are both correct.
   (e) We need more information about the coordinate system and the object.

10. Imagine that you stretch a spring 3 cm, and then another 3 cm. Do you do more, less, or the same amount of work stretching it the second 3 cm compared with the first?
    (a) The same work   (b) Less work   (c) More work

11. Two small spheres of putty, A and B, hang from the ceiling on massless strings of equal length. Sphere A is raised to the side so its string is horizontal to the ground. It is released, swings down, and collides with sphere B (initially at rest). The spheres stick together and swing upward along a circular path to a maximum height on the other side. Which of the following principles must be used to determine this final height?
    I. The work-energy equation
    II. The impulse-momentum equation
    (a) I only   (b) II only   (c) Both I and II
    (d) Either I or II but not both   (e) Neither I nor II

12. Two identical stones, A and B, are thrown from a cliff from the same height and with the same initial speed. Stone A is thrown vertically upward, and stone B is thrown vertically downward. Which of the following statements best explains which stone has a larger speed just before it hits the ground, assuming no effects of air friction?
    (a) Both stones have the same speed; they have the same change in U_g and the same K_i.
    (b) A, because it travels a longer path
    (c) A, because it takes a longer time interval
    (d) A, because it travels a longer path and takes a longer time interval
    (e) B, because no work is done against gravity

Conceptual Questions

13. Is energy a physical phenomenon, a model, or a physical quantity? Explain your answer.

14. Your friend thinks that the escape speed should be greater for more massive objects than for less massive objects. Provide a physics-based argument for his opinion. Then provide a counterargument for why the escape speed is independent of the mass of the object.

15. Suggest how you can measure the following quantities: work done by the force of friction, the power of a motor, the kinetic energy of a moving car, and the elastic potential energy of a stretched spring.

16. How can satellites stay in orbit without any jet propulsion system? Explain using work-energy ideas.

17. Why does the Moon have no atmosphere, but Earth does?

18. What will happen to Earth if our Sun becomes a black hole?

19. In the equation $U_g = mgy$, the gravitational potential energy is directly proportional to the distance of the object from a planet. In the equation $U_g = -G\dfrac{m_p m}{r}$, it is inversely proportional. How can you reconcile those two equations?

Problems

Below, BIO indicates a problem with a biological or medical focus. Problems labeled EST ask you to estimate the answer to a quantitative problem rather than derive a specific answer. Problems marked with ✐ require you to make a drawing or graph as part of your solution. Asterisks indicate the level of difficulty of the problem. Problems with no * are considered to be the least difficult. A single * marks moderately difficult problems. Two ** indicate more difficult problems.

6.1 Work and energy

1. Jay fills a wagon with sand (about 20 kg) and pulls it with a rope 30 m along the beach. He holds the rope 25° above the horizontal. The rope exerts a 20-N tension force on the wagon. How much work does the rope do on the wagon?

2. You have a 15-kg suitcase and (a) slowly lift it 0.80 m upward, (b) hold it at rest to test whether you will be able to move the suitcase without help in the airport, and then (c) lower it 0.80 m. What work did you do in each case? What assumptions did you make to solve this problem?

3. * You use a rope to slowly pull a sled and its passenger 50 m up a 20° incline, exerting a 150-N force on the rope. (a) How much work will you do if you pull parallel to the hill? (b) How much work will you do if you exert the same magnitude force while slowly lowering the sled back down the hill and pulling parallel to the hill? (c) How much work did Earth do on the sled for the trip in part (b)?

4. A rope attached to a truck pulls a 180-kg motorcycle at 9.0 m/s. The rope exerts a 400-N force on the motorcycle at an angle of 15° above the horizontal. (a) What is the work that the rope does in pulling the motorcycle 300 m? (b) How will your answer change if the speed is 12 m/s? (c) How will your answer change if the truck accelerates?

5. You lift a 25-kg child 0.80 m, slowly carry him 10 m to the playroom, and finally set him back down 0.80 m onto the playroom floor. What work do you do on the child for each part of the trip and for the whole trip? List your assumptions.

6. A truck runs into a pile of sand, moving 0.80 m as it slows to a stop. The magnitude of the work that the sand does on the truck is 6.0×10^5 J. (a) Determine the average force that the sand exerts on the truck. (b) Did the sand do positive or negative work? (c) How does the average force change if the stopping distance is doubled? Indicate any assumptions you made.

6.2 and 6.3 Energy is a conserved quantity and Quantifying gravitational potential and kinetic energies

7. A 5.0-kg rabbit and a 12-kg Irish setter have the same kinetic energy. If the setter is running at speed 4.0 m/s, how fast is the rabbit running?

8. * EST Estimate your average kinetic energy when walking to physics class. What assumptions did you make?

9. * A pickup truck (2268 kg) and a compact car (1100 kg) have the same momentum. (a) What is the ratio of their kinetic energies? (b) If the same horizontal net force were exerted on both vehicles, pushing them from rest over the same distance, what is the ratio of their final kinetic energies?

10. * When does the kinetic energy of a car change more: when the car accelerates from 0 to 10 m/s or from 30 m/s to 40 m/s? Explain.

11. * When exiting the highway, a 1100-kg car is traveling at 22 m/s. The car's kinetic energy decreases by 1.4×10^5 J. The exit's speed limit is 35 mi/h. Did the driver reduce the car's speed enough? Explain.

12. * You are on vacation in San Francisco and decide to take a cable car to see the city. A 5200-kg cable car goes 360 m up a hill inclined 12° above the horizontal. What can you learn about the energy of the system from this information? Answer quantitatively. What assumptions did you need to make?

13. * Flea jump A 5.4×10^{-7}-kg flea pushes off a surface by extending its rear legs for a distance of slightly more than 2.0 mm, consequently jumping to a height of 40 cm. What physical quantities can you determine using this information? Make a list and determine three of them.

14. * Roller coaster ride A roller coaster car drops a maximum vertical distance of 35.4 m. (a) Determine the maximum speed of the car at the bottom of that drop. (b) Describe any assumptions you made. (c) Will a car with twice the mass have more or less speed at the bottom? Explain.

15. * BIO EST Heart pumps blood The heart does about 1 J of work while pumping blood into the aorta during each heartbeat. (a) Estimate the work done by the heart in pumping blood during a lifetime. (b) If all of that work was used to lift a person, to what height could an average person be lifted? Indicate any assumptions you used for each part of the problem.

16. * **Wind energy** Air circulates across Earth in regular patterns. A tropical air current called the Hadley cell carries about 2×10^{11} kg of air per second past a cross section of Earth's atmosphere while moving toward the equator. The average air speed is about 1.5 m/s. (a) What is the kinetic energy of the air that passes the cross section each second? (b) About 1×10^{20} J of energy was consumed in the United States in 2005. What is the ratio of the kinetic energy of the air that passes toward the equator each second and the energy consumed in the United States each second?

17. * BIO **Bone break** The tibia bone in the lower leg of an adult human will break if the compressive force on it exceeds about 4×10^5 N (we assume that the ankle is pushing up). Suppose that you step off a chair that is 0.40 m above the floor. If landing stiff-legged on the surface below, what minimum stopping distance do you need to avoid breaking your tibias? Indicate any assumptions you made in your answer to this question.

18. * EST BIO **Climbing Mt. Everest** In 1953 Sir Edmund Hillary and Tenzing Norgay made the first successful ascent of Mt. Everest. How many slices of bread did each climber have to eat to compensate for the increase of the gravitational potential energy of the system climbers-Earth? (One piece of bread releases about 1.0×10^6 J of energy in the body.) Indicate all of the assumptions used. Note: The body is an inefficient energy converter—see the reading passage at the end of this section.

6.4 Quantifying elastic potential energy

19. * A door spring is difficult to stretch. (a) What maximum force do you need to exert on a relaxed spring with a 1.2×10^4-N/m spring constant to stretch it 6.0 cm from its equilibrium position? (b) How much does the elastic potential energy of the spring change? (c) Determine its change in elastic potential energy as it returns from the 6.0-cm stretch position to a 3.0-cm stretch position. (d) Determine its elastic potential energy change as it moves from the 3.0-cm stretch position back to its equilibrium position.

20. * You compress a spring by a certain distance. Then you decide to compress it further so that its elastic potential energy increases by another 50%. What was the percent increase in the spring's compression distance?

21. * A moving car has 40,000 J of kinetic energy while moving at a speed of 7.0 m/s. A spring-loaded automobile bumper compresses 0.30 m when the car hits a wall and stops. What can you learn about the bumper's spring using this information? Answer quantitatively and list the assumptions that you made.

22. * The force required to stretch a slingshot by different amounts is shown in the graph in **Figure P6.22**. (a) What is the spring constant of the sling? (b) How much work does a child need to do to stretch the sling 15 cm from equilibrium?

Figure P6.22

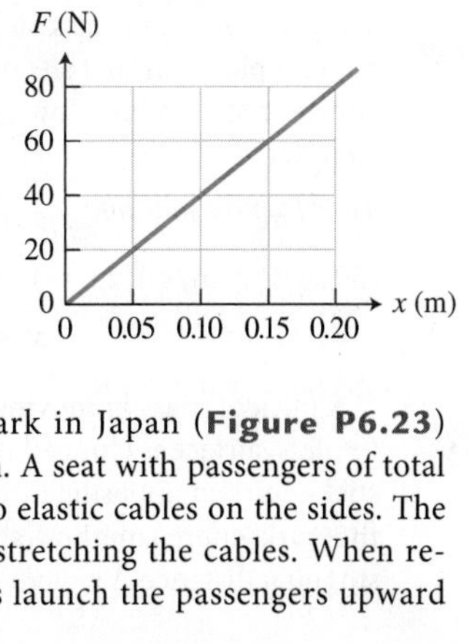

F (N)

[Graph showing a straight line from origin. Vertical axis marked 20, 40, 60, 80. Horizontal axis x (m) marked 0.05, 0.10, 0.15, 0.20.]

23. ** **Inverse bungee jump** The Ejection Seat at Lake Biwa Amusement Park in Japan (**Figure P6.23**) is an inverse bungee system. A seat with passengers of total mass 160 kg is connected to elastic cables on the sides. The seat is pulled down 12 m, stretching the cables. When released, the stretched cables launch the passengers upward above the towers to a height of about 30 m above their starting position at the ground. What can you learn about the mechanical properties of the cables using this information? Answer quantitatively. Assume that the cables are vertical.

Figure P6.23

6.5 Friction and energy conversion

24. * Jim is driving a 2268-kg pickup truck at 20 m/s and releases his foot from the accelerator pedal. The car eventually stops due to an effective friction force that the road, air, and other things exert on the car. The friction force has an average magnitude of 800 N. (a) Make a list of the physical quantities you can determine using this information and determine three of them. Specify the system and the initial and final states. (b) Would a heavier car travel farther before stopping or stop sooner? Identify the assumptions in your answer.

25. * A 1100-kg car traveling at 24 m/s coasts through some wet mud in which the net horizontal resistive force exerted on the car from all causes (mostly the force exerted by the mud) is 1.7×10^4 N. Determine the car's speed as it leaves the 18-m-long patch of mud.

26. * After falling 18 m, a 0.057-kg tennis ball has a speed of 12 m/s (the ball's initial speed is zero). Determine the average resistive force of the air in opposing the ball's motion. Solve the problem for the ball as your system and then repeat for a ball-Earth system. Are the answers the same or different?

27. * A water slide of length l has a vertical drop of h. Abby's mass is m. An average friction force of magnitude f opposes her motion. She starts down the slide at initial speed v_i. Use work-energy ideas to develop an expression for her speed at the bottom of the slide. Then evaluate your result using unit analysis and limiting case analysis.

28. ** / You are pulling a crate on a rug, exerting a constant force on the crate $\vec{F}_{Y \text{ on } C}$ at an angle θ above the horizontal. The crate moves at constant speed. Represent this process using a motion diagram, a force diagram, a momentum bar chart, and an energy bar chart. Specify your choice of system for each representation. Make a list of physical quantities you can determine using this information.

29. ** A 900-kg car initially at rest rolls 50 m down a hill inclined at an angle of 5.0°. A 400-N effective friction force opposes its motion. How fast is the car moving at the bottom? What distance will it travel on a similar horizontal surface at the bottom of the hill? Will the distance decrease or increase if the car's mass is 1800 kg?

30. * A car skids 18 m on a level road while trying to stop before hitting a stopped car in front of it. The two cars barely touch. The coefficient of kinetic friction between the first car and the road is 0.80. A policewoman gives the driver a ticket for exceeding the 35 mi/h speed limit. Can you defend the driver in court? Explain.

6.6 Skills for analyzing processes using the work-energy principle

31. * In a popular new hockey game, the players use small launchers with springs to move the 0.0030-kg puck. Each spring has a 120-N/m spring constant and can be compressed up to 0.020 m. What can you determine about the motion of the puck using this information? Make a list of quantities and determine their values.

32. * A 500-m-long ski slope drops at an angle of 6.4° relative to the horizontal. (a) Determine the change in gravitational potential energy of a 60-kg skier-Earth system when the skier goes down this slope. (b) If 20% of the gravitational potential energy change is converted into kinetic energy, how fast is the skier traveling at the bottom of the slope?

33. * A Frisbee gets stuck in a tree. You want to get it out by throwing a 1.0-kg rock straight up at the Frisbee. If the rock's speed as it reaches the Frisbee is 4.0 m/s, what was its speed as it left your hand 2.8 m below the Frisbee? Specify the system and the initial and final states.

34. A driver loses control of a car, drives off an embankment, and lands in a canyon 6.0 m below. What was the car's speed just before touching the ground if it was traveling on the level surface at 12 m/s before the driver lost control?

35. * You are pulling a box so it moves at increasing speed. Compare the work you need to do to accelerate it from 0 m/s to speed v to the work needed to accelerate it from speed v to the speed of $2v$. Discuss whether your answer makes sense. How many different situations do you need to consider?

36. * A cable lowers a 1200-kg elevator so that the elevator's speed increases from zero to 4.0 m/s in a vertical distance of 6.0 m. What is the force that the cable exerts on the elevator while lowering it? Specify the system, its initial and final states, and any assumptions you made. Then change the system and solve the problem again. Do the answers match?

37. ** EST **Hit by a hailstone** A 0.040-kg hailstone the size of a golf ball (4.3 cm in diameter) is falling at about 16 m/s when it reaches Earth's surface. Estimate the force that the hailstone exerts on your head—a head-on collision. Indicate any assumptions used in your estimate. Note that the cheekbone will break if something exerts a 900-N or larger force on the bone for more than 6 ms. Is this hailstone likely to break a bone?

38. * BIO **Froghopper jump** Froghoppers may be the insect jumping champs. These 6-mm-long bugs can spring 70 cm into the air, about the same distance as the flea. But the froghopper is 60 times more massive than a flea, at 12 mg. The froghopper pushes off for about 4 mm. What average force does it exert on the surface? Compare this to the gravitational force that Earth exerts on the bug.

39. * / **Bar chart Jeopardy 1** Invent in words and with a sketch a process that is consistent with the qualitative work-energy bar chart shown in **Figure P6.39**. Then apply in symbols the generalized work-energy principle for that process.

Figure P6.39

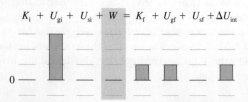

$$K_i + U_{gi} + U_{si} + W = K_f + U_{gf} + U_{sf} + \Delta U_{int}$$

40. * / **Bar chart Jeopardy 2** Invent in words and with a sketch a process that is consistent with the qualitative work-energy bar chart shown in **Figure P6.40**. Then apply in symbols the generalized work-energy principle for that process.

Figure P6.40

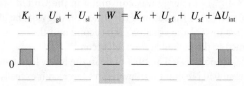

$$K_i + U_{gi} + U_{si} + W = K_f + U_{gf} + U_{sf} + \Delta U_{int}$$

41. * / **Equation Jeopardy 1** Construct a qualitative work-energy bar chart for a process that is consistent with the equation below. Then invent in words and with a sketch a process that is consistent with both the equation and the bar chart.

$$(1/2)(400\,\text{N/m})(0.20\,\text{m})^2 = (1/2)(0.50\,\text{kg})v^2 + (0.50\,\text{kg})(9.8\,\text{m/s}^2)(0.80\,\text{m})$$

42. * / **Equation Jeopardy 2** Construct a qualitative work-energy bar chart for a process that is consistent with the equation below. Then invent in words and with a sketch a process that is consistent with both the equation and the bar chart.

$$(120\,\text{kg})(9.8\,\text{m/s}^2)(100\,\text{m})\sin 53°$$
$$= (1/2)(120\,\text{kg})(20\,\text{m/s})^2 + f_k(100\,\text{m})$$

43. * **Evaluation 1** Your friend provides a solution to the following problem. Evaluate his solution. Constructively identify any mistakes he made and correct the solution. Explain possible reasons for the mistakes.

The problem: A 400-kg motorcycle, including the driver, travels up a 10-m-long ramp inclined 30° above the paved horizontal surface holding the ramp. The cycle leaves the ramp at speed 20 m/s. Determine the cycle's speed just before it lands on the paved surface.

Your friend's solution:

$$(1/2)(400\,\text{kg})(20\,\text{m/s}) = (400\,\text{kg})(9.8\,\text{m/s}^2)(10\,\text{m}) + (1/2)(400\,\text{kg})v^2$$
$$= -13.2\,\text{m/s}$$

44. * **Evaluation 2** Your friend provides a solution to the following problem. Evaluate her solution. Constructively identify any mistakes she made and correct the solution. Explain possible reasons for the mistakes.

The problem: Jim (mass 50 kg) steps off a ledge that is 2.0 m above a platform that sits on top of a relaxed spring of force constant 8000 N/m. How far will the spring compress while stopping Jim?

Your friend's solution:

$$(50\,\text{kg})(9.8\,\text{m/s}^2)(2.0\,\text{m}) = (1/2)(8000\,\text{N/m})x$$
$$x = 0.25\,\text{m}$$

45. * A puck of mass m moving at speed v_i on a horizontal, frictionless surface is stopped in a distance Δx because a hockey stick exerts an opposing force of magnitude F on it. (a) Using the work-energy method, show that $F = mv_i^2/2\Delta x$. (b) If the stopping distance Δx increases by 50%, by what percent does

the average force needed to stop the puck change, assuming that m and v_i are unchanged? Justify your result.

46. ** A rope exerts an 18-N force while lowering a 20-kg crate down a plane inclined at 20° (the rope is parallel to the plane). A 24-N friction force opposes the motion. The crate starts at rest and moves 10 m down the plane. Make a list of the physical quantities you can determine using this information and determine three of them. Specify the system and the initial and final states of the process.

6.7 Collisions: Putting it all together

47. ** You fire an 80-g arrow so that it is moving at 80 m/s when it hits and embeds in a 10-kg block resting on ice. (a) What is the velocity of the block and arrow just after the collision? (b) How far will the block slide on the ice before stopping? A 7.2-N friction force opposes its motion. Specify the system and the initial and final states for (a) and (b).

48. ** You fire a 50-g arrow that moves at an unknown speed. It hits and embeds in a 350-g block that slides on an air track. At the end, the block runs into and compresses a 4000-N/m spring 0.10 m. How fast was the arrow traveling? Indicate the assumptions that you made and discuss how they affect the result.

49. ** To confirm the results of Problem 48, you try a new experiment. The 50-g arrow is launched in an identical manner so that it hits and embeds in a 3.50-kg block. The block hangs from strings. After the arrow joins the block, they swing up so that they are 0.50 m higher than the block's starting point. How fast was the arrow moving before it joined the block?

50. ** A 1060-kg car moving west at 16 m/s collides with and locks onto a 1830-kg stationary car. (a) Determine the velocity of the cars just after the collision. (b) After the collision, the road exerts a 1.2×10^4-N friction force on the car tires. How far do the cars skid before stopping? Specify the system and the initial and final states of the process.

51. ** Jay rides his 2.0-kg skateboard. He is moving at speed 5.8 m/s when he pushes off the board and continues to move forward in the air at 5.4 m/s. The board now goes forward at 13 m/s. Determine Jay's mass and the change in the internal energy of the system during this process.

52. ** A 36-kg child is moving on a 2.0-kg skateboard at speed 6.0 m/s when she comes to a ledge that is 1.2 m above the surface below. Just before reaching the ledge, she pushes off the board. The board leaves the ledge moving horizontally and lands 8.0 m horizontally from the edge of the ledge. Make a list of the physical quantities describing the motion of the child after leaving the ledge and determine two of them. Describe any assumptions you made.

53. ** BIO Falcons While perched on an elevated site, a peregrine falcon spots a flying pigeon. The falcon dives, reaching a speed of 90 m/s (200 mi/h). The falcon hits its prey with its feet, stunning or killing it, then swoops back around to catch it in mid-air. Assume that the falcon has a mass of 0.60 kg and hits a 0.20-kg pigeon almost head-on. The falcon's speed after the collision is 60 m/s in the same direction. (a) Determine the final speed of the pigeon immediately after the hit. (b) Determine the internal energy produced by the collision. (c) Why does the falcon strike its prey with its feet and not head-on?

54. * When you play billiards, can you predict the velocities of the billiard balls after a collision if you know the velocity of a moving ball before the collision? Assume that the collision is head-on and elastic and that rotational motion can be ignored.

55. * A block of mass m_1 moving at speed v toward the west on a frictionless surface has an elastic head-on collision with a second, stationary block of mass m_2. Determine expressions for the final velocity of each block.

56. ** A 4.0-kg block moving at 2.0 m/s toward the west on a frictionless surface has an elastic head-on collision with a second 1.0-kg block traveling east at 3.0 m/s. Determine the final velocity of each block. (b) Determine the kinetic energy of each block before and after the collision. Note: The block with the least initial kinetic energy actually gains energy and the one with the most loses an equal amount. This is analogous to what happens when cool air comes into contact with warm air. The cool air warms (its molecules speed up) and the warm air cools (its molecules slow down).

6.8 Power

57. (a) What is the power involved in lifting a 1.0-kg object 1.0 m in 1.0 s? (b) While lifting a 10-kg object 1.0 m in 0.50? (c) While lifting the 10-kg object 2.0 m in 1.0 s? (d) While lifting a 20-kg object 1.0 m in 1.0 s?

58. * A fire engine must lift 30 kg of water a vertical distance of 20 m each second. What is the amount of power needed for the water pump for this fire hose?

59. * BIO Internal energy change while biking You set your stationary bike on a high 80-N friction-like resistive force and cycle for 30 min at a speed of 8.0 m/s. Your body is 10% efficient at converting chemical energy in your body into mechanical work. (a) What is your internal chemical energy change? (b) How long must you bike to convert 3.0×10^5 J of chemical potential while staying at this speed? (This amount of energy equals the energy released by the body after eating three slices of bread.)

60. * BIO Tree evaporation A large tree can lose 500 kg of water a day. (a) How much work does the tree need to do to lift the water 8.0 m? (b) If the loss of water occurs over a 12-h period, what is the average power in watts needed to provide this increase in gravitational energy in the water-Earth system?

61. * Climbing Mt. Mitchell An 82-kg hiker climbs to the summit of Mount Mitchell in western North Carolina. During one 2.0-h period, the climber's vertical elevation increases 540 m. Determine (a) the change in gravitational potential energy of the climber-Earth system and (b) the power of the process needed to increase the gravitational energy.

62. * BIO EST Sears stair climb The fastest time for the Sears Tower (now Willis Tower) stair climb (103 flights, or 2232 steps) is about 20 min. (a) Estimate the mechanical power in watts for a top climber. Indicate any assumptions you made. (b) If the body is 20% efficient at converting chemical energy into mechanical energy, approximately how many joules and kilocalories of chemical energy does the body expend during the stair climb? Note: 1 food calorie = 1 kilocalorie = 4186 J.

63. * BIO EST Exercising so you can eat ice cream You curl a 5.5-kg (12 lb) barbell that is hanging straight down in your hand up to your shoulder. (a) Estimate the work that your hand does in lifting the barbell. (b) Estimate the average mechanical power of the lifting process. Indicate any assumptions used in making the estimate. (c) Assuming the efficiency described at the end of Problem 62, how many times would you have to lift the barbell in order to burn enough calories to

use up the energy absorbed by eating a 300-food-calorie dish of ice cream? (Problem 62 provides the joule equivalent of a food calorie.) List the assumptions that you made.

64. ** **BIO** **Salmon move upstream** In the past, salmon would swim more than 1130 km (700 mi) to spawn at the headwaters of the Salmon River in central Idaho. The trip took about 22 days, and the fish consumed energy at a rate of 2.0 W for each kilogram of body mass. (a) What is the total energy used by a 3.0-kg salmon while making this 22-day trip? (b) About 80% of this energy is released by burning fat and the other 20% by burning protein. How many grams of fat are burned? One gram of fat releases 3.8×10^4 J of energy. (c) If the salmon is about 15% fat at the beginning of the trip, how many grams of fat does it have at the end of the trip?

65. * **EST** Estimate the maximum horsepower of the process of raising your body mass as fast as possible up a flight of 20 stair steps. Justify any numbers used in your estimate. The only energy change you should consider is the change in gravitational potential energy of the system you-Earth.

66. * A 1600-kg car smashes into a shed and stops. The force that the shed exerts on the car as a function of a position at the car's center is shown in **Figure P6.66**. How fast was the car traveling just before hitting the shed?

Figure P6.66

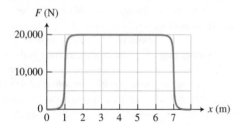

67. ** **/** Suppose the car in Problem 66 was moving at twice the speed and was stopped in the same manner and in the same distance, only now by a more solidly constructed shed. Draw a new graph for the force that this new shed exerted on the car as a function of the car's position. Include the appropriate numbers on the force axis and on the distance axis.

6.9 Improving our model of gravitational potential energy

68. At what distance from Earth is the gravitational potential energy of a spaceship-Earth system reduced to half the energy of the system before the launch?

69. * **Possible escape of different air molecule types** (a) Determine the ratio of escape speeds from Earth for a hydrogen molecule (H_2) and for an oxygen molecule (O_2). The mass of the oxygen is approximately 16 times that of the hydrogen. (b) In the atmosphere, the average random kinetic energy of hydrogen molecules and oxygen molecules is the same. Determine the ratio of the average speeds of the hydrogen and the oxygen molecules. (c) Based on these two results, give one reason why our atmosphere lacks hydrogen but retains oxygen.

70. Determine the escape speed for a rocket to leave Earth's Moon.

71. Determine the escape speed for a rocket to leave the solar system.

72. If the Sun were to become a black hole, how much would it increase the gravitational potential energy of the Sun-Earth system?

73. * A satellite moves in elliptical orbit around Earth, which is one of the foci of the elliptical orbit. (a) The satellite is moving faster when it is closer to Earth. Explain why. (b) If the satellite moves faster when it is closer to Earth, is the energy of the satellite-Earth system constant? Explain.

74. * Determine the maximum radius Earth's Moon would have to have in order for it to be a black hole.

General Problems

75. ** **EST** You wish to try bungee jumping, but want to make sure it is safe. The brochure provided at the ticket office says that the cord holding the jumper is initially 12 m long and has a spring constant of 160 N/m. The tower from which you plan to jump appears to be 10 floors high. Should you try the bungee jump? Explain your answer.

76. * Pose a problem involving work-energy ideas with real numbers. Then solve the problem choosing two different systems and discuss whether the answers were different.

77. ** Your dormitory has a nice balcony that looks over a pond in the grass below. You attach a 16-m-long rope to the limb of a tall tree beside the pond and pull it to the balcony so that it makes a 53° angle from the vertical. You hold the rope while standing on the balcony and then swing down. (a) How fast are you moving at the lowest point in your swing? Specify the system and the initial and final states. List the assumptions that you made. (b) How strong should the rope be to withstand your adventure (use your own mass for calculations)?

78. ** **Bungee jump at Squaw Valley** At the Squaw Valley ski area, 2500 m above sea level (**Figure P6.78**), a bungee jumper falls over a 152-m cliff above Lake Tahoe. Choose numbers for quantities, and make and solve a problem related to this bungee system.

Figure P6.78

79. ** **Six Flags roller coaster** A loop-the-loop on the Six Flags Shockwave roller coaster has a 10-m radius (**Figure P6.79**). The car is moving at 24 m/s at the bottom of the loop. Determine the force exerted by the seat of the car on an 80-kg rider when passing inverted at the top of the loop.

Figure P6.79

80. ** **Designing a ride** You are asked to help design a new type of loop-the-loop ride. Instead of rolling down a long hill to generate the speed to go around the loop, the 300-kg cart starts at rest (with two passengers) on a track at the same level as the bottom of the 10-m radius loop. The cart is pressed against a compressed spring that, when released, launches the cart along the track around the loop. Choose a spring of the appropriate spring constant to launch the cart so that the downward force exerted by the track on the cart as it passes the top of the loop is 0.2 times the force that Earth exerts on the cart. The spring is initially compressed 6.0 m.

81. **BIO EST Impact extinction** 65 million years ago over 50% of all species became extinct, ending the reign of dinosaurs and opening the way for mammals to become the dominant land vertebrates. A theory for this extinction, with considerable supporting evidence, is that a 10-km-wide 1.8×10^{15}-kg asteroid traveling at speed 11 km/s crashed into Earth. Use this information and any other information or assumptions of your choosing to (a) estimate the change in velocity of Earth due to the impact; (b) estimate the average force that Earth exerted on the asteroid while stopping it; and (c) estimate the internal energy produced by the collision (a bar chart for the process might help). By comparison, the atomic bombs dropped on Japan during World War II were each equivalent to 15,000 tons of TNT (1 ton of TNT releases 4.2×10^9 J of energy).

82. ** Newton's cradle is a toy that consists of several metal balls touching each other and suspended on strings (**Figure P6.82**). When you pull one ball to the side and let it strike the next ball, only one ball swings out on the other side. When you use two balls to hit the others, two balls swing out. Can you account for this effect using your knowledge about elastic collisions?

Figure P6.82

83. ** **Design of looping roller coaster** You are an engineer helping to design a roller coaster that carries passengers down a steep track and around a vertical loop. The coaster's speed must be great enough when at the top of the loop so that the rider stays in contact with the cart and the cart stays in contact with the track. Riders can withstand acceleration no more than a few "g's", where one "g" is 9.8 m/s². What are some reasonable values for the physical quantities you can use in the design of the ride? For example, one consideration is the height at which the cart and rider should start so that they can safely make it around the loop and the radius of the loop.

Reading Passage Problems

BIO Metabolic Rate Energy for our activities is provided by the chemical energy of the foods we eat. The absolute value of the rate of conversion of this chemical energy into other forms of energy ($\Delta E/\Delta t$) is called the metabolic rate. The metabolic rate depends on many factors—a person's weight, physical activity, the efficiency of bodily processes, and the fat-muscle ratio. **Table 6.8**

lists the metabolic rates of people under several different conditions and in several different units of measure: 1 kcal = 1000 calories = 4186 J. Dieticians call a kcal simply a Cal. A piece of bread provides about 70 kcal of metabolic energy.

In 1 hour of heavy exercise a 68-kg person metabolizes 600 kcal − 90 kcal = 510 kcal more energy than when at rest. Typically, reducing kilocalorie intake by 3500 kcal (either by burning it in exercise or not consuming it in the first place) results in a loss of 0.45 kg of body mass (the mass is lost through exhaling carbon dioxide—the product of metabolism).

84. Why is the metabolic rate different for different people?
    (a) They have different masses.
    (b) They have different body function efficiencies.
    (c) They have different levels of physical activity.
    (d) All of the above

85. A 50-kg mountain climber moves 30 m up a vertical slope. If the muscles in her body convert chemical energy into gravitational potential energy with an efficiency of no more than 5%, what is the chemical energy used to climb the slope?
    (a) 7 kcal   (b) 3000 J
    (c) 70 kcal   (d) 300,000 J

86. If 10% of a 50-kg rock climber's total energy expenditure goes into the gravitational energy change when climbing a 100-m vertical slope, what is the climber's average metabolic rate during the climb if it takes her 10 min to complete the climb?
    (a) 8 W   (b) 80 W   (c) 500 W
    (d) 700 W   (e) 800 W

87. **EST** A 68-kg person wishes to lose 4.5 kg in 2 months. Estimate the time that this person should spend in moderate exercise each day to achieve this goal (without altering her food consumption).
    (a) 0.4 h   (b) 0.9 h   (c) 1.4 h
    (d) 1.9 h   (e) 2.4 h

88. **EST** A 68-kg person walks at 5 km per hour for 1 hour a day for 1 year. Estimate the *extra* number of kilocalories of energy used because of the walking.
    (a) 40,000 kcal   (b) 47,000 kcal
    (c) 88,000 kcal   (d) 150,000 kcal

89. Suppose that a 90-kg person walks for 1 hour each day for a year, expending 50,000 extra kilocalories of metabolic energy (in addition to his normal resting metabolic energy use). What approximately is the person's mass at the end of the year, assuming his food consumption does not change?
    (a) 57 kg   (b) 61 kg   (c) 64 kg
    (d) 66 kg   (e) 67 kg

Table 6.8 Energy usage rate during various activities.

Type of activity	$\Delta E/\Delta t$ (watts)	$\Delta E/\Delta t$ (kcal/h)	$\Delta E/\Delta t$ (kcal/day)
45-kg person at rest	80	70	1600
68-kg person at rest	100	90	2100
90-kg person at rest	120	110	2600
68-kg person walking 3 mph	280	240	5800
68-kg person moderate exercise	470	400	10,000
68-kg person heavy exercise	700	600	14,000

Figure 6.14 A kangaroo hopping.

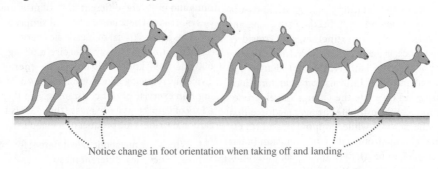

Notice change in foot orientation when taking off and landing.

Figure 6.15 The kangaroo stores elastic potential energy in its muscle and tendon when landing and uses some of that energy to make the next hop.

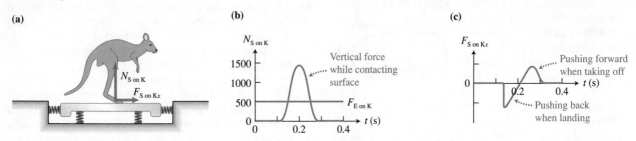

(a)

(b)

(c)

BIO Kangaroo hopping Hopping is an efficient method of locomotion for the kangaroo (see **Figure 6.14**). When the kangaroo is in the air, the Earth-kangaroo system has a combination of gravitational potential energy and kinetic energy. When the kangaroo lands, its Achilles tendons and the attached muscles stretch—a form of elastic potential energy. This elastic potential energy is used along with additional muscle tension to launch the kangaroo off the ground for the next hop. In the red kangaroo, more than 50% of the total energy used during each hop is recovered elastic potential energy. This is so efficient that the kangaroo's metabolic rate actually decreases slightly as its hopping speed increases from 8 km/h to 25 km/h.

The horizontal and vertical force components exerted by a firm surface on a kangaroo's feet while it hops are shown in **Figure 6.15a**. The vertical force $N_{\text{S on K}}$ (Figure 6.15b) varies: when the kangaroo is not touching the surface S, the force is zero; when it is pushing off, the force is about three times the gravitational force that Earth exerts on the kangaroo. The surface exerts a backward horizontal force ($F_{\text{S on K}x}$) on the kangaroo's foot while it lands and a forward horizontal force as it pushes off for the next hop (Figure 6.15c), similar to what happens to a human foot when landing in front of the body and when pushing off for another step when behind the body.

90. Why is hopping an energy-efficient mode of transportation for a kangaroo?
    (a) There is less resistance since there is less contact with the ground.
    (b) The elastic energy stored in muscles and tendons when landing is returned to help with the next hop.
    (c) The kangaroo has long feet that cushion the landing.
    (d) The kangaroo's long feet help launch the kangaroo.
91. Why does the horizontal force exerted by the ground on the kangaroo change direction as the kangaroo lands and then hops forward?

(a) The backward force when it lands prevents it from slipping, and the forward force when taking off helps propel it forward.
(b) One horizontal force is needed to help stop the kangaroo's fall and the other to help launch its upward vertical hop.
(c) Both forces oppose the kangaroo's motion, but one looks like it is forward because the kangaroo is moving fast.
(d) The kangaroo is not an inertial reference frame, and the forward force is not real.
(e) All of the above

92. Which answer below is closest to the vertical impulse that the ground exerts on the kangaroo while it takes off?
    (a) Zero (b) +50 N · s (c) +150 N · s
    (d) −50 N · s (e) −150 N · s

93. Which answer below is closest to the vertical impulse due to the gravitational force exerted on the kangaroo by Earth during the short time interval while it takes off?
    (a) 0 (b) +50 N · s (c) +150 N · s
    (d) −50 N · s (e) −150 N · s

94. Suppose the net vertical impulse on the 50-kg kangaroo due to all external forces was +100 N · s. Which answer below is closest to its vertical component of velocity when it leaves the ground?
    (a) +2.0 m/s (b) +3.0 m/s (c) +4.0 m/s
    (d) +8.0 m/s (e) +10 m/s

95. Which answer below is closest to the vertical height above the ground that the kangaroo reaches if it leaves the ground traveling with a vertical component of velocity of 2.5 m/s?
    (a) 0.2 m (b) 0.3 m (c) 0.4 m
    (d) 0.6 m (e) 0.8 m

Extended Bodies at Rest

7

Why is it best to lift heavy objects with your knees bent and the object near your body?

Why are doorknobs located on the side of the door opposite the hinges?

Why must your biceps muscle exert about seven times more force on your forearm to lift a barbell than the force that Earth exerts on it?

Back pain is a major health problem. In 2009 in the United States, the medical costs related to back pain were over $80 billion a year, about the same as the yearly cost of treating cancer. Eighty percent of people will suffer back pain at some point, usually beginning between the ages of 30 and 50. Back pain often results from incorrect lifting, which compresses the disks in the lower back, causing nerves to be pinched. Understanding the physics principles underlying lifting can help us develop techniques that minimize this compression and prevent injuries.

Be sure you know how to:

- Define the point-like model for an object (Section 1.2).
- Draw a force diagram for a system (Section 2.1).
- Use the component form of Newton's second law (Section 3.5).

So far in this book we have primarily been modeling objects as point-like with no internal structure. This method is appropriate when the shapes of objects do not affect the consequences of their interactions with each other—for example, a car screeching to a halt, an elevator moving up a shaft, or an apple falling into a pile of leaves. However, objects in general and the human body in particular are extraordinarily complex, with many internal parts that rotate and move relative to one another. To study the body and other complex structures, we need to develop a new way of modeling objects and of analyzing their interactions.

7.1 Extended and rigid bodies

In earlier chapters we focused on situations in which real objects that have nonzero dimensions could be reasonably modeled as point-like. Although any real object is extended in three dimensions, we assumed that an object's size and internal structure were not important for understanding the phenomena. When we analyzed the motion of such extended objects as cars, we neglected their parts (the turning wheels, the moving engine parts, and so forth) that moved with respect to each other internally even though the car as a whole moved in a straight line. We also neglected the size of the car. When the car made a turn, we considered the radius of the turn the same for all parts of the car. Put another way, when we model an object as a point-like, we ignore the fact that different parts of the same object move differently. Such motion—when an object moves as a whole from one location to another, without turning—is called **translational motion.**

Active Learning Guide >

In this chapter we will be analyzing more complex objects when they are at rest. We'd like, for example, to understand how the contemporary dancers in **Figure 7.1** manage to maintain their unusual balance. How large is the force exerted by the woman's hand on the man's foot? Where should each force be exerted and how large should their magnitudes be in order for the dancers to remain stable? We certainly cannot answer these questions by modeling the man and woman, who are extended objects, as point-like objects. We need a new model for extended objects and a new method for analyzing the forces that objects exert on each other. Our first task is to develop this new model for objects.

Rigid bodies

Notice that at the instant shown in the photo the various parts of the dancers' bodies are not moving with respect to each other. They are acting as a single rigid object. This observation motivates a new model for an extended object, in which the size of the object is not zero (it is not a point-like object), but

Figure 7.1 The point-like model of an object is not useful when we try to analyze the balance of these dancers.

parts of the object do not move with respect to each other. In physics, this model is called a **rigid body**.

> **Rigid body** A rigid body is a model of a real extended object. When we model an extended object as a rigid body, we assume that the object has a nonzero size but the distances between all parts of the object remain the same (the size and shape of the object do not change).

Many bones in your body can be reasonably modeled as rigid bodies, as can many everyday objects—buildings, bridges, streetlights, and utility poles. In this chapter we investigate what conditions are necessary for a rigid body to remain at rest.

Center of mass

Let's start with some simple experiments. Place a piece of thin, flat cardboard on a very smooth table. If we consider the cardboard to be a point-like object, on a force diagram the upward normal force that the table exerts on the cardboard will balance the downward gravitational force that Earth exerts on the cardboard (**Figure 7.2a**)—the cardboard will not accelerate. Now, we place the cardboard on a very small surface—like the eraser of a pencil. The cardboard tilts and falls off (Figure 7.2b). Does this result mean that the eraser cannot exert the same upward force on the cardboard that the table did, or is there some other explanation? The model of the point-like object cannot explain the tilting, since point-like objects do not tilt. Perhaps we need to model the cardboard as a rigid body. Before we do this, let us learn a little more about rigid bodies in Observational Experiment **Table 7.1**.

Figure 7.2 The cardboard is stable in (a) but not in (b). The place where the supporting force is exerted matters.

(a)

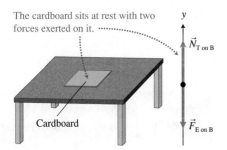

The cardboard sits at rest with two forces exerted on it.

(b)

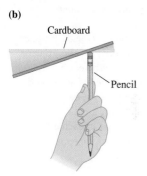

The cardboard tips if supported on the bottom off to the side.

OBSERVATIONAL EXPERIMENT TABLE

7.1 Pushing a board so that it moves without turning.

 VIDEO 7.1

Observational experiment	Analysis
We push with a pencil eraser at a point on the edge of a heart-shaped piece of flat cardboard so that the cardboard moves on a smooth surface *without* turning. We then exert the force on a different point and push the cardboard in a new direction so that it again moves without turning. We also push against the cardboard at points where it turns as it moves. 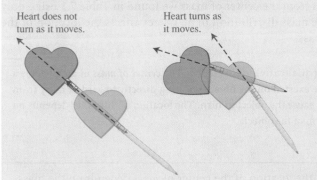	We draw lines across the top of the cardboard from the places and in the directions that the cardboard did not turn as it moved. 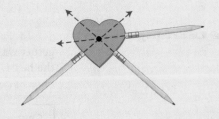

Pattern

1. All of the lines along which we had to push to move the cardboard without turning pass through a common point on the cardboard.
2. Pushing at the same locations in other directions causes the cardboard to turn as it moves.

Active Learning Guide ›

The analysis we did in Observational Experiment Table 7.1 indicates that a rigid body possesses a special point. If a force exerted on that object points directly toward or away from that point, the object will not turn. We call this point the object's **center of mass**. It appears that the center of mass is located approximately at the geometric center of the board (the geometric center is a point that in some sense is at the center of the object). In Testing Experiment **Table 7.2** we will test this idea that the center of mass of a rigid body is the point that we must push toward (or pull away from) in order to make an object move only translationally (as a whole, without turning).

TESTING EXPERIMENT TABLE

7.2   Where is the center of mass?

VIDEO 7.2

Testing experiment	Prediction	Outcome
Place a heavy object (the dark circle) on the heart-shaped cardboard used in Table 7.1(somewhere other than on the board's center of mass) and repeat the experiments, each time pushing at different locations along the edges, such that in each experiment the cardboard moves *without* turning.	If the idea is correct, and we push the cardboard so that in each experiment it moves but does not turn, then all of the lines along which the forces are exerted in all experiments should cross at one point: the new center of mass. Because we have added the heavy object, we expect it to be at a different location than in Table 7.1.	The lines along which the forces are exerted all cross at one point. This point is located somewhere between the old center of mass and the position of the added object.

Cardboard heart moves without turning.

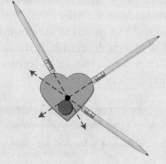

Conclusion

The outcome is consistent with the prediction—it supports the idea of a single center of mass in a rigid body or group of bodies.

Active Learning Guide ›

In Table 7.2, the center of mass of the heart and the added object together was different from the center of mass we found in Table 7.1 using only the heart. Thus, the mass distribution of an object affects the location of the object's center of mass.

> **Center of mass (qualitative definition)** The *center of mass* of an object is a point where a force exerted on the object pointing directly toward or away from that point, will not cause the object to turn. The location of this point depends on the mass distribution of the object.

> **TIP** Although the location of the center of mass depends on the mass distribution of the object, the mass of the object is not necessarily evenly distributed around the center of mass. We will learn more about the properties of the center of mass; we just want to caution you from taking the name of this point literally.

Where is the gravitational force exerted on a rigid body?

At the beginning of the chapter we found that we could not balance cardboard on the eraser of a pencil. Why did the cardboard fall off? Imagine drawing the forces exerted on the cardboard. The cardboard interacts with two objects: the eraser and Earth. The eraser exerts an upward normal force on the cardboard at the point where it touches it. Earth exerts a downward gravitational force on every part of the cardboard. Is it possible to simplify the situation and to find one location at which we can assume that the entire gravitational force is exerted on the cardboard?

Let us go back to the heart-shaped cardboard from Observational Experiment Table 7.1. If we place the eraser exactly under the cardboard's center of mass, the cardboard does not tip over and fall (**Figure 7.3**). If the cardboard does not tip, it means that all forces exerted on it, including the force exerted by Earth and the force exerted by the eraser, pass through the center of mass. The normal force exerted by the eraser passes through the place where it contacts the cardboard—below the center of mass.

Earth exerts a small force on each small part of the board, but we can assume that the total force is exerted exactly at the center of mass. That is why sometimes the object's center of mass is called the object's **center of gravity**.

When we model something as a point-like object, we model it as if all of the object's mass is located at its center of mass. Likewise, we can apply what we know about translational motion for point-like objects to rigid bodies, as long as we apply the rules to their centers of mass.

Review Question 7.1 You have an oval framed painting. How do you determine where you should insert a single nail into the frame so that the painting is correctly oriented in both vertical and horizontal directions?

7.2 Torque: A new physical quantity

We learned in the previous section that the turning effect of an individual force depends on where and in which direction the force is exerted on an object. The translational acceleration of the object's center of mass is still determined by Newton's second law, independently of where the force is exerted. In this section we will learn about the turning ability of a force that an object exerts on a rigid body.

Axis of rotation

When objects turn around an axis, physicists say that they undergo **rotational motion**. The axis may be a fixed physical axis, such as the hinge of a door, or it may not, as in the case of a spinning top. In this chapter we will focus on the conditions under which objects that could potentially rotate do not do so.

Consider a door. When you push on the doorknob perpendicular to the door's surface ($\vec{F}_2$ in **Figure 7.4**), it rotates easily about the door hinges. We call the imaginary line passing through the hinges the **axis of rotation**. You know from experience that pushing a door at or near the axis of rotation ($\vec{F}_1$ in Figure 7.4) is not as effective as pushing the doorknob. You also know that the harder you push near the knob, the more rapidly the door starts moving. Lastly, pushing on the outside edge of the door toward the axis of rotation ($\vec{F}_3$ in Figure 7.4) does not move the door at all.

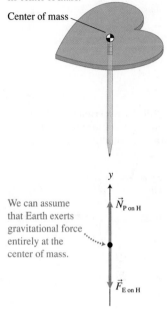

Figure 7.3 The eraser head can support the heart-shaped cardboard if the eraser is placed under the heart's center of mass.

The heart does not tip if supported under its center of mass.

Center of mass

We can assume that Earth exerts gravitational force entirely at the center of mass.

$\vec{N}_{\text{P on H}}$

$\vec{F}_{\text{E on H}}$

TIP When multiple forces are exerted on a rigid body, the center of mass of the rigid body accelerates translationally according to Newton's second law

$$\vec{a} = \frac{\Sigma \vec{F}}{m}$$

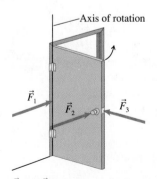

Figure 7.4 Different forces have different effects in turning a door about its axis of rotation.

Axis of rotation

$\vec{F}_1$ $\vec{F}_2$ $\vec{F}_3$

$\vec{F}_1$ and $\vec{F}_3$ do not rotate the door, whereas $\vec{F}_2$ moves it easily.

These observations suggest that three factors affect the turning ability of a force: (1) the place where the force is exerted, (2) the magnitude of the force, and (3) the direction in which the force is exerted. Next, let's construct a quantitative expression for this turning ability.

The role of position on the turning ability of a force

In order to construct a physical quantity that characterizes the turning ability of a force, we need to perform experiments where we exert measured forces at measured positions on a rigid body. We will take a 0.10-kg meter stick and suspend it at its center of mass from spring scale 2 (**Figure 7.5**); $\vec{F}_2$ is the force exerted on the meter stick by scale 2 at the point of suspension. Spring scale 1 pulls perpendicularly on the stick, exerting a downward force $\vec{F}_1$ at different locations on the left side, and scale 3 exerts a downward perpendicular force $\vec{F}_3$ at different locations on the right side. Earth exerts a 1.0-N force $\vec{F}_{\text{E on M}}$ on its center of mass, which is the point of suspension.

When either scale 1 or scale 3 pulls alone on the stick, the stick rotates (when scale 1 pulls on the stick, it rotates counterclockwise; when 3 pulls, it rotates clockwise). If we pull equally on scales 1 and 3 ($\vec{F}_1 = \vec{F}_3$) but the scales are located at different distances from the axis of rotation, the stick rotates (**Figure 7.6a**). We can use trial and error to find the combinations where the scales can be placed and pulled such that the stick does not rotate. For example, if scale 1 is twice as far from the axis of rotation and pulls half as hard as scale 3, the stick remains in equilibrium (Figure 7.6b). Similarly, if scale 1 is three times farther from the axis of rotation and exerts one-third of the force compared to scale 3, the stick remains in equilibrium (Figure 7.6c). When the stick does not rotate and does not move translationally, it is in a state of **static equilibrium.**

> **Static equilibrium** An object is said to be in static equilibrium when it *remains* at rest (does not undergo either translational or rotational motion) with respect to a particular observer in an inertial reference frame.

If we explore more situations in which the stick is not rotating, we find that when we have two springs pulling perpendicular to the stick, the stick remains in static equilibrium when the product of the magnitude of the force

Figure 7.5 An experiment to determine a condition necessary for multiple forces to balance a meter stick.

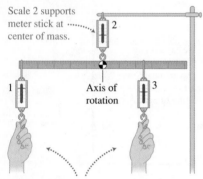

Scale 2 supports meter stick at center of mass.

We pull down on scales 1 and 3, exerting different forces at different places.

Active Learning Guide >

Figure 7.6 (a) The meter stick does not balance even though equal downward forces are exerted on each side. (b) and (c) A greater force on one side nearer the pivot point balances a smaller force on the other side farther from the pivot point.

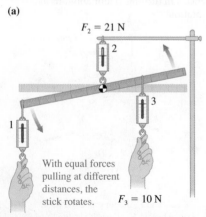

(a)

$F_2 = 21$ N

With equal forces pulling at different distances, the stick rotates.

$F_3 = 10$ N

$F_1 = 10$ N

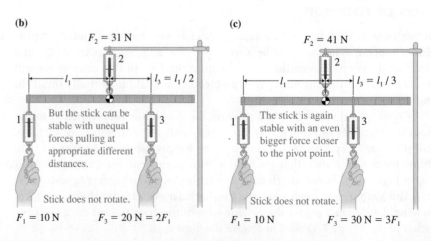

(b)

$F_2 = 31$ N

l_1 $l_3 = l_1 / 2$

But the stick can be stable with unequal forces pulling at appropriate different distances.

Stick does not rotate.

$F_1 = 10$ N $F_3 = 20$ N $= 2F_1$

(c)

$F_2 = 41$ N

l_1 $l_3 = l_1 / 3$

The stick is again stable with an even bigger force closer to the pivot point.

Stick does not rotate.

$F_1 = 10$ N $F_3 = 30$ N $= 3F_1$

exerted by the scale on the stick (F_1 or F_2 in our experiment) and the distance between where the force is exerted and the axis of rotation (l_1 or l_2 in our experiment) is the same for both forces:

$$F_1 l_1 = F_2 l_2$$

In other words, the turning ability of the force on the left cancels the turning ability of the force on the right, and the object is in static equilibrium.

The role of magnitude on the turning ability of a force

In addition, we notice that independently of whether the stick rotated or not, the reading of scale 2, F_2, supporting the stick always equals the sum of the readings of scales 1 and 3, $F_1 + F_3$, plus the magnitude of the force exerted by Earth on the stick $F_{\text{E on S}}$. In other words, in all cases the sum of the forces exerted on the stick was zero ($\Sigma F_y = 0$). This finding is consistent with what we know from Newton's laws—an object does not accelerate translationally if the sum of the forces exerted on it is zero. If it is originally at rest and does not accelerate translationally, then it remains at rest.

However, as we have seen from the experiment with the cardboard on the eraser, this sum-of-forces-equals-zero rule does not guarantee the rotational stability of rigid bodies (see Figure 7.2). Even when the sum of the forces exerted on the cardboard was zero, it could still start turning.

Another simple experiment helps illustrate this idea. Take a book with a glossy cover, place it on a smooth table (to minimize friction), and push it, exerting the same magnitude, oppositely directed force on each of two corners (as shown in **Figure 7.7a**). The force diagram in Figure 7.7b shows that the net force exerted on the book is zero—there is no translational acceleration. However, the book starts turning. The forces exerted by Earth and the table on the book pass through the book's center of mass; thus they do not cause turning. But the forces that you exert on the corners of the book do cause it to turn. Notice that these forces are of the same magnitude and are exerted at the same distance from the center of mass. You can imagine that there is an invisible axis of rotation passing through the center of mass perpendicular to the desk's surface. You would think that the turning effect caused by each force around this imaginary axis is the same; thus the two turning effects should cancel, as they did for scales 1 and 3 in the experiment described earlier. However, this does not happen. The fact that the book turns tells us that not only are the magnitude and placement of the force on the object important to describe the turning ability of the force, but the direction in which this force causes turning (for example, clockwise or counterclockwise around an axis of rotation) is also important.

By convention, physicists call counterclockwise turning about an axis of rotation positive and clockwise turning negative. So far, this is what we know about the new quantity that characterizes the turning ability of a force:

(a) It is equal to the product of the magnitude of the force and the distance the force is exerted from the axis of rotation.

(b) It is positive when the force tends to turn the object counterclockwise and negative when the force tends to turn the object clockwise.

(c) When one force tends to rotate an object counterclockwise and the other force tends to rotate an object clockwise, their effects cancel if $(F_{\text{counterclockwise}} l_1) + (-F_{\text{clockwise}} l_2) = 0$. In this case the object does not rotate.

Let's apply what we have devised so far to check whether this new quantity is useful for explaining other situations. Here is another simple experiment that you can do at home. Place a full milk carton or something of similar mass into a grocery bag. The bag with the milk carton by itself is not too heavy and

Figure 7.7 The turning effect of a force must depend on more than F and l.

(a)

The equal and opposite forces that you exert on the book at its corners cause the book to rotate.

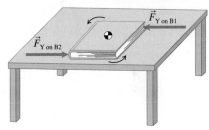

(b)

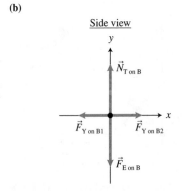

Figure 7.8 Holding a bag at the end of a stick is more difficult when the stick is horizontal than when the stick is tilted up.

(a)

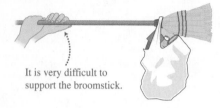

It is very difficult to support the broomstick.

(b)

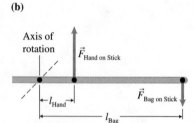

Axis of rotation

$\vec{F}_{\text{Hand on Stick}}$

$\vec{F}_{\text{Bag on Stick}}$

l_{Hand}

l_{Bag}

Note: $F_{\text{H on S}}l_{\text{H}} - F_{\text{B on S}}l_{\text{B}} = 0$

(c)

Tilting the broomstick up makes it easier to hold.

(d)

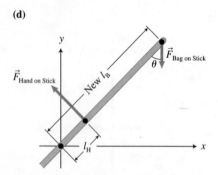

y

$\vec{F}_{\text{Hand on Stick}}$

New l_{B}

θ

$\vec{F}_{\text{Bag on Stick}}$

l_{H}

x

Active Learning Guide ›

you can easily lift it with one hand. Now hang the grocery bag from the end of a broomstick (**Figure 7.8a**). Try to support it by holding only the handle end of the broomstick with your hands close together. It is very difficult. Why?

The broomstick, the object of interest, can turn around an axis through the hand that is closest to you. The bag exerts a force on the broomstick far from this axis of rotation. This means that l_{Bag}, a quantity that we must find in order to determine the turning ability of the force exerted by the bag on the broomstick, is very large (Figure 7.8b). Your other hand, which is very close to the axis of rotation (distance l_{Hand}), must balance the effect of the bag. But since l_{Hand} is so small, the force your hand exerts must be very large. The outcome of this experiment agrees with what we learned so far about the turning ability of a force.

Holding the broomstick perpendicular to your body is quite difficult. However, if you hold the broomstick at an angle above the horizontal (Figure 7.8c), you find that the bag becomes easier to support. Why? Perhaps this has to do with the direction the force is exerted relative to the broomstick (see Figure 7.8d).

The role of angle on the turning ability of a force

Our current mathematical model of the physical quantity that characterizes the turning ability of the force ($\pm Fl$) takes into account the direction in which an exerted force can potentially rotate an object (clockwise or counterclockwise) but does not take into account the actual direction of the force. However, our experiment with the broomstick indicates that the angle at which we exert a force relative to the broomstick affects the turning ability of the force. We know from experience that pushing on a door on its outside edge directly toward the hinges does not cause it to rotate. The direction of the push must matter. How can we improve our model for the physical quantity to take the direction of the force into account?

To investigate this question we can change our experiment with the meter stick slightly by making scale 1 pull on the stick at an angle other than 90° (**Figure 7.9**). Scale 2, on the far right end of the meter stick, 0.50 m from the suspension point, will exert a constant force of 10.0 N downward at a 90° angle. Scale 1 on the far left end of the stick will pull at different angles θ so that the meter stick remains horizontal. The results are shown in **Table 7.3**. In all cases the stick is horizontal—therefore, the turning ability of the force on the right is balanced by the turning ability of the force on the left.

Using the data in the table, we see the effects of the magnitude and the angle of force $\vec{F}_1$ on its turning ability: the smaller the angle between the direction of the force and the stick, the larger the magnitude of the force that is necessary to produce the same turning ability. Thus we find that there are four factors that affect the turning ability of a force: (1) the direction (counterclockwise or clockwise) that the force can potentially rotate the object; (2) the magnitude of the force F, (3) the distance l of the point of application of the force from the axis of rotation, and (4) the angle θ that the force makes relative to a line from the axis of rotation to the point of application of the

Table 7.3 Magnitude, location, and direction of force and its turning ability.

Magnitude of $\vec{F}_1$	Distance to the axis of rotation	Angle θ between $\vec{F}_1$ and the stick	Turning ability produced by $\vec{F}_3$
10.0 N	0.50 m	90°	$-(10.0\,\text{N})(0.50\,\text{m}) = -5.0\,\text{N} \cdot \text{m}$
12.6 N	0.50 m	53°	$-(10.0\,\text{N})(0.50\,\text{m}) = -5.0\,\text{N} \cdot \text{m}$
14.2 N	0.50 m	45°	$-(10.0\,\text{N})(0.50\,\text{m}) = -5.0\,\text{N} \cdot \text{m}$
20.0 N	0.50 m	30°	$-(10.0\,\text{N})(0.50\,\text{m}) = -5.0\,\text{N} \cdot \text{m}$

force. If we combine these four factors, the physical quantity characterizing the turning ability of a force takes a form such as:

$$\pm Flf(\theta)$$

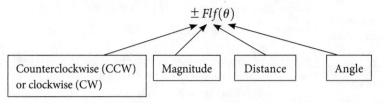

where $f(\theta)$ is some function of the angle θ.

Consider the last row of Table 7.3. The force exerted by scale 1 is 0.50 m from the axis of rotation ($l = 0.50$ m), and the scale exerts a 20-N force F_1 on the stick at a 30° angle relative to the stick. This force produces the counterclockwise $+5.0$ N·m effect needed to balance the -5.0 N·m clockwise effect of scale 2. What value would $f(\theta)$ have to be to get this rotational effect?

$$+5.0\,\text{N}\cdot\text{m} = (20.0\,\text{N})(0.5\,\text{m})\,f(30°)$$

or $f(30°)$ must be 0.50. Recall that $\sin 30° = 0.50$. Maybe the function $f(\theta)$ is the sine function, that is, $\tau = lF\sin\theta$. Is this consistent with the other rows in Table 7.3?

$$+(10.0\,\text{N})(0.5\,\text{m})(\sin 90°) = +5.0\,\text{N}\cdot\text{m}\,(1.00) = +5.0\,\text{N}\cdot\text{m}$$
$$+(12.6\,\text{N})(0.5\,\text{m})(\sin 53°) = +6.3\,\text{N}\cdot\text{m}\,(0.80) = +5.0\,\text{N}\cdot\text{m}$$
$$+(14.2\,\text{N})(0.5\,\text{m})(\sin 45°) = +7.1\,\text{N}\cdot\text{m}\,(0.71) = +5.0\,\text{N}\cdot\text{m}$$

This expression ($\tau = Fl\sin\theta$) is the mathematical definition of the new physical quantity that characterizes the ability of a force to turn (rotate) a rigid body. This physical quantity is called a **torque.** The symbol for torque is τ, the Greek letter tau.

Torque τ produced by a force The torque produced by a force exerted on a rigid body about a chosen axis of rotation is

$$\tau = \pm Fl\sin\theta \qquad (7.1)$$

where F is the magnitude of the force, l is the magnitude of the distance between the point where the force is exerted on the object and the axis of rotation, and θ is the angle that the force makes relative to a line connecting the axis of rotation to the point where the force is exerted (see **Figure 7.10**).

Figure 7.10 illustrates the method for calculating the turning ability (torque) due to a particular force. In this case, we are calculating the torque due to the force that the slanted rope exerts on the end of a beam that supports a load hanging from the beam. The torque is positive if the force has a counterclockwise turning ability about the axis of rotation, and negative if the force has a clockwise turning ability. The SI unit for torque is newton·meter, N·m (the British system unit is lb·ft).

TIP Notice that the units of torque (N·m) are the same as the units of energy (N·m = J). Torque and energy are very different quantities. We will always refer to the unit of torque as newton·meter (N·m) and the unit of energy as joule (J).

Figure 7.9 An experiment to determine the angle dependence of the turning ability caused by a force.

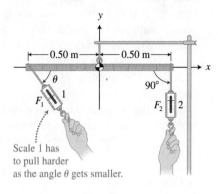

Scale 1 has to pull harder as the angle θ gets smaller.

Figure 7.10 A method to determine the torque (turning ability) produced by a force.

(a)

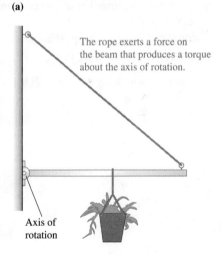

The rope exerts a force on the beam that produces a torque about the axis of rotation.

Axis of rotation

(b)

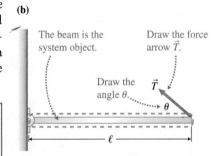

The beam is the system object.

Draw the force arrow $\vec{T}$.

Draw the angle θ.

Write an expression for the distance ℓ from the axis of rotation to the place the force is exerted.

QUANTITATIVE EXERCISE 7.1 **Rank the magnitudes of the torques**

Suppose that five strings pull one at a time on a horizontal beam that can pivot about a pin through its left end, which is the axis of rotation. The magnitudes of the forces exerted by the strings on the beam are either T or $T/2$. Rank the magnitudes of the torques that the strings exert on the beam, listing the largest magnitude torque first and the smallest magnitude torque last. Indicate if any torques have equal magnitudes. Try to answer the question before looking at the answer below.

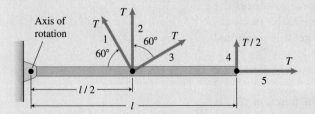

Represent mathematically A mathematical expression for the torque produced by each force is shown below. To understand why each torque is positive, imagine in what direction each string would turn the beam about the axis of rotation, if that were the only force exerted on it. You will see that each string tends to turn the beam counterclockwise (except string 5).

Torque due to string 1: $\tau_1 = +T(l/2)\sin 60° = +0.43\ Tl.$

Torque due to string 2: $\tau_2 = +T(l/2)\sin 90° = +0.50\ Tl.$

Torque due to string 3: $\tau_3 = +T(l/2)\sin 150° = +0.25\ Tl.$

Torque due to string 4: $\tau_4 = +(T/2)l\sin 90° = +0.50\ Tl.$

Torque due to string 5: $\tau_5 = +Tl\sin 0° = 0.$

Solve and evaluate Notice that the angle used for the torque for the force exerted by rope 3 was 30° and not 60°—the force makes a 30° angle relative to a line from the pivot point to the place where the string exerts the force on the beam. String 5 exerts a force parallel to the line from the pivot point to the place where it is exerted on the beam; as a result, torque 5 is zero. The rank order of the torques is $\tau_2 = \tau_4 > \tau_1 > \tau_3 > \tau_5$.

Try it yourself: Determine the torque caused by the cable pulling horizontally on the inclined drawbridge shown below. The force that the cable exerts on the bridge is 5000 N, the bridge length is 8.0 m, and the bridge makes an angle of 50° relative to the vertical support for the cable system.

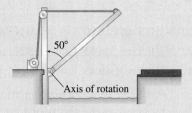

Answer: $\tau = +(5000\ \text{N})(8.0\ \text{m})\sin 40° = +26,000\ \text{N} \cdot \text{m}.$ Note that we did not use 50° in our calculation. Why?

TIP To decide the sign of the torque that a particular force exerts on a rigid body about a particular axis, pretend that a pencil is the rigid body. Hold it with two fingers at a place that represents the axis of rotation (**Figure 7.11**) and exert a force on the pencil representing the force whose torque sign you wish to determine. Does that force cause the pencil to turn counterclockwise (a + torque) or clockwise (a − torque) about the axis of rotation?

Figure 7.11 A method to determine the sign of a torque.

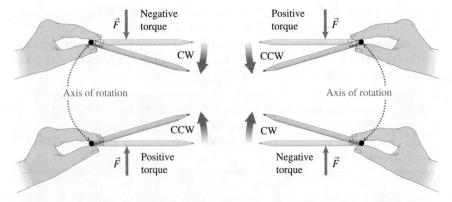

EXAMPLE 7.2 A painter on a ladder

A 75-kg painter stands on a 6.0-m-long 20-kg ladder tilted at 53° relative to the ground. He stands with his feet 2.4 m up the ladder. Determine the torque produced by the normal force exerted by the painter on the ladder for two choices of axis of rotation: (a) an axis parallel to the base of the ladder where it touches the ground and (b) an axis parallel to the top ends of the ladder where it touches the wall of the house.

Sketch and translate A sketch of the situation with the known information is shown below. The ladder is our system.

Find the torque that the painter exerts on the ladder about two different axes of rotation.

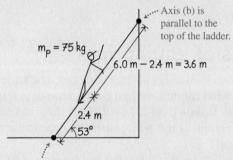

Axis (b) is parallel to the top of the ladder.

$m_p = 75$ kg

6.0 m − 2.4 m = 3.6 m

2.4 m

53°

Axis (a) is parallel to the base of the ladder.

Simplify and diagram Four objects interact with the ladder: Earth, the painter's feet, the wall, and the ground. Our interest in this problem is only in the torque that the painter's feet exert on the ladder. Thus, we diagram for the ladder and the downward normal force $\vec{N}_{P \text{ on } L}$ that the painter's feet exert on the ladder. The diagrams are for the two different axes of rotation.

$\vec{N}_{P \text{ on } L}$ causes a clockwise rotation about the axis of rotation at the bottom of the ladder (negative torque).

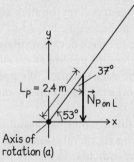

$L_p = 2.4$ m

37°

$\vec{N}_{P \text{ on } L}$

53°

Axis of rotation (a)

Represent mathematically We use Eq. (7.1) for the torque caused by the force that the painter's feet exert on the ladder (for simplicity, we write the magnitude of the force as N with no subscripts): $\tau = \pm Nl \sin \theta$.

Solve and evaluate (a) For the first calculation, we choose the axis of rotation at the place where the feet of the ladder touch the ground. The torque produced by the normal force exerted by the painter's feet on the ladder is

$$\tau = -Nl \sin \theta = -(m_p g)l \sin \theta$$
$$= -(75 \text{ kg})[(2.4 \text{ m})(9.8 \text{ N/kg})]\sin 37°$$
$$= -1100 \text{ N} \cdot \text{m}$$

Note that the normal force exerted by the painter's feet on the ladder has the same magnitude as the force mg that Earth exerts on the painter, but it is not the same force, as it is exerted on a different object and is a normal force and not a gravitational force. That normal force tends to rotate the ladder clockwise about the axis of rotation (a negative torque). The force makes a 37° angle relative to a line from the axis of rotation to the place where the force is exerted.

(b) We now choose the axis of rotation parallel to the wall at the top of the ladder where it touches the wall. The torque produced by the normal force exerted by the painter's feet on the ladder is

$$\tau = +Fl \sin \theta = +(m_p g)l \sin \theta$$
$$= +(75 \text{ kg})[(3.6 \text{ m})(9.8 \text{ N/kg})]\sin 143°$$
$$= +1600 \text{ N} \cdot \text{m}$$

$\vec{N}_{P \text{ on } L}$ causes a counterclockwise rotation about the axis of rotation at the top of the ladder (a positive torque).

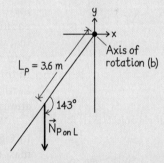

$L_p = 3.6$ m

143°

$\vec{N}_{P \text{ on } L}$

Axis of rotation (b)

When we choose the axis of rotation at the top of the ladder, the downward force exerted by the painter's feet on the ladder tends to rotate the ladder counterclockwise about the axis at the top (a positive torque). This force makes a 143° angle with respect to a line from the axis of rotation to the place where the force is exerted.

Note that the torque depends on where we place the axis of rotation. You cannot do torque calculations without carefully defining the axis of rotation.

Try it yourself: Write an expression for the torque produced by the upward normal force exerted by the floor on the ladder about the same two axes: (a) at the base of ladder and (b) at the top of the ladder.

(continued)

Answer: (a) The torque will be 0 at the bottom, because the normal force passes through the axis. (b) About the axis at the top of the ladder, the torque is $\tau = -(6.0\ \text{m})N_{\text{Floor on Ladder}} \sin 37°$. Note that the floor tends to push the ladder clockwise about an axis through the top of the ladder. The normal force makes a 37° angle relative to a line from the top axis to the place where the force is exerted.

Review Question 7.2 Give an example of a situation in which (a) a torque produced by a force is zero with respect to one choice of axis of rotation but not zero with respect to another; (b) a force is not exerted at the axis of rotation, but the torque produced by it is zero anyway; and (c) several forces produce nonzero torques on an object, but the object does not rotate.

7.3 Conditions of equilibrium

We can combine our previous knowledge of forces and our new knowledge of torque to determine under what conditions rigid bodies remain in static equilibrium, that is, at rest. Recall that on page 234 we defined static equilibrium as a state in which an object *remains* at rest with respect to a particular observer in an inertial reference frame.

It is possible for an object to be at rest briefly. For example, a ball thrown upward stops for an instant at the top of its flight, but it does not remain at rest. Thus, the word *remains* is important in the expression "remains at rest"—the object has to stay where it is. The words "with respect to an observer in an inertial reference frame" are also an important part of the definition of static equilibrium. Recall from the chapter on Newtonian mechanics (Chapter 2) that if an observer is not in an inertial reference frame, an object can accelerate with respect to the observer even if the sum of the forces exerted on it is zero. In this chapter we will only consider observers who are at rest with respect to Earth, since that is the most common point of view for observing real-life situations involving static equilibrium.

We again suspend the same 0.1-kg meter stick from spring scale 2, as shown in **Figure 7.12**. However, the suspension point is no longer at the center of mass of the meter stick. You and your friend again pull on the stick at different positions with spring scales 1 and 3. When pulled as described in Observational Experiment **Table 7.4** on the next page, the stick does not rotate. Pulling at other positions while exerting the same forces, or pulling at the same positions while exerting different forces, causes the stick to rotate. We need to find a pattern in the combinations of forces and torques exerted on the meter stick that keep the stick in static equilibrium.

For most situations that we analyze in this chapter, we assume that the objects rotate in the x-y plane and that the axis of rotation goes through the origin of the coordinate system and is perpendicular to the x-y plane.

Figure 7.12 Multiple objects exert forces on a meter stick.

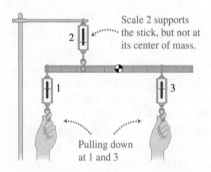

Scale 2 supports the stick, but not at its center of mass.

Pulling down at 1 and 3

OBSERVATIONAL EXPERIMENT TABLE

7.4 Meter stick in static equilibrium.

VIDEO 7.4

Observational experiment	Analysis

Experiment 1. Three spring scales and Earth exert forces on a meter stick at locations shown below. Examine the forces and torques exerted on the stick. Choose the axis of rotation at the place where the string from scale 2 supports the stick. This choice determines the distances in the torque equations for each force.

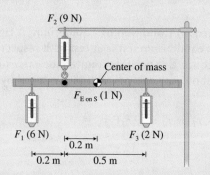

$\Sigma F_y = (-6.0\,\text{N}) + 9.0\,\text{N} + (-1.0\,\text{N}) + (-2.0\,\text{N}) = 0$

Counterclockwise torques:

$\tau_1 = (F_1)(l_1) = (6.0\,\text{N})(0.20\,\text{m}) = 1.2\,\text{N} \cdot \text{m}$

Clockwise torques:

$\tau_2 = (F_2)(l_2) = (9.0\,\text{N})(0) = 0$

$\tau_E = -(F_{\text{E on S}})(l_{\text{CM}}) = -(1.0\,\text{N})(0.2\,\text{m}) = -0.2\,\text{N} \cdot \text{m}$

$\tau_3 = -(F_3)(l_3) = -(2.0\,\text{N})(0.50\,\text{m}) = -1.0\,\text{N} \cdot \text{m}$

$\Sigma\tau = \tau_1 + \tau_2 + \tau_E + \tau_3$
$\quad = +1.2\,\text{N} \cdot \text{m} + 0 - 0.2\,\text{N} \cdot \text{m} - 1.0\,\text{N} \cdot \text{m} = 0$

Experiment 2. Three spring scales and Earth exert forces on a meter stick at locations shown below. Examine the forces and torques exerted on the stick. Choose the axis of rotation at the place where the string from scale 2 supports the stick.

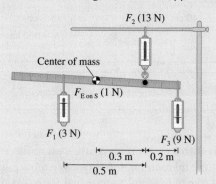

$\Sigma F_y = -3.0\,\text{N} + (-1.0\,\text{N}) + 13.\,\text{N} + (-9.0\,\text{N}) = 0$

Counterclockwise:

$\tau_1 = (F_1)(l_1) = (3.0\,\text{N})(0.50\,\text{m}) = 1.5\,\text{N} \cdot \text{m}$

$\tau_E = (F_{\text{E on S}})(l_{\text{CM}}) = (1.0\,\text{N})(0.3\,\text{m}) = 0.3\,\text{N} \cdot \text{m}$

Clockwise:

$\tau_2 = (F_2)(l_2) = (13\,\text{N})(0) = 0$

$\tau_3 = -(F_3)(l_3) = -(9.0\,\text{N})(0.20\,\text{m}) = -1.8\,\text{N} \cdot \text{m}$

$\Sigma\tau = \tau_1 + \tau_E + \tau_2 + \tau_3$
$\quad = +1.5\,\text{N} \cdot \text{m} + 0.3\,\text{N} \cdot \text{m} + 0 - 1.8\,\text{N} \cdot \text{m} = 0$

Patterns

- In both cases the net force exerted on the meter stick in the vertical direction is zero: $\Sigma F_y = 0$.
- In both cases the sum of the torques exerted on the meter stick equals zero: $\Sigma\tau = 0$.

The first pattern in both experiments is familiar to us. It is simply Newton's second law applied to the vertical axis of the meter stick for the case of zero translational acceleration. Because the sum of the vertical forces exerted on the meter stick is zero, there is no vertical acceleration. We had no horizontal forces, so the meter stick could not accelerate horizontally. The second pattern shows that in the experiments presented in the table in both cases the net torque is zero and the meter stick does not start turning. We will learn in the next chapter that these are examples of rigid bodies with zero rotational acceleration.

TIP Notice that you cannot determine the torque produced by a force without specifying the point at which the force is exerted on the object relative to the axis of rotation.

‹ Active Learning Guide

We can now state the conditions of static equilibrium mathematically as follows.

Condition 1. Translational (Force) Condition of Static Equilibrium An object modeled as a rigid body is in translational static equilibrium with respect to a particular observer if it is at rest with respect to that observer and the components of the sum of the forces exerted on it in the perpendicular x- and y-directions are zero:

$$\Sigma F_{\text{on O}x} = F_{1\,\text{on O}x} + F_{2\,\text{on O}x} + \cdots + F_{n\,\text{on O}x} = 0 \qquad (7.2x)$$

$$\Sigma F_{\text{on O}y} = F_{1\,\text{on O}y} + F_{2\,\text{on O}y} + \cdots + F_{n\,\text{on O}y} = 0 \qquad (7.2y)$$

The subscript n indicates the number of forces exerted by external objects on the rigid body.

Condition 2. Rotational (Torque) Condition of Equilibrium A rigid body is in turning or rotational static equilibrium if it is at rest with respect to the observer and the sum of the torques $\Sigma\tau$ (positive counterclockwise torques and negative clockwise torques) about any axis of rotation produced by the forces exerted on the object is zero:

$$\Sigma\tau = \tau_1 + \tau_2 + \cdots + \tau_n = 0 \qquad (7.3)$$

> **TIP** Remember that all the gravitational forces exerted by Earth on the different parts of the rigid body can be combined into a single gravitational force being exerted on the center of mass of the rigid body.

EXAMPLE 7.3 Testing the conditions of static equilibrium

Place the ends of a standard meter stick on two scales, as shown below. The scales each read 0.50 N. From this, we infer that the mass of the meter stick is about 0.10 kg (the gravitational force that Earth exerts on the meter stick would be $(0.10\ \text{kg})(9.8\ \text{N/kg}) = 1.0\ \text{N}$). Predict what each scale will read if you place a 5.0-kg brick 40 cm to the right of the left scale.

Sketch and translate A labeled sketch of the situation is shown below. We choose the stick as the system of interest and use a standard x-y coordinate system. We choose the axis of rotation at the place where the left scale touches the stick. By doing this, we are making the torque produced by the normal force exerted by the left scale on the stick zero—that force is exerted exactly at the axis of rotation. With this choice, we remove one of the unknown quantities from the torque condition of equilibrium and will be able to use that condition to find the force exerted by the right scale on the meter stick.

Predict the scale readings after a 5.0-kg brick is placed at the 40-cm mark.

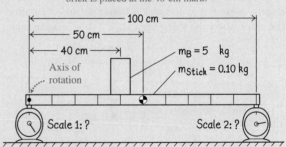

Simplify and diagram We model the meter stick as a rigid body with a uniform mass distribution (its center of mass is at the midpoint of the stick). We model the brick as a point-like object, and assume that the scales push up on the stick at the exact ends of the stick. We then draw a force diagram showing the forces exerted on the stick by Earth, the brick, and each of the scales. As noted, the left end of the stick has been chosen as the axis of rotation.

Analyzing the situation with the axis of rotation on the left side of the meter stick

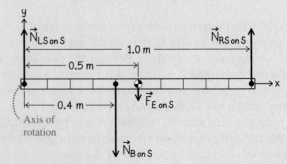

Represent mathematically According to the conditions of static equilibrium, the sum of the forces exerted on the meter stick should equal zero, as should the sum of the torques around the axis of rotation. The gravitational force exerted by Earth and the normal force exerted by the brick on the stick both have clockwise turning ability around the axis of rotation and produce negative torques. The force exerted by the right scale on the stick has counterclockwise turning ability and produces a positive torque. The force exerted by the left scale on the stick produces zero torque since it is exerted at the axis of rotation. The two conditions of equilibrium are then the following:

Translational (force) condition ($\Sigma F_y = 0$):

$$(-F_{\text{E on S}}) + (-N_{\text{B on S}}) + N_{\text{RS on S}} + N_{\text{LS on S}} = 0$$

Rotational (torque) condition ($\Sigma \tau = 0$):

$$-F_{\text{E on S}}(0.50\,\text{m}) - N_{\text{B on S}}(0.40\,\text{m}) + N_{\text{RS on S}}(1.00\,\text{m}) = 0$$

Since none of the forces have x-components, we didn't apply the x-component form of the force condition of equilibrium.

Earth exerts a downward gravitational force on the brick of magnitude:

$$F_{\text{E on B}} = m_{\text{B}}g = (5.0\,\text{kg})(9.8\,\text{N/kg}) \approx 50\,\text{N}$$

Thus, the stick must exert a balancing 50-N upward normal force on the brick. According to Newton's third law ($\vec{N}_{\text{B on S}} = -\vec{N}_{\text{S on B}}$), the brick must exert a downward 50-N normal force $\vec{N}_{\text{B on S}}$ on the stick.

Solve and evaluate We have two equations with two unknowns ($N_{\text{RS on S}}$ and $N_{\text{LS on S}}$). We first use the torque equilibrium condition to determine the magnitude of the force exerted by the right scale on the stick:

$$-[(0.10\,\text{kg})(10\,\text{N/kg})](0.50\,\text{m})$$
$$-[(5.0\,\text{kg})(10\,\text{N/kg})](0.40\,\text{m}) + N_{\text{RS on S}}(1.00\,\text{m}) = 0$$

or

$$N_{\text{RS on S}} = 20.5\,\text{N}$$

We can use this result along with the force equilibrium condition equation to determine the magnitude of the force exerted by the left scale on the stick.

$$-(0.10\,\text{kg})(10\,\text{N/kg}) - (5.0\,\text{kg})(10\,\text{N/kg})$$
$$+ 20.5\,\text{N} + N_{\text{LS on S}} = 0$$

or

$$N_{\text{LS on S}} = 30.5\,\text{N}$$

These predictions make sense because the sum of these two upward forces equals the sum of the two downward forces that Earth and the brick exert on the meter stick. Also, the force on the left end is greater because the brick

is positioned closer to it, which sounds very reasonable. Performing this experiment, we find that the outcome matches the predictions.

Using a different axis of rotation Remember that we had the freedom to choose whatever axis of rotation we wanted. Let's try it again with the axis of rotation at 40 cm from the left side, the location of the brick. See the force diagram below. The force condition of equilibrium will not change since it does not depend on the choice of the axis of rotation:

$$(-F_{\text{E on S}}) + (-N_{\text{B on S}}) + N_{\text{RS on S}} + N_{\text{LS on S}} = 0$$

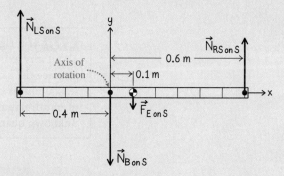

The axis of rotation is now at the 0.4-m position.

The torque condition will change:

$$[-F_{\text{E on S}}(0.10\,\text{m})] + [-N_{\text{LS on S}}(0.40\,\text{m})]$$
$$+ N_{\text{RS on S}}(0.60\,\text{m}) = 0$$

Now we have two unknowns in each of the two equations and have to solve them simultaneously to determine the unknowns. This will be somewhat harder than when we chose the axis of rotation at the left end of the stick. Let's solve the force condition equation for $N_{\text{RS on S}}$ and substitute the result into the torque condition equation:

$$N_{\text{RS on S}} = F_{\text{E on S}} + N_{\text{B on S}} - N_{\text{LS on S}}$$
$$\Rightarrow -F_{\text{E on S}}(0.10\,\text{m}) - N_{\text{LS on S}}(0.40\,\text{m})$$
$$+ (F_{\text{E on S}} + N_{\text{B on S}} - N_{\text{LS on S}})(0.60\,\text{m}) = 0$$

Combining the terms with $N_{\text{LS on S}}$ on one side, we get

$$N_{\text{LS on S}}(0.40\,\text{m} + 0.60\,\text{m}) = F_{\text{E on S}}(0.50\,\text{m})$$
$$+ N_{\text{B on S}}(0.60\,\text{m})$$
$$= (1.0\,\text{N})(0.50\,\text{m}) + (50\,\text{N})(0.60\,\text{m}) = 30.5\,\text{N} \cdot \text{m}$$

or $N_{\text{LS on S}} = 30.5\,\text{N}$. Substituting back into the force condition equation, we find that

$$N_{\text{RS on S}} = 1.0\,\text{N} + 50.0\,\text{N} - 30.5\,\text{N} = 20.5\,\text{N}$$

These are the same results we obtained from the original choice of the axis of rotation. The choice of the axis

(continued)

of rotation does not affect the results. This makes sense, in the same way that choosing a coordinate system does not affect the outcome of an experiment. The concepts of axes of rotation and coordinate systems are mental constructs and should not affect the outcome of actual experiments.

Try it yourself: A uniform meter stick with a 50-g object on it is positioned as shown below. The stick extends 30 cm over the edge of the table. If you push the stick so that it extends slightly further over the edge, it tips over. Use this result to determine the mass of the meter stick.

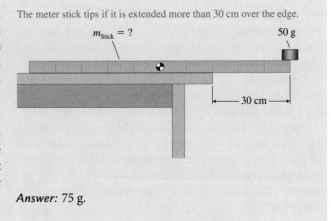

The meter stick tips if it is extended more than 30 cm over the edge.

$m_{Stick} = ?$ 50 g

30 cm

Answer: 75 g.

TIP If a rigid body is in static equilibrium, the sum of the torques about *any* axis of rotation is zero. It is often helpful in problem solving to place the axis at the place on the rigid body where the force you know least about is exerted. Then that force drops out of the second condition of equilibrium and you can use that equation to solve for some other unknown quantity.

Review Question 7.3 How do we choose the location of the axis of rotation when we are applying the conditions of equilibrium for a rigid body?

7.4 Center of mass

Many extended bodies are not rigid—the human body is a good example. A high jumper crossing the bar is often bent into an inverted U shape (**Figure 7.13**). Why does this shape allow her to jump higher? At the moment shown in the photo, her legs, arms, and head are below the bar as the trunk of the body passes over the top. As each part of the body passes over the bar, the rest of the body is at a lower elevation so that her center of mass is always slightly below the bar. The high jumper does not have to jump as high because she is able to reorganize her body's shape so that her center of mass passes *under* or at least not significantly over the bar.

Without realizing it, we change the position of our center of mass with respect to other parts of the body quite often. Try this experiment. Sit on a chair with your back straight and your feet on the floor in front of the chair (see **Figure 7.14a**). Without using your hands, try to stand up; you cannot. No matter how hard you try, you cannot raise yourself to standing from the chair if your back is vertical.

Why can't you stand? The center of mass of an average person when sitting upright is near the front of the abdomen. Consider what a force diagram for the experiment would look like (Figure 7.14b), assuming somehow you managed to lift yourself slightly off the chair—again keeping the back straight. Earth exerts a downward force at your center of mass, and the floor exerts an upward normal force on your feet. The torques caused by these two forces about any axis of rotation causes you to rotate back onto the chair.

To stand, you must tilt forward in the clockwise direction and move your feet back under the chair (Figure 7.14c). This shifts your center of mass

Figure 7.13 Where is the jumper's center of mass?

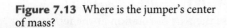

Figure 7.14 Getting out of a chair without using your hands.

(a)

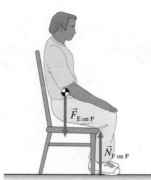

You sit on a chair with back straight and feet on the floor.

(b)

$\vec{F}_{E \, on \, P}$

$\vec{N}_{F \, on \, P}$

With your back straight as you lift yourself from the chair, the normal force exerted by the floor on your feet causes a counterclockwise torque about the center of mass. You fall backward.

(c)

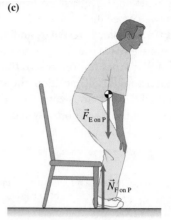

$\vec{F}_{E \, on \, P}$

$\vec{N}_{F \, on \, P}$

Bending forward so that your center of mass is in front of the floor's normal force causes a clockwise torque so that you can stand.

forward so that the downward gravitational force exerted on you by Earth is in front of the upward normal force exerted by the floor on your feet. Now, the torques caused by these two forces allow you to rotate forward and stand without touching the chair seat.

Calculating center of mass

How do we know that the center of mass of a sitting person is near the abdomen? In Section 7.1 we determined the location of an object's center of mass by investigating the directions along which one needs to push the object so it does not turn while being pushed on a flat smooth surface. When pushing in this way at different locations on the object, we found that lines drawn along the directions of these pushing forces all intersected at one point: the center of mass. This is a difficult and rather impractical way to find the center of mass of something like a human. Another method that we investigated consisted of balancing the object on a pointed support. This is also not very practical with respect to humans. Is there a way to predict where an object's center of mass is without pushing or balancing it? Our goal here is to develop a theoretical method that will allow us to determine the location of the center of mass of a complex object, such as the uniform seesaw with two apples shown in **Figure 7.15**. In the next example, we start with an object that consists of three other objects: two people of different masses and a uniform seesaw whose supporting fulcrum (point of support) can be moved (Figure 7.15). To determine the location of the center of mass of a system involving two people and a uniform beam, we find a place for the fulcrum to support the seesaw and the two people so that the system remains in static equilibrium.

Figure 7.15 Where is the center of mass of this seesaw?

Lighter apple   Heavier apple

EXAMPLE 7.4 Supporting a seesaw with two people

Find an expression for the position of the center of mass of a system that consists of a uniform seesaw of mass m_1 and two people of masses m_2 and m_3 sitting at the ends of the seesaw beam ($m_2 > m_3$).

Sketch and translate The figure on the next page shows a labeled sketch of the situation. The two people are represented as blocks. We choose the seesaw and two blocks as the system and construct a mathematical equation that lets us calculate the center of mass of that system. We place the x-axis along the seesaw with its

(continued)

origin at some arbitrary position on the left side of the seesaw. The center of mass of the seesaw beam is at x_1 and the two blocks rest at x_2 and x_3. At what *position x* should we place the fulcrum under the seesaw so that the system does not rotate—so that it remains in static equilibrium? At this position, the sum of all torques exerted on the system is zero. This position is the center of mass of the three-object system.

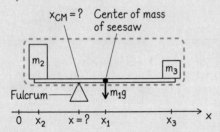

Where is the center of mass of the three-object system?

Simplify and diagram We model the seesaw as a rigid body and model each of the people blocks as point-like objects. Assume that the fulcrum does not exert a friction force on the seesaw. As you can see in the force diagram, we know the locations of all forces except the fulcrum force. We will calculate the unknown position x of the fulcrum so the seesaw with two people on it balances—so that it satisfies the second condition of equilibrium.

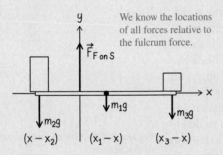

We know the locations of all forces relative to the fulcrum force.

Represent mathematically Apply the torque condition of equilibrium with the axis of rotation going through the unknown fulcrum position x. Then determine the torques around this axis produced by the forces exerted on the system. The gravitational force exerted by Earth on the center of mass of the seesaw has magnitude m_1g, is exerted a distance $x_1 - x$ from the axis of rotation, and has clockwise turning ability. The gravitational force exerted by Earth on block 2 has magnitude m_2g, is exerted a distance $x - x_2$ from the axis of rotation, and has counterclockwise turning ability. The gravitational force exerted by Earth on block 3 has magnitude m_3g, is exerted a distance $x_3 - x$ from the axis of rotation, and has clockwise turning ability. The torque condition of equilibrium for the system becomes

$$m_2g(x - x_2) - m_1g(x_1 - x) - m_3g(x_3 - x) = 0$$

Solve and evaluate Divide all terms of the equation by the gravitational constant g and collect all terms involving x on one side of the equation to get

$$m_2x + m_1x + m_3x = m_2x_2 + m_1x_1 + m_3x_3$$

or

$$x(m_2 + m_1 + m_3) = m_2x_2 + m_1x_1 + m_3x_3$$

Divide both sides of the equation by $(m_2 + m_1 + m_3)$ to obtain an expression for the location of the center of mass of the three-object system:

$$x = \frac{m_1x_1 + m_2x_2 + m_3x_3}{m_1 + m_2 + m_3}$$

Let's evaluate this result. The units of x are meters. Next check some limiting cases to see if the result makes sense. Imagine that there are no people sitting on the seesaw ($m_2 = m_3 = 0$). In this case, $x = \frac{m_1x_1}{m_1} = x_1$. The center of mass of the seesaw-only system is at the center of mass of the seesaw x_1, as it should be since we assumed its mass was uniformly distributed. Finally, if we increase the mass of one of the people on the seesaw, the location of the center of mass moves closer to that person.

Try it yourself: Where is the center of mass of a 3.0-kg, 2.0-m-long uniform beam with a 0.5-kg object on the right end and a 1.5-kg object on the left?

Answer: 0.8 m from the left end of the beam.

In Example 7.4 we arrived at an expression for the location of the center of mass of a three-object system where all objects were located along one straight line. We can apply the same method to a system whose masses are distributed in a two-dimensional x-y plane. For such a two-dimensional system, we get the following:

Center of mass (quantitative definition) If we consider an object as consisting of parts 1, 2, 3, . . . n whose centers of masses are located at the coordinates $(x_1, y_1); (x_2, y_2); (x_3, y_3); \ldots (x_n, y_n)$, then the center of mass of this whole object is at the following coordinates:

$$x_{cm} = \frac{m_1 x_1 + m_2 x_2 + m_3 x_3 + \ldots + m_n x_n}{m_1 + m_2 + m_3 + \ldots + m_n}$$

$$y_{cm} = \frac{m_1 y_1 + m_2 y_2 + m_3 y_3 + \ldots + m_n y_n}{m_1 + m_2 + m_3 + \ldots + m_n} \qquad (7.4)$$

Using the above equation for an object with a continuous mass distribution is difficult and involves calculus. For example, suppose you wanted to determine the x_{cm} and y_{cm} center-of-mass locations of the leg depicted in **Figure 7.16**. You would need to subdivide the leg into many tiny sections and insert the mass and position of each section into Eq. (7.4), and then add all the terms in the numerator and denominator together to determine the center of mass of the leg. For example, section 7 would contribute a term $m_7 x_7$ in the numerator of the x_{cm} equation and a term m_7 in the denominator. Usually, you will be given the location of the center of mass of such continuous mass distributions.

Knowledge of the center of mass helps you answer many questions: Why if you walk with a heavy backpack do you fall more easily? Why does a baseball bat have an elongated, uneven shape? Why do ships carry heavy loads in the bottom rather than near the top of the ship? Or why does a person lifting a barbell tilt backward (as in **Figure 7.17**)?

Mass distribution and center of mass

The term "center of mass" is deceiving. It might make you think that the center of mass of an object is located at a place where there is an equal amount of mass on each side. However, this is not the case. Consider again a uniform seesaw (see **Figure 7.18a**). Imagine that the mass of the seesaw is 10 kg, the length is 4.0 m, the mass of the person on the left end (m_1) is 70 kg, and the mass of the person on the right end (m_3) is 30 kg. Where is the center of mass of this two-person seesaw system and how much mass is on the left side and the right side of the center of mass?

To find the center of mass, we need a coordinate system with an origin. The origin can be anywhere. We put it at the location of the more massive person on the left side. The center of mass is then

$$x_{cm} = \frac{m_1 x_1 + m_2 x_2 + m_3 x_3}{m_1 + m_2 + m_3} = \frac{(70\ kg \cdot 0\ m) + (10\ kg \cdot 2.0\ m) + (30\ kg \cdot 4.0\ m)}{70\ kg + 10\ kg + 30\ kg}$$

$$= 1.3\ m$$

Figure 7.16 Finding the center of mass of a continuous mass distribution.

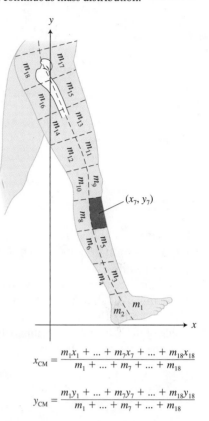

$$x_{CM} = \frac{m_1 x_1 + \ldots + m_7 x_7 + \ldots + m_{18} x_{18}}{m_1 + \ldots + m_7 + \ldots + m_{18}}$$

$$y_{CM} = \frac{m_1 y_1 + \ldots + m_7 y_7 + \ldots + m_{18} y_{18}}{m_1 + \ldots + m_7 + \ldots + m_{18}}$$

Figure 7.17 Why is the weight lifter tilted backward?

Figure 7.18 The masses on each side of a system's center of mass are unequal.

(a)

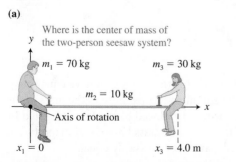

Where is the center of mass of the two-person seesaw system?

$m_1 = 70$ kg $m_3 = 30$ kg

$m_2 = 10$ kg

Axis of rotation

$x_1 = 0$ $x_3 = 4.0$ m

(b)

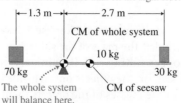

Note that there is more mass on the left side of the center of mass than on the right side.

1.3 m   2.7 m

CM of whole system

10 kg

70 kg   30 kg

CM of seesaw

The whole system will balance here.

The seesaw would balance about a fulcrum located 1.3 m from m_1 and 2.7 m from m_2 (Figure 7.18b). We see that the masses are not equal. The mass on the left side of the center of mass is much greater than on the right side—70 kg versus 40 kg. The larger mass on the left is a shorter distance from the center of mass than the smaller masses on the right, which on average are farther from the center of mass. In other words, the masses on the right and on the left of the center of mass are not equal! However, the product of mass and distance on each side balances out, causing torques of equal magnitude. We could rename the center of mass as "the center of torque" to reflect the essence of the concept, but since this is not the term used in physics, we will continue to use the term center of mass.

Let's use another example to test this idea that the mass on the left and on the right side of the center of mass is not necessarily equal.

CONCEPTUAL EXERCISE 7.5 Balancing a bread knife

You balance a bread knife by laying the flat side across one finger, as shown below. Where is the center of mass of the knife? How does the mass of the knife on the left side of the balance point compare to the mass of the knife on the right side of the balance point?

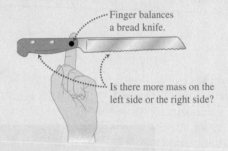
Finger balances a bread knife.

Is there more mass on the left side or the right side?

Sketch and translate We choose the knife as the system and orient a vertical y-axis upward.

Simplify and diagram Two objects exert forces on the knife: the finger exerts an upward normal force and Earth exerts a downward gravitational force. Since the knife is rotationally stable, these two forces must both pass through the axis of rotation at the location of the finger. If not, there would be an unbalanced torque exerted on the knife and it would tip. You can assume that the gravitational force is exerted at the knife's center of mass. Thus, the center of mass must be directly above the finger.

y

$\vec{N}_{F\,on\,K}$

$\vec{F}_{E\,on\,K}$

To determine how the mass on the right side of the balance point compares to the mass on the left side, we model the knife as two small spheres of mass m_1 and mass m_2, connected by a massless rod. The spheres are located at the center of mass of the respective sides of the knife. When the knife is balanced, there is less distance between sphere 1 and the balance point than between sphere 2 and the balance point. Thus, the mass of the handle end must be greater than the mass of the cutting end.

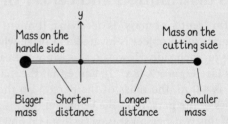

Mass on the handle side
Mass on the cutting side

Bigger mass   Shorter distance   Longer distance   Smaller mass

Try it yourself: A barbell has a 10-kg plate on one end and a 5-kg plate on the other end. Where is the center of mass of the barbell, which is also the balance point for the barbell? Ignore the mass of the 1.0-m-long rod that connects the plates on the ends.

Answer: The center of mass is 0.33 m from the 10-kg end and 0.67 m from the 5-kg end.

In Conceptual Exercise 7.5 we saw that for the torques to have equal magnitudes, the mass of the shorter handle end must be greater than the mass of the longer cutting end. In summary, the torques produced by the forces exerted by Earth on the objects on each side of the center of mass have equal magnitudes, but the masses of the objects themselves are not necessarily equal.

7.5 Skills for analyzing situations using equilibrium conditions

We often use the equations of equilibrium to determine one or two unknown forces if all other forces exerted on an object of interest are known. Consider the muscles of your arm when you lift a heavy ball or push down on a desktop (**Figure 7.19**). When you hold a ball in your hand, your biceps muscle tenses and pulls up on your forearm in front of the elbow joint. When you push down with your hand on a desk, your triceps muscle tenses and pulls up on a protrusion of the forearm behind the elbow joint. The equations of equilibrium allow you to estimate these muscle tension forces—see the next example, which describes a general method for analyzing static equilibrium problems. The right side of the table applies the general strategies to the specific problem provided.

< Active Learning Guide

Figure 7.19 Muscles in the upper arm lift and push down on the forearm.

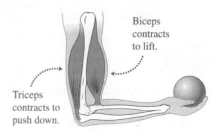

Biceps contracts to lift.

Triceps contracts to push down.

PROBLEM-SOLVING STRATEGY **Applying Static Equilibrium Conditions**

EXAMPLE 7.6 **Use the biceps muscle to lift**
Imagine that you hold a 6.0-kg lead ball in your hand with your arm bent. The ball is 0.35 m from the elbow joint. The biceps muscle attaches to the forearm 0.050 m from the elbow joint and exerts a force on the forearm that allows it to support the ball. The center of mass of the 12-N forearm is 0.16 m from the elbow joint. Estimate the magnitude of (a) the force that the biceps muscle exerts on the forearm and (b) the force that the upper arm exerts on the forearm at the elbow.

Sketch and translate

- Construct a labeled sketch of the situation. Include coordinate axes and choose an axis of rotation.
- Choose a system for analysis.

We choose the axis of rotation to be where the upper arm bone (the humerus) presses on the forearm at the elbow joint. This will eliminate from the torque equilibrium equation the unknown force that the upper arm exerts on the forearm.

    We choose the system of interest to be the forearm and hand.

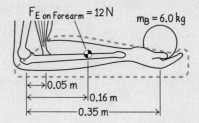

$F_{\text{E on Forearm}} = 12\,\text{N}$ $m_B = 6.0\,\text{kg}$

0.05 m
0.16 m
0.35 m

(continued)

Simplify and diagram

- Decide whether you will model the system as a rigid body or as a point-like object.

- Construct a force diagram for the system. Include the chosen coordinate system and the axis of rotation (the origin of the coordinate system).

Model the system as a rigid body and draw a force diagram for the forearm and hand.

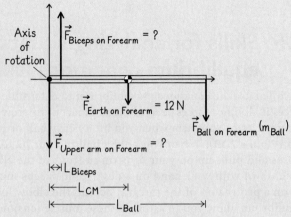

Represent mathematically

- Use the force diagram to apply the conditions of equilibrium.

$$\Sigma\tau = 0$$

$$(F_{\text{UA on FA}})(0) + (F_{\text{Biceps on FA}})(L_{\text{Biceps}}\sin 90°) + (-F_{\text{E on FA}}(L_{\text{CM}}\sin 90°))$$
$$+ (-F_{\text{Ball on FA}}(L_{\text{Ball}}\sin 90°)) = 0$$

$$\Sigma F_y = 0$$

$$(-F_{\text{UA on FA}}) + F_{\text{Biceps on FA}} + (-F_{\text{E on FA}}) + (-F_{\text{Ball on FA}}) = 0$$

Solve and evaluate

- Solve the equations for the quantities of interest.

- Evaluate the results. Check to see if their magnitudes are reasonable and if they have the correct signs and units. Also see if they have the expected values in limiting cases.

Substitute $\sin 90° = 1.0$ and rearrange the torque equation to find $F_{\text{Biceps on FA}}$.

$$F_{\text{Biceps on FA}} = [(F_{\text{E on FA}})(L_{\text{CM}}) + (F_{\text{Ball on FA}})(L_{\text{Ball}})]/L_{\text{Biceps}}$$
$$= [(12\,\text{N})(0.16\,\text{m}) + (59\,\text{N})(0.35\,\text{m})]/(0.050\,\text{m})$$
$$= 450\,\text{N}$$

Use the force equation to find $F_{\text{UA on FA}}$:

$$F_{\text{UA on FA}} = F_{\text{Biceps on FA}} - F_{\text{E on FA}} - F_{\text{Ball on FA}}$$
$$= 450\,\text{N} - 12\,\text{N} - 59\,\text{N} = 380\,\text{N}$$

The 450-N force exerted by the biceps on the forearm is much greater than the 59-N force exerted by the ball on the forearm. This difference occurs because the force exerted by the biceps is applied much closer to the axis of rotation than the force exerted by the lead ball.

If the center of mass of the forearm were farther from the elbow, the biceps would have to exert an even larger force.

Try it yourself: How would the force exerted by the biceps on the forearm change if the biceps were attached to the forearm farther from the elbow?

Answer: A longer L_{Biceps} in the torque equilibrium equation would mean that the biceps muscle would need to exert a smaller force on the forearm when lifting something.

Applying the conditions of equilibrium in more complex situations

In the previous example the forces exerted on the system were exerted at right angles to the system. What if this is not the case? Consider the next example.

EXAMPLE 7.7 Lifting a drawbridge

A drawbridge across the mouth of an inlet on the coastal highway is lifted by a cable to allow sailboats to enter the inlet. You are driving across the 16-m-long drawbridge when the bridge attendant accidentally activates the bridge. You abruptly stop the car 4.0 m from the end of the bridge. The cable makes a 53° angle with the horizontal bridge. The mass of your car is 1000 kg and the mass of the bridge is 4000 kg. Estimate the tension force that the cable exerts on the bridge as it slowly starts to lift the bridge.

Sketch and translate We sketch the situation below and choose the bridge as the system. We place the axis of rotation where the drawbridge connects by a hinge to the roadway at the left side of the bridge—a good choice, as we have no information about that force.

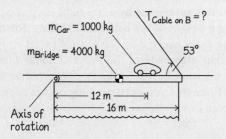

Simplify and diagram We model the car as a point-like object and the bridge as a rigid body with uniform mass distribution. The latter assumption means that Earth exerts a gravitational force on the center of the bridge. The bridge has just started to rise, so it is still approximately horizontal. Since it moves very slowly, we will assume that it is in static equilibrium. As we can see from the force diagram, four objects exert forces on the bridge. (1) The hinges on the left side exert a force $\vec{F}_{\text{H on B}}$ that is unknown in magnitude and direction. (2) Earth exerts a $(4000\,\text{kg})(9.8\,\text{N/kg}) = 39{,}200\,\text{N}$ gravitational force $\vec{F}_{\text{E on B}}$ on the center of the bridge. (3) The car pushes down on the bridge, exerting a $(1000\,\text{kg})(9.8\,\text{N/kg}) = 9800\,\text{N}$ force $\vec{F}_{\text{Car on B}}$ 4.0 m from the right side of the bridge. (4) The cable exerts an unknown force $\vec{T}_{\text{Cable on B}}$ on the right edge of the bridge at a 53° angle above the horizontal.

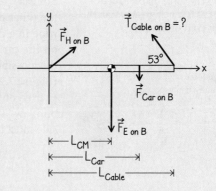

Represent mathematically Since four objects exert forces on the bridge, the torque condition of equilibrium will include four torques produced by these forces:

$$F_{\text{H on B}}(0) + (-F_{\text{E on B}}L_{\text{CM}} \sin 90°) + (-F_{\text{Car on B}} L_{\text{Car}} \sin 90°)$$
$$+ T_{\text{Cable on B}} L_{\text{Cable}} \sin 53° = 0.$$

Substitute $\sin 90° = 1.0$ and $\sin 53° = 0.80$ and rearrange the above to find an expression to determine the unknown tension force that the cable exerts on the bridge:

$$T_{\text{Cable on B}} = \frac{F_{\text{E on B}}L_{\text{CM}} + F_{\text{Car on B}}L_{\text{Car}}}{L_{\text{Cable}} \sin 53°}$$

Solve and evaluate Substitute the following values into the above equation: $L_{\text{CM}} = 8.0\,\text{m}$, $L_{\text{Car}} = 12.0\,\text{m}$, $L_{\text{Cable}} = 16.0\,\text{m}$, $F_{\text{E on B}} = 3.92 \times 10^4\,\text{N}$, and $F_{\text{Car on B}} = 9800\,\text{N}$. This yields

$$T_{\text{Cable on B}} = \frac{(3.92 \times 10^4\,\text{N})(8.0\,\text{m}) + (9800\,\text{N})(12.0\,\text{m})}{(16\,\text{m})\sin 53°}$$
$$= 34{,}000\,\text{N}$$

The unit is correct. The value of 34,000 N is reasonable given that the bridge holds the 9800-N car near the free end and that Earth exerts a force of 39,000 N on the bridge in its middle at its center of mass.

Try it yourself: What force would the cable have to exert on the bridge if your car was 4.0 m from the hinged end of the bridge instead of 4.0 m from the free end?

Answer: 28,000 N. This makes sense since the car is exerting a smaller torque when it is closer to the axis of rotation.

"Magnifying" a force

Our knowledge of equilibrium conditions allows us to understand how one can increase or decrease the turning ability of a force by exerting the force in a different location or in a different direction. Consider a situation when you need to get a car out of a rut in snow or mud. You know from experience that it is

Figure 7.20 The force that the rope exerts on the car is much greater than the force that you exert on the rope.

(a)

To exert a large force on a stuck car, push from the side on a tautly tied rope.

(b)

The force you exert on the rope is much less than the tension force exerted by the rope.

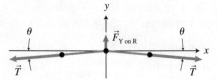

not easy. One way to do it is tie a rope to the front of the car and *tightly* wrap the other end around a tree (see **Figure 7.20a**). Then, push the middle of the rope in a direction perpendicular to the rope. The force that the rope exerts on the car is much greater than the force that you exert on the rope. How is this possible? Draw a force diagram for the short section of rope that you push (Figure 7.20b). For that small section of rope to remain stationary, the forces exerted on it must balance. Let's apply the y-component form of Newton's second law. The sum of the y-components of the tension forces pulling on each side must balance the force you exert on the rope:

$$-2T \sin \theta + (+F_{Y\text{ on R}}) = 0$$

$$\Rightarrow T = \frac{F_{Y\text{ on R}}}{2 \sin \theta}$$

If the angle θ of the rope's deflection is small (that's why you need the rope to be tight at the beginning), then $\sin \theta$ is also small and the rope tension T will be large.

Magnifying a force can have advantages or disadvantages, depending on the situation. For example, when you wear a backpack, the shoulder straps often rest on the trapezius muscles, which run across the top of your shoulders to your neck. If the backpack is not supported by a hip belt, each strap has to support approximately half the weight of the backpack, pulling down on the trapezius muscle. The tension force exerted by the muscle on its connection points at each end is somewhat greater than the perpendicular downward push that the strap exerts on the muscle, just as the force exerted by the rope on the car is greater than the force that you exert pushing on the rope. Carrying a heavy backpack can lead to injury.

EXAMPLE 7.8   The impact of carrying a heavy backpack

Assume that you are carrying a backpack with several books in it for a combined mass of 10 kg (Earth exerts about a 100-N gravitational force on it). This means that each of the two straps pulling on the trapezius muscle exerts a force of about 50 N on the muscle. This causes the muscle to deflect about 6° from the horizontal on each side of the strap. Estimate the force that the trapezius muscle exerts on its connecting points on the neck and shoulder (similar to the connections of the rope to the tree and the car in the previous example).

Sketch and translate Our sketch of the situation is shown below. Choose the section of muscle under the strap as the system of interest.

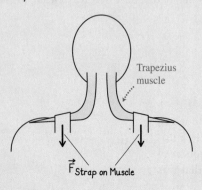

Simplify and diagram The force diagram shows the pull of the muscle tissue at an angle of 6° above the horizontal and the 50-N downward force exerted by the strap on that section of muscle.

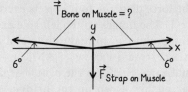

Represent mathematically We can apply the same equation we used to get the car out of the rut:

$$T_{\text{Bone on Muscle}} = \frac{F_{\text{Strap on Muscle}}}{2 \sin \theta}$$

Solve and evaluate Substituting the known information in the above equation, we get

$$T_{\text{Bone on Muscle}} = \frac{50 \text{ N}}{2 \sin 6°} = 240 \text{ N (or 54 lb)}$$

This force is almost 2.5 times larger than the force exerted by Earth on the backpack.

Try it yourself: A 70-kg tightrope walker stands in the middle of a tightrope that deflects upward 5.0° on each side of where he stands. Determine the force that each half of the rope exerts on a short section of rope beneath the walker's feet.

Answer: 3900 N.

Review Question 7.5 Earth exerts a 100-N force on a person's backpack. This force results in the person's trapezius muscle exerting a 240-N force on the bones it attaches to. How is this possible?

7.6 Stability of equilibrium

Often, objects can remain in equilibrium for a long time interval—you can sit comfortably for a long time on a living room couch without tipping. But sometimes equilibrium is achieved for only a short time interval—think of sitting on a chair and tilting it backward too far onto its rear legs.

Equilibrium and tipping objects

You have probably observed that it is easier to balance and avoid falling while standing in a moving bus or subway train if you spread your feet apart in the direction of motion. By doing this you are increasing the **area of support**, the area of contact between the object and the surface it is supported by. To understand area of support, in Observational Experiment **Table 7.5** we model a person riding the subway as a rigid body and consider the torques produced by the forces exerted by the floor on her feet.

OBSERVATIONAL EXPERIMENT TABLE

7.5   Stability of equilibrium.

Observational experiment	Analysis
Stationary train with respect to the observer on the platform: (a) Person A stands with feet together. (b) Person B stands with feet apart. You, the observer, are on the ground watching the train.	*Stationary train:* The forces and torques balance about an axis of rotation between their feet.

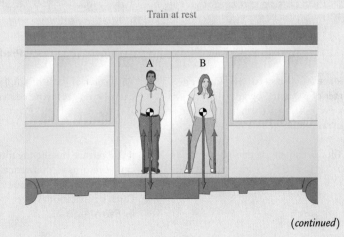

Train at rest

(continued)

Observational experiment	Analysis
Train accelerates to right with respect to the observer on the platform: Both people tend to remain stationary with respect to the platform (Newton's first law) and the train moves out from under them. They both tilt to the left. Consider the torques around an axis of rotation through the left foot of each person (as seen by the platform observer).	*Accelerating train:* (a) The force exerted by Earth on A has counterclockwise turning ability and causes A to tip over. 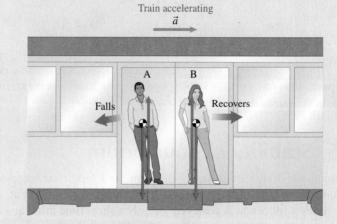 (b) The force exerted by Earth on B has clockwise turning ability and causes B to recover from the tilt without falling.

Patterns

Standing on stationary train: The net torque exerted on both people about any point is zero.

Standing on accelerating train:

PERSON A: With the feet together, the gravitational force exerts a counterclockwise torque and the person falls. The gravitational force points outside the area of support provided by the feet.

PERSON B: With the feet apart, the gravitational force exerts a clockwise torque and the person recovers. Note that the gravitational force points between the feet.

The patterns we observed in Observational Experiment Table 7.5 lead us to a tentative rule about tipping:

> For an object in static equilibrium, if a vertical line passing through the object's center of mass is within the object's area of support, the object does not tip. If the line is not within the area of support, the object tips.

If this is a general rule, then we can use it to predict the angle at which an object with a known center of mass will tip over (see Testing Experiment **Table 7.6**).

TESTING EXPERIMENT TABLE

7.6   Testing our tentative rule about tipping.

VIDEO 7.6

Testing experiment	Prediction	Outcome
Place a full soda can (typical diameter of 6 cm and height of 12 cm) on a flat but rough surface. Its center of mass is at its geometric center.	1. The center of mass of a full soda can is at its geometric center.	
Tilt the can a little and release it.	2. If you release the slightly tilted can, it returns to the vertical position because of the torque due to Earth's gravitational force exerted by Earth.	The can returns to the vertical position.

| Tilt the can at larger and larger angles. Predict the angle at which the can will tip. | 3. If you tilt the can more and more, eventually the line defined by the gravitational force passes over the support point for the can at its bottom edge. Tilting the can more than this critical angle causes it to tip over. 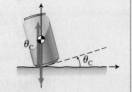
 4. *Prediction:* For a can with a diameter of 6 cm and a height of 12 cm, this angle will be $\theta_C = \tan^{-1}(6\ \text{cm}/12\ \text{cm}) = 27°$ | The outcome matches the prediction. |
| Repeat the experiment, but this time use an open can of corn syrup with the same mass as the soda. The can will only be about half full. Predict the angle at which the can will tip. | The center of mass should now be about one-fourth of the way from the bottom (if we neglect the mass of the can). The critical tilt angle should now be $\theta_C = \tan^{-1}(6\ \text{cm}/6\ \text{cm}) = 45°$ 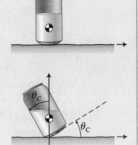 | The outcome matches the prediction within experimental uncertainty (about 1°). |

Conclusion

The outcomes are consistent with the predictions. We've increased our confidence in the tipping rule: For an object to be in static equilibrium, the line defined by the gravitational force exerted by Earth must pass within the object's area of support. If it is not within the area of support, the object tips over.

We now know that an object will tip if it is tilted so that the gravitational force passes beyond its area of support. If the area of support is large or if the center of mass is closer to the ground, more tipping is possible without the object falling over—it is more stable. This idea is regularly used in building construction. Tall towers (like the Eiffel Tower) have a wide bottom and a narrower top. The Leaning Tower of Pisa does not tip because a vertical line through its center of mass passes within the area of support (**Figure 7.21**). The same rule explains why you need to keep your feet apart when standing on a subway train entering or leaving a station (Observational Experiment Table 7.5).

> **The equilibrium of a system is stable against tipping if the vertical line through its center of mass passes through the system's area of support.**

Equilibrium and rotating objects

Objects that can rotate around a fixed axis can also have either stable or unstable equilibria. Consider a ruler with several holes in it. If you hang the ruler on a nail using a hole near one end, it hangs as shown in **Figure 7.22a**. If you pull the bottom of the ruler to the side and release it, the ruler swings back and forth with decreasing maximum displacement from the equilibrium position, but eventually hangs straight down. This equilibrium position is called *stable* because the ruler always tries to return to that position if free to rotate. However, if you turn the ruler 180° so that the axis of rotation is at the bottom of the ruler (see Figure 7.22b), it can stay in this position only if very carefully balanced. If disturbed, the ruler swings down and never returns. In this case the ruler is in *unstable* equilibrium.

Figure 7.21 Because $F_{\text{E on Tower}}$ passes through the base of the tower, the Leaning Tower of Pisa does not tip over.

$F_{\text{E on Tower}}$

Figure 7.22 Stable, unstable, and neutral equilibrium.

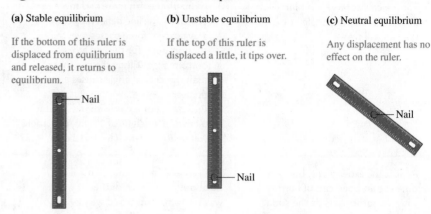

(a) Stable equilibrium

If the bottom of this ruler is displaced from equilibrium and released, it returns to equilibrium.

Nail

(b) Unstable equilibrium

If the top of this ruler is displaced a little, it tips over.

Nail

(c) Neutral equilibrium

Any displacement has no effect on the ruler.

Nail

The stability or lack of stability can be understood by considering the torque around the axis of rotation due to the force exerted by Earth on the object. In both positions a and b, the net force exerted on the ruler by Earth and by the nail (a normal force) is zero. The difference is that in the first case the center of mass is *below the axis of rotation*; in the second case it is *above the axis of rotation*. Torques produced by gravitational forces tend to lower the center of mass of objects. In other words, if it is possible for the object to rotate so that its center of mass becomes lower, it will tend to do so.

If we hang the ruler using the center hole (Figure 7.22c), it remains in whatever position we leave it. Both the normal force exerted by the nail and the gravitational force exerted by Earth produce zero torques. This is called *neutral* equilibrium.

> The equilibrium of a system is stable against rotation if the center of mass of the rotating object is below the axis of rotation.

CONCEPTUAL EXERCISE 7.9   Balancing a pencil

Is it possible to balance the pointed tip of a pencil on your finger?

Sketch and translate The figure to the right shows a sketch of the situation.

Simplify and diagram The forces exerted on the pencil are shown below. The tip of the pencil is the axis of rotation. When the pencil is tilted by only a small angle, the line defined by the gravitational force exerted by Earth on the pencil is not within the area of support of the pencil (which is just the pencil tip). This equilibrium is unstable. Having the center of mass *above* the axis of rotation leads to this instability. To make it stable, we need to lower the center of mass to below the axis of rotation. We can do it by attaching a pocketknife to the pencil, as shown below. Notice that the massive part of the knife is below the tip of the pencil and so is

the center of mass of the system. Then when the pencil tilts, the torque due to the gravitational force brings it back to the equilibrium position.

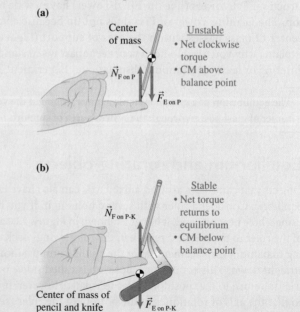

Axis of rotation

(a)

Center of mass

$\vec{N}_{\text{F on P}}$

$\vec{F}_{\text{E on P}}$

Unstable
• Net clockwise torque
• CM above balance point

(b)

$\vec{N}_{\text{F on P-K}}$

Stable
• Net torque returns to equilibrium
• CM below balance point

Center of mass of pencil and knife

$\vec{F}_{\text{E on P-K}}$

TIP Always try to understand new situations in terms of ideas we have already discussed. The trick of balancing the pencil can be understood with the rules we expressed above: (1) the equilibrium is most stable when the center of mass of the system is in the lowest possible position or, equivalently, (2) when the gravitational potential energy of the system has the smallest value.

The rules we have learned about equilibrium and stability have many applications, including circus tricks. Think about where the center of mass is located for the bicycle and the two people shown in **Figure 7.23**. Another application involves vending machines.

Center of mass is below balance point. Bicycle will not tip.

Figure 7.23 Balancing a bicycle on a high wire may not be as dangerous as it looks.

EXAMPLE 7.10 Tipping a vending machine

According to the U.S. Consumer Product Safety Commission, tipped vending machines caused 37 fatalities between 1978 and 1995 (2.2 deaths per year). Why is tipping vending machines so dangerous? A typical vending machine is 1.83 m high, 0.84 m deep, and 0.94 m wide and has a mass of 374 kg. It is supported by a leg on each of the four corners of its base.
(a) Determine the horizontal pushing force you need to exert on its front surface 1.5 m above the floor in order to just lift its front feet off the surface (so that it will be supported completely by its back two feet).
(b) At what critical angle would it fall forward?

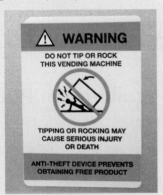

Sketch and translate (a) See the sketch of the situation below. The axis of rotation will go through the back support legs of the vending machine.

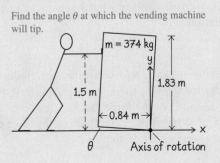

Simplify and diagram Model the vending machine as a rigid body. Three objects exert forces on the vending machine (shown in the side view force diagram below): the person exerting normal force $\vec{N}_{P \text{ on } M}$ on the machine, Earth exerting gravitational force $\vec{F}_{E \text{ on } M}$, and the floor exerting normal force $\vec{N}_{F \text{ on } M}$ and static friction force $\vec{f}_{s \text{ F on } M}$ on the machine.

Represent mathematically (a) We use the torque condition of equilibrium to analyze the force needed to tilt the vending machine until the front legs are just barely off the floor. The normal force exerted by the person has a clockwise turning ability while the gravitational force has counterclockwise turning ability. The force exerted by the floor on the back legs does not produce a torque, since it is exerted at the axis of rotation.

$$-N_{P \text{ on } M}L_P \sin \theta_P + F_{E \text{ on } M}L_E \sin \theta_E + N_{F \text{ on } M}(0)$$
$$+ f_{s \text{ F on } M}(0) = 0$$

Tilt not shown since vending machine is just barely off the floor.

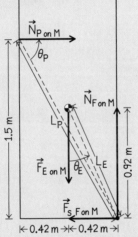

(continued)

(b) We can apply the analysis in Testing Experiment Table 7.6 for this situation:

$$\theta_C = \tan^{-1}\left(\frac{\text{depth}}{\text{height}}\right)$$

Solve and evaluate (a) Using the torque equation, we can find the normal force that the person needs to exert on the vending machine to just barely lift its front off the floor:

$$-N_{\text{P on M}}L_P\left(\frac{1.5 \text{ m}}{L_P}\right) + (3700 \text{ N})L_E\left(\frac{0.42 \text{ m}}{L_E}\right)$$
$$+ N_{\text{F on M}}(0) + f_{\text{s F on M}}(0) = 0$$

or $N_{\text{P on M}} = 1000$ N or 220 lb.

(b) We find that the critical tipping angle is

$$\theta_{\text{tipping}} = \tan^{-1}\left(\frac{\text{depth}}{\text{height}}\right) = \tan^{-1}\left(\frac{0.42 \text{ m}}{0.92 \text{ m}}\right) = 25°$$

Both answers seem reasonable.

Try it yourself: Determine how hard you need to push against the vending machine to keep it tilted at a 25° angle above the horizontal.

Answer: 0 N. A vertical line through the center of mass passes through the axis of rotation, so that the net torque about that axis is zero. However, this equilibrium is unstable and represents a dangerous situation.

The chance of being injured by a tipped vending machine is small since a large force must be exerted on it to tilt it up, and it must be tilted at a fairly large angle before it reaches an unstable equilibrium. A more common danger is falling bookcases in regions subject to earthquakes. The base of a typical 2.5-m-tall bookcase is less than 0.3 m deep. The shelves above the base that are filled with books are the same size as the base. If tilted by just

$$\theta_C = \tan^{-1}\left(\frac{\text{depth}}{\text{height}}\right) \approx \tan^{-1}\left(\frac{0.15 \text{ m}}{1.25 \text{ m}}\right) \approx 7°$$

the bookcase can tip over. In earthquake-prone regions, people often attach a bracket to the top back of the bookcase and then anchor it to the wall.

Review Question 7.6 Why is a ball hanging by a thread in stable equilibrium, while a pencil balanced on its tip is in unstable equilibrium?

7.7 Static equilibrium: Putting it all together

In this section we'll apply our understanding of static equilibrium to analyze three common situations: standing on your toes, lifting a heavy object, and safely climbing a ladder.

Standing on your toes

Most injuries to the Achilles tendon occur during abrupt movement, such as jumps and lunges. However, we will analyze what happens to your Achilles tendon in a less stressed situation.

EXAMPLE 7.11 **Standing with slightly elevated heel**

Suppose you stand on your toes with your heel slightly off the ground. In order to do this, the larger of the two lower leg bones (the tibia) exerts a force on the ankle joint where it contacts the foot. The Achilles tendon simultaneously exerts a force on the heel, pulling up on it in order for the foot to be in static equilibrium. What is the magnitude of the force that the tibia exerts on the ankle joint? What is the magnitude of the force that the Achilles tendon exerts on the heel?

Sketch and translate First we sketch the foot with the Achilles tendon and the tibia. We choose the foot as the system of interest. Three forces are exerted on the foot: the tibia is pushing down on the foot at the ankle joint; the floor is pushing up on the ball of the foot and the toes; and the Achilles tendon is pulling up on the heel. We choose the axis of rotation as the place where the tibia presses against the foot.

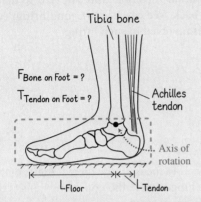

Simplify and diagram Model the foot as a very light rigid body. The problem says that the foot is barely off the ground, so we will neglect the angle between the foot and the ground and consider the foot horizontal. The gravitational force exerted on the foot by Earth is quite small compared with the other forces that are being exerted on it, so we will ignore it. A force diagram for the foot is shown below. When you are standing on the ball and toes of both feet, the floor exerts an upward force on each foot equal to half the magnitude of the gravitational force that Earth exerts on your entire body: $F_{\text{Floor on Foot}} = \dfrac{m_{\text{Body}}g}{2}$. The Achilles tendon pulls up on the heel of the foot, exerting a force $T_{\text{Tendon on Foot}}$. The tibia bone in the lower leg pushes down on the ankle joint exerting a force $F_{\text{Tibia on Foot}}$.

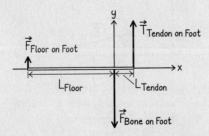

Represent mathematically Let's apply the conditions of equilibrium to this system. Note that the distance from the toes to the joint L_{Toe} is somewhat longer than the

distance from the joint to the Achilles tendon attachment point L_{Tendon}. The torque condition of equilibrium becomes

$$+\left[T_{\text{Tendon on Foot}}(L_{\text{Tendon}})\right] + F_{\text{Bone on Foot}}(0)$$
$$-\frac{m_{\text{Body}}g}{2}(L_{\text{Toe}}) = 0$$
$$\Rightarrow T_{\text{Tendon on Foot}} = \frac{m_{\text{Body}}g}{2}\left(\frac{L_{\text{Toe}}}{L_{\text{Tendon}}}\right)$$

Now, apply the y-scalar component of the force condition of equilibrium:

$$\Sigma F_y = T_{\text{Tendon on Foot}} + (-F_{\text{Bone on Foot}}) + \frac{m_{\text{Body}}g}{2} = 0$$

$$\Rightarrow F_{\text{Bone on Foot}} = T_{\text{Tendon on Foot}} + \frac{m_{\text{Body}}g}{2}$$

Solve and evaluate The distance from the place where the bone contacts the foot to where the floor contacts the foot is about 5 times longer than the distance from the bone to where the tendon contacts the foot. Consequently, the force that the Achilles tendon exerts on the foot is about

$$T_{\text{Tendon on Foot}} = \frac{m_{\text{Body}}g}{2}\left(\frac{L_{\text{Toe}}}{L_{\text{Tendon}}}\right) = \frac{m_{\text{Body}}g}{2}(5) = \frac{5}{2}m_{\text{Body}}g$$

or two and a half times the gravitational force that Earth exerts on the body. Using $g = 10$ N/kg for a 70-kg person, this force will be about 1750 N. That's a very large force for something as simple as standing with your heel slightly elevated! The force exerted on the joint by the leg bone would be

$$F_{\text{Bone on Foot}} = T_{\text{Tendon on Foot}} + \left(\frac{m_{\text{Body}}g}{2}\right)$$
$$= 1750 \text{ N} + 350 \text{ N} = 2100 \text{ N}$$

This force is three times the weight of the person! The forces are much greater when moving. Thus, every time you lift your foot to walk, run, or jump, the tendon tension and joint compression are several times greater than the gravitational force that Earth exerts on your entire body.

Try it yourself: Estimate the increase in the magnitude of the force exerted by the Achilles tendon on the foot of the person in this example if his mass were 90 kg instead of 70 kg.

Answer: An increase of at least 500 N (110 lb).

Figure 7.24 A bad way to lift. Lifting in this position causes considerable back muscle tension and disk compression in the lower back.

> Active Learning Guide >

Lifting from a bent position

Back problems often originate with improper lifting techniques—a person bends over at the waist and reaches to the ground to pick up a box or a barbell. In **Figure 7.24**, the barbell pulls down on the woman's arms far from the axis of rotation of her upper body about her hip area. This downward pull causes a large clockwise torque on her upper body. To prevent her from tipping over, her back muscles must exert a huge force on the backbone, thus producing an opposing counterclockwise torque. This force exerted by the back muscles compresses the disks that separate vertebrae and can lead to damage of the disks, especially in the lower back. We can use the equilibrium equations to estimate the forces and torques involved in such lifting.

EXAMPLE 7.12 Lifting incorrectly from a bent position

Estimate the magnitude of the force that the back muscle in the woman's back in Figure 7.24 exerts on her backbone and the force that her backbone exerts on the disks in her lower back when she lifts an 18-kg barbell. The woman's mass is 55 kg. Model the woman's upper body as a rigid body.

- The back muscle attaches two-thirds of the way from the bottom of her $l = 0.60$-m-long backbone and makes a 12° angle relative to the horizontal backbone.
- The mass of her upper body is $M = 33$ kg centered at the middle of the backbone and has uniform mass distribution. The axis of rotation is at the left end of the backbone and represents one of the disks in the lower back.

Sketch and translate The figure below is our mechanical model of a person lifting a barbell. We want to estimate the magnitudes of the force $T_{\text{M on B}}$ that the back muscle exerts on the backbone and the force $F_{\text{D on B}}$ that the disk in the lower back exerts on the backbone. The force that the disk exerts on the bone is equal in magnitude to the force exerted by the backbone on the disk. The upper body (including the backbone) is the system of interest, but we consider the back muscle to be external to the system

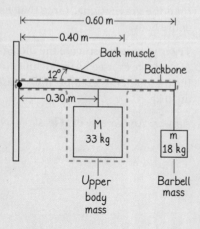

since we want to focus on the force it exerts on the backbone. The hinge where the upper body meets the lower body is the axis of rotation.

Simplify and diagram We next draw a force diagram for the upper body. The gravitational force that Earth exerts on the upper body $F_{\text{E on B}}$ at its center of mass is $Mg = (33\,\text{kg})(9.8\,\text{N/kg}) = 323\,\text{N}\,(73\,\text{lb})$. The barbell exerts a force on the upper body equal to $mg = (18\,\text{kg})(9.8\,\text{N/kg}) = 176\,\text{N}\,(40\,\text{lb})$. Because of our choice of axis of rotation, the force exerted by the disk on the upper body $F_{\text{D on B}}$ will not produce a torque. The gravitational force exerted by Earth on the upper body and the force that the barbell exerts on the upper body have clockwise turning ability, while the tension force exerted by the back muscles on the upper body has counterclockwise turning ability.

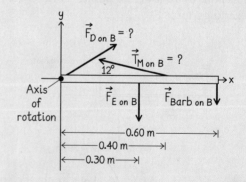

Represent mathematically The torque condition of equilibrium for the upper body is

$$\Sigma\tau = +(F_{\text{D on B}})(0) + [-(Mg)(l/2)\sin 90°]$$
$$+ (T_{\text{M on B}})(2l/3)\sin 12° + [-(F_{\text{Barb on B}})(l)\sin 90°]$$
$$= 0$$

The *x*- and *y*-component forms of the force condition of equilibrium for the backbone are

$$\Sigma F_x = F_{\text{D on B}x} + (-T_{\text{M on B}} \cos 12°) = 0$$
$$\Sigma F_y = F_{\text{D on B}y} + T_{\text{M on B}} \sin 12° + (-mg) + (-Mg) = 0$$

where $F_{\text{D on B}x}$ and $F_{\text{D on B}y}$ are the scalar components of the force that the disk exerts on the upper body.

Solve and evaluate We can solve the torque equation immediately to determine the magnitude of the force that the back muscle exerts on the backbone:

$$T_{\text{M on B}} = \frac{(Mg)(l/2)(\sin 90°) + (mg)(l) \sin 90°}{(2l/3)(\sin 12°)}$$

Note that the backbone length *l* in the numerator and denominator of all of the terms in this equation cancels out. Thus,

$$T_{\text{M on B}} = \frac{(Mg)(1/2)(\sin 90°) + (mg)(1) \sin 90°}{(2/3)(\sin 12°)}$$
$$= \frac{(323\ \text{N})(0.50)(1.0) + (176\ \text{N})(1)(1.0)}{(0.667\ \text{m})(0.208)}$$
$$= 2440\ \text{N}\ (550\ \text{lb})$$

We then find $F_{\text{D on B}x}$ from the *x*-component force equation:

$$F_{\text{D on B}x} = +T_{\text{M on B}} \cos 12°$$
$$= +(2440\ \text{N}) \cos 12°$$
$$= +2390\ \text{N}$$

and $F_{\text{D on B}y}$ from the *y*-component force equation:

$$F_{\text{D on B}y} = +Mg + mg - T_{\text{M on B}} \sin 12°$$
$$= +323\ \text{N} + 176\ \text{N} - (2440\ \text{N})(\sin 12°)$$
$$= -8\ \text{N}$$

Thus, the magnitude of $F_{\text{D on B}}$ is

$$F_{\text{D on B}} = \sqrt{(2390\ \text{N})^2 + (-8\ \text{N})^2} = 2390\ \text{N}\ (540\ \text{lb})$$

The direction of $\vec{F}_{\text{D on B}}$ can be determined using trigonometry:

$$\tan \theta = \frac{F_{\text{D on B}y}}{F_{\text{D on B}x}} = \frac{-8\ \text{N}}{2390\ \text{N}} = -0.0033$$

or $\theta = 0.19°$ below the horizontal. We've found that the back muscles exert a force more than four times the gravitational force that Earth exerts on the person and that the disks of the lower back are compressed by a comparable force.

Try it yourself: Suppose that a college football lineman stands on top of a 1-inch-diameter circular disk. How many 275-lb linemen, one on top of the other, would exert the same compression force on the disk as that exerted on the woman's disk when she lifts the 40-lb barbell?

Answer: The magnitude of the force exerted on the vertebral disk is equivalent to two linemen ($2 \times 275\ \text{lb} = 550\ \text{lb}$) standing on the 1-inch-diameter disk.

To lift correctly, keep your back more vertical with the barbell close to your body, as in **Figure 7.25**. Bend your knees and lift with your legs. With this orientation, the back muscle exerts one-third of the force that is exerted when lifting incorrectly. The disks in the lower back undergo one-half the compression they would experience from lifting incorrectly.

Using a ladder safely

Each year about 25 people per 100,000 experience serious falls from ladders. Why are ladders sometimes dangerous? Consider the physics of one aspect of ladder use.

Figure 7.25 A better way to lift things.

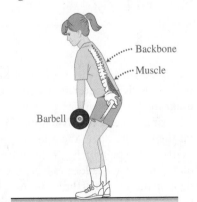

Backbone

Muscle

Barbell

EXAMPLE 7.13 Don't let your ladder slip

Any time you have to climb a ladder, you want the ladder to remain in static equilibrium. At what angle should a 60-kg painter place his ladder against the wall in order to climb two-thirds of the way up the ladder and have the ladder remain in static equilibrium? The ladder's mass is 10 kg and its length is 6.0 m. The exterior wall of the house is very smooth, meaning that it exerts a negligible friction force on the ladder. The coefficient of static friction between the floor and the ladder is 0.50.

Sketch and translate We've sketched the situation below. If the ladder is tilted at too large an angle from the vertical, everyday experience indicates that it will slide down the wall. The static friction force exerted by the floor on the ladder is what is preventing the ladder from sliding, so one way to look at the situation is to ask, What is the maximum angle relative to the wall that the ladder can have before the static friction force is insufficient to keep the ladder in static equilibrium? We choose the ladder and the painter together as the system of interest.

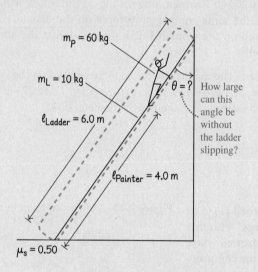

Simplify and diagram Model the ladder as a rigid body and the painter as a point-like object. A force diagram for the ladder-painter system includes the following forces: the gravitational force exerted by Earth on the ladder $\vec{F}_{\text{E on L}}$, the gravitational force that Earth exerts on the painter $\vec{F}_{\text{E on P}}$, the normal force of the wall on the top of the ladder $\vec{N}_{\text{W on L}}$, the normal force of the floor on the bottom of the ladder $\vec{N}_{\text{F on L}}$, and the static friction force of the floor on the bottom of the ladder $\vec{f}_{s\text{F on L}}$. We place the axis of rotation where the ladder touches the floor. Rather than determine the center of mass of the system, we have kept the gravitational forces exerted by Earth on the ladder and painter separate.

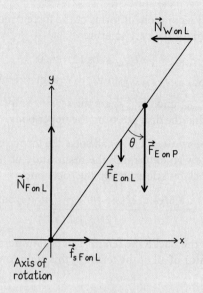

Represent mathematically Use the force diagram to apply the conditions of equilibrium. Since the forces in this situation point along more than one axis, we use both components of the force condition of equilibrium.

y-component force equation:
$$\Sigma F_y = N_{\text{F on L}} + (-m_{\text{Ladder}}g) + (-m_{\text{Painter}}g) = 0$$

x-component force equation:
$$\Sigma F_x = (-N_{\text{W on L}}) + (f_{s\text{ F on L}}) = 0$$

We can insert the expression for the maximum static friction force ($f_{s\text{ Max F on L}} = \mu_s N_{\text{F on L}}$) into the x-component force equation to get

$$N_{\text{W on L}} = \mu_s N_{\text{F on L}}$$

From the y-component equation, we get

$$N_{\text{F on L}} = m_{\text{Ladder}}g + m_{\text{Painter}}g$$

Combining the last two equations, we get

$$N_{\text{W on L}} = \mu_s(m_{\text{Ladder}} + m_{\text{Painter}})g$$

The torque equation is

$$\left[-F_{\text{E on L}}\left(\frac{l_{\text{Ladder}}}{2}\right)\sin\theta\right] + \left[-F_{\text{E on P}}\left(\frac{2l_{\text{Ladder}}}{3}\right)\sin\theta\right]$$
$$+ \left[N_{\text{W on L}}l_{\text{Ladder}}\sin(90° - \theta)\right] = 0$$
$$\Rightarrow -m_{\text{Ladder}}g\frac{l_{\text{Ladder}}}{2}\sin\theta - m_{\text{Painter}}g\frac{2l_{\text{Ladder}}}{3}\sin\theta$$
$$+ \mu_s(m_{\text{Ladder}} + m_{\text{Painter}})gl_{\text{Ladder}}\cos\theta = 0$$

Solve and evaluate We can cancel gl_{Ladder} out of each term of the torque equation and then solve what remains for θ:

$$\Rightarrow \frac{m_{Ladder}}{2}\sin\theta + \frac{2m_{Painter}}{3}\sin\theta$$

$$- \mu_s(m_{Ladder} + m_{Painter})\cos\theta = 0$$

$$\Rightarrow \left(\frac{m_{Ladder}}{2} + \frac{2m_{Painter}}{3}\right)\frac{\sin\theta}{\cos\theta} - \mu_s(m_{Ladder} + m_{Painter}) = 0$$

$$\Rightarrow \left(\frac{m_{Ladder}}{2} + \frac{2m_{Painter}}{3}\right)\tan\theta = \mu_s(m_{Ladder} + m_{Painter})$$

$$\Rightarrow \tan\theta = \frac{\mu_s(m_{Ladder} + m_{Painter})}{\dfrac{m_{Ladder}}{2} + \dfrac{2m_{Painter}}{3}}$$

$$= \frac{6\mu_s(1 + m_{Painter}/m_{Ladder})}{3 + 4m_{Painter}/m_{Ladder}}$$

$$= \frac{6(0.50)(1 + 6)}{3 + 4(6)} = \frac{7}{9}$$

$$\Rightarrow \theta = 38°$$

This seems reasonable and similar to what we observe in real life when someone uses a ladder. Some ladders come with a warning not to have the angle exceed 15°. As a limiting case, if the coefficient of static friction were zero, then our result says the angle would also have to be zero. This makes sense; without friction between the ground and the ladder, the ladder would need to be perfectly vertical or it would slip.

Try it yourself: If the exterior wall of the house were not smooth, how would the above method need to be changed?

Answer: The wall will exert an upward friction force on the ladder. We would have to take this force into account for both force and torque conditions of equilibrium.

Review Question 7.7   When lifting an object, the muscles in the body have to exert significantly larger forces than Earth exerts on the object being lifted. Use your understanding of torque and static equilibrium to explain this phenomenon.

Summary

Words	Pictorial and physical representations	Mathematical representation
Center of mass The gravitational force that the Earth exerts on an object can be considered to be exerted entirely on the object's center of mass. An external force pointing directly toward or away from the center of mass of a free object will not cause the object to turn or rotate. (Sections 7.1 and 7.4)	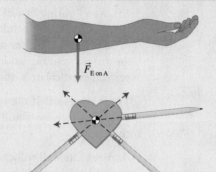	$$x_{cm} = \frac{m_1 x_1 + m_2 x_2 + m_3 x_3 + \cdots + m_n x_n}{m_1 + m_2 + m_3 + \cdots + m_n}$$ $$y_{cm} = \frac{m_1 y_1 + m_2 y_2 + m_3 y_3 + \cdots + m_n y_n}{m_1 + m_2 + m_3 + \cdots + m_n}$$ Eq. (7.4)
A **torque** τ around an *axis of rotation* is a physical quantity characterizing the turning ability of a force with respect to a particular axis of rotation. The torque is positive if the force tends to turn the object counterclockwise and negative if it tends to turn the object clockwise about the axis of rotation. (Section 7.2)	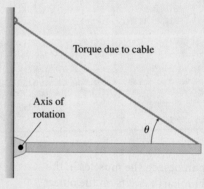 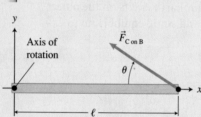	$\tau = \pm Fl \sin \theta$ Eq. (7.1)

Static equilibrium is a state in which a rigid body is at rest and remains at rest both translationally and rotationally. (Section 7.3)	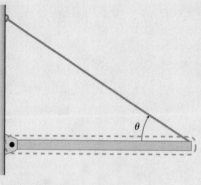
	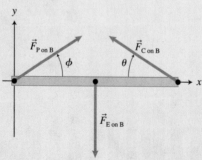
		Translational (force) condition
		
		$\Sigma F_x = F_{1 \text{ on } Bx} + \cdots + F_{n \text{ on } Bx} = 0$
		                       Eq. (7.2x)
		$\Sigma F_y = F_{1 \text{ on } By} + \cdots + F_{n \text{ on } By} = 0$
		                       Eq. (7.2y)
		
		Rotational (torque) condition
		
		$\Sigma \tau = \tau_1 + \tau_2 + \cdots + \tau_n = 0$ Eq. (7.3)

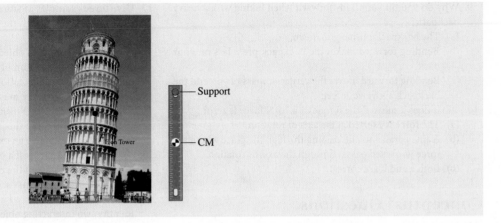

The **equilibrium** of a system is stable against tipping line through its center of mass is within the system's area of support.

The **equilibrium** of a system is stable against rotating of mass of the rotating object is below the axis of rotation. (Section 7.6)

E on Tower

Support

CM

 For instructor-assigned homework, go to MasteringPhysics.

Questions

Multiple Choice Questions

1. A falling leaf usually flutters while falling. However, we have learned that the force that Earth exerts on an object is exerted at its center of mass and thus should not cause rotational motion. How can you resolve this contradiction?
   (a) A leaf is not a rigid body and the rule does not apply.
   (b) There are other forces exerted on the leaf as it falls besides the force exerted by Earth.
   (c) Some forces were not taken into account when we defined the center of mass.

2. You have an irregularly shaped object. To find its center of mass you can do which of the following?
   (a) Find a point where you can put a fulcrum to balance it.
   (b) Push the object in different directions and find the point of intersection of the lines of action of the forces that do not rotate the object.
   (c) Separate the object into several regularly shaped objects whose center of masses you know, and use the mathematical definition of the center of mass to find it.
   (d) All of the above
   (e) a and b only

3. A hammock is tied with ropes between two trees. A person lies in it. Under what circumstances are its ropes more likely to break?
   (a) If stretched tightly between the trees
   (b) If stretched loosely between the trees
   (c) The ropes always have equal likelihood to break.

4. Where is the center of mass of a donut?
   (a) In the center of the hole
   (b) Uniformly distributed throughout the donut
   (c) Cannot be found

5. Which of the following objects is at rest but not in static equilibrium?
   (a) An object thrown upward at the top of its flight
   (b) A person sitting on a chair
   (c) A person in an elevator moving down at constant speed

6. Which of the following objects is not at rest but in equilibrium?
   (a) An object thrown upward at the top of its flight
   (b) A person sitting on a chair
   (c) A person in an elevator moving down at constant speed

7. A physics textbook lies on top of a chemistry book, which rests on a table. Which force diagram below best describes the forces exerted by other objects on the chemistry book (**Figure Q7.7**)?

Figure Q7.7

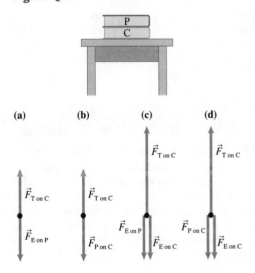

(a)   (b)   (c)   (d)

8. What does it mean if the torque of a force is positive?
   (a) The object exerting the force is on the right side of the axis of rotation.
   (b) The object exerting the force is on the left side of the axis of rotation.
   (c) The force points up.
   (d) The force points down.
   (e) None of these choices is necessarily correct.

9. Why do you tilt your body forward when hiking with a heavy backpack?
   (a) The backpack pushes you down.
   (b) Bending forward makes the backpack press less on your back.
   (c) Bending forward moves the center of mass of you and the backpack above your feet.
10. What does it mean if the torque of a 10-N force is zero?
   (a) The force is exerted at the axis of rotation.
   (b) A line parallel to and passing through the place where the force is exerted passes through the axis of rotation.
   (c) Both a and b are correct.

Conceptual Questions

11. Is it possible for an object not to be in equilibrium when the net force exerted on it by other objects is zero? Give an example.
12. Explain the meaning of torque so that a friend not taking physics can understand.
13. ✏ Something is wrong with the orientation of the ropes shown in **Figure Q7.13**. Use the first condition of equilibrium for the hanging pulley to help explain this error, and then redraw the sketch as you would expect to see it.
14. What are the two conditions of equilibrium? How do you know that these conditions create a state of equilibrium?
15. Describe an observational experiment that you will conduct so that a friend can discover the second condition of equilibrium.
16. Describe how you would test whether the second condition of equilibrium is correct. What is the difference between your answer to this question and the previous question?

Figure Q7.13

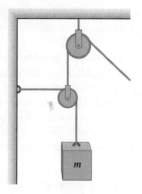

17. Give three examples of situations in which an object is starting to turn even though the sum of the forces exerted on the object is zero.
18. Give an example of an object that is accelerating translationally even though the sum of the torques of the forces exerted on it about an axis of rotation is zero.
19. The force that the body muscles exert on bones that are used to lift various objects is usually five to ten times greater than the gravitational force that Earth exerts on the object being lifted. Explain and give an example.
20. When a person jumps off a boat, what happens to the location of the center of mass of the person-boat system?
21. ✏ A ladder leans against a wall. Construct a force diagram showing the direction of all forces exerted on the ladder. Identify two interacting objects for each force.
22. ✏ Using a crowbar, a person can remove a nail by exerting little force, whereas pulling directly on the nail requires a large force to remove it (you probably can't). Why? Draw a sketch to support your answer.
23. Is it more difficult to do a sit-up with your hands stretched in front of you or with them behind your head? Explain.
24. Sit on a chair with your feet straight down at the front of the chair. Keeping your back perpendicular to the floor, try to stand up without leaning forward. Explain why it is impossible to do it.
25. Can you balance the tip of a wooden ruler vertically on a fingertip? Why is it so difficult? Design a method to balance the ruler on your fingertip. Describe any extra material(s) you will use.
26. Try to balance a sharp wooden pencil on your fingertip, point down. [*Hint:* A small pocketknife might help by lowering the center of mass of the system.]
27. Design a device that you can use to successfully walk on a tight rope.
28. Why do tightrope walkers carry long, heavy bars?
29. Explain why it is easier to keep your balance while jumping on two feet than while hopping on one.
30. A carpenter's trick to keep nails from bending when they are pounded into a hard material is to grip the center of the nail with pliers. Why does this help?
31. Why does a skier crouch when going downhill?

Problems

Below, BIO indicates a problem with a biological or medical focus. Problems labeled EST ask you to estimate the answer to a quantitative problem rather than develop a specific answer. Problems marked with ✏ require you to make a drawing or graph as part of your solution. Asterisks indicate the level of difficulty of the problem. Problems with no * are considered to be the least difficult. A single * marks moderately difficult problems. Two ** indicate more difficult problems.

7.2 Torque: A new physical quantity

1. Determine the torques about the axis of rotation P produced by each of the four forces shown in **Figure P7.1**. All forces have magnitudes of 120 N and are exerted a distance of 2.0 m from P on some unshown object O.

Figure P7.1

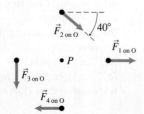

2. *EST Your hand holds a liter of milk while your arm is bent at the elbow in a 90° angle. Estimate the torque caused by the milk on your arm about the elbow joint. Indicate all numbers used in your calculations. This is an estimate, and your answer may differ by 10 to 50% from the answers of others.

3. *EST **Body torque** You hold a 4.0-kg computer. Estimate the torques exerted on your forearm about the elbow joint caused by the downward force exerted by the computer on the forearm and the upward 340-N force exerted by the biceps muscle on the forearm. Ignore the mass of the arm. Indicate any assumptions you make.

4. Three 200-N forces are exerted on the beam shown in **Figure P7.4**. (a) Determine the torques about the axis of rotation on the left produced by forces $\vec{F}_{1 \text{ on B}}$ and $\vec{F}_{2 \text{ on B}}$. (b) At what distance from the axis of rotation must $\vec{F}_{3 \text{ on B}}$ be exerted to cause a torque that balances those produced by $\vec{F}_{1 \text{ on B}}$ and $\vec{F}_{2 \text{ on B}}$?

Figure P7.4

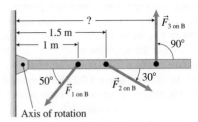

Axis of rotation

5. *A 2.0-m-long, 15-kg ladder is resting against a house wall, making a 30° angle with the vertical wall. The coefficient of static friction between the ladder feet and the ground is 0.40, and between the top of the ladder and the wall the coefficient is 0.00. Make a list of the physical quantities you can determine or estimate using this information and calculate them.

7.3 Conditions of equilibrium

6. Three friends tie three ropes in a knot and pull on the ropes in different directions. Adrienne (rope 1) exerts a 20-N force in the positive x-direction, and Jim (rope 2) exerts a 40-N force at an angle 53° above the negative x-axis. Luis (rope 3) exerts a force that balances the first two so that the knot does not move. (a) Construct a force diagram for the knot. (b) Use equilibrium conditions to write equations that can be used to determine $F_{L \text{ on } Kx}$ and $F_{L \text{ on } Ky}$. (c) Use equilibrium conditions to write equations that can be used to determine the magnitude and direction of $\vec{F}_{L \text{ on } K}$.

7. Adrienne from Problem 6 now exerts a 100-N force $\vec{F}_{A \text{ on } K}$ that points 30° below the positive x-axis and Jim exerts a 150-N force in the negative y-direction. How hard and in what direction does Luis now have to pull the knot so that it remains in equilibrium?

8. *Kate joins Jim, Luis, and Adrienne in the rope-pulling exercise described in the previous two problems. This time, they tie four ropes to a ring. The three friends each pull on one rope, exerting the following forces: $\vec{T}_{1 \text{ on } R}$ (50 N in the positive y-direction), $\vec{T}_{2 \text{ on } R}$ (20 N, 25° above the negative x-axis), and $\vec{T}_{3 \text{ on } R}$ (70 N, 70° below the negative x-axis). Kate pulls rope 4, exerting a force $\vec{T}_{4 \text{ on } R}$ so that the ring remains in equilibrium. (a) Construct a force diagram for the ring. (b) Use the first condition of equilibrium to write two equations that can be used to determine $T_{4 \text{ on } Rx}$ and $T_{4 \text{ on } Ry}$ (c) Solve these equations and determine the magnitude and direction of $\vec{T}_{4 \text{ on } R}$.

9. You hang a light in front of your house using an elaborate system to keep the 1.2-kg light in static equilibrium (see **Figure P7.9**). What are the magnitudes of the forces that the ropes must exert on the knot connecting the three ropes if $\theta_2 = 37°$ and $\theta_3 = 0°$?

Figure P7.9

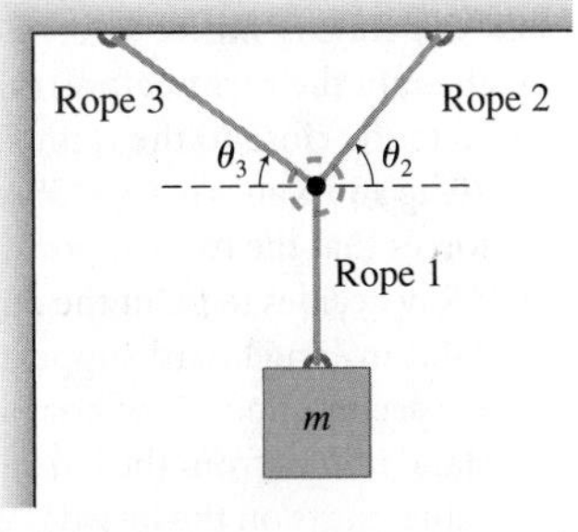

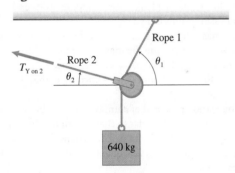

10. *Find the values of the forces the ropes exert on the knot if you replace the light in Problem 7.9 with a heavier 12-kg object and the ropes make angles of $\theta_2 = 63°$ and $\theta_3 = 45°$ (see Figure P7.9).

11. / Redraw Figure P7.9 with $\theta_2 = 50°$ and $\theta_3 = 0°$. Rope 2 is found to exert a 100-N force on the knot. Determine m and the magnitudes of the forces that the other two ropes exert on the knot.

12. *Determine the masses m_1 and m_2 of the two objects shown in **Figure P7.12** if the force exerted by the horizontal cable on the knot is 64 N.

Figure P7.12

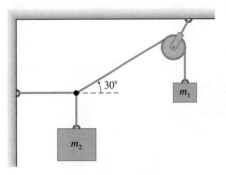

13. ***Lifting an engine** You work in a machine shop and need to move a huge 640-kg engine up and to the left in order to slide a cart under it. You use the system shown in **Figure P7.13**. How hard and in what direction do you need to pull on rope 2 if the angle between rope 1 and the horizontal is $\theta_1 = 60°$?

Figure P7.13

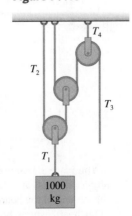

14. * / **More lifting** You exert a 630-N force on rope 2 in the previous problem (Figure P7.13). Write the two equations (x and y) for the first condition of equilibrium using the pulley as the object of interest for a force diagram. Calculate θ_1 and θ_2. You may need to use the identity $(\sin\theta)^2 + (\cos\theta)^2 = 1$.

15. * / **Even more lifting** A pulley system shown in **Figure P7.15** will allow you to lift heavy objects in the machine shop by exerting a relatively small force. (a) Construct a force diagram for each pulley. (b) Use the equations of equilibrium and the force diagrams to determine T_1, T_2, T_3, and T_4.

Figure P7.15

16. **Tightrope walking** A tightrope walker wonders if her rope is safe. Her mass is 60 kg and the length of the rope is about 20 m. The rope will break if its tension

exceeds 6700 N. What is the smallest angle at which the rope can bend up from the horizontal on either side of her to avoid breaking?

17. **Lifting patients** An apparatus to lift hospital patients sitting at the sides of their beds is shown in **Figure P7.17**. At what angle above the horizontal does the rope going under the pulley bend while supporting the 78-kg person hanging from the pulley?

Figure P7.17

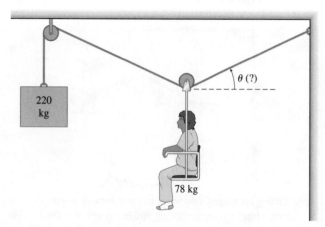

18. A mutineer on Captain Bligh's ship is made to "walk the plank." The plank, which extends 3.0 m beyond its support, will break if subjected to a torque greater than 3300 N·m. Will the sailor break the plank before stepping off its end? Explain. What assumptions did you make?

19. Brett (mass 70 kg) sits 1.2 m from the fulcrum of a uniform seesaw. (a) Determine the magnitude of the torque exerted by him on the seesaw. (b) At what distance from the fulcrum on the other side should 54-kg Dawn sit so that the seesaw is horizontal?

20. * You stand at the end of a uniform diving board a distance *d* from support 2 (similar to that shown in **Figure P7.20**). Your mass is *m*. What can you determine from this information? Make a list of physical quantities and show how you will determine them.

Figure P7.20

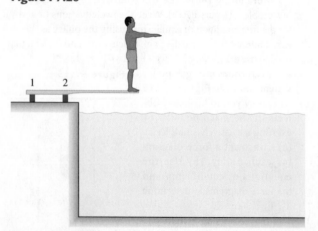

21. * You place a 3.0-m-long board across a chair to seat three physics students at a party at your house. If 70-kg Dan sits on the left end of the board and 50-kg Tahreen on the right end of the board, where should 54-kg Komila sit to keep the board stable? What assumptions did you make?

22. After dinner (see Problem 7.21), two guests decide to use the same 3.0-m-long 5.0-kg board as a seesaw, using a small bench as a fulcrum. An 82-kg man sits on one end and a 64-kg woman sits on the other end. Where should the bench be located so that the board balances?

23. **Car jack** You've got a flat tire. To lift your car, you make a homemade lever (see **Figure P7.23**). A very light 1.6-m-long handle part is pushed down on the right side of the fulcrum and a 0.050-m-long part on the left side supports the back of the car. How hard must you push down on the handle so that the lever exerts an 8000-N force to lift the back of the car?

Figure P7.23

24. * **Mobile** You are building a toy mobile, copying the design shown in **Figure P7.24**. Object A has a 1.0-kg mass. What should be the mass of object B? The numbers in Figure P7.24 indicate the relative lengths of the rods on each side of their supporting cords.

Figure P7.24

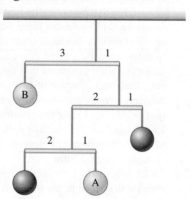

25. * ⫻ **Another mobile** You are building a toy mobile similar to that shown in Figure P7.24 but with different dimensions and replacing the objects with cups. The bottom rod is 20 cm long, the middle rod is 15 cm long, and the top rod is 8 cm long. You put one penny in the bottom left cup, three pennies in the bottom right cup, eleven pennies in the middle right cup, and five pennies in the top left cup. (a) Draw a force diagram for each rod. (b) Determine the cord attachment points and lengths on each side for each rod. (c) What assumptions did you make in order to solve the problem?

26. *EST Kate is sitting on a 1.5-m-wide porch swing. Because of the rain the night before, the left side of the swing is wet, and Kate sits close to the right side. The mass of the swing seat is 10 kg and Kate's mass is 55 kg. Estimate the magnitudes of the forces that the two supporting cables exert on the swing.

27. *Ray decides to paint the outside of his uncle's house. He uses a 4.0-m-long board supported by cables at each end to paint the second floor. The board has a mass of 21 kg. Ray (70 kg) stands 1.0 m from the left cable. What are the forces that each cable exerts on the board?

28. *The fulcrum of a uniform 20-kg seesaw that is 4.0 m long is located 2.5 m from one end. A 30-kg child sits on the long end. Determine the mass a person at the other end would have to be in order to balance the seesaw.

29. *A 2.0-m-long uniform beam of mass 8.0 kg supports a 12.0-kg bag of vegetables at one end and a 6.0-kg bag of fruit at the other end. At what distance from the vegetables should the beam rest on your shoulder to balance? What assumptions did you make?

30. *A uniform beam of length l and mass m supports a bag of mass m_1 at the left end, another bag of mass m_2 at the right end, and a third bag m_3 at a distance l_3 from the left end ($l_3 < 0.5l$). At what distance from the left end should you support the beam so that it balances?

7.4 Center of mass

31. *An 80-kg person stands at one end of a 130-kg boat. He then walks to the other end of the boat so that the boat moves 80 cm with respect to the bottom of the lake. (a) What is the length of the boat? (b) How much did the center of mass of the person-boat system move when the person walked from one end to the other?

32. EST Two people (50 kg and 75 kg) holding hands stand on rollerblades 1.0 m apart. (a) Estimate the location of their center of mass. (b) The two people push off each other and roll apart so the distance between them is now 4.0 m. Estimate the new location of the center of mass. What assumptions did you make?

33. *You have a disk of radius R with a circular hole of radius r cut a distance a from the center of the disk. Where is the disk's center of mass?

34. *A person whose height is 1.88 m is lying on a light board placed on two scales so that scale 1 is under the person's head and scale 2 is under the person's feet. Scale 1 reads 48.3 kg and scale 2 reads 39.3 kg. Where is the center of mass of the person?

35. **EST Estimate the location of the center of mass of the person described in the previous problem when he bends over and touches the floor with his hands.

36. *A seesaw has a mass of 30 kg, a length of 3.0 m, and fulcrum beneath its midpoint. It is balanced when a 60-kg person sits on one end and a 75-kg person sits on the other end. Locate the center of mass of the seesaw. Where is the center of mass of a uniform seesaw that is 3.0 m long and has a mass of 30 kg if two people of masses 60 kg and 75 kg sit on its ends?

37. **Find the center of mass of an L-shaped object. The vertical leg has a mass of m_a of length a and the horizontal leg has a mass of m_b of length b. Both legs have the same width w, which is much smaller than a or b.

38. **You have a 10-kg table with each leg of mass 1.0 kg—total mass 14 kg. If you place a 5.0-kg pot of soup in the back right corner of the table, where is the table's center of mass?

7.5 Skills for analyzing situations using equilibrium conditions

39. *Using biceps to hold a child A man is holding a 16-kg child using both hands with his elbows bent in a 90° angle. The biceps muscle provides the positive torque he needs to support the child. Determine the force that each of his biceps muscles must exert on the forearm in order to hold the child safely in this position. Ignore the triceps muscle and the mass of the arm.

40. *BIO Using triceps to push a table A man pushes on a table exerting a 20-N downward force with his hand. Determine the force that his triceps muscle must exert on his forearm in order to balance the upward force that the table exerts on his hand. Ignore the biceps muscle and the mass of the arm. If you did not ignore the mass of the arm, would the force you determined be smaller or larger? Explain.

41. *BIO Using biceps to hold a barbell Find the force that the biceps muscle shown in Figure P7.26 exerts on the forearm when you lift a 16-kg barbell with your hand. Also determine the force that the bone in the upper arm (the humerus) exerts on the bone in the forearm at the elbow joint. The mass of the forearm is about 5.0 kg and its center of mass is 16 cm from the elbow joint. Ignore the triceps muscle.

42. **Leg support** A person's broken leg is kept in place by the apparatus shown in **Figure P7.42**. If the rope pulling on the leg exerts a 120-N force on it, how massive should be the block hanging from the rope that passes over the pulley?

Figure P7.42

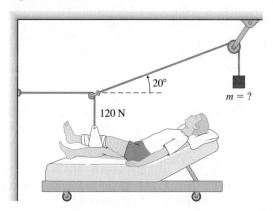

43. *BIO Hamstring You are exercising your hamstring muscle (the large muscle in the back of the thigh). You use an elastic cord attached to a hook on the wall while keeping your leg in a bent position (**Figure P7.43**). Determine the magnitude of the tension force $\vec{T}_{H \text{ on } L}$ exerted by the hamstring muscles on the leg and the magnitude of compression force $\vec{F}_{F \text{ on } B}$ at the knee joint that the femur exerts on the calf bone. The cord exerts a 20-lb force $\vec{F}_{C \text{ on } F}$ on the foot.

Figure P7.43

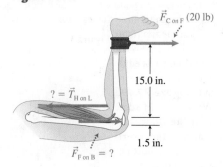

44. *You decide to hang a new 10-kg flowerpot using the arrangement shown in **Figure P7.44**. Can you use a slanted rope attached from the wall to the end of the beam if that rope breaks when the tension exceeds 170 N? The mass of the beam is not known but it looks light.

Figure P7.44

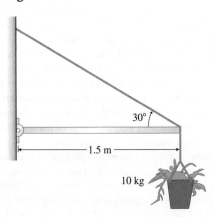

45. *You decide to hang another plant from a 1.5-m-long 2.0-kg horizontal beam that is attached by a hinge to the wall on the left. A cable attached to the right end goes 37° above the beam to a connecting point above the hinge on the wall. You hang a 100-N pot from the beam 1.4 m away from the wall. What is the force that the cable exerts on the beam?

46. **The plant in the hanging pot described in Problem 45 grows, and the pot and plant now have mass 12 kg. Determine the new force that the cable exerts on the beam and the force that the wall hinge exerts on the beam (its x- and y-components and the magnitude and direction of that force).

47. **Now you decide to change the way you hang the pot described in Problems 45 and 46. You orient the beam at a 37° angle above the horizontal and orient the cable horizontally from the wall to the end of the beam. The beam still holds the 2.0-kg pot and plant hanging 0.1 m from its end. Now determine the force that the cable exerts on the beam and the force that the wall hinge exerts on the beam (its x- and y-components and the magnitude and direction of that force).

48. **Diving board** The diving board shown in Figure P7.20 has a mass of 28 kg and its center of mass is at the board's geometrical center. Determine the forces that support posts 1 and 2 (separated by 1.4 m) exert on the board when a 60-kg person stands on the end of the board 2.8 m from support post 2.

49. **A uniform cubical 200-kg box sits on the floor with its bottom left edge pressing against a ridge. The length L of a side of the box is 1.2 m. Determine the least force you need to exert horizontally at the top right edge of the box that will cause its bottom right edge to be slightly off the floor, as shown in **Figure P7.49**. (Note: With the right edge slightly off the floor, the ground and ridge exert their forces on the bottom left edge of the box.)

Figure P7.49

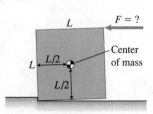

50. *If the force F shown in Figure P7.49 is 840 N and the bottom right edge of the box is slightly off the ground, what is the mass of the cubical box of side 1.2 m?

51. We know from the second condition of equilibrium that if two different magnitude forces are exerted on the same object, their rotational effects can cancel if their torques are the same magnitude but opposite sign. For example, you can lift a heavy boulder by exerting a force much smaller than the weight of the boulder. Design an experiment where you can lift a 100-kg rock by exerting a downward 100-N push (it is much easier to push than to pull).

52. *What mechanical work must you do to lift a log that is 3.0 m long and has a mass of 100 kg from the horizontal to a vertical position? [Hint: Use the work-energy principle.]

7.6 Stability of equilibrium

53. * / A 70-g meter stick has a 30-g piece of modeling clay attached to the end. Where should you drill a hole in the meter stick so that you can hang the stick horizontally in equilibrium on a nail in the wall? Draw a picture to help explain your decision.

54. **You are trying to tilt a very tall refrigerator (2.0 m high, 1.0 m deep, 1.4 m wide, and 100 kg) so that your friend can put a blanket underneath to slide it out of the kitchen. Determine the force that you need to exert on the front of the refrigerator at the start of its tipping. You push horizontally 1.4 m above the floor.

55. *You have an Atwood machine with two blocks each of mass m attached to the ends of a string of length l. The string passes over a frictionless pulley down to the blocks hanging on each side. While pulling down on one block, you release it. Both blocks continue to move at constant speed, one up and the other down. Is the system still in equilibrium? Find the vertical component of the center of mass of the two-block system. Indicate all of your assumptions and the coordinate system used.

56. *EST You stand sideways in a moving train. Estimate how far apart you should keep your feet so that when the train accelerates at 2.0 m/s² you can still stand without holding anything. List all your assumptions.

7.7 Static equilibrium: Putting it all together

57. **BIO **Lift with bent legs** You injure your back at work lifting a 420-N radiator. To understand how it happened, you model your back as a weightless beam (**Figure P7.57**), analogous

Figure P7.57

(a)

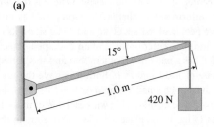

(b)

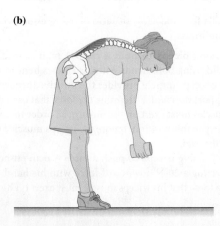

to the backbone of a person in a bent position when lifting an object. (a) Determine the tension force that the horizontal cable exerts on the beam (which is analogous to the force the back muscle exerts on the backbone) and the force that the wall exerts on the beam at the hinge (which is analogous to the force that a disk in the lower back exerts on the backbone). (b) Why do doctors recommend lifting objects with the legs bent?

58. **Determine the tension force that the horizontal cable exerts on the beam in Figure P7.57, but with the beam tilted at 30° rather than 15°. The cable remains horizontal.

59. **Determine the tension force that the horizontal cable exerts on the beam in **Figure P7.59** if the horizontal cable is moved down (compared to Figure P7.57) so that its right end attaches to the beam 0.30 m from its top right end. The cable remains horizontal and the beam is tilted at 15°.

Figure P7.59

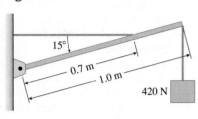

420 N

60. *BIO **Barbell lift I** A woman lifts a 3.6-kg barbell in each hand with her arm in a horizontal position at the side of her body and holds it there for 3 s (see **Figure P7.60**). What force does the deltoid muscle in her shoulder exert on the humerus bone while holding the barbell? The deltoid attaches 13 cm from the shoulder joint and makes a 13° angle with the humerus. The barbell in her hand is 0.55 m from the shoulder joint, and the center of mass of her 4.0-kg arm is 0.24 m from the joint.

Figure P7.60

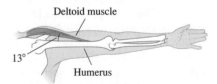

Deltoid muscle

13°

Humerus

61. **BIO **Barbell lift II** Repeat the previous problem with a 7.2-kg barbell. Determine both the force that the deltoid exerts on the humerus and the force that the lifter's shoulder joint exerts on her humerus.

62. *BIO **Facemask penalty** The head of a football running back (see **Figure P7.62**) can be considered as a lever with the vertebra at the bottom of the skull as a fulcrum (the axis of rotation). The center of mass is about 0.025 m in front of the axis of rotation. The torque caused by the force that Earth exerts on the 8.0-kg head/helmet is balanced by the torque caused by the downward forces exerted by a complex muscle system in the neck. That muscle system includes the trapezius and levator scapulae muscles, among others (effectively 0.057 m from the axis of rotation). (a) Determine the magnitude of the force exerted by the neck muscle system pulling down to balance the torque caused by the force exerted by Earth on the head. (b) If an opposing player exerts a downward 180-N (40-lb) force on the facemask, what muscle force would these neck muscles now need to exert to keep the head in equilibrium?

Figure P7.62

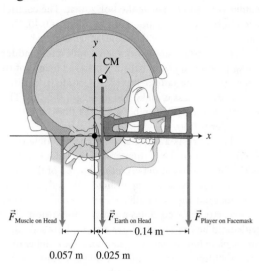

$\vec{F}_{Muscle\ on\ Head}$ $\vec{F}_{Earth\ on\ Head}$ $\vec{F}_{Player\ on\ Facemask}$

0.14 m

0.057 m   0.025 m

General Problems

63. **Design two experiments to determine the mass of a ruler, using different methods. Your available materials are the ruler, a spring, and a set of objects of standard mass: 50 g, 100 g, and 200 g. One of the methods should involve your knowledge of static equilibrium. After you design and perform the experiment, decide whether the two methods give you the same or different results.

64. *A board of mass m and length l is placed on a horizontal tabletop. The coefficient of static friction between the board and the table is μ. How far from the edge of the tabletop can one extend the board before it falls off?

65. ****Tightrope walker** A 60-kg 1.6-m-tall tightrope walker stands on a tightrope. (a) In his hands he holds a 10-kg 2.0-m-long horizontal bar. At each end of the bar are two 5.0-kg balls hanging from 0.50-m-long strings. How much does this apparatus lower his center of mass? (b) How long should the strings be so that the center of mass of the walker-bar-hanging ball system is at the level of the rope? Indicate all assumptions made for each part of the problem.

66. ****Lecturing on a beam** A 70-kg professor sits on a 20-kg beam while lecturing (**Figure P7.66**). A rope attached to the end of the beam passes over a pulley and down to a harness that wraps around the professor (the professor is supported partly by the beam and partly by the harness). The professor sits in the middle of the beam. Determine the force that the rope exerts on the harness and professor and the force that the beam exerts on the professor.

Figure P7.66

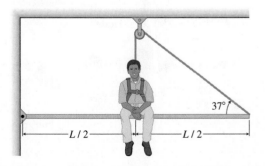

$L/2$ $L/2$ 37°

67. **A 70-kg person stands on a 6.0-m-long 50-kg ladder. The ladder is tilted 60° above the horizontal. The coefficient of friction between the floor and the ladder is 0.40. How high can the person climb without the ladder slipping?

68. **What is a safe angle between a wall and a ladder for a 60-kg painter to climb two-thirds of the height of the ladder without the ladder leaving the state of equilibrium? The ladder's mass is 10 kg and its length is 6.0 m. The coefficient of static friction between the floor and the feet of the ladder is 0.50.

69. **A ladder rests against a wall. The coefficient of static friction between the bottom end of the ladder and the floor is μ_1; the coefficient between the top end of the ladder and the wall is μ_2. At what angle should the ladder be oriented so it does not slip and both coefficients of friction are 0.50?

70. **Every rope or cord has a maximum tension that it can withstand before breaking. Investigate how a ski lift works and explain how it can safely move a large number of passengers of different mass uphill during peak hours, without the cord that carries the chairs breaking.

Reading Passage Problems

BIO **Muscles work in pairs** Skeletal muscles produce movements by pulling on tendons, which in turn pull on bones. Usually, a muscle is attached to two bones via a tendon on each end of the muscle. When the muscle contracts, it moves one bone toward the other. The other bone remains in nearly the original position. The point where a muscle tendon is attached to the stationary bone is called the *origin*. The point where the other muscle tendon is attached to the movable bone is called the *insertion*. The origin is like the part of a door spring that is attached to the doorframe. The insertion is similar to the part of the spring that is attached to the movable door.

During movement, bones act as levers and joints act as axes of rotation for these levers. Most movements require several skeletal muscles working in groups, because a muscle can only exert a pull and not a push. In addition, most skeletal muscles are arranged in opposing pairs at joints. Muscles that bring two limbs together are called flexor muscles (such as the biceps muscle in the upper arm in **Figure 7.26**). Those that cause the limb to extend outward are called extensor muscles (such as the triceps muscle in the upper arm). The flexor muscle is used when you hold a heavy object in your hand; the extensor muscle can be used, for example, to extend your arm when you throw a ball.

Figure 7.26 Muscles often come in flexor-extensor pairs.

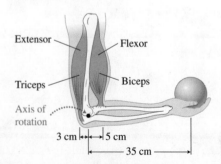

71. You hold a 10-lb ball in your hand with your forearm horizontal, forming a 90° angle with the upper arm (Figure 7.26). Which type of muscle produces the torque that allows you to hold the bell?
    (a) Flexor muscle in the upper arm
    (b) Extensor muscle in the upper arm.
    (c) Flexor muscle in the forearm
    (d) Extensor muscle in the forearm

72. In Figure 7.26, how far in centimeters from the axis of rotation are the forces that the ball exerts on the hand, that the biceps exerts on your forearm, and that the upper arm exerts on your forearm at the elbow joint?
    (a) 0, 5, 35   (b) 35, 5, 0   (c) 35, 5, 3
    (d) 35, 5, −3   (e) 30, 5, 0

73. Why is it easier to hold a heavy object using a bent arm than a straight arm?
    (a) More flexor muscles are involved.
    (b) The distance from the joint to the place where gravitational force is exerted by Earth on the object is smaller.
    (c) The distance from the joint to the place where force is exerted by the object on the hand is smaller.
    (d) There are two possible axes of rotation instead of one.

74. Why are muscles arranged in pairs at joints?
    (a) Two muscles can produce a bigger torque than one.
    (b) One can produce a positive torque and the other a negative torque.
    (c) One muscle can pull on the bone and the other can push.
    (d) Both a and b are true.

BIO **Improper lifting and the back** A careful study of human anatomy allows medical researchers to use the conditions of equilibrium to estimate the internal forces that body parts exert on each other while a person lifts in a bent position (**Figure 7.27a**). Suppose an 800-N (180-lb) person lifts a 220-N (50-lb) barbell in a bent position, as shown in Figure 7.27b. The cable (the back muscle) exerts a tension force $\vec{T}_{\text{M on B}}$ on the backbone and the support at the bottom of the beam (the disk in the lower back) exerts a compression force $\vec{F}_{\text{D on B}}$ on the backbone. The backbone in turn exerts the same magnitude force on the 2.5-cm-diameter fluid-filled disks in the lower backbone. Such disk compression can cause serious back problems. A force diagram of this situation is shown in Figure 7.27c. The magnitude of the gravitational force $\vec{F}_{\text{E on B}}$ that Earth exerts on the center of mass of the upper stomach-chest region is 300 N. Earth exerts a 380-N force on the head, arms, and 220-N barbell held in the hands. Using the conditions of equilibrium, we estimate that the back muscle exerts a 3400-N (760-lb) force $\vec{T}_{\text{M on B}}$ on the backbone and that the disk in the lower back exerts a 3700-N (830-lb) force $\vec{F}_{\text{D on B}}$ on the backbone. This is like supporting a grand piano on the 2.5-cm-diameter disk.

75. Rank in order the magnitudes of the distances of the four forces exerted on the backbone with respect to the joint (see Figure 7.27c), with the largest distance listed first.
    (a) 1 > 3 > 2 > 4   (b) 4 > 2 = 3 > 1
    (c) 4 > 3 > 2 > 1   (d) 2 > 3 > 1 > 4
    (e) 1 > 2 > 3 > 4

76. Rank in order the magnitudes of the torques caused by the four forces exerted on the backbone (see Figure 7.27c), with the largest torque listed first.
    (a) 1 > 2 > 3 > 4   (b) 2 = 3 > 1 > 4
    (c) 3 > 2 > 1 > 4   (d) 2 > 1 > 3 > 4
    (e) 1 = 2 = 3 = 4

Figure 7.27 Analysis of a person's backbone when lifting from a bent position.

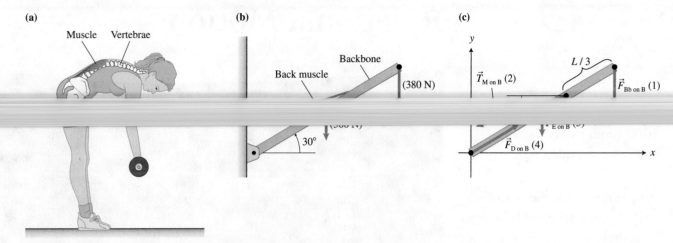

(a)

Muscle   Vertebrae

(b)

Backbone

Back muscle

(380 N)

30°

(c)

y

$L/3$

$\vec{T}_{\text{M on B}}$ (2)

$\vec{F}_{\text{Bb on B}}$ (1)

E on B

$\vec{F}_{\text{D on B}}$ (4)

x

77. What are the signs of the torques caused by forces 1, 2, 3, and 4, respectively, about the origin of the coordinate system shown in Figure 7.27c?
   (a) $+, +, +, +$ (b) $-, +, -, 0$
   (c) $+, -, +, 0$ (d) $-, -, -, 0$
   (e) $+, -, +, -$

78. Which expression below best describes the torque caused by force $F_{\text{D on B}} = F_{\text{Bb on B}}$, the force that the disk in the lower back exerts on the backbone of length L?
   (a) 0   (b) $F_4(2L/3)\sin 12°$
   (c) $F_4(L)\cos 30°$ (d) $-F_4(2L/3)\sin 12°$
   (e) $-F_4(L)\cos 30°$

79. Which expression below best describes the torque caused by force $F_3 = F_{\text{E on B}}$, the force that Earth exerts on the upper body at its center of mass for the backbone of length L?
   (a) 0   (b) $F_3(2L/3)\sin 12°$
   (c) $F_3(L/2)\cos 30°$ (d) $-F_3(2L/3)\sin 12°$
   (e) $-F_3(L/2)\cos 30°$

80. Which expression below best describes the torque caused by force $F_2 = T_{\text{M on B}}$ exerted by the muscle on the backbone?
   (a) 0   (b) $F_2(2L/3)\sin 12°$
   (c) $F_2(L)\cos 30°$ (d) $-F_2(2L/3)\sin 12°$
   (e) $-F_2(L)\cos 30°$

8 Rotational Motion

How can a star rotate 1000 times faster than a merry-go-round?

Why is it more difficult to balance on a stopped bike than on a moving bike?

How is the Moon slowing Earth's rate of rotation?

Be sure you know how to:

- Draw a force diagram for a system (Section 2.1).
- Determine the torque produced by a force (Section 7.2).
- Apply conditions of static equilibrium for a rigid body (Section 7.3).

In 1967, a group of astrophysicists from Cambridge University in England was looking for quasars using an enormous radio telescope. Jocelyn Bell, a physics graduate student, operated the radio telescope and analyzed the nearly 30 meters of printed radio telescope data that were collected daily. Bell noticed a series of regular radio pulses in the midst of a lot of receiver noise. It looked like somebody was sending a radio message, turning the signal on and off every 1.33 seconds. This is an incredibly small time for astronomical objects. At first, Bell's advisor Anthony Hewish believed that they had found signals from extraterrestrial life. This idea received considerable support from the scientific community, although it eventually proved to be an incorrect assertion. The group had, in fact, emit radio signals every second or so. The study of rotational motion explains how pulsars can emit signals so rapidly. In later chapters we will learn the mechanism behind them.

In the last chapter, we learned about the torque that a force produces on a rigid body. However, we only analyzed rigid bodies that were in static equilibrium—they remained at rest. In many cases, however, objects do not remain at rest when torques are exerted—they rotate. For example, the human leg rotates slightly around the hip joint while a person walks, and a car tire rotates around the axle as the car moves. In this chapter, we will learn how to describe, explain, and predict such motions.

8.1 Rotational kinematics

In order to understand the motion of rotating rigid bodies, we will follow the same strategy that we used for linear motion. We start by investigating how to describe rotational motion and then develop rules that explain how forces and torques cause objects to rotate in the way they do. Ultimately, this investigation will enable us to understand many aspects of the natural and human-made world, from why a bicycle is so stable when moving to how the gravitational pull of the Moon slows the rotation of Earth.

One of the simplest examples of a rotating rigid body is a disk. Suppose you stand at a lab bench with a rotating disk on top of it. You wish to describe the counterclockwise motion of the disk quantitatively. This is trickier than it might seem at first. When we investigated the motion of point-like objects, we did not have to specify which part of the object we were describing, since the object was located at a single point. With a rigid body, there are infinitely many points to choose from. For example, imagine that you place small coins at different locations on the disk, as shown in **Figure 8.1a**. As the disk turns, you observe that the direction of the velocity of each coin changes continually (see the coins on the outer edge of the disk in Figure 8.1a). In addition, a coin that sits closer to the edge moves faster and covers a longer distance during a particular time interval than a coin closer to the center (Figure 8.1b). This means that different parts of the disk move not only in different directions, but also at different speeds relative to you.

On the other hand, there are similarities between the motions of different points on a rotating rigid body. Perhaps we can find physical quantities that have the same value regardless of which point we consider. In Figure 8.1c, we see that during a particular time interval, all coins at the different points on the rotating disk turn through the same angle. Perhaps we should describe the rotational position of a rigid body using an angle.

Rotational (angular) position θ

Consider again a disk that rotates on a lab bench about a fixed point. The axis of rotation passes through the center of and is perpendicular to the disk

Figure 8.1 Top views comparing the velocities of coins traveling on a rotating disk.

(a)

The direction of the velocity $\vec{v}$ for each coin changes continually.

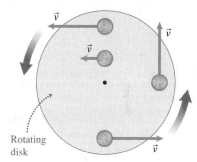

(b)

Coins at the edge travel farther during Δt than those near the center. The speed v will be greater for coins near the edge than for coins near the center.

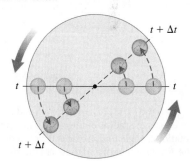

(c)

All coins turn through the same angle in Δt, regardless of their position on the disk.

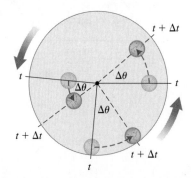

Figure 8.2 The rotational position (also called the angular position) of a point on a rotating disk. The units of rotational position can be either degrees or radians.

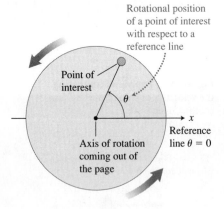

Rotational position of a point of interest with respect to a reference line

Point of interest

θ

Axis of rotation coming out of the page

Reference line $\theta = 0$

x

(**Figure 8.2**). A fixed line *perpendicular* to the axis of rotation (like the positive x-axis in Figure 8.2) is used as a reference line. We can draw another line on the disk from the axis of rotation to a point of interest, for example, to a coin sitting on the rotating disk. The angle θ in the counterclockwise direction between the reference line and the line to the point of interest is the **rotational position** (or **angular position**) of the point of interest. The observer is stationary beside the lab bench and looking down on the disk.

> **Rotational position** θ The rotational position θ of a point on a rotating object (sometimes called the angular position) is defined as an angle in the counterclockwise direction between a reference line (usually the positive x-axis) and a line drawn from the axis of rotation to that point. The units of rotational position can be either degrees or radians.

Units of rotational position

Two units are commonly used to indicate the rotational position θ of a point on a rotating object: degrees and radians. The degree (°) is the most familiar. There are 360° in a circle. If a point on the turning object is at the top of the circle, its position is 90° from a horizontal, positive x-axis. When at the bottom of the circle, its position is 270° or, equivalently, $-90°$.

The unit for rotational position that is most useful in physics is the **radian**. It is defined in terms of the two lengths shown in **Figure 8.3**. The arc length s is the distance in the counterclockwise direction along the circumference of the circle from the positive x-axis to the position of a point on the circumference of the rotating object. The other length is the radius r of the circle. The angle θ in units of radians (rad) is the ratio of s and r:

$$\theta \text{ (in radians)} = \frac{s}{r} \tag{8.1}$$

Figure 8.3 The rotational position θ in radians is the ratio of the arc length s and the radius r.

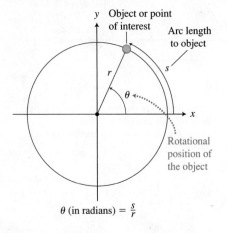

y Object or point of interest

Arc length to object

r

s

θ

x

Rotational position of the object

$\theta \text{ (in radians)} = \frac{s}{r}$

Note that the radian unit has no dimensions; it is the ratio of two lengths. We can multiply by the radian unit or remove the radian unit from an equation with no consequence. If we put the unit **rad** in the equation, it is usually because it is a reminder that we are using radians for angles.

> **TIP** From Eq. (8.1) we see that the arc length for a 1-rad angle equals the radius of the circle. For example, the 1-rad angle shown in **Figure 8.4** is the ratio of the 2-cm arc length and the 2-cm radius and is simply 1. If you use a calculator to work with radians, make sure it is in the radian mode.

Figure 8.4 The arc length for a 1-rad angle equals the radius of the circle. The 1-rad rotational angle in this case is the ratio of the 2-cm arc length and the 2-cm radius.

A 1-rad rotational position has equal arc length s and radius r.

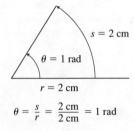

$s = 2 \text{ cm}$

$\theta = 1 \text{ rad}$

$r = 2 \text{ cm}$

$\theta = \frac{s}{r} = \frac{2 \text{ cm}}{2 \text{ cm}} = 1 \text{ rad}$

One complete rotation around a circle corresponds to a change in arc length of $2\pi r$ (the circumference of the circle) and a change in rotational position of

$$\theta \text{ (one complete rotation)} = \frac{s}{r} = \frac{2\pi r}{r} = 2\pi$$

Thus, there are 2π radians in one circle. We can now relate the two rotational position units:

$$360° = 2\pi \text{ rad}$$

We can use this equation to convert between degrees and radians.

We can use Eq. (8.1) to find the arc length s if the radius and rotational position θ are known:

$$s = r\theta \text{ (for } \theta \text{ in radians only)}$$

For example, if a car travels 2.0 rad (that is, from $\theta_0 = 0$ rad to $\theta = 2.0$ rad) around a highway curve of radius 100 m, the car travels a distance along the arc equal to

$$s = r(\theta - \theta_0) = (100 \text{ m})(2.0 \text{ rad} - 0) = 200 \text{ m}$$

We dropped the radian unit in the answer because angles measured in radians are dimensionless.

TIP You cannot calculate arc length using $s = r\theta$ when θ is measured in degrees. You must first convert θ to radians.

QUANTITATIVE EXERCISE 8.1 An old-fashioned watch

Your analog watch with hour and minute hands reads 3:30. What is the rotational position in radians of each of these hands? Use a reference line from the axis of rotation through the 12:00 position. Assume (contrary to reality) that the hour hand points directly at "3".

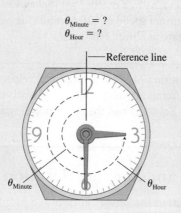

$\theta_{\text{Minute}} = ?$
$\theta_{\text{Hour}} = ?$

Reference line

θ_{Minute}
θ_{Hour}

Represent mathematically and solve The rotational position of the hour hand is the angle in the counterclockwise direction from the reference line to the 3:00 hour hand. The rotational position of the hour hand is three-quarters of the way around the clock from 12:00 going in the counterclockwise direction. So $\theta_{\text{Hour}} = +(3/4)2\pi = +3\pi/2$ radians. The rotational position of the minute hand is the angle in the counterclockwise direction from the reference line to the 6:00 minute hand. The positive counterclockwise position of the minute hand is $\theta_{\text{Minute}} = +(1/2)2\pi = +\pi$ radians.

Try it yourself: Your watch reads 6:15. What is the rotational position in radians from a 12:00 reference line to each hand of the clock? What assumption did you make?

Answer: $+\pi$ for the hour hand (assumption: it is still pointing at 6:00) and $+3\pi/2$ for the minute hand.

Rotational (angular) velocity ω

When we were investigating the motion of a point-like object along a single axis, we defined the translational velocity of that object as the rate of change of its linear position. Thus, it seems natural to define the **rotational (angular) velocity** ω of a rigid body as the rate of change of each point's rotational position. Because all points on the rigid body rotate through the same angle in the same period of time (see Figure 8.1c), each point on the rigid body has the same rotational velocity. This means we can just refer to the rotational velocity of the rigid body itself, rather than to any specific point within it.

Rotational velocity ω The average rotational velocity (sometimes called angular velocity) of a turning rigid body is the ratio of its change in rotational position $\Delta\theta$ and the time interval Δt needed for that change (see **Figure 8.5**):

$$\omega = \frac{\Delta\theta}{\Delta t} \qquad (8.2)$$

The sign of ω (omega) is positive for counterclockwise turning and negative for clockwise turning, as seen looking along the axis of rotation. *Rotational (angular) speed* is the magnitude of the rotational velocity. The most common units for rotational velocity and speed are radians per second (rad/s) and revolutions per minute (rpm).

Figure 8.5 Each point on a rigid body has the same rotational velocity ω.

Rotational velocity ω:
$$\omega = \frac{(\theta + \Delta\theta) - \theta}{(t + \Delta t) - t} = \frac{\Delta\theta}{\Delta t}$$

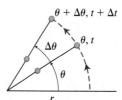

Rotational velocity is independent of the radius.

TIP Rotational velocity is the same for all points of a rotating rigid body. It is independent of the distance of a chosen point on the rigid body from the axis of rotation.

To distinguish the rotational velocity from the familiar velocity that characterizes the linear motion of an object, the latter is called linear velocity. When an object rotates, each point of the object has linear velocity. If you examine Figure 8.1a you see that the linear velocity vectors are tangent to the circle. Thus the linear velocity of a point on a rotating object is sometimes called **tangential velocity.**

The **revolution** is a familiar unit from everyday life. One revolution (rev) corresponds to a complete rotation about a circle and equals 360°. The revolution is not a unit of rotational position. It is a unit of *change* in rotational position $\Delta\theta$. Revolutions are usually used to indicate change in rotational position per unit time. For example, a motor that makes 120 complete turns in 1 min is said to have a rotational speed of 120 revolutions per minute (120 rpm). Automobile engines rotate at about 2400 rpm.

> **TIP** The definition of average rotational velocity or rotational speed becomes the instantaneous values of these quantities if you consider a small time interval in Eq. (8.2) and the corresponding small change in the rotational position.

Figure 8.6 Three rotational motion diagrams and the corresponding signs of the rotational accelerations.

(a)

$\Delta\theta$ is constant
ω is constant
$\alpha = 0$

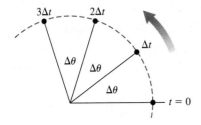

(b)

$\Delta\theta$ is increasing
ω is positive (counterclockwise) and increasing
$\alpha > 0$

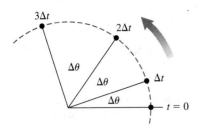

(c)

$\Delta\theta$ is decreasing
ω is positive (counterclockwise) and decreasing
$\alpha < 0$

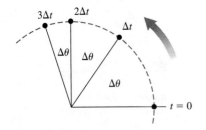

Rotational (angular) acceleration α

When we investigated the linear motion of a point-like object along a single axis, we developed the physical quantity acceleration to describe the object's change in velocity. This was translational acceleration, as it described the changing velocity of the object while moving from one position to another. We could apply the same translational acceleration idea to the center of mass of a rigid body that is moving as a whole from one position to another. But usually we are interested in the rate of change of the rigid body's rotational velocity, that is, its **rotational acceleration.** In other words, when the rotation rate of a rigid body increases or decreases, it has a nonzero rotational acceleration.

> **Rotational acceleration α** The average rotational acceleration α (alpha) of a rotating rigid body (sometimes called angular acceleration) is its change in rotational velocity $\Delta\omega$ during a time interval Δt divided by that time interval:
>
> $$\alpha = \frac{\Delta\omega}{\Delta t} \qquad (8.3)$$
>
> The unit of rotational acceleration is $(\text{rad/s})/\text{s} = \text{rad/s}^2$.

Figure 8.6 shows motion diagrams for three different types of rotational motion. Let's consider these rotational motion diagrams and try to develop a rule for how the sign of the rotational acceleration relates to the rotational velocity for a counterclockwise-turning disk. Note that when the disk's rotational velocity is constant (the lengths of arcs between the dots are the same), its rotational acceleration is zero (Figure 8.6a). When its counterclockwise rotational velocity (positive) is increasing (note that in Figure 8.6b the arcs between the dots increase in length), its rotational acceleration has the same sign (positive) as the rotational velocity. If the disk's counterclockwise rotational velocity (positive) is decreasing (note that in Figure 8.6c the arcs between the dots are shrinking), its rotational acceleration has the opposite sign (negative). Similarly, when a disk is rotating clockwise (negative rotational velocity) faster and faster, its rotational acceleration has the same sign (negative). If the disk's clockwise rotational velocity (negative) is decreasing, its rotational acceleration has the opposite sign (positive).

What could we conclude about signs of Earth's rotational velocity and acceleration if we were looking down on Earth from above the North Pole? The rotational velocity would have a positive sign (turning counterclockwise), and because the rotational velocity is constant, the rotational acceleration would be zero. (We will learn later that Earth's rotational velocity is not exactly constant.)

Relating translational and rotational quantities

Are there mathematical connections between physical quantities describing the rotational motion of a rigid body and the translational motion of different points on the body? Recall that the rotational position θ of a point on a turning object depends on the radial distance r of that point from the axis of rotation and the length s measured along the arc connecting that point to the reference axis (see Figure 8.3):

$$s = r\theta \tag{8.1}$$

If the angle changes by $\Delta\theta$, the distance of the point of the object along the arc changes by Δs, so that

$$\Delta s = r\,\Delta\theta$$

A similar relation exists between the speed of a point on a turning object and the rotational velocity of the turning object. An analogous relation also exists between the magnitude of acceleration of a point on a turning object and the rotational acceleration of the turning object. Suppose, for example, that a point on the object changes rotational position by $\Delta\theta$ in a time interval Δt. Its rotational velocity is $\omega = \Delta\theta/\Delta t$. The change in arc length is Δs along its circular path, and its tangential speed (the speed of the object tangent to the circle, sometimes called linear speed) is $v_t = \Delta s/\Delta t$. Substituting for Δs, we get

$$v_t = \frac{\Delta s}{\Delta t} = \frac{r\,\Delta\theta}{\Delta t} = r\left(\frac{\Delta\theta}{\Delta t}\right) = r\omega \tag{8.4}$$

Notice that while the rotational speed of all points of the same rigid body is the same, the tangential (linear) speed of different points increases as their distance from the axis of rotation increases. A similar relationship can be derived that relates that point's acceleration a_t tangent to the circle and its rotational acceleration α:

$$a_t = \frac{\Delta v_t}{\Delta t} = \frac{r\,\Delta\omega}{\Delta t} = r\left(\frac{\Delta\omega}{\Delta t}\right) = r\alpha \tag{8.5}$$

The signs of the rotational position and velocity are positive for counterclockwise turning, and the signs of the translational position and velocity are also positive for counterclockwise motion.

> **TIP** You get the familiar translational quantities for motion along the circular path by multiplying the corresponding angular rotational quantities by the radius r of the circle.

To visualize this relationship, imagine five people (the point objects in **Figure 8.7**) holding on to a long stick that can rotate horizontally about a vertical pole to which it is attached on one end. These people hold on to the stick as it completes a full circle. The person closest to the pole moves the slowest, the next person moves a little faster, and the one at the free end has to almost run to keep the stick in his hands. At a particular time, all of them have the same rotational position θ and the same rotational velocity ω. However, the linear distances and speeds are larger for the people farther from the axis of rotation (larger values of r).

> **TIP** The sign of the rotational acceleration is the same as the sign of the rotational velocity when the object rotates increasingly faster. The rotational acceleration has the opposite sign if the object is rotating increasingly slower.

Figure 8.7 A top-view diagram of five people (represented by dots) holding on to a stick that rotates about a fixed pole. Note that the speed v of each person depends on the person's distance from the axis of rotation.

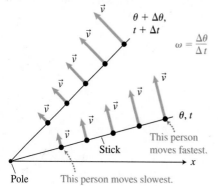

Top view

Five people (the dots) hold a stick that rotates about a fixed pole.

People sometimes play crack the whip while ice skating. The skaters hold hands and skate in a line. Then the person on one end of the line stops and the others continue moving but are pulled toward the stopped person. The line now moves in a circle with the farthest person going very fast, like the end of a whip.

Black holes are an extreme case of rotational motion. Black holes form when some stars at the end of their lives collapse, forming small, very dense objects. If the star was spinning when it was young, it continues to spin when it becomes a black hole, only much faster. The matter near the outer edge of the black hole is usually extremely hot gas that orbits the black hole with a tangential speed near the speed of light. The figure below is an artist's rendition of what such a black hole might look like, if we could see it.

QUANTITATIVE EXERCISE 8.2 Orbiting a black hole

Black hole GRS 1915+105 in the constellation Aquila (the Eagle) is about 35,000 light-years from Earth. It was formed when the core of a star with about 14 solar masses (mass of 14 times the mass of our Sun) collapsed. The boundary of the black hole, called the event horizon, is a sphere with radius about 25 km. Surrounding the black hole is a stable gaseous cloud with an innermost 30-km-radius stable circular orbit. This cloud moves in a circle about the black hole about 970 times per second. Determine the tangential speed of matter in this innermost stable orbit.

Black hole

Gaseous cloud just outside black hole

Represent mathematically We model the cloud as a rotating disk with radius $r = 30,000$ m; to find the rotational speed we convert the rotations in revolutions per second into radians per second: $\omega = (970 \text{ rev/s})(2\pi \text{ rad/rev}) = 6000 \text{ rad/s}$. We need to find the tangential speed v_t of the particles of gas in orbit around the black hole, which is related to the radius r of the circular orbit and the rotational speed ω of the matter:

$$v_t = r\omega$$

Solve and evaluate The speed of the matter in the innermost stable orbit will be

$$v_t = r\omega = (30,000 \text{ m})(6000 \text{ rad/s}) = 1.8 \times 10^8 \text{ m/s}$$

This is slightly more than half the speed of light! Actually, the physics we have developed is only moderately applicable in this environment of extreme gravitational forces and high speeds. Our answer is about 20% high compared to a more sophisticated analysis done using Einstein's theory of general relativity. Nevertheless, using ideas from uniform circular motion, we estimate that the radial acceleration of matter moving in that circular orbit is about $v^2/r \approx 10^{11} g$. It's not a good place to visit.

Try it yourself: You ride a carnival merry-go-round and are 4.0 m from its center. A motion detector held by your friend next to the merry-go-round indicates that you are traveling at a tangential speed of 5.0 m/s. What are the rotational speed and time interval needed to complete one revolution on the merry-go-round?

Answer: $\omega = 1.3 \text{ rad/s}$; $T = 5.0 \text{ s}$.

Rotational motion at constant acceleration

Earlier, we developed equations that related the physical quantities t, x, v, and a, which we used to describe the translational motion of a point-like object along a single axis with constant acceleration. Similar equations relate the rotational kinematics quantities t, θ, ω, and α, assuming the rotational acceleration is constant. We're not going to develop them based on observations in the way we did for the equations of translational motion, since the process will be

Table 8.1 Equations of kinematics for translational motion with constant acceleration and the analogous equations for rotational motion with constant rotational acceleration.

Translational motion	Rotational motion	
$v_x = v_{0x} + a_x t$	$\omega = \omega_0 + \alpha t$	(8.6)
$x = x_0 + v_{0x}t + \dfrac{1}{2}a_x t^2$	$\theta = \theta_0 + \omega_0 t + \dfrac{1}{2}\alpha t^2$	(8.7)
$2a_x(x - x_0) = v_x^2 - v_{0x}^2$	$2\alpha(\theta - \theta_0) = \omega^2 - \omega_0^2$	(8.8)

nearly the same. Instead, we will rely on the connections we have seen between the translational and rotational quantities. The analogous rotational motion equations are provided in **Table 8.1** along with the corresponding translational motion equations. Because the quantities that describe motion depend on the choice of the reference frame, always note the location of the observer in a particular situation.

For rotational motion, θ_0 is an object's rotational position at time $t_0 = 0$; ω_0 is the object's rotational velocity at time $t_0 = 0$; θ and ω are the rotational position and rotational velocity at some later time t; and α is the object's constant rotational acceleration during the time interval from time zero to time t. The sign of the rotational position is positive for counterclockwise θ and negative for clockwise θ from the reference axis. The sign of the rotational velocity ω depends on whether the object is rotating counterclockwise ($+$) or clockwise ($-$). The sign of the rotational acceleration α depends on how the rotational velocity is changing; α has the same sign as ω if the magnitude of ω is increasing and the opposite sign of ω if ω's magnitude is decreasing.

Review Question 8.1 Visualize an ice skater rotating faster and faster in a clockwise direction. What are the signs of rotational velocity and rotational acceleration? Then the skater starts slowing down. What are the signs of rotational velocity and acceleration now?

8.2 Torque and rotational acceleration

What causes a rigid body to have a particular rotational acceleration? When we investigated translational motion we learned that the acceleration of a point-like object was determined by its interactions with other objects, that is, forces that objects in the environment exerted on it. Perhaps there is an analogous way to think about what causes rotational acceleration. In the last chapter, we learned that the net torque produced by forces exerted on a system had to equal zero for the object to remain in static equilibrium, to not start rotating. What happens when the net torque isn't zero? Let's investigate this. In Observational Experiment **Table 8.2**, we perform four experiments with a bicycle that rests upside down so its front tire can rotate freely (**Figure 8.8**).

Figure 8.8 The front tire of an inverted bicycle, used for the experiments in Observational Experiment Table 8.2.

OBSERVATIONAL EXPERIMENT TABLE

8.2   Turning effects of forces exerted on a bicycle tire.

VIDEO 8.2

Observational experiment	Analysis
Experiment 1: Your bike sits upside down. You push on the front tire toward the axle. The tire does not turn.	A force is exerted on the tire, but the torque produced by the force about the axis of rotation (the axle) is zero. It has no effect on the tire's rotation.

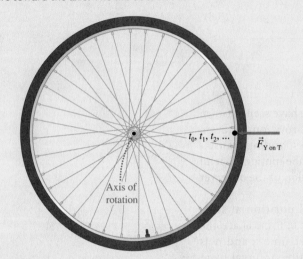

Experiment 2: You push lightly and continuously on the outside of the tire in a counterclockwise (ccw) direction tangent to the tire. As you continue to push, the tire rotates ccw faster and faster.	Your pushing causes a force and a ccw torque. The tire has increasing positive rotational velocity and a positive rotational acceleration.

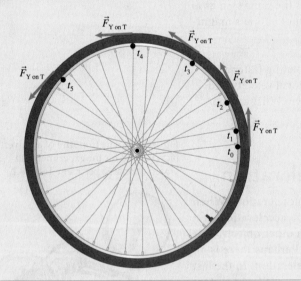

Experiment 3: You release the spinning tire and watch it. The tire continues rotating ccw at a constant rate.

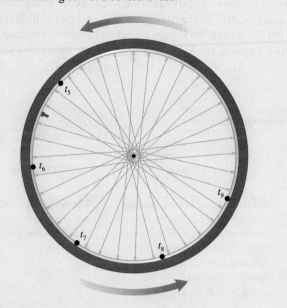

When the force and torque are zero, there is no change in rotation rate. The rotational velocity is constant and the rotational acceleration is zero.

Experiment 4: With the tire still rotating ccw fast, you gently and continuously push clockwise (cw) against the tire. The rotational speed decreases.

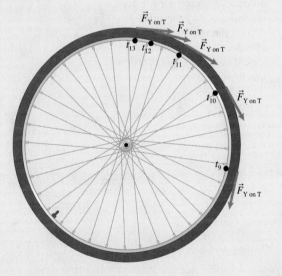

Pushing opposite the rotation causes a force and a cw torque and a decreasing ccw rotation rate. There is a clockwise (negative) rotational acceleration.

Patterns

- An external force that produces a *zero torque* on the tire does not change the tire's rotation rate. If the tire is at rest, it remains at rest.
- When there are no external forces exerting torques on a tire, its rotational velocity remains constant.
- An external force that produces a *nonzero torque* on the tire:
  - causes the tire to turn faster and faster if in the direction the tire is turning; and
  - causes the tire to turn slower and slower if opposite the direction it is turning.

The experiments in Table 8.2 indicate that a zero torque has no effect on the rotational motion of a rigid body. However, a nonzero torque does cause a change. If the torque is in the same direction as the direction of rotation of the rigid body, its rotational speed increases. If the torque is in the opposite direction, the rigid body's rotational speed decreases. These patterns are similar to the patterns that we found for translational motion.

Our goal is to determine what physical quantities cause rotational acceleration of an extended object. There really are just two possibilities: (1) the sum of the forces (net force) exerted on the object or (2) the net torque caused by the forces. Testing Experiment **Table 8.3** will help us determine which (if either) of these quantities might affect rotational acceleration.

TESTING EXPERIMENT TABLE

8.3 Testing two hypotheses explaining rotational acceleration.

Testing experiment	Prediction	Outcome
You have a cylinder that can rotate on an axle. Half of the cylinder's thickness has a large radius (R) and the other half has a smaller radius r. The axle is fixed. When the cylinder rotates, both parts rotate. **Experiment 1.** Wrap a string around the part of the cylinder with the small radius r and pull the string, exerting force F on the string. **Experiment 2.** Wrap a string around the part of the cylinder with the large radius R and pull the string, exerting the same force F on the string. In which experiment will the cylinder have a greater rotational acceleration?	*Prediction based on the hypothesis that rotational acceleration depends on the net force (sum of the forces):* Since the forces exerted on the cylinders are equal, the rotational acceleration should be the same in both experiments. *Prediction based on the hypothesis that rotational acceleration depends on the net torque:* Since the string pulling the large-radius cylinder produces a greater torque than the string pulling the small-radius cylinder, the cylinder of radius R in Experiment 2 should have the larger rotational acceleration.	The cylinder's rotational acceleration is greater in Experiment 2 than in Experiment 1.

Conclusions

- The outcome *does not* match the prediction based on the hypothesis that the *net force* exerted on an object causes its rotational acceleration.
- The outcome *does* match the prediction based on the hypothesis that the *net torque* exerted on an object causes its rotational acceleration.

Notice that the force exerted on the wheel in the first experiment in Table 8.2 produced a nonzero force and a zero torque on the wheel, and the wheel did not turn. Therefore, that experiment already disproves the hypothesis that the net force exerted on a rigid body causes its rotational acceleration. Additionally, the experiments in Table 8.3 disprove the force hypothesis and support the following provisional rule:

Changes in rotational velocity Rotational acceleration depends on net torque. The greater the net torque, the greater the rotational acceleration.

   Notice how this is similar to what we learned when studying translational motion. A nonzero net force (sum of the forces) needs to be exerted on an object to cause its velocity to change. The greater the net force, the greater the translational acceleration of the object.

   Remember that in all of the experiments we have performed so far in this chapter, rotational motion was all that was possible. Each rigid body was rotating about a fixed axis through its center of mass. If the rigid body were not held fixed, then a change in both translational and rotational motion could occur. The translational acceleration of the center of mass of such an object is determined by Newton's second law $\vec{a} = \Sigma\vec{F}/m$. The rotational acceleration around its center of mass will be determined by the ideas we will investigate over the next several sections.

Review Question 8.2 How do we know that rotational acceleration of an object depends on the torque, not the net force exerted on it?

8.3 Rotational inertia

We have found that a nonzero net torque will cause an object's rotational velocity to change. What other quantities might affect rotational acceleration? We know that the translational acceleration of an object is directly proportional to the net force exerted on it and inversely proportional to its mass. It seems that mass should also somehow affect an object's rotational acceleration. However, a simple experiment allows us to see that the mass alone is not the answer. Lay a broom on a hard smooth floor. First try to increase the broom's rotational speed about a vertical axis by turning it with one hand holding the broomstick far from the broom head. Then, do it again, holding the broom in the middle nearer the broom head. It is much easier to increase the rotational speed holding it in the middle. You are turning the same mass, but the location of that mass from the axis of rotation seems to make a difference. We explore this idea in Observational Experiment **Table 8.4**.

OBSERVATIONAL EXPERIMENT TABLE

8.4   Effect of mass distribution on rotational acceleration.

Observational experiment	Analysis
Two cylinders of the same radius and the same mass can rotate around an axis passing through their centers. One cylinder is solid and made of wood, and the other is hollow and made of iron. You wrap a string around each cylinder and pull equally hard on the strings wrapped around each cylinder. Both cylinders start accelerating, but the solid cylinder has a greater rotational acceleration than the hollow cylinder.	If we assume that the force and the torque exerted on the cylinders are the same, then the only difference is the distribution of mass of the cylinders. Most of the mass of the hollow cylinder is located far from the axis of rotation. The mass of the solid cylinder is distributed more uniformly, some of it closer to the axis of rotation.

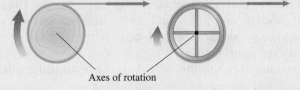

Axes of rotation

Pattern

The same torque produces smaller rotational acceleration if the mass of the rigid body is distributed farther from the axis of rotation.

Figure 8.9 Rotational acceleration decreases if the mass is farther from the axis of rotation, as in (b).

(a) Blocks close to axis of rotation

(b) Blocks far from axis

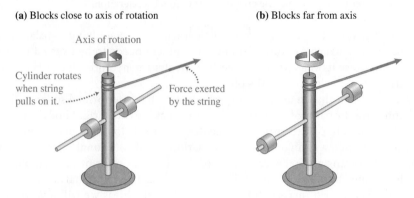

The pattern found in Table 8.4 is consistent with our simple broom rotation experiment. *If* it is correct and we change the distribution of the mass of a rigid body while keeping the total mass the same, *then* the rotational acceleration of the object due to a particular torque should decrease when the mass is moved farther away from the axis of rotation. Let's use this statement as a prediction of the outcome of a new experiment.

We use an apparatus that consists of metal blocks that can be moved on a metal rod so that they are different distances from the axis of rotation—see **Figure 8.9**. If we pull each string exerting the same force (represented by the solid red arrow), will the rotational acceleration be greater for the arrangement shown in Figure 8.9a or for that in Figure 8.9b? According to the above explanation, we predict that the rotational acceleration will be greater for arrangement (a) than arrangement (b) since the mass is nearer the axis of rotation in (a) than in (b). When we try the experiment, we find that the rate of rotation for (a) does change faster than for (b), consistent with the pattern we found in Table 8.4.

Thus, another important factor that affects the rotational acceleration of a rigid body is the *distribution* of mass of the rotating object. The closer the mass of the object to the axis of rotation, the easier it is to change its rotational motion. We call the physical quantity characterizing the location of the mass relative to the axis of rotation the **rotational inertia** (also known as the **moment of inertia**) of the object. Rotational inertia I depends on both the total mass of the object and the distribution of that mass about its axis of rotation. For objects of the same mass, the more mass that is located near the axis of rotation, the smaller the object's rotational inertia will be. Likewise, the more mass that is located farther away from the axis of rotation, the greater the object's rotational inertia will be. For objects of different mass but the same mass distribution, the more massive object has more rotational inertia. The higher the rotational inertia of an object, the harder it is to change its rotational motion. In summary, this quantity is the rotational equivalent of mass.

A common example of the effect of mass distribution on the change of the rotational velocity of an object involves swinging a baseball bat. When you "choke up" on the bat, that is, hold it farther up on the handle, you move the mass of the bat closer to your hands, and the bat becomes easier to swing. By distributing the mass so that more of it is located near the axis of rotation, you decrease the bat's rotational inertia.

We have now found two factors that affect the rotational acceleration of an object:

- The rotational inertia of the object
- The net torque produced by forces exerted on the object

In the next two sections we will investigate quantitatively how these two factors affect the rotational acceleration of an object.

A solid wooden ball and a smaller solid metal ball have equal mass (the metal ball is smaller because it is much denser than wood). Both can rotate on an axis going through their centers. You exert a force on each that produces the same torque about the axis of rotation. Which sphere's rotational motion will change the least? Explain.

8.4 Newton's second law for rotational motion

Let's see if we can construct a quantitative relationship between rotational acceleration, net torque, and rotational inertia. We start with a simple example of rotational motion: a small block attached to a light stick that can move on a smooth surface in a circular path (**Figure 8.10**). The axis of rotation passes through a pin at the other end of the stick. After we analyze this case, we will generalize the result to the rotation of an extended rigid body. You push the block with your finger, exerting a small force $\vec{F}_{\text{F on B}}$ on the block tangent to the circular path. This push causes a torque, which in turn causes the block and stick's rotational velocity about the pin to increase.

The torque produced by the force $\vec{F}_{\text{F on B}}$ is

$$\tau = rF_{\text{F on B}} \sin\theta = rF_{\text{F on B}} \sin 90° = rF_{\text{F on B}}$$

Since the block is small, we can reasonably model it as a point-like object. This allows us to apply Newton's second law. Since the mass of the block is much larger than the mass of the stick, we assume that the stick has no mass. The finger exerts a force of constant magnitude pushing lightly in a direction tangent to the block's circular path. Thus, the tangential component of Newton's second law for the block is

$$a_{\text{t}} = \frac{F_{\text{F on B}}}{m_{\text{B}}}$$

There is a mathematical way to get the torque produced by the pushing force. Rearrange the above equation to get

$$m_{\text{B}} a_{\text{t}} = F_{\text{F on B}}$$

Then, multiplying both sides of the equation by r, the radius of the circular path:

$$m_{\text{B}}\, r\, a_{\text{t}} = rF_{\text{F on B}}$$

Recall from Eq. (8.5) that $a_{\text{t}} = r\alpha$. Thus,

$$m_{\text{B}}\, r\, (r\alpha) = rF_{\text{F on B}}$$

The right side of this equation equals the torque τ caused by $\vec{F}_{\text{F on B}}$:

$$(m_{\text{B}} r^2)\, \alpha = \tau$$

$$\Rightarrow \alpha = \frac{\tau}{m_{\text{B}} r^2}$$

Examine the above equation and compare it to Newton's second law for the same object—the block not only acquires translational acceleration a_{t} due to the force exerted on it by the finger, but also acquires rotational acceleration around

Figure 8.10 A top view of an experiment to relate torque and rotational acceleration.

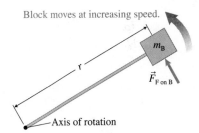

Block moves at increasing speed.

Your finger (not shown) pushes the block, causing its rotational acceleration.

the axis caused by the torque produced by that same force. This rotational acceleration is directly proportional to the torque produced by the force and inversely proportional to the mass of the block times the square of the distance between the block and the axis of rotation. The latter makes sense—we found experimentally that the farther the mass of the object is from the axis of rotation, the harder it is to change its rotational velocity. Thus the denominator in the equation above is an excellent candidate for the rotational inertia of the block about the pin (the axis of rotation in this situation).

In the above thought experiment, there was just a single force exerted on the object producing a torque about the axis of rotation. More generally, there could be several forces producing torques. It's reasonable that we should add the torques produced by all forces exerted on the object to determine its rotational acceleration:

$$\alpha = \frac{1}{m_B r^2}\Sigma\tau = \frac{1}{m_B r^2}(\tau_1 + \tau_2 + \ldots) \tag{8.9}$$

where $\tau_1, \tau_2, \ldots$ are the torques produced by forces $\vec{F}_{1\,\text{on O}}, \vec{F}_{2\,\text{on O}}, \ldots$ exerted on the object.

Analogy between translational motion and rotational motion

Notice how similar Eq. (8.9)

$$\alpha = \frac{1}{m_B r^2 \Sigma\tau}$$

is to Newton's second law for translational motion

$$\vec{a} = \frac{1}{m\Sigma\vec{F}}$$

When the same forces are exerted on a point-like object, we can describe its motion using two acceleration-type quantities—translational and rotational acceleration. The translational acceleration is determined by Newton's second law, and the rotational acceleration is determined by Eq. (8.9), called **Newton's second law for rotational motion.** There is a strong analogy between each of the three quantities in the two equations (see **Table 8.5**):

For translational motion, mass is a measure of an object's translational inertia—the tendency for its motion to not change. For the rotational motion of a point-like object, the object's mass times the square of its distance r from the axis of rotation (mr^2) is a measure of the object's rotational inertia—the tendency for its rotational motion to not change. In summary, the quantity mr^2 is the rotational inertia I of a point-like object of mass m around the axis that is the distance r from the location of the object.

For translational motion, the net force $\Sigma\vec{F}$ exerted on an object of interest by other objects causes that object's velocity to change—it has a translational acceleration ($\vec{a} = \Delta\vec{v}/\Delta t$). For rotational motion, the net torque $\Sigma\tau$ produced by forces exerted on the object causes its rotational velocity to change—it has a rotational acceleration ($\alpha = \Delta\omega/\Delta t$).

Table 8.5 Analogy between translational and rotational quantities in Newton's second law.

	Translational motion	Rotational motion
Inertia of a point-like object	m	mr^2
Cause of acceleration	$\Sigma\vec{F}$	$\Sigma\tau$
Acceleration	$\vec{a}$	α

EXAMPLE 8.3   Pushing a rollerblader

A 60-kg rollerblader holds a 4.0-m-long rope that is loosely tied around a metal pole. You push the rollerblader, exerting a 40-N force on her, which causes her to move increasingly fast in a counterclockwise circle around the pole. The surface she skates on is smooth, and the wheels of her rollerblades are well oiled. Determine the tangential and rotational acceleration of the rollerblader.

Sketch and translate We sketch the situation as shown below. We choose the rollerblader as the system object of interest.

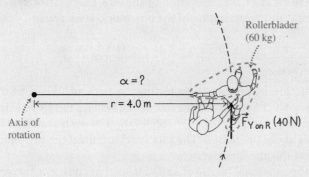

Simplify and diagram Since the size of the rollerblader is small compared to the length of the rope, we can model her as a point-like object. The figure below shows a force diagram for the rollerblader (viewed from above). Her tangential acceleration has no vertical component along a vertical axis (which would extend up and out of the page in the figure). The upward normal force $\vec{N}_{F \text{ on } R}$ that the floor exerts on her balances the downward gravitational force $\vec{F}_{E \text{ on } R}$ exerted by Earth on her (these forces are not shown in the figure). The tension force exerted by the rope on the rollerblader $\vec{T}_{Rope \text{ on } R}$ points directly toward the axis of rotation, so that force produces no torque.

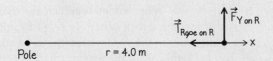

Represent mathematically From the force diagram we conclude that your pushing $\vec{F}_{Y \text{ on } R}$ on the rollerblader is the only force that produces a nonzero torque $\tau = rF_{Y \text{ on } R} \sin 90° = rF_{Y \text{ on } R}$, where r is the radius of the rollerblader's circular path.

Use Newton's second law in the tangential direction to determine the rollerblader's tangential acceleration:

$$a_t = \frac{1}{m_R}\Sigma F_t = \frac{F_{Y \text{ on } R}}{m_R}$$

Use Newton's second law for rotational motion [Eq. (8.9)] to determine the rollerblader's rotational acceleration:

$$\alpha = \frac{1}{m_R r^2}\Sigma\tau = \frac{1}{m_R r^2}\tau_{Y \text{ on } R} = \frac{1}{m_R r^2}(rF_{Y \text{ on } R}) = \frac{F_{Y \text{ on } R}}{m_R r}$$

Solve and evaluate For the tangential acceleration:

$$a_t = \frac{F_{Y \text{ on } R}}{m_R} = \frac{40 \text{ N}}{60 \text{ kg}} = 0.67 \text{ m/s}^2$$

For the rotational acceleration:

$$\alpha = \frac{F_{Y \text{ on } R}}{m_R r} = \frac{40 \text{ N}}{(60 \text{ kg})(4.0 \text{ m})} = 0.17 \text{ rad/s}^2$$

Let's check the units; note that $\frac{\text{N}}{\text{kg m}} = \frac{\text{kg} \cdot \text{m}}{\text{s}^2 \cdot \text{kg m}} = \frac{1}{\text{s}^2}$. Remember that the radian is not an actual unit. It is dimensionless. It is just a reminder that this is the angle unit appropriate for these calculations. So the units are correct, and the magnitudes for both results are reasonable. The rollerblader would have a rotational velocity of 0.17 rad/s after 1 s, 0.34 rad/s after 2 s, and so forth—the rotational velocity increases 0.17 rad/s each second.

Try it yourself: Suppose you exerted the same force on your friend, but the friend is holding an 8.0-m-long rope instead of a 4.0-m-long rope. How will this affect the rotational acceleration?

Answer: The tangential acceleration will not change, but the rotational acceleration will be half as large. The torque will be doubled because the distance from the axis of rotation to the point where the force is applied will be doubled. But the rotational inertia (mr^2) in the denominator will be quadrupled because of the r^2. The combination will be a reduction of the rotational acceleration by half—0.08 rad/s².

Newton's second law for rotational motion applied to rigid bodies

We know that the mass of an object composed of many small objects with masses m_1, m_2, m_3, etc. is the sum of the masses of its parts: $m = m_1 + m_2 + m_3 + \ldots$. Mass is a scalar quantity and therefore is always positive. The rotational inertia

Figure 8.11 Top views comparing the effect of rotational inertia on rotational acceleration.

(a)

The rotational inertia I of the two-block system should be twice that of the one-block system.

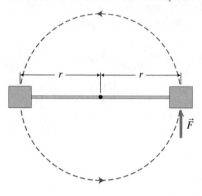

(b)

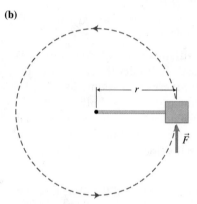

Thus, we predict that with equal torque and using $\alpha = \tau/I$,

$$\alpha_{\text{One block}} = 2\alpha_{\text{Two blocks}}$$

of a point-like object with respect to some axis of rotation is a scalar quantity. Thus, it is reasonable to think that the same rule applies to rigid bodies: the rotational inertia of a rigid body about some axis of rotation is the sum of the rotational inertias of the individual point-like objects that make up the rigid body.

To test this idea, we can use it to calculate the rotational inertia of a lightweight stick with a block attached to each end, as shown in **Figure 8.11a**. The axis of rotation is at the middle of the stick. If our reasoning is correct, the rotational inertia of this two-block rigid body should be twice the rotational inertia of a single block at the end of a stick that is half the length (Figure 8.11b):

$$I_{\text{Two block}} = 2I_{\text{One block}}.$$

If this is correct, and we exert the same torque ($\tau = rF \sin 90° = rF$) on both rigid bodies, the rotational acceleration of the two-block–stick system should be half the rotational acceleration of the one-block–stick system:

$$\alpha_{\text{Two block}} = \frac{\tau}{I_{\text{Two block}}} = \frac{\tau}{2I_{\text{One block}}} = \frac{1}{2}\left(\frac{\tau}{I_{\text{One block}}}\right) = \frac{1}{2}\alpha_{\text{One block}}$$

We check this prediction by performing a testing experiment. We exert the same force on the two-block system and the one-block system, thus producing the same torque, and measure the angular acceleration of each system. The outcome matches the above prediction. The rotational inertia of the two-block system is twice that of the one-block system.

It appears that the rotational inertia of a rigid body that consists of several point-like parts located at different distances from the axis of rotation is the sum of the mr^2 terms for each part:

$$I = m_1 r_1^2 + m_2 r_2^2 + \ldots$$

Let's apply this idea.

QUANTITATIVE EXERCISE 8.4

Use what you learned about rotational inertial to write an expression for the rotational inertia of the rigid body shown below. Each of the four blocks has mass m. They are connected with lightweight sticks of equal length $L/4$.

| $\leftarrow$——— L ———$\rightarrow$ |
| $\leftarrow L/4 \rightarrow \leftarrow L/4 \rightarrow \leftarrow L/4 \rightarrow \leftarrow L/4 \rightarrow$ |

$I = ?$

m m m m

Axis of rotation

Represent mathematically Each block of mass m contributes differently to the rotational inertia of the system. The farther the block from the axis of rotation, the greater its contribution to the rotational inertia of the system of blocks. We can add the rotational inertia of each block of mass m about the axis of rotation:

$$I = m(L/4)^2 + m(2L/4)^2 + m(3L/4)^2 + m(4L/4)^2$$

Solve and evaluate When added together, the rotational inertia of the four-block system is $I = 1.88\ mL^2$. Each block contributes to the system's rotational inertia. However, blocks farther from the axis of rotation contribute much more than those near the axis. In fact, the block at the right side of the rod contributes more to the rod's rotational inertia (mL^2) than the other three blocks combined ($0.88\ mL^2$).

Try it yourself: Calculate the rotational inertia of the same system altered so that the axis of rotation passes perpendicular through the rod halfway between the two central blocks.

Answer: $0.313\ mL^2$.

Calculating rotational inertia

We can calculate the rotational inertia of a rigid body about a specific axis of rotation in about the same way we determined the rotational inertia of the four-block system in Quantitative Exercise 8.4, by adding the rotational inertias of each part of the entire system. However, for most rigid bodies, the parts are not separate objects but are instead parts in a continuous distribution of mass—like in a door or baseball bat. In such a case we break the continuous distribution of mass into a very large number of very small pieces and add the rotational inertias for all of the pieces. Consider the person's leg shown in **Figure 8.12**, which we model as a rigid body if none of the joints bend. For example, mass element 7 contributes an amount $m_7 r_7^2$ to the rotational inertia of the leg. The rotational inertia of the whole leg is then

$$I = m_1 r_1^2 + m_2 r_2^2 + \ldots + m_7 r_7^2 + \ldots + m_{18} r_{18}^2 \qquad (8.10)$$

All r's in the above are from the same axis of rotation. The rotational inertia would be different if we chose a different axis of rotation. There are other ways to do the summation process in Eq. (8.10); often it is done using integral calculus, and sometimes I is determined experimentally.

Table 8.6 gives the rotational inertias of some common objects for specific axes of rotation. Notice the coefficients in front of the mR^2 and mL^2 expressions for the objects of different shapes. The value of the coefficient is determined by

Figure 8.12 Add the mr^2 of all the small parts to find the rotational inertia I of the leg.

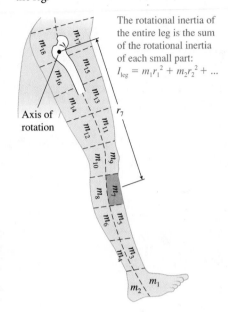

The rotational inertia of the entire leg is the sum of the rotational inertia of each small part:
$$I_{\text{leg}} = m_1 r_1^2 + m_2 r_2^2 + \ldots$$

Table 8.6 Expressions for the rotational inertia of standard shape objects.

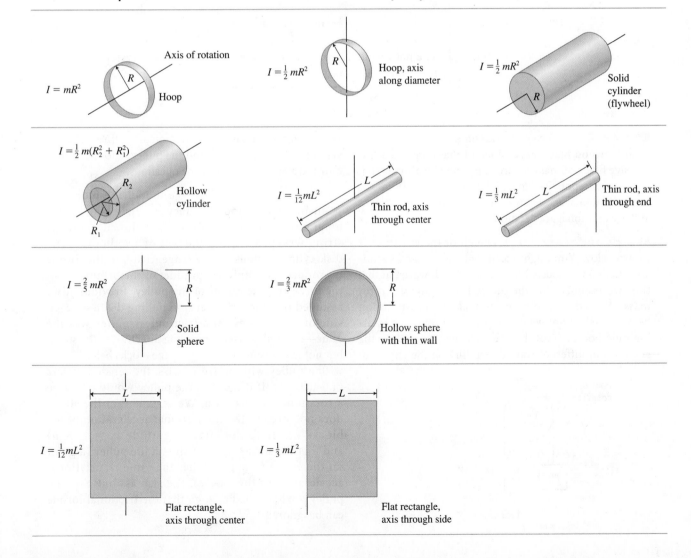

the mass distribution inside the object and the location of the axis of rotation. The closer the mass is to the axis of rotation, the less effect it has on the rotational inertial of the object. The same object has different rotational inertias for different axes of rotation.

We can now rewrite the rotational form of Newton's second law in terms of the rotational inertia of the rigid body.

Rotational form of Newton's second law One or more objects exert forces on a rigid body with rotational inertia I that can rotate about some axis. The sum of the torques $\Sigma\tau$ due to these forces about that axis causes the object to have a rotational acceleration α:

$$\alpha = \frac{1}{I}\Sigma\tau \qquad (8.11)$$

TIP By writing Newton's second law in the form

$$\vec{a}_S = \frac{1}{m_S}\Sigma\vec{F}_{\text{on S}} = \frac{\vec{F}_{O_1 \text{ on S}} + \vec{F}_{O_2 \text{ on S}} + \ldots + \vec{F}_{O_n \text{ on S}}}{m_S}$$

we see the cause-effect relationship between the net force $\Sigma\vec{F}_{\text{on S}}$ exerted on the system and the system's resulting translational acceleration $\vec{a}_S$. The same idea is seen in Eq. (8.11), only applied to the rotational acceleration:

$$\alpha = \frac{1}{I}\Sigma\tau = \frac{\tau_1 + \tau_2 + \ldots + \tau_n}{I}$$

The net torque $\Sigma\tau$ produced by forces exerted on the system causes its rotational acceleration α.

EXAMPLE 8.5 Atwood machine

In an Atwood machine, a block of mass m_1 and a less massive block of mass m_2 are connected by a string that passes over a pulley of mass M and radius R. What are the translational accelerations a_1 and a_2 of the two blocks and the rotational acceleration α of the pulley?

Sketch and translate A sketch of the situation is shown below. You might recall that we analyzed a similar situation previously (in Section 3.3). However, at that time we assumed that the pulley had negligible (zero) mass. We had no choice but to make that assumption because we had not yet developed the physics for rotating rigid bodies. Now that we have, we can analyze the situation in different ways depending on the choice of

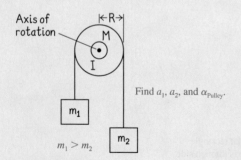

Axis of rotation

Find a_1, a_2, and α_{Pulley}.

$m_1 > m_2$

system. We will analyze the situation using three separate systems: block 1, block 2, and the pulley, and then combine the analyses to answer the questions.

Simplify and diagram We model the blocks as point-like objects and the pulley as a rigid body. Force diagrams for all three objects are shown below. A string wrapped around the rim of a pulley (or any disk/cylinder) pulls purely tangentially, so the torque it produces is simply the product of the magnitude of the force and the radius of the pulley. Previously, we assumed that the string tension pulling down on each side of the massless and frictionless pulley was the same—the pulley just changed the direction of the string but not the tension it exerted on the blocks below. Now, with a pulley with nonzero mass, the tension differs on each side. If it did not, the pulley would not have a rotational acceleration. We assume that the string does not stretch (the translational acceleration of the blocks will have the same magnitude $a_1 = a_2 = a$) and that the string does not slip on the pulley (a point on the edge of the pulley has the same translational acceleration as the blocks). We also assume that the pulley's axle is oiled enough that the frictional torque can be ignored.

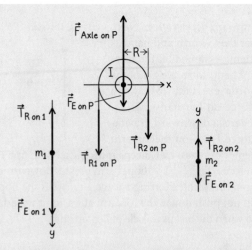

The translational acceleration of the hanging objects is due to the difference between the gravitational force that Earth exerts on them and the tension force that the string exerts on them. The rotational acceleration of the pulley is due to a nonzero net torque produced by the two tension forces exerted on the pulley.

We consider the pulley to be similar to a solid cylinder. Then according to Table 8.6, its rotational inertia around the axis that passes through its center is $I = \frac{1}{2}MR^2$, where R is the radius of the pulley and M is its mass.

Represent mathematically The force diagrams help us apply Newton's second law in component form for the two blocks and the rotational form for the pulley. The coordinate systems used in each case are shown. We choose the coordinate systems so the translational accelerations of both blocks are positive.

Block on the left: $m_1 a = +m_1 g + (-T_{R1\ on\ 1})$

Block on the right: $m_2 a = -m_2 g + T_{R2\ on\ 2}$

The pulley: $T_{R1\ on\ P}R + (-T_{R2\ on\ P}R) = I\alpha = \left(\dfrac{MR^2}{2}\right)\left(\dfrac{a}{R}\right)$

$\Rightarrow (T_{R1\ on\ 1})R - (T_{R2\ on\ 2})R = \dfrac{MRa}{2}$

$\Rightarrow T_{R1\ on\ 1} - T_{R2\ on\ 2} = \dfrac{Ma}{2}$

We now have three equations with three unknowns—the two tension forces exerted by the rope on the pulley and the magnitude of the acceleration a of the blocks. We can write expressions for $T_{R1\ on\ 1}$ and $T_{R2\ on\ 2}$ using the first two equations. We then have

$$T_{R1\ on\ 1} = m_1 g - m_1 a$$
$$T_{R2\ on\ 2} = m_2 g + m_2 a$$

After substituting these expressions for the rope forces into the pulley equation, we get

$$(m_1 g - m_1 a) - (m_2 g + m_2 a) = Ma/2$$

Solve and evaluate This equation can be rearranged to get an expression for the translational acceleration of the blocks:

$$a = \frac{m_1 - m_2}{\frac{1}{2}M + m_1 + m_2}g$$

Notice that if we neglect the mass of the pulley, the acceleration becomes

$$a = \frac{m_1 - m_2}{m_1 + m_2}g,$$

a larger acceleration than with the pulley (and consistent with the result we got in Chapter 3). Thus, the massive pulley decreases the acceleration of the blocks, which makes sense. If the pulley mass M is much heavier than the masses of the hanging objects m_1 and m_2, the acceleration becomes very small—it is almost like the blocks are hanging from a fixed massive object and not moving at all. We can find the rotational acceleration of the pulley by dividing the translational acceleration by the radius of the pulley:

$$\alpha = \frac{a}{R} = \left(\frac{m_1 - m_2}{\frac{1}{2}M + m_1 + m_2}\right)\frac{g}{R}$$

Try it yourself: Determine the translational acceleration of the blocks and the rotational acceleration of the pulley for the following given information: $m_1 = 1.2$ kg, $m_2 = 0.8$ kg, $M = 1.0$ kg, and $R = 0.20$ m.

Answer: $a = 1.6$ m/s^2 and $\alpha = 7.8$ rad/s^2.

Notice the calculations of the translational acceleration of the blocks in Example 8.5. The acceleration is equal to the sum of the forces divided by an effective total mass of the system.

$$a = \frac{m_1 - m_2}{\frac{1}{2}M + m_1 + m_2}g$$

Here you see that the pulley only contributes half of its mass due to its mass distribution. We can say that the pulley's "effective mass" is $M/2$, because of how its mass is distributed.

EXAMPLE 8.6   Throwing a bottle

A woman tosses a 0.80-kg soft drink bottle vertically upward to a friend on a balcony above. At the beginning of the toss, her forearm rotates upward from the horizontal so that the hand exerts a 20-N upward force on the bottle. Determine the force that her biceps exerts on her forearm during this initial instant of the throw. The mass of her forearm is 1.5 kg and its rotational inertia about the elbow joint is 0.061 kg·m². The attachment point of the biceps muscle is 5.0 cm from the elbow joint, the hand is 35 cm away from the elbow, and the center of mass of the forearm/hand is 16 cm from the elbow.

Sketch and translate A sketch of the situation is shown below. There is no information given about the kinematics of the process (for example, no way to directly determine the rotational acceleration of her arm). How can we use the rotational form of Newton's second law to determine the unknown force that the biceps muscle exerts on the woman's forearm during the throw? We do know the force her hand exerts on the bottle and the bottle's mass. So, in the first part of the problem, we can first use the translational form of Newton's second law with the bottle as the system to find the bottle's vertical acceleration at the beginning of the throw. We can then use this acceleration to find the angular acceleration of the arm and then finally use the rotational form of Newton's second law to find the force that the biceps muscle exerts on her arm during the throw. For this second part of the problem, we choose the lower arm and hand as the system of interest. The axis of rotation is at the elbow joint between the upper arm and the forearm.

Forearm tossing a bottle

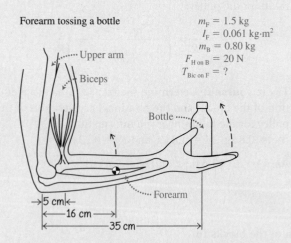

$m_F = 1.5$ kg
$I_F = 0.061$ kg·m²
$m_B = 0.80$ kg
$F_{H \text{ on } B} = 20$ N
$T_{Bic \text{ on } F} = ?$

Simplify and diagram The figure at top right is a force diagram for the bottle as a system. Earth exerts a downward 7.8-N gravitational force $\vec{F}_{E \text{ on } B}$ on the bottle and the woman's hand exerts an upward 20-N normal force $\vec{N}_{H \text{ on } B}$ on the bottle. Since these forces do not cancel, the bottle has an

initial upward acceleration. Next, consider the forearm and hand as the system. Assume that the forearm and hand form a rigid body. The bottle exerts a downward 20-N force on her hand $\vec{N}_{B \text{ on } H}$. Earth exerts a downward gravitational force $\vec{F}_{E \text{ on } F}$ on the forearm at its center of mass. Her biceps muscle exerts an upward tension force $\vec{T}_{Bic \text{ on } F}$. The upper arm presses down on the forearm at the joint, exerting a force $\vec{F}_{UA \text{ on } F}$. If the upper arm did not push down, the forearm at the joint would fly upward when the biceps muscle pulled up on it.

Bottle is system

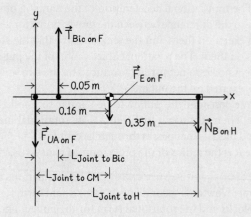

Forearm is system

Represent mathematically We first analyze the bottle's motion to determine its translational acceleration; then determine the rotational acceleration of the forearm and hand system; and finally apply the rotational form of Newton's second law to find the force that the biceps needs to exert on the system to cause this rotational acceleration. Consider the initial instant of the bottle's upward trip. The y-component form of Newton's second law applied to the bottle can be used to determine the vertical acceleration $a_{B \, y}$ for the bottle:

$$a_{B \, y} = \frac{N_{H \text{ on } B \, y} + F_{E \text{ on } B \, y}}{m_B} = \frac{N_{H \text{ on } B} + (-m_B g)}{m_B}$$

The rotational acceleration of the forearm/hand system at that instant is related to the vertical acceleration of the bottle:

$$\alpha_F = \frac{a_{B \, y}}{r}$$

where r is the distance from the axis of rotation to the hand. The magnitude of the force that the biceps muscle exerts on the forearm ($T_{Bic \text{ on } F}$) can be determined using the rotational form of Newton's second law applied to the forearm/hand. Notice that here the system consists of different parts joined together.

$$T_{\text{Bic on F}}L_{\text{Joint to Bic}} + (-F_{\text{E on F}}L_{\text{Joint to CM}})$$
$$+ (-N_{\text{B on H}}L_{\text{Joint to H}}) = I_F\alpha_F$$

Solve and evaluate We now use the known values of the quantities to solve the problem:

$$a_{B\,y} = \frac{N_{\text{H on B}} - m_B g}{m_B} = \frac{20\,\text{N} - (0.80\,\text{kg})(9.8\,\text{N/kg})}{(0.80\,\text{kg})}$$
$$= 15.2\,\text{m/s}^2$$

$$\alpha_F = \frac{a_{B\,y}}{r} = \frac{(15.2\,\text{m/s}^2)}{(0.35\,\text{m})} = +43.4\,\text{rad/s}^2$$

$$T_{\text{Bic on F}}(0.05\,\text{m}) - [(1.5\,\text{kg})(9.8\,\text{N/kg})](0.16\,\text{m})$$
$$- (20\,\text{N})(0.35\,\text{m}) = (0.061\,\text{kg}\cdot\text{m}^2)(43.4\,\text{rad/s}^2)$$

Solving the above equation, we find that $T_{\text{Bic on F}} = 240\,\text{N} = 54\,\text{lb}$, a reasonable magnitude for this force.

Try it yourself: Determine the force that the woman's biceps exerts on her forearm during the initial instant of a vertical toss of a 100-g rubber ball if she is exerting a 10-N force on the ball.

Answer: 430 N.

Review Question 8.4 How is Newton's second law for rotational motion similar to Newton's second law for translational motion?

8.5 Rotational momentum

Earlier in this textbook (Chapters 5 and 6), we constructed powerful principles for momentum and energy that allowed us to analyze complex processes that involved translational motion. Is it possible to find analogous principles for the rotational (angular) momentum and rotational energy of extended bodies? Consider the rotational inertia involved in the Observational Experiment **Table 8.7** following experiments.

OBSERVATIONAL EXPERIMENT TABLE

8.7 **Observations concerning rotational motion.**

Observational experiment	Analysis
(a) A figure skater initially spins slowly with a leg and two arms extended. Then she pulls her leg and arms close to her body and her spinning rate increases dramatically. 	*Initial situation:* Large rotational inertia *I* and small rotational speed *ω*. *Final situation:* Smaller rotational inertia *I* and larger rotational speed *ω*.

(continued)

Observational experiment	Analysis
(b) A man sitting on a chair that can spin with little friction initially holds barbells far from his body and spins slowly. When he pulls the barbells close to his body, the spinning rate increases dramatically. 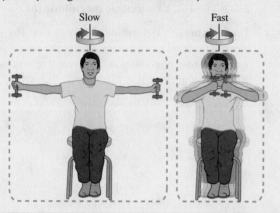	*Initial situation:* Large rotational inertia *I* and small rotational speed *ω*. *Final situation:* Smaller rotational inertia *I* and larger rotational speed *ω*.

Pattern

■ There are no external forces exerted on either person—no torques.
■ As the mass distribution of the system moves closer to the axis of rotation, the system's rotational inertia *I* decreases and the system's rotational speed *ω* increases (even though the net torque on the system is zero).

For each experiment in Table 8.7, the rotational inertia *I* of the spinning person decreased (the mass moved closer to the axis of rotation). Simultaneously, the rotational speed *ω* of the person increased. When *I* increases and *ω* decreases (or vice versa), *Iω* remains constant. We propose tentatively that when the rotational inertia *I* of an extended body in an isolated system decreases, its rotational speed *ω* increases, and vice versa.

Let's test this idea qualitatively in Testing Experiment **Table 8.8**.

TESTING EXPERIMENT TABLE

8.8 Testing the idea that *ω* increases if *I* decreases.

Testing experiment		Prediction	Outcome
A puck is tied to a string that passes through a hole in the center of an air table. As the puck moves, the string is pulled down through the table, decreasing the radius of the circular path of the puck.	Top view	As *r* decreases, the puck's rotational inertia $I = mr^2$ decreases. According to the proposed rule, the puck's rotational speed *ω* should increase—it should take less time to complete one rotation.	We observe that the rotational speed increases—it takes less time to complete one rotation around the post.

Conclusion

This result supports the idea that *ω* increases if the rotational inertia *I* of an isolated system decreases.

Rotational momentum is constant for an isolated system

Note that I is the rotational analogue of the mass m of a point-like object and ω is the rotational analogue of the translational velocity $\vec{v}$. The linear momentum of an object is the product of its mass m and its velocity $\vec{v}$. Let's propose that a turning object's rotational momentum L (analogous to linear momentum $\vec{p} = m\vec{v}$) is defined as

$$L = I\omega$$

In the chapter on linear momentum (Chapter 5) we derived a relationship [Eq. (5.4)] between the net force exerted on an object and the change in its linear momentum:

$$\Sigma\vec{F}(t_f - t_i) = \vec{p}_f - \vec{p}_i \qquad (5.4)$$

where $\vec{p} = m\vec{v}$. Torque τ is analogous to force $\vec{F}$. Thus, using the analogy between rotational and translational motion, we write

$$\Sigma\tau(t_f - t_i) = L_f - L_i$$

where $L = I\omega$. If a system with one rotating body is isolated, then the external torque exerted on the object is zero. In such a case, the rotational momentum of the object does not change ($0 = L_f - L_i$), and the object's rotational momentum is constant ($L_f = L_i$), or

$$I_i\omega_i = I_f\omega_f$$

Note that this is consistent with our tentative qualitative rule. If the final value of one quantity (I or ω) increases for an isolated system, then the other quantity must decrease. Now, let's try a quantitative test of this proposed rule—similar to the qualitative test done in Testing Experiment Table 8.8.

EXAMPLE 8.7 Puck on a string

Attach a 100-g puck to a string and let the puck glide in a counterclockwise circle (positive rotational momentum) on a horizontal, frictionless air table. The other end of the string passes through a hole at the center of the table. You pull down on the string so that the puck moves along a circular path of radius 0.40 m. It completes one revolution in 4.0 s. If you pull harder on the string so the radius of the circle slowly decreases to 0.20 m, what is the new period of revolution?

Sketch and translate A sketch of the situation is shown below. We choose the puck as the system and place the axis of rotation at the center of the table where the string passes through the hole.

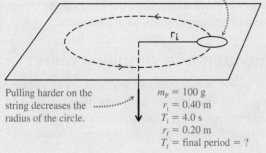

A puck glides in a circular path at the end of a string.

Pulling harder on the string decreases the radius of the circle.

$m_P = 100 \text{ g}$
$r_i = 0.40 \text{ m}$
$T_i = 4.0 \text{ s}$
$r_f = 0.20 \text{ m}$
$T_f = \text{final period} = ?$

Simplify and diagram   Side view
We model the puck as a point-like object. The force diagram to the right shows the situation viewed from the side of the air table looking from the back of the puck along its direction of motion. The tension force exerted by the string on the puck $\vec{F}_{S\,on\,P}$ passes through the axis of rotation and therefore produces zero torque. The upward normal force that the air table exerts on the puck $\vec{N}_{T\,on\,P}$ balances the downward gravitational force $\vec{F}_{E\,on\,P}$ of Earth on the puck, so they have zero net effect on the rotational motion of the puck. Also, the net torque due to these two forces is zero. So the net torque produced by all forces exerted on the puck is zero. According to the rule we are testing, this means that the rotational momentum of the puck should be constant, even though the radius of its circular path is decreasing.

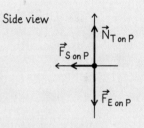

Represent mathematically Since the net external torque on the system is zero, the initial and final rotational momenta of the puck should be equal ($L_f = L_i$). Thus,

$$I_f\omega_f = I_i\omega_i$$

(continued)

The puck travels once around the circle (2π rad) in a time interval of one period T. Thus, the puck's rotational speed is

$$\omega = \frac{2\pi}{T}$$

The rotational inertia of the puck around the axis of rotation is

$$I = mr^2$$

Combining these last three equations, we get

$$(mr_i^2)\left(\frac{2\pi}{T_i}\right) = (mr_f^2)\left(\frac{2\pi}{T_f}\right)$$

Dividing each side by common terms, we get

$$\frac{r_f^2}{T_f} = \frac{r_i^2}{T_i}$$

Now multiply both sides of the equation by $T_f T_i$ and divide by r_i^2 to obtain an expression for T_f:

$$T_f = \frac{r_f^2 T_i}{r_i^2}$$

Solve and evaluate Insert the known quantities to get

$$T_f = \frac{r_f^2 T_i}{r_i^2} = \frac{(0.20\,\text{m})^2(4.0\,\text{s})}{(0.40\,\text{m})^2} = 1.0\,\text{s}$$

Remember that this is a testing experiment. When the experiment is performed, we find that the time interval with the reduced radius is very close to 1.0 s. The outcome of the experiment is consistent with the prediction.

Try it yourself: An 80-kg roller skater holds a rope that loops around a thick metal post, causing him to skate in a circular path. It takes him 8.0 s to complete one rotation around the pole. He starts pulling himself inward along the rope so that the radius of his motion decreases from 2.0 m to 1.0 m. Determine his rotational speed once he has pulled himself inward to a radius of 1.0 m.

Answer: 3.14 rad/s.

Rotational momentum and rotational impulse We now have a quantitative relation between rotational momentum $L = I\omega$ and rotational impulse.

$$L_i + \Sigma\tau\Delta t = L_f \qquad (8.12)$$

The initial rotational momentum of a turning object plus the product of the net external torque exerted on the object and the time interval during which it is exerted equals the final rotational momentum of the object.

  If the net torque that external objects exert on the turning object is zero, or if the torques add to zero, then the rotational momentum L of the turning object remains constant:

$$L_f = L_i \text{ or } I_f\omega_f = I_i\omega_i \qquad (8.13)$$

To explain most of the applications of torque and rotational momentum in this book, we account for their directions using positive or negative signs. A torque is positive if it tends to rotate the object counterclockwise and negative if it tends to rotate the object clockwise about the axis of rotation. A body rotating counterclockwise has positive rotational momentum and if rotating clockwise has negative rotational momentum.

TIP Rotational momentum is sometimes called angular momentum.

Rotational momentum of a shrinking object

At the beginning of the chapter, we described the discovery of pulsars, astronomical objects that rotate very quickly and emit repetitive radio signals with a very small time interval between them. The signals from the first discovered pulsar had a period of approximately 1.33 s. Astronomers could not explain at first how pulsars could rotate so rapidly. Most stars, including our Sun, rotate very much the way Earth does, usually taking several days to complete one rotation (about a month for our Sun).

However, as a star's core collapses and its mass moves closer to the axis of rotation, its rotational velocity increases because its rotational momentum is constant (assuming the star does not interact with any other objects). How much does a star need to shrink so that its period of rotation becomes seconds instead of days?

EXAMPLE 8.8 A pulsar

Imagine that our Sun ran out of nuclear fuel and collapsed. What would its radius have to be in order for its period of rotation to be the same as the pulsar described above? The Sun's current period of rotation is 25 days.

Sketch and translate First, sketch the process. The Sun is the system. We can convert the present period of rotation of the Sun into seconds ($T_i = 25$ days $= 2.16 \times 10^6$ s). Its mass is $m = 2.0 \times 10^{30}$ kg and its radius is $R_i = 0.70 \times 10^9$ m. After the Sun collapses, its period of rotation will be $T_f = 1.33$ s. What will be its radius?

In its initial state the Sun has a large I and a small ω.

$$\omega_i = \frac{2\pi}{25 \text{ days}}$$

In its final state the Sun has a small I and a large ω.

$$\omega_f = \frac{2\pi}{1.33 \text{ s}}$$

$m = 2.0 \times 10^{30}$ kg

$R_i = 0.70 \times 10^9$ m

$R_f = ?$

Simplify and diagram Assume that the Sun is a sphere with its mass distributed uniformly. Assume also that it does not lose any mass as it collapses.

Represent mathematically Now, apply the principle of rotational momentum conservation (Eq. 8.14) to the Sun's collapse:

$$I_i \omega_i = I_f \omega_f$$

From Table 8.6 we find that the rotational inertia of a sphere rotating around an axis passing through its center is

$$I = \left(\frac{2}{5}\right) mR^2$$

The rotational velocity of an object is

$$\omega = \frac{\Delta\theta}{\Delta t} = \frac{2\pi}{T}$$

where T is the period for one rotation. Combining the above three equations, we get

$$\left(\frac{2}{5} mR_i^2\right)\frac{2\pi}{T_i} = \left(\frac{2}{5} mR_f^2\right)\frac{2\pi}{T_f}$$

Dividing by the $2/5$, 2π, and m on each side of the equation, we get

$$\frac{R_i^2}{T_i} = \frac{R_f^2}{T_f}$$

Solve and evaluate Multiply both sides of the above by T_f and take the square root:

$$R_f = \sqrt{\frac{R_i^2 T_f}{T_i}}$$

$$= \sqrt{\frac{(0.70 \times 10^9 \text{ m})^2 (1.33 \text{ s})}{2.16 \times 10^6 \text{ s}}}$$

$$= 5.5 \times 10^5 \text{ m}$$

$$= 550 \text{ km}$$

Although this is much smaller than the radius of Earth, models of stellar evolution actually do predict that the Sun's core will eventually shrink to this size and possibly smaller.

Try it yourself: When massive stars explode, the collapse can shrink their radii to about 10 km. What would be the period of rotation of such a star if it originally had a mass twice the mass of the Sun, a radius that was 1.3 times the Sun's radius, and the same initial period of rotation as the Sun (25 days)?

Answer: 2.6×10^{-4} s.

Figure 8.13 Using the right-hand rule to determine the direction of an object's rotational momentum $\vec{L}$.

Circle fingers in direction of rotation. Thumb points in the direction of rotational momentum.

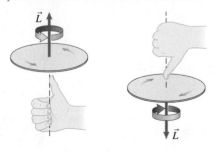

Figure 8.14 A bicycle that is not moving is in unstable equilibrium.

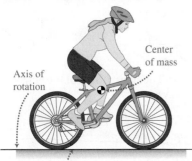

Axis of rotation is below center of mass—unstable.

Figure 8.15 Rotating bicycle tires have rotational momentum that stabilizes the bicycle.

Vector nature of torque, rotational velocity, and rotational momentum

We have not yet considered the vector nature of torque, rotational velocity, and rotational momentum. However several important applications depend on an understanding of the vector nature of these quantities. The vector direction of both rotational velocity and rotational momentum can be determined using **a right-hand rule**.

Right-hand rule for determining the direction of rotational velocity and rotational momentum Curl the four fingers of your right hand in the direction of rotation of the turning object. Your thumb, held perpendicular to the fingers, then points in the direction of both the object's rotational velocity and rotational momentum (**Figure 8.13**). To determine the **vector direction** of the torque that a force produces on an object (as opposed to the clockwise/counterclockwise way of describing it) about an axis of rotation, first imagine that the object is at rest and that the torque you are interested in is the only torque exerted on the object. Next, curl the fingers of your right hand in the direction that the torque would make the object rotate. Your thumb, held perpendicular to the fingers, shows the direction of this torque.

Bicycling We can use the vector nature of torque and rotational momentum to understand why a bicycle is much more stable when moving fast—especially if the bicycle has massive tires. Consider an axis of rotation parallel to the ground that passes through the two contact points of the tires with the ground. This axis of rotation is below the center of mass of the stationary bike—an unstable equilibrium (**Figure 8.14**). When the bicycle is moving quickly, the rotating tires (and therefore the bicycle + rider system) have considerable rotational momentum, which will change only when an unbalanced torque is exerted on the system. When a bicycle is moving on a smooth road, the rotational velocity and the rotational momentum vectors are perpendicular to the plane of rotation of the bike tires (**Figure 8.15**), and they are large due to the rapid rotation of the tires.

When the bike + rider system is balanced, the gravitational force exerted by Earth on the system produces no torque since that force points directly at the axis of rotation. If the rider's balance shifts a bit, or the wind blows, or the road is uneven, the system will start tilting. As a result, the gravitational force exerted on the system will produce a torque. However, since the rotational momentum of the system is large, this torque does not change its direction by much right away, but it takes only several tenths of a second for the torque to change the rotational momentum significantly. This is enough time for an experienced rider to make corrections to rebalance the system. The faster the person is riding the bike, the greater the rotational momentum of the system and the more easily she/he can keep the system balanced.

Gyroscopes Guidance systems for spaceships rely on the constancy of rotational momentum in isolated systems to help them maintain their chosen course. Once the ship is pointed in the desired direction, one or more heavy gyroscopes starts rotating. The gyroscope is a wheel whose axis of rotation keeps the ship oriented in the chosen direction. The gyroscope is similar to the rotating bicycle tires that help keep a rider upright without tipping or changing direction. Gyroscopes are also used in cameras to prevent them from vibrating or moving while the camera lens is open.

After a playground merry-go-round is set in motion, its rotational speed decreases noticeably if another person jumps on it. However, if a person riding the merry-go-round steps off, the rotational speed seems not to change at all. Explain.

8.6 Rotational kinetic energy

We are familiar with the kinetic energy $(1/2)mv^2$ of a single particle moving along a straight line or in a circle. It would be useful to calculate the kinetic energy of a rotating body—like Earth. Doing so would allow us to use the work-energy approach to solving problems involving rotation. Let's start by deriving an expression for the rotational kinetic energy of a single particle of mass m moving in a circle of radius r at speed v. According to the kinematics in Section 8.1, its linear speed v and rotational speed ω are related:

$$v = r\omega$$

Thus, the kinetic energy of this particle moving in a circle can be written as

$$K_{\text{rotational}} = \frac{1}{2}mv^2 = \frac{1}{2}m(r\omega)^2 = \frac{1}{2}(mr^2)\omega^2 = \frac{1}{2}I\omega^2$$

where $I = mr^2$ is the rotational inertia of a particle moving a distance r from the center of its circular path. The expression for the translational kinetic energy $(1/2)mv^2$ of a particle is similar to the rotational version, which involves the product of a mass-like term I and the square of a speed-like term ω. Can we use the expression $\frac{1}{2}I\omega^2$ for the rotational kinetic energy of a rotating rigid body?

To test this idea, consider a solid sphere of known radius R and mass m that can rotate freely on an axis. We wrap a string around the sphere and pull the string with a force probe exerting a constant force of a known magnitude so that the sphere starting at rest completes 5.0 revolutions (**Figure 8.16a**). After we stop pulling, we measure the rotational speed ω of the sphere. But before measuring it, we predict its value using this expression for rotational kinetic energy. If we choose the sphere as the system, the string is the only external object that exerts a force that causes a nonzero torque on the sphere. This string force does work, which changes the sphere's kinetic energy from zero to some new value (Figure 8.16b). Thus, the initial rotational kinetic energy of the sphere (zero) plus the work done by the string on the sphere during these five turns equals the final rotational kinetic energy of the sphere:

$$K_i + W = K_f$$

The string pulls parallel to the displacement of the edge of the sphere during the entire time. We can use the expression for the rotational kinetic energy under test to predict the magnitude of the final rotational speed:

$$K_i + W = K_f$$

$$0 + F_{\text{String on Sphere}}(5 \cdot 2\pi R)\cos 0 = \frac{1}{2}I\omega^2$$

where $5 \cdot 2\pi R$ is the distance the string is pulled—five circumferences of the sphere. The above leads to a prediction of the final rotational speed:

$$\omega_f = \sqrt{\frac{F_{\text{String on Sphere}} \cdot 20\pi R}{I}}$$

Figure 8.16 A string pulls a solid sphere.

(a)

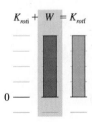

Axis of rotation

$R = 0.10\ \text{m}$

$m = 10\ \text{kg}$

5.0 N

(b)

Work done by the string causes the rotational kinetic energy of the sphere to increase.

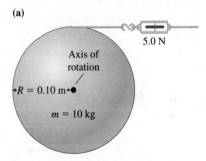

$K_{\text{roti}} + W = K_{\text{rotf}}$

0

From Table 8.6 we know that the rotational inertia of a solid sphere of radius R and mass m is $I = (2/5)mR^2$ (the axis passes through the center of the sphere) where $m = 10$ kg and $R = 0.10$ m. Thus, the rotational inertia is

$$I = (2/5)mR^2 = (2/5)(10\text{ kg})(0.10\text{ m})^2 = 0.040\text{ kg} \cdot \text{m}^2$$

We pull the string so that it exerts a 5.0-N force on the edge of the sphere. Thus, the sphere's final speed should be

$$\omega_f = \sqrt{\frac{(5.0\text{ N})(20\,\pi)(0.10\text{ m})}{(0.040\text{ kg} \cdot \text{m}^2)}} = \left(28\,\frac{\text{rad}}{\text{s}}\right)\left(\frac{1\text{ rev}}{2\pi\text{ rad}}\right) = 4.5\,\frac{\text{rev}}{\text{s}}$$

When we measure the final angular velocity, it is about 4.5 rev/s.

Let's test our idea for the mathematical expression of rotational kinetic energy in Testing Experiment **Table 8.9**.

TESTING EXPERIMENT TABLE

8.9 Testing the expression for rotational kinetic energy.

Testing experiment	Prediction	Outcome
A solid cylinder and a hoop of the same radius and mass start rolling at the top of an inclined plane. Which object reaches the bottom of the plane first? What is the ratio of their speeds at the bottom? 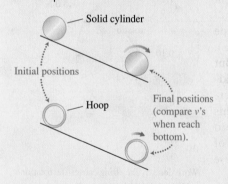	Both Earth-object systems start with the same gravitational potential energy. As they roll, both acquire translational and rotational kinetic energies. We can represent the process in a work-energy bar chart. The bar chart helps us construct a mathematical description.	The speeds, measured with a motion detector, are consistent with the prediction.

Initial positions

Solid cylinder

Hoop

Final positions (compare v's when reach bottom).

$U_{gi} = K_{tf} + K_{rf}$

$$mgy_i = \frac{1}{2}mv_f^2 + \frac{1}{2}I\omega_f^2$$

The rotational speed ω and translational speed v are related:

$$\omega = v/r$$

Substitute in the energy equation and then rearrange:

$$mgy_i = \frac{1}{2}\left(m + \frac{I}{r^2}\right)v_f^2$$

$$v_f = \sqrt{\frac{2mgy_i}{\left(m + \frac{I}{r^2}\right)}}$$

At the bottom of the inclined plane, the object with greater rotational inertia I will have the least translational speed v_f.
$I_{cylinder} = (1/2)mr^2$ and $I_{hoop} = mr^2$. Thus, $I_{hoop} > I_{cylinder}$. Thus, the hoop should be moving more slowly.

Calculations: Insert the expressions for the rotational inertia of the solid cylinder and of the hoop to find their final speeds:

Cylinder $\quad v_f = \sqrt{\dfrac{4}{3}gy_i}$

Hoop $\quad\quad v_f = \sqrt{gy_i}$

The ratio of their final speeds is $\sqrt{\dfrac{4}{3}}$, with the solid cylinder moving faster.

Conclusion

The outcome of the experiment supports the expression for the rotational kinetic energy of a rigid body:

$$K_{\text{rotational}} = \frac{1}{2}I\omega^2$$

This is the second testing experiment involving rotational kinetic energy in which the outcome matched the prediction. Given this support for our prediction and the lack of counterevidence, we will use this mathematical expression for a rigid body's rotational kinetic energy.

Rotational kinetic energy The rotational kinetic energy of an object with rotational inertia I turning with rotational speed ω is

$$K_{\text{rotational}} = \frac{1}{2}I\omega^2 \qquad\qquad (8.14)$$

TIP When you encounter a new physical quantity, always check whether its units make sense. In this particular case the units for I are $\text{kg}\cdot\text{m}^2$ and the units for ω^2 are $1/s^2$. Thus, the unit for kinetic energy is $\text{kg}\cdot\text{m}^2/s^2 = (\text{kg}\cdot\text{m}/s^2)\text{m} = \text{N}\cdot\text{m} = \text{J}$, the correct unit for energy.

Flywheels for storing and providing energy

You stop your car at a stoplight. Before stopping, the car had considerable kinetic energy; after stopping, the kinetic energy is zero. It has been converted to internal energy due to friction in the brake pads. Unfortunately, this thermal energy cannot easily be converted back into a form that is useful. Is there a way to convert that translational kinetic energy into some other form of energy that would help the car regain translational kinetic energy when the light turns green?

Efforts are under way to use the rotational kinetic energy of flywheels (rotating disks) for this purpose. Instead of rubbing a brake pad against the wheel and slowing it down, the braking system would, through a system of gears or through an electric generator, convert the car's translational kinetic energy into the rotational kinetic energy of a flywheel. As the car's translational speed decreases, the flywheel's rotational speed increases. This rotational kinetic energy could then be used to help the car start moving, rather than relying entirely on the chemical potential energy of gasoline.

EXAMPLE 8.9 Flywheel rotational speed

A 1600-kg car traveling at 20 m/s approaches a stop sign. If it could transfer all of its translational kinetic energy to a 0.20-m-radius, 20-kg flywheel while stopping, what rotational speed would the flywheel acquire?

Sketch and translate A sketch of the situation is shown below. The system of interest will be the car, including the flywheel.

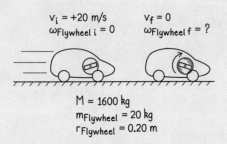

$v_i = +20$ m/s
$\omega_{Flywheel\,i} = 0$
$v_f = 0$
$\omega_{Flywheel\,f} = ?$

$M = 1600$ kg
$m_{Flywheel} = 20$ kg
$r_{Flywheel} = 0.20$ m

Simplify and diagram The process is represented in the figure below with a bar chart. The initial energy of the system is the car's translational kinetic energy; the final energy is the flywheel's rotational kinetic energy. Assume that the flywheel is a solid disk with rotational inertia of $(1/2)mr^2$ (see Table 8.6).

Braking converts the car's initial kinetic energy K_{ti} into the flywheel's final rotational kinetic energy K_{rf}—saved for future use.

$K_{ti} = K_{rf}$

0

Represent mathematically Use the bar chart to help construct an energy conservation equation:

$$K_{\text{translational i}} = K_{\text{rotational f}}$$

$$\frac{1}{2}Mv_i^2 = \frac{1}{2}I\omega_f^2$$

Multiplying both sides of the equation by 2 and dividing by I, we get

$$\omega_f^2 = \frac{Mv_i^2}{I}$$

The rotational inertia of the disk (a solid cylinder) is $I_{\text{cylinder}} = (1/2)mr^2$. Thus,

$$\omega_f^2 = \frac{Mv_i^2}{I} = \frac{Mv_i^2}{\frac{1}{2}mr^2} = \frac{2Mv_i^2}{mr^2}$$

Solve and evaluate To find the rotational speed, take the square root of both sides of the above equation:

$$\omega_f = \frac{v_i}{r}\sqrt{\frac{2M}{m}} = \frac{20\text{ m/s}}{0.20\text{ m}}\sqrt{\frac{2(1600\text{ kg})}{(20\text{ kg})}} = 1300\text{ rad/s}$$

$$= 200\text{ rev/s} = 12{,}000\text{ rpm}$$

Try it yourself: Could you store more energy in a rotating hoop or in a rotating solid cylinder, assuming they have the same mass, radius, and rotational speed?

Answer: The hoop has a greater rotational inertia and therefore would have a greater rotational kinetic energy at the same rotational speed.

Rolling versus sliding

Our knowledge of rotational kinetic energy helps explain a very simple but rather mysterious experiment that you can perform at home. For this experiment, you need two identical plastic water bottles. Fill one of them with snow. Pack the snow tightly. Fill the other one with water so that the mass of the water-filled bottle is the same as the mass of the snow-filled bottle. Place one of them at the top of an inclined plane (**Figures 8.17a** and b) and let it roll down. Then repeat the experiment with the other bottle. You observe an interesting effect. When the snow-filled bottle rolls down, it rotates and the solid snow inside rotates with the bottle. The bottle with water rotates, too, but the water inside does not rotate (Figure 8.17b). Thus, in effect the water slides down the incline and does not roll. Which bottle will win a race down the inclined plane if they travel side by side? Think about your prediction before you read on.

When the snow-filled bottle rolls down, it rotates as a solid cylinder, acquiring rotational kinetic energy in addition to translational kinetic energy.

(a)

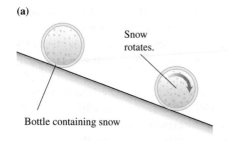

Snow rotates.

Bottle containing snow

(b)

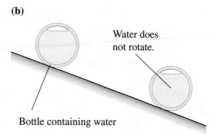

Water does not rotate.

Bottle containing water

Figure 8.17 Bottles with solid snow and liquid water race down identical inclines.

(c)

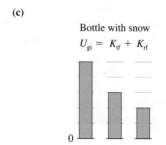

Bottle with snow
$U_{gi} = K_{tf} + K_{rf}$

0

The U_{gi} of the snow bottle converts to both translational and rotational kinetic energies.

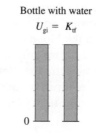

Bottle with water
$U_{gi} = K_{tf}$

0

The water does not rotate so all of its U_{gi} is converted to translational K—it wins the race.

In the case of the water-filled bottle, only the bottle rotates. The water inside just translates. Rolling is a combination of translation and rotation, whereas sliding involves only translation. The energy bar charts help us understand the energy transformations during the process (Figure 8.17c). The water-filled bottle has no rotational kinetic energy and a larger translational kinetic energy at the end of the plane. It should reach the bottom first. It's a fun experiment—try it and see if the outcome matches the prediction!

Review Question 8.6 Will a can of watery chicken noodle soup roll slower or faster down an inclined plane than an equal-mass can of thick, sticky English clam chowder?

8.7 Rotational motion: Putting it all together

We can use our knowledge of rotational motion to analyze a variety of phenomena that are part of our world. In this section, we consider two examples—the effect of the tides on the period of Earth's rotation (the time interval for 1 day) and the motion of bowling (also called pitching) in the sport of cricket.

Tides and Earth's day

The level of the ocean rises and falls by an average of 1 m twice each day, a phenomenon known as the tides. Many scientists, including Galileo, tried to explain this phenomenon and suspected that the Moon was a part of the answer. Isaac Newton was the first to explain how the motion of the Moon actually creates tides. He noted that at any moment, different parts of Earth's surface are at different distances from the Moon and that the distance from a given location on Earth to the Moon varied as Earth rotated. As illustrated in **Figure 8.18**, point A is closer to the Moon than the center of Earth or point B are, and therefore the gravitational force exerted by the Moon on point A is greater than the gravitational force exerted on point B. Due to the difference

Figure 8.18 The ocean bulges on both sides of Earth along a line toward the Moon.

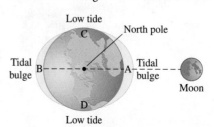

Low tide

North pole

C

Tidal bulge B - - - - - - - A Tidal bulge

Moon

D

Low tide

in forces, Earth elongates along the line connecting its center to the Moon's. This makes water rise to a high tide at point A and surprisingly also at B. The water "sags" a little at points C and D, forming low tides at those locations. When the Sun is aligned with the Moon and Earth, the bulging is especially pronounced.

As the solid Earth rotates beneath the tidal bulges, it attempts to drag the bulges with it. A large amount of friction is produced, which converts the rotational kinetic energy of Earth into internal energy. The time interval needed for Earth to complete one turn on its axis increases by 0.0016 s every 100 years. In other words, the Earth day is slowly getting longer. In a very long time, Earth will stop turning relative to the Moon and an unmoving tidal bulge will face toward and away from the Moon. This "tidal locking" has already occurred on the Moon (although it is solid, the same principle applies), which is why on Earth we only see one side of the Moon. It rotates around its axis with the same period as it moves around Earth.

Let's use our new understanding of rotational dynamics to estimate the friction force the tides exert on Earth, causing Earth's rotation rate to decrease.

EXAMPLE 8.10 Tides slow Earth's rotation

Estimate the effective tidal friction force exerted by ocean water on Earth that causes a 0.0016-s increase in Earth's rotation time every 100 years.

Sketch and translate The situation is already sketched in Figure 8.18. The solid Earth will be the system; the water covering most of its surface is considered an external object for this estimate. Earth rotates counterclockwise, taking 24 hours for one revolution (one period), as seen looking down on the North Pole. Remember that when an object rotates counterclockwise, its angular velocity is considered to be positive. Since Earth's rotation is gradually slowing, the tidal friction force is producing a negative torque, opposite the positive sign of the rotational velocity. We need to find the magnitude of the force that would increase the time of one revolution by 0.0016 s in 100 years.

Simplify and diagram Assume that the solid Earth is a sphere covered uniformly with water on its surface. Assume also that the frictional force exerted by the water on Earth is constant in magnitude and exerted at the equator. The force exerted by the tidal bulge on Earth produces a torque that opposes Earth's rotation.

$$\omega$$

$T_i = 24\ \text{h}$
$T_f = 24\ \text{h} + 0.0016\ \text{s}$
$\Delta t = 100\ \text{years}$
$F_{\text{T on E}} = ?$

North Pole

R_E

Tidal friction force exerted by ocean water on the rest of Earth

$\vec{F}_{\text{T on E}}$

Represent mathematically To estimate the friction force exerted by the oceans on Earth, we need to determine the torque produced by that force. We can use the rotational form of Newton's second law if we can determine the rotational inertia of Earth and its rotational acceleration. The rotational acceleration of Earth is

$$\alpha = \frac{\Delta\omega}{\Delta t} = \frac{\omega_f - \omega_i}{\Delta t} = \frac{\left(+\dfrac{2\pi}{T_f}\right) - \left(+\dfrac{2\pi}{T_i}\right)}{\Delta t}$$

$$= \frac{2\pi}{\Delta t}\left(\frac{1}{T_f} - \frac{1}{T_i}\right)$$

where $T_i = 24\ \text{h}\left(\dfrac{3600\ \text{s}}{1\ \text{h}}\right) = 86{,}400\ \text{s}$, $T_f = T_i + \Delta T = 86{,}400\ \text{s} + 0.0016\ \text{s}$, and $\Delta t = 100\ \text{years}\left(\dfrac{365\ \text{days}}{1\ \text{year}}\right)\left(\dfrac{86{,}400\ \text{s}}{1\ \text{day}}\right) = 3.15 \times 10^9\ \text{s}$.

Because T_f and T_i are so close, your calculator will likely evaluate the rotational acceleration of Earth to be zero. To deal with this, we put the equation into another form.

$$\alpha = \frac{2\pi}{\Delta t}\left(\frac{1}{T_f} - \frac{1}{T_i}\right) = \frac{2\pi}{\Delta t}\left(\frac{T_i - T_f}{T_i T_f}\right)$$

$$= \frac{2\pi}{\Delta t}\left(\frac{-\Delta T}{T_i(T_i + \Delta T)}\right) = \frac{2\pi}{\Delta t}\left(\frac{-\Delta T}{T_i^2 + T_i\Delta T}\right)$$

Now look at the two terms in the denominator, T_i^2 and $T_i\Delta T$. Because ΔT is so small, the second term is much less than the first term. This means the second term can be dropped without affecting the result in any significant way. Thus,

$$\alpha = -\frac{2\pi\Delta T}{\Delta t T_i^2}$$

We can use the rotational form of Newton's second law to get an alternative expression for Earth's rotational acceleration:

$$\alpha = \frac{1}{I}\Sigma\tau = \frac{1}{\frac{2}{5}mR_E^2}(F_{T\,on\,E}R_E \sin 90°)$$

The R_E in the numerator is the distance from the axis of rotation to the surface where the friction force is exerted, the radius of Earth. The R_E in the denominator is also the radius of Earth.

$$\alpha = \frac{1}{\frac{2}{5}mR_E^2}(F_{T\,on\,E}R_E \sin 90°) = \frac{5F_{T\,on\,E}}{2mR_E}$$

Setting the two expressions for the magnitude of the rotational acceleration equal to each other, we get

$$\frac{2\pi\Delta T}{\Delta t T_i^2} = \frac{5F_{T\,on\,E}}{2mR_E}$$

Solve and evaluate Solve the previous equation for the force of the tides on Earth:

$$
\begin{aligned}
F_{T\,on\,E} &= \frac{4\pi mR_E\Delta T}{5\Delta t T_i^2} \\
&= \frac{4\pi(5.97 \times 10^{24}\,\text{kg})(6.38 \times 10^6\,\text{m})(0.0016\,\text{s})}{5(3.15 \times 10^9\,\text{s})(86,400\,\text{s})^2} \\
&= 6.5 \times 10^9\,\text{N}
\end{aligned}
$$

The magnitude of this friction force seems big, and it is. But when exerted on an object of such large mass ($5.97 \times 10^{24}\,\text{kg}$), the effect is extremely tiny. By comparison, the gravitational force that the Sun exerts on Earth is $3.5 \times 10^{22}\,\text{N}$.

Try it yourself: Estimate the time interval in years that it will take for Earth's rotation to change from 24 hours to 27 days.

Answer: Using Eq. (8.6), we find $t = (\omega - \omega_0)/\alpha \approx -\omega_0/\alpha = (-7 \times 10^{-5}\,\text{rad/s})/(-4 \times 10^{-22}\,\text{rad/s}^2) = 2 \times 10^{17}\,\text{s}$, almost 10 billion years.

Cricket bowling

In the game of cricket, a bowler (comparable to the pitcher in baseball) delivers the ball toward the batsman in an overhand motion with an almost straight arm, swinging the arm in a vertical circle (see **Figure 8.19**, below). The record speed for a cricket ball pitch is about the same as the top speed for a baseball pitch—about 44 m/s (100 mph).

EXAMPLE 8.11 Cricket ball pitch

Estimate the average force that the bowler's hand exerts on the cricket ball (mass 0.156 kg) during the pitch. The bowler's body is moving forward at about 4 m/s and the ball leaves his hand at 40 m/s relative to the bowler's torso and at 44 m/s relative to the ground.

Sketch and translate The situation is shown in Figure 8.19. The ball will be the system of interest. We analyze the situation from the point of view of the bowler. Just before starting the pitch, the speed of the ball with respect to the bowler is 0 m/s. Just as the ball leaves the bowler's hand, it travels horizontally at 40 m/s with respect to the bowler. What is the average force that the bowler exerts on the ball to cause this velocity change?

Simplify and diagram Assume that the bowler's arm (estimated length 1.0 m) completes one full circle (2π rad) during a pitch. We model the ball as a point-like object and assume the force exerted by the bowler is the only force producing a significant torque on it. The axis of rotation is the bowler's shoulder.

Figure 8.19 A cricket bowler's arm pushes the ball around a circular path during a pitch.

(continued)

Represent mathematically To estimate the force exerted on the ball by the bowler's arm, we use the rotational form of Newton's second law, which will require determining the ball's rotational inertia and its rotational acceleration. Since the ball is being modeled as a point-like object, its rotational inertia is $I = mr^2$. The rotational form of Newton's second law becomes

$$\alpha = \frac{1}{I}\Sigma\tau = \frac{1}{mr^2}(F_{\text{Bowler on Ball}}r)$$

$$\Rightarrow F_{\text{Bowler on Ball}} = mr\alpha$$

Next, we use rotational kinematics to determine the rotational acceleration of the ball:

$$F_{\text{Bowler on Ball}} = mr\alpha = mr\left(\frac{\omega_f^2 - \omega_i^2}{2(\theta_f - \theta_i)}\right)$$

$$= mr\left(\frac{\left(\frac{v}{r}\right)^2 - 0}{2(2\pi)}\right) = \frac{mv^2}{4\pi r}$$

Solve and evaluate We can now substitute the known information in the above to determine the force that the bowler's hand exerts on the ball:

$$F_{\text{Bowler on Ball}} = \frac{(0.156\,\text{kg})(40\text{m/s})^2}{4\pi(1.0\,\text{m})} = 20\,\text{N}$$

The answer has the correct units and has a reasonable magnitude, although perhaps smaller than you might have expected. Remember that this is the force that the bowler's hand exerts on the ball. The muscles in his shoulder exert a much greater force on the arm during this throwing motion. The mass of a human arm is about 3–4 kg, whereas the mass of the cricket ball is 0.156 kg. Thus, most of the bowler's effort goes into swinging his own arm. The ball is just along for the ride!

Let's check some limiting cases. If the final velocity of the ball were zero, then the above equation indicates that the bowler would need to exert zero force on the ball. This makes sense. If the bowler's arm were longer, then according to the above equation, he would need to exert less force on the ball; it gets up to the desired speed more easily at the end of a long arm.

Try it yourself: If you could exert a 20-N push on a 0.156-kg ball along a straight line that is $2\pi(1.0\,\text{m})$ long and the ball started at rest, how fast would it be traveling at the end of this push? Explain.

Answer: 40 m/s, the same speed as the cricket ball in the last example.

Review Question 8.7 How can you explain the increasing length of a day on Earth?

Summary

Words	Pictorial and physical representations	Mathematical representation
Rotational kinematics The rotational motion of a rigid body can be described using quantities similar to those for translational motion—rotational position θ, rotational velocity ω, and rotational acceleration α. (Section 8.1)	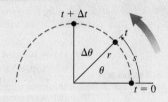	■ Rotational position (in radians) $\theta = s/r$ Eq. (8.1) ■ Rotational velocity (in rad/s) $\omega = \Delta\theta/\Delta t$ Eq. (8.2) ■ Rotational acceleration (in rad/s^2) $\alpha = \Delta\omega/\Delta t$ Eq. (8.3)
Rotational inertial is the physical quantity equal to the sum of the mr^2 terms for each part of an object and depends on the distribution of mass relative to an axis of rotation. (Section 8.3)	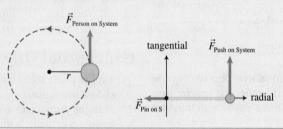	$I = \Sigma mr^2$ Eq. (8.10)
Rotational dynamics A rigid body's rotational acceleration equals the net torque produced by forces exerted on the body divided by its rotational inertia. (Section 8.4)		$\alpha = \dfrac{\Sigma\tau}{I}$ Eq. (8.11)
Rotational momentum L is the product of the rotational inertia I of an object and its rotational velocity ω, positive for counterclockwise rotation and negative for clockwise rotation. For an isolated system (zero net torque exerted on it), the rotational momentum of the system is constant. (Section 8.6)	$I_i\omega_i$ $I_f\omega_f$	$L = I\omega$ Eq. (8.13) For isolated system, $I_i\omega_i = I_f\omega_f$ Eq. (8.14)
Rotational kinetic energy $K_{\text{rotational}}$ of a rigid body is energy due to the rotation of the object about a particular axis. This is another form of kinetic energy that is included in the work-energy principle. (Section 8.7)	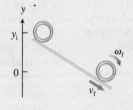	$K_{\text{rotational}} = \dfrac{1}{2}I\omega^2$ Eq. (8.15)

For instructor-assigned homework, go to
MasteringPhysics.

Questions

Multiple Choice Questions

1. Is it easier to open a door that is made of a solid piece of wood or a door of the same mass made of light fiber with a steel frame?
   (a) Wooden door
   (b) Fiber door with a steel frame
   (c) The same difficulty
   (d) Not enough information to answer

2. You push a child on a swing. Why doesn't the child continue in a vertical loop over the top of the swing?
   (a) The torque of the force that Earth exerts on the child pulls him back.
   (b) The swing does not have enough kinetic energy when at the bottom.
   (c) The swing does not have enough rotational momentum.
   (d) All of the above are correct.

3. In terms of the torque needed to rotate your leg as you run, would it be better to have a long calf and short thigh, or vice versa?
   (a) Long calf and short thigh
   (b) Short calf and long thigh
   (c) Does not matter

4. Suppose that two bicycles have equal overall mass, but one has thin lightweight tires while the other has heavier tires made of the same material. Why is the bicycle with thin tires easier to accelerate?
   (a) Thin tires have less area of contact with the road.
   (b) With thin tires, less mass is distributed at the rims.
   (c) With thin tires, you don't have to raise the large mass of the tire at the bottom to the top.

5. When riding a 10-speed bicycle up a hill, a cyclist shifts the chain to a larger-diameter gear attached to the back wheel. Why is this gear preferred to a smaller gear?
   (a) The torque exerted by the chain on the gear is larger.
   (b) The force exerted by the chain on the gear is larger.
   (c) You pedal more frequently to travel the same distance.
   (d) Both a and c are correct.

6. A meter stick is supported horizontally at each end by your fingers. A heavy object rests on one end of the stick. If you remove your fingers under the end that holds the object, that end of the meter stick falls faster than the object. Why?
   (a) The object is heavier than the meter stick.
   (b) The rotational inertia of the object is larger than that of a meter stick.
   (c) The acceleration of certain parts of the meter stick can be greater than 9.8 m/s^2.
   (d) Both b and c are correct.

7. You have a raw egg and a cooked egg. If you exert the same torque on each of them for the same period of time, which one spins faster in the end?
   (a) The raw egg
   (b) The cooked egg
   (c) They spin at the same speed.
   (d) The whole process is random and unpredictable.

8. If you turn on a coffee grinding machine sitting on a smooth tabletop, what do you expect it to do?
   (a) Start rotating in the same direction as the blades rotate
   (b) Start rotating in the direction opposite the blade rotation
   (c) Grind the coffee without any rotation of the machine

9. If you could accelerate from zero to 4 m/s when running anywhere on Earth's surface, where and in which direction would you run to increase the length of the day most?
   (a) East-West at the poles   (b) East-West at the equator
   (c) West-East at the poles   (d) West-East at the equator

10. The Mississippi River carries sediment from higher latitudes toward the equator. How does this affect the length of the day?
    (a) Increases the day
    (b) Decreases the day
    (c) Does not affect the day
    (d) There is no relation between the mass distribution and the length of the day.

Conceptual Questions

11. Explain your choices for Questions 1–10 (your instructor will choose which ones).

12. If all the people on Earth took elevators to the tops of high buildings in their communities, estimate the effect of this on the length of the day. Explain.

13. A spinning raw egg, if stopped momentarily and then released by the fingers, will resume spinning. Explain. Will this happen with a hard-boiled egg? Explain.

14. Compare the magnitude of Earth's rotational momentum about its axis to that of the Moon about Earth. The tides exert a torque on Earth and Moon so that eventually they rotate with the same period. The object with the greatest rotational momentum will experience the smallest percent change in the period of rotation. Will Earth's solar day increase more than the moon's period of rotation decreases? Explain.

15. You lay a pencil on a smooth desk (ignore sliding friction). You push the pencil, exerting a constant force first directly at its center of mass and then close to the tip of the pencil. In both cases, the force is exerted perpendicular to the body of the pencil. If the forces that you exert on the pencil are exactly the same in magnitude and direction, in which case is the translational acceleration of the pencil greater in magnitude?

16. If you watch the dive of an Olympic diver, you note that she continues to rotate after leaving the board. However, her center of mass follows a parabolic curve. Explain why.

17. Explain why you do not tip over when riding a bicycle but do tip when stationary at a stoplight.

18. Sometimes a door is not attached properly and it will open by itself or close by itself. But it will never do both. Why?

Problems

Below, BIO indicates a problem with a biological or medical focus. Problems labeled EST ask you to estimate the answer to a quantitative problem rather than develop a specific answer. Problems marked with ∕ require you to make a drawing or graph as part of your solution. Asterisks indicate the level of difficulty of the problem. Problems with no * are considered to be the least difficult. A single * marks moderately difficult problems. Two ** indicate more difficult problems.

8.1 Rotational kinematics

1. The sweeping second hand on your wall clock is 20 cm long. What is (a) the rotational speed of the second hand, (b) the translational speed of the tip of the second hand, and (c) the rotational acceleration of the second hand?

2. You find an old record player in your attic. The turntable has two readings: 33 rpm and 45 rpm. What do they mean? Express these quantities in different units.

3. * Consider again the turntable described in the last problem. Determine the magnitudes of the rotational acceleration in each of the following situations. Indicate the assumptions you made for each case. (a) When on and rotating at 33 rpm, it is turned off and slows and stops in 60 s. (b) When off and you push the play button, the turntable attains a speed of 33 rpm in 15 s. (c) You switch the turntable from 33 rpm to 45 rpm, and it takes about 2.0 s for the speed to change. (d) In the situation in part (c), what is the magnitude of the average tangential acceleration of a point on the turntable that is 15 cm from the axis of rotation?

4. You step on the gas pedal in your car, and the car engine's rotational speed changes from 1200 rpm to 3000 rpm in 3.0 s. What is the engine's average rotational acceleration?

5. You pull your car into your driveway and stop. The drive shaft of your car engine, initially rotating at 2400 rpm, slows with a constant rotational acceleration of magnitude 30 rad/s^2. How long does it take for the drive shaft to stop turning?

6. An old wheat-grinding wheel in a museum actually works. The sign on the wall says that the wheel has a rotational acceleration of 190 rad/s^2 as its spinning rotational speed increases from zero to 1800 rpm. How long does it take the wheel to attain this rotational speed?

7. **Centrifuge** A centrifuge at the same museum is used to separate seeds of different sizes. The average rotational acceleration of the centrifuge according to a sign is 30 rad/s^2. If starting at rest, what is the rotational velocity of the centrifuge after 10 s?

8. * **Potter's wheel** A fly sits on a potter's wheel 0.30 m from its axle. The wheel's rotational speed decreases from 4.0 rad/s to 2.0 rad/s in 5.0 s. Determine (a) the wheel's average rotational acceleration, (b) the angle through which the fly turns during the 5.0 s, and (c) the distance traveled by the fly during that time interval.

9. * During your tennis serve, your racket and arm move in an approximately rigid arc with the top of the racket 1.5 m from your shoulder joint. The top accelerates from rest to a speed of 20 m/s in a time interval of 0.10 s. Determine (a) the magnitude of the average tangential acceleration of the top of the racket and (b) the magnitude of the rotational acceleration of your arm and racket.

10. * An ant clings to the outside edge of the tire of an exercise bicycle. When you start pedaling, the ant's speed increases from zero to 10 m/s in 2.5 s. The wheel's rotational acceleration is 13 rad/s^2. Determine everything you can about the motion of the wheel and the ant.

11. * The speedometer on a bicycle indicates that you travel 60 m while your speed increases from 0 to 10 m/s. The radius of the wheel is 0.30 m. Determine three physical quantities relevant to this motion.

12. * You peddle your bicycle so that its wheel's rotational speed changes from 5.0 rad/s to 8.0 rad/s in 2.0 s. Determine (a) the wheel's average rotational acceleration, (b) the angle through which it turns during the 2.0 s, and (c) the distance that a point 0.60 m from the axle travels.

13. **Mileage gauge** The odometer on an automobile actually counts axle turns and converts the number of turns to miles based on knowledge that the diameter of the tires is 0.62 m. How many turns does the axle make when traveling 10 miles?

14. **Speedometer** The speedometer on an automobile measures the rotational speed of the axle and converts that to a linear speed of the car, assuming the car has 0.62-m-diameter tires. What is the rotational speed of the axle when the car is traveling at 20 m/s (45 mph)?

15. * **Ferris wheel** A Ferris wheel starts at rest, acquires a rotational velocity of ω rad/s after completing one revolution and continues to accelerate. Write an expression for (a) the magnitude of the wheel's rotational acceleration (assumed constant), (b) the time interval needed for the first revolution, (c) the time interval required for the second revolution, and (d) the distance a person travels in two revolutions if he is seated a distance l from the axis of rotation.

16. * You push a disk-shaped platform on its edge 2.0 m from the axle. The platform starts at rest and has a rotational acceleration of 0.30 rad/s^2. Determine the distance you must run while pushing the platform to increase its speed at the edge to 7.0 m/s.

17. * **EST** Estimate what Earth's rotational acceleration would be in rad/s^2 if the length of a day increased from 24 h to 48 h during the next 100 years.

8.4 Newton's second law for rotational motion

18. A turntable turning at rotational speed 33 rpm stops in 50 s when turned off. The turntable's rotational inertia is 1.0×10^{-2} kg·m^2. How large is the resistive torque that slows the turntable?

19. A 0.30-kg ball is attached at the end of a 0.90-m-long stick. The ball and stick rotate in a horizontal circle. Because of air resistance and to keep the ball moving at constant speed, a continual push must be exerted on the stick, causing a 0.036-N·m torque. Determine the magnitude of the resistive force that the air exerts on the ball opposing its motion. What assumptions did you make?

20. * **Centrifuge** A centrifuge with a 0.40-kg·m^2 rotational inertia has a rotational acceleration of 100 rad/s^2 when the power is turned on. (a) Determine the minimum torque that the motor supplies. (b) What time interval is needed for the centrifuge's rotational velocity to increase from zero to 5000 rad/s?

21. * **Airplane turbine** What is the average torque needed to accelerate the turbine of a jet engine from rest to a rotational velocity of 160 rad/s in 25 s? The turbine's rotating parts have a 32-kg·m^2 rotational inertia.

22. * / The solid two-part pulley in **Figure P8.22** initially rotates counterclockwise. Two ropes pull on the pulley as shown. The inner part has a radius of 1.5*a*, and the outer part has a radius of 2.0*a*. (a) Construct a force diagram for the pulley with the origin of the coordinate system at the center of the pulley. (b) Determine the torque produced by each force (including the sign) and the resultant torque exerted on the pulley. (c) Based on the results of part (b), decide on the signs of the rotational velocity and the rotational acceleration.

Figure P8.22

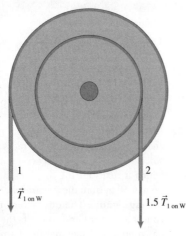

$\vec{T}_{1 \text{ on W}}$ 1 2 $1.5\,\vec{T}_{1 \text{ on W}}$

23. * The flywheel shown in Figure P8.22 is initially rotating clockwise. Determine the relative force that the rope on the right needs to exert on the wheel compared to the force that the left rope exerts on the wheel in order for the wheel's rotational velocity to (a) remain constant, (b) increase in magnitude, and (c) decrease in magnitude. The outer radius is 2.0*a* compared to 1.5*a* for the inner radius.

24. * The flywheel shown in Figure P8.22 is initially rotating in the clockwise direction. The force that the rope on the right exerts on it is 1.5*T* and the force that the rope on the left exerts on it is *T*. Determine the ratio of the maximum radius of the inner circle compared to that of the outer circle in order for the wheel's rotational speed to decrease.

25. * A pulley such as that shown in **Figure P8.25** has rotational inertia $10 \text{ kg} \cdot \text{m}^2$. Three ropes wind around different parts of the pulley and exert forces $T_{1 \text{ on W}} = 80 \text{ N}$, $T_{2 \text{ on W}} = 100 \text{ N}$, and $T_{3 \text{ on W}} = 50 \text{ N}$. Determine (a) the rotational acceleration of the pulley and (b) its rotational velocity after 4.0 s. It starts at rest.

Figure P8.25

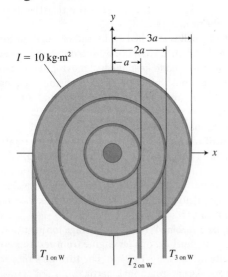

$I = 10 \text{ kg} \cdot \text{m}^2$

y

3*a*
2*a*
a
x

$T_{1 \text{ on W}}$ $T_{2 \text{ on W}}$ $T_{3 \text{ on W}}$

26. * / **Equation Jeopardy 1** The equation below describes a rotational dynamics situation. Draw a sketch of a situation that is consistent with the equation and construct a word problem for which the equation might be a solution. There are many possibilities.

$$-(2.2 \text{ N})(0.12 \text{ m}) = \left[(1.0 \text{ kg})(0.12 \text{ m})^2 \right] \alpha$$

27. * / **Equation Jeopardy 2** The equation below describes a rotational dynamics situation. Draw a sketch of a situation that is consistent with the equation and construct a word problem for which the equation might be a solution. There are many possibilities.

$$-(2.0 \text{ N})(0.12 \text{ m}) + (6.0 \text{ N})(0.06 \text{ m})$$
$$= \left[(1.0 \text{ kg})(0.12 \text{ m})^2 \right] \alpha$$

28. Determine the rotational inertia of the four balls shown in **Figure P8.28** about an axis perpendicular to the paper and passing through point A. The mass of each ball is *m*. Ignore the mass of the rods to which the balls are attached.

Figure P8.28

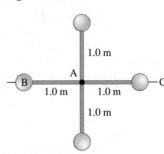

1.0 m

B A C
1.0 m 1.0 m

1.0 m

29. Repeat the previous problem for an axis perpendicular to the paper through point B.

30. Repeat the previous problem for axis BC, which passes through two of the balls.

31. **Merry-go-round** A mechanic needs to replace the motor for a merry-go-round. What torque specifications must the new motor satisfy if the merry-go-round should accelerate from rest to 1.5 rad/s in 8.0 s? You can consider the merry-go-round to be a uniform disk of radius 5.0 m and mass 25,000 kg.

32. * A small 0.80-kg train propelled by a fan engine starts at rest and goes around a circular track with a 0.80-m radius. The fan air exerts a 2.0-N force on the train. Determine (a) the rotational acceleration of the train and (b) the time interval needed for it to acquire a speed of 3.0 m/s. Indicate any assumptions you made.

33. * The train from the previous problem is moving along the rails at a constant rotational speed of 5.4 rad/s (the fan has stopped). Determine the time interval that is needed to stop the train if the wheels lock and the rails exert a 1.8-N friction force on the train.

34. * **Motor** You wish to buy a motor that will be used to lift a 20-kg bundle of shingles from the ground to the roof of a house. The shingles are to have a $1.5\text{-m}/\text{s}^2$ upward acceleration at the start of the lift. The very light pulley on the motor has a radius of 0.12 m. Determine the minimum torque that the motor must be able to provide.

35. * A thin cord is wrapped around a grindstone of radius 0.30 m and mass 25 kg supported by bearings that produce negligible friction torque. The cord exerts a steady 20-N tension force on the grindstone, causing it to accelerate from rest to 60 rad/s in 12 s. Determine the rotational inertia of the grindstone.

36. ** / A string wraps around a 6.0-kg wheel of radius 0.20 m. The wheel is mounted on a frictionless horizontal axle at the top of an inclined plane tilted 37° below the horizontal.

The free end of the string is attached to a 2.0-kg block that slides down the incline without friction. The block's acceleration while sliding down the incline is 2.0 m/s². (a) Draw separate force diagrams for the wheel and for the block. (b) Apply Newton's second law (either the translational form or the rotational form) for the wheel and for the block. (c) Determine the rotational inertia for the wheel about its axis of rotation.

37. * Elena, a black belt in tae kwon do, is experienced in breaking boards with her fist. A high-speed video indicates that her forearm is moving with a rotational speed of 40 rad/s when it reaches the board. The board breaks in 0.0040 s and her arm is moving at 20 rad/s just after breaking the board. Her fist is 0.32 m from her elbow joint and the rotational inertia of her forearm is 0.050 kg · m². Determine the average force that the board exerts on her fist while breaking the board (equal in magnitude to the force that her fist exerts on the board). Ignore the gravitational force that Earth exerts on her arm and the force that her triceps muscle exerts on her arm during the break.

38. ** / **Like a yo-yo** Sam wraps a string around the outside of a 0.040-m-radius 0.20-kg solid cylinder and uses it like a yo-yo (**Figure P8.38**). When released, the cylinder accelerates downward at (2/3) g. (a) Draw a force diagram for the cylinder and apply the translational form of Newton's second law to the cylinder in order to determine the force that the string exerts on the cylinder. (b) Determine the rotational inertia of the solid cylinder. (c) Apply the rotational form of Newton's second law and determine the cylinder's rotational acceleration. (d) Is your answer to part (c) consistent with the application of $a = r\alpha$, which relates the cylinder's linear acceleration and its rotational acceleration? Explain.

Figure P8.38

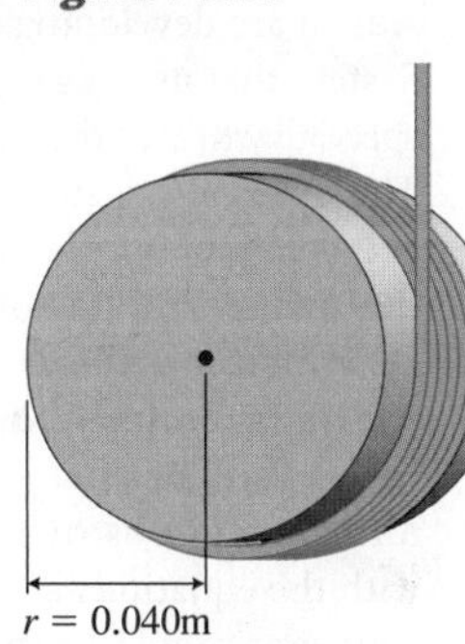

$r = 0.040$m

39. ** / **Fire escape** A unique fire escape for a three-story house is shown in **Figure P8.39**. A 30-kg child grabs a rope wrapped around a heavy flywheel outside a bedroom window. The flywheel is a 0.40-m-radius uniform disk with a mass of 120 kg. (a) Make a force diagram for the child as he moves downward at increasing speed and another for the flywheel as it turns faster and faster. (b) Use Newton's second law for translational motion and the child force diagram to obtain an expression relating the force that the rope exerts on him and his acceleration. (c) Use Newton's second law for rotational motion and the flywheel force diagram to obtain an expression relating the force the rope exerts on the flywheel and the rotational acceleration of the flywheel. (d) The child's acceleration a and the flywheel's rotational acceleration α are related by the equation $a = r\alpha$, where r is the flywheel's radius. Combine

Figure P8.39

this with your equations in parts (b) and (c) to determine the child's acceleration and the force that the rope exerts on the wheel and on the child.

40. ** / An Atwood machine is shown in Example 8.5. Use $m_1 = 0.20$ kg, $m_2 = 0.16$ kg, $M = 0.50$ kg, and $R = 0.10$ m. (a) Construct separate force diagrams for block 1, for block 2, and for the solid cylindrical pulley. (b) Determine the rotational inertia of the pulley. (c) Use the force diagrams for blocks 1 and 2 and Newton's second law to write expressions relating the unknown accelerations of the blocks. (d) Use the pulley force diagram and the rotational form of Newton's second law to write an expression for the rotational acceleration of the pulley. (e) Noting that $a = R\alpha$ for the pulley, use the three equations from parts (c) and (d) to determine the magnitude of the acceleration of the hanging blocks.

41. ** / A physics problem involves a massive pulley, a bucket filled with sand, a toy truck, and an incline (see **Figure P8.41**). You push lightly on the truck so it moves down the incline. When you stop pushing, it moves down the incline at constant speed and the bucket moves up at constant speed. (a) Construct separate force diagrams for the pulley, the bucket, and the truck. (b) Use the truck force diagram and the bucket force diagram to help write expressions in terms of quantities shown in the figure for the forces $T_{1\text{ on Truck}}$ and $T_{2\text{ on Bucket}}$ that the rope exerts on the truck and that the rope exerts on the bucket. (c) Use the rotational form of Newton's second law to determine if the tension force $T_{1\text{ on Pulley}}$ that the rope on the right side exerts on the pulley is the same, greater than, or less than the force $T_{2\text{ on Pulley}}$ that the rope exerts on the left side.

Figure P8.41

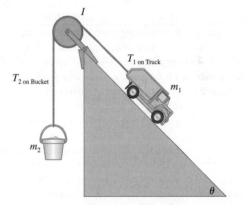

8.5 Rotational momentum

42. (a) Determine the rotational momentum of a 10-kg disk-shaped flywheel of radius 9.0 cm rotating with a rotational speed of 320 rad/s. (b) With what magnitude rotational speed must a 10-kg solid sphere of 9.0 cm radius rotate to have the same rotational momentum as the flywheel?

43. **Ballet** A ballet student with her arms and a leg extended spins with an initial rotational speed of 1.0 rev/s. As she draws her arms and leg in toward her body, her rotational inertia becomes 0.80 kg · m² and her rotational velocity is 4.0 rev/s. Determine her initial rotational inertia.

44. * A 0.20-kg block moves at the end of a 0.50-m string along a circular path on a frictionless air table. The block's initial rotational speed is 2.0 rad/s. As the block moves in the circle, the string is pulled down through a hole in the air table at the

axis of rotation. Determine the rotational speed and tangential speed of the block when the string is 0.20 m from the axis.

45. * / **Equation Jeopardy 3** The equation below describes a process. Draw a sketch representing the initial and final states of the process and construct a word problem for which the equation could be a solution.

$$\left(\frac{2}{5}mR^2\right)\left(\frac{2\pi}{30 \text{ days}}\right) = \left[\frac{2}{5}m\left(\frac{R}{100}\right)^2\right]\left(\frac{2\pi}{T_f}\right)$$

46. * A student sits motionless on a stool that can turn friction-free about its vertical axis (total rotational inertia I). The student is handed a spinning bicycle wheel, with rotational inertia I_{wheel}, that is spinning about a vertical axis with a counterclockwise rotational velocity ω_0. The student then turns the bicycle wheel over (that is, through 180°). Estimate, in terms of ω_0, the final rotational velocity acquired by the student.

47. * **Neutron star** An extremely dense neutron star with mass equal to that of the Sun has a radius of about 10 km—about the size of Manhattan Island. These stars are thought to rotate once about their axis every 0.03 to 4 s, depending on their size and mass. Suppose that the neutron star described in the first sentence rotates once every 0.040 s. If its volume then expanded to occupy a uniform sphere of radius 1.4×10^8 m (most of the Sun's mass is in a sphere of this size) with no change in mass or rotational momentum, what time interval would be required for one rotation? By comparison, the Sun rotates once about its axis each month.

8.6 Rotational kinetic energy

48. Determine the change in rotational kinetic energy when the rotational velocity of the turntable of a stereo system increases from 0 to 33 rpm. Its rotational inertia is $6.0 \times 10^{-3} \text{ kg} \cdot \text{m}^2$.

49. A grinding wheel with rotational inertia I gains rotational kinetic energy K after starting from rest. Determine an expression for the wheel's final rotational speed.

50. ** **Flywheel energy for car** The U.S. Department of Energy had plans for a 1500-kg automobile to be powered completely by the rotational kinetic energy of a flywheel. (a) If the 300-kg flywheel (included in the 1500-kg mass of the automobile) had a $6.0 \text{-kg} \cdot \text{m}^2$ rotational inertia and could turn at a maximum rotational speed of 3600 rad/s, determine the energy stored in the flywheel. (b) How many accelerations from a speed of zero to 15 m/s could the car make before the flywheel's energy was dissipated, assuming 100% energy transfer and no flywheel regeneration during braking?

51. * The rotational speed of a flywheel increases by 40%. By what percent does its rotational kinetic energy increase? Explain your answer.

52. ** **Rotating student** A student sitting on a chair on a circular platform of negligible mass rotates freely on an air table at initial rotational speed 2.0 rad/s. The student's arms are initially extended with 6.0-kg dumbbells in each hand. As the student pulls her arms in toward her body, the dumbbells move from a distance of 0.80 m to 0.10 m from the axis of rotation. The initial rotational inertia of the student's body (not including the dumbbells) with arms extended is $6.0 \text{ kg} \cdot \text{m}^2$, and her final rotational inertia is $5.0 \text{ kg} \cdot \text{m}^2$. (a) Determine the student's final rotational speed. (b) Determine the change of kinetic energy of the system consisting of the student together with the two dumbbells. (c) Determine the change in the kinetic energy of the system consisting of the two dumbbells alone without the student. (d) Determine the change of kinetic energy of the system consisting of student alone

without the dumbbells. (e) Compare the kinetic energy changes in parts (b) through (d).

53. * A turntable whose rotational inertia is $1.0 \times 10^{-3} \text{ kg} \cdot \text{m}^2$ rotates on a frictionless air cushion at a rotational speed of 2.0 rev/s. A 1.0-g beetle falls to the center of the turntable and then walks 0.15 m to its edge. (a) Determine the rotational speed of the turntable with the beetle at the edge. (b) Determine the kinetic energy change of the system consisting of the turntable and the beetle. (c) Account for this energy change.

54. * Repeat the previous problem, only assume that the beetle initially falls on the edge of the turntable and stays there.

55. * **Water turbine** A Verdant Power water turbine (a "windmill" in water) turns in the East River near New York City. Its propeller is 2.5 m in radius and spins at 32 rpm when in water that is moving at 2.0 m/s. The rotational inertia of the propeller is approximately $3.0 \text{ kg} \cdot \text{m}^2$. Determine the kinetic energy of the turbine and the electric energy in joules that it could provide in 1 day if it is 100% efficient at converting its kinetic energy into electric energy.

56. * **Flywheel energy** Engineers at the University of Texas at Austin are developing an Advanced Locomotive Propulsion System that uses a gas turbine and perhaps the largest high-speed flywheel in the world in terms of the energy it can store. The flywheel can store 4.8×10^8 J of energy when operating at its maximum rotational speed of 15,000 rpm. At that rate, the perimeter of the rotor moves at approximately 1,000 m/s. Determine the radius of the flywheel and its rotational inertia.

57. * / **Equation Jeopardy 4** The equations below represent the initial and final states of a process (plus some ancillary information). Construct a sketch of a process that is consistent with the equations and write a word problem for which the equations could be a solution.

$$(80 \text{ kg})(9.8 \text{ N/kg})(16 \text{ m}) = \frac{1}{2}(80 \text{ kg})v_f^2 + \frac{1}{2}(240 \text{ kg} \cdot \text{m}^2)\omega_f^2$$

$$v_f = (0.40 \text{ m})\omega_f$$

58. ** A bug of a known mass m stands at a distance d cm from the axis of a spinning disk (mass m_d and radius r_d) that is rotating at f_i revolutions per second. After the bug walks out to the edge of the disk and stands there, the disk rotates at f_f revolutions per second. (a) Use the information above to write an expression for the rotational inertia of the disk. (b) Determine the change of kinetic energy in going from the initial to the final situation for the total bug-disk system.

59. ** **Merry-go-round** A 40-kg boy running at 4.0 m/s jumps tangentially onto a small stationary circular merry-go-round of radius 2.0 m and rotational inertia $20 \text{ kg} \cdot \text{m}^2$ pivoting on a frictionless bearing on its central shaft. (a) Determine the rotational velocity of the merry-go-round after the boy jumps on it. (b) Find the change in kinetic energy of the system consisting of the boy and the merry-go-round. (c) Find the change in the boy's kinetic energy. (d) Find the change in the kinetic energy of the merry-go-round. (e) Compare the kinetic energy changes in parts (b) through (d).

60. ** Repeat the previous problem with the merry-go-round initially rotating at 1.0 rad/s in the same direction that the boy is running.

61. ** Repeat the previous problem with the merry-go-round initially rotating at 1.0 rad/s opposite the direction that the boy was running before he jumped on it.

62. * **Another merry-go-round** A carnival merry-go-round has a large disk-shaped platform of mass 120 kg that can rotate

about a center axle. A 60-kg student stands at rest at the edge of the platform 4.0 m from its center. The platform is also at rest. The student starts running clockwise around the edge of the platform and attains a speed of 2.0 m/s relative to the ground. (a) Determine the rotational velocity of the platform. (b) Determine the change of kinetic energy of the system consisting of the platform and the student.

63. ** A rough-surfaced turntable mounted on frictionless bearings initially rotates at 1.8 rev/s about its vertical axis. The rotational inertia of the turntable is 0.020 kg·m². A 200-g lump of putty is dropped onto the turntable from 0.0050 m above the turntable and at a distance of 0.15 m from its axis of rotation. The putty adheres to the surface of the turntable. (a) Find the initial kinetic energy of the turntable. (b) What is the final rotational speed of the system (the lump of putty and turntable)? (c) What is the final linear speed of the lump of putty? Find the change in kinetic energy of (d) the turntable, (e) the putty, and (f) the putty-turntable combination. How do you account for your answers?

8.7 Rotational dynamics: Putting it all together

64. * **Stopping Earth's rotation** Suppose that Superman wants to stop Earth so it does not rotate. He exerts a force on Earth $\vec{F}_{S\,on\,E}$ at Earth's equator tangent to its surface for a time interval of 1 year. What magnitude force must he exert to stop Earth's rotation? Indicate any assumptions you make when completing your estimate.

65. * BIO **Triceps and darts** Your upper arm is horizontal and your forearm is vertical with a 0.010-kg dart in your hand (**Figure P8.65**). When your triceps muscle contracts, your forearm initially swings forward with a rotational acceleration of 35 rad/s². Determine the force that your triceps muscle exerts on your forearm during this initial part of the throw. The rotational inertia of your forearm is 0.12 kg·m² and the dart is 0.38 m from your elbow joint. You triceps muscle attaches 0.03 m from your elbow joint.

Figure P8.65

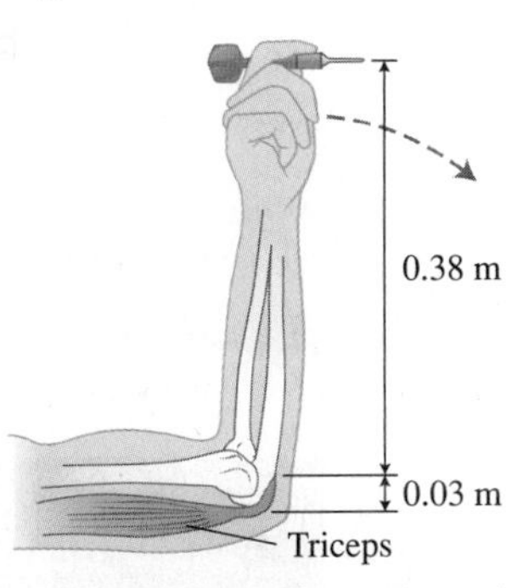

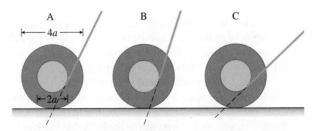

66. * BIO **Bowling** At the start of your throw of a 2.7-kg bowling ball, your arm is straight behind you and horizontal (**Figure P8.66**). Determine the rotational acceleration of your arm if the muscle is relaxed. Your arm is 0.64 m long, has a rotational inertia of 0.48 kg·m², and has a mass of 3.5 kg with its center of mass 0.28 m from your shoulder joint.

Figure P8.66

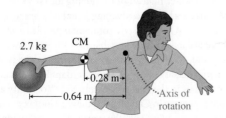

67. ** BIO **Leg lift** You are doing one-leg leg lifts (**Figure P8.67**) and decide to estimate the force that your iliopsoas muscle

exerts on your upper leg bone (the femur) when being lifted (the lifting involves a variety of muscles). The mass of your entire leg is 15 kg, its center of mass is 0.45 m from the hip joint, and its rotational inertia is 4.0 kg·m², and you estimate that the rotational acceleration of the leg being lifted is 35 rad/s². For calculation purposes assume that the iliopsoas attaches to the femur 0.10 m from the hip joint. Also assume that the femur is oriented 15° above the horizontal and that the muscle is horizontal. Estimate the force that the muscle exerts on the femur.

Figure P8.67

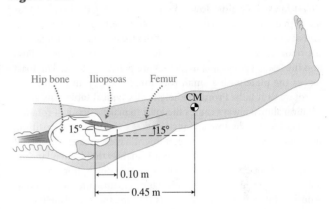

General Problems

68. * BIO EST **Punting a football** Estimate the tangential acceleration of the foot and the rotational acceleration of the leg of a football punter during the time interval that the leg starts to swing forward in an arc until the instant just before the foot hits the ball. Indicate any assumptions that you make and be sure that your method is clear.

69. * EST Estimate the average rotational acceleration of a car tire as you leave an intersection after a light turns green. Discuss the choice of numbers used in your estimate.

70. ** EST **Door on fingers** Estimate the average force that a car door exerts on a person's fingers if the door is closed when the fingers are in the door opening. Justify all assumptions you make.

71. ** **Yo-yo trick** A yo-yo rests on a horizontal table. The yo-yo is free to roll but friction prevents it from sliding. When the string exerts one of the following tension forces on the yo-yo (shown in **Figure P8.71**), which way does the yo-yo roll? Try the problem for each force: (a) $\vec{T}_{A\,S\,on\,Y}$; (b) $\vec{T}_{B\,S\,on\,Y}$; and (c) $\vec{T}_{C\,S\,on\,Y}$. [Hint: Think about torques about a pivot point where the yo-yo touches the table.]

Figure P8.71

72. ** EST **Running to change time interval of day** At present, the motion of people on Earth is fairly random; the number moving east equals the number moving west, etc. Assume that we could get all of Earth's inhabitants lined up along the

land at the equator. If they all started running as fast as possible toward the west, estimate the change in the length of a day. Indicate any assumptions you made.

73. * EST **White dwarf** A star the size of our Sun runs out of nuclear fuel and, without losing mass, collapses to a white dwarf star the size of our Earth. If the star initially rotates at the same rate as our Sun, which is once every 25 days, determine the rotation rate of the white dwarf. Indicate any assumptions you make.

Reading Passage Problems

Toast lands jelly-side down You are preparing the breakfast table with coffee and a plate of toast. While setting the plate down, you accidently tilt it slightly so that the toast slides off and falls to the floor. It always seems to land jelly-side down. Newton's laws and the laws of rotational dynamics are partly to blame. The toast leaves the plate with a small rotational velocity and continues to rotate as it falls. From the height of a typical table, this small rotation almost always causes the toast to make a one-half rotation and land jelly-side down. If it started at a higher initial position, like your elbow when standing, it might have a greater number of rotations and land jelly-side up; but if slightly tilted when landing it might bounce and flip over with jelly-side down. If it first lands jelly-side down, it does not bounce but remains stuck to the floor. Another interesting observation is that the toast lands jelly-side down only if it slides slowly off the plate. If it is shoved from the table with high velocity, it can land either way.

74. What is the force that provides the torque that causes the toast to rotate?
    (a) The normal force exerted by the plate on the trailing edge of the toast
    (b) The force due to air resistance exerted by the air on the toast
    (c) The gravitational force exerted by Earth on the toast when partly off the plate
    (d) The centripetal force of the toast's rotation
    (e) The answer depends on the choice of axis.

75. The toast is more likely to fall on the jelly side if it makes how many revolutions?
    (a) 0.5 revolutions   (b) 0.8 revolutions
    (c) 0.9 revolutions   (d) No revolutions

76. What does the number of revolutions that the toast sliding off the plate will make before it touches the floor depend on?
    (a) The amount of jelly
    (b) The height of its starting position
    (c) The length of the toast

77. Why does toast have a better chance of landing jelly-side up if it is quickly shoved off the plate or table?
    (a) It falls faster than if slowly slipping from the plate and does not have time to rotate.
    (b) It moves in a parabolic path.
    (c) The torque due to the gravitational force about the trailing edge of the toast as it leaves the plate has very little time to change the toast's rotational momentum.
    (d) The hand probably gives it an extra twist and the toast makes a full rotation instead of a half rotation.

78. The length of the toast is about 0.10 m and the mass is about 0.050 kg. Which answer below is closest to the torque about the trailing edge of the toast due to the force that Earth exerts on the toast when its trailing edge is just barely on the plate and the rest is off the plate?
    (a) Zero   (b) 0.0025 N·m
    (c) 0.005 N·m   (d) 0.025 N·m
    (e) 0.05 N·m

Tidal energy Tides are now used to generate electric power in two ways. In the first, huge dams can be built across the mouth of a river where it exits to the ocean. As the ocean tide moves in and out of this tidal basin or estuary, the water flows through tunnels in the dam (see **Figure 8.20**). This flowing water turns turbines in the tunnels that run electric generators. Unfortunately, this technique works best with large increases in tides—a 5-m difference between high and low tide. Such differences are found at only a small number of places. Currently, France is the only country that successfully uses this power source. A tidal basin plant in France, the La Rance station, makes 240 megawatts of power—enough energy to power 240,000 homes. Damming tidal basins can have negative environmental effects because of reduced tidal flow and silt buildup. Another disadvantage is that they can only generate electricity when the tide is flowing in or out, for about 10 hours each day.

Figure 8.20 Dams built across tidal basins can generate electric power.

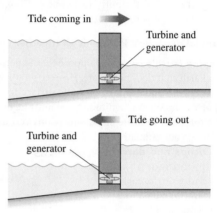

Tide coming in

Turbine and generator

Tide going out

Turbine and generator

As the tide rises and falls, water passes through the turbine, which runs a generator.

Figure 8.21 Turbines harness the energy of moving water.

A second method for collecting energy from the tidal flow (as well as all water flow) is to place turbines directly in the water—like windmills in moving water instead of in moving air (see **Figure 8.21**). These water turbines have the advantages that they are much cheaper to build, they do not have the environmental problems of a tidal basin, and there are many more suitable sites for such water flow energy farms. Also, the energy density of flowing water is about 800 times the energy density of dry air flow. Verdant Power is developing turbine prototypes in the East River near New York City and in the Saint Lawrence Seaway in Canada, and they are looking at other sites in the Puget Sound and all over the world. The worldwide potential for hydroelectric power is about 25 terawatts = 25×10^{12} J/s—enough to supply the world's energy needs.

79. If the La Rance tidal basin station in France could produce power 24 hours a day, which answer below is closest to the daily amount of energy in joules that it could produce?
   (a) 240 J
   (b) 240 × 10^6 J
   (c) 6 × 10^9 J
   (d) 2.5 × 10^10 J
   (e) 2 × 10^13 J

80. Suppose a tidal basin is 5 m above the ocean at low tide and that the area of the basin is 4 × 10^7 sq m (about 4 miles by 4 miles). Which answer below is closest to the gravitational potential energy change if the water is released from the tidal basin to the low-tide ocean level? The density of water is 1000 kg/m³. [Hint: The level does not change by 5 m for all of the water.]
   (a) 5 × 10^8 J
   (b) 5 × 10^11 J
   (c) 1 × 10^12 J
   (d) 5 × 10^12 J
   (e) 1 × 10^13 J

81. The La Rance tidal basin can only produce electricity when what is occurring?
   (a) Water is moving into the estuary from the ocean.
   (b) Water is moving into the ocean from the estuary.
   (c) Water is moving in either direction.
   (d) The moon is full.
   (e) The moon is full and directly overhead.

82. Why do water turbines seem more promising than tidal basins for producing electric energy?
   (a) Turbines are less expensive to build.
   (b) Turbines have less impact on the environment.
   (c) There are many more locations for turbines than for tidal basins.
   (d) Turbines can operate 24 hours/day versus for only 10 hours/day for tidal basins.
   (e) All of the above

83. Why do water turbines have an advantage over air turbines (windmills)?
   (a) Air moves faster than water.
   (b) The energy density of moving water is much greater than that of moving air.
   (c) Water turbines can float from one place to another, whereas air turbines are fixed.
   (d) All of the above
   (e) None of the above

84. Which of the following is a correct statement about water turbines?
   (a) Water turbines can operate only in moving tidal water.
   (b) Water turbines can produce only a small amount of electricity.
   (c) Water turbines have not had a proof of concept.
   (d) Water turbines cause significant ocean warming.
   (e) None of the above are correct statements.

9

Gases

Why does a plastic bottle left in a car overnight look crushed on a chilly morning?

How hard is air pushing on your body?

How long will the Sun shine?

Be sure you know how to:

- Draw force diagrams (Section 2.1).
- Use Newton's second and third laws to analyze interactions of objects (Section 2.8).
- Use the impulse-momentum principle (Section 5.3).

When you inflate the tires of your bicycle in a warm basement in winter, they tend to look a bit flat when you take the bike outside. The same happens to a basketball—you need to pump it up before playing outside on a cold day. A plastic bottle half full of water left in a car looks crushed on a chilly morning. What do all those phenomena have in common, and how do we explain them?

When we studied energy (in Chapters 6 and 8), we found that in many processes some of the mechanical energy of a system transforms into internal energy, resulting in a change in the temperature of the interacting objects. One of the goals of this chapter is to investigate the connection between temperature and internal energy. It turns out that the key to this connection lies in understanding the internal structure of matter.

9.1 Structure of matter

When we look at objects that surround us, we do not see their internal structure—water in a cup looks homogeneous, and the cup itself looks like one piece of material. However, they must be made up of *something*. What are the building blocks of matter? To begin to answer this question, consider a simple observational experiment.

Imagine that you dip a cotton ball in rubbing alcohol and wipe it across a piece of paper (**Figure 9.1**). The wet alcohol strip disappears gradually, with the edges of the strip disappearing first. You observe the same behavior when you wipe water or acetone on the paper, except that the water strip disappears more slowly and the acetone strip disappears more quickly. This phenomenon is the same one we observe with wet clothes and puddles as they dry.

Since the alcohol disappeared gradually, it is reasonable to suggest that it is made of "pieces" too small to be seen. If the alcohol were composed of one piece, it would be gone all at once. However, the model of small pieces does not explain *how* the disappearance occurs. Let's try to construct some possible *mechanisms* that explain how the alcohol disappears (these mechanisms would also be applicable to the drying of water or acetone). Three of the many possible mechanisms are described below:

Mechanism 1. The little pieces of liquid move to the *inside* of the paper and are still there, even though the paper looks dry.

Mechanism 2. The air surrounding the paper somehow pulls the liquid pieces out of the paper.

Mechanism 3. The pieces of liquid are moving—they bump into each other and slowly bump each other out of the paper one by one.

All of these mechanisms seem viable. Testing Experiment **Table 9.1** will help us decide if we can rule out any of them.

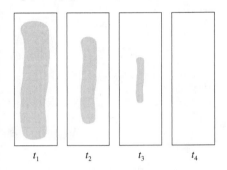

Figure 9.1 A disappearing moist strip on a piece of paper.

t_1 t_2 t_3 t_4

< **Active Learning Guide**

< **Active Learning Guide**

TESTING EXPERIMENT TABLE

9.1 Testing various mechanisms for the drying of wet objects.

 VIDEO 9.1

Testing experiment 1	Prediction based on each mechanism	Outcome
Weigh a dry strip of paper on a sensitive scale. Record its mass. Then moisten the paper with alcohol and record the mass of the wet paper. What happens to the mass of the strip when it dries? 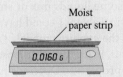 Moist paper strip	**Mechanism 1.** If the alcohol pieces go "inside" the paper, then the mass of the paper after it dries should be greater than it was before it was made wet and exactly the same as it was just after it was made wet. **Mechanism 2.** If the air pieces absorb and carry away the alcohol pieces, then after the paper dries, it should have the same mass as it had before it was made wet. **Mechanism 3.** If the moving alcohol pieces bump each other out of the paper, then after the paper dries it should have the same mass as it had before it was made wet.	After the paper dries, its mass is the same as before it was made wet.

(continued)

Conclusion 1

The alcohol pieces left the paper. Mechanism 1 is disproved by this experiment, but Mechanisms 2 and 3 are consistent with it. This leaves us with only two possible mechanisms.

Testing experiment 2	Prediction based on each mechanism	Outcome
Moisten two strips of paper. Place one inside a sealed glass jar attached to a vacuum pump. Place the other strip outside the jar. Pump the air from the jar. Moist paper / Vacuum	**Mechanism 2.** If the air pulls the alcohol pieces out of the strip, then the strip inside the vacuum jar with little air should dry more slowly than the strip outside the jar. **Mechanism 3.** If the moving alcohol pieces bump each other out of the paper, then the strip inside the vacuum jar should not dry more slowly than the strip outside the jar.	The paper inside the evacuated jar dries faster.

Conclusion 2

Mechanism 2 is disproved by this experiment. Mechanism 3 is not disproved by it.* Now we have only one mechanism that has not been disproved.

Testing experiment 3	Prediction	Outcome
Add a droplet of colored alcohol to a glass of clear alcohol. What happens to the food coloring? Food coloring	Mechanism 3. If the small pieces of clear alcohol are moving and bumping each other, then they should bump the colored pieces and cause them to spread.	The color slowly spreads throughout the clear alcohol.

Conclusion 3

Mechanism 3 is supported by this experiment.

*Note that while Mechanism 3 was not disproved in Experiment 2, the mechanism does not explain why the strip in the vacuum jar actually dried faster.

Based on these experiments, it is reasonable to assume that alcohol and other liquids are composed of smaller objects, called **particles,** which move randomly in all directions. These particles need empty space between them so that particles of other materials can move between them, as happened in Experiment 3 in Table 9.1. This model of the internal structure of alcohol can be used to explain many other phenomena that we encounter—the way some liquids mix or smells spread. In fact, experiments such as those described in Table 9.1 could have led the Greek philosopher Democritus (460–370 B.C.) to the *atomistic* model. *Atomos* in Greek means *indivisible*. According to Democritus, matter was composed of small, indivisible pieces with different shapes and properties. Different substances were made of different combinations of these pieces. Democritus also suggested that the pieces were separated by tiny regions of completely empty space. Democritus proclaimed, "There is nothing in the world but atoms and empty space." We see that Democritus's views are in line with our reasoning.

In 1827, a Scottish botanist named Robert Brown used a microscope to observe the random movement of pollen granules in a droplet of water. To explain his observations, Brown used Mechanism 3—that water itself is composed of particles smaller than the granules and these particles move randomly between frequent collisions (**Figure 9.2a**). The water particles randomly hit the pollen granules from all directions and caused random changes in the position of a granule (Figure 9.2b). This experiment supported the model of water composed of invisible particles that were in continual random motion.

In 1905, Albert Einstein constructed a quantitative model to describe the phenomenon observed by Brown, called *Brownian motion*. He predicted the average distance that a granule of a certain size would move in a given time interval. Einstein's model predicted that the distance depended on the temperature of the liquid in a specific way. Later experiments by Jean Perrin were consistent with Einstein's predictions. Physicists consider Perrin's experiments strong support for the *particle structure of matter*.

We now know that **atoms** are the smallest objects that still retain the chemical properties of a particular element (hydrogen, oxygen, carbon, iron, gold, etc.). Atoms of these various elements combine to build solids, liquids, and gases. A **molecule** is a certain combination of atoms that bond together in a particular arrangement. Molecules may consist of two atoms (such as oxygen, O_2, and nitrogen, N_2), three atoms (such as water—two hydrogen atoms and one oxygen atom, H_2O), or many atoms. Protein molecules can consist of thousands of atoms. Atoms are composed of fundamental particles—protons, neutrons, and electrons. We will discuss these fundamental particles later in the book (Chapters 27 and 28). For now, we will use the word *particle* to indicate an object approximately the size of a molecule or smaller.

The particle model explains how we can smell things even when we are not near them. Suppose we open a bottle of perfume while standing in the middle of a room. Several minutes later, people all over the room can smell the perfume. According to the particle model, the little particles of perfume leave the bottle and gradually disperse, eventually arriving at our nostrils. Since everyone in the room eventually smells the perfume, the perfume particles must move in all directions. However, it takes time to smell the perfume if you are far away from the bottle. Why is that? Perfume particles leaving the bottle move quickly but collide with air particles along the way and reach your nose only after many collisions.

If air is composed of tiny particles, and those particles have mass, then air must have mass. We can test this hypothesis with another simple experiment. Take two rigid metal cylinders—one that has had the air evacuated with a vacuum pump, and the other filled with air. Then weigh them. If the hypothesis that air has mass is correct, the evacuated jar should weigh less than the unevacuated jar. The cylinder filled with air is indeed heavier than the evacuated cylinder (**Figure 9.3**). However, the difference in masses is small (8 g difference for cylinders of about $6 \times 10^{-3} \text{m}^3$ (1.5 gallons).

Gases, liquids, and solids

We know from experience that gases are easy to compress, while liquids and solids are almost incompressible (**Figure 9.4**). The particle model helps us explain this difference: we assume that matter in all states is composed of small particles, but the amount of empty space between the particles is different in solids, liquids, and gases. In solids and liquids, the particles are closely packed (almost no empty space between them), while in gases the particles are packed more loosely (lots of empty space).

Gases tend to occupy whatever volume is available. If you take the air that fills a small cylinder and move it to a much larger cylinder, the air also fills the

Figure 9.2 (a) Water particles moving randomly. (b) Their collisions with a pollen granule cause its random motion.

(a)

The shaded water particle moves randomly due to collisions with other particles.

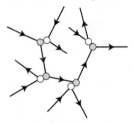

(b)

Water molecules collide and cause the pollen granule to move in random directions.

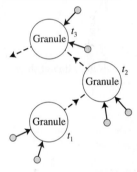

‹ Active Learning Guide

Figure 9.3 Air has mass.

The cylinder with air has more mass than the evacuated cylinder.

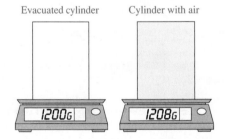

Figure 9.4 A gas can be compressed, but a liquid cannot.

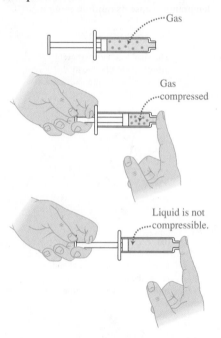

Gas

Gas compressed

Liquid is not compressible.

Figure 9.5 Although gases and liquids are both fluids, they do not share all the same properties.

(a)

The same gas completely fills a different volume.

(b)

The liquid volume remains the same regardless of the container.

Active Learning Guide ›

Figure 9.6 Ideal gas model. Gas particles are modeled as point-like objects that interact only when colliding with each other or with the walls of their containers.

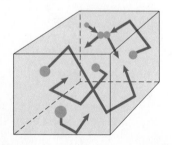

large cylinder (**Figure 9.5a**). In contrast, if we move the liquid filling a small container to a much larger container, the liquid volume remains the same independent of the container's shape (Figure 9.5b). Solids maintain not only their volume but also their shape.

In order for particles to form a substance, the particles must somehow be attracted to each other. We will learn later (in Chapters 14 and 27) that although the nature of this attractive force between particles is different from their gravitational attraction for each other, there are some similarities between the two types of attraction. For instance, in both cases the particles are attracted to each other at a distance, without any direct contact. The behavior of gases indicates that particles in gases must be so far from each other that such attractive forces are very small or even negligible. In solids the particles are close together, and these new attractive forces between them are strong. The attractive forces between particles in a liquid are stronger than for gas. In solids they are even stronger.

Ideal gas model

From the observed behavior of gases and our explanations based on the particle nature of matter, we can conclude that gases will be the easiest to study, as we can neglect the very small forces between the particles. We start by constructing a simplified model of a gas as a system.

In this model, we assume that the average distance between the particles (molecules or atoms) is much larger than the size of the particles. We assume that the gas particles are point-like objects that only interact with each other and with any surfaces in their containers during collisions (**Figure 9.6**), similar to the impulsive interactions of a billiard ball with a side of a pool table. In our model the particles do not attract each other at a distance the way they do in solids and liquids. We also assume that the motions of these particles obey Newton's laws. Thus, between collisions the particles move in straight lines at constant velocity and only change their velocities during collisions. Altogether, these simplifying assumptions make up the **ideal gas model**.

> **Ideal gas model** A model of a system in which gas particles are considered point-like and only interact with each other and the walls of their container through collisions. This model also assumes that the particles and their interactions are accurately described using Newton's laws.

Ideal in this context does not mean perfect; it means *simplified*. This is a simplified model with certain assumptions. Whether or not this model can be used to represent a real gas remains to be seen. Only testing experiments can resolve that issue.

How useful is this new model of a gas? As with our previous models of objects (point-like objects, rigid bodies), we will use it to describe and explain known phenomena and then to predict new phenomena. However, so far our model is only qualitative. We need to devise physical quantities to represent the features and behavior of the model. Only then can we use it to develop descriptions, explanations, and predictions of new phenomena involving gases. This process will then allow us to construct a mathematical description of an ideal gas and predict its behavior.

Review Question 9.1 Use the particle model to explain how moist objects dry out.

9.2 Pressure, density, and the mass of particles

In this section we will identify several new physical quantities that are useful for describing the properties of the model of an ideal gas. These quantities are also applicable to the properties of real gases, liquids, and solids.

Pressure

Imagine you are holding an air-filled balloon. Try to crush it a little bit. You feel the balloon resisting the crushing, as if something inside it pushes back on your fingers. How can we explain this "resistance" of the air inside the balloon? As we will learn later in this section, the particles of matter (atoms and molecules) are rather small. As air particles move randomly in space, they eventually collide with the solid surfaces of any objects in that space. In each of these collisions, the particle exerts an impulsive force on the object—like a tennis ball hitting a practice wall (see **Figure 9.7a**). However, when a huge number of particles bombard a solid surface at a constant rate, these collisions collectively exert an approximately constant force on the object (Figure 9.7b). This impulsive force must be what we feel when we are trying to squeeze the balloon. Notice that we have now constructed a *model of a process* that explains how the motion of the particles of gas inside the balloon accounts for the observational evidence—the apparent resistance of the balloon to squeezing. This ideal process that we imagined serves as a mechanism for what we observe.

We can test this process model with a simple experiment. Consider that when you inflate a balloon, its surface expands outward. Our model actually predicts this outcome because as we blow, we add air particles to the interior of the balloon; thus there are more particles inside colliding with the walls. This greater collision rate results in a larger outward average force on each part of the balloon's surface, causing it to expand outward.

On the other hand, there is also air outside the balloon, and particles of this air hit the outside of the balloon walls. In Testing Experiment **Table 9.2**, we investigate what will happen if we decrease the number of particles outside the balloon.

Figure 9.7 Impulsive forces during collisions cause an approximately constant force against a wall.

(a)

Each ball or particle exerts an impulsive force on the wall during a collision.

(b)

Many particles hitting the wall cause a near-constant force.

TESTING EXPERIMENT TABLE

9.2 Testing the model of moving gas particles pushing on the surface.

Testing experiment	Prediction based on a model of gas particle motion	Outcome
Place a partially inflated balloon inside a vacuum jar. Seal the jar. 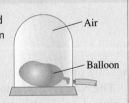 What happens to the balloon's shape when you start pumping air out of the jar?	As we remove air particles from outside the balloon, the collisions of particles on the outside of the balloon are less frequent and exert less force on each part of the balloon's outer surface. The collision rate of the particles inside the balloon does not decrease. Therefore, the balloon should expand.	As air outside of the balloon is removed, the balloon expands. 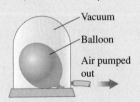

Conclusion

The model of air consisting of moving particles colliding with objects exposed to the air has not been disproved. The results of this experiment support the model.

In normal situations, an extremely large number of gas particles collide each second with the surface that is in contact with the gas. For example, about 10^{23} particles of air collide each second with each square centimeter of your skin. Making a force diagram for the skin by including these individual particle collision forces is not practical. A different approach is needed.

As we have discussed, although each particle collision is impulsive, the forces are so small and so frequent that the force exerted by the gas on the walls of the container can be modeled as a single constant force. The force also depends on the area of that surface—the bigger the area, the more particles push on it during collisions. Thus, instead of using force to describe gas processes, we use a new force-like physical quantity called **pressure**. As we will see (in Chapters 10 and 11), this quantity is useful for describing the behavior of liquids and solids as well.

> **Pressure P** Pressure is a physical quantity equal to the magnitude of the perpendicular component of the force $F_\perp$ that a gas, liquid, or solid exerts on a surface divided by the contact area A over which that force is exerted:
>
> $$P = \frac{F_\perp}{A} \qquad (9.1)$$

The SI unit of pressure is the pascal (Pa), where $1\,\mathrm{Pa} = 1\,\mathrm{N/m^2}$. In British units pressure is measured in pounds per square inch (psi). You will find a complete list of pressure units on the inside front cover of the text.

Pressure is easy to visualize when you think about two solid surfaces that contact each other. Compare the magnitude of force you exert on soft snow when wearing street shoes versus the magnitude of force that you exert on the snow when wearing snowshoes. The magnitude of force that you exert on the snow is the same, but the corresponding pressure is not. The snowshoes decrease the pressure you exert on the snow, since the force is spread over a much larger area. Similarly, when you decrease the area over which the force is spread, the pressure increases. Scissors and knives increase pressure on a surface because they decrease the area over which they exert a force.

Consider the concept of pressure inside a gas or liquid. The air particles hitting one wall in a room collectively exert an average force on the wall, and therefore a pressure. However, the air particles in the center of the room are not interacting with the wall at all, at least not for the moment. Yet, the air there has a pressure as well. If a table were placed in the center of the room, the air particles would exert an average force, and therefore a pressure, on the top, bottom, and sides of the table. Thus, air has a pressure whether or not a solid object is present.

Measuring pressure

Many instruments are used to measure pressure. An aneroid barometer is used to measure gas pressure directly. The barometer contains a small aneroid cell (**Figure 9.8**). Inside, the cell has almost no air. A lever is attached to the cell's moveable wall. As the outside air pressure on this wall changes, the cell thickness changes, and the lever causes a pointer needle on the aneroid barometer to change, indicating the outside air pressure. Measurements show that the pressure of the atmospheric air at sea level is on average $10^5\,\mathrm{N/m^2}$, or $10^5\,\mathrm{Pa}$. This atmospheric pressure defines yet another unit of pressure, called an atmosphere: $1.0\,\mathrm{atm} = 1.0 \times 10^5\,\mathrm{Pa} = 1.0 \times 10^5\,\mathrm{N/m^2}$.

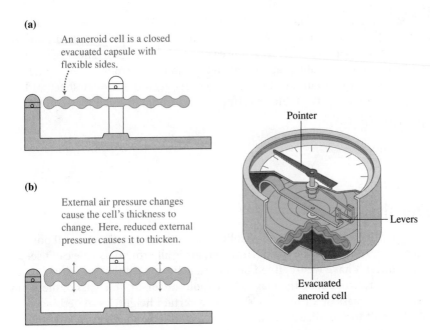

(a)

An aneroid cell is a closed evacuated capsule with flexible sides.

(b)

External air pressure changes cause the cell's thickness to change. Here, reduced external pressure causes it to thicken.

Pointer

Levers

Evacuated aneroid cell

Figure 9.8 An aneroid barometer consists of an evacuated aneroid cell that gets thicker or thinner depending on outside air pressure.

Another way to measure gas pressure is to compare the pressure of a gas in contact with a gauge to the atmospheric pressure. For example, when you use a tire gauge to measure the air pressure in a car tire, you are comparing the air pressure inside the tire to that of the atmosphere outside the tire. The pressure in the tire is called **gauge pressure.** If the pressure in a container is 1.0 atm, its gauge pressure is zero, because there is no difference between the pressure inside the container and outside of it.

Gauge pressure P_{gauge} Gauge pressure is the difference between the pressure in some container and the atmospheric pressure outside the container:

$$P_{gauge} = P - P_{atm} \qquad (9.2)$$

where $P_{atm} = 1.0 \text{ atm} = 1.0 \times 10^5 \text{ N/m}^2$.

We are continually immersed in a fluid—the gaseous atmosphere. What is the magnitude of the force that the air exerts on one side of your body?

QUANTITATIVE EXERCISE 9.1 The force that air exerts on your body

Estimate the total force that the air exerts on the front side of your body, assuming that the pressure of the atmosphere is constant.

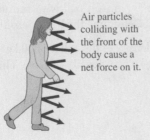

Air particles colliding with the front of the body cause a net force on it.

Represent mathematically To calculate the total force exerted by the air on the front of your body, multiply your front surface area by 10^5 N/m^2, the atmospheric air pressure. To estimate the surface area of the front of your body, model your body as a rectangular box—your height times your width. Assume a height of 1.8 m and a width of 0.3 m. Use Eq. (9.1) to estimate the force:

$$F_{A \text{ on } F} = P \cdot A$$

Solve and evaluate $F_{A \text{ on } F} = P \cdot A = (10^5 \text{ N/m}^2)[(1.8 \text{ m})(0.3 \text{ m})] \approx 5 \times 10^4 \text{ N}$. That's $5 \times 10^4 \text{ N} \times 0.22 \text{ lb/N} = 1.1 \times 10^4 \text{ lb}$, or 10,000 lb. So why aren't you immediately thrown backward? [*Hint:* Think about the force that the air behind you is exerting on you as well.]

(continued)

Try it yourself: What is the minimum force that you must exert to lift a 10 cm × 20 cm, 10-kg rubber sheet that is stuck flat to a tabletop? Assume that there is no air below the sheet.

Answer: $F_{Y \text{ on } S} = (P_{atm} \cdot A) + mg = (10^5 \text{ N/m}^2)[(0.10 \text{ m})(0.20 \text{ m})] + (0.10 \text{ kg})(9.8 \text{ N/kg}) = 2000 \text{ N},$

or about 450 lb. Since it is not really this difficult (although it is not easy!) to lift a rubber sheet off the table, under real circumstances there must be a small amount of air trapped between the rubber sheet and the table, exerting an additional upward force on the sheet.

Density

The quantity *mass* helps describe solid objects that have discrete real boundaries—a person, a car, or a ball. However, air is all around us. We can't see it, and it doesn't have well-defined boundaries. If we used the quantity mass to describe air, would it be the mass of one molecule of air, the mass of the air in a room, the mass of air over the street to a certain height, or something else? It's difficult to visualize air as a macroscopic object. For gases, a much more useful physical quantity is the mass of one unit of volume—**density**.

Density measures the mass of one cubic meter of a substance. For example, at sea level and 0°C the mass of 1.0 m³ of air is 1.3 kg. We say that the density of air is 1.3 kg/m³. If we had 2.0 m³ of air at sea level, its mass would be 2.6 kg. Its density is still 1.3 kg/m³, since

$$\frac{2.6 \text{ kg}}{2.0 \text{ m}^3} = 1.3 \text{ kg/m}^3.$$

TIP Density is different from mass. Air in a room has a particular mass and density. If you divide the room into two equal parts using a screen, the mass of air in each part will be half the total mass, but the density in each part will remain the same.

Density ρ The density ρ (lowercase Greek letter "rho") of a substance or of an object equals the ratio of the mass *m* of a volume *V* of the substance (for example, air or water) divided by that volume *V*:

$$\rho = \frac{m}{V} \tag{9.3}$$

The unit of density is kg/m³.

QUANTITATIVE EXERCISE 9.2 The density of a person

Estimate the density of a person.

Represent mathematically Assume the following about the person: mass is 80 kg; dimensions are 1.8 m tall, 0.3 m wide, and 0.1 m thick; and volume is $V \approx 1.8 \text{ m} \times 0.3 \text{ m} \times 0.1 \text{ m} = 0.054 \text{ m}^3$.

The person's density is given by Eq. (9.3):

$$\rho = \frac{m}{V}$$

Solve and evaluate Substitute the person's mass and volume into the above to get

$$\rho = \frac{80 \text{ kg}}{0.054 \text{ m}^3} = 1500 \text{ kg/m}^3$$

This is the correct unit for density. In a later chapter, we will decide if this is a reasonable estimate for the density of a human.

Try it yourself: An iron ball with radius 5.0 cm has a mass of 2.0 kg. Determine the ball's density.

Answer: The density of the ball is 3800 kg/m³, much less than the 7860-kg/m³ density of iron. The ball must be hollow.

Mass and size of particles

In 1811, an Italian scientist named Avogadro proposed that equal volumes of different types of gas, when at the same temperature and pressure, contain the same number of gas particles. Using Avogadro's hypothesis, scientists could determine the relative masses of different types of particles by comparing the masses of equal volumes of the gases. Presently, scientists use Avogadro's number N_A to indicate the number of atoms or molecules present in 22.4 L ($22.4 \times 10^3 \text{ cm}^3$) of any gas at $0°$ C and standard atmospheric pressure.

The mass in grams of any substance that has exactly Avogadro's number of particles is equal to the atomic mass. For example, the atomic mass of molecular hydrogen is 2, that of molecular oxygen is 32, and that of lead is 207; therefore, 2 g of molecular hydrogen, 32 g of molecular oxygen, and 207 g of lead all have the same number of particles—exactly $N_A = 6.02 \times 10^{23}$. This number of particles is called a **mole** (**Figure 9.9**).

> **Avogadro's number and the mole** Avogadro's number $N_A = 6.02 \times 10^{23}$ particles is called a mole. The number of particles in a mole is the same for all substances and is the number of particles whose total mass equals the atomic mass of that substance.

The mass of one mole of particles of any substance is called **molar mass.** One mole of hydrogen (H_2) has a mass of 2.0 g, one mole of oxygen (O_2) has a mass of 32 g, and one mole of lead has a mass of 207 g.

We can now easily determine the mass of a single gas particle of any substance (for example, hydrogen, oxygen, or nitrogen) by dividing the molar mass of the substance by 6.02×10^{23}, the number of particles in one mole of the substance:

$$m_{\text{particle}} = \frac{m_{\text{mole}}}{N_A}$$

We find that air (typically 70% N_2 with $m_{N_2} = 28$ g, 29% O_2 with $m_{O_2} = 32$ g, and small percentages of other gases) has a molar mass of about 29 g. Thus, the mass per air particle is approximately

$$m_{\text{air particle}} = \frac{29 \times 10^{-3} \text{ kg}}{6.02 \times 10^{23} \text{ air particles}} = 4.8 \times 10^{-26} \text{ kg/air particle}$$

In addition to a particle's mass, its size is also important. The size of particles was estimated much later than the mass—in the 1860s by Josef Loschmidt. Loschmidt found the linear size of the particles that made up gases, liquids, and solids to be about $d \approx 10^{-9}$ m $= 10^{-7}$ cm $= 1$ nm. Contemporary methods indicate that nitrogen and oxygen are about 0.3 nm.

Figure 9.9 A mole (6×10^{23} particles) of helium (4 g of helium in the balloon), water (18 g), and salt (58 g).

‹ Active Learning Guide

EXAMPLE 9.3 **The average distance between air particles**

What is the average separation between nearby gas particles in the air, and how does it compare to the size of the particles themselves?

Sketch and translate Sketch the situation as a mole of air divided into cubes as shown at the right, with one particle located at the center of each cube. The diameter of an individual particle is d; the average distance between them is D.

We visualize the mole of gas divided into equal-sized cubes of volume D^3, each containing one particle of diameter d.

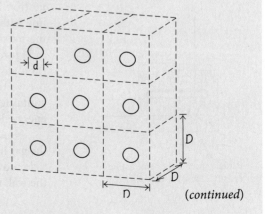

(continued)

Simplify and diagram Assume standard conditions, with one mole of gas particles occupying a volume of $22.4 \times 10^3 \, cm^3$. Now, imagine this volume divided equally between the particles (the cubes in the figure) and each particle being contained in a cube whose volume is D^3. We can think of the distance between the particles as the side of this cube D.

Represent mathematically We determine the volume of the cube corresponding to each particle and then take the cube root of that volume to estimate the average distance between them.

$$V_{\text{per particle}} = \frac{V_{\text{one mole}}}{N_{\text{particles in a mole}}}$$

$$D = \sqrt[3]{V_{\text{per particle}}} = \sqrt[3]{\frac{V_{\text{one mole}}}{N_{\text{particles in a mole}}}}$$

Solve and evaluate

$$D = \sqrt[3]{\frac{V_{\text{one mole}}}{N_{\text{particles in a mole}}}} = \sqrt[3]{\frac{22.4 \times 10^3 \, cm^3}{6.02 \times 10^{23} \, \text{particles}}}$$

$$= \sqrt[3]{3.72 \times 10^{-20} \, cm^3/\text{particle}}$$

$$= \sqrt[3]{37.2 \times 10^{-21} \, cm^3/\text{particles}}$$

$$= 3.34 \times 10^{-7} \, cm$$

Recall that the average diameter of a single particle of air is $3 \times 10^{-8} \, cm$. Thus, the approximate distance between the particles of air is on average $\dfrac{3.34 \times 10^{-7} \, cm}{3 \times 10^{-8} \, cm} \approx 10$ times the size of the particles ($D/d = 10$). This is a lot of empty space. A macroscopic analogy would be 20 people lying down and spread uniformly over the area of a football field.

Try it yourself: Estimate the distance between water molecules in liquid water and compare this distance to their dimensions. The density of water is $1.0 \times 10^3 \, kg/m^3$. The molar mass of water is $18 \times 10^{-3} \, kg/\text{mole}$.

Answer: The volume occupied by one mole of water is about

$$V_{\text{one mole}} \approx \frac{(18 \, g)(10^{-3} \, kg/g)}{1000 \, kg/m^3} = 18 \times 10^{-6} \, m^3$$

The volume occupied by one molecule is

$$V_{\text{one molecule}} \approx \frac{18 \times 10^{-6} \, m^3}{6 \times 10^{23}} = 3 \times 10^{-29} \, m^3$$

The distance between particles is $D_{\text{between water molecules}} \approx 3 \times 10^{-8} \, cm$. This is just a little larger than the size of the molecules, which means there is very little space between the water molecules.

Review Question 9.2 The distance between air particles is very small—about $3 \times 10^{-7} \, cm$. How can we say that there is considerable empty space in air?

9.3 Quantitative analysis of ideal gas

We can use the quantities pressure, density, and the mole to construct a mathematical description of an ideal gas that will allow us to make predictions about new phenomena.

To start, we make a few more simplifying assumptions. First, in addition to modeling the particles (atoms or molecules) as point-like objects whose motion is governed by Newton's laws, assume that the particles do not collide with each other—they only collide with the walls of the container, exerting pressure on the walls (in other words, they move like the model depicted in Figure 9.6 but with no particle collisions). This is a reasonable assumption for a gas of low density. Second, assume that the collisions of particles with the walls are elastic. This makes sense, as the pressure of the gas in a closed container remains constant, which would not happen if the particles' kinetic energy decreased during inelastic collisions.

Now, let's construct a mathematical description of an ideal gas. Imagine the gas inside a cubic container with sides of length L (see **Figure 9.10**). A particle moves at velocity $\vec{v}$ with respect to a vertical wall. When it hits the wall, the wall exerts a force on the particle that causes it to reverse direction. Since

Figure 9.10 A gas particle bouncing back and forth between the walls of a container.

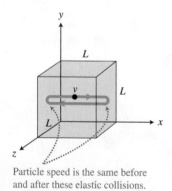

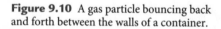

Particle speed is the same before and after these elastic collisions.

the kinetic energy of the particle is the same before and after the collision, the same is true for its speed. The wall exerts a force on the particle, and the particle in turn exerts an equal-magnitude and oppositely directed force on the wall (Newton's third law).

Let's use impulse-momentum ideas to analyze the particle–wall collision. During the collision with the right wall, it exerts a normal force on the particle in the negative x-direction. Before the collision, the particle has a positive x-component of velocity v_{xi}. After the collision, the particle has a negative x-component of velocity v_{xf}. The impulse-momentum equation gives

$$mv_{xi} + F_{\text{W on P }x}\Delta t = mv_{xf}$$

Because the particle's speed is the same before and after the collision, $v_{xf} = -v_{xi}$. In addition, because of Newton's third law, $F_{\text{W on P }x} = -F_{\text{P on W }x}$. Thus,

$$mv_{xi} + (-F_{\text{P on W }x}\Delta t) = -mv_{xi}$$
$$2mv_{xi} = F_{\text{P on W }x}\Delta t$$

In the above equation, $F_{\text{P on W }x}$ is the impulsive force that the particle exerts on the wall during the very short time interval Δt that the particle is actually touching and colliding with the wall. How can we determine an average effect of these impulsive collisions in order to determine the pressure of the gas on the wall? We rewrite the right side of the equation as the product of the *average force* exerted by the particle on the wall from one collision to the next, multiplied by the time that *passes between collisions*. Note that this average force is much smaller than the impulsive force, since most of the time the particle is flying through the container and is not in contact with the wall (**Figure 9.11**). However, the time interval between collisions is longer than the impulsive time interval. The product of the big impulsive force and short time interval equals the product of the small average force and the long time interval between collisions.

Looking at the x-component of the motion of the particle, we see that the time interval between collisions with the wall is $\Delta t_{\text{between collisions}} = (2L/v_{xi})$ since the particle must travel a distance $2L$ in the x-direction before colliding with the wall again. So, our equation becomes

$$2mv_{xi} = F_{\text{P on W }x}\Delta t_{\text{collision}} = \overline{F_{\text{P on W }x}}\Delta t_{\text{between collisions}} = \overline{F_{\text{P on W }x}}\frac{2L}{v_{xi}}$$

The bar above the force in the last two expressions indicates that it is the *average force* exerted over the time interval between collisions. Multiplying by v_{xi} and dividing by 2 gives

$$mv_{xi}^2 = \overline{F_{\text{P on W }x}}L$$

To relate this microscopic relationship to macroscopic quantities, such as the pressure of the gas and its volume, multiply both sides of the equation by N, the number of particles of gas in the container. Because the particles do not all move with the same speed, we also replace the quantity v_{xi}^2 (the square of the x-component of the velocity of an individual particle) with its average value for all the particles:

$$Nm\overline{v_x^2} = \left(N\frac{\overline{F_{\text{P on W }x}}}{L^2}\right)L^3$$

We have also multiplied the right side by L^2/L^2. The term in parentheses is the pressure exerted by the gas on the wall (the force exerted by all N particles divided by the wall area L^2.) The L^3 outside the bracket is the volume occupied by the gas. Thus, the above equation becomes

$$Nm\overline{v_x^2} = PV$$

Figure 9.11 A method to find the average force of particles colliding with the container wall.

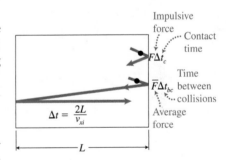

The impulsive force exerted by a particle against the wall during the short contact time interval equals the average force exerted by the particle against the wall during the long time interval between collisions: $F\Delta t_c = \overline{F}\Delta t_{bc}$.

Figure 9.12 The average velocity squared $\overline{v^2}$ is the sum of the x, y, and z average velocity squared components.

(a)

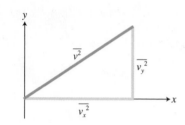

(b)

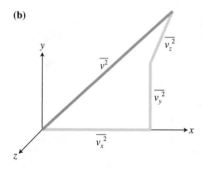

Note that this equation refers only to the motion of the particles along the x-direction. The particles also move in the y- and z-directions. Those equations are

$$Nm\overline{v_y^2} = PV$$
$$Nm\overline{v_z^2} = PV$$

To make an equation that simultaneously takes into account the motion of the particles in all three directions, we just add these three equations:

$$Nm(\overline{v_x^2} + \overline{v_y^2} + \overline{v_z^2}) = 3PV$$

The quantity $(\overline{v_x^2} + \overline{v_y^2} + \overline{v_z^2})$ is the sum of the averages of the squares of the speeds of the particles in the three directions. Think of the meaning of this sum. We are adding squared x-, y-, and z-components of some vector. If we worked in two dimensions, the sum of the squared x- and y-components of some vector would be the square of the magnitude of that vector (the Pythagorean theorem— **Figure 9.12a**). In our three dimensional situation, $(\overline{v_x^2} + \overline{v_y^2} + \overline{v_z^2})$, this sum must be the average of the squared speed $(\overline{v_x^2} + \overline{v_y^2} + \overline{v_z^2}) = \overline{v^2}$, as depicted in Figure 9.12b. Using this new average of the squared speed, we can write

$$Nm\overline{v^2} = 3PV$$

The left side of the equation is the average kinetic energy of the particles if we multiply by ½. After multiplying by ½ and rearranging the equation, we get

$$PV = \frac{2}{3}N\left(\frac{1}{2}m\overline{v^2}\right) = \frac{2}{3}N\overline{K}$$

Dividing by V, we get

$$P = \frac{2}{3}\left[\frac{N\left(\frac{1}{2}m\overline{v^2}\right)}{V}\right] \tag{9.4}$$

TIP Every time you derive a new equation, ask yourself whether it makes sense. In Eq. (9.4), the pressure is proportional to the average squared speed of the particles. This means that doubling the speed of all the particles quadruples the pressure. Try to explain this dependence before you read on.

When the particles have high speed, they (1) hit the walls of the container more frequently and (2) exert a greater force during the collisions. Both factors lead to a greater pressure. Thus, it is the speed squared and not just the speed of the particles that affects the pressure.

Suppose the number of particles N and the particle average speed squared $\overline{v^2}$ remains the same, but the volume of the box decreases. Then the pressure must increase. This seems reasonable. The smaller the container, the more frequently particles collide with the walls, and the greater the pressure. A unit analysis with a little manipulation indicates that both sides of Eq. (9.4) have the units $kg/(m \cdot s^2)$, so the equation checks out from the point of view of dimensional analysis.

QUANTITATIVE EXERCISE 9.4 How fast do they move?
Estimate the average speed of air particles at standard conditions, when the air is at atmospheric pressure ($1.0 \times 10^5 \, N/m^2$), and one mole of the air particles (6.02×10^{23} molecules) occupies 22.4 L or $22.4 \times 10^{-3} \, m^3$. Although air is composed of many types of particles, we will assume that the air particles have an average mass $m_{air} = 4.8 \times 10^{-26} \, kg/particle$.

Represent mathematically We can estimate the average speed using Eq. (9.4):

$$P = \frac{2}{3}\left[\frac{N\left(\frac{1}{2}m\overline{v^2}\right)}{V}\right]$$

Solve and evaluate Multiply both sides by 3, divide by Nm_p, and take the square root to get

$$\overline{v} \approx \sqrt{\frac{3PV}{Nm_p}}$$

Inserting the appropriate values gives

$$\overline{v} \approx \sqrt{\frac{3PV}{Nm_p}} = \sqrt{\frac{3(1.0 \times 10^5\,\text{N/m}^2)(22.4 \times 10^{-3}\,\text{m}^3)}{(6.02 \times 10^{23})(4.8 \times 10^{-26}\,\text{kg})}}$$
$$= 480\,\text{m/s}$$

This number seems too high. We know that it takes several minutes for the smell of perfume to propagate across a room. How could it take so long if the perfume molecules move at hundreds of meters per second (about 1000 mph)? We will find out in the next section.

Try it yourself: The pressure in a diver's full oxygen tank is about $4 \times 10^7\,\text{N/m}^2$. What happens to the average speed of the particles when some of the oxygen is used up and the pressure in the tank drops to half this value? What assumptions did you make?

Answer: If we assume that the only change is in the number of particles in the tank, then their average speed should stay the same. However, when a gas expands (as this one does when the valve to the tank is opened), the gas inside the tank has to push outward against the environment in order to leave the tank. This requires energy, which lowers the average kinetic energy (and therefore the speed) of the particles inside.

Time interval between collisions

We estimated that the average speed of air particles is $v = 480\,\text{m/s}$, or about 1000 mph. Then why does it take 5 to 10 minutes for the smell of perfume to travel across a room? Remember that we estimated that the average distance between particles in a gas at normal conditions is about $D = 3.3 \times 10^{-7}$ cm. While deriving our mathematical descriptions of the ideal gas model, we assumed that the particles do not collide with each other. Perhaps the gas particles actually do collide with neighboring particles and change direction due to each collision, thus making little progress in crossing the room. More detailed estimates show that particles collide about 10^9 times a second under typical atmospheric conditions. They change direction at each collision, and even though they are moving very fast, their migration from one place to another is very slow.

What's next?

So far, we have said nothing about the temperature of the gas. Could there be a connection between the temperature of a gas and the average kinetic energy of its particles? We consider this idea next.

Review Question 9.3 In the expression $PV = \frac{1}{3}N(m_p\overline{v^2})$, pressure is proportional to the particle mass times the average squared speed of the particles. Thus, doubling the average speed of particles leads to quadrupling the pressure. Explain why this makes sense.

9.4 Temperature

As with other physical quantities, temperature can be measured. A common way to measure temperature is with a liquid thermometer. A liquid thermometer consists of a narrow tube connected to a bulb at the bottom (**Figure 9.13**). The bulb and part of the tube are filled with a liquid that expands predictably when heated and shrinks when cooled. To calibrate a thermometer, one marks the height of the liquid at the freezing and boiling conditions of water and then divides this interval by a set number of degrees. On the Celsius scale,

Temperature conversions

$$T_F = (9/5)T_C + 32°$$

$$T_C = (5/9)T_F - 32°$$

$$T_K = T_C + 273.15°$$

Figure 9.13 Thermometers calibrated for (a) Celsius scale, (b) Fahrenheit scale, and (c) absolute (Kelvin) scale.

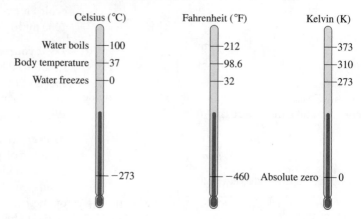

100 C° separates the boiling point and freezing point of water. On the Fahrenheit scale, 180 F° separates these same points (212 °F for boiling and 32 °F for freezing).

What does temperature really quantify?

We measure the temperature of an object indirectly by measuring the changing volume of a liquid that contacts the object. But what is actually different about an object at higher temperature compared to an object at lower temperature? Consider the experiments in Observational Experiment **Table 9.3**.

OBSERVATIONAL EXPERIMENT TABLE

9.3   Connecting properties of matter to its temperature.

 VIDEO 9.3

Observational experiment	Analysis
A beach ball is soft when in a very cold garage. When taken into a warm house, it becomes firmer and bouncier.	Assuming that the air pressure inside and outside the house is the same, the particles of air inside the ball must exert a higher pressure when the gas inside the ball gets warmer.
A balloon filled with air shrinks if placed in ice water.	Assuming that the air pressure in the room stays the same, the particles of air inside the balloon must exert a lower pressure when the gas inside the balloon gets colder.

Pattern

Changing the temperature of the gas seems to change its pressure.

Active Learning Guide ›

How can we explain this pattern? In the experiment with the ball taken into a warm room, we can hypothesize that the ball expands because the impulses of the particles against the inside walls are larger when the gas is warm then when it is cold. This would happen if the particles were moving faster. If so, they would also collide with the walls more frequently. In the balloon experiment the particles seem to exert a smaller impulse on the walls of the balloon when the gas is cooler. This would happen if the particles were moving slower. If so, they would also collide less frequently. Based on this reasoning, we can hypothesize that the temperature of a gas is related to the speed of the random motion of its particles. Is the temperature of the gas related to any other properties of the gas (the pressure or volume of the gas, or perhaps how many particles comprise the gas and how massive they are)?

Let's do more observational experiments, this time conducted in a physics laboratory. We place three different gases in three containers of different but known volumes. A pressure gauge measures the pressure inside each container. The number N of particles (atoms or molecules) in each container is determined by measuring the mass of the gas m_{gas} in each container and then calculating

$$N = \frac{m_{gas}}{m_{molar\ mass}} N_A,$$

where N_A is Avogadro's number. Each container, with known V, P, and N for each gas, is placed first in an ice water bath and then in boiling water, as depicted in **Figure 9.14**. Notice that the volume, pressure, and the number of particles in each container are different, but the temperature of the matter in the three containers is the same. Collected data show the following pattern: independently of the type of gas in a container, the ratio PV/N is identical for all of the gases in the containers when they are at the same temperature:

$$\frac{P_N V_N}{N_N} = \frac{P_O V_O}{N_O} = \frac{P_{He} V_{He}}{N_{He}}$$

The ratio for the gases in the containers at $T_1 = 0\,°C$ is smaller than that for the gases in the containers at $T_2 = 100\,°C$.

From these experiments we can conclude that if you have any amount of a particular type of gas and know its pressure, volume, and the number of particles, then the ratio PV/N only depends on the temperature of the gas. Maybe it equals the temperature? Consider the units of this quantity:

$$\frac{(N/m^2)m^3}{particle} = \frac{N \cdot m}{particle} = J/particle$$

The joule is a unit of energy, not temperature! Perhaps gas particles at the same temperature have the same average energy per particle. Remember that in the ideal gas model, the particles do not have any potential energy between them. This suggests that temperature is related to the average kinetic energy per particle of the gas.

Can we mathematically relate the energy per particle of the gas molecules to the temperature of the gas? The simplest relationship is a direct proportionality to the temperature:

$$\frac{PV}{N} = kT \tag{9.5}$$

where k is a proportionality constant whose value we need to determine. Notice that this relationship immediately leads to a difficulty. The kinetic energy per particle is always a positive number. But in the Celsius and Fahrenheit scales, temperatures can have negative values. So the particle energy cannot be proportional to temperature if measured using either of those scales.

Absolute (Kelvin) temperature scale and the ideal gas law

We need a scale in which the zero point is the lowest possible temperature. That way, all temperatures will be positive. This lowest possible temperature can be found by applying Eq. (9.5) to measurements with a container of gas at two different reference temperatures—the freezing and boiling temperatures of water. The data in **Table 9.4** were collected when a constant-volume metal container

Figure 9.14 The ratio PV/N seems to depend only on the gas temperature.

Ice water 0 °C

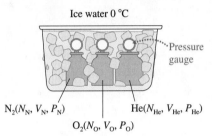

$N_2(N_N, V_N, P_N)$ He(N_{He}, V_{He}, P_{He})

$O_2(N_O, V_O, P_O)$

The ratio of $\frac{PV}{N}$ is the same for different gases if in the same temperature bath.

Boiling water 100 °C

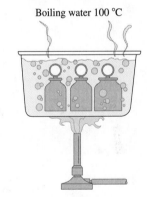

Table 9.4 PV/N for one mole of gas in a 22.4-L container at two different temperatures.

Conditions in the bath	Pressure	Volume	$\dfrac{PV}{N} = kT$
Ice water (T)	$1.013 \times 10^5\,\text{N/m}^2$	$22.42 \times 10^{-3}\,\text{m}^3$	$3.773 \times 10^{-21}\,\text{J}$
Boiling water ($T + 100$)	$1.384 \times 10^5\,\text{N/m}^2$	$22.42 \times 10^{-3}\,\text{m}^3$	$5.154 \times 10^{-21}\,\text{J}$

with 1 mol of nitrogen ($N = N_\text{A} = 6.02 \times 10^{23}$ particles) was placed in baths at two different temperatures. If we assume that the ratio PV/N is proportional to the absolute temperature of the gas, we can find the coefficient of proportionality.

We now have two equations with two unknowns: the constant k and the water temperature T on the new scale.

$$3.773 \times 10^{-21}\,\text{J} = kT$$
$$5.154 \times 10^{-21}\,\text{J} = k(T + 100)$$

We subtract the first equation from the second to get $k = 1.38 \times 10^{-23}\,\text{J/degree}$. The freezing temperature T of the water is then

$$T = \frac{3.773 \times 10^{-21}\,\text{J}}{k} = \frac{3.773 \times 10^{-21}\,\text{J}}{1.381 \times 10^{-23}\,\text{J/degree}} = 273.2 \text{ degrees}.$$

On this new scale, water freezes at $T = 273.2$ degrees above the lowest possible temperature. The lowest possible temperature on the new scale is 0 and on the Celsius scale should be $-273.2\,°\text{C}$. Considerable modern research has refined this value to $-273.15\,°\text{C}$. This temperature scale is called the absolute temperature scale or the Kelvin scale (because it was invented by William Thomson (Lord Kelvin) in 1848). Temperatures are described in kelvin (see the right scale in Figure 9.13). Temperature intervals on the Kelvin scale are the same as on the Celsius scale: a change in temperature of $1\,\text{C}°$ is equivalent to a change in temperature of 1 K. Celsius temperatures are related to kelvin as follows:

$$T_\text{K} = T_\text{C} + 273.15 \tag{9.6}$$

In other words, a temperature of 273.15 K is equivalent to $0\,°\text{C}$.

The constant k that we determined using the data in Table 9.4 is called Boltzmann's constant after the German physicist Ludwig Boltzmann (1844–1906):

$$k = 1.38 \times 10^{-23}\,\text{J/K}$$

We can now rewrite Eq. (9.5) using the absolute temperature T and the value of the constant k:

$$PV = NkT \tag{9.7}$$

Equation (9.7) is called **the ideal gas law**. It is also commonly used in a slightly different form. Rather than referring to the number of particles that comprise the gas (typically an extremely large number), we refer to the number n of moles of the gas. Since one mole has Avogadro's number of particles, $N = nN_\text{A}$. Substituting this into Eq. (9.7), we get

$$PV = nN_\text{A}kT$$

> **TIP** Note that Eq. (9.7) implies that when the absolute temperature of the ideal gas is zero, its pressure must be zero.

The product of the two constants $N_A k$ is another constant called the **universal gas constant** R. Equation (9.7) then becomes

$$PV = nRT$$

where $R = N_A k = \left(6.02 \times 10^{23} \frac{\text{particles}}{\text{mole}}\right)\left(1.38 \times 10^{-23} \frac{\text{J}}{\text{K}}\right) = 8.3 \frac{\text{J}}{\text{K} \cdot \text{mole}}$

This is the more common form of the ideal gas law.

Ideal gas law For an ideal gas, the quantities pressure P, volume V, number of particles N, temperature T (in kelvins), and Boltzmann's constant $k = 1.38 \times 10^{-23}$ J/K are related in the following way:

$$PV = NkT \qquad (9.7)$$

The law can also be written in terms of the number of moles of particles n, and the universal gas constant $R = 8.3 \dfrac{\text{J}}{\text{K} \cdot \text{mole}}$:

$$PV = nRT \qquad (9.8)$$

Temperature and particle motion

Let us look back at what we have done so far. First, we found the relation between the pressure and volume of an ideal gas and the average kinetic energy of the particles that comprise it, $PV = \frac{2}{3}N\overline{K}$ [Eq. (9.4)]. This was a reasonable finding: the faster the particles move inside the gas, the more often and the harder they hit the walls. The particle mass and the number of them per unit volume N/V also affect the pressure.

Next, we found that the product of the pressure and volume of a gas is related to the temperature of the gas, $PV = NkT$ [Eq. (9.7)]. We can now connect the average kinetic energy of the gas particles to the absolute temperature of the gas. Rearrange Eqs. (9.4) and (9.5) so they each have PV/N on the left side. Insert the average kinetic energy $\overline{K} = \frac{1}{2}m\overline{v^2}$ in the right side of Eq. (9.4) and then set the right sides of the two equations equal to each other to get

$$\overline{K} = \frac{3}{2}kT \qquad (9.9)$$

The temperature of a gas is an indication of the average random translational kinetic energy of the particles in the ideal gas. Note that temperature is an indication of not only the particle's speed but also its kinetic energy—the mass of the particle also matters. One implication of this discovery is that when you have a mixture of particles of different gases in one container (for example, in air there are nitrogen molecules, oxygen molecules, carbon dioxide molecules, etc.), the lighter molecules move faster than the heavier ones, though each species of particles has the same average kinetic energy (since each species will have the same temperature once the gas has mixed together thoroughly).

Before we move to an example, let's think more about temperature. Imagine that you have two metal containers with identical gases that have been sitting in the same room for a long time. One container is large and the other one is small. Which one has a higher temperature? Since the average kinetic energy per particle is the same in each container, the temperatures of the two gases are the same. However, the total kinetic energy of the particles in the large container is larger because it contains more particles.

Imagine another scenario: You have two containers with the same type of gas. In one container the gas is hot and in the other it is cold. What will happen if you mix those two gases together? The faster moving particles of the hot gas will collide with the slower moving particles of the cold gas. If we use the laws of momentum and kinetic energy conservation (assuming that collisions are elastic), we find that following a collision, the faster moving particle on average is moving slower than before the collision, and the slower particle is on average moving faster than before. Eventually, the particles of the two gases have the same average kinetic energy and therefore the same temperature. Physicists say that the gases are in *thermal equilibrium*.

> **TIP** Only when temperature is measured in kelvins can Eq. (9.9) be used to calculate the average kinetic energy of the particles.

QUANTITATIVE EXERCISE 9.5 Speed of air particles

Estimate the average speed of air particles in a typical room. Air consists of particles whose average molar mass is 29×10^{-3} kg/mole.

Represent mathematically The temperature in an average room is about 20 °C or 293 K. The average kinetic energy of each particle at temperature T is

$$\overline{K} = \frac{1}{2}m_{\mathrm{p}}\overline{v^2} = \frac{3}{2}kT$$

The mass m_{p} of one particle is the mass M of a mole of that type of particle divided by the number of particles in one mole N_{A}:

$$m_{\mathrm{p}} = \frac{M}{N_{\mathrm{A}}}$$

Solve and evaluate Combining the last two equations, we find the square root of the average speed squared, called the **root-mean-square speed** (the rms speed) of the air particles. We'll use this as our estimate of the average speed of the particles:

$$\sqrt{\overline{v^2}} = \sqrt{\frac{3kT}{m_{\mathrm{p}}}} = \sqrt{\frac{3kTN_{\mathrm{A}}}{M}}$$

Inserting the appropriate values gives

$$\sqrt{\overline{v^2}} = \sqrt{\frac{3(1.38 \times 10^{-23}\,\mathrm{J/K})(293\,\mathrm{K})(6.02 \times 10^{23}\,\mathrm{particles/mole})}{29 \times 10^{-3}\,\mathrm{kg/mole}}}$$

$$= 500\,\mathrm{m/s}$$

This speed is close to the speed calculated in Quantitative Exercise 9.4. Let's check the units:

$$\sqrt{\overline{v^2}} = \sqrt{\frac{\left(\frac{\mathrm{J}}{\mathrm{K}}\right)(\mathrm{K})\left(\frac{1}{\mathrm{mole}}\right)}{\left(\frac{\mathrm{kg}}{\mathrm{mole}}\right)}}$$

$$= \sqrt{\frac{\mathrm{J}}{\mathrm{kg}}} = \sqrt{\frac{\mathrm{N} \cdot \mathrm{m}}{\mathrm{kg}}}$$

$$= \sqrt{\frac{\left(\mathrm{kg} \cdot \frac{\mathrm{m}}{\mathrm{s}^2}\right) \cdot \mathrm{m}}{\mathrm{kg}}} = \frac{\mathrm{m}}{\mathrm{s}}$$

The units are consistent.

Try it yourself: What happens to the average speed of the molecules in the room when the temperature drops by one-half?

Answer: It depends on which scale we use to measure the temperature. If we use the Celsius scale, the temperature would be 10 °C, (283 K on the Kelvin scale) and the rms speed would be about 490 m/s. If we use the Kelvin scale, the temperature would be 147 K and the rms speed would be 350 m/s.

Review Question 9.4 If there is a mixture of different molecules in a container, which ones have a higher average speed, the more massive molecules or the less massive molecules?

9.5 Testing the ideal gas law

In order to determine if the ideal gas law describes the behavior of real gases, we will use the law to predict the outcomes of some testing experiments. If the predictions match the outcomes, we gain confidence in the ideal gas law. In the experiments below we will keep one of the variables (T, V, or P) constant and predict the relation between the two other variables. Processes in which T, V, or P are constant are called **isoprocesses.** The three types of isoprocesses we will investigate are **isothermal** (T = constant), **isochoric** (V = constant) and **isobaric** (P = constant).

‹ Active Learning Guide

TESTING EXPERIMENT TABLE

9.5 Does the ideal gas law apply to real gases?

Testing experiment	Prediction	Outcome
Experiment 1: Isothermal process. n moles of gas are in a variable volume V container that is held in an ice bath at constant 0 °C (273 K) temperature T. How does the pressure of the gas change as we change the volume of the container? We push the piston slowly so that the temperature of the gas is always the same as the ice bath. Isothermal process: constant n and T 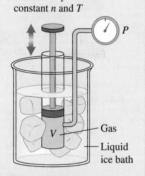 Gas — Liquid ice bath	According to the ideal gas law $PV = nRT$, during a constant temperature process, the product of PV should remain constant. We predict that as the volume decreases, the pressure will increase so that the product remains constant.	Data collected: $V\,(m^3)$ $P\,(N/m^2)$ 3.0×10^{-4} 2.0×10^5 6.0×10^{-4} 1.0×10^5 9.0×10^{-4} 0.67×10^5 The product of volume and pressure remains constant in all experiments.
Experiment 2: Isochoric process. n moles of gas and the gas volume V are kept constant. The container is placed in different-temperature baths. How does the gas pressure change as the temperature changes? Isochoric process: constant n and V 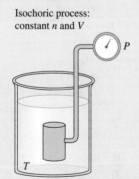	According to the ideal gas law $PV = nRT$, during a constant volume process, the ratio $\dfrac{P}{T} = \dfrac{nR}{V}$ should remain constant. We predict that the pressure should increase in proportion to the temperature.	Data collected: $T\,(K)$ $P\,(N/m^2)$ 300 1.0×10^5 400 1.3×10^5 500 1.7×10^5 The ratio of pressure and temperature is constant in all experiments.
Experiment 3: Isobaric process. n moles of gas and the gas pressure P are held constant, as a piston in the gas container can move freely up and down keeping the pressure constant. How does the gas volume change as the temperature changes? Isobaric process: constant n and P 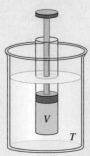	According to the ideal gas law $PV = nRT$, during a constant pressure process, the ratio $\dfrac{V}{T} = \dfrac{nR}{P}$ should remain constant. We predict that the volume should increase in proportion to the temperature.	Data collected: $T\,(K)$ $V\,(m^3)$ 300 3.0×10^{-4} 400 4.0×10^{-4} 500 5.0×10^{-4} The ratio of volume and temperature remains constant in all experiments.

(continued)

Conclusion

The outcomes of all three experiments are consistent with the predictions.

- In the first experiment, the product of pressure and volume remains constant, as predicted.
- In the second experiment, the pressure increases in direct proportion to the temperature, as predicted.
- In the third experiment, the volume increases in direct proportion to the temperature, as predicted.

The outcomes of the experiments in Testing Experiment **Table 9.5** were consistent with predictions based on the ideal gas law, giving us increased confidence that the law applies to real gases (however, we cannot say that we proved it). A summary of gas processes (some of which are not isoprocesses) is provided in **Table 9.6**.

Reflection on the process of construction of knowledge

Let's pause here and reflect on the process through which we arrived at the mathematical version of the ideal gas law. The first step was to construct a

Table 9.6 A summary of ideal gas law processes.

Name	Constant quantities	Changing quantities	Equation	Graphical representation
Isothermal	N or n, T	P, V	$PV = \text{constant}$ $P_1V_1 = P_2V_2$	
Isobaric	N or n, P	V, T	$\dfrac{V}{T} = \text{constant}$ $\dfrac{V_1}{T_1} = \dfrac{V_2}{T_2}$	
Isochoric	N or n, V	P, T	$\dfrac{P}{T} = \text{constant}$ $\dfrac{P_1}{T_1} = \dfrac{P_2}{T_2}$	
[No name]	N or n	P, V, T	$\dfrac{PV}{T} = \text{constant}$ $\dfrac{P_1V_1}{T_1} = \dfrac{P_2V_2}{T_2}$	
[No name]		P, V, T, N or n	$\dfrac{PV}{NT} = k$ $\dfrac{PV}{nT} = R$	

simplified model of a system that could represent a real gas—the ideal gas model. This involved making assumptions about the internal structure of gases. This model was based on some observations and also on the knowledge of particle motion and interactions developed earlier—Newton's laws of motion. We used this model to devise a mathematical description of the behavior of gases, the ideal gas law. We then tested its applicability to real gases by using it to predict how macroscopic quantities describing the gas (temperature, pressure, volume, and the amount of gas) would change during specific processes (isothermal, isobaric, and isochoric) and used the ideal gas law to construct equations that described those processes. These predictions were consistent with the outcomes of the new testing experiments.

The process of constructing the knowledge described above looks relatively smooth and straightforward—observe, simplify, explain, test. However, in real physics, knowledge construction is not that simple and straightforward. For example, the isoprocesses mentioned above were known in physics long before the ideal gas model was constructed. They were discovered in the 17th and 18th centuries and carried the names of the people who discovered them through patterns found in observational experiments. The relation $PV =$ constant for constant temperature processes is called **Boyle's law** and was discovered experimentally by Robert Boyle in 1662. The relation $V/T =$ constant for constant pressure processes is called **Charles's law** and was discovered by Jacques Charles in 1787, though the work was published by Joseph Gay-Lussac only in 1802. Gay-Lussac also discovered the relation $P/T =$ constant, now called **Gay-Lussac's law**. When these relations were discovered empirically, there was no explanation for why gases behaved in these ways. The explanations arrived much later via the ideal gas model. The real process of knowledge construction is often more complicated and nonlinear than how it is presented in a textbook. The skills you are learning by constructing knowledge through experimentation will prepare you for those more complicated situations.

Applications of the ideal gas law

The ideal gas law has numerous everyday applications. Try to explain the following phenomena using the ideal gas law. Clearly state your assumptions.

- A sealed, half-full bottle of water shrinks when placed in the refrigerator.
- A container full of nitrogen explodes if the temperature rises too high.
- A bubble of gas expands as it rises from the bottom of a lake.
- Air rushes into your lungs when your diaphragm, a dome-shaped membrane, contracts.

Below we consider in greater detail several of these phenomena.

Breathing During inhalation, our lungs absorb oxygen from the air, and during exhalation, they release carbon dioxide, a metabolic waste product. Yet the lungs have no muscle to push air in or out. The muscle that makes inhaling and exhaling possible, the diaphragm, is not part of the lungs. The diaphragm is a large dome-shaped muscle that separates the rib cage from the abdominal cavity (see **Figure 9.15**).

The diaphragm works like a bellows. As the diaphragm contracts and moves down from the base of the ribs, the volume of the chest cavity and lungs increases. If we assume that for a brief instant both the temperature and the number of particles are constant (an isothermal process),

$$P = \frac{nRT}{V},$$

Figure 9.15 The diaphragm is a large dome-shaped muscle that separates the rib cage from the abdominal cavity. Relaxing or contracting the diaphragm changes the volume of the lungs and chest cavity.

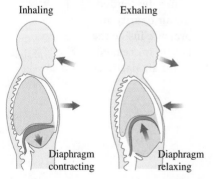

Inhaling   Exhaling

Diaphragm contracting   Diaphragm relaxing

Diaphragm contraction and relaxation cause air to enter and leave the lungs.

then the pressure in the cavity and in the lungs will decrease. The pressure of the outside air is greater than that of the inside. Because of this, the outside air at normal atmospheric pressure enters the mouth or nostrils and fills the lungs with fresh new air (an increase $\Delta n > 0$ of the amount of air). To exhale, the opposite occurs. The diaphragm relaxes and rises, decreasing the volume of the chest cavity and the lungs, thus increasing the pressure inside (again assuming that for a brief instant both the temperature and the number of particles are constant),

$$P = \frac{nRT}{V}$$

Because the inside pressure is higher than the outside pressure, air is then forced out of the lungs.

Your Water Bottle on an Airplane The behavior of an empty plastic water bottle on an airplane is another example of an isothermal process.

CONCEPTUAL EXERCISE 9.6 A shrunk bottle
On a flight from New York to Los Angeles you drink most of a bottle of water and then store it in your seat pocket. Just before landing, you take it out again as the flight attendant is collecting trash. The bottle has changed shape; it looks like someone crushed it. How can we explain this shape change?

Sketch and translate The bottle's volume decreased, though the temperature in the cabin didn't change much during the flight.

Simplify and diagram If we assume the bottle was perfectly sealed, and the temperature of the gas inside the bottle remained constant, then we can model the process as an isothermal process, represented graphically in the figure below. Although the cabin is pressurized, at higher elevations air pressure and air density inside the cabin are slightly less than at lower elevations. You closed the bottle when it was filled with low-density air at high elevation. As the plane descended, the air density and pressure inside the cabin increased. More air particles

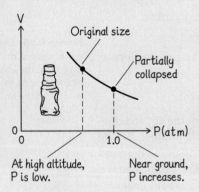

were hitting the outside walls of the empty bottle than were hitting the inside. The higher pressure from outside partially crushed the bottle. If you open the cap after landing, the bottle will pop back to its original shape.

Try it yourself: A process is represented on a graph in the figure below. Describe the process in words. Assume the mass to be constant.

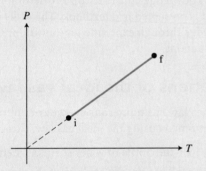

Answer: The graph represents an isochoric process. The graph line, if extended, passes through the origin; thus the pressure is directly proportional to temperature. The gas was in a sealed container. The gas container was first placed in a water bath at low temperature and then transferred to a water bath at high temperature. The volume of the gas is constant. The pressure increases as the particles move faster and collide with the walls of the container more often.

Comparing different gases

How does the type of gas that is in a container affect the pressure of the gas? Consider the next conceptual exercise.

CONCEPTUAL EXERCISE 9.7 Analyzing two types of gas

You have two containers, each with pistons of equal mass that can move up and down depending on the pressure of the gas below. Each container has a nozzle that allows you to add gases to the containers. Container 1 holds nitrogen (molar mass M_1); we label its volume V_1. Container 2 (V_2) holds helium (molar mass $M_2 < M_1$). The volumes are the same ($V_1 = V_2$), and the containers sit in the same room. Determine everything you can about the situation.

Piston

Nozzle to insert gas

Same mass piston

Nozzle to insert gas

$V_1 = V_2$
$T_1 = T_2$
$M_1 > M_2$
$P_{external} = P_{Atm}$

Sketch and translate We sketch the situation in the figure to the right. Quantities that we can try to compare between the two containers are the following:

(a) the relative pressure inside the containers
(b) the average kinetic energy of a particle of each type of gas
(c) the mass of individual gas particles
(d) the rms speeds of each particle
(e) the number of particles in each container
(f) the mass of the gas in each container
(g) the density of the gas in the containers.

Simplify and diagram Assume that the ideal gas model accurately describes the gases.

(a) Consider the pressures inside the containers by analyzing the identical moveable pistons above the gases. Force diagrams for each piston are shown at right. Earth exerts a downward force $\vec{F}_{E\ on\ P}$, on the piston, the atmospheric gas above a piston pushes down on the piston $\vec{F}_{Atm\ on\ P}$, and the gases inside the containers push up on their pistons $\vec{F}_{N_2\ on\ P}$ or $\vec{F}_{He\ on\ P}$. As the pistons are not

accelerating, the net force exerted on each is zero. As the downward forces exerted by Earth and by the outside atmospheric air on each piston are the same, the upward forces exerted by the gases in each container on the piston must also be the same. Since the surface area of each piston is the same, the pressure of the gas inside each container must be the same, $P_{N_2} = P_{He}$.

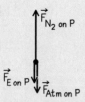

The downward forces on each piston are the same, so the upward force is also the same.

(b) Since the temperature of a gas depends only on the average kinetic energy of the particles, and the temperature of the two gases is the same, the average kinetic energy of each particle is the same:

$$\frac{3}{2}kT = \frac{1}{2}m_{N_2\ in\ 1}\overline{v_{N_2\ in\ 1}^2} = \frac{1}{2}m_{He\ in\ 2}\overline{v_{He\ in\ 2}^2}.$$

(c) and (d) Since the N_2 particles in gas 1 have a higher molar mass than the He particles in gas 2, the N_2 molecular mass $m = M_1/N_A$ is also the higher mass ($m_{N_2\ in\ 1} > m_{He\ in\ 2}$). Using this result and that from (b), we find that $\overline{v_{N_2\ in\ 1}^2} < \overline{v_{He\ in\ 2}^2}$. The more massive particles move slower.

(e) The pressure, volume, and temperature are the same for each gas. Thus, according to the ideal gas law, both gases have the same number of particles and the same number of moles of gas in their containers ($PV/kT = N$; $PV/RT = n$).

(f) and (g) Since the mass of a nitrogen molecule is greater than the mass of a helium atom, and there are equal numbers of particles in each container, the total mass of the gas in the nitrogen container must be greater than the total mass of the gas in the helium container ($M_{N_2\ in\ 1} > M_{He\ in\ 2}$). Since the volumes of the containers are equal, the density $\rho = M/V$ of the gas in the nitrogen container must be greater than that in the helium container ($\rho_{N_2\ in\ 1} > \rho_{He\ in\ 2}$).

Try it yourself: Why are the gases at the same temperature?

Answer: Both containers sit in the same room temperature environment.

Review Question 9.5 What is the difference between the following two equations: $PV = 1/3\ Nm\overline{v^2}$ and $PV = nRT$?

9.6 Speed distribution of particles

In our previous analysis, we found that the average molecular kinetic energy of gas particles depends on the temperature of the gas

$$\overline{K} = \frac{1}{2}m\overline{v^2} = \frac{3}{2}kT \tag{9.9}$$

Consequently, the root-mean-square speed of a gas atom or molecule (the rms speed) is

$$v_{\text{rms}} = \sqrt{\overline{v^2}} = \sqrt{\frac{3kT}{m}} \tag{9.10}$$

We found that for air molecules at room temperature, the average root-mean-square speed was about 500 m/s. When we derived relationships such as these, we assumed for simplicity that gas particles do not collide with each other, just with the walls of a container. However, we know that if gas particles did not collide, the smells of food and perfume would spread at hundreds of meters per second—almost instantly. The smells spread slowly, so the particles must be colliding. What happens if we no longer ignore collisions of particles with each other?

Maxwell speed distribution

In 1860 James Clerk Maxwell included the collisions of the particles in his calculations involving an ideal gas. This inclusion led to the following prediction: at a particular temperature, the collisions of gas particles with each other cause a very specific distribution of speeds. When we were deriving Eq. (9.4) we assumed that the speeds of the particles were different, but we did not have any idea of why they were different. Maxwell's work explained this variability of speeds by the collisions of the particles with each other. Consider the yellow ^{20}Ne line (neon atoms) in **Figure 9.16**. On the vertical axis, we plot the percentage of particles that have a particular speed. According to Maxwell, a certain percentage of the particles should have speeds around 100 m/s, more around 200 m/s, the most around 500 m/s, and fewer at higher speeds. Very few particles have extremely low or extremely high speeds. Most should have intermediate speeds. The most probable speed is at the highest point on the curve in Figure 9.16. Surprisingly, the root-mean-square speed of the particles at a particular temperature is the same as that predicted by the model with no collisions—Eq. (9.10).

Figure 9.16 The Maxwell particle speed distributions at a particular temperature for four gases.

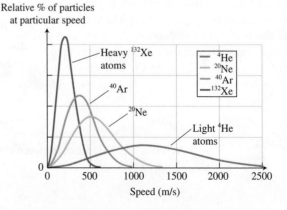

To test Maxwell's ideas, one needs to actually measure the speeds of atoms and molecules at a particular temperature and then compare the distribution of speeds to the calculated distribution curve shown in Figure 9.16. The task—measuring the speeds of objects that are 10^{-10} m in diameter—seems almost impossible. However, the problem was tackled and successfully solved by German physicist Otto Stern in 1920, many years after the development of Eq. (9.10). Stern's experiments led to a whole field of study called molecular beam spectroscopy.

An apparatus such as that shown in **Figure 9.17** is used in molecular beam spectroscopy. A gas is heated to some predetermined temperature. A small fraction of the rapidly moving gas particles leaves the container through slit A. Some of these particles pass through slits B and C, forming a narrow beam of particles that hits a rapidly rotating drum with a slit D. The particles can only enter the drum as slit D passes along the line from slits A to C. The particles that enter the drum travel across it to the other side, where they are detected by a sensitive film that produces a mark (a dot) when hit by a particle. Fast-moving particles hit the film almost directly across from the slit, whereas slow-moving particles hit the film somewhat later, as the drum has rotated farther. After the drum completes one rotation, a new group of particles enters the drum. Thus, even if the beam has only a few particles hitting slit D per rotation, after many rotations a denser pattern develops on the film.

The density of the number of particles hitting a particular part of the film indicates the relative speed of those particles. Thus, you can make a graph of the darkening of the film (the relative number of particles hitting a part of the film) versus the position on the film (the speed of particles hitting that part of the film). The pattern can be used to calculate the average particle speed squared. The experiment can be repeated multiple times with the gas at a different temperature each time. The determined average speed squared and gas temperature are consistent with Eq. (9.10). The measured speed distribution patterns in Stern's experiments matched the Maxwell predicted distributions perfectly (Figure 9.16). Finally, note that all of the speed distributions in Figure 9.16 were for different types of gas particles at the same temperature. Note that the less massive particles moved at higher speeds—in complete agreement with experimental results. These results provided strong support for the kinetic theory of gases.

Therefore, we can say that the ideal gas model is a productive model for describing gases. A combination of the model, the ideal gas law, and all of the testable predictions and testing experiments is called **kinetic molecular theory**—a theory that describes and explains the behavior of gases based on their particle structure.

Figure 9.17 An apparatus to measure the speed distribution of particles in a gas.

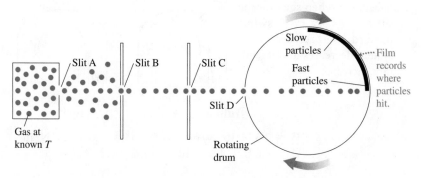

Limitations of the ideal gas law

Isoprocesses were discovered empirically with experiments long before the model of ideal gas was constructed. However, the fact that the ideal gas law, derived from the ideal gas model, predicts those empirical results is evidence that the law itself and the ideal gas model on which the law is based are quite reliable. Nevertheless, the model and the gas law have their limitations.

For real gases such as air, the measurements of pressure and volume at the conditions of normal pressure (1.0×10^5, as elsewhere) and temperature (room temperature) are consistent with predictions of the ideal gas law. But at very high pressures or very low temperatures, real measurements differ from those predictions (for example, the model of ideal gas does not predict that one can turn a gas into a liquid). The ideal gas law describes gases accurately only over certain temperature and pressure ranges.

Review Question 9.6 What is the role of Stern's molecular beam experiment in the development of kinetic molecular theory?

9.7 Skills for analyzing processes using the ideal gas law

In this section, we adapt our problem-solving strategy to analyze gas processes. A general strategy for analyzing such processes is described on the left side of the table in Example 9.8 and illustrated on the right side for a specific process.

PROBLEM-SOLVING STRATEGY   Applying the Ideal Gas Law

EXAMPLE 9.8   Scuba diver returns to the surface
The pressure of the air in a scuba diver's lungs when she is 15 m under the water surface is $2.5 \times 10^5 \, \text{N/m}^2$ (equal to the pressure of the water at that depth), and the air occupies a volume of 4.8 L. Determine the volume of the air in the diver's lungs when she reaches the surface, where the pressure is $1.0 \times 10^5 \, \text{N/m}^2$.

Sketch and translate

- Sketch the process. Choose a system and characterize the initial and final states of the system.

- Describe the process (a word description of the changes in the system between the initial and final states) in terms of macroscopic quantities (pressure, volume, temperature, moles of gas).

The gas inside the diver's lungs is the system. The initial state of the system is when the diver is underwater; the final state is when she is at the surface. As the diver swims upward, the pressure of the system decreases and its volume increases.

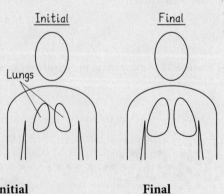

Initial	Final
$P_i = 2.5 \times 10^5 \, \text{N/m}^2$	$P_f = 1.0 \times 10^5 \, \text{N/m}^2$
$V_i = 4.8 \, \text{L} = 4.8 \times 10^{-3} \, \text{m}^3$	$V_f = ?$

Simplify and diagram

- Decide if the system can be modeled as an ideal gas.

- Decide which macroscopic quantities remain constant and which do not.

- If helpful, draw P vs. V, P vs. T, and/or V vs. T graphs to represent the process.

Assume that the gas inside the lungs can be modeled as an ideal gas. Assume that the diver does not exhale, which means that the moles of gas remain constant. Assume the temperature of the gas is constant at body temperature. A pressure-versus-volume graph for the process is shown below.

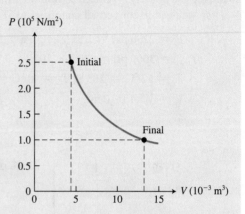

Represent mathematically

Use your sketch or the graphs to help construct a mathematical description.

$$P_i V_i = P_f V_f$$

$$\Rightarrow V_f = \frac{P_i V_i}{P_f}$$

Solve and evaluate

- Solve for the unknowns.

- Evaluate the answer: is it reasonable? (For example, evaluate the magnitude of the answer, its units, and how the solution behaves in limiting cases.)

$$V_f = \frac{P_i V_i}{P_f} = \frac{(2.5 \times 10^5 \, \text{N/m}^2)(4.8 \times 10^{-3} \, \text{m}^3)}{(1.0 \times 10^5 \, \text{N/m}^2)} = 12 \times 10^{-3} \, \text{m}^3$$

We found the lung volume to be 12 L, much larger than seems possible. Is the answer realistic? Remember that we assumed that the mass of the gas inside remains constant—the person does not exhale. As we see from the solution, it is important for divers to exhale as they are ascending and the gas expands. If not, they can suffer severe internal damage.

Try it yourself: How many moles of gas are in the diver's lungs and how many should she exhale so the final volume is only 6 L instead of 12? Assume that the gas temperature is 37 °C.

Answer: 0.47 moles; 0.23 moles.

TIP In some gas processes, notice that the graph lines for isobaric processes (constant pressure) pass through the origin in V-versus-T graphs and those for isochoric (constant volume) processes pass through the origin in P-versus-T graphs.

EXAMPLE 9.9 Will the container burst?

A 100-L oxygen tank filled at night was left outside in the sunshine the next day. When it was filled, the pressure was 2250 psi (pounds per square inch is another unit for pressure, where 1 psi $= 6.89 \times 10^3$ Pa) and the temperature was 12 °C. Is it dangerous to leave the container outside when the temperature is 40 °C, if the warning on the container says "pressure not to exceed 3000 psi?" Explain the process microscopically.

Sketch and translate Label a diagram with known quantities for the initial and final states, as shown in the figure below. The gas inside the tank is the system. The initial state is when it was filled at night. The final state is

(continued)

during the hot day. We need to find the final pressure P_f. First, we need to convert all quantities to SI units.

$$P_i = 2250 \, \text{psi} = 1.55 \times 10^7 \, \text{Pa}$$
$$P_{max} = 3000 \, \text{psi} = 2.07 \times 10^7 \, \text{Pa}$$
$$T_i = 12 \, °C = 285 \, \text{K}; \; T_f = 40 \, °C = 313 \, \text{K}$$
$$V_i = V_f = 0.10 \, \text{m}^3$$

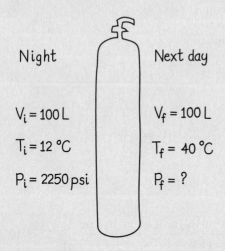

Night

$V_i = 100 \, L$

$T_i = 12 \, °C$

$P_i = 2250 \, psi$

Next day

$V_f = 100 \, L$

$T_f = 40 \, °C$

$P_f = ?$

Simplify and diagram Can we model the gas in the container as an ideal gas? The pressure starts at $P_i = 1.55 \times 10^7 \, \text{Pa}$ and is not to exceed $P_{max} = 2.07 \times 10^7 \, \text{Pa}$, which is more than 200 times atmospheric pressure. The particles of the gas are much closer together than at atmospheric pressure; thus the ideal gas model might not be applicable. However, we do not have another mathematical model to use, so we will use the ideal gas law, keeping in mind that our estimate of the final pressure might not be reasonable. The tank's volume remains constant, so we can use the mathematical description of an isochoric process.

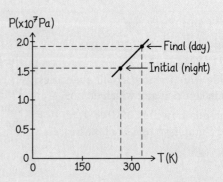

Represent mathematically For an isochoric process,

$$\frac{P_i}{T_i} = \frac{P_f}{T_f}$$

Solve and evaluate Solving for P_f and inserting the appropriate values gives

$$P_f = \frac{P_i T_f}{T_i} = \frac{(1.55 \times 10^7 \, \text{Pa})(313 \, \text{K})}{285 \, \text{K}} = 1.70 \times 10^7 \, \text{Pa}$$

We are still under the limit, but getting close. Taking into account that we are not sure that the ideal gas law applies to the gas inside, we should remember to put the tank in the shade!

Try it yourself: How many moles of gas are in the tank when its temperature is 12 °C and when it is 40 °C?

Answer: 650 moles; n does not change—it's a sealed container.

> **TIP** Notice that if you continue the graph line in the above Example, it passes through the origin, as the pressure is proportional to the absolute temperature.

Review Question 9.7 Why is it helpful to know whether the mass of the gas is constant during a particular process?

9.8 Thermal energy, the Sun, and diffusion: Putting it all together

The ideal gas model applies to many real-world situations involving gases, provided that the gas particles are far from each other compared to their own sizes. Let's consider some of these situations.

Thermal energy of air

Gas consisting of many atoms and molecules at normal temperature possesses thermal energy—the random kinetic energy of the constituent particles. How does this thermal energy compare to the gravitational potential energy of the system consisting of Earth and its atmosphere? To answer these questions we could estimate the thermal energy of a cup of air. A cup of air at $27\,°C$ contains about 10^{22} molecules. Each molecule has some kinetic energy; the thermal energy of these molecules is the sum of their individual average kinetic energies:

$$U_{\text{thermal}} = N\left(\frac{3}{2}kT\right)$$

Or, using the provided information,

$$U_{\text{thermal}} = N\left(\frac{3}{2}\right)kT = (10^{22})\frac{3}{2}(1.38 \times 10^{-23}\,\text{J/K})[(273 + 27)\text{K}] = 60\,\text{J}$$

How high would we need to lift the air molecules in the cup so that the gravitational potential energy of the gas-Earth system equals the thermal energy of the gas? If we assume that the zero level of gravitational potential energy is at Earth's surface, then the gravitational potential energy of the system once the gas has been lifted to a height h is $U_g = m_{\text{total}}gh$, where m_{total} is the total mass of the gas. Thus,

$$U_{\text{thermal}} = N\left(\frac{3}{2}kT\right) = m_{\text{total}}gh$$

$$\Rightarrow h = \frac{N\left(\frac{3}{2}kT\right)}{m_{\text{total}}g}$$

The mass of the air molecules in the cup is

$$m_{\text{total}} = Nm_{\text{single particle}} = N\frac{M_{\text{mole}}}{N_A}$$

(the mass of one particle is its molar mass divided by the number of particles N_A in one mole). Inserting the numbers, we get

$$h = \frac{U_{\text{thermal}}}{m_{\text{total}}g} = \frac{U_{\text{thermal}}}{N\frac{M_{\text{mole}}}{N_A}g} = \frac{60\,\text{J}}{\left[(10^{22})\frac{(29 \times 10^{-3}\,\text{kg})}{(6.02 \times 10^{23})}\right](9.8\,\text{N/kg})} = 13{,}000\,\text{m}$$

The thermal energy present in that cup of air is equivalent to the work needed to lift that air 13 km! This gives you an idea of the very significant amount of thermal energy in the atmosphere.

How long will the Sun shine?

The technique that we used to determine the thermal energy of gases can be used to help analyze many kinds of systems, such as automobile engines, Earth's atmosphere, and the Sun. The mass of the Sun is about 2×10^{30} kg, its radius is about 7×10^8 m, and the temperature of the Sun's surface is about 6×10^3 K. Its core temperature is about 10^7 K. Every second, the Sun radiates about 4×10^{26} J as visible light and other forms of radiation. How long will our Sun shine if its particles possess *only* thermal energy?

We first need to determine how much thermal energy the Sun possesses. Then we can divide this amount of energy by the amount it loses every second from the emission of visible light and other forms of radiation to find the number of seconds the Sun can shine if its thermal energy is the only source of this radiative energy. Although in reality there are some complications with modeling the material of the Sun as an ideal gas, we will do so in this case for simplicity.

To estimate the thermal energy of the Sun, we need to know the number of particles and the average temperature of these particles. The Sun consists mostly of hydrogen atoms. To find the number of hydrogen atoms in the Sun, we estimate the number of moles of hydrogen gas and then multiply by the number of particles in one mole (note that 1 mole of hydrogen has a mass of $1 \text{ g} = 10^{-3} \text{ kg}$):

$$N = \frac{m_{\text{Sun}}}{m_{\text{hydrogen atom}}} = \frac{m_{\text{Sun}}}{M_{\text{molar mass hydrogen}}/N_A}$$

$$= \frac{(2 \times 10^{30} \text{ kg})}{(10^{-3} \text{ kg/mole})/(6 \times 10^{23} \text{ particles/mole})}$$

$$= \frac{(2 \times 10^{30} \text{ kg})}{(1.7 \times 10^{-27} \text{ kg/particle})}$$

$$= 12 \times 10^{56} \text{ particles}$$

At the very high temperatures within the Sun, each hydrogen atom separates into two smaller particles called an electron and a proton—subjects of later study. This separation of hydrogen atoms doubles the number of particles to 24×10^{56}.

These particles do not spread out into space or collapse in toward the center of the Sun because two competing forces remain in balance. All parts of the Sun exert a gravitational force on all other parts. If we select a small volume inside the Sun as the system of interest and add the forces that all other particles of the Sun exert on the system (see **Figure 9.18**), the net gravitational force points toward the center of the Sun. There is another force exerted on the system. It is the pressure force exerted by other particles on the system. This force points outward and balances the gravitational force. As long as these forces balance each other, the Sun is in equilibrium and does not expand or collapse.

Now consider the total thermal energy of the Sun's particles for two extreme cases. A lower bound for the Sun's thermal energy assumes its temperature throughout equals its surface temperature. An upper bound for the Sun's thermal energy assumes its temperature throughout equals its core temperature.

$$U_{\text{thermal min}} = \frac{3}{2} NkT_{\text{min}}$$

$$= \frac{3}{2}(24 \times 10^{56} \text{ particles})(1.38 \times 10^{-23} \text{ J/K})(6 \times 10^3 \text{ K})$$

$$= 3 \times 10^{38} \text{ J}$$

$$U_{\text{thermal max}} = \frac{3}{2} NkT_{\text{max}}$$

$$= \frac{3}{2}(24 \times 10^{56} \text{ particles})(1.38 \times 10^{-23} \text{ J/K})(10^7 \text{ K})$$

$$= 5 \times 10^{41} \text{ J}$$

Both numbers are so large that they imply that the Sun will shine for a long time. The life expectancy of the Sun in seconds (based on the thermal energy alone) equals the total thermal energy divided by the energy radiated

Figure 9.18 Gas in the Sun is in equilibrium due to two forces.

This small volume of gas (the system) is kept in equilibrium by the gravitational force and pressure force exerted on it by other particles in the Sun.

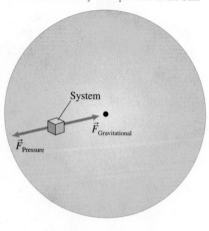

per second, $R = 4 \times 10^{26}$ J/s. We can convert our result to years, noting that there are about 3×10^7 s in 1 year.

$$\Delta t_{min} = \frac{E_{min}}{R} = \frac{3 \times 10^{38} \text{ J}}{4 \times 10^{26} \text{ J/s}} = 7.5 \times 10^{11} \text{ s} = \frac{7.5 \times 10^{11} \text{ s}}{3 \times 10^7 \text{ s/year}} = 2 \times 10^4 \text{ years}$$

$$\Delta t_{max} = \frac{E_{max}}{R} = \frac{5 \times 10^{41} \text{ J}}{4 \times 10^{26} \text{ J/s}} = 1.2 \times 10^{15} \text{ s} = \frac{1.2 \times 10^{15} \text{ s}}{3 \times 10^7 \text{ s/year}} = 4 \times 10^7 \text{ years}$$

The maximum possible lifetime for the Sun if it simply converts its thermal energy into light and other forms of radiation is 4×10^7 years, or 40 million years. Yet we know that the geological age of Earth is 4.5×10^9 years—4.5 billion years! Earth cannot be older than the Sun; thus this simple estimate suggests that either the Sun's material cannot be modeled as an ideal gas or some source of energy other than the thermal energy of the Sun's particles has been supporting its existence for a time interval equal to the age of Earth. As the ideal gas model explains many phenomena occurring on the Sun, it means that there must be another energy source within the Sun that far exceeds the thermal energy present. (We will learn about it in Chapter 28.)

Diffusion

Molecules spread from regions of high concentration to regions of low concentration due to their random motion. This process is called **diffusion.** The explanation of diffusion follows from the ideal gas model. Diffusion plays an important role in biological processes. Oxygen is carried by the hemoglobin in the blood from the heart to the tiny capillary vessels spread throughout the body. Oxygen diffuses from the oxygen-rich blood inside the capillaries to the oxygen-poor cells near the capillaries. Some of these cells may be muscle fibers. Since a muscle fiber needs oxygen to twitch, the action of muscles may be limited by the rate of oxygen diffusion into the fiber. Thus, diffusion limits the rate of some processes in our bodies.

Review Question 9.8 How do we know that the Sun's thermal energy is not the main source of the light energy it produces?

Summary

Words	Pictorial and physical representations	Mathematical representation
Pressure P The perpendicular component of the force F that another object or that a gas or liquid exerts perpendicular to a surface of area A divided by that area. (Section 9.2)	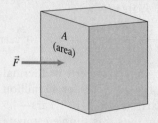	$P = \dfrac{F_\perp}{A}$ Eq. (9.1)
Density ρ The mass m of a substance divided by the volume V that the substance occupies. (Section 9.2)	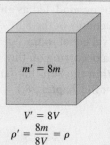 $\rho = \dfrac{m}{V}$ $\rho' = \dfrac{8m}{8V} = \rho$	$\rho = \dfrac{m}{V}$ Eq. (9.3)
Moles n and **Avogadro's number** N_A A mole of any type of particle equals Avogadro's number of that type of particle. (Section 9.3)	≈ 54 g iron (1 mole)	$N_A = 6.02 \times 10^{23}$ particles/mole
Temperature T and **temperature scales** Temperature measures how hot or cool a substance is. When measured in kelvins, the temperature is directly proportional to the average random kinetic energy of a particle in that gas. (Section 9.4)		The average kinetic energy per atom or molecule in a gas is $$\bar{K} = \frac{3}{2}kT$$ $T_F = (9/5)T_C + 32°$ $T_C = (5/9)(T_F - 32°)$ $T_K = T_C + 273.15°$ Eq. (9.6)
Thermal energy U_{thermal} The random kinetic energy of *all* atoms and molecules in a system. (Section 9.8)		$U_{\text{thermal}} = N\left(\dfrac{3}{2}kT\right)$
Ideal gas model and ideal gas law The ideal gas model is a simplified model of gas in which atoms/molecules are considered to be point-like objects that obey Newton's laws. They only interact with each other and with the walls of the container during collisions exerting pressure. The ideal gas law relates the macroscopic quantities of such a gas. (Sections 9.1, 9.3, and 9.4)		$P = \dfrac{2}{3}\left[\dfrac{N\left(\frac{1}{2}m\overline{v^2}\right)}{V}\right]$ Eq. (9.4) $PV = NkT$ Eq. (9.7) $PV = nRT$ Eq. (9.8)

Gas Processes (Examples)

Constant pressure (isobaric) process A container with a frictionless plunger is filled with gas. The air outside is at constant pressure, thus the pressure inside the container is constant. (Section 9.5)	**Microscopic** The molecules inside collide with container walls at different speeds and varying frequency. If gas warms, the particles collide harder and more often, thus causing the gas to expand.	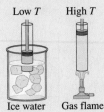	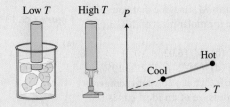	Assume that N, n, and P are constant. Then, $$\frac{V_1}{T_1} = \frac{V_2}{T_2}$$
Constant volume (isochoric) process A closed oxygen tank sits outside on a sunny summer day. Its volume is constant. (Section 9.5)	**Microscopic** As gas warms, the molecules inside move faster and collide with walls more often, thus exerting greater pressure on walls.		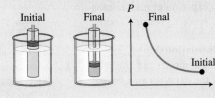	Assume that N, n, and V are constant. Then, $$\frac{P_1}{T_1} = \frac{P_2}{T_2}$$
Constant temperature (isothermal) process A closed plastic bottle shrinks as an airplane descends. The temperature inside the bottle is always equal to the temperature outside. (Section 9.5)	**Microscopic** As pressure increases, the collisions of air molecules against the outside of the bottle become more frequent, causing the bottle volume to decrease.			Assume that N, n, and T are constant. Then, $$P_1V_1 = P_2V_2$$

 For instructor-assigned homework, go to MasteringPhysics.

Questions

Multiple-Choice Questions

1. What experimental evidence rejects the explanation that wet clothes become dry because the air absorbs the water?
   (a) The clothes dry faster if you blow air across them.
   (b) They do not dry if you put the wet clothes in a plastic bag.
   (c) The clothes dry faster under a vacuum jar with the air pumped out.

2. What is the difference between the words particle, molecule, and atom?
   (a) A particle is bigger than a molecule or an atom.
   (b) Particles can be microscopic and macroscopic, while atoms and molecules are only microscopic.
   (c) Molecules are made of atoms; both can be called particles.
   (d) All are correct.
   (e) Both b and c are correct.

3. You have a basketball filled with gas. Which method below changes its volume because of a mass change of the gas inside?
   (a) Put it into a refrigerator.   (b) Squeeze it.
   (c) Pump more gas into it.   (d) Hold it under water.
   (e) Leave it in the sunshine.

4. Choose the quantities describing the air inside a bike tire that do *not* change when you pump the tire.
   (a) Mass   (b) Volume
   (c) Density   (d) Pressure
   (e) Particle mass   (f) Particle concentration

5. Which answer below does *not* explain the decrease in size of a basketball after you take it outside on a cold day?
   (a) The pressure inside the ball decreases.
   (b) The temperature of the gas inside the ball decreases.
   (c) The volume of the ball decreases.
   (d) The number of gas particles inside the ball decreases.

6. What causes balloons filled with air or helium to deflate as time passes?
   (a) The elasticity of the rubber decreases.
   (b) The temperature inside decreases.
   (c) The pressure outside the balloon increases.
   (d) The gas from inside diffuses into the atmosphere.
7. From the list below, choose the assumption that we did not use in deriving the ideal gas law.
   (a) Gas particles can be treated as objects with zero size.
   (b) The particles do not collide with each other inside the container.
   (c) The particles collide partially inelastically with the walls of the container.
   (d) The particles obey Newton's laws.
8. You have a mole of oranges of mass M. Imagine that you split each orange in half. What will be the molar mass of this pile of half oranges in kilograms?
   (a) $M/2$ (b) $M/(6.02 \times 10^{23})$
   (c) $2M/(6.02 \times 10^{23})$ (d) $(M/2)/(6.02 \times 10^{23})$
9. How did physicists come to know that at a constant temperature and constant mass, the pressure of an ideal gas is inversely proportional to its volume?
   (a) They could have conducted an experiment maintaining the gas as described above and made a pressure-versus-volume graph.
   (b) They could have derived this relationship using the equations describing the ideal gas model and the relationship between the speed of the particles and the gas temperature.
   (c) Both a and b are correct.
10. A gas is in a sealed container with a heavy top that is free to move. When the gas is heated, the top moves up, causing the volume of the gas to increase. Which equation below best describes this process?
    (a) $V_{initial}/T_{initial} = V_{final}/T_{final}$
    (b) $P_{initial}V_{initial} = P_{final}V_{final}$
    (c) $P_{initial}V_{initial} = nRT_{final}$
    (d) $P_{initial}V_{initial} = NkT_{final}$
11. A gas is in a sealed container with a heavy top that is free to move. With constant external pressure pushing down on the top, the top moves up, causing the volume of the gas to increase. Which graph below best represents this process (**Figure Q9.11**)?

Figure Q9.11

(a)   (b)   (c)

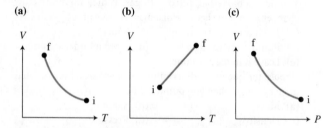

12. A completely closed rigid container of gas is taken from the oven and placed in ice water. Which graph at top right does not represent this process (**Figure Q9.12**)?

Figure Q9.12

(a)   (b)   (c)   (d)

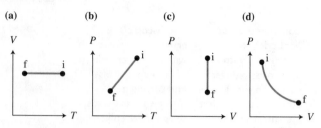

(e) None of them do.
13. Which graph below does not represents a process described by the equation $V_{initial}/T_{initial} = V_{final}/T_{final}$ (**Figure Q9.13**)?

Figure Q9.13

(a)   (b)   (c)   (d)

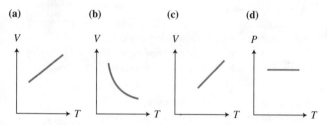

(e) They all do.
14. What does contracting and relaxing the diaphragm allow?
    (a) Greater O_2 absorption by humans and mammals compared to other species
    (b) Greater pressure difference between air in the lungs and outside air
    (c) The only way to facilitate air intake into and out of the lungs
    (d) a and b
    (e) a, b, and c
15. When is air is inhaled into the lungs?
    (a) When lung muscles cause them to expand
    (b) When the diaphragm contracts
    (c) When the chest expands
    (d) b and c
    (e) a, b, and c
16. What happens if the volume of the lungs increases?
    (a) The air pressure in the lungs increases.
    (b) The air pressure in the lungs decreases.
    (c) The amount n of air in the lungs increases.
    (d) a and c can occur simultaneously.
    (e) b and c can occur simultaneously.

Conceptual Questions

17. (a) How do we know if a real gas can be described as an ideal gas? (b) How can you decide if air in the physics classroom can be described using the ideal gas model?
18. Imagine that you are an astronaut on a space station. What happens if you open a perfume bottle in outer space?
19. Why does it hurt to walk barefoot on gravel?
20. In the magic trick in which a person lies on a bed of nails, why doesn't the person get hurt by the nails?

21. Why does food cook faster in a pressure cooker than in an open pot?
22. What does it mean if the density of a gas is 1.29 kg/m³?
23. How many oranges would you have if you had two moles of oranges?
24. Imagine that you have an unknown gas. What experiments do you need to do and what real equipment do you need to determine the mass of one molecule of this gas?
25. One mole of chicken feathers is spread uniformly over the surface of Earth. *Estimate* the thickness of this layer of feathers. Justify any assumptions you made in your calculations. The radius of Earth is about 6400 km.
26. How many molecules are there in 1 g of air at normal conditions? If these molecules were distributed uniformly on Earth's surface, *estimate* the number that would be under your feet right now. The radius of Earth is about 6400 km.

27. Describe how temperature and one degree are defined on the Celsius scale.
28. Describe how temperature and one degree are defined on the Kelvin scale.
29. Why does sugar dissolve faster in hot tea than in cold water?
30. (a) Describe experiments that were used to test the predictions of the molecular kinetic theory. (b) What experiments revealed its limitations?
31. Give three examples of diffusion that are important for human life.
32. Why do very light gases such as hydrogen not exist in Earth's atmosphere but do exist in the atmospheres of giant planets such as Uranus and Saturn?
33. Why does the Moon have no atmosphere?

Problems

Below, BIO indicates a problem with a biological or medical focus. Problems labeled EST ask you to estimate the answer to a quantitative problem rather than develop a specific answer. Problems marked with ✐ require you to make a drawing or graph as part of your solution. Asterisks indicate the level of difficulty of the problem. Problems with no * are considered to be the least difficult. A single * marks moderately difficult problems. Two ** indicate more difficult problems.

9.2 Pressure, density, and the mass of particles

1. * EST A water molecule in a glass of water has 10^{14} collisions with other molecules every second. Estimate the number of years a college football player would have to play (24 hours a day) to have the same number of collisions. Explain all of your assumptions.
2. What are the molar masses of molecular and atomic hydrogen, helium, oxygen, and nitrogen? What are their molecular masses?
3. EST Estimate the number of hydrogen atoms in the Sun. The mass of the Sun is 2×10^{30} kg. About 70% of it by mass is hydrogen, 30% is helium, and there is a negligible amount of other elements.
4. * The average particle density in the Milky Way galaxy is about one particle per cubic centimeter. Express this number in SI units (kg/m³). Indicate any assumptions you made.
5. * (a) What is the concentration (number per cubic meter) of the molecules in air at normal conditions? (b) What is the average distance between molecules compared to the dimensions of the molecules? (c) Can you consider air to be an ideal gas? Explain your answer.
6. * EST Estimate the number of collisions that one molecule of air in the physics classroom experiences every second. List all the assumptions that you made.
7. What is the mass of a water molecule in kilograms? What is the mass of an average air molecule in kilograms?
8. You find that the average gauge pressure in your car tires is about 35 psi. How many newtons per square meter is it? What is gauge pressure?
9. BIO **Forced vital capacity** Physicians use a machine called a spirometer to measure the maximum amount of air a person

can exhale (called the forced vital capacity). Suppose you can exhale 4.8 L. How many kilograms of air do you exhale? What assumptions did you make to answer the question? How do these assumptions affect the result?
10. A container is at rest with respect to a desk. Inside the container a particle is moving horizontally at a speed v with respect to the desk. It collides with a vertical wall of the container elastically and rebounds. Qualitatively, determine the direction and the speed of the particle if the wall is (a) at rest with respect to the desk; (b) moving in the same direction as the particle at a speed smaller than the particle's; and (c) moving in the direction opposite to the motion of the particle at a smaller speed.

9.3 Quantitative analysis of ideal gas

11. * **Hitting tennis balls against a wall** A 0.058-kg tennis ball, traveling at 25 m/s, hits a wall, rebounds with the same speed in the opposite direction, and is hit again by another player, causing the ball to return to the wall at the same speed. The ball returns to the wall once every 0.60 s. (a) Determine the force that the ball exerts on the wall averaged over the time between collisions. State the assumptions that you made. (b) If 10 people are practicing against a wall with an area of 30 m², what is the average pressure of the 10 tennis balls against the wall?
12. * Friends throw snowballs at the wall of a 3.0 m × 6.0 m barn. The snowballs have mass 0.10 kg and hit the wall moving at an average speed of 6.0 m/s. They do not rebound. Determine the average pressure exerted by the snowballs on the wall if 40 snowballs hit the wall each second. Which problem, this or the previous problem, resembles the actions of the molecules of an ideal gas hitting the walls of their container?
13. * A ball moving at a speed of 3.0 m/s with respect to the ground hits a stationary wall at a 30° angle with respect to the surface of the wall. Determine the direction and the magnitude of the velocity of the ball after it rebounds. Explain carefully what physics principles you used to find the answer. What assumptions did you make? How will the answer change if one or more of them are not valid?
14. **Oxygen tank for mountains** Consider an oxygen tank for a mountain climbing trip. The mass of one molecule of

oxygen is 5.3×10^{-26} kg. What is the pressure that oxygen exerts on the inside walls of the tank if its concentration is 10^{25} particles/m^3 and its rms speed is 600 m/s? What assumptions did you make?

15. You have five molecules with the following speeds: 300 m/s, 400 m/s, 500 m/s, 450 m/s, and 550 m/s. What is their average speed?

16. What is the rms speed of the molecules in the previous problem? Is it different from the average speed?

17. Two gases in different containers have the same concentration and same rms speed. The mass of a molecule of the first gas is twice the mass of a molecule of the second gas. What can you say about their pressures? Explain.

18. * You are hiking up a mountain. About halfway up you pass through a cloud and become moist from cloud water. How can this water be at such a high elevation? To answer this question, compare the molar masses and densities of dry air and humid air. Explain. List all of your assumptions.

19. * / BIO **Breathing** You are breathing heavily while hiking up the mountain. To inhale, you expand your diaphragm and lungs. Explain, using your knowledge of gas pressure, why this mechanical movement leads to the air flowing into your nose or mouth. Support your reasoning with diagrams if necessary.

20. * **Oxygen tank for mountain climbing** An oxygen container that one can use in the mountains has a 90-min oxygen supply at a speed of 6 L/min. Determine everything you can about the gas in the container. Make reasonable assumptions.

9.4 Temperature

21. You are cooking dinner in the mountains. At 7000 feet, water boils at 92.3 °C. Convert the boiling temperature to °F and suggest two possible reasons why the boiling temperature is lower at this elevation than at sea level.

22. Your temperature, when taken orally, is 98.6 °F. When taken under your arm, it's 36.6 °C. Are these results consistent? Explain.

23. On top of Mount Everest, the temperature is −19 °C in July. Being a physicist, you determine by how many degrees Celsius one needs to change the air temperature to double the average kinetic energy of its molecules. Explain your reasoning.

24. Air consists of many different molecules, for example, N_2, O_2, H_2O, and CO_2. Which molecules are the fastest on average? The slowest on average? Explain.

25. What is the average kinetic energy of a particle of air at standard conditions?

26. Air is a mixture of molecules of different types. Compare the rms speeds of the molecules of N_2, O_2, and CO_2 at standard conditions. What assumptions did you make?

27. * How many moles of air are in a regular 1-L water bottle when you finish drinking the water? What assumptions did you make? How do these assumptions affect your result?

28. * At approximately what temperature does the average random kinetic energy of a N_2 molecule in an ideal gas equal the macroscopic translational kinetic energy of a copper atom in a penny that is dropped from the height of 1.0 m?

29. ** A molecule moving at speed v_1 collides head-on with a molecule of the same mass moving at speed v_2. Compute the speeds of the molecules after the collision. What assumptions did you make? How does the answer to this problem explain why the mixing of hot and cold gases causes the cold gas to become warmer and the hot gas to become cooler?

30. **Balloon flight** For a balloon ride, the balloon must be inflated with helium to a volume of 1500 m^3 at sea level. The balloon will rise to an altitude of about 12 km, where the temperature

is about −52 °C and the pressure is about 20 kPa. How much helium should be put into the balloon? What assumptions did you make?

31. * BIO **Ears pop** The middle ear has a volume of about 6.0 cm^3 when at a pressure of 1.0×10^5 N/m^2 (1.0 atm). Determine the volume of that same air when the air pressure is 0.83×10^5 N/m^2, as it is at an elevation of 1500 m above sea level (assume the air temperature remains constant). If the volume of the middle ear remains constant, some air will have to leave as the elevation increases. That is why ears "pop."

32. * Even the best vacuum pumps cannot lower the pressure in a container below 10^{-15} atm. How many molecules of air are left in each cubic centimeter in this "vacuum?" Assume that the temperature is 273 K.

33. **Pressure in interstellar space** The concentration of particles (assume neutral hydrogen atoms) in interstellar gas is 1 particle/cm^3, and the average temperature is about 10 K. What is the pressure of the interstellar gas? How does it compare to the best vacuum that can be achieved on Earth (see the previous problem)?

9.5 Testing the ideal gas law; 9.6 Speed distribution of particles; 9.7 Skills for analyzing processes using the ideal gas law

34. * Describe experiments to determine if each of the three gas isoprocess laws works. The experiments should be ones that you could actually carry out.

35. * The following data were collected for the temperature and volume of a gas. Can this gas be described by the ideal gas model? Explain how you know.

Temperature (°C)	Volume (ml)
11	95.0
25	100.0
47	107.5
73	116.0
159	145.1
233	170.0
258	177.9

36. * Describe a mechanical model of Stern's molecular beam experiment.

37. * Explain the microscopic mechanisms for the relation of macroscopic variables for an isothermal process, an isobaric process, and an isochoric process.

38. * **Scuba diving** The pressure of the air in a diver's lungs when he is 20 m under the water surface is 3.0×10^5 N/m^2, and the air occupies a volume of 4.8 L. How many moles of air should he exhale while moving to the surface, where the pressure is 1.0×10^5 N/m^2?

39. * BIO EST **Alveoli surface area** Estimate the size of the surface area of a single alveolus in the lungs.

40. * When surrounded by air at a pressure of 1.0×10^5 N/m^2, a basketball has a radius of 0.12 m. Compare its volume at this condition with the volume that it would have if you take it 15 m below the water surface where the pressure is 2.5×10^5 N/m^2. What assumptions did you make?

41. ** / You have gas in a container with a movable piston. The walls of the container are thin enough so that its temperature stays the same as the temperature of the surrounding medium. You have baths of water of different temperatures, different objects that you can place on top of the piston, etc.

(a) Describe how you could make the gas undergo an isothermal process so that the pressure inside increases by 10%, then undergo an isobaric process so that the new volume decreases by 20%, and finally undergo an isochoric process so that the temperature increases by 15%. (b) Represent all processes in P-versus-T, V-versus-T, and P-versus-V graphs. (c) What are the new pressure, volume, and temperature of the gas?

42. * **Bubbles** While snorkeling, you see air bubbles leaving a crevice at the bottom of a reef. One of the bubbles has a radius of 0.060 m. As the bubble rises, the pressure inside it decreases by 50%. Now what is the bubble's radius? What assumptions did you make to solve the problem?

43. ** ∕ **Diving bell** A cylindrical diving bell, open at the bottom and closed at the top, is 4 m tall. Scientists fill the bell with air at the pressure of $1.0 \times 10^5 \, \text{N/m}^2$. The pressure increases by $1.0 \times 10^5 \, \text{N/m}^2$ for each 10 m that the bell is lowered below the surface of the water. If the bell is lowered 30 m below the ocean surface, how many meters of air space are left inside the bell? Why doesn't water enter the entire bell as it goes under water? Draw several sketches for this problem.

44. * **Mount Everest** (a) Determine the number of molecules per unit volume in the atmosphere at the top of Mount Everest. The pressure is $0.31 \times 10^5 \, \text{N/m}^2$, and the temperature is $-30\,^\circ\text{C}$. (b) Determine the number of molecules per unit volume at sea level, where the pressure is $1.0 \times 10^5 \, \text{N/m}^2$ and the temperature is $20\,^\circ\text{C}$.

45. * EST **Breathing on Mount Everest** Using the information from Problem 44, estimate how frequently you need to breathe on top of Mount Everest to inhale the same amount of oxygen as you do at sea level. The pressure is about one-third the pressure at sea level.

46. **Capping beer** You would like to make homemade beer, but you are concerned about storing it. Your beer is capped into a bottle at a temperature of $27\,^\circ\text{C}$ and a pressure of $1.2 \times 10^5 \, \text{N/m}^2$. The cap will pop off if the pressure inside the bottle exceeds $1.5 \times 10^5 \, \text{N/m}^2$. At what maximum temperature can you store the beer so the gas inside the bottle does not pop the cap? List the assumptions that you made.

47. * **Car tire** With a tire gauge, you measure the pressure in a car tire as $2.1 \times 10^5 \, \text{N/m}^2$. How can this be if you know that absolute pressure in the tire is three times higher than atmospheric? The tire looks okay. What's the deal?

48. * **Car tire dilemma** Imagine a car tire that contains 5.1 moles of air when at a gauge pressure of $2.1 \times 10^5 \, \text{N/m}^2$ (the pressure above atmospheric pressure) and a temperature of $27\,^\circ\text{C}$. The temperature increases to $37\,^\circ\text{C}$, the volume decreases to 0.8 times the original volume, and the gauge pressure decreases to $1.6 \times 10^5 \, \text{N/m}^2$. Can these measurements be correct if the tire did not leak? If it did leak, then how many moles of air are left in the tire?

49. There is a limit to how much gas can pass through a pipeline, because the pipes can only tolerate so much pressure on the walls. To increase the amount of gas going through the pipeline, engineers decide to cool the gas (to reduce its pressure). Suggest how much they should lower the temperature of the gas if they want to increase the mass per unit time by 1.5 times.

50. Explain how you know that the volume of one mole of gas at standard conditions is 22.4 L.

51. * At what pressure is the density of $-50\,^\circ\text{C}$ nitrogen gas (N_2) equal to 0.10 times the density of water?

52. * In the morning, the gauge pressure in your car tires is 35 psi. During the day, the air temperature increases from 20 °C to 30 °C and the pressure increases to 36.5 psi. By how much did the volume of one of the tires increase? What assumptions did you make?

53. **Equation Jeopardy 1** The equation below describes a process. Construct a word problem for a process that is consistent with the equations (there are many possibilities). Provide as much detailed information as possible about your proposed process.

$$\frac{1.2 \times 10^5 \, \text{N/m}^2}{293 \, \text{K}} = \frac{2.0 \times 10^5 \, \text{N/m}^2}{T}$$

54. * **Equation Jeopardy 2** The equation below describes a process. Construct a word problem for a process that is consistent with the equations (there are many possibilities). Provide as much detailed information as possible about your proposed process.

$$\Delta n = \frac{(0.67 \times 10^5 \, \text{N/m}^2)(0.60 \times 10^{-6} \, \text{m}^3)}{(8.3 \, \text{J/mole} \cdot \text{K})(303 \, \text{K})}$$
$$- \frac{(1.00 \times 10^5 \, \text{N/m}^2)(0.60 \times 10^{-6} \, \text{m}^3)}{(8.3 \, \text{J/mole} \cdot \text{K})(310 \, \text{K})}$$

55. ** The P-versus-T graph in **Figure P9.55** describes a cyclic process comprising four hypothetical parts. (a) What happens to the pressure of the gas in each part? (b) What happens to the temperature of the gas in each part? (c) What happens to the volume of the gas in each part? (d) Explain each part microscopically. (e) Use the information from (a)–(c) to represent the same parts in P-versus-V and V-versus-T graphs. [*Hint:* It helps to align the P-versus-V graph *beside* the P-versus-T graph using the same P values on the ordinate (vertical axes) and to place the V-versus-T graph *below* the P-versus-T graph using the same T values on the abscissa (horizontal axes). This helps keep the same scale for the variables.]

Figure P9.55

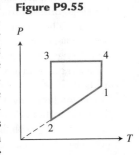

56. ** The V-versus-T graph in **Figure P9.56** describes a cyclic process comprising four hypothetical parts. (a) What happens to the pressure of the gas in each parts? (b) What happens to the temperature of the gas in each parts? (c) What happens to the volume of the gas in each parts? (d) Explain each part microscopically. (e) Use the information from (a)–(c) to represent the same process in a P-versus-T graph (below the V-versus-T graph) and a P-versus-V graph (beside the P-versus-T graph). See the hint in the previous problem about the graph alignments.

Figure P9.56

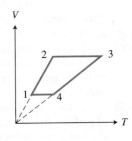

9.8 Thermal energy, the Sun, and diffusion: Putting it all together

57. ** EST **Sun's life expectancy** (a) Estimate the average kinetic energy of the particles in the Sun. Assume that it is made of atomic hydrogen and that its average temperature is 100,000 K. The mass of the Sun is 2×10^{30} kg. (b) For how long would the Sun shine using this energy if it radiates 4×10^{26} W/s? Is your answer reasonable? Explain.

58. * The temperature of the Sun's atmosphere near the surface is about 6000 K, and the concentration of atoms is about 10^{15} particles/m^3. What is the average pressure and density of its atmosphere? What assumptions did you make to solve the problem?

59. ** A gas that can be described by the ideal gas model is contained in a cylinder of volume V. The temperature of the gas is T. The particle mass is m, and the molar mass is M. Write an expression for the total thermal energy of the gas. Now, imagine that the exact same gas has been placed in a container of volume $2V$. What happens to its pressure? What happens to its temperature? What happens to its thermal energy?

60. * BIO EST **Breathing and metabolism** We need about 0.7 L of oxygen per minute to maintain our resting metabolism and about 2 L when standing and walking. Estimate the number of breaths per minute for a person to satisfy this need when resting and when standing and walking. What assumptions did you make? Remember that oxygen is about 21% of the air.

61. ** In 1896, Lord Rayleigh showed that a mixture of two gases of different atomic masses could be separated by allowing some of it to diffuse through a porous membrane into an evacuated space. Rayleigh proposed that the molecules of lighter gas diffuse through the membrane faster, leaving the heavier gas behind in the original space. He described this "separation factor" as equal to $\sqrt{m_2/m_1}$, where m_1 is the molecular weight of a lighter gas and m_2 is the molecular weight of a heavier gas. Give a reason why the separation factor depends on the square root of the ratio of the molecular masses.

62. * **Equation Jeopardy 3** The three equations below describe a physical situation. Construct a word problem for a situation that is consistent with the equations (there are many possibilities). Provide as much detailed information as possible about the situation.

$$m = (1.3 \text{ kg/m}^3)(3.0 \text{ m} \times 5.0 \text{ m} \times 2.0 \text{ m})$$

$$N = m/[(29 \times 10^{-3} \text{ kg})/(6.0 \times 10^{23} \text{ particles})]$$

$$U_{thermal} = N(3/2)(1.38 \times 10^{-23} \text{ J/K})(273 \text{ K})$$

General Problems

63. * **No H_2 in Earth atmosphere** Explain why Earth has almost no free hydrogen in its atmosphere.

64. * **No atmosphere on the Moon** Why does the Moon have no atmosphere? Explain.

65. * **Different planet compositions** Explain why planets closer to the Sun have low concentrations of light elements, but are relatively abundant in giant planets such as Jupiter, Uranus, and Saturn, which are far from the Sun.

66. ** EST **Density of our galaxy** Estimate the average density of particles in our galaxy, assuming that the most abundant element is atomic hydrogen. There are about 10^9 stars in the galaxy and the size of the galaxy is about 10^5 light-years. A light-year is the distance light travels in 1 year moving at a speed of 3×10^8 m/s. What do you need to assume about the stars in order to answer this question?

67. * BIO **Breathing** Observe yourself breathing and count the number of times you inhale per second. During each breath you probably inhale about 0.50 L of air. How many oxygen molecules do you inhale if you are at sea level?

68. ** **Car engine** During a compression stroke of a cylinder in a diesel engine, the air pressure in the cylinder increases from 1.0×10^5 N/m^2 to 50×10^5 N/m^2, and the temperature increases from 26 °C to 517 °C. Using this information, how would you convince your friends that knowledge about ideal gases can help explain how hot gases burned in the car engine affect the motion of the car?

69. ** EST How can the pressure of air in your house stay constant during the day if the temperature rises? Estimate the volume of your house and the number of moles of air that leave the house during the daytime. Assume that nighttime temperature and daytime temperature differ by about 10 °C. List all other assumptions that are necessary to answer the question.

70. ** **Tell-all problem** Tell everything you can about the process described by the pressure-versus-volume graph shown in **Figure P9.70**.

Figure P9.70

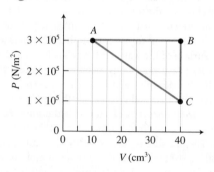

71. ** Two massless, frictionless pistons are inside a horizontal tube opened at both ends. A 10-cm-long thread connects the pistons. The cross-sectional area of the tube is 20 cm^2. The pressure and temperature of gas between the pistons and the outside air are the same and are equal to $P = 1.0 \times 10^5$ N/m^2 and $T = 24$ °C. At what temperature will the thread break if it breaks when the tension reaches 30 N?

72. * A closed cylindrical container is divided into two parts by a light, movable, frictionless piston. The container's total length is 100 cm. Where is the piston located when one side is filled with nitrogen (N_2) and the other side with the same mass of hydrogen (H_2) at the same temperature?

Reading Passage Problems

BIO **Vascular wall tension and aortic blowout** The walls of blood vessels contain varying amounts of elastic fibers that allow the vessels to expand and contract as the pressure and amount of fluid inside vary (these fibers are more prevalent in the aorta and large arteries than in the small arterioles and capillaries). These fibers in the cylindrical walls produce a wall tension T, defined as

$$T = \frac{F}{L}$$

where L is the length of an imaginary cut parallel to the axis of the vessel and F is the magnitude of the force that each side of the cut must exert on the other side to hold the two sides together.

Three forces are exerted on a short section of wall fiber—the system. (1) The fluid inside pushes outward, due to fluid pressure from inside $P_{\text{inside fluid pushing out}} = P_{out}$; (2) fluid outside the vessel pushes inward, due to fluid pressure from outside $P_{\text{outside fluid pushing in}} = P_{in}$; and (3) the wall next to the system exerts wall tension T on each side of that wall. The pressure difference across the wall $\Delta P = P_{out} - P_{in}$, the wall tension T, and the radius R of the cylindrical vessel are related by Laplace's law:

$$\Delta P = \frac{T}{R} \quad \text{or} \quad T = \Delta P \cdot R$$

The inward gauge pressure P_{in} of tissue surrounding the vessels is approximately zero. Thus, the pressure difference $\Delta P = P_{out} - P_{in} = P_{vessel} - 0 = P_{vessel}$ is the gauge pressure in the blood vessel. We can now estimate the wall tension for different types of vessels. The tension in the aorta is approximately

$$T = \Delta P \cdot R \approx (100 \text{ mm Hg})\left(\frac{133 \text{ N/m}^2}{1 \text{ mm Hg}}\right)(1.3 \times 10^{-2} \text{ m})$$

$$= 170 \text{ N/m}$$

Using similar reasoning, the wall tension in the low-pressure, very small radius capillaries is about 0.016 N/m—about 0.0003 times the tension needed to tear a facial tissue. Because such little tension is needed to hold a capillary together, its wall can be very thin, allowing easy diffusion of various molecules across the wall.

The walls of a healthy aorta can easily provide the tension needed to support the increased blood pressure when it fills with blood from the heart during each heartbeat. However, aging and various medical conditions may weaken the aortic wall in a short section, and increased blood pressure can cause it to stretch. The weakened wall bulges outward; this is called an aortic aneurism. The increased radius causes increased tension, which can increase bulging. This cycle can result in a rupture to the aorta: an aortic blowout.

73. Why is the wall tension in capillaries so small?
    (a) There are so many capillaries.
    (b) Their radii are so small.
    (c) The outward pressure of the blood inside is so small.
    (d) b and c
    (e) a, b, and c

74. Which answer below is closest to the wall tension in a typical arteriole of radius 0.15 mm and 60 mm Hg blood pressure?
    (a) 0.001 N/m   (b) 0.01 N/m
    (c) 0.1 N/m   (d) 1 N/m
    (e) 10 N/m

75. According to Laplace's law, elevated blood pressure in an artery should cause the wall tension in the artery to do what?
    (a) Increase   (b) Remain unchanged
    (c) Decrease   (d) Impossible to decide

76. As a person ages, the fibers in arteries become less elastic and the wall tension increases. According to Laplace's law, this will cause the blood pressure to do what?
    (a) Increase   (b) Remain unchanged
    (c) Decrease   (d) Impossible to decide

77. Aortic blowout occurs when part of the wall of the aorta becomes weakened. What does this cause?
    (a) A bulge and increased radius of the aorta when the blood pressure inside increases
    (b) An increased radius of the aorta, which causes increased tension in the wall
    (c) An increased tension in the aorta, which causes the radius to increase
    (d) a and b
    (e) a, b, and c

BIO **Portable hyperbaric chamber** In 1997, a hiking expedition was stranded for 38 days on the Tibetan side of Mount Everest at altitudes over 5200 m. For 30 days the party was stranded above 6500 m in severe weather conditions. While at altitudes over 8000 m, 10 climbers suffered acute mountain sickness. A 37-year-old climber with acute pulmonary edema (buildup of fluid in the lungs) was treated with a portable Gamow bag (see **Figure 9.19**), named after its inventor, Igor Gamow. The Gamow bag is a windowed cylindrical portable hyperbaric chamber constructed of nonpermeable nylon that requires constant pressurization with a foot pump attached to the bag. The climber enters the bag, and a person outside pumps air into the bag so that the air pressure inside is somewhat higher than the outside pressure.

The bag and pump have a 6.76-kg mass. The volume of the inflated bag is 0.476 m³. The maximal bag pressure is $0.14 \times 10^5 \text{ N/m}^2$ above the air pressure at the site where it is used. In the 1997 climb, with the temperature at $-20 \,^\circ\text{C}$, the bag was filled in about 2 min with 10–20 pumps per minute. This raised the pressure in the bag to $0.58 \times 10^5 \text{ N/m}^2$ (equivalent to an elevation of 4400 m) instead of the actual outside pressure of $0.43 \times 10^5 \text{ N/m}^2$ at the 6450-m elevation. The treatment lasted for 2 h, with the person inhaling about 15 times/min at about 0.5 L/inhalation, and was successful—the pulmonary edema disappeared.

Figure 9.19 A Gamow bag, used to help climbers with acute altitude sickness.

78. What is closest to the volume of the Gamow bag?
    (a) 50 L   (b) 100 L   (c) 200 L   (d) 500 L   (e) 1000 L

79. What is closest to the temperature at the 6450-m elevation on the day described in the problem?
    (a) 37 K   (b) 253 K   (c) 20 K   (d) -20 K   (e) 273 K

80. What is closest to the number n of gram-moles of air in the filled bag when at 4400 m?
    (a) 3 g · moles   (b) 10 g · moles
    (c) 13 g · moles   (d) 110 g · moles
    (e) 170 g · moles

81. What is closest to the number n of gram-moles of air in the bag if at the 6450-m pressure?
    (a) 3 g · moles   (b) 10 g · moles
    (c) 13 g · moles   (d) 110 g · moles
    (e) 170 g · moles

82. Estimate the fraction of the air in the Gamow bag that an occupant would inhale in 1 h, assuming no replacement of the air.
    (a) 0.01   (b) 0.1   (c) 0.3   (d) 0.5   (e) 1.0

10

Static Fluids

How can a hot air balloon travel for hours in the sky?

Why is it dangerous for scuba divers to ascend to the surface quickly from a deep-sea dive?

Why does a 15-g nail sink in water and a cargo ship float?

Be sure you know how to:

- Draw a force diagram for a system of interest (Section 1.6).
- Apply Newton's second law in component form (Section 3.5).
- Define pressure (Section 9.2).

The first balloon flight occurred in 1731 in Russia, in a balloon filled with smoke. Fifty years later the brothers Montgolfier used heated air to fill a balloon in France. That ride carried a sheep, a duck, and a chicken in the balloon's basket. Both smoke and hot air are less dense than cold air. By carefully balancing the density of the balloon and its contents with the density of air, the pilot can control the force that the outside cold air exerts on the balloon. What do pressure, volume, mass, and temperature have to do with this force?

In the previous chapter, we constructed the ideal gas model and used it to explain the behavior of gases. That model was built on the knowledge of the particle nature of matter and a Newtonian analysis of the motion of those particles. The temperature of the gas and the pressure it exerted on surfaces played an important role in the phenomena we analyzed. We ignored the effect of the gravitational force exerted by Earth on the gas particles. This simplification was reasonable, since in most of the processes we analyzed the gases had little mass and occupied a relatively small region of space, like in a piston. In this chapter, our interest expands to include phenomena in which the force exerted by Earth plays an important role. We will confine the discussion to static fluids—fluids that are not moving.

10.1 Density

We are familiar with the concept of density (from Chapter 9). To find the density of an object or a substance, determine its mass and volume and then calculate the ratio of the mass and volume:

$$\rho = \frac{m}{V} \tag{10.1}$$

Archimedes (Greek, 287–212 B.C.) discovered how to determine the density of an object of irregular shape. First determine its mass using a scale. Then determine its volume by submerging it in a graduated cylinder with water (**Figure 10.1**). Finally, divide the mass in kilograms by the volume in cubic meters to find the density in kilograms per cubic meter. Using this method we find that the density of an iron nail is $7800 \, \text{kg/m}^3$—relatively large. A gold coin has an even larger density—$19{,}320 \, \text{kg/m}^3$. The universe, though, contains much denser objects. For example, the rapidly spinning neutron star known as a pulsar (discussed in Chapter 8) has a density of approximately $10^{18} \, \text{kg/m}^3$. **Table 10.1** lists the densities of various solids, liquids, and gases.

Figure 10.1 Measuring the density of an irregularly shaped object.

1. Measure mass of object.
2. Place the object in water in a graduated cylinder.
3. Measure the volume change of the water. Volume change of water = volume of object.
4. Density = $\rho = m/V$

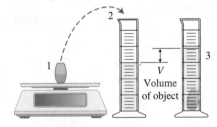

QUANTITATIVE EXERCISE 10.1
Ping-pong balls with different densities
Saturn has the lowest density of all the planets in the solar system ($M_{\text{Saturn}} = 5.7 \times 10^{26} \, \text{kg}$ and $V_{\text{Saturn}} = 8.3 \times 10^{23} \, \text{m}^3$). The average density of a neutron star is $10^{18} \, \text{kg/m}^3$. Compare the mass of a ping-pong ball filled with material from Saturn with that of the same ball filled with material from a neutron star. An empty ping-pong ball has a 0.037-m diameter (0.020-m radius) and a 2.7-g mass.

Represent mathematically To find the mass of a ping-pong ball filled with a particular material, we add the mass of the ball alone and the calculated mass of the material inside:

$$m_{\text{filled ball}} = m_{\text{ball}} + m_{\text{material}}$$

where

$$m_{\text{material}} = \rho_{\text{material}} V_{\text{ball}}$$

(continued)

The density of the neutron star is given, and the density of Saturn can be found using the operational definition $\rho_{\text{Saturn}} = \dfrac{m_{\text{Saturn}}}{V_{\text{Saturn}}}$. The interior of the ping-pong ball is a sphere of volume

$$V_{\text{sphere}} = \frac{4}{3}\pi R^3$$

Assume that the plastic shell of the ball has negligible volume. The mass of either filled ball is

$$m_{\text{filled ball}} = m_{\text{ball}} + m_{\text{material}} = m_{\text{ball}} + \rho_{\text{material}} V_{\text{ball}}$$
$$= m_{\text{ball}} + \rho_{\text{material}}(4/3)\pi R_{\text{ball}}^3$$

Solve and evaluate For the neutron-star-filled ball:

$$m_{\text{neutron star ball}} = (0.003 \text{ kg}) + (10^{18}\text{ kg/m}^3)\frac{4}{3}\pi(0.020 \text{ m})^3$$
$$= 3.4 \times 10^{13} \text{ kg}$$

For the Saturn-filled ball:

$$m_{\text{Saturn ball}} = m_{\text{ball}} + \left(\frac{m_{\text{Saturn}}}{V_{\text{Saturn}}}\right)\left(\frac{4}{3}\pi R_{\text{ball}}{}^3\right)$$

$$= 0.003 \text{ kg} + \left(\frac{5.7 \times 10^{26}\text{ kg}}{8.3 \times 10^{23}\text{ m}^3}\right)\left(\frac{4}{3}\pi(0.020 \text{ m})^3\right)$$

$$= 0.003 \text{ kg} + 0.023 \text{ kg} = 0.026 \text{ kg}$$

The material from Saturn has less mass than an equal volume of water. The ball filled with the material from a neutron star has a mass of more than a billion tons!

Try it yourself: The mass of the ping-pong ball filled with soil from Earth's surface is 0.050 kg. What is the density of the soil?

Answer: 1400 kg/m^3.

Table 10.1 Densities of various solids, liquids, and gases.

Solids		Liquids		Gases	
Substance	Density (kg/m³)	Substance	Density (kg/m³)	Substance	Density (kg/m³)
Aluminum	2700	Acetone	791	Dry air 0° C	1.29
Copper	8920	Ethyl alcohol	789	10° C	1.25
Gold	19,300	Methyl alcohol	791	20 °C	1.21
Iron	7860	Gasoline	726	30 °C	1.16
Lead	11,300	Mercury	13,600	Helium	0.178
Platinum	21,450	Milk	1028–1035	Hydrogen	0.090
Silver	10,500	Seawater	1025	Oxygen[2]	1.43
Bone	1700–2000	Water 0 °C	999.8		
Brick	1400–2200	3.98 °C	1000.00		
Cement	2700–3000	20 °C	998.2		
Clay	1800–2600	Blood plasma	1030		
Glass	2400–2800	Blood whole[1]	1050		
Ice	917				
Balsa wood	120				
Oak	600–900				
Pine	500				
Planet Earth	5515				
Moon	3340				
Sun	1410				
Universe (average)	10^{-26}				
Pulsar	10^{11}–10^{18}				

[1]Densities of liquids are at 0 °C unless otherwise noted.
[2]Densities of gases are at 0 °C and 1 atm unless otherwise noted.

Density and floating

Understanding density allows us to pose questions about phenomena that we observe almost every day. For example, why does oil form a film on water? If you pour oil into water or water into oil, they form layers (see **Figure 10.2a**). Independently of which fluid is poured first—the layer of oil is always on top of the water. The density of oil is less that the density of water. If you pour corn syrup and water into a container, the corn syrup forms a layer below the water (Figure 10.2b); the density of corn syrup is 1200 kg/m^3, greater than the density of water. Why, when mixed together, is the lower density substance always on top of the higher density substance?

Similar phenomena occur with gases. Helium-filled balloons accelerate upward in air while air-filled balloons accelerate (slowly) downward. The mass of helium atoms is much smaller than the mass of any other molecules in the air. (Recall that at the same pressure and temperature, atoms and molecules of gas have the same concentration; because helium atoms have much lower mass, their density is lower.) The air-filled balloon must be denser than air. The rubber with which the skin of any balloon is made is denser than air. We can disregard the slight compression of the gas by the balloon, because even though it increases the density of the gas, the effect is the same for both the air and helium in the balloons.

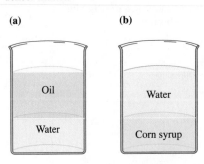

Figure 10.2 Less dense matter floats on denser matter.

(a) Oil / Water

(b) Water / Corn syrup

Solid water floats in liquid water

The solid form of a particular substance is almost always denser than the liquid form of the substance, with one very significant exception: liquid water and solid ice. Since ice floats on liquid water, we can assume that the density of ice is less than that of water. This is in fact true: the density of water changes slightly with temperature and is the highest at 4 °C: 1000 kg/m³. The density of ice is 917 kg/m³. Ice has a lower density because in forming the crystal structure of ice, water molecules spread apart. The fact that water expands when it forms ice is important for life on Earth (see the second Reading Passage at the end of this chapter). Fish and plants living in lakes survive cold winters in liquid water under a shield of ice and snow at the surface. Water absorbed in the cracks of rocks freezes and expands in the winter, cracking the rock. Over the years, this process of liquid water absorption, freezing, and cracking eventually converts the rock into soil.

Why do denser forms of matter sink in less dense forms of matter? We learned (in Chapter 9) that the quantity *pressure* describes the forces that fluids exert on each other and on the solid objects they contact. Let us investigate whether pressure explains, for example, why a nail sinks in water or why a hot air balloon rises.

‹ Active Learning Guide

Review Question 10.1 How would you determine the density of an irregularly shaped object?

10.2 Pressure exerted by a fluid

We know that as gas particles collide with the walls of the container in which they reside, they exert pressure. In fact, if you place any object inside a gas, the gas particles exert the same pressure on the object as the gas exerts on the walls of the container. Do liquids behave in a similar way? In the last chapter we learned that the particles in a liquid are in continual random motion, somewhat similar to particles in gases.

Let's conduct a simple observational experiment. Take a plastic water bottle and poke several small holes at the same height along its perimeter. Close the holes with tacks, fill the bottle with water, open the cap, remove the tacks, and observe what happens (**Figure 10.3**). Identically-shaped parabolic streams of water shoot out of the holes. The behavior of the water when the tacks are removed is analogous to a person leaning on a door that is suddenly opened from the other side—the person falls through the door. Evidently, the water inside must push out perpendicular to the wall of the bottle, just as gas pushes out perpendicular to the wall of a balloon. Due to their similar behaviors, liquids and gases are often studied together and are collectively referred to as fluids. In addition, since the four streams are identically shaped, the pressure at all points at the same depth in the fluid is the same.

Figure 10.3 Arcs of water leaving holes at the same level in a bottle.

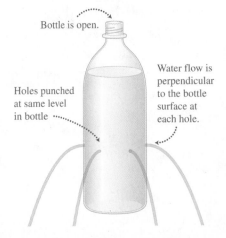

Bottle is open.

Holes punched at same level in bottle

Water flow is perpendicular to the bottle surface at each hole.

Figure 10.4 Pascal's first law: Increasing the pressure of a fluid at one location causes a uniform pressure increase throughout the fluid.

(a)

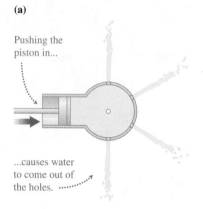

Pushing the piston in...

...causes water to come out of the holes.

(b)

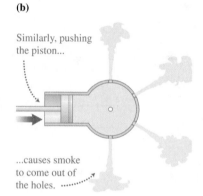

Similarly, pushing the piston...

...causes smoke to come out of the holes.

Pascal's first law

Many practical applications involve situations in which an external object (for example, a piston) exerts a force on a particular part of a fluid. What happens to the pressure at other places inside the fluid? To investigate how the pressure at different points in a fluid are related, we use a special instrument that consists of a round glass bulb with holes in it connected to a glass tube with a piston on the other side (**Figure 10.4a**). When we fill the apparatus with water and push the piston, water comes out of all of the holes, not only those that align with the piston. When we fill the apparatus with smoke and push the piston, we get the same result (Figure 10.4b). The liquid and the gas behave similarly.

How can we explain this observation? Pushing the piston in one direction caused a greater pressure in the fluid close to the piston. It seems that almost immediately the pressure throughout the fluid increased as well, as the fluid was pushed out of *all* of the holes in the bulb in the same way. This phenomenon was first discovered by French scientist Blaise Pascal in 1653 and is called Pascal's first law.

> **Pascal's first law** An increase in the pressure of a static, enclosed fluid at one place in the fluid causes a uniform increase in pressure throughout the fluid.

The above experiment describes Pascal's first law macroscopically. We can also explain Pascal's first law at a microscopic level. Particles inside a container move randomly in all directions. When we push harder on one of the surfaces of the container, the fluid compresses near that surface. The molecules near that surface collide more frequently with their neighbors farther from the surface. They in turn collide more frequently with their neighbors. The extra pressure exerted at the one surface quickly spreads and soon there is increased pressure throughout the fluid.

Glaucoma

Pascal's first law can help us understand a common eye problem—glaucoma. A clear fluid called aqueous humor fills two chambers in the front of the eye (**Figure 10.5**). In a healthy eye, new fluid is continually secreted into these chambers while old fluid drains from the chambers through sinus canals. A person with glaucoma has closed drainage canals. The buildup of fluid causes increased pressure throughout the eye, including at the retina and optic nerve, which can lead to blindness. Ophthalmologists diagnose glaucoma by measuring the pressure at the front of the eye. The eye pressure of a person with glaucoma is about 3000 N/m² above atmospheric pressure.

Active Learning Guide ›

Figure 10.5 Glaucoma is an increase in intraocular pressure, caused by blockage of the ducts that normally drain aqueous humor from the eye.

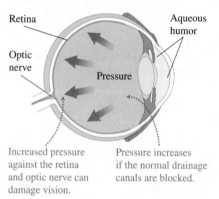

Retina

Aqueous humor

Optic nerve

Pressure

Increased pressure against the retina and optic nerve can damage vision.

Pressure increases if the normal drainage canals are blocked.

Hydraulic lift

One of the technical applications of Pascal's first law is a hydraulic press, a form of simple machine that converts small forces into larger forces, or vice versa. Automobile mechanics use hydraulic presses to lift cars, and dentists and barbers use them to raise and lower their clients' chairs. The hydraulic brakes of an automobile are also a form of hydraulic press. Most of these devices work on the simple principle illustrated in **Figure 10.6**, although the actual devices are usually more complicated in construction.

In Figure 10.6, a downward force $\vec{F}_{1\,\text{on}\,L}$ is exerted by piston 1 (with small area A_1) on the liquid. This piston compresses a liquid (usually oil) in the lift. The pressure in the fluid just under piston 1 is

$$P_1 = \frac{F_{1\,\text{on}\,L}}{A_1}$$

Because the pressure changes uniformly throughout the liquid, the pressure under piston 2 is also $P = F_{1\,on\,L}/A_1$, assuming the pistons are at the same elevation. Since piston 2 has a greater area A_2 than piston 1, the liquid exerts a greater upward force on piston 2 than the downward force on piston 1:

$$F_{L\,on\,2} = PA_2 = \left(\frac{F_{1\,on\,L}}{A_1}\right)A_2 = \left(\frac{A_2}{A_1}\right)F_{1\,on\,L} \qquad (10.2)$$

Since A_2 is greater than A_1, the lift provides a significantly greater upward force $F_{L\,on\,2}$ on piston 2 than the downward push of the smaller piston 1 on the liquid $F_{1\,on\,L}$.

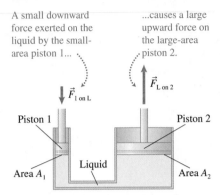

Figure 10.6 Schematic of a hydraulic lift.

A small downward force exerted on the liquid by the small-area piston 1...

...causes a large upward force on the large-area piston 2.

Piston 1   Piston 2

Area A_1 Liquid   Area A_2

‹ Active Learning Guide

EXAMPLE 10.2 Lifting a car with one hand

A hydraulic lift similar to the press described above has a small piston with surface area 0.0020 m² and a larger piston with surface area 0.20 m². Piston 2 and the car placed on piston 2 have a combined mass of 1800 kg. What is the minimal force that piston 1 needs to exert on the fluid to slowly lift the car?

Sketch and translate The situation is similar to that shown in Figure 10.6. We need to find $F_{1\,on\,L}$ so the fluid exerts a force great enough to support the mass of the car and piston 2. The hydraulic lift Eq. (10.2) should then allow us to determine $F_{1\,on\,L}$.

Simplify and diagram Assume that the levels of the two pistons are the same and that the car is being lifted at constant velocity. Use the force diagram for the car and piston 2 (see below diagram) and Newton's second law to determine $F_{L\,on\,2}$. Note that the force that the liquid exerts

Liquid under the large-area piston 2 supports the car and piston.

$\vec{F}_{1\,on\,L}$

$\vec{F}_{L\,on\,Car + Piston\,2}$

$\vec{F}_{E\,on\,Car + Piston\,2}$

Piston 1

Piston 2

Liquid

$A_1 = 0.0020\ \text{m}^2$ $A_2 = 0.20\ \text{m}^2$
$F_{1\,on\,L} = ?$ $m_{Car + Piston\,2} = 1800\ \text{kg}$

on the large piston 2 $F_{L\,on\,2}$ is equal in magnitude to the force that piston 2 and the car exert on the liquid $F_{2\,on\,L}$, which equals the downward gravitational force that Earth exerts on the car and piston: $F_{2\,on\,L} = m_{Car+piston}g$.

Represent mathematically We rewrite the hydraulic lift Eq. (10.2) to determine the unknown force:

$$F_{1\,on\,L} = \left(\frac{A_1}{A_2}\right)F_{2\,on\,L} = \left(\frac{A_1}{A_2}\right)m_{Car+piston}g$$

Solve and evaluate

$$F_{1\,on\,L} = \frac{(0.0020\ \text{m}^2)}{(0.20\ \text{m}^2)}\,[(1800\ \text{kg})(9.8\,\text{N/kg})] = 180\ \text{N}.$$

That is the force equal to lifting an object of mass 18 kg, which is entirely possible for a person. The units are also consistent.

Try it yourself: If you needed to lift the car about 0.10 m above the ground, what distance would you have to push down on the small piston? Specify the assumptions you made.

Answer: 10 m. Since liquids are essentially incompressible, the total volume of the liquid won't change significantly. The downward pushing distance (10 m) times the area of piston 1 equals the upward lifting distance (0.10 m) times the area of piston 2—the volume changes are equal. The small piston must push down significantly farther than the large piston will rise.

Review Question 10.2 If you poke many small holes in a closed toothpaste tube and squeeze it, the paste comes out equally from all holes. Why?

10.3 Pressure variation with depth

Pascal's first law states that an increase in the pressure in one part of an enclosed fluid results in an increase at all other parts of the fluid. Does that mean that the pressure is the same throughout a fluid—for example, in a vertical column of fluid? To test this hypothesis, consider another experiment with a water bottle. This time, poke holes vertically along one side of the bottle. Place tacks in the holes, and fill the bottle with water. Leave the cap off. If the pressure is the same throughout the liquid, when the tacks are removed the water should come out of each hole in the same arcs, as shown in **Figure 10.7a**.

Figure 10.7 Water seems to be pushed harder from holes deeper in the water.

(a) Predicted   **(b)** Observed

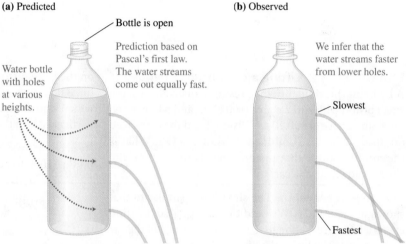

Bottle is open

Prediction based on Pascal's first law. The water streams come out equally fast.

Water bottle with holes at various heights.

We infer that the water streams faster from lower holes.

Slowest

Fastest

Active Learning Guide ›

However, when the tacks are removed, we observe that the water squirts out in a wider arc at the bottom than at the top (Figure 10.7b). Should we abandon Pascal's first law now because the prediction based on it did not match the outcome of the experiment? In such cases, scientists do not immediately throw out the principle but first examine the additional assumptions that were used to make the prediction. In our first experiment with the water bottle, we did not consider the impact of poking holes at different heights. Maybe this was an important factor in the experiment. Let's investigate this in Observational Experiment **Table 10.2**.

OBSERVATIONAL EXPERIMENT TABLE

10.2   How does the location of the holes affect the streams leaving the holes?

Observational experiment	Analysis
Experiment 1. Place two tacks on each side of a plastic bottle, one hole above the other, and fill the bottle with water above the top tack. Remove the tacks. Water comes out on the left and right and the stream from the lower holes shoots farther.	There must be greater pressure inside than outside. The pressure must be greater at the bottom holes than at the top holes.

Observational experiment	Analysis
Experiment 2. Repeat Experiment 1 but this time fill the bottle with water to the same distance above the bottom tack as it was filled above the top tack in Experiment 1. Remove the tacks. The stream comes out the bottom holes with the same arc as it came out of the top holes in Experiment 1.	The total water depth seems not to matter, just the height of the water above the hole.
Experiment 3. Repeat Experiment 1 using a thinner bottle with the water level initially the same distance above the top tack as it was in Experiment 1. Remove the tacks. The water streams are identical to those in Experiment 1.	Because the water comes out in exactly the same arc in a bigger bottle and in a smaller bottle when the water level above the top tack is at the same height, we can conclude that the mass of the water in the bottle does not affect the pressure.

Patterns

The stream shape at a particular level:
■ Depends on the height of the water above the hole.
■ Is the same in different directions at the same level.
■ Does not depend on the amount of liquid (volume or mass) above the hole (just the height of the water above the hole).
■ Does not depend on the amount (mass or volume) or depth of the water below the hole.

From the patterns above we reason that the pressure of the liquid at the hole depends only on the *height* of the liquid above the hole, and not on the *mass* of the liquid above. We also see that the pressure at a given depth is the same in all directions. This is consistent with the experience you have when you dive below the surface of the water in a swimming pool, lake, or ocean. The pressure on your ears depends only on how deep you are below the surface. Pascal's first law fails to explain this pressure variation at different depths below the surface. Now we need to understand why the pressure varies with depth and to devise a rule to describe this variation quantitatively.

Why does pressure vary at different levels?

To explain the variation of pressure with depth we can use an analogy of stacking ten books on a table (**Figure 10.8a**). Imagine that each book is a layer of water in a cylindrical tube (see Figure 10.8b). Consider the pressure (force per unit area) from above on the top surface of each book. The only force exerted on the top surface of the top book is due to air pushing down from above. However, there are in effect two forces exerted on the top surface of the second book: the force that the top book exerts on it (equal in magnitude to the weight of the book) plus the force exerted by the air on the top book. The top surface of the bottom book in the stack must balance the force exerted by the nine books above it (equal in magnitude to the weight of nine books) plus the

Figure 10.8 Pressure increases with depth.

(a)

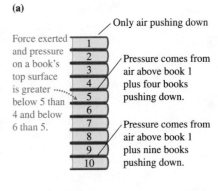

Force exerted and pressure on a book's top surface is greater below 5 than 4 and below 6 than 5.

Only air pushing down
Pressure comes from air above book 1 plus four books pushing down.
Pressure comes from air above book 1 plus nine books pushing down.

(b)

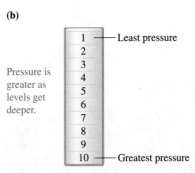

Pressure is greater as levels get deeper.

Least pressure
Greatest pressure

Figure 10.9 Water coming from holes on the side of a bottle behaves differently if the bottle is closed at the top.

(a)

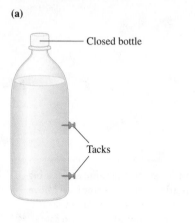

Closed bottle

Tacks

(b)

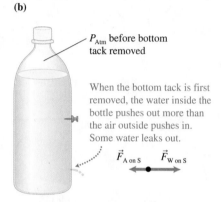

P_{Atm} before bottom tack removed

When the bottom tack is first removed, the water inside the bottle pushes out more than the air outside pushes in. Some water leaks out.

$\vec{F}_{A \text{ on } S}$ $\vec{F}_{W \text{ on } S}$

(c)

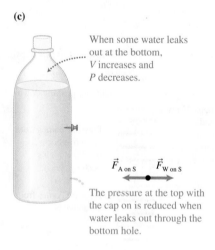

When some water leaks out at the bottom, V increases and P decreases.

$\vec{F}_{A \text{ on } S}$ $\vec{F}_{W \text{ on } S}$

The pressure at the top with the cap on is reduced when water leaks out through the bottom hole.

(d)

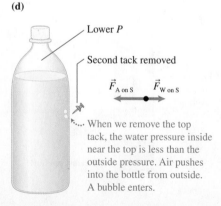

Lower P

Second tack removed

$\vec{F}_{A \text{ on } S}$ $\vec{F}_{W \text{ on } S}$

When we remove the top tack, the water pressure inside near the top is less than the outside pressure. Air pushes into the bottle from outside. A bubble enters.

pressure force exerted by the air on the top book. So the pressure increases on the top surface of each book in the stack as we go lower in the stack.

Similar reasoning applies for the liquid-filled tube divided into a number of imaginary thin layers in Figure 10.8b. Air pushes down on the top layer. The second layer balances the weight of the top layer plus the force exerted by the air pushing down on the top layer, and so on. The pressure is lowest at the top of the fluid and greatest at the bottom.

Note that, at each layer, the pressure is the same in all directions. If we could take a pressure sensor and place it inside the container of water, the readings of the sensor would be the same independent of the orientation of the sensor as long as its depth remains the same.

This helps us understand why a sealed empty water bottle on an airplane collapses as the plane descends from a higher elevation to a lower elevation. Even in a pressurized cabin, air pressure at the lower elevation is slightly greater than at higher elevation, and the increased pressure crushes the bottle.

We can test this "layer model" by making one small change in the experiment that involves the bottle with holes in it. So far we have kept the cap of the bottle open. What happens if we repeat the experiment, but this time with the cap closed? Let's start with a closed bottle full of water with two tacks, one above the other on one side. (**Figure 10.9a**) What does the model predict will happen if we remove the bottom tack from the hole? If we consider a tiny portion of water at the bottom hole as our system of interest, then the acceleration of that system can be either out of the bottle, into the bottle, or zero. There are two forces exerted on the water portion (we will call it the system S); the force exerted by the outside air pushing inward $\vec{F}_{A \text{ on } S}$ and the force of the inside water pushing outward $\vec{F}_{W \text{ on } S}$ (Figure 10.9b). If one of the forces is smaller, then the water will either accelerate outward and we will see the bottle leaking, or it will accelerate inward and we will see bubbles of air coming in.

What does the model predict will happen? Closing the bottle traps a little bit of air above the water. The air is at atmospheric pressure since the bottle was originally open. Inside the water at the position of the bottom tack, in addition to the pressure due to air in the bottle above pushing down on the water, there is additional pressure from the water pressing downward. At the level of the lower tack, the force that the water exerts on the water at the hole (the system) should be greater than the force exerted by the air outside the hole on that same water. If you remove that tack, the water should accelerate outward.

However, when water comes out of the hole, the volume available to the air at the top of the closed bottle increases. Because the bottle is closed, there is no way for air to enter the bottle. As a result, the air expands to fill the larger volume and the air pressure decreases. This pressure decrease spreads throughout the fluid, decreasing the magnitude of the force exerted by the water on the system. As water drains out, eventually the forces exerted by the water and the air on the system balance (Figure 10.9c). Thus, the model predicts that some water will come out of the bottle when we remove one tack, but then the leaking will soon stop. This is exactly what happens when we perform the experiment.

Now, what happens if you remove the higher tack while the bottom hole remains open? Try to predict the outcome before continuing. After you make your prediction, look at Figure 10.9d. Remember that water flow from the bottom tack stops when the inside pressure at the lower tack equals the outside air pressure pushing in. Thus the pressure in the water at the level of the higher opening will be less than the air pressure outside. Thus, the outside air will exert a greater inward force on water at the higher opening than the outward force that water in the bottle exerts on that same portion. Air flows into the top hole, forming bubbles, which float to the top of the closed bottle. The water pressure increases and more water squirts out of the lower hole, reducing pressure at the top of the bottle again. More air comes in the upper hole, and another squirt of water comes out of the bottom hole. The bottle drains in small squirts from the bottom hole. Is this what you predicted?

How can we quantify pressure change with depth?

We know that pressure increases with depth, but does the pressure depend on the depth linearly, quadratically, or in some other way? Consider the shaded cylinder C of water shown in **Figure 10.10a** as our system of interest. The walls on opposite sides of the cylinder push inward, exerting equal-magnitude and oppositely directed forces—the forces exerted by the sides cancel. What about the forces exerted by the water above and below? If the pressure at elevation y_2 is P_2 and the cross-sectional area of the cylinder is A, then the fluid above pushes down, exerting a force of magnitude $F_{\text{fluid above on C}} = P_2 A$ (Figure 10.10b). Similarly, fluid from below the shaded section of fluid at elevation y_1 exerts on the cylinder an upward force of magnitude $F_{\text{fluid below on C}} = P_1 A$. Earth exerts a third force on the shaded cylinder $\vec{F}_{\text{E on C}}$ equal in magnitude to $m_C g$, where m_C is the mass of the liquid in the cylinder. Since the liquid is not accelerating, these three forces add to zero. Choosing the y-axis pointing up, we have

$$\Sigma F_y = (-F_{\text{fluid above on C}}) + F_{\text{fluid below on C}} + (-m_C g) = 0$$

Substituting the earlier expressions for the forces, we have

$$(-P_2 A) + P_1 A + (-m_C g) = 0$$

The mass of the fluid in the shaded cylinder is the product of the fluid's density and the volume of the cylinder:

$$m_C = \rho_{\text{fluid}} V = \rho_{\text{fluid}} [A(y_2 - y_1)]$$

Substituting this expression for the mass in the above expression for the forces, we get

$$-P_2 A + P_1 A - \rho_{\text{fluid}} [A(y_2 - y_1)] g = 0$$

Divide by the common A in all of the terms and rearrange to get

$$P_1 = P_2 + \rho_{\text{fluid}}(y_2 - y_1)g$$

This is Pascal's second law. As we see, pressure varies linearly with depth.

Pascal's second law—variation of pressure with depth The pressure P_1 in a static fluid at position y_1 can be determined in terms of the pressure P_2 at position y_2 as follows:

$$P_1 = P_2 + \rho_{\text{fluid}}(y_2 - y_1)g, \tag{10.3}$$

where ρ_{fluid} is the fluid density, assumed constant throughout the fluid, and $g = 9.8 \text{ N/kg}$. The positive y-direction is up.

TIP When using Pascal's second law [Eq. (10.3)], picture the situation and be sure to include a vertical y-axis that points upward and has a defined origin, or zero point. Then choose the two points of interest and identify their vertical y-positions relative to the axis. This lets you relate the pressures at those two points.

In order to test whether the pressure of a fluid depends on the depth and not on the mass of the fluid, Blaise Pascal filled a barrel with water and inserted a long, narrow vertical tube into the water from above. He then sealed the barrel (see **Figure 10.11**). He predicted that when he filled the tube with water, the barrel would burst even though the mass of water in the thin tube was small, because the pressure of the water in the barrel would depend on the height, not the mass, of the water column. The barrel burst, matching Pascal's prediction: thus supporting the idea that the height of the liquid above, not the amount of water above, determined the pressure.

Figure 10.10 Using Newton's second law to determine how fluid pressure changes with the depth in the fluid.

(a)

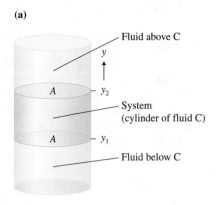

(b)

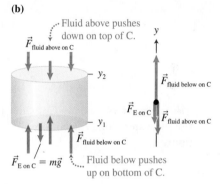

‹ Active Learning Guide

Figure 10.11 Pascal tests his second law.

Pascal burst a barrel by filling a long, narrow tube with water.

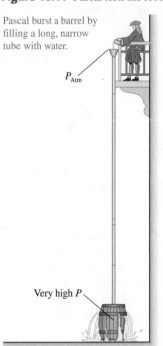

Equation (10.3) applies to any static fluid, whether liquid or gas. This explains why air pressure is greater at the bottom of a mountain than at the top.

CONCEPTUAL EXERCISE 10.3 Pascal's paradox

Blaise Pascal came up with a paradoxical situation involving the apparatus shown below. When he poured water into the apparatus, the water level was the same in all parts of the apparatus despite differences in the shapes of the apparatus in different parts and the mass of water in each part. Explain.

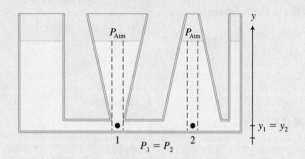

Sketch and translate Consider the horizontal portion of water along the bottom of the apparatus (see the figure above).

Simplify and diagram Since this water is stationary, the pressure at all points along the bottom must be the same. Consider in particular the water at positions 1 and 2. These pressures can be equal only if the columns of water above points 1 and 2 are the same height. Thus, each part of Pascal's unusually shaped device fills to the same level.

Try it yourself: Use your understanding of pressure to explain why our ears pop when we climb a mountain or during takeoff in an airplane.

Answer: When we begin climbing a mountain, the air pressure inside P_{inside} the eardrum and the air pressure P_1 of outside air pushing in on the eardrum are equal. At the top of the mountain, the air pressure P_2 outside is less than the air pressure P_{inside} inside. We can equalize these pressures by venting a little air from the middle ear. That's the popping sensation. The same thing happens when you are in an ascending airplane.

QUANTITATIVE EXERCISE 10.4 Pop your ears

If your ears did not pop, then what would be the net force exerted by the inside and outside air on your eardrum at the top of a 1000-m-high mountain? You start your hike from sea level. The area of your eardrum is 0.50 cm². The density of air at sea level at standard conditions is 1.3 kg/m³. Assume the air density remains constant during the hike. The situation at the start at $y_1 = 0$ and at the end of the hike at $y_2 = 1000$ m is sketched below.

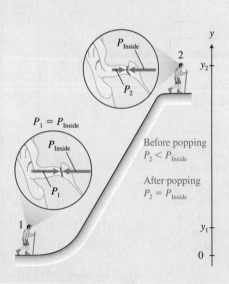

Represent mathematically We use an upward pointing vertical y-axis with the origin at sea level. Assume that the air pressure inside the eardrum remains constant at its sea level value $P_{inside} = P_1$. The air pressure difference between the top of the mountain and sea level would be

$$P_2 - P_1 = \rho(y_1 - y_2)g$$

When at 1000 m above sea level, the outside air exerts a lower pressure P_2 on the eardrum. So the net force exerted by the air on the drum is the pressure difference of air pushing out $P_{inside} = P_1$ and air pushing in P_2 times the area of the eardrum: $F_{net\ air\ on\ drum} = (P_1 - P_2)A$. This pressure difference can be determined using Pascal's second law.

Solve and evaluate

$$P_1 - P_2 = \rho(y_2 - y_1)g$$
$$= (1.3\ \text{kg/m}^3)(1000\ \text{m} - 0)(9.8\ \text{N/kg})$$
$$= +0.13 \times 10^5\ \text{N/m}^2$$

This is 0.13 atm!

$$F_{net\ air\ on\ drum} = (P_1 - P_2)A$$
$$= (0.13 \times 10^5\ \text{N/m}^2)(0.5\ \text{cm}^2)$$
$$\times (1\ \text{m/100\ cm})^2$$
$$= +0.6\ \text{N}$$

The net force is exerted outward and is about half the gravitational force that Earth exerts on an apple. No wonder it can hurt until you get some air out of that middle ear!

Try it yourself: Determine the difference in water pressure on your ear when you are 1.0 m underwater compared to when you are at the surface. The density of water is 1000 kg/m³.

Answer: +9800 N/m² greater pressure when under the water, or 0.1 atm.

Review Question 10.3 Pascal's first law says that an increase in pressure in one part of an enclosed liquid results in an increase in pressure throughout all parts of that liquid. Why then does the pressure differ at different heights?

10.4 Measuring atmospheric pressure

We can now use Pascal's second law to develop a method for measuring atmospheric air pressure.

Measuring atmospheric pressure

During Galileo's time (1600s), suction pumps were used to lift drinking water from wells and to remove water from flooded mines. The suction pump was like a long syringe. The pump consisted of a piston in a long cylinder that pulled up water (**Figure10.12a**). Such pumps could lift water a maximum of 10.3 m. Why 10.3 meters?

Evangelista Torricelli (1608–1647), one of Galileo's students, hypothesized that the pressure of the air in the atmosphere could explain the limit to how far water could be lifted. Torricelli did not know of Pascal's second law, which was published in the year that Torricelli died. However, it is possible that Torricelli's work influenced Pascal. Let's analyze the situation shown in Figure 10.12b.

Consider the pressure at three places: point 1, at the water surface in the pool outside the cylinder; point 2, at the same elevation only inside the cylinder; and point 3, in the cylinder 10.3 m above the pool water level. The pressure at point 1 is atmospheric pressure. The pressure at point 2, according to Pascal's second law, is also atmospheric pressure, since it is at the same level as point 1. To get the water to the 10.3-m maximum height, the region above the water surface inside the cylinder and under the piston must be at the least possible pressure—essentially a vacuum. Thus, we assume that the pressure at point 3 is zero.

Now we will use Eq. (10.3) with $y_3 - y_2 = 10.3$ m to predict the pressure P_2:

$$P_2 = P_3 + \rho_{water}(y_3 - y_2)g = 0 + (1.0 \times 10^3 \text{ kg/m}^3)(10.3 \text{ m} - 0)(9.8 \text{ N/kg})$$
$$= 1.01 \times 10^5 \text{ N/m}^2$$

This number is exactly the value of the atmospheric pressure that we encountered in our discussion of gases (in Chapter 9). The atmospheric pressure pushing down on the water outside the tube can push water up the tube a maximum of 10.3 m if there is a vacuum (absence of any matter) above the water in the tube.

At Torricelli's time the value of normal atmospheric pressure was unknown, so the huge number that came out of this analysis surprised Torricelli. Not believing the result, he tested it using a different liquid—mercury. Mercury is 14 times denser than water ($\rho_{Hg} = 13{,}600$ kg/m³); hence, the column of mercury should rise only 1/14 times as high in an evacuated tube. However,

Figure 10.12 A piston pulled up a cylinder causes water to rise to a maximum height of 10.3 m.

(a)

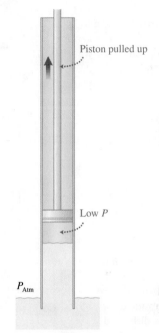

Piston pulled up

Low P

P_{Atm}

Atmospheric pressure P_{Atm} pushes water up the cylinder.

(b)

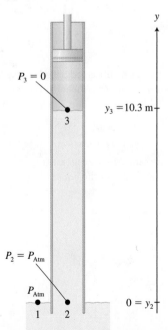

$P_3 = 0$

$y_3 = 10.3$ m

3

$P_2 = P_{Atm}$

P_{Atm}

$0 = y_2$

1   2

instead of using a piston to lift mercury, Torricelli devised a method that guaranteed that the pressure at the top of the column was about zero. Consider Testing Experiment **Table 10.3**.

TESTING EXPERIMENT TABLE

10.3   Testing Torricelli's hypothesis using mercury.

Testing experiment	Prediction based on Torricelli's hypothesis that atmospheric pressure limits the height of the liquid in a suction pump	Outcome
■ Torricelli filled a long glass tube closed at one end with mercury. 	Mercury should start leaking from the tube into the dish. When it leaks, it leaves an empty evacuated space at the top of the tube. It will leak until the height of the mercury column left in the tube produces the same pressure as the atmosphere at the bottom of the column at position 1. The height of the mercury in the tube should be $$y_2 - y_1 = \frac{P_1 - P_2}{\rho_{mercury}g}$$ $$= \frac{(1.01 \times 10^5\,\text{N/m}^2 - 0)}{(13.6 \times 10^3\,\text{kg/m}^3)(9.8\,\text{N/kg})}$$ $$= 0.76\,\text{m}$$	Torricelli observed some mercury leaking from the tube and then the process stopped. He measured the height of the remaining mercury to be $0.76\,\text{m} = 760\,\text{mm}$, in agreement with the prediction.
■ He put his finger over the open end and placed it upside down in a dish filled with mercury. He then removed his finger. 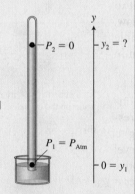 Predict what he observed based on the hypothesis that atmospheric pressure limits the height of the liquid in a suction pump.		

Conclusion

The outcome of the experiment was consistent with the prediction based on Torricelli's hypothesis that atmospheric pressure limits the height of the liquid being lifted in a suction pump. Thus the hypothesis is supported by evidence.

Torricelli also used his understanding of pressure and fluids to predict that in the mountains, where the atmospheric pressure is lower, the height of the mercury column should be lower. Experiments have shown that the mercury level indeed decreases at higher elevation. These experiments supported the explanation that atmospheric air pushes the liquids upward into the tubes.

Torricelli's apparatus with the mercury tube became a useful device for measuring atmospheric pressure—called a barometer. However, since mercury is toxic, the Torricelli device has since been replaced by the aneroid barometer (described in Chapter 9).

> **TIP** We now understand why pressure is often measured and reported in mm Hg and why atmospheric pressure is 760 mm Hg. The atmospheric pressure ($101,000 \, \text{N/m}^2$) can push mercury of density ($13,600 \, \text{kg/m}^3$) 760 mm up a column.

Diving bell

Our understanding of atmospheric pressure allows us to explain many simple experiments that lead to important practical applications. For example, have you ever submerged a transparent container upside down under water? If you do, you will see that at first little water enters the container; then as the inverted container is pushed deeper into the water, more water enters the container. One practical application of this phenomenon is a diving bell—a large bottomless chamber lowered under water with people and equipment inside. Divers use the diving bell to take a break and refill on oxygen.

‹ Active Learning Guide

EXAMPLE 10.5 Diving bell

The bottom of a 4.0-m-tall cylindrical diving bell is at an unknown depth underwater. The pressure of the air inside the bell is 2.0 atm (it was 1 atm before the bell entered the water). The average density of ocean water is slightly higher than fresh water, $\rho_{\text{ocean water}} = 1027 \, \text{kg/m}^3$. How high is the water inside the bell and how deep is the bottom of the bell under the water?

Sketch and translate We sketch the situation below. We want to find the height h of the water in the bell and the depth d of the bottom of the bell under the water.

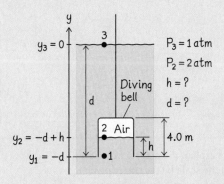

Simplify and diagram Assume that Boyle's law (PV is constant for a constant temperature gas) applies to the air inside the bell and use it to relate the state of the air inside the bell before it is submerged to its state after being submerged. This will let us find the ratio of its volumes

before and after, and from that we can determine the height of the water inside the bell. Once we have that result, we can use Pascal's second law to determine how far the bottom of the bell is under the water.

Represent mathematically Use Boyle's law for the process of submerging the bell, where the initial state is just before the bell starts to be submerged in the water and the final state is when submerged to some unknown depth. We have:

$$P_i V_i = P_f V_f$$

Next, use Pascal's second law to determine the depth of the diving bell. The water surface $y_3 = 0$ will be the origin, with the positive direction pointing upward. The bottom of the bell will be at $y_1 = -d$. Compare the pressure at the ocean surface P_3 to the pressure at the water surface inside the bell P_2.

$$P_2 = P_3 + \rho(y_3 - y_2)g$$

Solve and evaluate The air volume inside the diving bell is half of what it was before entering the water.

$$V_f = \frac{P_i}{P_f}V_i = \frac{1.0 \, \text{atm}}{2.0 \, \text{atm}}V_i = \frac{1}{2}V_i$$

Thus, the bell is half full of water, which means the height of the water level inside is

$$h = \frac{1}{2}(4.0 \, \text{m}) = 2.0 \, \text{m}$$

(continued)

Rearrange Pascal's second law to solve for the position $y_2 = -d + h$ of the water level inside the bell:

$$y_2 = y_3 - \frac{P_2 - P_3}{\rho g}$$

$$= 0 - \frac{(2.02 \times 10^5 \, \text{Pa} - 1.01 \times 10^5 \, \text{Pa})}{(1027 \, \text{kg/m}^3)(9.8 \, \text{N/kg})}$$

$$= 0 - 10.0 \, \text{m}$$

But $y_2 = -d + h = -d + 2.0 \, \text{m}$. Set these two expressions for y_2 equal to each other:

$$y_2 = -d + 2.0 \, \text{m} = 0 - 10.0 \, \text{m}$$

Thus, $d = 12.0 \, \text{m}$. The position of the bottom of the bell is 12.0 m below the ocean's surface.

Try it yourself: Suppose the air pressure in the bell is 3.0 atm. Now how high is water in the bell and how deep is the bottom of the bell?

Answer: The water is 2.7 m high in the bell (1.3 m of air at top) and the bottom is 23 m below the water surface.

Review Question 10.4 What does it mean if atmospheric pressure is 760 mm of mercury?

10.5 Buoyant force

Active Learning Guide >

Pascal's first law tells us that pressure changes in one part of a fluid result in pressure changes in other parts. Pascal's second law describes how the pressure in a fluid varies depending on the depth in the fluid. Do these laws explain why some objects float and others sink? Consider Observational Experiment **Table 10.4.**

OBSERVATIONAL EXPERIMENT TABLE

10.4   Effect of depth of submersion on a steel block suspended in water.

VIDEO 10.4

Observational experiment	Analysis
Experiment 1. Hang a 1.0-kg block from a spring scale. The force that the scale exerts on the block balances the downward force that Earth exerts on the block ($mg = (1.0 \, \text{kg})(9.8 \, \text{N/kg}) = 9.8 \, \text{N}$).  9.8 N	Force diagram for the block B: 
Experiment 2. Lower the block into a container of water, so it is partially submerged. The water level rises. The reading of the scale decreases. 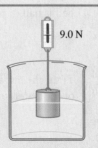 9.0 N	We explain the decreased reading on the scale by the water pushing upward a little on the block. 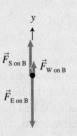

Observational experiment		Analysis
Experiment 3. Lower the same block into the container of water only to the point where the block is completely submerged. As the water level rises, the reading of the scale decreases.	8.0 N	The upward force exerted by the water increases.
Experiment 4. Lower the block into the container of water so that the block is completely submerged near the bottom. The water level and the reading of the scale do not change.	8.0 N	The upward force exerted by the water does not change once it is completely submerged.

Patterns

We notice two effects:

1. The level of the water in the container rises as more of the block is submerged in the water.
2. The scale reading decreases as more of the block is submerged. The water exerts an upward force on the block. The magnitude of this force depends on how much of the block is submerged. After it is totally submerged, the force does not change, even though the depth of submersion changes.

In Table 10.4, after the block is completely submerged, the scale reads 8.0 N instead of 9.8 N. Evidently, the water exerts a 1.8-N upward force on the block. What is the mechanism responsible for this force?

The magnitude of the force of fluid on a submerged object

Consider *only* the fluid forces exerted on the block shown in **Figure 10.13a**. The fluid pushes inward on the block from all sides, including the top and the bottom. The forces exerted by the fluid on the vertical sides of the block cancel, since the pressure at a specific depth is the same magnitude in all directions.

What about the fluid pushing down on the top and up on the bottom of the block? The pressure is greater at elevation y_1 at the bottom of the block than at elevation y_2 at the top surface of the block. Consequently, the force exerted by the fluid pushing up on the bottom is greater than the force exerted by the fluid pushing down on the top of the block. Arrows in Figure 10.13b represent the forces that the fluid exerts on the top and bottom of the block. The vector sum of these two fluid forces always points up and is called a buoyant force $\vec{F}_{\text{F on O}}$ (fluid on object).

To calculate the magnitude of the upward buoyant force $F_{\text{F on O}}$ exerted by the fluid on the block, we use Eq. (10.3) to determine the upward pressure P_1 of the fluid on the bottom surface of the block compared to the downward pressure P_2 of the fluid on the top surface (see Figure 10.13):

$$P_1 = P_2 + \rho_{\text{fluid}}(y_2 - y_1)g$$

Figure 10.13 A fluid exerts an upward buoyant force on the block.

(a)

The force $\vec{F}_{\text{FB on B}}$ exerted by the fluid below on the block is greater than the force $\vec{F}_{\text{FA on B}}$ exerted by the fluid above.

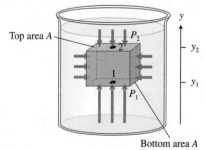

(b)

The upward force of the fluid on the bottom surface is greater than the downward force of the fluid on the top surface.

Active Learning Guide ➤

The magnitudes of the forces exerted by the fluid on the top and on the bottom of the object are the products of the pressure P and the area A of the top and bottom surfaces of the block:

$$P_1 A = P_2 A + \rho_{\text{fluid}}(y_2 - y_1)Ag$$

or

$$F_1 = F_2 + \rho_{\text{fluid}}(y_2 - y_1)Ag$$

The volume of the block is

$$V = A(y_2 - y_1)$$

where A is the cross-sectional area of the block and $(y_2 - y_1)$ is its height. Substitute this volume into the above force equation and rearrange it to get an expression for the magnitude of the total upward **buoyant force** $F_{\text{F on O}}$ that the fluid exerts on the block (the object of interest):

$$F_{\text{F on O}} = F_1 - F_2 = \rho_{\text{fluid}} gV$$

Note that for a totally submerged block, V is the volume of the block. However, when it is partially submerged, V is the volume of the submerged part.

We can now understand the results of the experiments in Table 10.4. When submerging the block further into the water, the scale reading decreased because the buoyant force was increasing. The scale reading stopped changing after the block was completely under the water. Once completely underwater, the submerged volume did not change; the upward buoyant force exerted by the fluid on the block remained constant. This equation is known as Archimedes' principle.

Active Learning Guide ➤

Archimedes' principle—the buoyant force A stationary fluid exerts an upward buoyant force on an object that is totally or partially submerged in the fluid. The magnitude of the force is the product of the fluid density ρ_{fluid}, the volume V_{fluid} of the fluid that is displaced by the object, and the gravitational constant g:

$$F_{\text{F on O}} = \rho_{\text{fluid}} V_{\text{fluid}} g \qquad (10.4)$$

TIP (1) If an object is completely submerged in the fluid (as in the case of the block), the volume used in Eq. (10.4) is just the volume of the object. However, if the object floats, the volume in the equation then equals the volume of space taken up by the object below the fluid's surface. (2) The derivation of Eq. (10.4) was for a solid cube, but the result applies to objects of any shape, though calculus is needed to establish that.

Review Question 10.5 Why does a fluid exert an upward force on an object submerged in it?

10.6 Skills for analyzing static fluid processes

In this section we adapt our problem-solving strategy to analyze processes involving static fluids. A general strategy is described on the left side of the table in Example 10.6 and illustrated on the right side for that specific problem.

PROBLEM-SOLVING STRATEGY **Analyzing Static Fluid Processes**

EXAMPLE 10.6 **Buoyant force exerted by air on a human**

Suppose your mass is 70.0 kg and your density is 970 kg/m³. If you could stand on a scale in a vacuum chamber on Earth's surface, the reading of the scale would be $mg = (70.0\text{ kg})(9.80\text{ N/kg}) = 686\text{ N}$. What will the scale read when you are completely submerged in air of density 1.29 kg/m³?

Sketch and translate

- Make a labeled sketch of the situation and choose the system of interest.
- Include all known information in the sketch and indicate the unknown(s) you wish to determine.

You are the system object.

The scale reads 686 N when in a vacuum. Your density is 970 kg/m³ and the density of air is 1.29 kg/m³. What does it read when you are submerged in air?

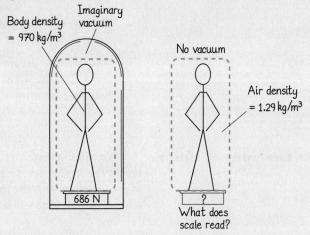

Simplify and diagram

- Indicate any assumptions you are making.
- Identify objects outside the system that interact with the system object(s).
- Construct a force diagram for the system, including a vertical coordinate axis. The buoyant force is just one of the forces included in the diagram.

Assume that the air density is uniform.

Three objects exert forces on you. Earth exerts a downward gravitational force $F_{\text{E on Y}} = mg = 686\text{ N}$. The air exerts an upward buoyant force $F_{\text{A on Y}} = \rho_{\text{air}} g V_{\text{you}}$. The scale exerts an unknown upward normal force of magnitude $N_{\text{S on Y}}$.

Represent mathematically

- Use the force diagram to help apply Newton's second law in component form.
- Use the expression for the buoyant force and the definitions of pressure and density if needed.

The y-component form of Newton's second law for your body with zero acceleration is (assuming the upward direction as positive)

$$0 = N_{\text{S on Y}} + F_{\text{A on Y}} + (-F_{\text{E on Y}})$$

or

$$N_{\text{S on Y}} = +F_{\text{E on Y}} - F_{\text{A on Y}}$$

The buoyant force that the air exerts on your body has magnitude $F_{\text{A on Y}} = \rho_{\text{air}} V_{\text{you}} g$. The volume of your body is $V_{\text{you}} = (m/\rho_{\text{body}})$. The magnitude of the buoyant force that the air exerts on you is

$$F_{\text{A on Y}} = \rho_{\text{air}}\left(\frac{m}{\rho_{\text{body}}}\right)g = mg\left(\frac{\rho_{\text{air}}}{\rho_{\text{body}}}\right)$$

(continued)

Thus the reading of the scale should be:

$$N_{\text{S on Y}} = mg - mg\left(\frac{\rho_{\text{air}}}{\rho_{\text{body}}}\right)$$

Solve and evaluate

- Insert the known information and solve for the desired unknown.

- Evaluate the final result in terms of units, reasonable magnitude, and whether the answer makes sense in limiting cases.

$$N_{\text{S on Y}} = +686\,\text{N} - (686\,\text{N})\frac{1.29\,\text{kg/m}^3}{970\,\text{kg/m}^3} = 685\,\text{N}$$

According to Newton's third law, the force that you exert on the scale $N_{\text{Y on S}}$ is equal in magnitude to the force the scale exerts on you.

The reading of the scale is actually 0.1% less when you step on the scale in air—not a big deal. We can usually neglect air's buoyant force. Notice that the atmospheric air pushes up on objects, and not down.

Try it yourself: What will the scale read if you weigh yourself in a swimming pool with your body completely submerged?

Answer: 0 N. Because you are less dense than water, the buoyant force exerted by the water on you will completely support you and the scale will not push upward on you at all.

Active Learning Guide >

We will use these strategies to analyze several more situations in the chapter. In the next example, it might not be obvious how to arrive at the answer to the question with the information provided. Following the suggested problem-solving routine will help you arrive at a solution.

EXAMPLE 10.7   Is the crown made of gold?

You need to determine if a crown is made from pure gold or some less valuable metal. From Table 10.1 you know that the density of gold is 19,300 kg/m³. You find that the force that a string attached to a spring scale exerts on the crown is 25.0 N when the crown hangs in air and 22.6 N when the crown hangs completely submerged in water.

Sketch and translate We draw a sketch of the situation and label the givens. If you could measure the mass m_C of the crown and its volume V_C, you could calculate the density $\rho_C = m_C/V_C$ of the crown—it should be 19,300 kg/m³. You can determine the mass of the crown easily from the measurement of the scale when the crown hangs in air. But how can you determine the volume from the given information? Crowns have irregular shapes, and it would be difficult to determine its volume by simple measurements and calculations.

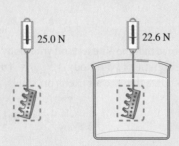

Simplify and diagram Let's just follow the recommended strategy and see what happens. First, we draw force diagrams for the crown hanging in air and again when hanging in water. When the crown is in air, the upward force exerted by the string attached to the spring scale balances the downward force exerted by Earth. We ignore the buoyant force that air exerts on the crown when hanging in air, since it will be very small in magnitude compared with the other forces exerted on the crown. When the crown is in water, the upward force exerted by the string (the force measured by the scale) and the upward buoyant force that the water exerts on the crown combine to balance the downward gravitational force that Earth exerts on the crown.

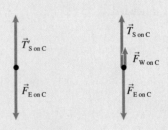

Represent mathematically Since the crown is in equilibrium, the forces exerted on it must add to zero in both cases. When the crown is hanging in air, the vertical component form of Newton's second law is

$$0 = \Sigma F_y = T'_{\text{S on C}} + (-F_{\text{E on C}})$$

where $T'_{\text{S on C}}$ is the 25.0-N string tension force exerted on the crown when it is suspended in air. Thus,

$$F_{\text{E on C}} = m_{\text{C}}g = 25.0\,\text{N}$$

or

$$m_{\text{C}} = \frac{25.0\,\text{N}}{9.8\,\text{N/kg}} = 2.55\,\text{kg}$$

The vertical component form of Newton's second law when the crown hangs in water becomes (the upward direction is positive):

$$\Sigma F_y = T_{\text{S on C}} + F_{\text{W on C}} + (-F_{\text{E on C}}) = 0$$

where $T_{\text{S on C}} = 22.6\,\text{N}$ is the magnitude of the string tension force, and the buoyant force that the water exerts on the crown is $F_{\text{W on C}} = \rho_{\text{W}}V_{\text{C}}g$. Substituting in the above, we get

$$T_{\text{S on C}} + \rho_{\text{W}}V_{\text{C}}g - m_{\text{C}}g = 0$$

Solve and evaluate We see now that the last equation can be used to determine the volume of the crown:

$$V_{\text{C}} = \frac{m_{\text{C}}g - T_{\text{S on C}}}{\rho_{\text{W}}g}$$

$$= \frac{25.0\,\text{N} - 22.6\,\text{N}}{(1000\,\text{kg/m}^3)(9.8\,\text{N/kg})} = 0.000245\,\text{m}^3$$

We now know the crown mass and volume and can calculate its density:

$$\rho = \frac{m}{V} = \frac{2.55\,\text{kg}}{0.000245\,\text{m}^3} = 10,400\,\text{kg/m}^3$$

Oops! Since $10,400\,\text{kg/m}^3$ is much less than the $19,300\,\text{kg/m}^3$ density of gold, the crown is not made of pure gold. The goldsmith must have combined the gold with some less expensive metal.

Try it yourself: What is the density of the crown if the scale reads 0 when submerged in water?

Answer: $1000\,\text{kg/m}^3$.

Review Question 10.6 Two objects have the same volume, but one is heavier than the other. When they are completely submerged in oil, on which one does the oil exert a greater buoyant force?

‹ **Active Learning Guide**

10.7 Buoyancy: Putting it all together

As we learned in Section 10.1, whether an object floats or sinks depends on its density relative to the density of the fluid. The reason for this lies in the interactions of the object with the fluid and Earth. Specifically, the magnitude of the buoyant force is $F_{\text{F on O}} = \rho_{\text{fluid}}V_{\text{submerged part}}g$, and the magnitude of the force exerted by Earth is $F_{\text{E on O}} = \rho_{\text{object}}V_{\text{object}}g$. These forces are exerted in the opposite directions. The relative magnitudes of the forces and consequently the relative magnitudes of the densities of the fluid and the object determine what happens to the object when placed in the fluid.

■ If the object's density is less than that of the fluid $\rho_{\text{object}} < \rho_{\text{fluid}}$, then $\rho_{\text{object}}V_{\text{object}}g < \rho_{\text{fluid}}V_{\text{object}}g$; the object floats *partially* submerged since the buoyant force can balance the gravitational force with less than the entire object below the surface of the fluid.

■ If the densities are the same $\rho_{\text{object}} = \rho_{\text{fluid}}$, then $\rho_{\text{object}}V_{\text{object}}g = \rho_{\text{fluid}}V_{\text{object}}g$; the sum of the forces exerted on the object is zero and it remains wherever it is placed totally submerged at any depth in the fluid.

■ If the object is denser than the fluid $\rho_{\text{object}} > \rho_{\text{fluid}}$, then $\rho_{\text{object}}V_{\text{object}}g > \rho_{\text{fluid}}V_{\text{object}}g$; the magnitude of the gravitational force is always greater than the magnitude of the buoyant force. The object sinks at increasing speed until it reaches the bottom of the container.

These cases show that by changing the density of an object relative to the density of the fluid, the object can be made to float or sink in the same fluid. In this section we investigate this phenomenon and its many practical applications.

How do submarines manage to sink and then rise in the water?

A submarine's density increases when water fills its compartments. With enough water in the compartments, the submarine's density is greater than that of the water outside, and it sinks. When the water is pumped out, leaving behind empty compartments, the submarine's density decreases. With enough air in the compartments, its density is less than that of the outside water, and the submarine rises toward the surface.

Building a stable ship

For years ships were made of wood. In the middle of the 17th century, people decided to try building metal ships. Many thought that this idea was absurd: Iron is denser than water and an iron boat would certainly sink. In 1787 British engineer John Wilkinson succeeded in building the first iron ship that did not sink. Since the middle of the 19th century, large ships have been made primarily of steel, which is less dense that iron but much denser than water. These ships can float because part of the volume of a ship is filled with air, which reduces the average density of the ship to a density lower than that of water.

EXAMPLE 10.8 Should we take this trip?

An empty life raft of cross-sectional area 2.0 m × 3.0 m has its top edge 0.36 m above the waterline. How many 75-kg passengers can the raft hold before water starts to flow over its edges? The raft is in seawater of density 1025 kg/m³.

Sketch and translate We make a sketch of the unloaded raft. As people get on the raft, it sinks deeper into the water, and the upward buoyant force increases until the raft reaches a maximum submerged volume, when the maximum number of people are on board. The maximum submerged volume is $V_{\text{submerged}} = 2.0\text{ m} \times 3.0\text{ m} \times 0.36\text{ m} = 2.16\text{ m}^3$. We need to determine the maximum buoyant force the seawater can exert on the raft and then decide how to convert this into the number of passengers the raft can hold. The raft and the passengers are our system of interest.

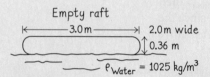

Empty raft
3.0 m
2.0 m wide
0.36 m
$\rho_{\text{Water}} = 1025$ kg/m³

Simplify and diagram We have no information about the mass of the raft; thus we will assume it is negligible. We draw a sketch of the filled raft and a force diagram for the raft with passengers (the system). The vertical

axis points up. There are two forces exerted on the system: the upward force exerted by the water $\vec{F}_{\text{W on S}}$ of magnitude $\rho_{\text{water}}gV_{\text{submerged}}$ and the downward force exerted by Earth $\vec{F}_{\text{E on S}}$. The magnitude of the force Earth exerts on N people is $N_{\text{people}}m_{\text{person}}g$. As the system is in equilibrium, the net force exerted on it is zero.

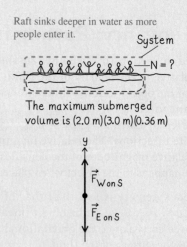

Raft sinks deeper in water as more people enter it.

System

N = ?

The maximum submerged volume is (2.0 m)(3.0 m)(0.36 m)

$\vec{F}_{\text{W on S}}$

$\vec{F}_{\text{E on S}}$

Represent mathematically Using the upward direction as positive, apply the vertical component form of Newton's second law:

$$\Sigma F_y = F_{\text{W on S}} + (-F_{\text{E on S}}) = 0$$

$$\rho_{\text{water}}gV_{\text{submerged}} - N_{\text{people}}m_{\text{person}}g = 0$$

Assuming that all people have the same mass, we find the number of people:

$$N_{people} = \frac{\rho_{water} V_{submerged} g}{m_{person} g} = \frac{\rho_{water} V_{submerged}}{m_{person}}$$

Solve and evaluate

$$N_{people} = \frac{\rho_{water} V_{submerged}}{m_{person}}$$

$$= \frac{(1025 \text{ kg/m}^3)(0.36 \text{ m} \times 2.0 \text{ m} \times 3.0 \text{ m})}{75 \text{ kg}}$$

$$= 29.5$$

The raft can precariously hold 29 passengers, which is a reasonable number. The number is inversely proportional to the mass of a person. This makes sense—the heavier the people, the fewer of them the raft should hold. The units, dimensionless, also make sense. We assumed that the raft has negligible mass. If we take the mass into account, the number of people will be smaller.

Try it yourself: Suppose that 10 people of average mass 80 kg entered the raft. Now how far would the water line be from the top of the raft?

Answer: The raft would sink 0.13 m into the water, and the water line would be 0.23 m below the top edge of the raft.

Making a ship or a raft float is only part of the challenge of building watercraft. Another problem is to maintain stable equilibrium for the ship, allowing it to right itself if it tilts to one side due to wind or rough seas. Refresh your knowledge of stable equilibrium (Section 7.6) before you read on.

Consider a floating bottle partially filled with sand. Earth exerts a gravitational force at the center of mass of the bottle (**Figure 10.14a**). The buoyant force exerted by the water on the bottle is effectively exerted at the geometrical center of the part of the bottle that is underwater, which equals the center of mass of the displaced water. If this point is above the center of mass of the bottle, then any slight tipping causes these forces to produce a torque that attempts to return the bottle to an upright position (see Figure 10.14b). However, if the geometrical center of the part of the bottle that is underwater is above the center of mass of the bottle, slight tipping causes the gravitational force to produce a torque that enhances the tipping—unstable equilibrium (Figure 10.14c).

So to build a good ship, make the average density of the ship less than the density of water, and make the ship's center of mass lower than the center of mass of the fluid the ship displaces. This is why ships have their cargo stored at the bottom.

Ballooning

Balloons used for transportation are filled with hot air. Why hot air? The density of 100 °C hot air is 0.73 times the density of 0 °C air. Thus, balloonists can adjust the average density of the balloon (the balloon's material, people, equipment, etc.) to match the density of air so that the balloon can float at any location in the atmosphere (up to certain limits). A burner under the opening of the balloon regulates the temperature of the air inside the balloon and hence its volume and density. This allows control over the buoyant force that the outside cold air exerts on the balloon. The same approach that we used in Example 10.8 allows us to predict that a balloon with radius of 5.0 m and a mass of 20.0 kg filled with hot air at the temperature of 100 °C can carry about 160 kg (three slim 53-kg people or two medium mass 80-kg people). This is not a heavy load.

At one time, hydrogen was used in closed balloons instead of air. The density of hydrogen is 1/14 times the density of air. Unfortunately, hydrogen can burn explosively in the presence of oxygen. The hydrogen-filled Hindenburg, a German

‹ Active Learning Guide

Figure 10.14 Making a bottle float with stable equilibrium.

(a)

Bottle partially filled with sand floats upright if the center of mass is below the geometrical center of bottle.

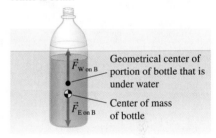

$\vec{F}_{W \text{ on } B}$ Geometrical center of portion of bottle that is under water

$\vec{F}_{E \text{ on } B}$ Center of mass of bottle

(b)

If the bottle is tipped, torques due to forces exerted on the bottle return it to the upright position.

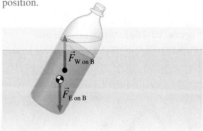

$\vec{F}_{W \text{ on } B}$

$\vec{F}_{E \text{ on } B}$

(c)

If this bottle tips slightly, torques due to forces exerted on the bottle cause it to overturn.

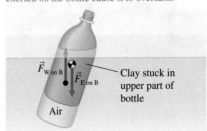

$\vec{F}_{W \text{ on } B}$ $\vec{F}_{E \text{ on } B}$ Clay stuck in upper part of bottle

Air

Table 10.5 The pressure of air and the pressure due to the oxygen in the air (called partial pressure) at different elevations.

Location	Elevation (m)	P_{air} (atm)	P_{oxygen} (atm)
Sea level	0	1.0	0.21
Mount Washington	1917	0.93	0.18
Pikes Peak	4300	0.59	0.12
Mount McKinley	6190	0.47	0.10
Mount Everest	8848	0.34	0.07
Jet travel	12,000	0.23	0.05

Figure 10.15 The Hindenburg explosion.

airship, caught fire and exploded in 1937, killing 36 people (see **Figure 10.15**). Balloonists then turned to helium—an inert gas that does not interact readily with other types of atoms.

Effects of altitude on humans

As we know, the pressure that a liquid exerts on the walls of a container and on any object submerged in it increases with depth. Water pressure is lower near the surface than deep underwater. Similarly, the pressure of atmospheric air decreases as one moves higher and higher in altitude—see **Table 10.5**.

Climbers and balloonists have to guard against altitude sickness, caused by the low pressure and lack of oxygen. Below 3000 m, there is little effect of altitude on performance. Between 3000 m and 4600 m, climbers experience compensated hypoxia—increased heart and breathing rates. Between 4600 m and 6000 m, manifest hypoxia sets in. Heart and breathing rates increase dramatically, and cognitive and sensory function and muscle control decline. Climbers may feel lethargy and euphoria and even experience hallucinations. Between 6000 m and 8000 m, climbers undergo critical hypoxia, characterized by rapid loss of muscular control, loss of consciousness, and possibly death.

These symptoms were exhibited clearly in the attempt on April 15, 1875 by three French balloon pioneers to set an altitude record. They carried bags of oxygen with them, but slowly lost the mental awareness needed to use the bags as their elevation increased. Instruments indicate that the balloon reached a maximum elevation of 8600 m twice. During the second time, two of the balloonists died. The third lost consciousness but survived.

Figure 10.16 External air pressure collapses an evacuated can.

When air was pumped out of the can, outside air pressure caused it to collapse.

Scuba diving

The sport of scuba diving depends on an understanding of fluid pressure and buoyant force to avoid cases of excess internal pressure, oxygen overload, and decompression sickness.

Air exerts a 50,000-N force (over 5 tons) on the surface of your body. Fortunately, fluids inside the body push outward and balance the force exerted by the outside air. For example, the pressure inside your lungs is approximately atmospheric pressure. What would happen if the fluid pressure on the inside remained constant while the pressure on the outside doubled or tripled? Would you be crushed, the way a can or barrel is crushed by outside air pressure when the air pressure inside the can is much lower than the pressure outside (**Figure 10.16**)? Scuba divers face this problem.

We can determine the pressure P_d at a depth d in the water compared to the atmospheric pressure at the water's surface using Pascal's second law [Eq. (10.3)]. We've defined the zero point to be at the depth d:

$$P_2 = P_1 + \rho(y_1 - y_2)g$$
$$= P_{\text{surface}} + \rho_{\text{water}}(d - 0)g$$
$$= (1 \times 10^5 \, \text{Pa}) + (1 \times 10^3 \, \text{kg/m}^3)d \, (10 \, \text{N/kg})$$
$$= (1 \times 10^5 \, \text{Pa}) + (1 \times 10^4 \, \text{Pa/m})d$$

At depth $d = 10$ m, the water pressure will be $2 \times 10^5 \, \text{N/m}^2$, or 2 atm. At 40 m below the water surface, the pressure is about $5 \times 10^5 \, \text{N/m}^2$, or 5 atm. This would surely be a problem for a scuba diver if the internal pressure were only 1 atm!

To avoid this problem, divers breathe compressed air. While moving slowly downward, a diver adjusts the pressure outlet from the compressed air tank in order to accumulate gas from the cylinder into her lungs and subsequently into other body parts, increasing the internal pressure to balance the increasing external pressure. If a diver returns too quickly to the surface, the great gas pressure in the lungs can force bubbles of gas into the bloodstream. These bubbles can behave like blood clots, blocking blood flow to the brain and possibly causing death. Blood vessels could rupture if the pressure difference between the inside and outside of the vessel is too great. Thus, a diver rises to the surface slowly so that pressure changes gradually and bubbles of gas do not form. This gradual process is called **decompression.**

When humans travel to dangerous environments (mountaintops, the deep sea, or outer space), physics intersects with human physiology. A careful understanding of gases, liquids, and the effects of changing pressures on the human body is needed to allow humans to survive in these places. As we explore the universe, we will need to learn to adapt to ever more challenging environments.

Review Question 10.7   A ship's waterline marks the maximum safe depth of the ship in the water when it has a full cargo. An empty ship is at the dock with its waterline somewhat above the water level. How could you estimate its maximum cargo?

Summary

Words	Pictorial and physical representations	Mathematical representation
Pressure P The ratio of the force F that a fluid exerts perpendicular to a surface of area A divided by that area. The pressure is caused by fluid particles colliding elastically with objects in contact with the fluid. (Section 10.2)	$\vec{F}_\perp$ A (area)	$P = \dfrac{F_\perp}{A}$ Eq. (9.1)
Density ρ The ratio of the mass m of a substance divided by the volume of that substance. (Section 10.1)	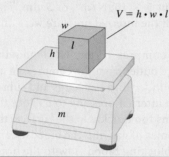 $V = h \cdot w \cdot l$	$\rho = \dfrac{m}{V}$ Eq. (10.1)
Pascal's first law—hydraulic press An increase in the pressure in one part of an enclosed fluid increases the pressure throughout the fluid. In a hydraulic press, a small force F_1 exerted on a small piston of area A_1 can cause a large force F_2 to be exerted on a large piston of area A_2. (Section 10.2)	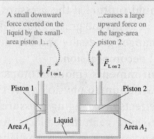 A small downward force exerted on the liquid by the small-area piston 1... ...causes a large upward force on the large-area piston 2. $\vec{F}_{\text{L on 1}}$ $\vec{F}_{\text{L on 2}}$ Piston 1 Piston 2 Liquid Area A_1 Area A_2	For a hydraulic press: $F_{\text{L on 2}} = PA_2 = (A_2/A_1)\,F_{\text{1 on L}}$ Eq. (10.2)
Pascal's second law—variation of pressure with depth On a vertical upward-pointing y-axis, the pressure of a fluid P_2 at position y_2 depends on the pressure P_1 at position y_1 and on the density of the fluid. (Section 10.3)	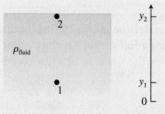 2 y_2 ρ_{fluid} 1 y_1 0	$P_1 = P_2 + \rho_{\text{fluid}}(y_2 - y_1)g$ Eq. (10.3)
Buoyant force A fluid exerts an upward-pointing buoyant force on an object totally or partially immersed in the fluid. The force depends on the density of the fluid and on the volume of the fluid displaced. (Section 10.5)	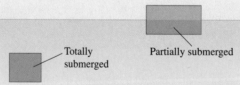 Totally submerged Partially submerged ρ fluid	$F_{\text{F on O}} = \rho_{\text{fluid}} V_{\text{fluid displaced}}$ Eq. (10.4)
Newton's second law Use the standard problem-solving strategies with this law (sketches, force diagrams, math descriptions) to find some unknown quantity. The problems often involve the buoyant force, pressure, and density. (Section 10.6)	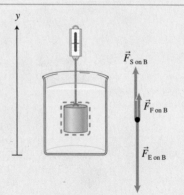 y $\vec{F}_{\text{S on B}}$ $\vec{F}_{\text{F on B}}$ $\vec{F}_{\text{E on B}}$	$a_y = \Sigma \dfrac{F_y}{m}$ Eq. (2.6)

Questions

Multiple Choice Questions

1. What does it mean if the pressure exerted by a fluid equals $10\ N/m^2$?
   (a) The fluid exerts a force of 10 N.
   (b) The surface area of the object is $1\ m^2$.
   (c) The ratio of the magnitude of the force component perpendicular to the surface area is $10\ N/m^2$.
   (d) All three choices are correct.

2. Rank in increasing order the pressure that the *italicized objects* exert on the surface.
   I. A person standing with *bare feet* on the floor
   II. A person in *skis* standing on snow
   III. A person in *rollerblades* standing on a road
   IV. A person in *ice skates* standing on ice
   (a) I, II, III, IV   (b) IV, III, I, II   (c) IV, III, II, I
   (d) III, II, IV, I   (e) II, I, III, IV

3. Choose a device that reduces the pressure caused by a force.
   (a) Scissors   (b) Knife   (c) Snowshoes
   (d) Nail   (e) Syringe

4. What does it mean if the density of a material equals $2000\ kg/m^3$?
   (a) The mass of the material is 2000 kg.
   (b) The volume of the material is $1\ m^3$.
   (c) The ratio of the mass of any amount of this material to the volume is equal to $2000\ kg/m^3$.

5. Is a material with a density of $10\ kg/m^3$ more or less dense than a material that causes a pressure of $10\ N/m^2$?
   (a) More   (b) Less   (c) The same
   (d) Not enough information to answer

6. If you hold a sheet of paper in your hands by the ends so it is oriented horizontally, what is the net force that the atmosphere exerts on the paper?
   (a) Downward   (b) Upward
   (c) Zero—the up and down pressure are equal and cancel

7. If you hold a cylinder vertically, what is the net force exerted by the atmospheric pressure on it?
   (a) Downward   (b) Upward   (c) Zero

8. How do we know that a fluid exerts an upward force on an object submerged in the fluid?
   (a) Fluid pushes on the object in all directions.
   (b) The reading of a scale supporting the object when submerged in the fluid is less than when not in the fluid.
   (c) The fluid pressure on the bottom of the object is greater than the pressure on the top.
   (d) Both (b) and (c) are correct.

9. When you suspend an object from a spring scale, it reads 15 N. Then you place the same object and scale under a vacuum jar and pump out the air. What happens to the reading of the scale?
   (a) It increases slightly.
   (b) It decreases slightly.
   (c) It says the same.
   (d) Don't have enough information to answer.

10. Why can't a suction pump lift water higher than 10.3 m?
    (a) Because it does not have the strength to pull up higher
    (b) Because the atmospheric pressure is equal to the pressure created by a 10.3-m-high column of water

(c) Because suction pumps are outdated lifting devices
(d) Because most suction cups have an opening to the bulb that is too narrow

11. If Torricelli had a wider tube in his mercury barometer, what would the height of the mercury column in the tube do?
    (a) Decrease   (b) Increase   (c) Stay the same

12. If Torricelli took his mercury barometer to the mountains, what would the height of the mercury column in the tube do?
    (a) Decrease   (b) Increase   (c) Stay the same

13. Two identical beakers with the same amount of water sit on the arms of an equal arm balance. A wooden block floats in one of them. What does the scale indicate?
    (a) The beaker with the block is heavier.
    (b) The beaker without the block is heavier.
    (c) The scale shows that both beakers weigh the same.

14. What would happen to the level of water in the oceans if all icebergs presently floating in the oceans melted?
    (a) The level would rise.
    (b) The level would stay the same.
    (c) The level would drop.

15. A piece of steel and a bag of feathers are suspended from two spring scales in a vacuum. Each scale reads 100 N. What happens when you repeat the experiment outside under normal conditions?
    (a) The scale with feathers reads more than the scale with steel.
    (b) The scale with feathers reads less that the scale with steel.
    (c) The scales have the same reading, but the reading is less than the reading in a vacuum.

16. A metal boat floats in a pool. What happens to the level of the water in the pool if the boat sinks?
    (a) It rises.   (b) It falls.   (c) It stays the same.

17. A helium balloon floats in a car with the windows closed. In which direction will the balloon move with respect to the ground when the car accelerates forward from a stop sign?
    (a) Opposite the direction of the car's acceleration
    (b) It will not move.
    (c) In the same direction as the car's acceleration

18. Will a boat float higher or lower in salt water than in fresh water?
    (a) Higher   (b) Lower   (c) The same

19. Three blocks are floating in oil as shown in **Figure Q10.19**. Which block has a higher density?
    (a) A   (b) B   (c) C
    (d) All blocks have the same density.

20. Three blocks are floating in oil as shown in Figure Q10.19. On which block does the oil exert a greater buoyant force?
    (a) A   (b) B   (c) C
    (d) The oil exerts the same force on all of them.

Figure Q10.19

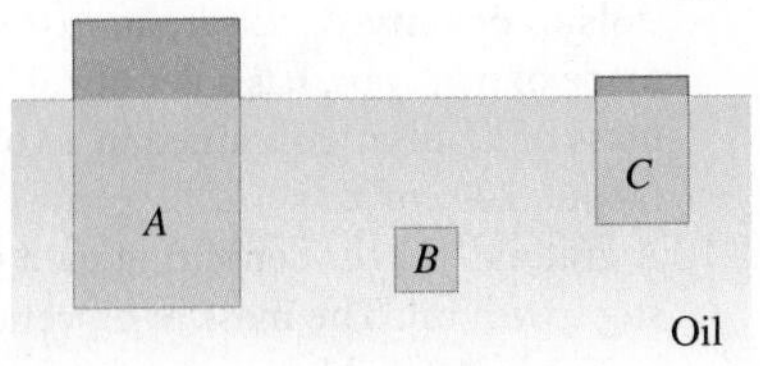

Conceptual Questions

21. Describe a method to measure the density of a liquid.
22. How can you determine the density of air?
23. Design an experiment to determine whether air has mass.
24. Does air exert an upward force or a downward force on an object submerged in the air? How can you test your answer experimentally?
25. What causes the pressure that air exerts on a surface that is in the air?
26. Why, when you fill a teapot with water, is the water always at the same level in the teapot and in the spout?
27. What experimental evidence supports Pascal's first law?
28. Fill a plastic cup to the very top with water. Put a piece of paper on top of the cup so that the paper covers the cup at the edges and is not much bigger that the surface of the cup. Turn the cup and paper upside down (practice over the sink first) and hold the bottom of the cup (now on the top). Why doesn't the water fall out of the cup?
29. Why does a fluid exert an upward force on an object submerged in the fluid?
30. Describe how you could predict whether an object will float or sink in a particular liquid without putting it into the liquid.

31. Why can you lift objects while in water that are too heavy to lift when in the air?
32. When placed in a lake, an object either floats on the surface or sinks. It does not float at some intermediate location between the surface and the bottom of the lake. However, a weather balloon floats at some intermediate distance between Earth's surface and the top of its atmosphere. Explain.
33. A flat piece of aluminum foil sinks when placed under water. Take the same piece and shape it so that it floats in the water. Explain why the method worked.
34. Ice floats in water in a beaker. Will the level of the water in the beaker change when the ice melts? Explain.
35. The density of ice at 0 °C is less than the density of water at 0 °C. Why is this very important for the existence of life on Earth?
36. How would you determine the density of an irregular-shaped unknown object if (a) it sinks in water and (b) it floats in water? List all the steps and explain the reasoning behind them.
37. Why do people sink in fresh water and in most seawater (if they do not make an effort to stay afloat) but do not sink in the Dead Sea?

Problems

Below, BIO indicates a problem with a biological or medical focus. Problems labeled EST ask you to estimate the answer to a quantitative problem rather than develop a specific answer. Problems marked with ✏ require you to make a drawing or graph as part of your solution. Asterisks indicate the level of difficulty of the problem. Problems with no * are considered to be the least difficult. A single * marks moderately difficult problems. Two ** indicate more difficult problems.

10.1 Density

1. BIO **Water in human body** About two-thirds of your body mass consists of water. Determine the volume of water in a 70-kg person.
2. * Determine the average density of Earth. What data did you use? What assumptions did you make?
3. * EST **Height of atmosphere** Use data for the normal pressure and the density of air near Earth's surface to estimate the height of the atmosphere, assuming it has *uniform* density. Indicate any additional assumptions you made. Are you on the low or high side of the real number?
4. EST A single-level home has a floor area of 200 m² with ceilings that are 2.6 m high. Estimate the mass of the air in the house.
5. * BIO A diet decreases a person's mass by 5%. Exercise creates muscle and reduces fat, thus increasing the person's density by 2%. Determine the percent change in the person's volume.
6. **Pulsar density** A pulsar, an extremely dense rotating star made of neutrons, has a density of 10^{18} kg/m³. Determine the mass of a pulsar contained in a volume the size of your fist (about 200 cm³).
7. A graduated cylinder sitting on a platform scale is filled in steps with oil. The mass of oil versus its volume is reported in **Table 10.6**. Make a mass-versus-volume graph for the oil and from the graph determine its density.

Table 10.6

Oil volume (m³)	Oil mass (kg)
5.00×10^{-5}	4.4×10^{-2}
10.0×10^{-5}	9.0×10^{-2}
15.0×10^{-5}	13.5×10^{-2}
20.0×10^{-5}	18.2×10^{-2}
25.0×10^{-5}	22.4×10^{-2}

8. * Use the graph lines in **Figure P10.8** to determine the densities of the three liquids in SI units. If you place them in one container, how will they position themselves? How does

Figure P10.8

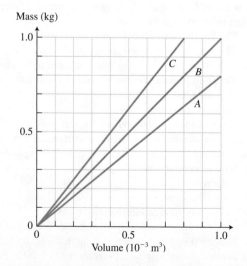

the density of each liquid change as its volume increases? As its mass decreases? Compare the masses of the three liquids when they occupy the same volume. Compare the volumes of the three liquids when they have the same mass.

9. * Imagine that you have gelatin cut into three cubes: the side of cube A is *a* cm long, the side of cube B is double the side of A, and the side of cube C is three times the side of A. Compare the following properties of the cubes: (a) density, (b) volume, (c) surface area, (d) cross-sectional area, and (e) mass.

10. Determine the density of the material whose mass-versus-volume graph line is shown in **Figure P10.10**. If you double the mass of this substance, what will happen to its density? What substance might this be?

Figure P10.10

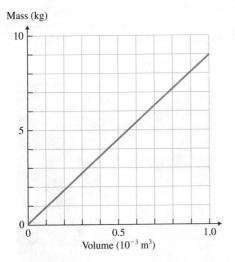

11. An object made of material A has a mass of 90 kg and a volume of 0.45 m³. If you cut the object in half, what would be the density of each half? If you cut the object into three pieces, what would be the density of each piece? What assumptions did you make?

12. You have a steel ball that has a mass of 6.0 kg and a volume of 3.0×10^{-3} m³. How can this be?

13. A material is made of molecules of mass 2.0×10^{-26} kg. There are 2.3×10^{29} of these molecules in a 2.0-m³ volume. What is the density of the material?

14. You compress all the molecules described in Problem 13 into 1.0 m³. Now what is the density of the material? What type of material could possibly behave this way?

15. * Bowling balls are heavy. However, some bowling balls float in water. Use available resources to find the dimensions of a bowling ball and explain why some balls float while others do not.

16. * **EST** Estimate the average density of a glass full of water and then the glass when the water is poured out (do not forget the air that now fills the glass instead of water).

10.2 Pressure exerted by a fluid

17. * Anita holds her physics textbook and complains that it is too heavy. Andrew says that her hand should exert no force on the book because the atmosphere pushes up on it and balances the downward pull of Earth on the book (the book's weight). Jim disagrees. He says that the atmosphere presses down on things and that is why they feel heavy. Who is correct? Approximately how large is the force that

the atmosphere exerts on the bottom of the book? Why does this force not balance the force exerted by Earth on the book?

18. **EST** **Force exerted by the air on your state** Estimate the force exerted by Earth's atmosphere on the state where you are taking your physics course.

19. * The air pressure in the tires of a 980-kg car is 3.0×10^5 N/m². Determine the average area of contact of each tire with the road.

20. * **EST** Estimate the pressure that you exert on the floor while wearing hiking boots. Now estimate the pressure under each heel if you change into high-heeled shoes. Indicate any assumptions you made.

21. **Hydraulic car lift** You are designing a hydraulic lift for a machine shop. The average mass of a car it needs to lift is about 1500 kg. What should be the specifications on the dimension of the pistons if you wish to exert a force on a smaller piston of not more than 500 N? How far down will you need to push the piston in order to lift the car 30 cm?

22. **Venus pressure and underwater pressure** Atmospheric pressure on Venus is 9.0×10^6 N/m². How deep underwater on Earth would you have to go to feel the same pressure?

23. **EST** **Force of air on forehead** Estimate the force that air exerts on your forehead. Describe the assumptions you made.

24. * A cylindrical iron plunger is held against the ceiling, and the air is pumped from inside it. A 72-kg person hangs by a rope from the plunger (**Figure P10.24**). List the quantities that you can determine about the situation and determine them. Make assumptions if necessary.

Figure P10.24

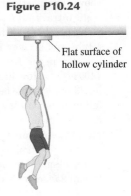

Flat surface of hollow cylinder

25. You have a rubber pad with a handle attached to it (**Figure P10.25**). If you press the pad firmly on a smooth table, it is impossible to lift it off the table. Why? What force would you need to exert on the handle to lift it? The surface area of the pad is 0.023 m².

26. * ✎ You vacuum up a small piece of paper on the floor. Draw a force diagram for the paper just as it is being lifted up into the vacuum cleaner.

Figure P10.25

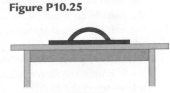

27. ✎ **EST** **Toy bow and arrow** A child's toy arrow has a suction cup on one end. When the arrow hits the wall, it sticks. Draw a force diagram for the arrow stuck on the wall and estimate the magnitudes of the forces exerted on it when it is in equilibrium. The mass of the arrow is about 10 g. Why are the words "suction cup" not appropriate?

10.3 Pressure variation with depth

28. **Pressure on the Titanic** The Titanic rests 4 km (2.5 miles) below the surface of the ocean. What physical quantities can you determine using this information?

29. You have three reservoirs (**Figure P10.29**). Rank the pressures at the bottom of each and explain your rankings. Then rank the net force that the water exerts on the bottom of each reservoir. Explain your rankings.

Figure P10.29

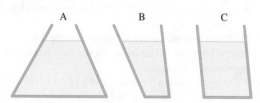

30. * A bucket filled to the top with water has a piece of ice floating in it. (a) Will the pressure on the bottom of the bucket change when the ice melts? Explain. (b)Will the level of water rise when the ice melts? Explain.

31. **Water reservoir and faucet** The pressure at the top of the water in a city's gravity-fed reservoir is $1.0 \times 10^5 \text{ N/m}^2$. Determine the pressure at the faucet of a home 42 m below the reservoir.

32. **Dutch boy saves Holland** An old story tells of a Dutch boy who used his fist to plug a 2.0-cm-diameter hole in a dike that was 3.0 m below sea level, thus preventing the flooding of part of Holland. What physical quantities can you determine from this information? Determine them.

33. **EST BIO Blood pressure** Estimate the pressure of the blood in your brain and in your feet when standing, relative to the average pressure of the blood in your heart of 1.3×10^4-N/m^2 above atmospheric pressure.

34. * **BIO Intravenous feeding** A glucose solution of density 1050 kg/m^3 is transferred from a collapsible bag through a tube and syringe into the vein of a person's arm. The pressure in the arm exceeds the atmospheric pressure by 1400 N/m^2. How high above the arm must the top of the liquid in the bottle be so that the pressure in the glucose solution at the needle exceeds the pressure of the blood in the arm? Ignore the pressure drop across the needle and tubing due to viscous forces.

35. * **Mountain climbing** Determine the change in air pressure as you climb from elevation of 1650 m at the timberline of Mount Rainier to its 4392-m summit, assuming an average air density of 0.82 kg/m^3. Will the real change be more or less than the one you calculated? Explain.

36. **EST BIO Giraffe raises head** Estimate the pressure change of the blood in the brain of a giraffe when it lifts its head from the grass to eat a leaf on an overhead tree. Without special valves in its circulatory system, the giraffe could easily faint when lifting its head.

37. * **EST Car in pond** Your car slides off an embankment into a pond. Estimate the force you must exert on the door to open it if the top of the door is 0.5 m below the surface. Describe in detail how you might escape without opening the door.

38. ✏ **Drinking through a straw** You are drinking water through a straw in an open glass. Select a small volume of water in the straw as a system and draw a force diagram for the water inside this volume that explains why the water goes up the straw.

39. * **More straw drinking** While you are drinking through the straw, the pressure in your mouth is 30 mm Hg below atmospheric pressure. What is the maximum length of a straw in an open glass that you can use to drink a fruit drink of density 1200 kg/m^3?

40. * Your office has a 0.020 m^3 cylindrical container of drinking water. The radius of the container is about 14 cm. When the container is full, what is the pressure that the water exerts on the sides of the container halfway down from the top? All the way down?

41. * **EST BIO Eardrum** Estimate the net force on your 0.5-cm^2 eardrum that air exerts on the inside and the outside after you drive from Denver, Colorado (elevation 1609 m) to the top of Pikes Peak (elevation 4301 m). Assume that the air pressure inside and out are balanced when you leave Denver and that the average density of the air is 0.80 kg/m^3. What other assumptions did you make?

42. **BIO Eardrum again** You now go snorkeling. What is the pressure on your eardrum when you are 2.4 m under the water, assuming the pressure was equalized before the dive?

43. Water and oil are poured into opposite sides of an open U-shaped tube. The oil and water meet at the exact center of the U at the bottom of the tube. If the column of oil of density 900 kg/m^3 is 16 cm high on one side, how high is the water on the other side?

44. * Examine the photo of Hoover Dam (**Figure P10.44**). What do you notice about its vertical structure? Explain why a dam is thicker at the bottom than at the top.

Figure P10.44

45. * A test tube of length L and cross-sectional area A is submerged in water with the open end down so that the edge of the tube is a distance h below the surface. The water goes up into the tube so its height inside the tube is l. Describe how you can use this information to decide whether the air that was initially in the tube obeys Boyle's law. List your assumptions.

10.4 Measuring atmospheric pressure

46. The reading of a barometer in your room is 780 mm Hg. What does this mean? What is the pressure in pascals?

47. How long would Torricelli's barometer have had to be if he had used oil of density 950 kg/m^3 instead of mercury?

48. Sometimes gas pressure is measured with a device called a *liquid manometer* (**Figure P10.48**). Explain how this

Figure P10.48

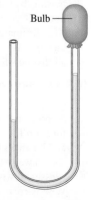

Bulb —

instrument can be used to measure the pressure of gas in a bulb attached to one of the tubes.

49. You use a liquid manometer with water to measure the pressure inside a rubber bulb. Before you squeeze the bulb, the water is at the same level in both legs of the tube. After you squeeze the bulb, the water in the opposite leg rises 20 cm with respect to the leg connected to the bulb (**Figure P10.49**). What is the pressure in the bulb? What assumptions did you make? How will the answer change if the assumptions are not valid?

Figure P10.49

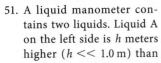

50. * In a mercury-filled manometer (**Figure P10.50**), the open end is inserted into a container of gas and the closed end of the tube is evacuated. The difference in the height of the mercury is 80 mm. The radius of the connecting tube is 0.50 cm. (a) Determine the pressure inside the container in newtons per square meter. (b) An identical manometer has a connecting tube that is twice as wide. If the difference in the height of the mercury is the same, then what is the pressure in the container?

Figure P10.50

51. A liquid manometer contains two liquids. Liquid A on the left side is h meters higher ($h \ll 1.0$ m) than the liquid on the right side. What do you know about those two liquids? What assumptions did you make? If the assumptions were not true, how would the answer be different?

52. Examine the reading of the manometer that you use to measure the pressure inside car tires. What are the units? Does the manometer measure the absolute pressure of the air inside the tires or gauge pressure? How do you know?

53. * Marjory thinks that the mass of a liquid above a certain level should affect the pressure at this level. Describe how you will test her idea.

10.5 Buoyant force

54. / Draw a force diagram for an object that is floating at the surface of a liquid.

55. / Draw a cubic object that is completely submerged in a fluid but not resting on the bottom of the container. Then draw arrows to represent the forces exerted by the fluid on the top, sides, and bottom of the object. Make the arrows the correct relative lengths. What is the direction of the total force exerted by the fluid on the object?

56. / Draw a force diagram for a helium balloon that you just released. Then draw a force diagram for an air-filled balloon that you just released.

57. / You are holding a brick that is completely submerged in water. Draw a force diagram for the brick. Why does it feel lighter in water than when you hold it in the air?

58. * This textbook says that the upward force that a fluid exerts on a submerged object is equal in magnitude to the product of the density of the fluid, the gravitational constant g, and the volume of the submerged part of the object. Where did this equation come from?

59. * **Design** This textbook says that the upward force that a fluid exerts on a submerged object is equal in magnitude to the product of the density of the fluid, the gravitational constant g, and the volume of the submerged part of the object. Design an experiment to test this expression, including a prediction about the outcome of the experiment.

60. * / You have four objects at rest, each of the same volume. Object A is partially submerged, and objects B, C, and D are totally submerged in the same container of liquid, as shown in **Figure P10.60**. Draw a force diagram for each object. Rank the densities of the objects from least to greatest and indicate whether any objects have the same density.

Figure P10.60

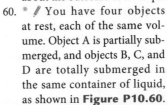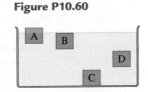

61. * **Does air affect what a scale reads?** A 60-kg woman with a density of 980 kg/m³ stands on a bathroom scale. Determine the reduction of the scale reading due to air.

62. * When analyzing a sample of ore, a geologist finds that it weighs 2.00 N in air and 1.13 N when immersed in water. What is the density of the ore? What assumptions did you make to answer the question? If the assumptions are not correct, how would the answer be different?

63. * A pin through a hole in the middle supports a meter stick. Two identical blocks hang from strings at an equal distance from the center so the stick is balanced. What happens to the stick if one block is submerged in water of density 1000 kg/m³ and the other block in kerosene of density 850 kg/m³?

64. ** / A meter stick is supported by a pin through a hole in the middle. (a) Two blocks made of the same material but different sizes hang from strings at different positions in such a way that the stick balances. What happens when the blocks hang entirely submerged in beakers of water? (b) Next you hang two blocks of different masses but the same volume at different positions so the stick balances. What happens when these blocks hang completely submerged in beakers of water? Support your answer for each part using force diagrams with arrows drawn with the correct relative lengths.

10.6 Skills for analyzing static fluid processes

65. **Goose on a lake** A 3.6-kg goose floats on a lake with 40% of its body below the 1000-kg/m³ water level. Determine the density of the goose.

66. * **Floating in seawater** A person of density of ρ_1 floats in seawater of density ρ_2. What fraction of the person's body is submerged? Explain.

67. * **Floating in seawater** A person of density of 980 kg/m³ floats in seawater of density 1025 kg/m³. What can you determine using this information? Determine it.

68. ** (a) Determine the force that a vertical string exerts on a 0.80-kg rock of density of 3300 kg/m³ when it is fully submerged in water of density 1000 kg/m³. (b) If the force exerted by the string supporting the rock increases by 12% when the rock is submerged in a different fluid, what is that fluid's density? (c) If the density of another rock of the same volume is 12% greater, what happens to the buoyant force the water exerts on it?

69. * **Snorkeling** A 60-kg snorkeler (including snorkel, mask, and other gear) displaces 0.058 m³ of water when 1.2 m under the surface. Determine the magnitude of the buoyant force exerted by the 1025-kg/m³ seawater on the person. Will the person sink or drift upward?

70. * A helium balloon of volume 0.12 m³ has a total mass (the helium plus the balloon) of 0.12 kg. Determine the buoyant force exerted on the balloon by the air if the air has density 1.13 kg/m³. Determine the initial acceleration of the balloon when released.

71. ** A bucket filled to the top with water has a piece of ice floating in it. Will the pressure on the bottom change when the ice melts? Justify your answer.

72. * / BIO **Protein sinks in water** A protein molecule of mass 1.1×10^{-22} kg and density 1.3×10^3 kg/m³ is placed in a vertical tube of water of density 1000 kg/m³. (a) Draw a motion diagram and a force diagram at the moment immediately after the molecule is released. (b) Determine the initial acceleration of the protein.

73. * How can you determine if a steel ball of known radius is hollow? List the equipment that you will need for the experiment, and describe the procedure and calculations. Can you determine how big the hollow part is if present in the ball?

74. ** **Crown composition** A crown is made of gold and silver. The scale reads its mass as 3.0 kg when in air and 2.75 kg when in water. Determine the masses of the gold and the silver in the crown. The density of gold is 19,300 kg/m³ and that of silver is 10,500 kg/m³.

10.7 Buoyancy: Putting it all together

75. * **Wood raft** Logs of density 600 kg/m³ are used to build a raft. What is the weight of the maximum load that can be supported by a raft built from 300 kg of logs?

76. ** A cylinder has radius R. How high should a column of liquid be so that the magnitude of the force averaged over the side wall surface area that the liquid exerts on the wall equals the magnitude of the force that the liquid exerts on the bottom surface of the cylinder? Explain.

77. **Standing on a log** A log is L long and d in diameter. What is the mass of a person who can stand on the log without getting her feet wet?

78. * **Ferryboat** A ferryboat is 12 m long and 8 m wide. Two cars, each of mass 1600 kg, ride on the boat for transport across the lake. How much farther does the boat sink into the water?

79. EST **Iceberg** Estimate the fraction of the volume of an iceberg that is underwater.

80. * **Life preserver** A life preserver is manufactured to support a 70-kg person with 20% of his volume out of the water. If the density of the life preserver is 100 kg/m³ and it is completely submerged, what must its volume be? List your assumptions.

General Problems

81. * Compare the density of water at 0 °C to the density of ice at 0 °C. Suggest possible explanations in terms of the molecular arrangements inside the liquid and solid forms of water that would account for the difference. If necessary, use extra resources to help answer the question.

82. * **Collapsing star** The radius of a collapsing star destined to become a pulsar decreases by 10% while at the same time 12% of its mass escapes. Determine the percent change in its density.

83. ** EST BIO **Syringe pressure** You are getting a flu shot. Estimate the average pressure of the fluid entering your arm during the shot. Indicate any assumptions you made. Compare this to the pressure of your shoes on the ground when standing.

84. Explain qualitatively and quantitatively how we drink through a straw. Make sure you can account for the water going up the length of the straw.

85. * **Deep dive** The *Trieste* research submarine traveled 10.9 km below the ocean surface while exploring the Mariana Trench in the South Pacific, the deepest place in the ocean. Determine the force needed to prevent a 0.10-m-diameter window on the side of the submarine from imploding. The density of the water is 1025 kg/m³.

86. ** **Bursting a wine barrel** Pascal placed a long 0.20-cm-radius tube in a wine barrel of radius 0.24 m. He sealed the barrel where the tube entered it. When he added wine of density 1050 kg/m³ to the tube so the column of wine was 8.0 m high, the cover of the barrel burst off the top of the barrel. What was the net force that caused the cover to come off?

87. * BIO **Lowest pressure in lungs** Experimentally determine the maximum distance you can suck water up a straw. Use this number to determine the pressure in your lungs above or below atmospheric pressure while you are sucking. Be sure to indicate any assumptions you made and show clearly how you reached your conclusion.

88. * **Landing on Venus** Atmospheric pressure on Venus is 92×10^5 N/m². Suppose that NASA is planning to land a 1.0-m-radius spherical research vehicle on Venus. (a) Determine the force on each square centimeter of its surface. (b) What is the buoyant force exerted by the atmosphere on the spherical vehicle?

89. ** You have an empty water bottle. Predict how much mass you need to add to it to make it float half-submerged. Then add the calculated mass and explain any discrepancy that you found. How did you make your prediction?

90. ** BIO **Flexible bladder helps fish sink or rise** A 1.0-kg fish of density 1025 kg/m³ is in water of the same density. The fish's bladder contains 10 cm³ of air. The bladder compresses to 4 cm³, reducing its volume by 6 cm³. Now what is the density of the fish? Will it sink or rise? Explain.

91. * **Plane lands on Nimitz aircraft carrier** When a 27,000-kg fighter airplane lands on the deck of the aircraft carrier Nimitz, the carrier sinks 0.25 cm deeper into the water. Determine the cross-sectional area of the carrier.

92. ** To determine the density of an object and an unknown liquid, it is first weighed in air, then in water, and then in an unknown liquid. The readings of the scale are T_1, T_2, and T_3, respectively. Suggest a method of using these data to determine the density of the object and of the liquid. Decide what additional equipment and measurements you would need to make to test whether the results of the first method are correct.

93. ** Two upward-moving balloons carry equal loads. The first balloon has an upward acceleration of $(g/3)$. The second balloon moves up at constant speed. The density of the gas inside both balloons is one-third the density of air. The volume of the first balloon is V_1. What is the volume of the second balloon? The masses of the balloons are the same.

94. Derive an equation for determining the unknown density of a liquid by measuring the magnitude of a force $T_{S \text{ on } O}$ that a string needs to exert on a hanging object of unknown mass m and density ρ to support it when the object is submerged in the liquid.

Reading Passage Problems

BIO **Free diving** So-called "no-limits" free divers slide to deep water on a weighted sled that moves from a boat down a vinyl-coated steel cable to the bottom of a dive site. The diver reaches depths where a soda can would implode. After reaching the target depth, the diver releases the sled and an air bag opens and brings the diver quickly back to the surface. The divers have no external oxygen supply— just lungs full of air at the start of the dive. In August 2002, Tanya Streeter of the Cayman Islands held the women's no-limits free dive record at 160 m. In 2005 Patrick Musimu set the men's record with a 209.6-m free dive in the Red Sea just off the Egyptian coast (the record was later broken by Herbert Nitsch of Austria).

Musimu's 2005 dive took 3 minutes 28 seconds. He began the dive with his 9-L lungs full of air. By the time he passed the 200-m mark, Musimu's lungs had contracted to the size of a tennis ball. His body transferred blood from his limbs to essential organs such as the heart, lungs, and brain. This "blood shift" occurs when mammals submerge in water. Blood plasma fills the chest cavity, especially the lungs. Without this adaptation, the lungs would shrink and press against the chest walls, causing permanent damage. When he reached his target, Musimu released the weighted segment of the specialized sled that had taken him down and opened an airbag, which began his return to the surface at an average speed of 3–4 m/s.

95. The pressure of the water when Musimu was 209.6 m below the surface was closest to which of the following?
    (a) 2 atm   (b) 3 atm   (c) 21 atm
    (d) 22 atm   (e) 200 atm

96. Assuming Musimu weighs 670 N (150 lb) and is 1.6 m tall, 0.30 m wide, and 0.15 m thick, which answer below is closest to the magnitude of the force that the deep water exerted on one side of his body?
    (a) 0   (b) 670 N (130 lb)
    (c) 15,000 N (3000 lb)   (d) 10^5 N (20,000 lb)
    (e) 10^6 N (200,000 lb)

97. Musimu's training allows him to hold up to 9 L = 9000 cm^3 of air when in a 1 atm environment. Which answer below is closest to the volume of that air if at pressure 22 atm?
    (a) 100 cm^3 (b) 200 cm^3 (c) 400 cm^3
    (d) 9000 cm^3 (e) 2 × 10^5 cm^3

98. As Musimu descends, the buoyant force that the water exerts on him
    (a) remains approximately constant.
    (b) increases a lot because the pressure is so much greater.
    (c) decreases significantly because his body is being compressed and made much smaller.
    (d) is zero for the entire dive.
    (e) There is not enough information to answer the question.

99. Why don't his lungs, heart, and chest completely collapse?
    (a) The return balloon helps counteract the external pressure.
    (b) There is no external force pushing directly on the organs.
    (c) The sled that helps him descend protects the front of his body.
    (d) Blood plasma moves from his extremities to his chest and the organs in it.
    (e) The air originally in the lungs is transferred to the vital organs.

100. Using the dimensions in Question 96, which answer below is closest to the buoyant force that the water exerts on Musimu (without his sled or his return balloon)? Assume that the density of water is 1000 kg/m^3.
    (a) 200 N   (b) 400 N   (c) 700 N
    (d) 1000 N   (e) 2 × 10^6 N

Lakes freeze from top down We all know that ice cubes float in a glass of water. Why? Virtually every substance contracts when it solidifies—the solid is denser than the liquid. If this happened to water, ice cubes would sink to the bottom of a glass, and ice sheets would sink to the bottom of a lake. Fortunately, this doesn't happen. Liquid water expands by 9% when it freezes into solid ice at 0 °C, from a liquid density of 1000 kg/m^3 to a solid density of 917 kg/m^3. Consequently, in the winter when the water in a lake freezes, the solid ice stays at the top, forming an ice sheet. Snow covering the icy surface forms a protective blanket that insulates the ice and water below and helps to keep the lake from completely freezing into a solid chunk of ice. Fish and lake plants below the ice survive during the winter.

The expansion of water when it freezes has another important environmental benefit: the so-called freeze-thaw effect on sedimentary rocks. Water is absorbed into cracks in these rocks and then freezes in cold weather. The solid ice expands and cracks the rock—like a wood cutter splitting logs. This continual process of liquid water absorption, freezing, and cracking releases mineral and nitrogen deposits into the soil and can eventually break the rock down into soil.

101. When is water denser?
    (a) When liquid at 0 °C.
    (b) When solid ice at 0 °C.
    (c) Water is always 1000 kg/m^3.
    (d) When it is near room temperature.

102. Why does water freeze from the top down?
    (a) The denser water at 0 °C sinks below the ice.
    (b) The less dense ice at 0 °C rises above the liquid water at 0 °C.
    (c) The solid ice is denser than the liquid, just like for metals.
    (d) a and b
    (e) a, b, and c

103. Using Newton's second law, expressions for buoyant force and other forces, and the densities of liquid and solid water at 0 °C, find the fraction of an iceberg or an ice cube that is under liquid water.
    (a) 0.84   (b) 0.88   (c) 0.92
    (d) 0.96   (e) 1.00

104. A swimming pool at 0 °C has a very large chunk of ice floating in it—like an iceberg in the ocean. When the ice melts, what happens to the level of the water at the edge of the pool?
    (a) It rises.   (b) It stays the same.
    (c) It drops.   (d) It depends on the size of the chunk.

105. Which of the following are benefits of the decrease in the density of water when it freezes?
    (a) Fish and plants can survive winters without being frozen.
    (b) Over time, soil is formed from sedimentary rocks.
    (c) Water pipes when frozen in the winter do not burst.
    (d) Two of the above three.
    (e) All of the first three.

11

Fluids in Motion

How does blood flow dislodge plaque from an artery?

Why can a strong wind cause the roof to blow off a house?

Why do people snore?

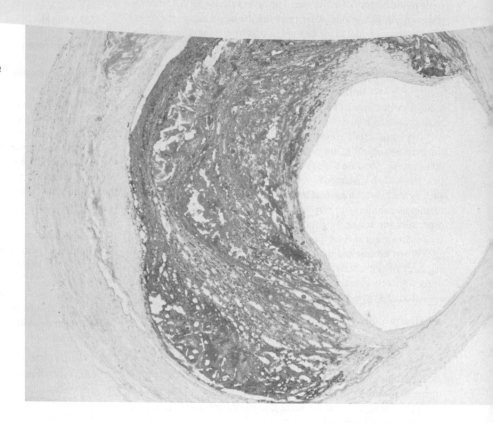

Be sure you know how to:

- Draw work-energy bar charts (Section 6.2).
- Apply the concept of pressure to explain the behavior of liquids (Section 10.2).
- Draw force diagrams and apply Newton's second law (Section 3.2).

Plaque (fatty deposits that accumulate on the walls of an artery) grows larger as cholesterol-laden blood flows by. As plaque on the artery wall grows, blood flows past at higher speed. The fast-flowing blood pulls on the plaque like a suction cup. If the blood is moving fast enough, it can completely remove the deposit from the wall. The plaque floats downstream until it reaches another narrow opening, where it may become lodged and completely stop blood flow. If this stoppage occurs in an artery in the heart, it can cause a heart attack by preventing blood from flowing to the cardiac muscles responsible for pumping the blood. In this chapter we will learn why blood flows faster through an artery clogged with plaque and why the fast-moving stream of blood tends to pull the plaque off the artery wall.

In the previous chapter, we investigated the behavior of static fluids. We learned that a fluid's pressure increases with depth beneath the surface of the fluid, that the physics of static fluids explains how a hydraulic lift works, and that a fluid exerts a buoyant force on objects partially or completely submerged in it. In all of these situations, however, the fluid was at rest. What happens when a gas or liquid moves across a surface—for example, when air moves across the roof of a house or when blood moves through a blood vessel? In this chapter, we will investigate and explain phenomena involving moving fluids—fluid dynamics.

Figure 11.1 Air blowing across the top of the ball causes it to float in the air.

11.1 Fluids moving across surfaces— Qualitative analysis

How does air blowing over the top of a beach ball, shown in **Figure 11.1**, lift and support the ball? To answer questions like this, we must compare the forces that stationary air exerts on a surface to the forces exerted on the surface by moving air. In Observational Experiment **Table 11.1**, we apply this analysis to three systems. We know the directions of the forces that external objects exert on these systems and can thus deduce the direction of the net force due to air pressure exerted on different sides of these objects.

OBSERVATIONAL EXPERIMENT TABLE

11.1 Does fluid pressure against a surface depend on the motion of the fluid across the surface?

 VIDEO 11.1

Observational experiment	Analysis
Experiment 1. We blow air through a straw between two lightbulbs hanging from strings. The bulbs move closer to each other. 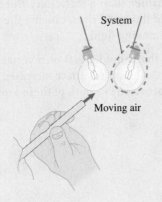	Choose the right bulb as the system. After it comes to equilibrium, the forces exerted on the bulb are balanced. The vertical component of the force exerted by the string balances the downward force exerted by Earth. The horizontal component of the force exerted toward the right by the string must balance the net force exerted by the air. Thus, the force exerted toward the left by the non-moving air on the right side of the bulb must be greater than the force exerted toward the right by the air moving on the left side of the bulb. The pressure of moving air must be lower than the pressure of nonmoving air. 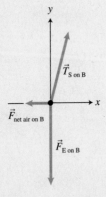

(continued)

Observational experiment	Analysis
Experiment 2. As a car with a canvas-covered trailer moves fast, the canvas bulges upward as though something is pushing up from below.	Choose part of the canvas as the system. Earth exerts a downward gravitational force on the canvas. In order for the canvas to bulge, the stationary air below the canvas and the moving air above must exert a net upward force. This can only happen if the air pressure pushing the canvas up is greater than the air pressure pushing the canvas down. Therefore, we can conclude that the air pressure on the side of moving air is less that on the side of stationary (with respect to the canvas) air.
Experiment 3. We blow a stream of air across the top of a large air-filled balloon. The moving air allows it to float with no other support.	Choose the floating balloon as the system. Earth exerts a downward gravitational force. To remain stationary, the net force caused by the air must point upward toward the moving-air side of the balloon. This can only happen if the air pressure pushing the balloon up is greater than the air pressure pushing the balloon down. Therefore, we can conclude that the air pressure on top of the balloon (the side of the moving air) is less than that on the bottom of the balloon (the side of the stationary air)

Pattern

Based on the analysis of each experiment using the force diagrams, we can infer that moving air exerts a smaller force on the system and consequently has a smaller pressure then stationary air.

In each of the three experiments in Table 11.1 the air exerted a net force on the object in the direction of the side past which the air was moving. This means that stationary air exerted a greater pressure on the object than the moving air. Two explanations might account for the observed pattern.

Explanation 1: Temperature. The pressure on one side of an object decreases because a moving fluid (such as air) is warmer than a stationary fluid; the warm fluid has a lower density than the cold fluid and rises, causing the pressure to decrease on the side with the warm, moving fluid.

Explanation 2: Fluid speed. The pressure that a fluid exerts on a surface decreases as the speed with which the fluid moves across the surface increases.

Let's devise an experiment that can disprove one or both of these explanations. See Testing Experiment **Table 11.2**.

TESTING EXPERIMENT TABLE

11.2   Testing the two explanations for why moving fluid exerts a lower pressure on a surface than stationary fluid.

VIDEO 11.2

Testing experiment	Prediction	Outcome
Experiment 1. What happens to a piece of paper held at the corners when you blow hard across the *top* surface of the paper? 	***Prediction based on Explanation 1:*** The moving air originating inside your body is warmer than the external air below the paper. The warm air above the paper rises, reducing the pressure on the top surface. The paper will rise. ***Prediction based on Explanation 2:*** There is less pressure on the top surface where the air is moving compared to the pressure from the stationary air below the paper. The greater pressure from below will cause the paper to rise.	The paper rises when blowing air across the top. 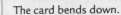
Experiment 2. What happens if you vigorously blow air through a straw *under* a card folded in an inverted-U shape? 	***Prediction based on Explanation 1:*** The card should bend up because the warm air from your breath is less dense than surrounding air and tends to rise. ***Prediction based on Explanation 2:*** The air moving under the card exerts less pressure on the bottom surface than the stationary air above the card—the card should bend down.	The card bends down. Moving air

Conclusion

- In Experiment 1, both explanations correctly predicted the outcome. Thus, neither explanation can be rejected based on this experiment.
- In Experiment 2, the outcome matches the prediction based on Explanation 2 and is opposite the prediction based on Explanation 1. We can reject Explanation 1 based on the outcome of this experiment.

    Our testing experiments did not reject Explanation 2. In addition, if we repeat Experiment 1 in Table 11.1, carefully observing what happens when we increase the speed at which air is moving, we notice that the angle of the threads supporting the bulbs increases. This means that the difference in the pressure of stationary and moving air increases with the increase of air speed. At this point, without evidence to the contrary, we can assume that as a fluid's speed increases, the pressure that the moving fluid exerts on the surface decreases. Explanation 2 is a qualitative version of a rule formulated in 1738 by Daniel Bernoulli and named in his honor.

◄ Active Learning Guide

> **Bernoulli's principle** The pressure that a fluid exerts on a surface *decreases* as the speed with which the fluid moves across the surface *increases*.

    In short, the magnitude of the speed of the fluid has a significant effect on the pressure—the greater the speed, the lower the pressure. Bernoulli's principle has important fluid-flow implications in biological systems—for example, in the flow of blood through blood vessels. The blood pressure against the wall of a vessel depends on how fast the blood is moving—pressure is lower when the blood is moving faster. Let's look at two other applications of Bernoulli's principle.

Figure 11.2 A clarinet reed closes when air starts moving across the top and opens when air flow over the reed stops.

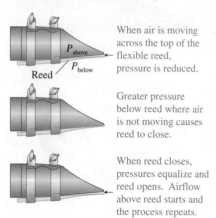

When air is moving across the top of the flexible reed, pressure is reduced.

Greater pressure below reed where air is not moving causes reed to close.

When reed closes, pressures equalize and reed opens. Airflow above reed starts and the process repeats.

Figure 11.3 Snoring occurs when the soft palate opens and closes due to the starting and stopping of air flow across it.

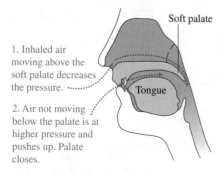

Soft palate

1. Inhaled air moving above the soft palate decreases the pressure.

Tongue

2. Air not moving below the palate is at higher pressure and pushes up. Palate closes.

3. When airflow stops, the pressures equalize and the soft palate reopens. The moving air causes the process to repeat.

4. The vibrating palate and air flow cause the snoring sound.

Figure 11.4 Flow rate is the volume of fluid that passes a cross section of a vessel in a given time interval.

(a) $t = 0$

A volume $V = lA$ of fluid flows past cross-section A in time interval Δt.

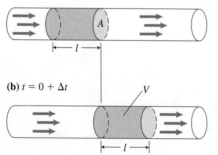

(b) $t = 0 + \Delta t$

Clarinet reed

A clarinet is a woodwind instrument. The musician induces sound by blowing air into the clarinet's mouthpiece, where air moves across the top of a reed (see **Figure 11.2**). Air pressure across the top of the reed decreases relative to the pressure below the reed, where the air is stationary. In response to the drop in pressure, the flexible reed rises and closes the mouthpiece of the clarinet. Once the mouthpiece is closed, the airflow stops and the pressure above and below the reed equalizes. The reed opens and air flows again across the top of the reed. Once again, pressure above the reed drops, the reed rises, and the mouthpiece closes. The rhythmic opening and closing of the reed initiates the sound heard from a clarinet.

Snoring

A snoring sound occurs when air moving through the narrow opening above the soft palate at the back of the roof of the mouth has lower pressure than nonmoving air below the palate (**Figure 11.3**). The normal air pressure below the soft palate, where the air is not moving, pushes the palate closed. When airflow stops, the pressures equalize and the passage reopens. The rhythmic opening and closing of the soft palate against the throat leads to the snoring sound (like the vibration of a clarinet reed).

Review Question 11.1 Why did the lightbulbs in the first observational experiment come together when the air was moving between them?

11.2 Flow rate and fluid speed

We have found qualitatively that the pressure of a fluid against a surface is lower when the fluid is moving across the surface than when it is static. In order to apply this idea quantitatively later, we need first to quantify the rate of fluid flow.

Perhaps you have taken a shower in which little water flowed from the showerhead. In physics, we would say the water flow rate was low. The **flow rate Q** is an important consideration in designing showerheads. A smaller flow rate will save water, but a larger flow rate will get you cleaner. Flow rate is defined as the volume V of fluid that moves through a cross section of a pipe divided by the time interval Δt during which it moved (see **Figure 11.4**):

$$Q = \frac{V}{\Delta t} \tag{11.1}$$

The SI unit of flow rate is m^3/s, but you may also see it as ft^3/s, ft^3/min, gallons/min, or any unit of volume divided by any unit of time interval. Notice that flow rate in m^3/s is different from the speed of the fluid v in m/s.

> **TIP** The symbols V, t, and Q are also used in other aspects of physics. For example, a lowercase v denotes speed, the capital letter T is used for temperature, and in future chapters we will use Q for two other unrelated quantities. Because these symbols are often used to indicate different quantities, it is important when working with equations to try to visualize their meaning with concrete images (for example, the volume of water flowing out of a faucet during 1 s).

You probably feel intuitively that the flow rate should be related to the speed of the moving fluid. To explore the relationship, consider Figure 11.4a. Over a certain time interval Δt the darkened volume of fluid passes a cross section of area A at some position along the pipe. Thus, after a time Δt, the back part of this fluid volume has in effect moved forward to the position shown in Figure 11.4b. The volume V of fluid in the darkened portion of the cylinder is the product of its length l and the cross-sectional area A of the pipe:

$$V = lA$$

Thus, the fluid flow rate is

$$Q = \frac{V}{\Delta t} = \frac{lA}{\Delta t} = \left(\frac{l}{\Delta t}\right)A$$

However, l is also the distance the fluid moves in a time interval Δt. Thus, $\frac{l}{\Delta t}$ is the average fluid speed v. Substituting $v = \frac{l}{\Delta t}$ into the above equation, we find that

$$Q = \frac{V}{\Delta t} = vA \qquad (11.2)$$

The flow rate is equivalent to the average fluid speed multiplied by the cross-sectional area of the vessel.

‹Active Learning Guide

QUANTITATIVE EXERCISE 11.1 Speed of blood flow in aorta

The heart pumps blood at an average flow rate of 80 cm^3/s into the aorta, which has a diameter of 1.5 cm. Determine the average speed of blood flow in the aorta.

Represent mathematically The flow rate can be determined by rearranging Eq. (11.2):

$$v = \frac{Q}{A}$$

where the cross-sectional area of the aorta is

$$A = \pi r^2 = \pi\left(\frac{d}{2}\right)^2$$

Solve and evaluate Combining the above two equations, we find that the average speed of blood flow in the aorta is

$$v = \frac{Q}{A} = \frac{Q}{\pi(d/2)^2} = \frac{(80 \text{ cm}^3/\text{s})}{\pi(1.5 \text{ cm}/2)^2} = 45 \text{ cm/s}$$

The unit is correct. The magnitude is reasonable—about half a meter each second.

Try it yourself: Determine the average speed of blood flow if the diameter is reduced from 1.5 cm to 1.0 cm—with the same flow rate.

Answer: 100 cm/s.

Notice that blood speed more than doubles when the aorta diameter decreases by 33%. Vessel diameter has a very significant effect on the flow rate of fluid through a vessel, including those in biological systems. The narrower the blood vessel, the faster the blood flows, increasing the risk of dislodging plaque. Likewise, the narrower the airway from the nose to the mouth, the faster the air moves and the more likely you are to snore. These effects depend on the speed of the fluid, like blood or air in different parts of a vessel or pipe in which the diameter changes from one section to another. Let's see if we can relate the speed of the fluid in one part of a vessel to another.

Continuity equation

Frequently, the radius of a single pipe varies from one part of the pipe to another. How does this affect the flow rate and the speed of the moving fluid in the different parts of the pipe? Consider blood flowing through the artery

Figure 11.5 The flow speed $v_2 > v_1$ depends on the cross-sectional area of pipe carrying the fluid.

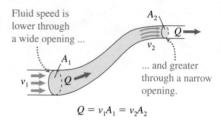

Fluid speed is lower through a wide opening ...

... and greater through a narrow opening.

$$Q = v_1 A_1 = v_2 A_2$$

shown in **Figure 11.5**. Is it possible for more fluid to flow through cross section 1 (labeled A_1 on the figure) than through cross section 2 (A_2)? In such a case, more fluid would be entering the region between 1 and 2 than was leaving that region. This scenario might occur if the region between 1 and 2 were expanding. It could also occur if the fluid were compressed so that it took up less space between 1 and 2. However, let's assume that the vessel does not change its shape and remember that liquids (as opposed to gases) are almost incompressible. This means that the amount of fluid entering at 1 must equal the amount of fluid leaving at 2. What must change is the speed of the fluid as it travels through the narrower part of the vessel. In a narrow section of the vessel (A_2) the speed will be greater in order to keep the flow rate constant. Thus, the flow rate past cross section 1 will equal that past cross section 2:

$$Q_1 = v_1 A_1 = v_2 A_2 = Q_2 \qquad (11.3)$$

where v_1 is the average speed of the fluid passing cross section A_1 and v_2 is the average speed of the fluid passing cross section A_2. Equation (11.3) is called the **continuity equation** and is used to relate the cross-sectional area and average speed of fluid flow in different parts of a rigid vessel carrying an incompressible fluid.

QUANTITATIVE EXERCISE 11.2 Blood flow speed

Blood normally flows at an average speed of about 10 cm/s in a large artery with a radius of about 0.30 cm. Assume that the radius of a small section of the artery is reduced by one-half because of *atherosclerosis*, a thickening of the arterial walls. Determine the speed of the blood as it passes through the constriction—point 2 in the figure below.

$v_1 = 10$ cm/s
$v_2 = ?$
$r_1 = 0.30$ cm 1●   2●
$r_2 = r_1/2$

Represent mathematically Point 1 indicates the wider part of the artery and point 2 is the constricted part. Assume that the cross-sectional areas are circular ($A = \pi r^2$). We can rearrange the continuity equation to find the average speed of blood flow past point 2 in terms of its speed past 1 and the two cross-sectional areas:

$$v_2 = \left(\frac{A_1}{A_2}\right) v_1$$

Solve and evaluate To find v_2 in terms of v_1, insert expressions for the two areas:

$$v_2 = \left(\frac{A_1}{A_2}\right) v_1 = \left(\frac{\pi r_1^2}{\pi r_2^2}\right) v_1 = \left(\frac{r_1}{r_2}\right)^2 v_1 = \left(\frac{r_1}{0.5 r_1}\right)^2 v_1$$
$$= 4 v_1$$

Thus, v_2 will be 40 cm/s if v_1 is 10 cm/s. The increase in speed will have a significant effect on the pressure that the blood exerts on the atherosclerotic material (called plaque) on the vessel wall.

Try it yourself: Suppose the radius of this small section of artery is reduced to one-third its normal value because of atherosclerosis. What now is the speed of blood flow past this constriction?

Answer: $9v_1$, or 90 cm/s.

Figure 11.6 Why is $v_2 > v_1$?

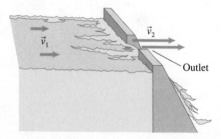

$\vec{v}_1$

$\vec{v}_2$

Outlet

Review Question 10.2 Why does water in a river flow more slowly just before a dam than it does while passing through the outlet of the dam (**Figure 11.6**)?

11.3 Causes and types of fluid flow

We've learned qualitatively that the fluid pressure against a surface decreases as the fluid speed increases. We have also learned that flow rate of a fluid through a pipe depends on the speed of the fluid and the cross-sectional area of the pipe. The next question is: What causes fluids to flow?

Fluid flow is caused by differences in pressure. When the pressure in one region of the fluid is lower than in another region, the fluid tends to flow from the higher pressure region toward the lower pressure region. For example, large masses of air in Earth's atmosphere move from regions of high pressure into regions of low pressure. Blood flows through the circulatory system from the arterial side, which has a gauge pressure of about 100 mm Hg, to the venous side, which has a gauge pressure of about 5 mm Hg.

What kinds of flow can occur? We're familiar with different kinds of fluid flow. There is the smooth flow that we see in a wide river and the more turbulent flow that we see as water rushes and swirls through a narrow channel. Studies of fluid flow in wind tunnels indicate that there are two primary kinds of flow: streamline (or laminar) flow and turbulent flow. In **streamline flow**, every particle of fluid that passes a particular point follows the same path as particles that preceded it. This is the smooth flow that we see in a wide river (see the fluid flowing smoothly in the wide tube in **Figure 11.7**). **Turbulent flow**, on the other hand, is characterized by agitated, disorderly motion. Instead of following a given path, the fluid forms whirlpool patterns called eddies, which come and go randomly, or sometimes become semi-stable (see **Figure 11.8**). Turbulent flow occurs when a fluid moves around objects and through pipes at high speed.

The force exerted by the fluid on objects is called the **drag force** and is somewhat greater during turbulent flow than during streamline flow. Designing a car so that air moves over it with streamline flow reduces the drag force that the air exerts on the car and improves gasoline mileage. Placing a curved dome above the cab of a truck deflects air up and over the trailer, reducing turbulent flow and increasing gas mileage by more than 10% (see **Figure 11.9**).

Review Question 11.3 Is it easier for the heart to pump blood if the flow of the blood through the blood vessels is streamline or if it is turbulent? Explain.

11.4 Bernoulli's equation

Earlier in this chapter we developed the qualitative version of Bernoulli's principle: the pressure of a fluid against a surface decreases as the speed of the fluid across the surface increases. A quantitative version of Bernoulli's principle relates the properties (pressure, speed) of a fluid at one position to its properties at another position.

To derive the quantitative version of Bernoulli's principle, called Bernoulli's equation, we again use the case of a fluid flowing through a pipe, as shown in **Figure 11.10a**. We assume that (1) the fluid is incompressible, (2) the fluid flows without friction, and (3) the flow is streamline. We can apply the work-energy equation to describe the behavior of the fluid as it moves a short distance along the vessel. Consider the system to be composed of the shaded volume of fluid pictured in Figures 11.10a and b and Earth.

Figure 11.10a shows us the initial state of the system. As the volume of fluid flows to the right, the fluid behind it at position 1 exerts a force of magnitude $F_1 = P_1 A_1$ to the right where P_1 is the fluid pressure against the left side of the volume, and A_1 is the cross-sectional area of the left side of the volume. Simultaneously, the fluid ahead of the system at position 2 exerts a force in the opposite direction of magnitude $F_2 = P_2 A_2$, where P_2 is the fluid pressure against the right side of the volume and A_2 is the cross-sectional area of the right side of the volume.

In Figure 11.10b, the shaded volume of fluid has moved to the right. Because the pipe is narrower at 2, the right side of the fluid at position 2 moves a greater distance than the left side of the same volume of fluid in the wider part

Figure 11.7 Streamline flow.

Streamline flow of water through a tube

Figure 11.8 Turbulent flow.

Turbulent flow of water through a narrow part of a tube

Eddies

Figure 11.9 A dome over the cab of a truck reduces the turbulent flow of air and increases the truck's fuel efficiency.

‹Active Learning Guide

Figure 11.10 Applying the work-energy equation to fluid flow.

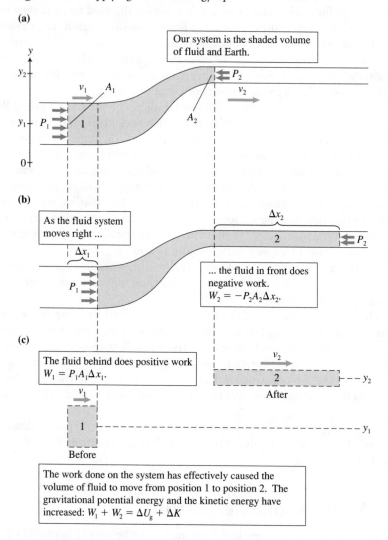

(a)

Our system is the shaded volume of fluid and Earth.

(b)

As the fluid system moves right ...

... the fluid in front does negative work.
$W_2 = -P_2 A_2 \Delta x_2$.

(c)

The fluid behind does positive work
$W_1 = P_1 A_1 \Delta x_1$.

After

Before

The work done on the system has effectively caused the volume of fluid to move from position 1 to position 2. The gravitational potential energy and the kinetic energy have increased: $W_1 + W_2 = \Delta U_g + \Delta K$

of the pipe at position 1. The net effect of the movement of the fluid a short distance to the right is summarized in Figure 11.10c. The volume of fluid initially at position 1 moving at speed v_1 has now been transferred, in effect, to position 2 where it moves at speed v_2. The volume of fluid stays constant, since we assume the fluid is incompressible. The fluid is now moving faster through the narrow tube at position 2 than it was earlier when moving through the wider tube at position 1, an increase in kinetic energy. The fluid at position 2 is at a higher elevation than when at position 1, thus the gravitational potential energy of the system increases. The energies changed as a result of the work done by the forces exerted by the fluid behind and ahead of the shaded volume. We can represent this quantitatively using the generalized work-energy equation [Eq. (6.3)].

$$(K_1 + U_{g1}) + W = (K_2 + U_{g2})$$

If we move the terms with the subscript 1 to the right side of the equation, we have

$$W = (K_2 - K_1) + (U_{g2} - U_{g1})$$

or

$$W = \Delta K + \Delta U_g \tag{11.4}$$

Let us now write expressions for each of the terms in the above equation.

Work Done Two forces are doing work on the system. The fluid behind the system exerts a force F_1 to the right on the left side of the system over a distance Δx_1. The fluid ahead of the system exerts a force F_2 to the left on the right side of the system over a distance Δx_2. (Figures 11.10a and b show fluid pressures; the corresponding forces have magnitudes $F_1 = P_1 A_1$ and $F_2 = P_2 A_2$). The force F_1 does positive work since it points in the direction of the motion of the system. The force F_2 does negative work since it points in the direction opposite the motion of the system. The total work done on the system is

$$W = F_1 \Delta x_1 \cos 0° + F_2 \Delta x_2 \cos 180°$$
$$= P_1 A_1 \Delta x_1 - P_2 A_2 \Delta x_2$$

The volume of fluid V that has moved from the left to the right is $V = A_1 \Delta x_1 = A_2 \Delta x_2$ since the fluid is incompressible. The preceding expression for work becomes

$$W = P_1 V - P_2 V = (P_1 - P_2) V$$

Change in Kinetic Energy The mass m of the system (the moving volume of fluid) is related to its density ρ and volume V:

$$m = \rho V$$

As the system moves from 1 to 2, its speed changes from v_1 to v_2. Thus, the kinetic energy change of the mass m of fluid shown in Figure 11.10c is

$$\Delta K = \frac{1}{2} m v_2^2 - \frac{1}{2} m v_1^2 = \frac{1}{2} \rho V v_2^2 - \frac{1}{2} \rho V v_1^2$$

Change in Gravitational Potential Energy The gravitational potential energy of the system has also changed, because the system has moved from elevation y_1 to elevation y_2. The change in gravitational potential energy is then

$$\Delta U_g = mg(y_2 - y_1) = \rho V g(y_2 - y_1)$$

We can now substitute the above three expressions into Eq. (11.4) to get

$$(P_1 - P_2) V = \left(\frac{1}{2} \rho V v_2^2 - \frac{1}{2} \rho V v_1^2 \right) + \rho V g(y_2 - y_1)$$

If the common V is canceled from each term, we find that

$$P_1 - P_2 = \frac{1}{2} \rho (v_2^2 - v_1^2) + \rho g(y_2 - y_1)$$

By dividing by V in that last step we have changed the units of each term in the equation from energy (measured in joules) to energy density (measured in joules per cubic meter). Energy density is the same as energy per unit of volume of the fluid and appears on the right side of the above equation. The left hand side represents the amount of work done on the fluid per unit volume of fluid.

Bernoulli's equation relates the pressures, speeds, and elevations of two points along a single streamline in a fluid:

$$P_1 - P_2 = \frac{1}{2} \rho (v_2^2 - v_1^2) + \rho g(y_2 - y_1) \qquad (11.5)$$

The equation can be rearranged into an alternate form:

$$\frac{1}{2} \rho v_1^2 + \rho g y_1 + P_1 = P_2 + \frac{1}{2} \rho v_2^2 + \rho g y_2 \qquad (11.6)$$

The sum of the kinetic and gravitational potential energy densities and the pressure at position 1 equals the sum of the same three quantities at position 2.

Bernoulli's equation describes the flow of a frictionless, nonturbulent, incompressible fluid. The equation is the quantitative equivalent of Bernoulli's principle, which we developed in Section 11.1.

Using Bernoulli bar charts to understand fluid flow

Bernoulli's equation looks fairly complex and might be difficult to use for visualizing fluid dynamics processes. However, since Bernoulli's equation is based on the work-energy principle, we can represent such processes using energy bar charts similar to the ones used in Chapter 6 (here the bars represent pressures and energy densities). The following Reasoning Skill box describes how to construct a fluid dynamics bar chart for a process. The procedure is illustrated for the following process: A fire truck pumps water through a big hose up to a smaller hose on the ledge of a building. Water sprays out of the smaller hose onto a fire in the building. Compare the pressure in the hose just after leaving the pump to the pressure at the exit of the small hose.

REASONING SKILL Constructing a bar chart for a moving fluid

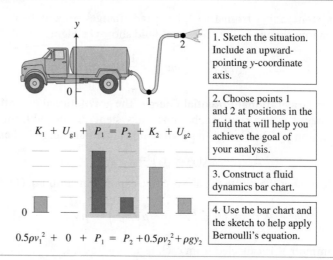

$$K_1 + U_{g1} + P_1 = P_2 + K_2 + U_{g2}$$

$$0.5\rho v_1^2 + 0 + P_1 = P_2 + 0.5\rho v_2^2 + \rho g y_2$$

1. Sketch the situation. Include an upward-pointing y-coordinate axis.

2. Choose points 1 and 2 at positions in the fluid that will help you achieve the goal of your analysis.

3. Construct a fluid dynamics bar chart.

4. Use the bar chart and the sketch to help apply Bernoulli's equation.

To start, first draw a sketch of the process. Then, choose positions 1 and 2 at appropriate locations in order to help answer the question. One of the positions might be a place where you want to determine the pressure in the fluid and the other position a place where the pressure is known. For the water pump-hose process, it is useful to choose position 1 at the exit of the pump (the location of the unknown pressure) and position 2 at the exit of the water from the small hose (at known atmospheric pressure).

Next, represent this process by placing bars of appropriate relative lengths on the chart (the absolute lengths are not known). It is often easiest to start by analyzing the gravitational potential energy density. Use a vertical y-axis with a well-defined origin to keep track of the gravitational potential energy densities. For the fire hose process, choose position 1 as the origin of the vertical coordinate system. The gravitational potential energy density at position 1 is then zero. The exit of the water from the small hose is at higher elevation; thus, there is a positive gravitational potential energy density bar for position 2 (in the bar chart we arbitrarily assign it one positive unit of energy density).

Next, consider the kinetic energy. The water flows from a wider hose at position 1 to a narrower hose at position 2. Thus the kinetic energy density at 2 is greater than at 1—thus the longer bar at position 2. We arbitrarily assume that the kinetic energy density bar for position 1 is one unit and for position 2 is three units.

Notice now that the total length of the bars on the right side of the chart is much higher than on the left side (the difference is three units). To account for the difference we need to consider the change in pressure. Since the fluid pressures at 1 and 2 are analogous to the work done on the system in an ordinary work-energy bar chart, P_1 and P_2 appear in the shaded box in the center of the bar chart, where work is represented. The difference in the pressure heights should account for the total difference in the energy densities. Thus, we draw the bar for P_1 three units higher than for P_2. The bar chart is now complete. We can use it to write a mathematical description for the process and solve for any unknown quantity.

Review Question 11.4 Compare and contrast work-energy bar charts and Bernoulli bar charts.

11.5 Skills for analyzing processes using Bernoulli's equation

In this section, we adapt our problem-solving strategy to analyze processes involving moving fluids. In this case, we describe and illustrate a strategy for finding the speed of water as it leaves a bottle. The general strategy is on the left side of the table and the specific process is on the right.

PROBLEM-SOLVING STRATEGY **Applying Bernoulli's Equation**

EXAMPLE 11.3 **Removing a tack from a water bottle**
What is the speed with which water flows from a hole punched in the side of an open plastic bottle? The hole is 10 cm below the water surface.

Sketch and translate

- Sketch the situation. Include an upward-pointing y-coordinate axis. Choose an origin for the axis.
- Choose points 1 and 2 at positions in the fluid where you know the pressure/speed/position or which involve the quantity you are trying to determine.
- Choose a system.

- Choose the origin of the vertical y-axis to be the location of the hole.
- Choose position 1 to be the place where the water leaves the hole and position 2 to be a place where the pressure, elevation, and water speed are known—at the water surface $y_2 = 0.10$ m and $v_2 = 0$. The pressure in Bernoulli's equation at both positions 1 and 2 is atmospheric pressure, since both positions are exposed to the atmosphere $(P_1 = P_2 = P_{\text{atm}})$.
- Choose Earth and the water as the system.

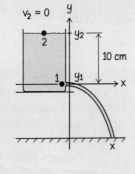

Simplify and diagram

- Identify any assumptions you are making. For example, can we assume flow without friction?
- Construct a Bernoulli bar chart.

- Assume that the fluid flows without friction.
- Assume that y_2 and y_1 stay constant during the process, since the elevation of the surface decreases slowly compared to the speed of the water as it leaves the tiny hole.
- Draw a bar chart that represents the process.

$$K_1 + U_{g1} + \boxed{P_1 = P_2} + K_2 + U_{g2}$$

(continued)

Represent mathematically

- Use the sketch and bar chart to help apply Bernoulli's equation.
- You may need to combine Bernoulli's equation with other equations, such as the equation of continuity $Q = v_1A_1 = v_2A_2$ and the definition of pressure $P = \dfrac{F}{A}$.

- We see from the sketch and the bar chart that the speed of the fluid at position 2 is zero (zero kinetic energy density) and that the elevation is zero at position 1 (zero gravitational potential energy density). Also, the pressure is atmospheric at both 1 and 2. Thus,

$$(1/2)\rho(0)^2 + \rho gy_2 + P_{atm} = P_{atm} + (1/2)\rho v_1^2 + \rho g(0)$$
$$\Rightarrow \rho gy_2 = (1/2)\rho v_1^2$$

Solve and evaluate

- Solve the equations for an unknown quantity.
- Evaluate the results to see if they are reasonable (the magnitude of the answer, its units, how the answer changes in limiting cases, and so forth).

- Solve for v_1

$$v_1 = \sqrt{2gy_2}$$

Substituting for g and y_2, and y_2, we find that

$$v_1 = \sqrt{2(9.8 \text{ m/s}^2)(0.10 \text{ m})} = 1.4 \text{ m/s}$$

- The unit m/s is the correct unit for speed. The magnitude seems reasonable for water streaming from a bottle (if we obtained 120 m/s it would be unreasonably high).

Active Learning Guide >

Try it yourself: In the above situation the water streams out of the bottle onto the floor a certain horizontal distance away from the bottle. The floor is 1.0 m below the hole. Predict this horizontal distance using your knowledge of projectile motion. [*Hint:* Use Eqs. (3.7) and (3.8).]

Answer: The equations yield a result of 0.63 m. However, if we were to actually perform this experiment with a tack-sized hole, the water would land short of our prediction because there is friction between a small hole and the water. In order to make the water land 0.63 m from the bottle, we must increase the diameter of the hole to about 3 mm. We discuss the effect of friction on fluid flow later in the chapter.

EXAMPLE 11.4 Drying your basement

After a rainstorm, your basement is filled with water to a depth of 0.10 m. The surface area of the basement floor is 150 m². A water pump with a short 1.2-cm-radius outlet pipe connects to a 0.90-cm-radius hose that in turn goes out the basement window to the ground outside, 2.4 m above the pump outlet pipe. The pump can remove water from the basement at a rate of 6.8 m³/hour (1800 gal/h). Determine the time interval needed to remove the water and determine the water pressure produced at the pump outlet pipe.

Sketch and translate A labeled sketch of the situation is shown at right. The system is Earth and the water in the basement, in the pump, and in the hose. The pump itself is not a part of the system. A vertical y-axis points upward with its origin at the pump outlet pipe. We can calculate the volume of water in the basement and use this volume and the flow rate $Q = 6.8$ m³/h to determine the time interval needed to remove the water.

The system is the shaded water and Earth.

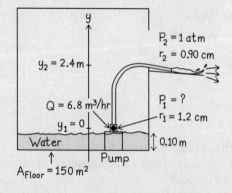

We use Bernoulli's equation to determine the pressure at the pump outlet pipe. Point 1 is at the outlet pipe in the basement at vertical position $y_1 = 0$ (the place where we want to determine the pressure). Position 2 is the exit point of the hose on the ground outside at vertical position $y_2 = +2.4$ m (we know that pressure $P_2 = P_{atm}$, since the fluid is exposed to the air). Since we know the fluid flow rate Q, we use $Q = Av$ to determine the speed of the fluid at the pump outlet pipe and at the exit of the hose (since the fluid flow rate must be the same at both positions).

Simplify and diagram Assume that the water is incompressible and that the water flows without friction or turbulence. Then draw a fluid dynamics bar chart to represent the process. We include the following energy density changes.

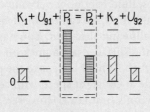

Kinetic energy density The fluid moves faster through the narrower hose at position 2 than through the wider pipe at position 1. Thus, the kinetic energy density K_2 is greater than K_1.

Gravitational potential energy density Position 2 is at a higher elevation than position 1. Thus, U_{g2} is greater than U_{g1}, which we know is zero, having chosen the origin of the vertical y-axis at position 1.

Pressure To balance the initial-final energy densities, the pressure P_1 at the outlet of the pump in the basement must be greater than the atmospheric pressure $P_2 = P_{atm}$ of the air as water leaves the hose at position 2.

Represent mathematically The volume V of water in the basement is the product of its area $A = 150$ m^2 and its depth $d = 0.10$ m; thus, $V = A \cdot d = (150$ m$^2)(0.10$ m$) = 15$ m^3. The time interval for pumping all the water out is

$$\Delta t = \frac{V}{Q}$$

The bar chart helps us apply Bernoulli's equation [Eq. (11.5) or (11.6)]. There is nonzero term in the equation for each nonzero bar in the chart. We get

$$\frac{1}{2}\rho v_1^2 + P_1 = P_2 + \frac{1}{2}\rho v_2^2 + \rho g y_2$$

Substitute for the pressure at point 2 and rearrange the equation to find P_1:

$$P_1 = P_{atm} + \frac{1}{2}\rho(v_2^2 - v_1^2) + \rho g y_2$$

The speeds v_1 and v_2 are determined using Eq. (11.2): $v_1 = Q/A_1$ and $v_2 = Q/A_2$.

Solve and evaluate The flow rate is 6.8 m^3/h. The time interval needed to pump all of the water out of the basement will be

$$\Delta t = \frac{V}{Q} = \frac{15\ \text{m}^3}{6.8\ \text{m}^3/\text{h}} = 2.2\ \text{h}$$

The unit for time interval is correct, and the magnitude is reasonable—usually pumping water from the basement takes hours, not seconds or years.

Next, use Bernoulli's equation to determine the pressure at the pump outlet. First use the expression for flow rate Eq. (11.2) to determine the speed of the water at positions 1 and 2. The flow rate is $Q = 6.8$ m^3/h$(1$h$/3600$ s$) = 0.00188$ m^3/s. Thus, the water speeds at positions 1 and 2 are

$$v_1 = \frac{Q}{A_1} = \frac{Q}{\pi r_1^2} = \frac{0.00188\ \text{m}^3/\text{s}}{\pi(0.012\ \text{m})^2} = 4.2\ \text{m/s}$$

$$v_2 = \frac{Q}{A_2} = \frac{Q}{\pi r_2^2} = \frac{0.00188\ \text{m}^3/\text{s}}{\pi(0.0090\ \text{m})^2} = 7.4\ \text{m/s}$$

The pressure P_1 is determined using Eq. (11.5):

$$P_1 = P_{atm} + \frac{1}{2}\rho(v_2^2 - v_1^2) + \rho g y_2$$

$$= (1.0 \times 10^5\ \text{N/m}^2) + \frac{1}{2}(1000\ \text{kg/m}^3)[(7.4\ \text{m/s})^2$$
$$- (4.2\ \text{m/s})^2] + (1000\ \text{kg/m}^3)(9.8\ \text{N/kg})(2.4\ \text{m})$$
$$= (1.0 \times 10^5\ \text{N/m}^2) + (0.19 \times 10^5\ \text{N/m}^2)$$
$$+ (0.24 \times 10^5\ \text{N/m}^2)$$
$$= 1.4 \times 10^5\ \text{N/m}^2$$

Note that 1 kg/m$\cdot$s$^2 = 1$ N/m^2. This would be a gauge pressure of 0.4×10^5 N/m^2 or 0.4 atm, certainly possible with a pump from most hardware stores.

Try it yourself: Assume that the hose from the pump to the outside has a radius of 0.50 cm instead of 0.90 cm. Now what is the speed of the water exiting the hose and the pressure at the pump outlet pipe?

Answer: 24 m/s (a hose with a smaller radius produces a somewhat higher water speed) and 4.0×10^5 N/m^2 (we need a huge increase in pressure because of the smaller radius hose).

Review Question 11.5   In Example 11.3 we said that the pressure was the same at two levels when we drew the bar chart. Doesn't the pressure in a fluid increase with depth?

11.6 Viscous fluid flow

In our previous discussions and examples in this chapter, we assumed that fluids flow without friction. That is, we assumed no interaction either between the fluid and the walls of the pipes they flow in, or between the layers of the fluid. However, in Example 11.3 we found that this assumption was not reasonable. In fact, for many processes, such as the transport of blood in the small vessels in our bodies, fluid friction is very important. When we cannot neglect this friction inside the fluid, we call the fluid **viscous.**

Consider the following situation. You have an object that can slide on a frictionless horizontal surface, say, a puck on smooth ice. You push the puck abruptly and then let go. What happens to the puck? Once in motion, the puck will continue to slide at constant speed with respect to the ice even if nothing else pushes it (Newton's first law). However, if there is friction between the contacting surfaces (there is a little sand in the ice), then the puck starts slowing down; for it to continue moving at constant speed, someone or something has to push it forward to balance the opposing friction force.

By analogy, if a fluid flows through a horizontal tube without friction, we would expect it to continue to flow at a constant rate with no additional forward pressure. But if friction is present, there must be greater pressure at the back of the fluid than at the front of the fluid. If this is the case, the force exerted on any volume of the fluid due to the forward pressure is greater than the force exerted on the same volume of the fluid due to the pressure in the opposite direction.

Factors that affect fluid flow rate

What factors affect the flow rate in the vessel with friction? What is the functional dependence of those factors? Let's think about the physical properties of the fluid and the vessel that can affect the flow rate. The following quantities might be important.

Pressure Difference The flow rate should depend on how hard the fluid is pushed forward, that is, on the difference between the fluid pressure pushing forward from behind and the fluid pressure pushing back from in front of the fluid, or $(P_1 - P_2)$.

Radius of the Tube The radius r of the tube carrying the fluid should affect the flow rate. From everyday experience we know that it is more difficult to push (a greater pressure difference) fluid through a tube of tiny radius than through a tube with a large radius.

Length of the Tube The length l of the tube might also affect the ease of fluid flow. A long tube offers more resistance to flow than a shorter tube.

Fluid Type Water flows much more easily than molasses does. Thus some property of a fluid that characterizes its "thickness" or "stickiness" should affect the flow.

Let's design an experiment to investigate exactly how the first three of these four factors (P, r, and l) affect the fluid flow rate Q. As shown in **Figure 11.11**, a pump that produces an adjustable pressure P_1 causes fluid to flow through tubes of different radii r and lengths l. We collect the fluid exiting the tube and measure the flow rate Q, which is the volume V of fluid leaving the tube in a certain time interval Δt divided by that time interval. The results of the experiments are reported in **Table 11.3**.

How is the flow rate affected by each of the three factors?

Pressure difference Looking at the first three rows of the table, we notice that the flow rate is proportional to the pressure difference ($Q \propto P_1 - P_2$).

Figure 11.11 How do $P_1 - P_2$, r, and l affect the flow rate Q?

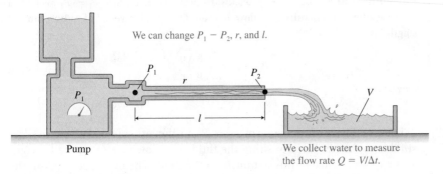

We can change $P_1 - P_2$, r, and l.

P_1 r P_2 V

P_1

l

Pump

We collect water to measure the flow rate $Q = V/\Delta t$.

Table 11.3 Different quantities affect the flow rate Q of fluid through a tube. The data are reported in relative units.

$P_1 - P_2$ (Pressure difference)	r (Radius)	l (Length)	Q (Flow rate)
1	1	1	1
2	1	1	2
3	1	1	3
1	2	1	16
1	3	1	81
1	1	2	0.5
1	1	3	0.33

Radius of the tube Looking at rows 1, 4, and 5, we notice that the flow rate increases rapidly as the radius increases. Doubling the radius causes the flow rate to increase by a factor of 16 (2^4). Tripling the radius causes the flow rate to increase by a factor of 81 (3^4). The flow rate is proportional to the fourth power of the radius of the tube ($Q \propto r^4$).

Length of the tube Looking at rows 1, 6, and 7, we notice that the flow rate decreases as the length of the tube increases. It is proportional to the inverse of the length ($Q \propto 1/l$).

These three relationships can be combined in a single equation:

$$Q \propto \frac{r^4(P_1 - P_2)}{l}$$

Viscosity and Poiseuille's law

In this experiment we did not investigate the fourth factor: the type of fluid. Under the same conditions, water flows faster than oil, which flows faster than molasses. If we use the same pressure difference to push different fluids through the same tube, we find that the fluids have different flow rates. The quantity by which we measure this effect on flow rate is called the **viscosity** η of the fluid. The flow rate is inversely proportional to viscosity:

$$Q \propto \frac{1}{\eta}$$

In 1840, using an experiment similar to that described above, French physician and physiologist Jean Louis Marie Poiseuille established a relationship between these physical quantities. However, instead of writing the flow rate in terms of the other four quantities, he wrote an expression for the pressure difference needed to cause a particular flow rate.

> **Poiseuille's law** The forward-backward pressure difference $P_1 - P_2$ needed to cause a fluid of viscosity η to flow at a rate Q through a vessel of radius r and length l is
>
> $$P_1 - P_2 = \left(\frac{8}{\pi}\right)\frac{\eta l}{r^4}Q \qquad (11.8)$$

TIP Notice that the pressure difference needed to cause a particular flow rate is proportional to the inverse of the fourth power of the radius of the vessel. If the radius of a vessel carrying fluid is reduced by a factor of 0.5, the pressure difference needed to cause the same flow rate must increase by $(1/0.5)^4 = 16$. We need 16 times the pressure difference to cause the same flow rate.

The pressure difference term $P_1 - P_2$ on the left side of Poiseuille's law is similar to the net force pushing the fluid. The flow rate Q on the far right side is a consequence of this net push on the fluid. The term before Q on the right side (the $\left(\frac{8}{\pi}\right)\frac{\eta l}{r^4}$ term) can be thought of as the resistance of the fluid to flow—the resistance is greater if the fluid has greater viscosity η, is greater for a longer vessel (greater l), and is far more resistive if the vessel through which the fluid flows has a smaller radius ($1/r^4$ is much greater for small r). This idea has many applications relative to the circulatory system—see the example a little later in this section.

From Poiseuille's law we can determine the unit for viscosity. To do this, we express the viscosity using other quantities in Eq. (11.8):

$$P_1 - P_2 = \left(\frac{8}{\pi}\right)\frac{\eta l}{r^4}Q \Rightarrow \eta = \frac{(P_1 - P_2)r^4\pi}{8Ql}$$

We use the latter equation to find the units for the viscosity. Remember that the units of pressure are

$$Pa = \frac{N}{m^2} = \frac{kg \cdot m}{s^2 \cdot m^2} = \frac{kg}{s^2 \cdot m}$$

and the units for flow rate are m^3/s. Using these units we get

$$\eta = \frac{(kg)(m^4)(s)}{(s^2 \cdot m)(m^3)(m)} = \frac{kg}{s \cdot m}.$$

We can also rewrite the last combination of units as $N \cdot s/m^2$. A list of viscosities of several fluids appears in **Table 11.4** using $N \cdot s/m^2$ for the units of η.

Table 11.4 Viscosities of some liquids and gases.

Substance	Viscosity η $(N \cdot s/m^2)$
Air (30 °C)	1.9×10^{-5}
Water vapor (30 °C)	1.25×10^{-5}
Water (0 °C)	1.8×10^{-3}
Water (20 °C)	1.0×10^{-3}
Water (40 °C)	0.66×10^{-3}
Water (80 °C)	0.36×10^{-3}
Blood, whole (37 °C)	4×10^{-3}
Oil, SAE No. 10	0.20

QUANTITATIVE EXERCISE 11.5 Blood flow through a narrow artery

Because of plaque buildup, the radius of an artery in a person's heart decreases by 40%. Determine the ratio of the present flow rate to the original flow rate if the pressure across the artery, its length, and the viscosity of blood are unchanged.

Represent mathematically In this exercise, we are interested in the change in the flow rate and not in the change in pressure. Consequently, we rearrange Poiseuille's law for the flow rate in terms of the other quantities:

$$Q = \left(\frac{\pi}{8}\right)\left(\frac{\Delta P}{\eta l}\right)r^4$$

If the radius decreases by 40%, the new radius is $100\% - 40\% = 60\%$ of the original. Thus the radius r of the vessel at the present time is related to the radius r_0 years earlier by the equation $r = 0.60 r_0$.

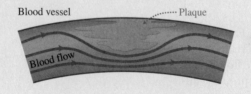

Blood vessel ⋯⋯ Plaque

Blood flow

Solve and evaluate The ratio of the flow rates is

$$\frac{Q}{Q_0} = \frac{\left(\frac{\pi}{8}\right)\left(\frac{\Delta P}{\eta l}\right) r^4}{\left(\frac{\pi}{8}\right)\left(\frac{\Delta P}{\eta l}\right) r_0^4} = \frac{r^4}{r_0^4} = \left(\frac{r}{r_0}\right)^4 = (0.60)^4 = 0.13$$

The flow rate is only 13% of the original flow rate! To compensate for such a dramatically reduced flow rate, the person's blood pressure will increase.

Try it yourself: Determine the reduction in flow rate, assuming a constant pressure difference, if the radius of the vessel is reduced 90% (to 0.10 times its original value). This is not an unusual reduction for people with high blood pressure.

Answer: $Q/Q_0 = 0.0001$, or 0.01% of its original value!

Limitations of Poiseuille's Law: Reynolds Number Poiseuille's law describes the flow of a fluid accurately only when the flow is streamline, or laminar. Experiments indicate that to determine when the flow is laminar or turbulent, one needs to determine what is called the **Reynolds number** R_e:

$$R_e = \frac{2\bar{v}r\rho}{\eta} \tag{11.9}$$

where $\bar{v}$ is the average speed of the fluid, ρ is its density, η is the viscosity, and r is the radius of the vessel that carries the fluid. Experiments show that if the Reynolds number is less than 2000, the fluid flow is laminar; if it is more than 3000, the flow is turbulent; and between 2000 and 3000 the flow is unstable and can be either laminar or turbulent.

Review Question 11.6 Describe some of the physics-related effects on the cardiovascular system of medication that lowers the viscosity of blood).

11.7 Applying fluid dynamics: Putting it all together

We can explain many phenomena using ideas from fluid dynamics. In this section, we'll analyze a roof being ripped off a house during a high-speed wind and the dislodging of plaque in an artery.

Blowing the roof off a house

You've no doubt seen images of roofs being blown from houses during tornadoes or hurricanes. How does that happen? On a windy day, the air inside the house is not moving, whereas outside the air is moving very rapidly. The air pressure inside the house is therefore greater than the air pressure outside, creating a net pressure against the roof and windows that pushes outward. If the net pressure becomes great enough, the roof and/or the windows will blow outward off of the house. In the following example, we do a quantitative estimate of the net force exerted by the inside and outside air on a roof.

EXAMPLE 11.6 Effect of high-speed air moving across the roof of a house

During a storm, air is moving at speed 45 m/s (100 mi/h) across the top of the 200-m² flat roof of a house. Estimate the net force exerted by the air pushing up on the inside of the roof and the outside air pushing down on the outside of the roof. Indicate any assumptions made in your estimate.

Sketch and translate The situation is shown below. We need to determine the pressure just above and below the roof.

What is the net force exerted by the inside and outside air on the roof?

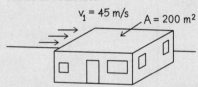

$v_1 = 45$ m/s $A = 200$ m²

Simplify and diagram A force diagram for the roof is shown below. The air above the house exerts a downward force on the roof $F_{1 \text{ on R}} = P_1 A$, where P_1 is the air pressure above the house and A is the area of the roof. The air inside the house pushes up, exerting a force on the roof $F_{2 \text{ on R}} = P_{\text{atm}} A$, where P_{atm} is the assumed atmospheric pressure of the stationary air inside the house. We assume the air is incompressible and flows without friction or turbulence and that the roof is fairly thin so that the air has approximately the same gravitational potential energy density at points 1 and 2.

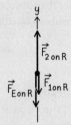

Represent mathematically With the y-axis oriented upward, the net force exerted by the air on the roof is

$$F_{\text{net Air}} = F_{2 \text{ on R}} - F_{1 \text{ on R}} = P_{\text{atm}} A - P_1 A$$
$$= (P_{\text{atm}} - P_1) A$$

We use Bernoulli's equation to find this pressure difference.

$$P_2 + \frac{1}{2}\rho v_2^2 + \rho g y_2 = \frac{1}{2}\rho v_1^2 + \rho g y_1 + P_1$$

$$\Rightarrow P_{\text{atm}} + 0 + \rho g y_2 = \frac{1}{2}\rho v_1^2 + \rho g y_1 + P_1$$

$$\Rightarrow P_{\text{atm}} - P_1 = \frac{1}{2}\rho v_1^2 + (\rho g y_1 - \rho g y_2) = \frac{1}{2}\rho v_1^2 + 0$$

We can now determine the net force exerted by the air on the roof.

$$F_{\text{net Air}} = (P_{\text{atm}} - P_1)A = \frac{1}{2}\rho v_1^2 A$$

Solve and evaluate

$$F_{\text{net Air}} = \frac{1}{2}\rho v_1^2 A$$
$$= \frac{1}{2}(1.3 \text{ kg/m}^3)(45 \text{ m/s})^2(200 \text{ m}^2)$$
$$= 2.6 \times 10^5 \text{ N}$$

The result is an upward net force that is enough to lift more than ten cars of combined mass 30,000 kg.

We were a little lax in applying Bernoulli's equation in Example 11.6. The equation relates the properties of a fluid at two points along the same streamline. A streamline does not flow between just below the roof and just above the roof. We could, with some more complex reasoning, use the equation correctly by considering two streamlines that start far from the house at the same pressure. One ends up in the house under its roof with the air barely moving. The other passes just above the roof with the air moving fast. We would get the same result in a somewhat more cumbersome manner.

Try it yourself: A 2.0 m × 2.0 m canvas covers a trailer. The trailer moves at 29 m/s (65 mi/h). Determine the net force exerted on the canvas by the air above and below it.

Answer: An upward force of 2000 N (about 500 lb). No wonder the canvas covering a truck trailer moving on a highway balloons outward.

TIP Remember: Hurricane winds do not tear the roofs from houses. The air inside the house actually pushes the roof upward from the house. Similarly, when you drink through a straw, the atmospheric air pushes down on the liquid in the glass, which then moves up into the lower pressure air in the straw.

Dislodging plaque

The physical principles of a roof being lifted from a house also explain how plaque can become dislodged from the inner wall of an artery. Consider the figure in Example 11.5. The plaque may block a considerable portion of the area where blood normally flows. Suppose the radius of the vessel opening is one-third its normal value because of the plaque. Then the area available for blood flow, proportional to r^2, is about one-ninth the normal value. The speed of flow in the narrowed portion of the artery will be about 9 times greater than in the unblocked part of the vessel. The kinetic energy density term in Bernoulli's equation is proportional to v^2 and therefore is 81 times greater in the constricted area than in the open part of the vessel.

Notice that in Bernoulli's equation, the sum of the gravitational potential energy density, the kinetic energy density, and the pressure at one location should equal the sum of the same three terms at some other location along a streamline in the blood. As blood speeds by the plaque, its kinetic energy density is 81 times greater, and consequently its pressure is much less than the pressure in the open vessel just before and just after the plaque. This pressure differential could cause the plaque to be pulled off the wall and tumble downstream, causing a blood clot (a process called thrombosis). Let's estimate the net force that the blood exerts on the plaque.

EXAMPLE 11.7 A clogged artery

Blood flows through the unobstructed part of a blood vessel at a speed of 0.50 m/s. The blood then flows past a plaque that constricts the cross-sectional area to one-ninth the normal value. The surface area of the plaque parallel to the direction of blood flow is about $0.60 \text{ cm}^2 = 6.0 \times 10^{-5} \text{ m}^2$. Estimate the net force that the fluid exerts on the plaque.

Sketch and translate A simplified sketch of the situation is shown below. Point 1 is above the plaque in stationary blood pooled in a channel where the plaque attaches to the artery wall. Point 2 is in the bloodstream below the plaque, where blood flows rapidly past it. The net force will depend on the differences in pressure at points 1 and 2. Thus, we need to first find the pressure difference. Since points 1 and 2 are not on the same streamline, we cannot automatically use Bernoulli's equation. But since the streamlines that do go through points 1 and 2 were side by side before they reached the plaque, each streamline will have the same $P + (1/2)\rho v^2 + \rho g y$ value. This means we can equate the $P + (1/2)\rho v^2 + \rho g y$ values at points 1 and 2.

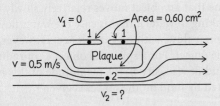

Simplify and diagram Assume for simplicity that the blood is nonviscous and flows with laminar flow without turbulence. Assume also that the vertical distance between points 1 and 2 is small ($y_1 - y_2 \approx 0$) and that the area of the stationary blood above the plaque is the same as the area where the blood moves below the plaque. A bar chart represents the process. The blood pressure at position 2 is less than at position 1 because the blood flows at high speed through the constricted artery, whereas it sits at rest in the channels at position 1 ($v_1 = 0$).

Represent mathematically Compare the two points using Bernoulli's equation:

$$P_1 - P_2 = \frac{1}{2}\rho\left(v_2^2 - v_1^2\right) + \rho g\left(y_2 - y_1\right)$$

$$= \frac{1}{2}\rho\left(v_2^2 - 0\right) + 0 = \frac{1}{2}\rho v_2^2$$

Since the cross-sectional area of the vessel at the location of the plaque is one-ninth its normal area (the radius is one-third the normal value), the blood must be flowing at nine times its normal speed. The net force exerted by the blood on the plaque downward and perpendicular to the direction the blood flows will be

$$F_{\text{net blood on P }y} = F_{\text{blood 1 on P}} + \left(-F_{\text{blood 2 on P}}\right)$$
$$= P_1 A - P_2 A = \left(P_1 - P_2\right)A$$
$$\Rightarrow F_{\text{net blood on P }y} = \frac{1}{2}\rho v_2^2 A$$

Solve and evaluate

$$F_{\text{net blood on P }y}$$
$$= \frac{1}{2}\left(1050 \text{ kg/m}^3\right)\left(9 \times 0.5 \text{ m/s}\right)^2\left(6.0 \times 10^{-5}\text{ m}^2\right)$$
$$= 0.64 \text{ N} \approx 0.6 \text{ N}.$$

This is about the weight of one-half of an apple pulling on this tiny plaque. In addition, an "impact" force caused by blood hitting the plaque's upstream side contributes to the risk of breaking the plaque off the side wall of the vessel. The loose plaque can then tumble downstream and block blood flow in a smaller vessel in the heart (causing a heart attack) or in the brain (causing a stroke).

Try it yourself: Air of density 1.3 kg/m³ moves at speed 10 m/s across the top surface of a clarinet reed that has an area of 3 cm². The air below the reed is not moving and is at atmospheric pressure. Determine the net force exerted on the reed by the air above and below it.

Answer: 0.02 N upward, toward the inside of the mouthpiece.

Figure 11.12 Measuring blood pressure

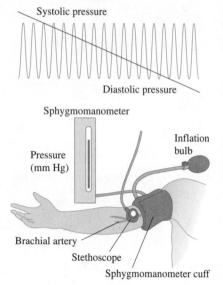

Systolic pressure

Diastolic pressure

Sphygmomanometer

Pressure (mm Hg)

Inflation bulb

Brachial artery

Stethoscope

Sphygmomanometer cuff

Measuring blood pressure

To measure a person's blood pressure, a nurse uses a device called a sphygmomanometer (see **Figure 11.12**). She places a cuff around the upper arm of a patient at about the level of the heart and places a stethoscope on the inside of the elbow above the brachial artery in the arm. The nurse then increases the gauge pressure in the cuff to about 180 mm Hg by pumping air into the cuff. The expanded cuff pushes on the brachial artery and stops blood flow in the arm. Then the nurse slowly releases the air from the cuff, decreasing the pressure of the air in it. When the pressure in the cuff is equal to the systolic pressure (120 mm Hg if the systolic blood pressure is normal), blood starts to squeeze through the artery past the cuff. The flow is intermittent and turbulent and causes a sound heard with the stethoscope. This turbulent sound continues until the cuff pressure decreases below the diastolic pressure (80 mm Hg for normal diastolic blood pressure). At that point the artery is continually open and blood flow is laminar and makes no sound. The systolic and diastolic pressure numbers together make up the blood pressure measurement.

Review Question 11.7 Why does a roof lift up off a house during a high-speed wind?

11.8 Drag force

So far in this chapter all of our analyses have focused on a moving fluid—air flowing past a roof or blood flowing through an artery. In Section 11.6 we were concerned with the friction-like resistance that occurs as these fluids move through a tube. Now we focus on solid objects moving through a fluid—for example, a swimmer moving through water, a skydiver falling through the air, or a car traveling through air. As you know from experience, the fluid in these and in other cases exerts a resistive **drag force** on the object moving through the fluid. So far we have been neglecting this force in our mechanics problems. Now we will not only learn how to calculate this force but also learn whether our assumptions were reasonable: for example, is the resistive drag force insignificant when people and objects fall from small and large heights?

Laminar Drag Force Imagine that an object moves relatively slowly through a fluid (for example, a rock sinking in water). In this case the water flows around the object in streamline laminar flow, with no turbulence. However, the fluid does exert a drag force on the object. For a spherical object O of radius r moving at speed v through a liquid of viscosity η, the magnitude of this nonturbulent drag force $F_{\mathrm{D\,F\,on\,O}}$ exerted by fluid on the object is given by the equation

$$F_{\mathrm{D\,F\,on\,O}} = 6\pi\eta r v \qquad (11.10)$$

This equation is called **Stokes's law**. Notice that the drag force is proportional to the speed of the object relative to the fluid and to the radius of the object.

Turbulent Drag Force A rock falls through air much faster than through water. In this case the motion of the air past the falling rock is turbulent and Eq. (11.10) does not apply. A different Reynolds number can be used to decide whether the flow of fluid past an object is laminar or turbulent:

$$R_e = \frac{v l \rho}{\eta} \qquad (11.11)$$

where v is the object's speed with respect to the fluid, l is the length of the object, and ρ and η are the density and viscosity of the fluid. When the Reynolds number is calculated using this equation, the threshold value for the laminar flow is 1. If the Reynolds number is more than 1, the flow is turbulent and we cannot use Stokes's law. In this case, a new equation for drag force applies:

$$F_{\text{D F on O}} \approx \frac{1}{2}C_{\text{D}}\rho A v^2 \qquad (11.12)$$

where ρ is the density of fluid, A is the cross-sectional area of the object as seen along its line of motion, and C_{D} is a dimensionless number called the *drag coefficient*. The drag coefficient depends on the shape of the object (the lower the number, the smaller the drag force and the more streamline the flow past the object).

Drag force exerted on a moving vehicle

Does Stokes's law apply to moving cars? At 60 mi/h (about 30 m/s), for a car about 2 m wide in air with density 1.3 kg/m³ and viscosity 2×10^{-5} N·s/m², the estimated Reynolds number will be

$$R_{\text{e}} = \frac{v l \rho}{\eta} = \frac{(30 \text{ m/s})(2 \text{ m})(1.3 \text{ kg/m}^3)}{(2 \times 10^{-5} \text{ N·s/m}^2)} \approx 4 \times 10^6.$$

This is much more than 1. We need to use Eq. (11.12) for the drag force.

QUANTITATIVE EXERCISE 11.8 Drag force exerted on a car

Estimate the magnitude of the drag force that air exerts on a compact car traveling at 27 m/s (60 mi/h). The drag coefficient C_{D} is approximately 0.5 for a well-designed car, and the air density is 1.3 kg/m³.

Represent mathematically The flow of air past the car is turbulent, and we use Eq. (11.12) to estimate the drag force that the air exerts on the car:

$$F_{\text{D F on O}} = \frac{1}{2}C_{\text{D}}\rho A v^2$$

Solve and evaluate We estimate that the cross-sectional area of a car is 2 m². Thus,

$$F_{\text{D F on O}} = \frac{1}{2}(0.5)(1.3 \text{ kg/m}^3)(2 \text{ m}^2)(27 \text{ m/s})^2 = 470 \text{ N}$$

or a force of about 100 lb. Designing cars to minimize drag force improves fuel economy.

Try it yourself: Estimate the drag force (without using a calculator) if the car's speed is 13 m/s, half the value in the quantitative exercise.

Answer: About one-fourth the answer above, or about 120 N.

Drag force exerted on a falling person

We can estimate whether the drag force exerted on a person falling from a building is significant. For example, assume that a 70-kg person accidentally falls from a second-floor window. The force exerted on him by Earth is about 700 N. The buoyant force exerted on the person by the air is about

$$F_{\text{B A on P}} = \rho_{\text{air}}g V_{\text{body}} \approx (1.3 \text{ kg/m}^3)(9.8 \text{ N/kg})\frac{(70 \text{ kg})}{(1000 \text{ kg/m}^3)} \approx 0.9 \text{ N}$$

where the person's volume is his mass divided by his density. We see that the buoyant force is very small. To estimate the drag force, assume that the drag coefficient is 1, that the person's cross-sectional area is about $(1.6 \text{ m})(0.3 \text{ m}) = 0.5 \text{ m}^2$, and that the air density is 1.3 kg/m³. The speed after falling freely a distance of 2 m is about 6 m/s ($v_{\text{f}}^2 = 2g(h_{\text{f}} - h_{\text{i}})$). Thus, our estimate for the drag force exerted by air on the person after falling 2 m is about

$$F_{\text{D F on P}} = \frac{1}{2}(1)(1.3 \text{ kg/m}^3)(0.5 \text{ m}^2)(6 \text{ m/s})^2 \approx 12 \text{ N}$$

> **TIP** Note that if the speed of a car increases by two times, the drag force exerted on it quadruples. Thus, because of air drag, when you increase your driving speed, you reduce your gas mileage.

Thus, for a person falling from small heights, the drag force is significant but small. What about falls from higher heights?

Terminal speed

As a skydiver falls through the air, her speed increases and the drag force that the air exerts on the diver also increases. Eventually, the diver's speed becomes so great that the resistive drag force that the air exerts on the diver equals the downward gravitational force that Earth exerts on the diver. The net force exerted on the diver is balanced, so the diver moves downward at a constant speed, known as **terminal speed**. Let's estimate the terminal speed for a skydiver.

EXAMPLE 11.9 Terminal speed of skydiver

Estimate the terminal speed of a 60-kg skydiver falling through air of density 1.3 kg/m^3, assuming a drag coefficient $C_D = 0.6$.

Sketch and translate The situation is sketched below. When the diver is moving at terminal speed, the forces that the air exerts on the diver and that Earth exerts on the diver balance—the net force is zero. We choose the diver as the system of interest with vertical y-axis pointing upward.

When the forces that the air and the Earth exert on the diver are equal in magnitude, the diver falls at constant terminal speed.

$m = 60\,kg$

$v_{Terminal} = ?$

Simplify and diagram A force diagram for the diver is shown to the right. Assume that the buoyant force that the air exerts on the diver is negligible in comparison to the other forces exerted on her and that the drag force involves turbulent airflow past the diver.

y

$\vec{F}_{D\,A\,on\,D}$

$\vec{F}_{E\,on\,D}$

Represent mathematically Use the force diagram to help apply Newton's second law for the diver:

$$ma_y = \Sigma F_y$$
$$\Rightarrow 0 = F_{D\,A\,on\,D} + (-F_{E\,on\,D})$$
$$\Rightarrow 0 = \frac{1}{2}C_D\rho A v_{terminal}^2 - mg$$

Solve and evaluate Solving for the diver's terminal speed gives

$$v_{terminal} = \sqrt{\frac{2mg}{C_D\rho A}}$$

All of the quantities in the above expression are known except the cross-sectional area of the diver along her line of motion. If we assume that she is 1.5 m tall and 0.3 m wide, her cross-sectional area is about 0.5 m^2. We find that her terminal speed is

$$v_{terminal} = \sqrt{\frac{2(60\,kg)(9.8\,N/kg)}{(0.6)(1.3\,kg/m^3)(0.5\,m^2)}} = 55\,m/s$$

The unit is correct. The magnitude seems reasonable— about 120 mi/h.

Try it yourself: Suppose she pulled her legs near her chest so she was more in the shape of a ball. How qualitatively would that affect her terminal speed? Explain.

Answer: Her cross-sectional area would be smaller, and according to the above equation, her terminal speed would be greater.

Review Question 11.8 When a skydiver falls downward at constant terminal speed, shouldn't the resistive drag force that the air exerts on the skydiver be a little less than the downward gravitational force that Earth exerts on the diver? If they are equal, shouldn't the diver stop falling? Explain.

Summary

Words	Pictorial and physical representations	Mathematical representation
Flow rate The flow rate Q of a fluid is the volume V of fluid that passes a cross section in a tube divided by the time interval Δt needed for that volume to pass. The flow rate also equals the product of the average speed v of the fluid and the cross-sectional area A of the vessel. (Section 11.2)	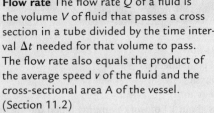	Flow rate Q: $$Q = \frac{V}{\Delta t} = vA \qquad \text{Eq. (11.2)}$$
Continuity equation If fluid does not accumulate, the flow rate into a region (position 1) must equal the flow rate out of the region (position 2). At position 1 the fluid has speed v_1 and the tube has cross-sectional area A_1. At position 2 the fluid has speed v_2 and the tube has cross-sectional area A_2. (Section 11.2)	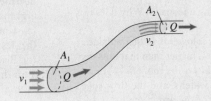	Continuity equation: $$Q = v_1 A_1 = v_2 A_2 \qquad \text{Eq. (11.3)}$$
Bernoulli's equation For a fluid flowing without friction or turbulence, the sum of the kinetic energy density $(1/2)\rho v^2$, the gravitational potential energy density $\rho g y$, and pressure P of the fluid is a constant. (Section 11.4)	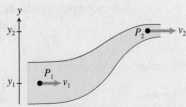	Bernoulli's equation: $$\frac{1}{2}\rho v_1^{\,2} + \rho g y_1 + P_1$$ $$= P_2 + \frac{1}{2}\rho v_2^{\,2} + \rho g y_2$$ $$\text{Eq. (11.6)}$$
Poiseuille's law For viscous fluid flow, the pressure drop $(P_1 - P_2)$ across a fluid of viscosity η flowing in a tube depends on the length l of the tube, its radius r, and the fluid flow rate Q. (Section 11.6)	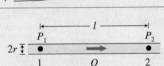	Poiseuille's law: $$P_1 - P_2 = \frac{8}{\pi}\frac{\eta l}{r^4}Q \quad \text{Eq. (11.8)}$$
Laminar drag force When a spherical object (like a balloon falling in air) moves slowly through a fluid, the fluid exerts a resistive drag force on the object that is proportional to the object's speed v. **Stokes's law** describes the force. (Section 11.8)	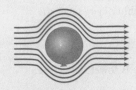	Laminar drag force (Stokes's law): $$F_D = 6\pi\eta r v \qquad \text{Eq. (11.10)}$$
Turbulent drag force For an object moving at faster speed through a fluid (like a tilted airplane wing), turbulence occurs and the resistive drag force is proportional to the square of the speed. (Section 11.8)		Turbulent drag force: $$F_D = \frac{1}{2}C_D\rho A v^2 \qquad \text{Eq. (11.12)}$$

* For instructor-assigned homework, go to **MasteringPhysics**.

Questions

Multiple Choice Questions

1. Two empty soda cans stand on a smooth tabletop. If you blow horizontally between the cans, what will they do?
   (a) Move further apart.
   (b) Move closer together.
   (c) Stay where they are.

2. A roof is blown off a house during a tornado. Why does this happen?
   (a) The air pressure in the house is lower than that outside.
   (b) The air pressure in the house is higher than that outside.
   (c) The wind is so strong that it blows the roof off.

3. A river flows downstream and widens, and the flow speed slows. As a result, the pressure of the water against a dock downstream compared to upstream will be
   (a) higher.   (b) lower.   (c) the same.

4. Air blowing past the window of a building causes the window to be blown
   (a) out.   (b) in.
   (c) It depends on the direction of the wind.

5. Why does the closed top of a convertible bulge when the car is riding along a highway?
   (a) The volume of air inside the car increases.
   (b) The air pressure is greater outside the car than inside.
   (c) The air pressure inside the car is greater than the pressure outside.
   (d) The air blows into the front part of the roof, lifting the back part.

6. How does Bernoulli's principle *help* explain air going up the chimney of a house?
   (a) Air blowing across the top of the chimney reduces the pressure above the chimney.
   (b) The air above the chimney attracts the ashes.
   (c) The hot ashes seek the cooler outside air.
   (d) The gravitational potential energy is lower above the chimney.

7. Why does cutting the end of an envelope and blowing air past the cut end cause the envelope to bulge open?
   (a) The blown air goes in and increases the pressure inside.
   (b) The blown air blowing past the outside exerts less pressure than air inside.
   (c) The person blowing tends to squish the edges of the envelope.

8. As a river approaches a dam, the width of the river increases and the speed of the flowing water decreases. What can explain this effect?
   (a) Bernoulli's equation
   (b) The continuity equation
   (c) Poiseuille's law

9. What is an incompressible fluid?
   (a) A law of physics
   (b) A physical quantity
   (c) A model of an object

10. What is viscous flow?
    (a) A physical phenomenon
    (b) A law of physics
    (c) A physical quantity

11. The heart does about 1 J of work pumping blood during one heartbeat. What is the immediate first and main type of energy that increases due to the heart's work?
    (a) Kinetic energy
    (b) Thermal energy
    (c) Elastic potential energy

12. You have a glass of water with a straw in it (see **Figure Q11.12**). The end of the straw is about 1 cm above the water. You now blow hard through a second straw so that air moves across the top of the first straw (do not blow into it). Predict what happens.
    (a) The water level in the straw goes down.
    (b) The water level in the straw goes up and possibly sprays out of the straw.
    (c) The water level in the straw is not affected by the air moving above it.

Figure Q11.12

Blow

Conceptual Questions

13. A hair dryer blowing air over a ping-pong ball will support it, as shown in **Figure Q11.13**. Construct a force diagram for the ball. Explain in terms of forces how the ball can remain in equilibrium.

Figure Q11.13

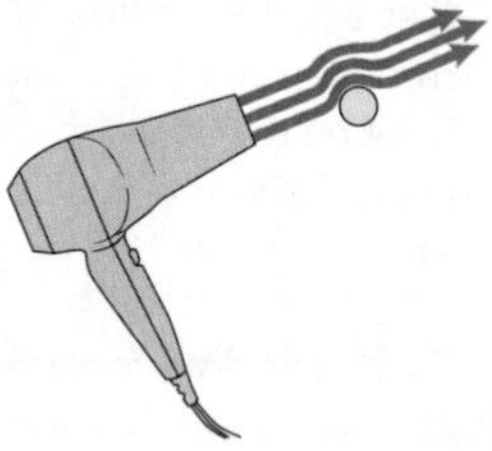

Air

14. You have two identical large jugs with small holes on the side near the bottom. One jug is filled with water and the other with liquid mercury. The liquid in each jug, sitting on a table, squirts out the side hole into a container on the floor. Which container, the one catching the water or the one catching the mercury, must be closer to the table in order to catch the fluid? Or should they be placed at the same distance? Which jug will empty first, or do they empty at the same time? Explain.

15. Why does much of the pressure drop in the circulatory system occur across the arterioles (small vessels carrying blood to the capillaries) and capillaries as opposed to across the much larger diameter arteries?

16. If you partly close the end of a hose with your thumb, the water squirts out farther. Give at least one explanation for why this phenomenon occurs.

Problems

Below, BIO indicates a problem with a biological or medical focus. Problems labeled EST ask you to estimate the answer to a quantitative problem rather than develop a specific answer. Asterisks indicate the level of difficulty of the problem. Problems with no * are considered to be the least difficult. A single * marks moderately difficult problems. Two ** indicate more difficult problems. Unless stated otherwise, assume in these problems that atmospheric pressure is $1.01 \times 10^5 \, \text{N/m}^2$ and that the densities of water and air are $1000 \, \text{kg/m}^3$ and $1.3 \, \text{kg/m}^3$, respectively.

11.1 and 11.2 Fluids moving across surfaces—qualitative analysis and Flow rate and fluid speed

1. **Watering plants** You water flowers outside your house. (a) Determine the flow rate of water moving at an average speed of 32 cm/s through a garden hose of radius 1.2 cm. (b) Determine the speed of the water in a second hose of radius 1.0 cm that is connected to the first hose.

2. **Irrigation canal** You live near an irrigation canal that is filled to the top with water. (a) It has a rectangular cross section of 5.0-m width and 1.2-m depth. If water flows at a speed of 0.80 m/s, what is its flow rate? (b) If the width of the stream is reduced to 3.0 m and the depth to 1.0 m as the water passes a flow-control gate, what is the speed of the water past the gate?

3. **Fire hose** During a fire, a firefighter holds a hose through which $0.070 \, \text{m}^3$ of water flows each second. The water leaves the nozzle at an average speed of 25 m/s. What information about the hose can you determine using these data?

4. The main waterline for a neighborhood delivers water at a maximum flow rate of $0.010 \, \text{m}^3/\text{s}$. If the speed of this water is 0.30 m/s, what is the pipe's radius?

5. * **BIO** **Blood flow in capillaries** The flow rate of blood in the aorta is $80 \, \text{cm}^3/\text{s}$. Beyond the aorta, this blood eventually travels through about 6×10^9 capillaries, each of radius 8.0×10^{-4} cm. What is the speed of the blood in the capillaries?

6. * **Irrigating a field** It takes a farmer 2.0 h to irrigate a field using a 4.0-cm-diameter pipe that comes from an irrigation canal. How long would the job take if he used a 6.0-cm pipe? What assumption did you make? If this assumption is not correct, how will your answer change?

11.4 Bernoulli's equation

Figure P11.7

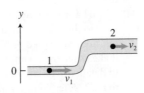

7. Represent the process sketched in **Figure P11.7** using a qualitative Bernoulli bar chart and an equation (include only terms that are not zero).

8. Represent the process sketched in **Figure P11.8** using a qualitative Bernoulli bar chart and an equation (include only terms that are not zero).

Figure P11.8

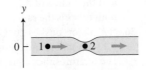

9. **Fluid flow problem** Write a symbolic equation (include only terms that are not zero) and draw a sketch of a situation that could be represented by the qualitative Bernoulli bar chart shown in **Figure P11.9** (there are many possibilities).

10. Repeat Problem 9 using the bar chart in **Figure P11.10**.

11. Repeat Problem 9 using the bar chart in **Figure P11.11**.

12. Repeat Problem 9 using the bar chart in **Figure P11.12**.

13. An application of Bernoulli's equation is shown below. Construct a qualitative Bernoulli bar chart that is consistent with the equation and draw a sketch of a situation that could be represented by the equation (there are many possibilities).
$$\rho g \, y_2 = 0.5 \rho v_1^2$$

14. Repeat Problem 13 using the equation below. The size of the symbols represents the relative magnitudes of the physical quantities at two points.
$$0.5\rho \, v_1^2 + (P_1 - P_2) = 0.5 \, \rho v_2^2$$

15. Repeat Problem 13 using the equation below. The size of the symbols represents the relative magnitudes of the physical quantities at two points.
$$0.5\rho \, v_1^2 + (P_1 - P_2) = 0.5 \, \rho v_2^2 + \rho g y_2$$

16. * **Wine flow from barrel** While visiting a winery, you observe wine shooting out of a hole in the bottom of a barrel. The top of the barrel is open. The hole is 0.80 m below the top surface of the wine. Represent this process in multiple ways (a sketch, a bar chart, and an equation) and apply Bernoulli's equation to a point at the top surface of the wine and another point at the hole in the barrel.

17. **Water flow in city water system** Water is pumped at high speed from a reservoir into a large-diameter pipe. This pipe connects to a smaller diameter pipe. There is no change in elevation. Represent the water flow from the large pipe to the smaller pipe in multiple ways—a sketch, a bar chart, and an equation.

Figure P11.9

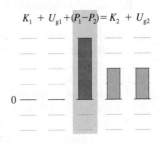

$$K_1 + U_{g1} + (P_1 - P_2) = K_2 + U_{g2}$$

Figure P11.10

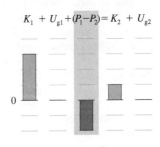

$$K_1 + U_{g1} + (P_1 - P_2) = K_2 + U_{g2}$$

Figure P11.11

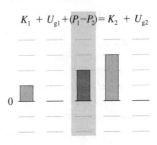

$$K_1 + U_{g1} + (P_1 - P_2) = K_2 + U_{g2}$$

Figure P11.12

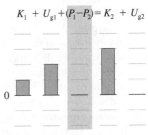

$$K_1 + U_{g1} + (P_1 - P_2) = K_2 + U_{g2}$$

11.5 Skills for analyzing processes using Bernoulli's equation

18. * The pressure of water flowing through a 0.060-m-radius pipe at a speed of 1.8 m/s is $2.2 \times 10^5 \, \text{N/m}^2$. What is (a) the flow rate of the water and (b) the pressure in the water after it goes up a 5.0-m-high hill and flows in a 0.050-m-radius pipe?

19. * **Siphoning water** You want to siphon rainwater and melted snow from the cover of an above-ground swimming pool. The cover is 1.4 m above the ground. You have a plastic hose of 1.0-cm radius with one end in the water on the pool cover and the other end on the ground. (a) At what speed does water exit the hose? (b) If you want to empty the pool cover in half the time, how much wider should the hose be? (c) How much faster does the water flow through this wider pipe?

20. * **Cleaning skylights** You are going to wash the skylights in your kitchen. The skylights are 8.0 m above the ground. You connect two garden hoses together—a 0.80-cm-radius hose to a 1.0-cm-radius hose. The smaller hose is held on the roof of the house and the wider hose is attached to the faucet on the ground. The pressure at the opening of the smaller hose is 1 atm, and you want the water to have the speed of 6.0 m/s. What should be the pressure at ground level in the large hose? What should be the speed?

21. ** **Community water system** A large city waterline pipe has radius 0.060 m and feeds ten smaller pipes, each of radius 0.020 m, that carry water to homes. The flow rate of water in each of the smaller pipes is to be $6.0 \times 10^{-3} \, \text{m}^3/\text{s}$, and the pressure is $4.00 \times 10^5 \, \text{N/m}^2$. The homes are 10.0 m above the main pipe. What is the average speed of the water in (a) a smaller pipe and (b) the main pipe? (c) What is the pressure in the main pipe?

22. * **BIO Blood flow in artery** Blood flows at an average speed of 0.40 m/s in a horizontal artery of radius 1.0 cm. The average pressure is $1.4 \times 10^4 \, \text{N/m}^2$ above atmospheric pressure (the gauge pressure). (a) What is the average speed of the blood past a constriction where the radius of the opening is 0.30 cm? (b) What is the gauge pressure of the blood as it moves past the constriction?

23. * **Window in wind** You are on the 48th floor of the Windom Hotel. The day is stormy, and you wonder whether the hotel is a safe place. According to the weather report, the air speed is 20 m/s; the size of the window in your room is 1.0 m × 2.0 m. (a) What is the difference in pressure between the inside and outside air? (b) What is the net force that the air exerts on the window (magnitude and direction)? (c) What assumption did you make to answer these questions?

24. * **Straw aspirator** A straw extends out of a glass of water by a height h. How fast must air blow across the top of the straw to draw water to the top of the straw?

25. * **Gate for irrigation system** You observe water at rest behind an irrigation dam. The water is 1.2 m above the bottom of a gate that, when lifted, allows water to flow under the gate. Determine the height h from the bottom of the dam that the gate should be lifted to allow a water flow rate of $1.0 \times 10^{-2} \, \text{m}^3/\text{s}$. The gate is 0.50 m wide.

11.6 Viscous fluid flow

26. * A 5.0-cm-radius horizontal water pipe is 500 m long. Water at 20 °C flows at a rate of $1.0 \times 10^{-2} \, \text{m}^3/\text{s}$. (a) Determine the pressure drop due to viscous friction from the beginning to the end of the pipe. (b) What radius pipe must you use if you

want to keep the pressure difference constant and double the flow rate?

27. **Fire hose** A volunteer firefighter uses a 5.0-cm-diameter fire hose that is 60 m long. The water moves through the hose at 12 m/s. The temperature outside is 20 °C. What is the pressure drop due to viscous friction across the hose?

28. **Another fire hose** The pump for a fire hose can develop a maximum pressure of $6.0 \times 10^5 \, \text{N/m}^2$. A horizontal hose that is 50 m long is to carry water of viscosity $1.0 \times 10^{-3} \, \text{N} \cdot \text{s/m}^2$ at a flow rate of $1.0 \, \text{m}^3/\text{s}$. What is the minimum radius for the hose?

29. * **Solar collector water system** Water flows in a solar collector through a copper tube of radius R and length l. The average temperature of the water is T °C and the flow rate is $Q \, \text{cm}^3/\text{s}$. Explain how you would determine the viscous pressure drop along the tube, assuming the water does not change elevation.

30. * **BIO Blood flow through capillaries** Your heart pumps blood at a flow rate of about 80 cm^3/s. The blood flows through approximately 9×10^9 capillaries, each of radius 4×10^{-4} cm and 0.1 cm long. Determine the viscous friction pressure drop across a capillary, assuming a blood viscosity of $4 \times 10^{-3} \, \text{N} \cdot \text{s/m}^2$.

11.7 Applying fluid dynamics: Putting it all together

31. * **BIO Flutter in blood vessel** A person has a 5200-N/m^2 gauge pressure of blood flowing at 0.50 m/s inside a 1.0-cm-radius main artery. The gauge pressure outside the artery is 3200 N/m^2. When using his stethoscope, a physician hears a fluttering sound farther along the artery. The sound is a sign that the artery is vibrating open and closed, which indicates that there must be a constriction in the artery that has reduced its radius and subsequently reduced the internal blood pressure to less than the external 3200-N/m^2 pressure. What is the maximum artery radius at this constriction?

32. * **BIO Effect of smoking on arteriole radius** The average radius of a smoker's arterioles, the small vessels carrying blood to the capillaries, is 5% smaller than those of a nonsmoker. (a) Determine the percent change in flow rate if the pressure across the arterioles remains constant. (b) Determine the percent change in pressure if the flow rate remains constant.

33. * **Roof of house in wind** The mass of the roof of a house is 2.1×10^4 kg and the area of the roof is 160 m^2. At what speed must air move across the roof of the house so that the roof is lifted off the walls? Indicate any assumptions you made.

34. * You have a U-shaped tube open at both ends. You pour water into the tube so that it is partially filled. You have a fan that blows air at a speed of 10 m/s. (a) How can you use the fan to make water rise on one side of the tube? Explain your strategy in detail. (b) To what maximum height can you get the water to rise? Note: You cannot touch the water yourself.

35. * Determine the ratio of the flow rate through capillary tubes A and B (that is, Q_A/Q_B). The length of A is twice that of B, and the radius of A is one-half that of B. The pressure across both tubes is the same.

36. * A piston pushes 20 °C water through a horizontal tube of 0.20-cm radius and 3.0-m length. One end of the tube is open and at atmospheric pressure. (a) Determine the force needed to push the piston so that the flow rate is 100 cm^3/s. (b) Repeat the problem using SAE 10 oil instead of water.

37. * Engineers use a venturi meter to measure the speed of a fluid traveling through a pipe (see **Figure P11.37**). Positions 1 and 2 are in pipes with surface areas A_1 and A_2, with A_1 greater than A_2, and are at the same vertical height. How can you determine the relative speeds at positions 1 and 2 and the pressure difference between positions 1 and 2?

Figure P11.37

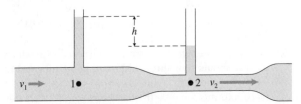

38. * How can you use the venturi meter system (see Problem 37) to determine whether viscous fluid needs an additional pressure difference to flow at the same speed as a nonviscous fluid?

11.8 Drag force

39. **Car drag** A 2300-kg car has a drag coefficient of 0.60 and an effective frontal area of 2.8 m². Determine the air drag force on the car when traveling at (a) 24 m/s (55 mi/h) and (b) 31 m/s (70 mi/h).

40. * EST **Air drag when biking** Estimate the drag force opposing your motion when you ride a bicycle at 8 m/s.

41. BIO **Drag on red blood cell** Determine the drag force on a red blood cell with a radius of 1.0×10^{-5} m and moving through 20 °C water at speed 1.0×10^{-5} m/s. (Assume laminar flow.)

42. * BIO EST **Protein terminal speed** A protein of radius 3.0×10^{-9} m falls through a tube of water with viscosity $\eta = 1.0 \times 10^{-3}$ N·s/m². Earth exerts a constant downward 3.0×10^{-22}-N force on the protein. (a) Use Stokes's law and the information provided to estimate the terminal speed of the protein. Assume no buoyant force is exerted on the protein. (b) How many hours would be required for the protein to fall 0.10 m?

43. * Earth exerts a constant downward force of 7.5×10^{-13} N on a clay particle. The particles settle 0.10 m in 820 min. Determine the radius of a clay particle. Assume no buoyant force is exerted on the clay particle. The viscosity of water is 1.0×10^{-3} N·s/m².

44. * A sphere falls through a fluid. Earth exerts a constant downward 0.50-N force on the sphere. The fluid exerts an opposing drag force on the fluid given by $F_D = 2v$, where F_D is in newtons if v is in meters per second. Determine the terminal speed of the sphere.

45. * **Terminal speed of balloon** A balloon of mass m drifts down through the air. The air exerts a resistive drag force on the balloon described by the equation $F_D = 0.03v^2$ where F_D is in newtons if v is in meters per second. What is the terminal speed of the balloon?

General Problems

46. ** EST A cooler filled with water has a hole of radius 0.40 cm at the bottom. The hole is originally closed with a plug. The cooler is about 1.0 m tall, and the bottom has area 0.4 m × 0.6 m. Determine the initial flow rate of water after removing the plug. Estimate how long it will take to empty the

cooler. What assumptions did you make? If they are not valid, will the real time be greater or smaller than the estimate?

47. ** Design an elevator-like device that can lift you to your dorm room by blowing air across the top surface of the elevator. Be sure to provide the details in your design and indicate any difficulties you might encounter.

48. ** BIO **Pressure needed for intravenous needle** A glucose solution of viscosity 2.2×10^{-3} N·s/m² and density 1030 kg/m³ flows from an elevated open bag into a vein. The needle into the vein has a radius of 0.20 mm and is 3.0 cm long. All other tubes leading to the needle have much larger radii, and viscous forces in them can be ignored. The pressure in the vein is 1000 N/m² above atmospheric pressure. (a) Determine the pressure relative to atmospheric pressure needed at the entrance of the needle to maintain a flow rate of 0.10 cm³/s. (b) To what elevation should the bag containing the glucose be raised to maintain this pressure at the needle?

49. ** **Viscous friction with Bernoulli** We can include the effect of viscous friction in Bernoulli's equation by adding a term for the thermal energy generated by the viscous retarding force exerted on the fluid. Show that the term to be added to Eq. (11.5) for flow in a vessel of uniform cross-sectional area A is

$$\frac{\Delta U_{Th}}{V} = \frac{4\pi \eta l v}{A}$$

where v is the average speed of the fluid of viscosity η along the center of a pipe whose length is l.

50. ** (a) Show that the work W done per unit time Δt by viscous friction in a fluid with a flow rate Q across which there is a pressure drop ΔP is

$$\frac{W}{\Delta t} = \Delta P Q = Q^2 R = \frac{\Delta P^2}{R}$$

where $R = 8\eta l/\pi r^4$ is called the *flow resistance* of the fluid moving through a vessel of radius r. (b) By what percentage must the work per unit time increase if the radius of a vessel decreases by 10% and all other quantities including the flow rate remain constant (the pressure does not remain constant)?

51. ** EST BIO **Thermal energy in body due to viscous friction** Estimate the thermal energy generated per second in a normal body due to the viscous friction force in blood as it moves through the circulatory system.

52. ** BIO **Essential hypertension** Suppose your uncle has hypertension that causes the radii of his 40,000 arterioles to decrease by 20%. Each arteriole initially was 0.010 mm in radius and 1.0 cm long. By what factor does the resistance $R = 8\eta l/\pi r^4$ to blood flow through an arteriole change because of these decreased radii? The pressure drop across all of the arterioles is about 60 mm Hg. If the flow rate remains the same, what now is the pressure drop change across the arteriole part of the circulatory system?

53. * **Parachutist** A parachutist weighing 80 kg, including the parachute, falls with the parachute open at a constant 8.5-m/s speed toward Earth. The drag coefficient $C_D = 0.50$. What is the area of the parachute?

54. A 0.20-m-radius balloon falls at terminal speed 0.40 m/s. If the drag coefficient is 0.50, what is the mass of the balloon?

55. ** **Terminal speed of skier** A skier going down a slope of angle θ below the horizontal is opposed by a turbulent drag

force that the air exerts on the skier and by a kinetic friction force that the snow exerts on the skier. Show that the terminal speed is

$$v_T = \left[\frac{2mg(\sin\theta - \mu\cos\theta)}{C_D \rho A} \right]^{1/2}$$

where μ is the coefficient of kinetic friction between the skis and the snow, ρ is the density of air, A is the skier's frontal area, and C_D is the drag coefficient.

56. ** A grain of sand of radius 0.15 mm and density 2300 kg/m³ is placed in a 20 °C lake. Determine the terminal speed of the sand as it sinks into the lake. Do not forget to include the buoyant force that the water exerts on the grain.

57. ** EST **Comet crash** On June 30, 1908, a monstrous comet fragment of mass greater than 10^9 kg is thought to have devastated a 2000-km² area of remote Siberia (this impact was called the Tunguska event). Estimate the terminal speed of such a comet in air of density 0.70 kg/m³. State all of your assumptions.

Reading Passage Problems

BIO **Intravenous (IV) feeding** A patient in the hospital needs fluid from a glucose nutrient bag. The glucose solution travels from the bag down a tube and then through a needle inserted into a vein in the patient's arm (**Figure 11.13a**). Your study of fluid dynamics makes you think that the bag seems a little low above the arm and the narrow needle seems long. You wonder if the glucose is actually making it into the patient's arm. What height should the bag (open at the top) be above the arm so that the glucose

Figure 11.13 (a) A glucose solution flowing from an open container into a vein. (b) The analysis of the needle in this system.

(a)

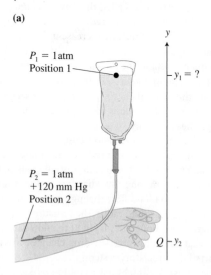

(b)

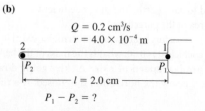

solution (density 1000 kg/m³ and viscosity 1.0×10^{-3} N·s/m²) drains from the open bag down the 0.6-m-long, 2.0×10^{-3}-m radius tube and then through the 0.020-m-long, 4.0×10^{-4}-m radius needle and into the vein? The gauge pressure in the vein in the arm is +930 N/m² (or 7 mm Hg). The nurse says the flow rate should be 0.20×10^{-6} m³/s (0.2 cm³/s).

58. Which answer below is closest to the speed with which the glucose should flow out of the end of the needle at position 2 in Figure 11.13b?
    (a) 0.0004 m/s   (b) 0.004 m/s   (c) 0.04 m/s
    (d) 0.4 m/s   (e) 4 m/s

59. Which answer below is closest to the speed with which the glucose should flow through the end of the tube just to the right of position 1 in Figure 11.13b?
    (a) 0.0002 m/s   (b) 0.002 m/s   (c) 0.02 m/s
    (d) 0.2 m/s   (e) 2 m/s

60. Assume that there is no resistive friction pressure drop across the needle (as could be determined using Poiseuille's law). Use the Bernoulli equation and the results from Problems 58 and 59 to determine which answer below is closest to the change in pressure between positions 1 and 2 ($P_1 - P_2$) in Figure 11.13b.
    (a) 8 N/m²   (b) 80 N/m²   (c) 800 N/m²
    (d) 8000 N/m²   (e) 80,000 N/m²

61. Now, in addition to the Bernoulli pressure change from position 1 to position 2 calculated in Problem 60, there may be a Poiseuille resistive friction pressure change across the needle from position 1 to position 2. Which answer below is closest to that pressure change?
    (a) 0.4 N/m²   (b) 4 N/m²   (c) 40 N/m²
    (d) 400 N/m²   (e) 4000 N/m²

62. The blood pressure in the vein at position 2 in Figure 11.13b at the exit of the needle into the blood is 930 N/m². Use this value and the results of Problems 60 and 61 to determine which answer below is closest to the gauge pressure at position 1 in the tube carrying the glucose to the needle.
    (a) 1010 N/m²   (b) 1410 N/m²   (c) 1980 N/m²
    (d) 2800 N/m²   (e) 4620 N/m²

63. Suppose that there is no Poiseuille resistive friction pressure decrease from the top of the glucose solution in the open bag (position 1 in Figure 11.13a) through the tube and down to position 2 near the entrance to the needle. Which answer below is closest to the minimum height of the top of the bag in order for the glucose to flow down from the tube and through the needle into the blood? Remember that the pressure at position 1 is atmospheric pressure, which is a zero gauge pressure.
    (a) 0.04 m   (b) 0.08 m   (c) 0.14 m
    (d) 0.27 m   (e) 0.60 m

64. Suppose there is a Poiseuille resistive friction pressure decrease from the top of the glucose solution (position 1 in Figure 11.13a) through the tube and down to position 2 near the entrance to the needle. How will this affect the placement of the bag relative to the arm?
    (a) The bag will need to be higher.
    (b) The bag can remain the same height above the arm.
    (c) The bag can be placed lower relative to the arm
    (d) Too little information is provided to answer the question.

BIO **The human circulatory system** In the human circulatory system, depicted in **Figure 11.14**, the heart's left ventricle pumps about 80 cm³ of blood into the aorta. The blood then moves into

Figure 11.14 A schematic representation of the circulatory system including the pressure variation across different types of vessels.

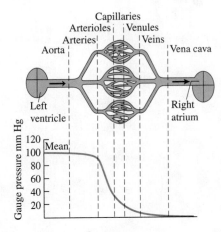

a larger and larger number of smaller radius vessels (aorta, arteries, arterioles, and capillaries). After the capillaries, which deliver nutrients to the body cells and absorb waste products, the vessels begin to combine into a smaller number of larger radius vessels (venules, small veins, large veins, and finally the vena cava). The vena cava returns blood to the heart (see **Table 11.5**).

The flow rate Q of blood through the arteries equals the flow rate through the arterioles, which equals the flow rate through the capillaries, and so forth. The average blood gauge pressure in the aorta is about 100 mm Hg. The pressure drops as blood passes through the different groups of vessels and is approximately 0 mm Hg when it returns to the heart at the vena cava.

A working definition of the resistance R to flow by a group of vessels is the ratio of the gauge pressure drop ΔP across those vessels divided by the flow rate Q through the vessels:

$$R = \frac{\Delta P}{Q} \qquad (11.13)$$

The gauge pressure drop across the whole system is (100 mm Hg − 0), and the total resistance is

$$R_{\text{total}} = \frac{\Delta P_{\text{total}}}{Q} = \frac{100 \text{ mm Hg}}{80 \text{ cm}^3/\text{s}} = 1.25 \frac{\text{mm Hg}}{\text{cm}^3/\text{s}}$$

Table 11.5 A rough description of the different types of vessels in the circulatory system.

Vessel type	Number of vessels	Approximate radius (mm)
Aorta	1	5
Large arteries	40	2
Smaller arteries	2400	0.4
Arterioles	40,000,000	0.01
Capillaries	1,200,000,000	0.004
Venules	80,000,000	0.02
Small veins	2400	1
Large veins	40	3
Vena cava	1	6

The gauge pressure drop across the whole system is the sum of the drops across each type of vessel:

$$\Delta P_{\text{total}} = \Delta P_{\text{aorta}} + \Delta P_{\text{arteries}} + \Delta P_{\text{arterioles}} + \Delta P_{\text{capillaries}}$$
$$+ \cdots + \Delta P_{\text{vena cava}}$$

Now rearrange and insert Eq. (11.13) into the above for the pressure drop across each part:

$$QR_{\text{total}} = QR_{\text{aorta}} + QR_{\text{arteries}} + QR_{\text{arterioles}} + QR_{\text{capillaries}}$$
$$+ \cdots + QR_{\text{vena cava}}$$

Canceling the common flow rate through each group of vessels, we have an expression for the total resistance of the circulatory system:

$$R_{\text{total}} = R_{\text{aorta}} + R_{\text{arteries}} + R_{\text{arterioles}} + R_{\text{capillaries}} + \cdots + R_{\text{vena cava}}$$

The measured gauge pressure drop across the arterioles is about 50 mm Hg, and the arteriole resistance is

$$R_{\text{arterioles}} = \frac{\Delta P_{\text{arterioles}}}{Q} = \frac{50 \text{ mm Hg}}{80 \text{ cm}^3/\text{s}} = 0.62 \frac{\text{mm Hg}}{\text{cm}^3/\text{s}}$$

or about 50% of the total resistance. The next most resistive group of vessels is the capillaries, at about 25% of the total resistance. These percentages vary significantly from person to person. A person with essential hypertension has arterioles and capillaries that are reduced in radius. The resistance to blood flow increases dramatically (the resistance has a $1/r^4$ dependence). The blood pressure has to be greater (for example, double the normal value) in order to produce reasonable flow to the body cells. Even with the increased pressure, the flow rate may still be lower than normal.

65. The capillaries typically produce about 25% of the resistance to blood flow. Which pressure drop below is closest to the pressure drop across the group of capillaries?
    (a) 5 mm Hg   (b) 15 mm Hg   (c) 25 mm Hg
    (d) 35 mm Hg   (e) 45 mm Hg

66. We found that the arteriole resistance to fluid flow was about 0.62 mm Hg/(cm³/s). By what factor would you expect the resistance of all the arterioles to change if the radius of each arteriole decreased by 0.8?
    (a) 1.3   (b) 1.6   (c) 2.4
    (d) 0.4   (e) 0.6

67. Why is the resistance to fluid flow through unobstructed arteries relatively small compared to resistance to fluid flow through the arterioles and capillaries?
    (a) The arteries are nearer the heart.
    (b) There are a relatively small number of arteries.
    (c) The artery radii are relatively large.
    (d) b and c
    (e) a, b, and c

68. The huge number of capillaries and venules is needed to
    (a) provide nutrients (such as O_2) and remove waste products from all of the body cells.
    (b) distribute water uniformly throughout the body.
    (c) reduce the resistance of the circulatory system.
    (d) b and c
    (e) a, b, and c

69. Which number below best represents the ratio of the resistance of a single capillary to the resistance of a single arteriole, assuming they are equally long?
    (a) 40   (b) 6   (c) 2.5
    (d) 0.4   (e) 0.026

12

First Law of Thermodynamics

How is the CO_2 concentration in the atmosphere related to the melting of glaciers?

How does sweating protect our bodies from overheating?

Why does the thermal energy of a spoon increase when you place it in a cup of hot tea even though no external objects do work on it?

Be sure you know how to:

- Identify a system and decide on the initial and final states of a process (Section 2.1).
- Draw an energy bar chart and use it to help apply the work-energy equation (Section 6.2).
- Apply your understanding of molecular motion to explain gas processes (Section 9.6).

Earth's average temperature is increasing. The 1980s were the warmest decade in recorded history. However, each decade since has beaten that record. Before the Industrial Revolution of the 1800s, the concentration of atmospheric CO_2 ranged from 200 ppm (parts per million) during relatively cold periods to 280 ppm during relatively warm periods. The CO_2 concentration is now over 370 ppm and is increasing at a rate of about 1.6 ppm/year. What is the connection between the concentration of CO_2 in Earth's atmosphere and the changes in its average global temperature? See one consequence in **Figure 12.1**.

In this chapter, we combine our understanding of the random motion of atoms and molecules (kinetic molecular theory) with the ideas of work and energy to describe and explain such phenomena as how clouds form, what is causing global climate change, and why athletes get overheated on hot summer days. In physics, these questions are part of the study of **thermodynamics**.

When we studied work and energy (Chapter 6), we learned that a system's energy could change when external objects do work on it. However in real life we know many examples of processes that can change the internal energy of a system without external objects exerting forces and doing work. Imagine for example that you put a room-temperature spoon into a cup of hot coffee. Our understanding of work-energy processes does not explain why the spoon gets hot. However, we can use thermodynamics to analyze such processes.

12.1 Internal energy and work in gas processes

The ideal gas model describes a gas consisting of atoms and molecules that are assumed to be identical point objects—particles. These particles do not interact with each other at a distance but do obey Newton's laws when colliding with each other and with the walls of their container. We determined (in Chapter 9) an expression for the energy of the particles in a container of such a gas.

Thermal energy of ideal gas

Imagine a bottle filled with air. The air is our system. What energy does this system possess? It does not have any gravitational potential energy (Earth is not in the system) and it has no organized kinetic energy. However, the gas consists of individual particles that are moving randomly.

Earlier (in Chapter 9) we reasoned that each gas particle has some average kinetic energy

$$\overline{K}_{particle} = \frac{3}{2}kT_K$$

due to its random motion. T is the absolute (kelvin) temperature of the gas. Consequently, the N molecules in the container have a total random kinetic energy (called **thermal energy**):

$$U_{thermal} = N\left(\frac{3}{2}kT_K\right)$$

Do gas molecules possess any potential energy? We assumed that ideal gas particles do not interact at a distance; thus, the system has no potential energy due to particle interactions. Therefore, in an ideal gas the total internal energy of the gas particles equals its thermal energy.

When we studied ideal gases (Chapter 9), we expressed the thermal energy of the gas in different ways. The number N of particles can also be written as the number of moles n of the gas times Avogadro's number of particles N_A in one mole ($N = nN_A$). Also, recall that $N_A k = R$, the universal gas constant and $T = T_K$ is the temperature in kelvins. Thus we can rewrite the equation as

$$U_{thermal} = N\left(\frac{3}{2}kT\right) = nN_A\left(\frac{3}{2}kT\right) = n\left(\frac{3}{2}N_A kT\right) = \frac{3}{2}nRT \quad (12.1)$$

Note that the thermal energy of this gas depends only on its absolute temperature and on the number of moles of gas. It does not depend on the volume it occupies.

Figure 12.1 Warming Earth causes glaciers to melt. Photos taken in 1941 and 2004 show the melting of Alaska's Muir glacier.

‹ Active Learning Guide

The thermal energy of a gas is called a **state function**. A state function is a property of a system that describes the system at a particular time. It is possible that we will need several state functions to completely describe the system.

TIP We distinguish between the thermal energy

$$U_{\text{thermal}} = N\left(\frac{3}{2}kT\right)$$

of all N particles in a container of gas and the average random kinetic energy

$$\overline{K} = \frac{3}{2}kT$$

of a single particle. Imagine a small and a large container of the same gas, both at the same temperature. The gas in the larger container has more thermal energy even though the average kinetic energy of individual molecules in both containers is the same.

EXAMPLE 12.1 Thermal energy of the gas in a room

Construct an expression that could be used to estimate the thermal energy of air at pressure P in a room of volume V. Use the expression to estimate the thermal energy of the air filling a small bedroom.

Sketch and translate We have an expression for the thermal energy of a container of gas in terms of its absolute temperature T and the number n of moles of particles (Eq. 12.1). We need to connect these quantities to the gas pressure P and the room volume V. The ideal gas law seems promising for relating these quantities.

Simplify and diagram Assume that the air in the room is an ideal gas, that the air pressure is $P_{\text{atm}} = 1.0 \times 10^5 \, \text{N/m}^2$, and that the dimensions of a small bedroom are $4 \, \text{m} \times 3 \, \text{m} \times 3 \, \text{m} = 36 \, \text{m}^3$.

Represent mathematically The thermal energy of the gas is given by Eq. (12.1):

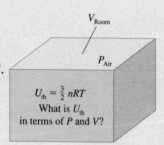

$U_{\text{th}} = \frac{3}{2} nRT$
What is U_{th} in terms of P and V?

$U_{\text{th}} = (3/2)nRT$. Using the ideal gas law, $PV = nRT$, we find that

$$U_{\text{th}} = \frac{3}{2}nRT = \frac{3}{2}PV$$

We can use this expression to estimate the thermal energy of the gas.

Solve and evaluate The thermal energy of the gas in a small bedroom is approximately:

$$U_{\text{th}} = \frac{3}{2}PV = \frac{3}{2}(10^5 \, \text{N/m}^2)(36 \, \text{m}^3)$$

$$= 54 \times 10^5 \, \text{N} \cdot \text{m} \approx 5 \times 10^6 \, \text{J}$$

The result is equivalent to the energy given off by a 100-J/s (100-watt) lightbulb during a 50,000-s (14-hour) time interval—a significant amount of energy! Notice that when estimating, we keep only one significant figure.

Try it yourself: What is the thermal energy of the air in an empty 1-L soft drink bottle? State your assumptions.

Answer: 150 J. We assumed that the gas pressure inside the bottle equaled atmospheric pressure and that the air inside the bottle could be modeled as an ideal gas.

Work done on a gas

Figure 12.2 The piston does work on the gas, changing its thermal energy.

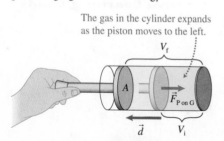

The gas in the cylinder expands as the piston moves to the left.

Gases play an important role in many common mechanical devices. For example, the explosion of a mixture of air and gasoline in the cylinder of a car's engine increases the gas pressure against the piston in that cylinder. The gas at high pressure pushes the piston outward, which causes the driveshaft to rotate and causes the car wheels to turn. How much work does the gas in the cylinder do during such an expansion?

Imagine that you have a gas (the system) at high pressure in a cylinder with a movable piston (**Figure 12.2**). You hold the piston and allow it to move slowly outward (to the left in Figure 12.2). The gas pushes to the left on the piston and the piston in turn pushes toward the right on the gas, exerting a force on the gas $\vec{F}_{\text{P on G}}$. The gas inside the piston expands slowly. What work is done by the force that the piston exerts on the gas? Assume that the piston moves outward a distance d, allowing the gas to expand from its initial volume V_i to its final larger volume V_f.

Recall that work is the product of the magnitude of the force exerted on an object, the object's displacement, and the cosine of the angle between the directions of the force and the displacement ($W = Fd\cos\theta$). This equation only works for a *constant* magnitude force. In our case, as the high-pressure gas expands, there is a decrease in the frequency of collisions of particles with the piston, and the pressure decreases. Thus, the piston exerts a *changing* force on the expanding gas.

To determine the work done by the piston on the gas, we could use calculus. However, it is also possible to determine an expression for the work done on the changing volume of gas by breaking the big volume change into many small increments and adding the work done during each small change. Let's start with a process in which the piston moves outward in small steps each of distance Δx starting at position x_i and ending at position x_f (**Figure 12.3a**). For each step the gas volume change is small, the gas pressure changes little, and the force that the piston exerts on the gas remains approximately constant. The gas pushes outward on the piston whether the piston is causing an expansion or contraction of the gas. The piston in turn exerts a force on the gas in the opposite direction (toward the right in Figure 12.3a). If the gas is expanding, then the piston in Figure 12.3a is moving left, and the work done by the piston on the gas for the Δx expansion is

$$W_{\text{Piston on Gas}} = F_{\text{P on G}}\Delta x \cos(180°) = -F_{\text{P on G}}\Delta x$$

The force that the piston exerts on the gas and the piston's displacement point in opposite directions—hence the minus sign.

The change in the volume of the gas when the piston moves out a distance Δx is $\Delta V = A\Delta x$ where A is the cross-sectional area of the piston. Rearranging the above, we find that $\Delta x = (\Delta V)/A$. Recall also the definition of pressure: $P = (F/A)$ or $F = P \cdot A$. Insert these expressions for F and Δx into the above expression for work to get

$$W_{\text{Piston on Gas}} = -F_{\text{P on G}}\Delta x = -(P \cdot A)\left(\frac{\Delta V}{A}\right) = -P\Delta V \qquad (12.2)$$

As the gas has expanded and has a larger volume, ΔV is positive. Since the piston is pushing opposite the expansion, it does negative work on the gas and the energy of the gas decreases. If instead the piston exerted a force that caused the gas to compress, the force would do positive work on the system (ΔV would be negative and it would cancel the negative sign in Eq. (12.2); the gas would have increased energy).

Remember also that this is the work done *on the gas* by the external force that the piston exerts on the gas during a small change in the gas volume. This work done by the piston on the gas during a small expansion of the gas is the negative of the area under the narrow dark-shaded column of the P-versus-V graph line in Figure 12.3b. This area is the height P of the narrow column times its width ΔV.

We split the large change in gas volume for the whole process into n tiny volume changes. For each small volume change, assume that the pressures are approximately constant (Figure 12.3b). Then add the work done during each of these volume changes to get the total work done by the piston on the gas:

$$W_{\text{Piston on Gas}} = (-P_1\Delta V_1) + (-P_2\Delta V_2) + \cdots + (-P_n\Delta V_n)$$

Each term is the negative of the area of a narrow column in Figure 12.3b. Thus, the work done during the expansion from volume V_i to volume V_f is the negative area under the entire graph line between those initial and final volumes.

Figure 12.3 Determine the work done on a changing volume of gas.

(a)

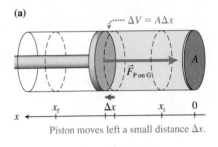

Piston moves left a small distance Δx.

(b)

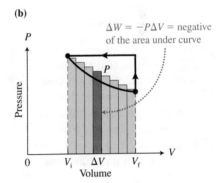

Figure 12.4 Work depends on the way in which the system changes from the initial to the final state.

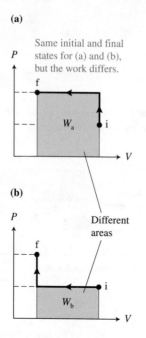

(a)

Same initial and final states for (a) and (b), but the work differs.

(b)

Different areas

TIP Notice that the thermal energy of the gas in a cylinder depends only on the temperature of the gas when in that state (and on the number of particles). Temperature and internal thermal energy are state functions.

Work depends on the process between the initial and final states

Many different processes can take a gas from an initial pressure P_i and volume V_i to a final pressure P_f and volume V_f (different curves or lines on a P-versus-V graph). The amount of work done on the gas can differ in each case (see the examples in **Figures 12.4a** and b). Note that in Figure 12.4a, to go from the initial to the final state, the gas first has a constant volume and increasing pressure and then a constant pressure and decreasing volume. On the other hand, in (b), the gas first has a constant-pressure volume decrease and then a constant-volume pressure increase to get to the same final state. Note that the work done for each process, represented by the shaded area under the graph lines, is different—less for the process in (b) than the process in (a), even though the initial and final states are the same. The work done on or by a gas depends on how the pressure and volume change in going from the initial to the final state. The work depends on the process that occurs in changing state and is *not* a state function. It does not describe a single state or the difference between the two states of the system.

Work done *by* a gas *on* its environment

The gas mixture in the cylinders of your automobile engine is part of a motor. In this case we are not interested in the work done by the environment (the piston) on the gas (the system) but are instead interested in the work done by that gas on the environment. When the gas in the cylinder expands, it does positive work on its environment. The environment in turn does the same magnitude negative work on the gas. The work that the piston does on the gas is equal in magnitude and opposite in sign to the work that the gas does on the piston—an example of Newton's third law. Thus, we can define work involving gas changes as follows:

> **Work done on or by a gas** The work done by the environment on a gas $W_{\text{Environment on Gas}}$ when a gas changes from volume V_i to V_f is the negative of the area under the P-versus-V graph for that process. The work is negative if the gas volume increases and positive if its volume decreases. The work done by the gas on the environment $W_{\text{Gas on Environment}}$ has the same magnitude but the opposite sign. If the pressure is constant, the work done by the environment on the gas is
>
> $$W_{\text{Environment on Gas}}(\text{constant pressure process}) = -P(V_f - V_i) = -P\Delta V \quad (12.2)$$
>
> and the work done by the gas on the environment is
>
> $$W_{\text{Gas on Environment}}(\text{constant pressure process}) = +P(V_f - V_i) = +P\Delta V \quad (12.3)$$

Review Question 12.1 Imagine that a balloon expands when brought from a cold garage into a warm room. The room is at atmospheric pressure and the change in the volume of the balloon is ΔV. What is the work that the air in the room does on the balloon during the process?

12.2 Two ways to change the energy of a system

Throughout this text, we have stressed the importance of defining a system—a carefully identified object or group of objects on which we focus our attention. Everything outside the system is the environment. A system is characterized by physical quantities, such as the system's internal energy, mass, volume, temperature, and number of particles. The interaction of the environment

Active Learning Guide >

with the system is defined in terms of the work done by the environment on the system. We devised the generalized work-energy equation [Eq. (6.3)] that summarizes these ideas:

$$U_i + W_{\text{Environment on System}} = U_f$$

Using this equation, we find that the total initial energy U_i of the system plus the work done on the system by the environment $W_{\text{Environment on System}}$ equals the total final energy U_f of the system. The energy of the system can be in many different forms, but the only way the total energy of the system can change is if the environment does work on the system.

$$(K_i + U_{gi} + U_{si} + \cdots) + W_{\text{E on S}} = (K_f + U_{gf} + U_{sf} + \cdots + \Delta U_{\text{internal}})$$

Are there situations where these ideas are inadequate? Consider Observational Experiment **Table 12.1**. To analyze these experiments we will use a rewritten version of the generalized work-energy equation:

$$W_{\text{Environment on System}} = U_f - U_i$$

OBSERVATIONAL EXPERIMENT TABLE

12.1   Using the work-energy principle to analyze two processes.

Observational experiment	Analysis
Experiment 1. Place a fixed amount of gas in a fixed-volume container. A flame below causes the gas to warm from temperature T_i to T_f.  Constant V	■ The temperature of the gas increased; hence its thermal energy increased: $\Delta U_{th} > 0$. ■ The work done on the gas was zero, because the gas volume did not change: $W = 0$. ■ Thus, $0 = \Delta U_{th}$. *According to the generalized work-energy equation, this process cannot occur—the left side is zero and the right side is a positive number.*
Experiment 2. Place a fixed amount of gas in a cylindrical enclosure with a piston that can move freely up or down in the cylinder, thus keeping the pressure exerted on the gas constant. A flame below causes the gas to expand and warm from temperature T_i to T_f. It takes longer to reach T_f than it did in Experiment 1. 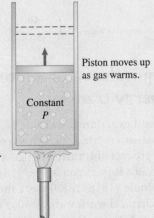 Piston moves up as gas warms. Constant P	■ The temperature of the gas increased; hence its thermal energy increased: $\Delta U_{th} > 0$. ■ The piston pushed down on the gas opposite the direction of its expansion; thus negative work was done on the gas ($W_{\text{Piston on Gas}} < 0$). ■ Thus, $-|W| = +|U_{th}|$. *According to the generalized work-energy equation, this process cannot occur—the left side is negative and the right side is positive.*

Pattern

Analyzing both experiments using the work-energy principle, we find that according to this principle the processes cannot possibly occur. However, since we observe these processes in real life, they must be possible. We conclude that either the principle is incorrect or it needs modification.

The generalized work-energy principle is unable to account for the energy changes in the two experiments in Table 12.1. What modification in the principle would help us explain what happened in those two experiments?

Particle motion explains the change in thermal energy of the gas

To reconcile the work-energy principle with the observational experiments described in Table 12.1, we need to assume that energy can be transferred to the gas by means other than work. As we know, the **temperature** of a substance is related to the random motion and vibration of the particles of which the substance is made. Consider Experiment 1. In a hot flame the particles are moving very quickly in random directions. They collide with the slower randomly moving particles in the cool cylinder containing the gas. The particles in that cool cylinder gain random kinetic energy and move faster and/or vibrate more. The particles have greater random motion, which corresponds to a higher temperature, so the cylinder gets warmer. Thus the energy of the hot flame was transferred to the cooler cylinder without any work done. The same mechanism transfers energy from the cylinder walls to the cool gas inside it. Overall, the energy from the hot flame is transferred to the cool gas. This new way of transferring energy from an object at one temperature (the hot flame) to an object at a different temperature (the cool gas) is called **heating**.

> **TIP** Note that we use the term *heating* rather than *heat*. We do this deliberately. Heating does not mean the actual increase of the temperature of the system; it means the process by which energy is transferred from a hot object to a cooler object (without any work being done).

> **Heating** Q is a physical quantity that characterizes a process of transferring energy from the environment to a system, which is at a different temperature. The environment does no work on the system.

The SI unit of heating is the joule. However, another common unit of heating is the calorie. One calorie is the amount of energy that must be transferred to 1 g of water to increase its temperature by 1 degree C. If you have a known mass of water that changes its temperature by a known number of degrees, you can determine how many calories of energy were transferred to this water through the process of heating. To do this you must multiply one calorie by the mass and the temperature difference; the answer will be in units of calories:

$$Q_{\text{Environment to Water}} = \left(1\,\frac{\text{cal}}{\text{g}\cdot{}^\circ\text{C}}\right)(m_{\text{water}})(\Delta T)$$

Testing the equivalence of heating and work as a means of energy transfer

Recall that in the second experiment in Table 12.1 negative work was done on the system, but the system's thermal energy increased. How do we explain this effect using our new process of transferring energy by heating? German physician and physicist Julius Robert von Mayer (1814–1878) and English brewer and physicist James Joule (1818–1889) were the first scientists to speculate that heating and mechanical work were similar processes in terms of energy transfer. The difference between the two processes, they hypothesized, is that for the energy to be transferred to a system through heating, there must be some temperature difference between the system and the environment. This is not the case with mechanical work. Both men performed experiments to test this idea.

First we will describe James Joule's testing experiment in Testing Experiment **Table 12.2.**

TESTING EXPERIMENT TABLE

12.2 Testing the equivalence of work and heating as a means for energy transfer.

Testing experiment	Prediction based on the idea under test	Outcome
Joule placed a paddle wheel in water. The wheel was connected with a string passing over pulleys tied to a heavy block. When the block went down, the paddle wheel rotated in the water. 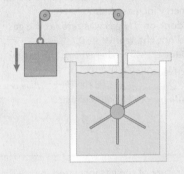	Our system will be Earth, block, paddle wheel, and water. The work W done in lifting a block of mass m a height h equals the gain in the gravitational potential energy of the system $\Delta U_g = m_{block}gh$ measured in joules. When the block is released and returns to its starting position, the paddle wheel turning in the water should cause the water to gain the same amount of thermal energy. We can predict the temperature change of the water assuming that the gravitational potential energy of the system is converted into the internal energy of the water: $$m_{block}gh = \left(1\frac{cal}{g\cdot{}^\circ C}\right)(m_{water})(\Delta T)$$ or $$\Delta T = \frac{m_{block}gh}{\left(1\frac{cal}{g\cdot{}^\circ C}\right)(m_{water})}.$$ To validate the assumption and minimize the transfer of energy to the container, pulleys, strings, and surrounding air, Joule covered the container in blankets, used light and smooth pulleys, and made other adjustments.	Joule lifted the block multiple times to get a bigger temperature increase and repeated the experiment many times. In all experiments the observed change in temperature of the water agreed with the predicted change.

Conclusion

The amount that the system's energy changes when external objects do mechanical work on it is the same as when external objects transferred the same amount of energy by heating.

Based on Table 12.2 we can conclude that the thermal energy of a system can be changed by two means: mechanical work and heating. Mechanical work and heating are both ways of transferring energy between the environment and the system. Joule found that 1.0 J of work done by the person lifting the block and transferred to the kinetic energy of the paddle wheel inside the liquid caused the same effect as the transfer of 0.24 cal through the process of heating to the same liquid. In independent experiments others observed the same effect:

$$1.0\ J = 0.24\ cal$$
$$1.0\ cal = 4.2\ J$$

Consequently, the amount of energy that we need to transfer to 1 kg of water to change its temperature by 1 °C is 1000 cal or 4200 J.

Review Question 12.2 Suppose that heating with an electric burner transfers 400 cal of energy to a pan of water. How many joules of mechanical energy would produce the same change in the water's thermal energy?

12.3 First law of thermodynamics

Now that we have established the equivalence of heating and work as mechanisms through which energy can be transferred between an environment and a system, we can incorporate the concept of heating into the generalized work-energy principle. Let's try now to explain the outcomes of the experiments in Table 12.1 by including a heating term in the equation:

$$W_{\text{Environment on System}} + Q_{\text{Environment to System}} = \Delta U_{\text{System}}$$

OBSERVATIONAL EXPERIMENT TABLE

12.3 Using heating to explain the experiments performed in Table 12.1.

Observational experiment		Analysis
Experiment 1. Place a fixed amount of gas in a fixed-volume container. A flame below causes the gas to warm from temperature T_i to T_f.	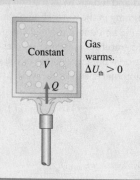 Constant V — Gas warms. $\Delta U_{\text{th}} > 0$ Q	■ The temperature of the gas increased; hence its thermal energy increased: $\Delta U_{\text{th}} > 0$. ■ The work done on the gas was zero, as the gas volume did not change: $W = 0$. ■ The hot flame transferred energy to the gas by heating: $Q_{\text{Flame to Gas}} > 0$. ■ Thus, $0 + Q_{\text{Flame to Gas}} = \Delta U_{\text{th}}$. *The equation is consistent with the experimental outcome.*
Experiment 2. Place a fixed amount of gas in a cylindrical enclosure with a piston that can move freely up or down in the cylinder, thus keeping the pressure exerted on the gas constant. A flame below causes the gas to expand and warm from temperature T_i to T_f. It takes longer to reach the final temperature than in Experiment 1.	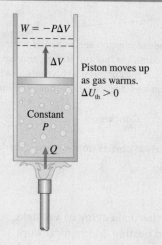 $W = -P\Delta V$ ΔV Constant P Q Piston moves up as gas warms. $\Delta U_{\text{th}} > 0$	■ The temperature of the gas increased; hence the thermal energy increased: $\Delta U_{\text{th}} > 0$. ■ The piston pushed down on the gas opposite the direction of its expansion, thus negative work was done on the gas ($W_{\text{Piston on Gas}} < 0$). ■ The hot flame transferred more energy to the gas by heating than in Experiment 1—the gas was exposed to the flame for a longer period of time. $Q_{\text{Flame to Gas}} > 0$. ■ Thus, $-|W| + Q_{\text{Flame to Gas}} = +\Delta U_{\text{th}}$. *The equation is consistent with the experimental outcome.*

Pattern

The new work-heating-energy change principle explains the outcomes of both experiments.

The work-heating-energy equation succeeds in explaining many phenomena that the old work-energy principle could not—such as the change of the temperature of water (the system) placed on a hot electric stove (the environment) and the change in the temperature of a hot hard-boiled egg (the system) placed in a bowl of cold water (the environment). In each of these cases, the energy of the system changes with no work done on it. But in all cases the system contacts a part of the environment that is *at a different temperature* than the system.

A note about temperature, thermal energy, and heating

We've learned that energy can only be transferred spontaneously through heating when the temperatures of the system and the environment are different. This is different than saying that their thermal energies must be different. Imagine a small spoon at room temperature placed in a bathtub filled with water at room temperature. Although the water in the bathtub has far more thermal energy than the water in the bathtub has far more thermal energy than the spoon does, there is no energy transfer from the water to the spoon.

The spoon and bathtub example highlights the difference between three physical quantities: temperature, thermal energy, and heating. Temperature is the physical quantity that quantifies the *average* random kinetic energy of the individual particles that comprise the object. Temperature does not depend on the number of particles in the object. Thermal energy is a physical quantity that quantifies the *total* random kinetic energy of all the particles. Adding more particles at the same temperature to the system increases the system's thermal energy. Heating is the physical quantity that quantifies the *process* through which some *amount* of thermal energy is transferred between the system and the environment when they are at different temperatures. If two systems have the same temperatures but different thermal energies, there is no transfer of energy between them through heating.

> **TIP** When the temperature of a system is lower than that of the environment, "positive heating" occurs—the system gains energy from the environment; when the temperature of the system is higher than the environment's, "negative heating" occurs—the system loses energy to the environment.

Quantitative analysis of a process involving heating

The physician Robert Mayer approached the idea of work-heating-energy from a different angle. His reasoning is used in the following example.

EXAMPLE 12.2 Quantitative analysis of heating a gas

It was known in Mayer's times that 1.0 g of air requires 0.17 cal of energy transferred through heating at *constant volume* for its temperature to increase by 1.0 °C and 0.24 cal of energy transferred through heating at *constant pressure* for the temperature to change by 1.0 °C. Are these data consistent with the work-heating-energy equation and the equivalence of work and heating?

Sketch and translate A sketch for each experiment is shown below. The system is the gas. In the first experiment an electric heater warms the gas (transfer of energy though heating) while the gas remains at constant volume. In the second experiment the heater again warms the gas but this time at constant pressure P. As the gas warms at constant pressure, it expands and pushes outward on a piston.

(a) Experiment 1. Constant volume heating

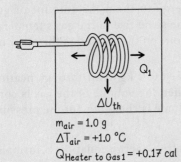

$m_{air} = 1.0$ g
$\Delta T_{air} = +1.0$ °C
$Q_{Heater\ to\ Gas\ 1} = +0.17$ cal

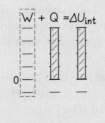

(b) Experiment 2. Constant pressure heating

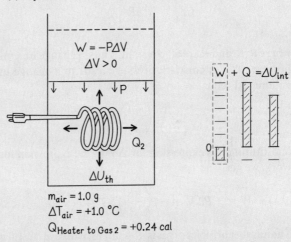

$m_{air} = 1.0$ g
$\Delta T_{air} = +1.0$ °C
$Q_{Heater\ to\ Gas\ 2} = +0.24$ cal

Simplify and diagram We assume that the gas in each experiment is an ideal gas and that the piston moves without friction. We can represent each experiment with an energy bar chart. Notice the Q bar next to the work bar; it indicates the energy transferred through heating to the system. The temperature increases by 1.0 °C in each experiment—the same internal energy change in both cases. In the second experiment, the piston does negative work on the gas (because the force exerted by the piston on the gas points opposite the direction of expansion of the gas). To balance the bar chart in the second experiment, the heating must be greater than in the first experiment.

Represent mathematically In the first experiment, the heating of the gas leads to a change in its internal energy:

$$0 + Q_{Heater\ to\ Gas\ 1} = \Delta U_{Thermal\ Gas\ 1}$$

(continued)

In the second experiment, in addition to heating the gas, the environment does negative work on the gas:

$$W_{\text{Environment on Gas 2}} + Q_{\text{Heater to Gas 2}} = \Delta U_{\text{Thermal Gas 2}}$$

Since both gases 1 and 2 undergo the same temperature change, the internal energy changes are equal ($\Delta U_{\text{Thermal Gas 1}} = \Delta U_{\text{Thermal Gas 2}}$). Consequently,

$$Q_{\text{Heater to Gas 1}} = W_{\text{Environment on Gas 2}} + Q_{\text{Heater to Gas 2}}$$

or

$$W_{\text{Environment on Gas 2}} = Q_{\text{Heater to Gas 1}} - Q_{\text{Heater to Gas 2}}$$

We could determine the work done by an independent method involving the definition of work and the ideal gas law. The work done by the piston on the gas in experiment 2 was $W_{\text{Environment on Gas 2}} = -P\Delta V$, where P was the constant pressure of the gas and ΔV was the volume change of the gas ($W_{\text{Environment on Gas 2}} < 0$, since pressure is always a positive number and the gas expands, thus its volume change is positive).

To find work, we need to determine ΔV. Using the ideal gas law for a constant pressure process, we can write

$$V = \left(\frac{nR}{P}\right)T,$$

where n, R, and P are all constant. Thus, a change in temperature ΔT at constant pressure leads to a change in volume ΔV equal to

$$\Delta V = \left(\frac{nR}{P}\right)\Delta T$$

Substituting this expression for ΔV in the expression for work, we get:

$$W = -P\Delta V = -P\left(\frac{nR}{P}\right)\Delta T = -nR\Delta T$$

The number of moles of gas is $n = m/M$, where $m = 1.0$ g is the mass of gas and $M = 29.0$ g/mole is the molecular mass of air.

$$n = \frac{1.0\text{ g}}{29.0\text{ g}/1.0\text{ mole}} = 0.0345\text{ moles}$$

Solve and evaluate We can now calculate the work W done by the environment on the gas in two ways. The first method gives us

$$W_{\text{Environment on Gas 2}} = Q_{\text{Heater to Gas 1}} - Q_{\text{Heater to Gas 2}}$$
$$= 0.17\text{ cal} - 0.24\text{ cal} = -0.07\text{ cal}$$

The second method gives us

$$W = -nR\Delta T = -(0.0345\text{ mole})\left(\frac{8.3\text{ J}}{\text{mole}\cdot{}^{\circ}\text{C}}\right)(1.0\,{}^{\circ}\text{C})$$
$$= -0.29\text{ J} = -0.29\text{ J}\left(\frac{0.24\text{ cal}}{1.0\text{ J}}\right) = -0.07\text{ cal}$$

The two independent methods of calculating the work produce consistent results. This strengthens our confidence in the new work-heating-energy equation.

Try it yourself: If in the experiment Mayer had 1.0 mole of gas, what would be the ratio of the heating needed to raise the temperature of this gas by one degree at constant pressure and the heating needed to raise the temperature of the same mole of gas by one degree at constant volume?

Answer: 5/3. The thermal energy increase in each gas is $(3/2)nR\Delta T$ and the negative work done on the constant pressure gas is $nR\Delta T$. Thus,

$$\frac{Q_{\text{Env. to Gas 2 (constant pressure)}}}{Q_{\text{Env. to Gas 1 (constant volume)}}} = \frac{\Delta U_{\text{th}} - (-P\Delta V)}{\Delta U_{\text{th}}}$$
$$= \frac{(3/2)nR\Delta T + nR\Delta T}{(3/2)nR\Delta T} = \frac{5}{3}$$

This ratio is different from what Mayer would have calculated, 0.24 cal/0.17 cal = 1.4 = 7/5. This is because the ideal gas law assumes molecules are point-like and have no internal structure. This assumption is only marginally reasonable for air in this situation.

Notice two important points here.

1. Providing the same amount of energy to the same amount of gas through heating *might not* lead to the same rise in the gas's temperature if in one experiment work was done on the gas and in the other no work was done.

2. The energy that needs to be transferred to 1 kg of air through heating to change its temperature by 1 °C when the volume of the gas is constant is 710 J and 1000 J when the volume is changing but the pressure remains constant.

What we have been calling the work-heating-energy equation is formally known as the **first law of thermodynamics**, a fundamental principle of physics.

First law of thermodynamics
Consider a process involving a system of interest. The system's internal energy change $(U_{\text{f}} - U_{\text{i}}) = \Delta U_{\text{System}}$ is equal to the amount of work W done on the system plus the amount of energy Q transferred to the system through the process of heating:

$$W_{\text{Environment on System}} + Q_{\text{Environment to System}} = \Delta U_{\text{System}}$$
$$(12.4a)$$

We can rewrite the first law by including all types of energy in a system:

$$(U_{gi} + K_i + U_{si}) + W + Q = (U_{gf} + K_f + U_{sf} + \Delta U_{int}) \qquad (12.4b)$$

Think of how you can warm your right hand (the system) after coming from a trip outside on a cool windy day. You can press your other hand against it and rub them together—transferring energy to your right hand by doing work on it (notice that both of your hands are at the same temperature). You can also submerge your right hand in warmer water—transferring energy to it through heating.

Review Question 12.3 How are the first law of thermodynamics and the work-energy equation similar? How are they different?

> **TIP** Heating, like work, is a physical quantity characterizing a process during which energy is transferred between a system and its environment. Both heating and work lead to a change in the energy of a system. Remember, heating and work are not things that reside within the system.

12.4 Specific heat

We need to transfer 1000 cal or 4200 J to 1 kg of water to change its temperature by 1 °C. Do all materials need the same amount of energy to change their temperature by 1 °C per unit mass? Consider the experiments in Observational Experiment **Table 12.4**.

OBSERVATIONAL EXPERIMENT TABLE

12.4 **Energy transferred to different materials to achieve the same temperature change.**

Observational experiment	Analysis
Experiment 1. A closed Styrofoam container holds 1.0 kg of water at 20 °C and a thermometer. A large pan of boiling water (at 100 °C) holds a 1.0 kg iron block. You transfer the block very quickly to the container with 20 °C water. You observe the temperature rising and then stabilizing at around 28 °C.	Although the masses of the water and iron block were the same, the water temperature increased by 8 °C whereas the iron block's temperature decreased by 100 °C − 28 °C = 72 °C. We assume that the water received 4200 J of energy for each 1 °C temperature change, or a total of 8(4200 J) = 3.4 × 10^4 J. The iron must have lost the same amount of energy. Thus the energy loss per 1 °C for the 1.0-kg iron block was $$\frac{3.4 \times 10^4\,J}{(1.0\,kg)(72\,°C)} = \frac{470\,J}{kg \cdot °C}$$
Experiment 2. You repeat Experiment 1, replacing the iron block with a 1.0 kg aluminum block. The final temperature is 34 °C.	The masses of the water and aluminum block were the same (1.0 kg), but the water temperature increased by 14 °C whereas the aluminum block's temperature decreased by 100 °C − 34 °C = 66 °C. We assume that the water received 4200 J of energy for each 1 °C temperature change, or a total of 14(4200 J) = 5.88 × 10^4 J. The aluminum must have lost the same amount of energy. Thus the energy loss per 1 °C for the 1.0 kg aluminum block was $$\frac{5.88 \times 10^4\,J}{(1.0\,kg)(66\,°C)} = \frac{890\,J}{kg \cdot °C}$$

Pattern

The same amount of energy added to or removed from objects of the same mass by heating or cooling leads to different changes of temperature of these objects. Alternatively, different amounts of energy are gained or lost for each 1 °C temperature change, depending on the type of material.

To explain the patterns in Table 12.4 we hypothesize that there is a physical quantity that characterizes how much additional energy 1 kg of a substance needs to change its temperature by 1°C, and this physical quantity is different for

Table 12.5 Specific heats (c) of various solid and liquid substances.

Solid Substances	c (J/kg · °C)	Liquid Substances	c (J/kg · °C)
Iron, steel	450	Water	4180
Copper	390	Methanol	2510
Aluminum	900	Ethanol	2430
Silica glass	840	Benzene	1730
Sodium chloride	860	Ethylene glycol	2420
Lead	130		
Ice	2090		
Wood	~1700		
Sand	~800		
Brick	~800		
Concrete	~900		
Human body	~3500		

different substances. As we found, the amount of energy is different for water, iron, and aluminum and is about

$$4200 \ \frac{J}{kg \cdot °C}, \ 470 \ \frac{J}{kg \cdot °C}, \text{ and } 890 \ \frac{J}{kg \cdot °C},$$

respectively. These values are the values of a physical quantity, called the **specific heat** of each substance.

> Specific heat c is the physical quantity equal to the amount of energy that needs to be added to 1 kg of a substance to increase its temperature by 1 °C. The symbol for specific heat is c and the units are $\frac{J}{kg \cdot °C}$. This energy is added through heating or work or both.

Table 12.5 lists specific heat capacities for several solid and liquid materials.

Water has the largest specific heat in Table 12.5. Notice that the specific heats of sand, bricks, and concrete are about one-fifth that of water. These materials have much greater temperature changes than water when equal masses of these materials absorb the same energy. This is one reason why the sand on a beach or the concrete beside a swimming pool feels so much hotter than the adjacent water on a sunny day. The sand and concrete have bigger temperature changes than the water when exposed to the same sunlight.

The physical quantity specific heat allows us to construct an equation that can be used to determine the total amount of energy ΔU that needs to be transferred to a mass m of some type of material to cause its temperature to change by ΔT. If the amount of energy equal to the specific heat c in joules (J) needs to be transferred to 1 kg of material to change its temperature by 1 °C, then to change the energy of m kg of the same material we need to transfer $c \cdot m$ J. If we now want to change the temperature by ΔT degrees instead of 1 degree, we need to transfer $c \cdot m\Delta T$ joules. Note that the unit of the product of these three quantities is the unit of energy:

$$\frac{J}{kg \cdot C°} \times kg \times °C = J$$

Therefore, we conclude that an amount of energy ΔU must be added to a substance of mass m and specific heat c to cause its temperature to change by ΔT:

$$\Delta U = cm\Delta T \tag{12.5}$$

TIP Every time you encounter a new physical quantity, think of its meaning. For example, what does it mean that the specific heat of ethanol is 2430 J/kg · °C? It means that we would have to transfer 2430 joules of energy to 1 kg of ethanol to change its temperature by 1 °C.

The change in the energy of the system can occur both due to the process of heating and due to external objects doing work on the system. This leads to an interesting consequence. Imagine that you are heating a gas in a rigid container. No work is done on the gas since it is not expanding, so the change in thermal energy of the gas is exactly equal to the heating. However, if the container is very flexible and there is atmospheric air outside, that atmospheric air does negative work on the system, reducing the change in thermal energy. To achieve the same change in temperature as in the previous process and thus the same change in thermal energy of the gas, one needs to transfer more energy through heating. This is exactly what happened in Example 12.2. Recall the statement of the problem: It was known in Mayer's times that 1.0 g of air requires 0.17 cal of energy transferred through heating at *constant volume* for its temperature to increase by 1.0 °C and 0.24 cal of energy transferred through heating at *constant pressure* for the temperature to change by 1.0 °C. In the first case there is no work done on the system, and thus there is less energy transferred through heating needed to change its temperature by one degree. If the work done on the gas is negative and its magnitude is exactly equal to the positive energy transferred through heating, the temperature of the gas and its thermal energy do not change. However, in most applications, we deal with solids and liquids whose volume changes very little. In this case the thermal energy changes only due to the process of heating, and thus $Q + W = Q + 0 = \Delta U = cm\Delta T$.

Specific heat is a useful quantity. For example, we can use it to determine how much thermal energy a car engine would need to gain by heating in order to warm it to the boiling temperature of the water in its cooling system. Normally, a radiator cooling system prevents the engine from getting too hot, but what if the radiator isn't functioning?

QUANTITATIVE EXERCISE 12.3 Heating a car engine

A car has a 164-kg mostly aluminum engine that when operating at highway speeds converts chemical potential energy into thermal energy at a rate of $P = 1.8 \times 10^5$ J/s (240 horsepower). Suppose the cooling system shuts down and all of the thermal energy continually generated by the engine remains in the engine. Determine the time interval for the engine's temperature to change from $T_i = 20\,°C$ to $T_f = 100\,°C$, the boiling temperature of water.

Represent mathematically Assume that the energy is transferred from the burning fuel to the engine through the mechanism of heating, which produces the temperature change in the engine:

$$Q_{\text{Fuel to Engine}} = c_{\text{Engine}} m_{\text{Engine}} (T_{\text{Engine, f}} - T_{\text{Engine, i}})$$

The burning fuel converts chemical energy into thermal energy at a rate of

$$P = \frac{\Delta U_{\text{thermal}}}{\Delta t} = \frac{\Delta Q}{\Delta t} = \frac{Q_{\text{Fuel to Engine}}}{\Delta t}$$

Thus, the time interval Δt needed for the burning fuel to warm the engine is

$$\Delta t = \frac{Q_{\text{Fuel to Engine}}}{P}$$

Solve and evaluate The engine is made mostly of aluminum with specific heat $c_{\text{aluminum}} = 900$ J/kg·°C. Thus,

$$Q_{\text{Fuel to Engine}} = \left(900\frac{\text{J}}{\text{kg}\cdot°C}\right)(164\,\text{kg})(100\,°C - 20\,°C)$$
$$= 1.2 \times 10^7\,\text{J}$$

The engine will reach boiling temperature in

$$\Delta t = \frac{Q_{\text{Fuel to Engine}}}{P}$$
$$= \frac{(1.2 \times 10^7\,\text{J})}{(1.8 \times 10^5\,\text{J/s})} = 67\,\text{s}$$

The unit is correct. The engine warms quickly. An operating cooling system is important!

Try it yourself: You hold two spoons of identical shape and volume in a flame. One spoon is aluminum and one is steel. Why does the temperature of the aluminum spoon change more quickly than that of the steel spoon, even though the specific heat of aluminum is about twice that of steel?

Answer: The density of aluminum is about one-third of the density of steel; hence the steel spoon is more massive.

Sharing energy through the process of heating when objects are in contact

Active Learning Guide >

If you have a fever, bathing in cool bath water lowers your temperature and raises the water temperature. If you are pulled from a cold lake, to warm up you can bathe in hot bath water—the hot water warms your body and your cool body cools the water. These changes in the thermal energies of two objects can be explained using the idea of energy transfer: the hot object loses thermal energy to the cold object through the process of heating, and the cold object gains energy through the process of heating from the hot object. We can summarize the above discussion as follows:

$$Q_{\text{Hot to Cold}} + Q_{\text{Cold to Hot}} = 0$$

It is important to note that when a cold object and a hot object are in contact, the energy transfers from the hot object to the cold object until their temperatures become the same. At that point the average random kinetic energies of their particles are the same and there is no more energy transfer. The objects are said to be in **thermal equilibrium**.

> **TIP** When object 1 at high temperature is in contact with object 2 at lower temperature, the process of energy transfer from 1 to 2 is still called heating, despite the fact that the temperature of object 1 goes down.

EXAMPLE 12.4 Warming the body of a hypothermic person

The kayak of a 70-kg man tips on a spring day and he falls into a cold stream. When he is rescued from the cold water, his body temperature is 33 °C (91.4 °F). You place him in 50 kg of warm bath water at temperature 41 °C (105.8 °F). What is the final temperature of the man and the water?

Sketch and translate Draw a sketch of the situation. Since the initial bath water temperature is higher than the man's initial temperature, the decrease in the thermal energy of the water equals the increase in the thermal energy of the man. The specific heats of water and of humans are listed in Table 12.5.

$$M_{\text{Man}} = 70 \text{ kg} \qquad M_{\text{Water}} = 50 \text{ kg}$$
$$T_{i\,\text{Man}} = 33 °C \qquad T_{i\,\text{Water}} = 41 °C$$
$$T_f = ?$$

Simplify and diagram Assume that no energy is transferred from the water and man to the surroundings. In addition, assume that chemical energy converted into thermal energy in the man's body is small and can be ignored during the short time interval in the water.

Represent mathematically The thermal energy change of the man equals the energy transferred from the water by heating to the man: $Q_{\text{Water to Man}} = m_{\text{Man}}c_{\text{Man}}(T_{f\,\text{Man}} - T_{i\,\text{Man}})$. The thermal energy change of the water equals the energy transferred by heating from the man to the water: $Q_{\text{Man to Water}} = m_{\text{Water}}c_{\text{Water}}(T_{f\,\text{Water}} - T_{i\,\text{Water}})$. We also know that

$T_{f\,\text{Water}} = T_{f\,\text{Man}} = T_f$ and that $Q_{\text{Man to Water}} + Q_{\text{Water to Man}} = 0$. Putting this all together, we get

$$m_{\text{Water}}c_{\text{Water}}(T_f - T_{i\,\text{Water}}) + m_{\text{Man}}c_{\text{Man}}(T_f - T_{i\,\text{Man}}) = 0$$

Notice that the first expression on the left side of the above equation will be negative, as the water's final temperature is less than its initial temperature. The above can now be solved for the final temperature (try to do it yourself before you look at the expression shown below).

Solve and evaluate Solving for the final temperature of the system, we get

$$
\begin{aligned}
T_f &= \frac{m_{\text{Water}}c_{\text{Water}}T_{i\,\text{Water}} + m_{\text{Man}}c_{\text{Man}}T_{i\,\text{Man}}}{m_{\text{Water}}c_{\text{Water}} + m_{\text{Man}}c_{\text{Man}}} \\
&= \frac{(50 \text{ kg})(4180 \text{ J/kg} \cdot °C)(41 °C) + (70 \text{ kg})(3470 \text{ J/kg} \cdot °C)(33 °C)}{(50 \text{ kg})(4180 \text{ J/kg} \cdot °C) + (70 \text{ kg})(3470 \text{ J/kg} \cdot °C)} \\
&= 36.7 °C
\end{aligned}
$$

The answer is reasonable: the final temperature is 3.7 °C above the initial temperature of the man and 4.3 °C below the water's initial temperature. The final temperature is 0.3 °C below the normal 37 °C body temperature.

Try it yourself: Fill a Styrofoam cup with 100 g of water at 90 °C and then place a 30-g aluminum spoon at 25 °C into it. The water cools to 86 °C and the spoon warms to 86 °C, the same temperature as the water. Use these results to estimate the specific heat of aluminum. What assumptions did you make?

Answer: 900 J/kg·°C. We assumed that the spoon is completely submerged in the water and no energy is transferred through heating to the cup or any other part of the environment.

Review Question 12.4 What experimental evidence indicates that the specific heat of aluminum or iron is much smaller that the specific heat of water?

12.5 Applying the first law of thermodynamics to gas processes

Let us now apply the first law of thermodynamics to several processes involving gases. We will represent the processes using words, graphs, bar charts, and equations and explain what happens using a microscopic approach (molecules and their motion) and a first law of thermodynamics approach (energy transfer to or from the system and its changes). Let's begin with a simple process.

Effect of a moving piston on the temperature of an enclosed gas

Suppose a gas is enclosed in a cylinder with a piston on the right end that can move, allowing the gas to expand or contract (see **Figure 12.5a**). How does the average speed of the gas particles and therefore the gas temperature change if the gas expands?

Consider a situation in which one particle collides with the piston. The particle will bounce off the piston, but because the piston is moving away from the particle, its speed will not on average be as great after the collision as it was before the collision, and the gas temperature will decrease slightly. Consider an extreme example of such a collision. Suppose the particle is moving at velocity +10 m/s with respect to Earth when it collides with a piston that is moving outward in the same direction as the particle, only at +9 m/s with respect to Earth (Figure 12.5b). Then with respect to the piston, the particle is moving towards the piston at +1m/s. Since the collision is elastic, the particle rebounds at −1 m/s (Figure 12.5c) with respect to the piston. With respect to Earth, then, the particle should be moving to the right at 8 m/s, because (−1 m/s) + (9 m/s) = 8m/s. Its speed has decreased by 20%—from 10 m/s to 8 m/s with respect to Earth, and it is therefore a much "cooler" particle.

In a real situation, the piston moves much more slowly than the particles and their speeds decrease less dramatically. But they do move somewhat more slowly after colliding with the expanding piston. As a result, the temperature of the gas decreases as the piston moves outward. Similarly, if the piston moves inward, the average speed of the particles after colliding with the piston will be greater than before the collision. In this case, the temperature of the gas increases as the piston moves inward.

Now that we have an understanding of this, let's move on to more complex processes.

Analyzing gas processes

In **Table 12.6**, we analyze various gas processes using the first law of thermodynamics using particle-based and energy-based explanations. In all of these processes the number of moles n of gas is assumed to be constant. Recall what we have learned about isoprocesses (in Chapter 9)—processes occurring in a gas of constant mass in which one of the following physical quantities also remained constant: P, T, or V. In addition, we will analyze a new type of process called an **adiabatic** process. In an adiabatic process no energy is transferred through heating ($Q = 0$).

< Active Learning Guide

Figure 12.5 Piston motion affects speed of gas particles. (a) The piston is moving right. (b) Particle speed of a gas particle before it hits a piston that is moving away. (c) The gas particle speed is lower after it hits the piston.

(a)

The piston can move either left or right but is moving toward the right at the moment.

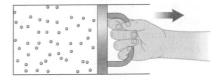

(b)

Before the collision, the particle is traveling at +10 m/s and moves toward the piston at 1 m/s relative to the piston.

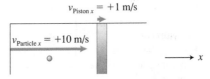

(c)

After the elastic collision, the particle moves away from the piston at 1 m/s, which is a velocity of +8 m/s to the right with respect to Earth.

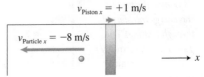

Table 12.6 Application of the first law to gas processes. In each process, the gas starts at position 1 (the initial state) on the P-versus-V graph and moves to position 2 (the final state).

Process	Graph	Particle-level explanation	Explanation and bar chart using the first law of thermodynamics
(a) *Isothermal process (constant temperature)* The gas is in a non-insulated container with a piston, all submerged in a large bath of water at constant temperature, which is the same as the temperature of the gas. Someone pulls the piston out very slowly.	P-V graph with point 1 at upper left, point 2 at lower right, curve labeled $T = \text{const}$	As the piston is pulled out, the gas molecules collide elastically with the piston and rebound at a slower speed than before the collision, which causes a temperature decrease. But the container is in contact with the water bath, which provides heating so the gas remains at constant temperature.	■ As the temperature of the gas stays constant, $\Delta U_{int} = 0$. ■ The external work done on the gas is negative ($W < 0$). ■ Transfer of energy through heating ($Q > 0$) is needed to make the internal energy change zero. ■ $-|W| + Q = 0$ bar chart: $W + Q = \Delta U_{int}$
(b) *Isochoric process (constant volume)* The gas is in a noninsulated container with a fixed piston. The container is placed in a bath of higher temperature. The temperature of the gas increases.	P-V graph with point 2 above point 1, vertical line	As the gas warms, its molecules move faster and faster, colliding with the walls of the container more often and more violently. The gas pressure increases.	■ Because the volume is constant, there is no work done on the gas ($W = 0$). ■ Energy is transferred to the gas through heating ($Q > 0$). ■ Transfer of energy through heating leads to an increase in the thermal energy of the gas ($\Delta U_{int} > 0$). ■ $0 + Q = +\Delta U_{int}$ bar chart: $W + Q = \Delta U_{int}$
(c) *Isobaric process (constant pressure)* Gas is in a noninsulated container that has a piston that can move freely up (or down), keeping the gas at constant pressure. The container is placed in a bath at higher temperature, causing the gas to expand at constant pressure.	P-V graph with point 1 at left, point 2 at right, horizontal line	The hot temperature bath warms the gas by heating. But the gas warms less because the piston moves outward, slowing the particles that hit it and also causing negative work to be done by the environment on the gas.	■ The environment does negative work on the gas ($W < 0$). ■ Energy is transferred through heating ($Q > 0$) and is greater than the negative work. ■ The thermal energy of the gas increases ($\Delta U_{int} > 0$). ■ $-|W| + Q = +\Delta U_{int}$ bar chart: $W + Q = \Delta U_{int}$
(d) *Adiabatic process (no energy transferred through heating)* The gas is in a thermally insulated container or is compressed very quickly.	P-V graph with point 2 at upper left, point 1 at lower right, curve	As the gas is being compressed, the molecules collide with a piston moving inward; the speed of the particles reflected off the incoming piston is greater than before, and the gas's temperature increases.	■ The environment does positive work on the gas ($W > 0$). ■ Since the process happens very quickly or the gas is in an insulated container, we assume that there is no transfer of energy through heating ($Q = 0$). ■ The internal thermal energy of the gas increases ($\Delta U_{int} > 0$). ■ $W + 0 = +\Delta U_{int}$ bar chart: $W + Q = \Delta U_{int}$

Skills for solving gas problems using the first law of thermodynamics

Everyday life offers many examples of isoprocesses and adiabatic processes, where one quantity T, P, V, or Q is constant (in an adiabatic process $Q = \text{const} = 0$). A useful strategy for analyzing such processes using the first law of thermodynamics is described on the left side of the table in Example 12.5 and illustrated for the problem in that example.

PROBLEM-SOLVING STRATEGY   **Using thermodynamics for gas law processes**

EXAMPLE 12.5 Hot air balloon

A burner heats $1.0\ \text{m}^3$ of air inside a small hot air balloon. Initially, the air is at $37\ °\text{C}$ and atmospheric pressure. Determine the amount of energy that needs to be transferred to the air through heating (in joules) to make it expand from $1.0\ \text{m}^3$ to $1.2\ \text{m}^3$.

Sketch and translate

- Make a labeled sketch of the process.
- Choose a system of interest and the initial and final states of the process.

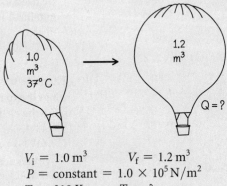

$$V_i = 1.0\ \text{m}^3 \qquad V_f = 1.2\ \text{m}^3$$
$$P = \text{constant} = 1.0 \times 10^5\ \text{N/m}^2$$
$$T_i = 310\ \text{K} \qquad T_f = ?$$
$$Q_{\text{Environment to System}} = ?$$

The system is the gas inside the balloon. The initial state is before the flame was turned on; the final state is after the balloon has partially expanded.

Simplify and diagram

- Decide whether you can model the system as an ideal gas.
- Decide whether the gas undergoes one of the isoprocesses.
- Determine whether you can ignore any interactions of the environment with the system.
- Draw the following representations if appropriate: a work-heating-energy bar chart and/or a P-versus-V graph.

- We can model the air as an ideal gas since conditions are close to normal atmospheric conditions.
- Assume that the gas expands slowly so that it remains at constant atmospheric pressure—an isobaric process. This also assumes that the skin of the balloon does not exert a significant additional inward pressure on the gas.

P-versus-V graph   **Q-W-ΔU bar chart**

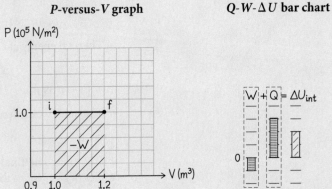

(continued)

Represent mathematically

- Use the bar chart to help apply the first law of thermodynamics:

$$U_{gi} + K_i + U_{si} + W + Q$$
$$= U_{gf} + K_f + U_{sf} + \Delta U_{int}.$$

Decide if the mechanical energy of the system is constant (kinetic, gravitational potential, or elastic).

- If the system can be modeled as an ideal gas, use, if needed, gas law equations such as $PV = nRT = NkT$.

- For this process, the initial and final gravitational potential and (non-random) kinetic energies of the gas are equal, and there are no springs in the system. Thus the first law equation becomes

$$W + Q = \Delta U_{int}$$

We need to determine the amount of energy transferred by heating:

$$Q = \Delta U_{int} - W.$$

The gas expands; thus the environment pushes in the direction opposite to the motion of the balloon walls and does negative work (area under P-versus-V graph line with a negative sign):

$$W = -P\Delta V$$

- To find the change in internal energy, we need to use the expression for the internal energy of the ideal gas $U_{int} = \frac{3}{2}nRT$. To find the change, we need to determine the number of moles of gas and then the change in temperature. We can use the ideal gas law to determine the number of moles of air initially in the 1.0 m³ of air:

$$n = P_i V_i / RT_i$$

Then use the ideal gas law to determine the final temperature of the air when it expands to the final volume: $T_f = P_f V_f / nR$. From this, determine the change in thermal energy of the air:

$$\Delta U_{int} = (3/2)nR(T_f - T_i)$$

- We can now put everything together into the equation

$$Q = \Delta U_{int} - W$$

Solve and evaluate

- Solve the equations for the unknown quantity.
- Evaluate the results to see if they are reasonable (the magnitude of the answer, its units, and how the solution changes in limiting cases).

The work done on the gas is:

$$W = -P\Delta V = -(1.0 \times 10^5 \, \text{N/m}^2)(1.2 \, \text{m}^3 - 1.0 \, \text{m}^3)$$
$$= -0.2 \times 10^5 \, \text{N} \cdot \text{m} = -2.0 \times 10^4 \, \text{J}$$

- The number of moles of air is

$$n = P_i V_i / RT_i = (1 \times 10^5 \, \text{N/m}^2)(1 \, \text{m}^3)/(8.3 \, \text{J/mole} \cdot \text{K})(310 \, \text{K})$$
$$= 39 \text{ moles}$$

- The final temperature of the air is

$$T_f = P_f V_f / nR$$
$$= (1 \times 10^5 \, \text{N/m}^2)(1.2 \, \text{m}^3)/(39 \, \text{mole})(8.3 \, \text{J/mole} \cdot \text{K}) = 371 \, \text{K}$$

- Thermal energy change of the gas is

$$\Delta U_{thermal} = (3/2)nR(T_f - T_i)$$
$$= (3/2)(39 \, \text{mole})(8.3 \, \text{J/mole} \cdot \text{K})[(371 - 310)\text{K}] = +3.0 \times 10^4 \, \text{J}$$

- The energy transferred to the gas through heating is:

$$Q = \Delta U_{thermal} - W = +3.0 \times 10^4 \, \text{J} - (-2.0 \times 10^4 \, \text{J}) = +5.0 \times 10^4 \, \text{J}$$

Evaluation: The final temperature is reasonable, as is the number of moles of gas. Notice that almost half of the energy transferred by heating is needed to compensate for negative work done on the system by the environment as the gas system expands.

Try it yourself: Air rushing out of the open valve of a pumped bicycle tire feels cool. Why?

Answer: This is an approximately adiabatic process ($Q = 0$) since it happens so rapidly. The gas is expanding, so the environment does negative work on the gas ($W < 0$). Thus, the internal energy change should also be negative ($W + 0 = \Delta U_{int}$). The gas cools as it expands adiabatically.

In the next example we analyze an isochoric process.

‹ Active Learning Guide

EXAMPLE 12.6 Temperature control in a domed stadium

You are the heating-cooling contractor for a new 68,000-person sports dome construction project. Estimate the air temperature change in the building caused by the metabolic processes of the spectators during a football game, assuming that the building is perfectly insulated and has no air conditioning. Identify any other assumptions that you make.

Sketch and translate A sketch of the situation is shown below. Choose the air in the dome as the system. If we assume that the volume of the air inside the dome remains constant and that no air escapes the dome, then no work is done on the air by the environment ($W = -P\Delta V = 0$ and $0 + Q = \Delta U_{th}$). The change in the internal energy (thermal energy) of the gas equals the energy transferred by heating from the spectators to the air in the dome (see the bar chart below):

$$Q_{\text{People to Air}} = \Delta U_{\text{th Air}}$$

(a)

$$\Delta T = ?$$

Approximate metabolic rate = 100 J/s = $Q_{\text{Person to Air}}$

(b)

$$\boxed{W} + \boxed{Q} = \Delta U_{int}$$

We can use Eq. (12.5) to estimate the air temperature change ΔT caused by this thermal energy change during the game:

$$\Delta T = \frac{\Delta U_{\text{th Air}}}{c_{\text{Air}} m_{\text{Air}}} = \frac{Q_{\text{People to Air}}}{c_{\text{Air}} m_{\text{Air}}}$$

To complete this task, we need to estimate $Q_{\text{People to Air}}$ and the mass of the air inside the dome m_{Air}. We also need to know the specific heat c_{Air} of the air at constant volume.

Simplify and diagram The stadium holds $N = 68,000$ spectators, each with a resting metabolic rate MR_{Person} of about 100 J/s (about 100 W). Assume the game lasts about 3 hours (notice that we do not worry about significant figures here, as we are doing an estimate). Also assume that there is no energy transfer between the stadium and the environment.

Represent mathematically The energy that an average person adds to the air during the game is

$$Q_{\text{Person to Air}} = MR_{\text{Person}} \Delta t$$

where $MR_{\text{Person}} = 100\,\text{W} = 100\,\text{J/s}$ is the metabolic rate (power output) of a resting person. Since about N people are at the game, the total energy added to the air through heating during the game is approximately

$$Q_{\text{People to Air}} = N Q_{\text{Person to Air}} = N \cdot MR_{\text{Person}} \Delta t$$

To find the temperature change we need to know the mass of the air and its specific heat. The mass of the air will be the density of air times the stadium volume:

$$m_{\text{Air}} = \rho_{\text{Air}} V_{\text{Stadium}}$$

Solve and evaluate The total energy added through the process of heating to the air by the people during the game is

$$Q_{\text{People to Air}} = N \cdot MR_{\text{Person}} \Delta t$$

$$= (68{,}000\,\text{persons})\left[(100\,\text{J/s})/\text{person}\right]\left(3\,\text{h}\,\frac{3600\,\text{s}}{1\,\text{h}}\right)$$

$$\approx 7 \times 10^{10}\,\text{J}$$

What is the mass of the air in the dome? The approximate inside volume V_{Stadium} of such a stadium is $(140\,\text{m} \times 130\,\text{m} \times 40\,\text{m}) \approx 7 \times 10^5\,\text{m}^3$. The density of air at normal conditions is $\rho_{\text{Air}} = 1.3\,\text{kg/m}^3$ and the specific heat at constant volume is about

(continued)

$\approx 700 \, \text{J/kg} \cdot \text{°C}$. Thus, our estimate of the mass of the air inside the stadium is

$$m_{\text{Air}} = \rho_{\text{Air}} V_{\text{Stadium}} = (1.3 \, \text{kg/m}^3)(7 \times 10^5 \, \text{m}^3)$$
$$= 9 \times 10^5 \, \text{kg} \approx 10^6 \, \text{kg}$$

We can now estimate the expected change in temperature:

$$\Delta T = \frac{Q_{\text{People to Air}}}{c_{\text{Air}} m_{\text{Air}}} \approx \frac{(7 \times 10^{10} \, \text{J})}{(700 \, \text{J/kg} \cdot \text{°C})(10^6 \, \text{kg})} \approx 100 \, \text{°C}$$

Wow! If the air starts at 20 °C (room temperature), the final temperature would be about 120 °C. This example illustrates the importance of having a cooling system that can remove energy from the stadium at the same rate that it is being added by the people. Next time you attend a crowded party, note the temperature in the room. It becomes warmer due to the 100 J/s heating effect of each person.

Try it yourself: If the volume of a tent is about 9 m³, estimate how much the air in the tent warms during an 8-hour night while two people sleep in it. What assumptions did you make? How reasonable are they?

Answer: 700 °C! We assumed that the air in the tent exchanges energy only with the two sleeping people (100 J/s each)—the result is unreasonable. A large amount of energy must transfer to the environment through the walls of the tent.

Figure 12.6 An experiment to study phase changes and temperature changes due to heating.

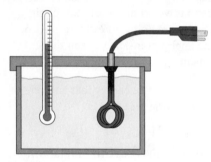

Active Learning Guide >

Figure 12.7 Effect of heating on ice, liquid water, and steam.

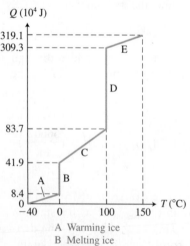

Review Question 12.5 Describe two situations in which the same amount of energy is supplied through heating to the same mass of gas at the same initial conditions, but the gas ends up at a different final temperature in each situation.

12.6 Changing state

We have found that providing energy to a system through heating or by doing work causes its temperature to change. Are there any processes in which the thermal energy of a system changes but the temperature of the system does not? Suppose you warm an ice cube in a Styrofoam cup so that some of it melts and becomes liquid water. Is the melted water warmer than the solid ice cube that remains? The transformation of ice into water is a **phase change**—a process during which a substance changes from one state to another. We are most familiar with the gas, liquid, and solid states. There are other states, such as plasma. Plasma is what fills neon lights and is the main state of matter in stars. We will learn about plasma in later chapters of the book. Let's consider an experiment involving the heating of water that starts as very cold solid ice.

Assume that you have 1.0 kg of water, which you place in a foam container with the top initially open (**Figure 12.6**). You insert a thermometer and an electric heating coil into the water. You place all of this in a freezer that is set at −40 °C. After the water freezes and reaches −40 °C, you place a foam cover over the container with the thermometer sticking out. You then remove the container from the freezer and turn on the electric coil. You observe the temperature change with time, as shown in the graph in **Figure 12.7**.

Warming from −40 °C to 0 °C

Between −40 °C and 0 °C, the contents of the container remain frozen (A in Figure 12.7). The energy transferred to the ice from the electric coil causes the ice to warm and eventually reach its melting temperature at 0 °C. Because you know the power output of the coil, you learn that it takes 2090 J to warm 1.0 kg of ice by 1.0 °C. In other words, it takes 8.4×10^4 J to warm 1.0 kg of ice from −40 °C to 0 °C. This means that the specific heat of ice is 2090 J/kg · °C. It takes less energy to warm ice by 1.0 °C than it does to warm the same amount of liquid water. So far the transfer of energy to the ice led to a change of its temperature—no surprises here.

Melting and freezing

At 0 °C the graph changes suddenly, becoming vertical (B in Figure 12.7). We observe that when the ice reaches 0 °C, it starts to melt, but the thermometer reading does not change. According to the graph, it takes 3.35×10^5 J of energy to completely melt 1.0 kg of ice. While this energy is added, the temperature of the system does not change. Why?

Remember that in a solid, neighboring molecules are close to each other and are attracted and bonded to each other—that is why solids keep their shape. Their potential energy of interaction is negative, meaning they are bound to one another much like the Earth-Moon system is bound by negative gravitational potential energy. The molecules vibrate and therefore have kinetic energy of random motion (though they do not fly around freely as they would in a gas). As the temperature of the solid rises toward its melting temperature, the vibrations of the neighboring water molecules become larger in amplitude and less ordered. When the solid reaches its melting temperature, the vibrations of neighboring molecules become so disruptive that the molecules begin to separate from one another and are no longer bound (**Figure 12.8**). There is still considerable interaction between neighboring molecules, but the interaction is less than in the rigid solid. Therefore, when melting or freezing occurs, it is the potential energy of particle interactions that changes. When we transfer energy to the solid material at the melting temperature, all of this energy goes into changing the potential energy of particle interactions, not the kinetic energy—thus the temperature does not change.

The energy needed to melt 1.0 kg of ice into water is called the **heat of fusion** of water and is given the symbol L_f ($L_{f\,water} = 3.35 \times 10^5$ J/kg). (The heat of fusion is sometimes called the *latent heat of fusion* to underscore that this energy does not lead to temperature change; the change is in a hidden or latent form.) When a liquid freezes, it releases the same amount of energy (in other words, the reaction occurs in the opposite direction). The heat of fusion differs for different materials; their melting-freezing temperatures also differ (see **Table 12.7**). The unit of heat of fusion is joules per kilogram.

> **Energy to melt or freeze** The energy in joules needed to melt a mass m of a solid at its melting temperature, or the energy released when a mass m of the liquid freezes at that same temperature, is
>
> $$\Delta U_{int} = \pm m L_f \qquad (12.6)$$
>
> L_f is the **heat of fusion** of the substance (see Table 12.7). The plus sign is used when the substance melts and the minus sign when it freezes.

Figure 12.8 Melting. Depiction of structural change as ice in (a) melts to form liquid water in (b).

(a)

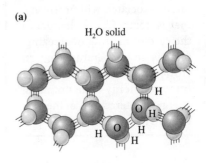

H₂O solid

(b)

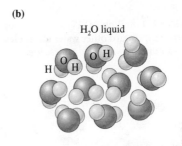

H₂O liquid

Table 12.7 Heats of fusion and vaporization.

Substance	Melting temperature (°C)	L_f, Heat of fusion at the melting temperature (J/kg)	Boiling temperature (°C)	L_v, Heat of vaporization at the boiling temperature (J/kg)
Water	0	3.35×10^5	100	2.256×10^6
Ethanol	−114	1.01×10^5	78	0.837×10^6
Hydrogen (H_2)	−259	0.586×10^5	−253	0.446×10^6
Oxygen (O_2)	−218	0.138×10^5	−183	0.213×10^6
Nitrogen (N_2)	−210	0.257×10^5	−196	0.199×10^6
Aluminum	660	3.999×10^5	2520	10.9×10^6
Copper	1085	2.035×10^5	2562	4.73×10^6
Iron	1538	2.473×10^5	2861	6.09×10^6

TIP The term heat of fusion is confusing, as the word *heat* is often associated with warming, that is, temperature change. In the case of melting or freezing, there is no temperature change. However, there is still change in internal energy—the potential energy of interactions of the particles in the system.

Figure 12.9 Boiling. (a) Water molecules are attracted toward other water molecules. (b) Fast moving molecules can escape.

(a)

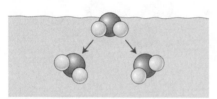

A water molecule at the surface of the water is prevented from leaving the water by its attraction to neighboring molecules.

(b)

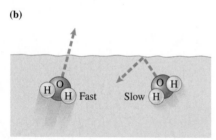

A fast-moving water molecule can escape from a liquid when it hits the surface. A slow-moving one cannot escape.

Warming liquid water from 0 °C to 100 °C

Once the ice has completely melted (C in Figure 12.7), further transfer of energy by heating causes the liquid water to warm. The graph in this region has a slope of 4180 J/°C and indicates that the specific heat of liquid water is 4180 J/kg·°C. With continued heating, the water temperature continues to rise.

Boiling and condensation

At 100 °C, the graph again changes abruptly (D in Figure 12.7). We observe that the water is boiling and steam is coming out. The temperature of the water remains at the boiling temperature until all of the liquid water boils into vapor. From the graph we see that when the water reaches 100 °C, we must add 2.256×10^6 J of energy to completely convert 1.0 kg of liquid water to gaseous water vapor (steam). Once again, energy has been added to the water, but its temperature does not change. Where does the energy go? For a molecule to leave the surface of a liquid, it must have enough kinetic energy to break away from the neighboring molecules, which are exerting attractive forces on it (**Figure 12.9a**). A fast-moving water molecule can escape, while a slow-moving one cannot (Figure 12.9b). Escape of the fast-moving molecules leaves the remaining slow-moving "cooler" molecules behind. A continuous supply of energy through heating when the substance is at the boiling temperature causes a continual number of water molecules to gain the energy needed to leave the water surface but does not change the water temperature until all molecules have become water vapor. The energy transferred to the liquid leads to a change in the potential energy component of the internal energy.

Liquids can evaporate at any temperature, not just at the boiling temperature. At any temperature some molecules have enough kinetic energy to escape the liquid. However, as they leave, the temperature of the liquid left behind decreases, since the average kinetic energy of the molecules left behind decreases. This is how we cool ourselves by evaporation on hot days.

The **heat of vaporization** of a substance L_v (sometimes called the *latent heat of vaporization*) is the energy needed to transform 1 kg of a liquid at its boiling temperature into its gaseous state at that same temperature. For water, $L_{v\,water} = 2.256 \times 10^6$ J/kg. When 1.0 kg of the gas turns into liquid (condenses) at its boiling temperature, it releases the same amount of energy. The heat of vaporization differs for different substances; their boiling temperatures also differ (see Table 12.7). The unit of heat of vaporization is joules per kilogram.

Energy to boil or condense The energy in joules needed to vaporize a mass m of a liquid at its boiling temperature, or the energy released when a mass m of a gas condenses at that same temperature is

$$\Delta U_{int} = \pm mL_v \tag{12.7}$$

L_v is the **heat of vaporization** of the substance (see Table 12.7). The plus sign is used when the substance vaporizes and the minus sign when it condenses.

The heat of fusion, melting temperature, heat of vaporization, and boiling temperature depend on the type of substance, as you can see in Table 12.7.

Notice that the values of the heats of fusion and vaporization are much larger than the specific heat of these substances. This means that much more energy is needed to change the *state* of a substance than to change the *temperature* of the substance by one degree. Also notice that the heats of vaporization are significantly larger than the heats of fusion; more energy is required to boil the same mass of the same substance than to melt it. Finally, notice that the values are provided for boiling temperatures. Liquids evaporate at temperatures below the boiling temperature; however, the energy needed to keep them at the same temperature during this evaporation is greater. For example, the heat of vaporization at the boiling temperature for water is $2.256 \times 10^6 (J/kg)$ whereas at body temperature (when sweat evaporates) the heat of vaporization is $L_{v \text{ water}} \approx 2.4 \times 10^6 \, J/kg$.

Warming gaseous water vapor above 100 °C

We cannot perform the next part of the experiment (E in Figure 12.7) with the apparatus shown in Figure 12.6 since the water vapor would escape. But the experiment could be performed if the container were airtight. In that case, if we continued to add energy to 1.0 kg of water vapor, we would find that its temperature increased 1 °C for each approximately 2000 J of additional energy added to it. This tells us that the specific heat of water vapor is approximately $c_{\text{water vapor}} = 2000 \, J/kg \cdot °C$.

Reversing the process

We now put the 1.0 kg of hot water vapor back in our −40 °C freezer and slowly remove thermal energy to reverse the process shown graphically in Figure 12.7. We find that the water first cools to its boiling temperature (100 °C for water). It then releases considerable energy as it condenses to form liquid water. Next, it cools to 0 °C as more energy is removed. The liquid water again releases considerable energy as it freezes. The ice finally cools to a final temperature of −40 °C.

QUANTITATIVE EXERCISE 12.7 Sweating in the Grand Canyon

Eugenia carries a 10-kg backpack on a hike down into the Grand Canyon and then back up. Sweat evaporates off her skin at a rate of 0.28 g/s. At what rate is thermal energy transferred from inside her body to the surface of her skin in order to provide the energy needed to vaporize the sweat on her skin at this rate? At body temperature the heat of vaporization for sweat is about $L_{v \text{ Sweat}} \approx 2.4 \times 10^6 \, J/kg$.

Represent mathematically The sweat on the skin will be the system. The energy needed to evaporate a small mass Δm of the sweat that is added to the sweat from inside Eugenia's body is $Q_{\text{Eugenia to Sweat}} = +\Delta m \cdot L_v$. Dividing each side by a small time interval Δt, we get

$$\frac{Q_{\text{Eugenia to Sweat}}}{\Delta t} = \frac{\Delta m_{\text{Sweat}} L_{v \text{ Sweat}}}{\Delta t} = \left(\frac{\Delta m_{\text{Sweat}}}{\Delta t} \right) L_{v \text{ Sweat}}$$

where the rate of sweating is $\frac{\Delta m_{\text{Sweat}}}{\Delta t} = 0.28 \, g/s = 0.28 \times 10^{-3} \, kg/s$.

Solve and evaluate We find the rate of energy transfer through evaporative heating is

$$\frac{Q_{\text{Eugenia to Sweat}}}{\Delta t} = \left(\frac{\Delta m_{\text{Sweat}}}{\Delta t} \right) L_{v \text{ Sweat}}$$
$$= (0.28 \times 10^{-3} \, kg/s)(2.4 \times 10^6 \, J/kg)$$
$$= 670 \, J/s = 670 \, W$$

This is the rate that energy is being transferred from inside her body to the moisture on her skin. Another way to think about this is that Eugenia's body is being cooled at a rate of 670 W!

The evaporation of sweat is a natural cooling mechanism that protects us from overheating. Increasing the rate of evaporation from the skin increases the body's cooling rate. A sick child's fever is often reduced by rubbing the skin with alcohol, since alcohol evaporates much faster than water.

Try it yourself: After swimming in a pool, you come out with 300 g of water absorbed in your swimsuit. How much energy will this water absorb from the environment to evaporate? How can you evaluate your answer?

Answer: $7.2 \times 10^5 \, J$. If we assume that the swimsuit dries only using the energy generated by your body at a rate of 100 J/s, it should take about 7200 s to dry, which is about 2 hours—this is a reasonable time. When the Sun is out, a swimsuit dries somewhat faster, so the absorption of energy from the Sun's radiation must be significant.

We can often solve problems in multiple ways. We could solve the following problem by treating the two objects (hot coffee and an ice cube) as two different systems. However, this time we will treat them as a single system. The hot object loses thermal energy as its temperature decreases, and the cooler object gains an equal amount of internal energy (potential and thermal) as it melts and warms.

EXAMPLE 12.8 Cooling hot coffee

You add 10 g of ice at temperature 0 °C to 200 g of coffee at 90 °C. Once the ice and coffee reach equilibrium, what is their temperature? Indicate any assumptions you made.

Sketch and translate A sketch of the process is shown below. We choose the coffee and ice as the system. The ice gains thermal energy and melts at 0 °C. Then the melted ice, now liquid water, warms to the final equilibrium temperature. Simultaneously, the coffee loses an equal amount of thermal energy as it cools until the mixture is at a single final equilibrium temperature.

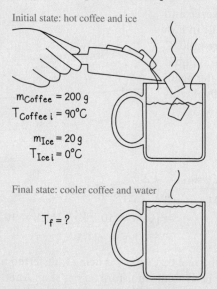

Initial state: hot coffee and ice

$m_{Coffee} = 200$ g
$T_{Coffee\,i} = 90°C$

$m_{Ice} = 20$ g
$T_{Ice\,i} = 0°C$

Final state: cooler coffee and water

$T_f = ?$

Simplify and diagram The coffee and ice are the system. Assume that the coffee-ice cube system is isolated; there is no energy transferred between the system and the environment, including the cup. Assume also that the coffee has the same specific heat as water, which is reasonable since coffee is mostly water. The process is

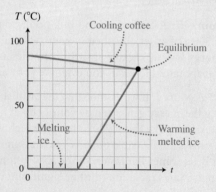

represented qualitatively in a temperature-versus-time (T-versus-t) graph using two lines for the ice and one line for the coffee. Notice that after some time, they reach the same thermal equilibrium temperature.

Represent mathematically The energy of the system is constant, so the internal energy change of the ice plus the thermal energy change of the coffee must add to zero.

$$\Delta U_{int} = \Delta U_{int\,Ice} + \Delta U_{th\,Coffee} = 0$$

The ice first melts, then warms. The coffee cools.

$$(m_{Ice}L_{Ice} + m_{Ice}c_{Water}(T_f - T_{Ice\,i}))$$
$$+ m_{Coffee}c_{Coffee}(T_f - T_{Coffee\,i}) = 0$$

Note that the system reaches equilibrium, so $T_{Ice\,f} = T_{Coffee\,f} = T_f$. Also, the initial temperature of the melted ice is $T_{Ice\,i} = 0$ °C.

Solve and evaluate Solve for the final temperature T_f:

$$T_f = \frac{m_{Coffee}c_{Coffee}T_{Coffee\,i} + m_{Ice}c_{Water}(0\,°C) - m_{Ice}L_{Ice}}{m_{Ice}c_{Water} + m_{Coffee}c_{Coffee}}$$

$$= \frac{(0.2\,\text{kg})(4180\,\text{J/kg}\cdot°C)(90\,°C) + 0 - (0.01\,\text{kg})(3.35 \times 10^5\,\text{J/kg})}{(0.01\,\text{kg})(4180\,\text{J/kg}\cdot°C) + (0.2\,\text{kg})(4180\,\text{J/kg}\cdot°C)}$$

$$= 82\,°C$$

This is a reasonable temperature considering the small amount of ice used.

Try it yourself: You pour 200 g water at 5.0 °C into a Styrofoam container and add 500 g of ice at −40 °C. What is the result? Describe the steps in solving the problem and determine the numerical answer.

Answer: Four possible outcomes are (a) all ice melts and the resulting water is above 0 °C; (b) all water freezes and the resulting ice is below 0 °C; (c) some but not all of the water freezes and the resulting mixture is at 0 °C; (d) some but not all ice melts and the mixture is at 0 °C. To decide which outcome is correct, first determine how much energy the water will release to cool to 0 °C and how much energy the ice will need to warm to 0 °C. Calculations show that the water will release 4.2×10^3 J and the ice will require 42×10^3 J. Thus cases (b) and (c) are possible. Final answer: the mixture will be at 0 °C and the amount of water that freezes is 110 g.

Review Question 12.6 Why are the units for specific heat and
heat of fusion/vaporization different?

12.7 Heating mechanisms

So far we have described the process of heating as a way for energy to be
transferred into or out of a system. However, we haven't discussed the energy transfer mechanisms by which heating occurs. These mechanisms
underlie and explain many phenomena. For example, how can we reduce
energy losses from our homes during the winter? Why is the "dry heat"
in Arizona less uncomfortable than a hot, humid day in Mississippi? Why
should we be concerned about climate change? These questions are related
to energy transfer mechanisms and, in particular, to the rate at which these
transfers occur.

We define heating rate H as the energy transfer by heating Q from one
object to another divided by the time interval Δt needed for that heating energy transfer:

$$\text{Heating rate} = H = \frac{Q}{\Delta t}$$

This transfer can occur through several mechanisms that we discuss below.

Conduction

You are boiling water for morning coffee at your campsite. You have two cups:
one cup is Styrofoam and the other aluminum. Which cup would you prefer
to hold just after filling it with hot water? The water inside both cups is at the
same temperature. Experience tells us that the Styrofoam cup will feel more
comfortable while the aluminum cup may burn our fingers.

Later, you cook scrambled eggs in a frying pan on the camp stove. You
start to remove the pan from the stove, but the handle is very hot, even though
the handle hasn't touched the stove's hot burner. How can you explain these
observations?

Recall that the temperature of a substance is related to the average random
kinetic energy of the particles of which it is made. The high-energy, high-speed
hot water molecules hit the inside wall of the aluminum cup or the Styrofoam
cup, making the particles on the inside wall of each cup vibrate faster. These
more energetic particles in the wall then bump into neighboring particles in
the wall, which in turn bump into their neighbors, which are closer to the outside wall. Thus, the thermal energy transfers from particle to particle from
the inside wall to the outside wall. Your fingers touch the outside wall. When
its surface is hot, the high-energy particles on the surface bump the particles
on the surface of your skin, causing them to move faster. The high-energy
vibrations of the particles on your skin can produce a burning sensation. The
process by which thermal energy is transferred through physical contact is
called **conductive heating** or **cooling**.

Why does the aluminum cup of hot water feel hotter than the Styrofoam
cup filled with equally hot water? Aluminum transfers thermal energy from
particle to particle quickly. Styrofoam transfers thermal energy more slowly.
We say that aluminum is a good thermal energy conductor and Styrofoam
is a poor thermal energy conductor. A quantity called **thermal conductivity K** characterizes the rate at which a particular material transfers thermal
energy (see **Table 12.8**). The larger the thermal conductivity, the faster thermal energy transfers from molecule to molecule in that material. You want to
use a material with low thermal conductivity to make a cup for holding hot

Why is the handle hot, even though it has
not touched the flame?

**Table 12.8 Thermal conductivity
of materials (at 25 °C).**

Material	K (W/m·°C)
Metals	
Aluminum	250
Brass	109
Copper	400
Stainless steel	16
Other solids	
Soil	0.6–4
Brick	1.31
Concrete, stone	1.7
Attic insulation	0.04
Glass, window	0.9
Ice (at 0 °C)	2.18
Styrofoam	0.03
Wood	0.04–0.14
Fat	0.2
Muscle	0.4
Bone	0.4
Other materials	
Air	0.024
Water	0.58

beverages. Likewise, you want to use a material with low thermal conductivity to insulate your home—to prevent thermal energy from leaving the house during the winter or to prevent it from entering during the summer.

> **Conductive heating/cooling** The rate of conductive heating $H_{\text{Conduction}}$ from the hot side of a material to the cool side is proportional to the temperature difference between the two sides ($T_{\text{Hot}} - T_{\text{Cool}}$), the thermal conductivity K of the material separating the sides, and the cross-sectional area A across which the thermal energy flows. The rate is inversely proportional to the distance l that the thermal energy travels.
>
> $$H_{\text{Conduction}} = \frac{Q_{\text{Hot to Cool}}}{\Delta t} = \frac{KA(T_{\text{Hot}} - T_{\text{Cool}})}{l} \qquad (12.8)$$

CONCEPTUAL EXERCISE 12.9 Snow on the roof

In the morning on a cloudy winter day your house with an attached garage has a light snow covering it. What might you expect to see 24 hours later, assuming that the temperature stays low and there is no new snow?

Sketch and translate The melting on different parts of the roof depends on the conductive thermal energy transfer through the roof from inside the house to the outside.

Simplify and diagram Assume that the energy delivered to the roof comes from inside the house and not from the Sun. Assume also that the insulation and roofing material is the same in all parts of the house (K is the same for the roof over all parts of the house). Now divide the house into two parts: the living quarters and the garage. Thermal energy transfer occurs from the warmer regions inside the house through the roof to cooler regions outside of the house. The conductive heating transfer rate depends on the temperature difference ($T_{\text{Hot inside}} - T_{\text{Cold outside}}$). The temperature difference is greater from in the living area to the outside than it is from the inside of the garage to the outside. Thus, the thermal energy transfer rate should be greater through the roof above the living quarters than it is through the roof over the garage. Because of this, there will be more snow melting above the living area than above the garage.

The photo below confirms this reasoning. The faster melting over the living area tells us that energy is being transferred out of the house more quickly than out of the garage.

Can we reduce the amount of energy we are releasing to the environment (and therefore keep more of the energy—and warmth—in the house)? To reduce energy transfer, we could slow the heating transfer rate by using lower thermal conductivity K insulation or thicker insulation. When we use thicker insulation [thereby increasing length l in Eq. (12.8)], the thermal energy has to travel a longer distance across the interface.

Try it yourself: Your skin temperature is about 35 °C. Imagine that you touch the wooden top of a desk in a room where the temperature is 20 °C. The desk feels comfortable. Then you touch the steel leg of the desk and it feels cold. Does it mean that the legs of the desk are at a lower temperature than the wooden top?

Answer: The legs and the desktop are at the same temperature—room temperature. When you touch each of them with your hand, which is somewhat warmer than these objects, thermal energy transfers from your hand to the objects. The larger thermal conductivity of steel makes this transfer occur at a higher rate, which is why the steel feels cooler to your hand.

The concept of thermal conductivity explains why a thermos keeps drinks hot or cold. The thermos depicted in **Figure 12.10** consists of two metal cups separated by a layer of low-density air (as close to a vacuum as we can get). Because the particles of air in the nearly evacuated space between the metal cups are very far from each other, they slowly transfer energy by conduction, resulting in a very small thermal conductivity. (In a perfect vacuum K would be zero, because there would be no particles to interact with each other.) Thus, if a liquid in the inner cup is cold, it stays cold; if it is warm, it stays warm.

Another application of the concept of thermal conductivity is fur that allows animals to get through cold winters. The fur consists of hair strands separated by air; the strands work almost like a thermos in preventing the conductive energy transfer from a warm body to a cold environment.

Figure 12.10 How a thermos bottle works. A thermos consists of an inner chamber surrounded by a vacuum.

Near vacuum

Evaporation

At the beginning of this section we asked, Why are the hot, dry summers in Arizona less uncomfortable than hot, humid summers elsewhere in the United States? The answer is that we more effectively transfer energy by evaporation in a dry climate. On a humid day water not only evaporates from the skin but also condenses on the skin from the gaseous water vapor in the air. When gaseous water vapor converts to liquid water (condenses), energy is released and returned to your body, raising your temperature. To stay cool, you want the rate of evaporation to be somewhat greater than the rate of condensation. In a "dry" climate with low humidity, not much water vapor condenses on your skin. You evaporate more water than condenses on you, and there is a net cooling effect. **Figure 12.11** indicates some typical evaporation rates under different circumstances.

Figure 12.11 Evaporation rates for different conditions.

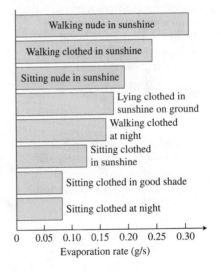

Evaporative heating/cooling Evaporative heating occurs when a gas condenses on a surface. Evaporative cooling occurs when a liquid evaporates from a surface. The evaporative energy transfer rate $H_{\text{Evaporation}}$ is proportional to the rate at which liquid mass evaporates or condenses $\Delta m/\Delta t$ on the surface and to the heat of vaporization L_v of the liquid at the temperature of the liquid:

$$H_{\text{Evaporation}} = \pm\left(\frac{\Delta m}{\Delta t}\right)L_v \tag{12.9}$$

We use the positive sign for condensation (energy transferred to the system) and the negative sign for evaporation (energy transferred out of the system).

Convection

Conduction as the heating mechanism works well in materials in which particles interact with each other. In gases, for example, the particles are so far apart that these interactions are almost nonexistent. In liquids the particles interact but much less than in a solid. How does thermal energy transfer in those materials? Consider the experiments in Observational Experiment **Table 12.9** on the next page.

OBSERVATIONAL EXPERIMENT TABLE

12.9 Convective heating/cooling.

Observational experiment	Analysis
Experiment 1. You light a candle and hold your hand near one side of it. Your hand does not feel much warmer than when it is far away from the candle.	The air near the flame quickly warms and expands. Since its density is now less than the density of the surrounding air, it rises and this warmed air contacts the skin of your hand directly above the candle. 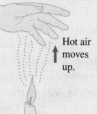 Hot air moves up.
However, if you place your hand above the candle, you can only hold it there for a short time interval before you risk burning yourself. Ouch!	

Experiment 2. You have the apparatus shown, placed over a flame. Potassium permanganate ($KMnO_4$) is a powdery substance of dark purple color that when placed in water turns it into a pink or purple color. You place some powder in a small spoon with holes and lower the spoon into the water above the flame.

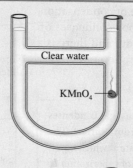

Clear water

$KMnO_4$

Colored water first spreads upward, then through the horizontal pipe, and then down the other vertical tube, and back to the area above the flame.

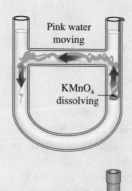

Pink water moving

$KMnO_4$ dissolving

Thermal conduction warms the water next to the flame. This warm water expands and rises. The warm water that moved up starts a *flow* of warm liquid that moves around the apparatus. As the pink water moves horizontally, it cools and sinks, replacing the water heated by the flame. The pink water is then heated again and the process repeats.

Pattern

In both processes the warmed fluid becomes less dense and rises. As the warm fluid rises, it is replaced by cooler fluid.

The processes described in Table 12.9 are called **natural convective heating**.

> **Natural convective heating/cooling** Natural convective heating/cooling is the process of energy transfer by a fluid moving in a vertical direction away from/toward the center of Earth. Fluids such as air or water rise because their density decreases as their temperature increases and sink because their density increases as their temperature decreases.

While convection can occur as a natural process, it can also be induced. For example, in a forced-air heating system, air in the furnace is warmed by a gas or electric heater. The warmed air is forced by a fan through ducts to different rooms in the house. Other ducts return cooler air back to the furnace to be warmed. In cars, burning gasoline makes the engine very hot. To cool it, a pump moves fluid (water or a coolant) from the radiator through hoses into and through the engine (**Figure 12.12**). Thermal energy is transmitted by conduction to the coolant from the engine. The energy is then carried away by the fluid (forced convective thermal energy transfer) through a radiator hose and back to the radiator, where outside air moves past the fins in the radiator and cools the fluid again. This process is called forced convective heating (or cooling). The forced-air heating system in a house and the cooling system in cars are examples of **forced convection**.

> **Forced convective heating/cooling** In forced convective heating, a hot fluid is forced to flow past a cooler object, transferring thermal energy to the object. The fluid moves away with less thermal energy. In forced convective cooling, a cool fluid moves past a warmer object (for example, a cold winter wind moving across your warm skin). The object transfers thermal energy to the cooler fluid. The now-warmed fluid carries the thermal energy away from the object.

Calculating the rate of convective energy transfer is a complex subject and depends on many factors. Often, the calculations are estimates and involve the use of tables of data assembled by energy transfer specialists. There are many practical examples of natural and forced convective heating and cooling.

A *convection oven* has advantages over a traditional oven. In a traditional oven, a thin layer of air surrounds the food. This air is at a temperature between that of the food and the hotter air in the rest of the oven. Thus, cooking occurs at a lower temperature than the temperature of air in the oven. In a convection oven, a fan (maybe more than one) at the back of the oven moves heated air around the cooking compartment. This causes fresh hot air to continually move past the food. Cooking time is reduced by about 20% and cooking temperature can be 10–20 °C lower than in a traditional oven. You get the same result with less energy use.

Radiation

Consider another phenomenon. You are sitting by a campfire on a cool evening. The air above the flames warms and rises with the smoke from the fire, an example of natural convective heating. You sit on a log at the side of the fire and are also warmed by it. The warming is not the result of the convection from the fire—the warm air is moving upward and does not move across your skin. You notice that the part of your body facing the flame is warmed, while the part of your body facing away from the flame remains cool. Why is that?

Consider a similar process. Outside Earth's atmosphere is an approximate vacuum with low concentrations of atomic particles. The particles are very far from each other and therefore essentially do not transfer energy by conduction. The Earth has remained warm enough for habitation because the Sun irradiates Earth in much the same way that a campfire warms the part of your body facing it. The Earth is warmer on the side facing the Sun and cooler on

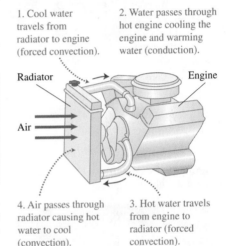

Figure 12.12 A car's engine is cooled mostly by forced convection.

1. Cool water travels from radiator to engine (forced convection).

2. Water passes through hot engine cooling the engine and warming water (conduction).

Radiator

Engine

Air

4. Air passes through radiator causing hot water to cool (convection).

3. Hot water travels from engine to radiator (forced convection).

the side facing away from it. Both the Sun and the campfire emit infrared radiation and some light that is the source of this thermal energy.

> **Radiative heating/cooling** All objects, hot and cool, emit different forms of radiation that can travel through a vacuum. The hotter the object, the greater the rate of radiation emission. Objects in the path of this radiation may absorb it, gaining energy.

Figure 12.13 A passive solar heating system.

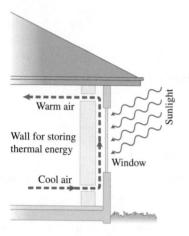

Passive solar heating is an important application of radiative heating (and other heating mechanisms). Passive solar buildings are constructed so that they have walls with large heat capacities for storing radiative energy (sunlight) that passes through windows during the daytime (**Figure 12.13**). The light energy absorbed by the wall transforms its energy into the thermal energy of the surface atoms and molecules in the wall. These warmer atoms and molecules transfer their energy by conduction to other atoms and molecules deeper in the wall. A well-designed wall has large mass m with a high specific heat c. As a result, the wall's temperature change

$$\Delta T = \frac{Q}{m \cdot c}$$

will be small during the day, because both m and c are large. The large amount of energy transferred to the wall by heating during the day is released slowly at night when it is needed. Energy is transferred to the building at night by natural convective heating as cool air moves up past the warm wall and into the building as well as from radiative heating from the wall into the building.

Radiation is also used for diagnostic methods in what is called *thermography*. One application is breast thermography, which detects differences in infrared radiation from warmer and cooler surfaces. Thermography detects "hot spots" caused by the increased metabolic activity in cancer cells and combines advanced digital technology with ultrasensitive infrared cameras for safe and early screening of breast cancer.

Thermography was also employed to take the picture in **Figure 12.14** of a house at night. The parts of the house that are radiating energy at the fastest rate are yellow. The next fastest radiating part is orange, then pink, purple, violet, and black. None of these regions are sources of visible light. All lights in the house are off. This radiation is being emitted simply because the house is not at zero absolute temperature.

Figure 12.14 An infrared photo of a house at night (no lights are on).

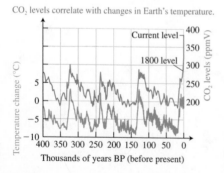

Review Question 12.7 What are the most effective ways of transferring energy through solids, liquids, gases, and vacuum?

12.8 Climate change and controlling body temperature: Putting it all together

The first law of thermodynamics and the various heating mechanisms help us understand two crucial phenomena—global climate change and body temperature control.

The greenhouse effect and climate change

Figure 12.15 Earth's average temperature and CO_2 concentration for the past 400,000 years.

Figure 12.15 shows a strong correlation between the CO_2 concentration in Earth's atmosphere and the average Earth temperature during the last 400,000 years. During the coldest periods, when much of Earth was covered with glaciers, the CO_2 levels were lowest. During warm periods, some parts of Earth became much dryer while other parts experienced flooding. During these

warm periods, the CO_2 levels were highest. Glaciers melted, leading to a rising ocean and coastal flooding. More frequent and severe weather events occurred (such as hurricanes, typhoons, droughts, and blizzards). Scientific evidence indicates that the concentration of CO_2 in our atmosphere plays a big part in these dramatic climate changes.

The cause of the increase in atmospheric CO_2 is a complex subject. Part of the increase is due to human activities, such as the increasing use of fossil fuels and the removal of forests and plant life, which absorb CO_2.

We can qualitatively investigate the role of CO_2 concentration in the Earth's changing temperature by applying the first law of thermodynamics to the system pictured in **Figure 12.16**. The system inside the dashed lines consists of Earth and its atmosphere. The Earth/atmosphere system gains energy through radiative heating from the Sun ($Q_{\text{Sun to Earth}} > 0$) and loses energy through radiation that it emits into space ($Q_{\text{Space to Earth}} < 0$):

$$Q_{\text{Sun to Earth}} + Q_{\text{Space to Earth}} = \Delta U_{\text{System}}$$

If we divide both sides of this equation by an arbitrary time interval Δt during which heating occurs, we get an equation that involves the heating rates and the rate of the system's energy change:

$$\frac{Q_{\text{Sun to Earth}}}{\Delta t} + \frac{Q_{\text{Space to Earth}}}{\Delta t} = \frac{\Delta U_{\text{System}}}{\Delta t}$$

or

$$H_{\text{Sun to Earth}} + H_{\text{Space to Earth}} = \frac{\Delta U_{\text{System}}}{\Delta t}$$

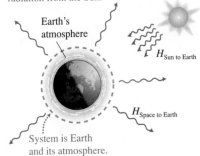

Figure 12.16 The Earth's temperature (including the atmosphere) depends on the balance of incoming and outgoing radiation.

Earth emits longer wavelength radiation into space. Earth absorbs shorter wavelength radiation from the Sun.

Earth's atmosphere

$H_{\text{Sun to Earth}}$

$H_{\text{Space to Earth}}$

System is Earth and its atmosphere.

The terms on the left are the radiative heating rates from the Sun to Earth ($H_{\text{Sun to Earth}} > 0$) and from Earth to space ($H_{\text{Space to Earth}} < 0$). If $H_{\text{Sun to Earth}} + H_{\text{Space to Earth}} = 0$, then the internal energy of the system (Earth and its atmosphere) remains constant. If $H_{\text{Sun to Earth}} + H_{\text{Space to Earth}} > 0$, the system's energy increases. If $H_{\text{Sun to Earth}} + H_{\text{Space to Earth}} < 0$, its energy decreases.

Which type of radiation is emitted by an object depends on the temperature of the object. Hot objects such as the Sun emit more energy as short-wavelength infrared radiation and visible light. Cooler objects such as Earth emit longer wavelength infrared radiation and very little short-wavelength infrared radiation and visible light. Carbon dioxide is a strong absorber of long-wavelength infrared radiation, but not a strong absorber of short-wavelength infrared radiation and visible light. This means that carbon dioxide does not absorb much of the radiation Earth receives from the Sun, but it does absorb considerable radiation emitted by Earth. Since the magnitude of $H_{\text{Space to Earth}}$ decreases as CO_2 concentration increases, then $H_{\text{Sun to Earth}} + H_{\text{Space to Earth}} > 0$ and the thermal energy of the Earth/atmosphere system increases, which manifests as a global temperature increase. Our simple model based on the first law of thermodynamics and the mechanisms of energy transfer predicts global warming if CO_2 concentration is increasing.

The concentration of CO_2 in Earth's atmosphere is now at the highest level in the last 400,000 years (Figure 12.15). The hottest years since Earth temperature data were first recorded in 1861 have all occurred since 1990, with each decade getting slightly warmer than the previous decades.

Controlling body temperature

A healthy human body maintains a central core temperature of about $37.0\,°C$ ($98.6\,°F$). To do so, the body must shed excess thermal energy that is converted from chemical energy due to metabolic processes. This is especially challenging during vigorous exercise, when the body's core temperature could increase by as much as $6\,°C$ in only 15 min, which could lead to convulsions or even brain damage.

Energy transfer is the key to maintaining a consistent temperature. We can analyze how the human body uses heating mechanisms of energy transfer by applying the first law of thermodynamics ($W_{\text{Environment on System}} + Q_{\text{Environment to System}} = \Delta U_{\text{System}}$) to the body (the system) during a short time interval Δt. Divide the quantities in the first law by the time interval during which the changes occurred:

$$\frac{W_{\text{Environment on System}}}{\Delta t} + \frac{Q_{\text{Environment to System}}}{\Delta t} = \frac{\Delta U_{\text{System}}}{\Delta t}$$

The second term in the above equation is the heating energy transfer rate ($Q_{\text{Environment to System}}/\Delta t$) from the environment to the system. We can separate Q into four separate terms, which represent the rates of conduction, convection, radiation, and evaporation (represented by H symbols with subscripts):

$$\frac{W_{\text{Environment on System}}}{\Delta t} + (H_{\text{Conduction}} + H_{\text{Convection}} + H_{\text{Radiation}} + H_{\text{Evaporation}})$$

$$= \frac{\Delta U_{\text{Thermal}}}{\Delta t} + \frac{\Delta U_{\text{Chemical}}}{\Delta t}$$

We also substituted the two types of internal energy change that can occur in a person's body: the rate of thermal energy change and the rate of chemical potential energy change. This latter term is the metabolic rate, the rate of energy released by the chemical reactions occurring in the body.

Now, let's apply the above equation to a real process. Suppose that a person runs at moderate speed on an indoor air-conditioned track. **Table 12.10** shows quantities that are representative of a real process, although the quantitative methods used to determine the numbers have not been developed in this book.

Consider each value in Table 12.10. Air exerts a drag force that opposes the runner's motion, doing work on the runner at a rate of -100 J/s. The cool air in the indoor facility passes across the runner's skin and causes a -300 J/s convective cooling rate as the skin loses thermal energy and the air gains it. The walls of the indoor facility are cooler than the runner's skin temperature, and consequently the runner absorbs energy transmitted by radiative heating from the walls at a lower rate than the body emits energy by radiative cooling from the skin (a net -75 J/s radiative energy transfer rate). Finally, the runner's evaporative energy transfer rate equals -325 J/s. Adding all these, we get -800 J/s. To keep the body temperature constant, the runner's metabolic rate must be 800 J/s. If the person's metabolic rate is more than 800 J/s, the person would need additional cooling in order to maintain a constant body core temperature.

Review Question 12.8 How does the concentration of CO_2 in the atmosphere contribute to the rise of the atmosphere's temperature?

Table 12.10 Work, heating, and energy changes in the body of a runner (moderate speed in an air-conditioned building).

Type of energy change	Work and heating rates (watts = J/s)
Drag force exerted by air (work)	-100
Conduction	≈ 0
Thermal energy transfer due to convection	-300
Transfer of energy by radiative cooling	-75
Cooling due to evaporation of sweat	-325

	Rate of Internal Energy Change (watts = J/s)
Thermal	≈ 0
Chemical (metabolic rate)	-800

Summary

Words	Pictorial and physical representations	Mathematical representation

Thermal energy U_{thermal} The random kinetic energy of the atoms and molecules in a substance. (Section 12.1)

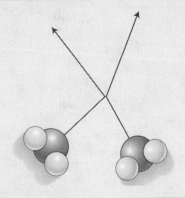

$$U_{\text{thermal}} = \frac{3}{2}nRT = \frac{3}{2}NkT_K$$
Eq. (12.1)

(Ideal gas, T in kelvins)

Work W The process of energy transfer from the environment to the system due to mechanical motion. For gases, instead of force and distance to describe work, we use the pressure of the environment against the system and the volume change of the system. For gases the work equals the negative of the area under a P-versus-V graph). (Section 12.1)

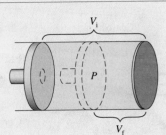
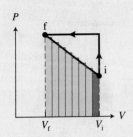

$$W_{\text{Environment on System}} = -P\Delta V$$
Eq. (12.2)

for constant P

Heating Q Process of energy transfer from the environment to the system due to differences in their temperatures. $Q > 0$ if energy is transferred into the system; $Q < 0$ if energy is transferred out of the system. (Section 12.2)

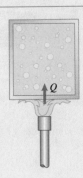

First law of thermodynamics A system's energy can change $\Delta U_{\text{int}} = (U_f - U_i)$ during a process due to the work W done on the system by the environment and/or due to the heating Q of the system by the environment. (Section 12.3)

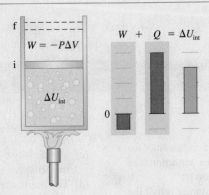

$W_{\text{Environment on System}}$
$+ Q_{\text{Environment to System}}$
$= \Delta U_{\text{System}}$ Eq. (12.4)

(continued)

Words	Pictorial and physical representations	Mathematical representation
Consequences of heating (no work done on the system) ■ *Temperature change* The system's temperature changes when there is no phase change occurring. ■ *Phase changes* System melts ($\Delta U > 0$) or freezes ($\Delta U < 0$) at constant temperature. System boils ($\Delta U > 0$) or condenses ($\Delta U < 0$) at constant temperature. (Sections 12.4 and 12.6)	Q T_{Change} (gas) Boil/condense T_{Change} (liquid) Melt/freeze T_{Change} (solid) t	$\Delta U = cm\Delta T$ Eq. (12.5) $Q = \pm mL_f$ Eq. (12.6) $Q = \pm mL_v$ Eq. (12.7)
Thermodynamic energy transfer mechanisms ■ *Conduction* $H_{Conduction}$ Energy transfer from particle to particle via contact. ■ *Convection* $H_{Convection}$ Energy transfer by particles moving from one place to another. ■ *Evaporation* $H_{Evaporation}$ Energy transfer due to evaporation/ condensation on surface. ■ *Radiation* $H_{Radiation}$ Energy transfer via absorption or emission of radiation. (Section 12.7)	Conduction ℓ A Heat T_2 T_1 Convection Warmer air Cool air Evaporation H O H O H H H Fast Slow Radiation R_{Earth} Earth R_{Sun} Sun R_{Sun} Earth's atmosphere R_{Earth}	$H = Q/\Delta t$ Conduction: $H_{Conduction} = KA(T_{Hot} - T_{Cold})/\ell$ Eq. (12.8) Convection: *No equation* Evaporation: $H_{Evaporation} = (\Delta m/\Delta t)L_v$ Eq. (12.9) Radiation: *No equation*

 For instructor-assigned homework, go to **MasteringPhysics.**

Questions

Multiple Choice Questions

1. An ideal gas in a container is separated with a divider into two identical half-size containers, each with the same amount of gas. Which of the following statements are correct?
   (a) The temperature of the gas in each container is half the previous temperature.

   (b) The mass of gas in each container is half the original mass.
   (c) The density of the gas in each container is half the original density.
   (d) The thermal energy of gas in each container is half the original thermal energy.

2. A container of gas has a movable piston, which you hold with your hand. You remove your hand and simultaneously place the container in a vacuum. Which of the following statements are correct? (More than one statement may be correct.)
   (a) The gas volume increases.
   (b) The internal energy of the gas does not change.
   (c) The temperature of the gas decreases.
   (d) The internal energy of the gas decreases.
3. A container of gas has a movable piston, which you hold with your hand. You remove your hand. Which of the following statements are always correct at the moment at which you release the piston? (More than one statement may be correct.)
   (a) The volume of the gas will increase.
   (b) The internal energy of the gas will not change.
   (c) The temperature of the gas will decrease.
   (d) The internal energy of the gas will decrease.
   (e) None of the above are always true.
4. Which of the following are possible means to change the energy of a system?
   (a) Do work on the system.
   (b) Heat the system.
   (c) Transfer energy without doing any work or heating.
5. Which gas has the greatest specific heat?
   (a) A gas in a constant volume container
   (b) The same gas in a constant pressure container with a movable piston
   (c) The specific heat is the same for both containers.
   (d) There is not enough information to answer the question.
6. The specific heat of water is 4200 J/°C · kg. Which of the following statements is true?
   (a) One kilogram of water stores 4200 J of energy when its temperature is 1 °C.
   (b) If you add 4200 J to water, its temperature will increase by 1 °C.
   (c) If you add 42,000 J to 1 kg of water, its temperature will increase by 10 °C.
7. How much heat is stored in 10 kg of water at 95 °C?
   (a) 42,000 J
   (b) 399,000 J
   (c) Zero; heat is not energy but is a process for transferring thermal energy.
8. We define the specific heat of a material as the energy that must be transferred to 1.0 kg of that material in order to cause it to warm 1.0 °C. What happens to the specific heat if we transfer twice that much energy?
   (a) The specific heat doubles.
   (b) The specific heat halves.
   (c) The specific heat does not change.
9. Dublin, Ireland and Edmonton, Alberta, Canada are at the same latitude (53.5°). The January and July average temperatures in Dublin are 5 °C and 16 °C, respectively; in Edmonton they are −12 °C and 21 °C, respectively. Why is the temperature variation so much greater in Edmonton than in Dublin?
   (a) Dublin and Edmonton are in different hemispheres.
   (b) Dublin is a bigger city.
   (c) Dublin is on the ocean.
10. Why do people sometimes wear fur coats in the winter in northern cities?
   (a) Fur reflects the sunlight.
   (b) Fur has low thermal conductivity.
   (c) Wearing a fur coat raises your temperature.

Conceptual Questions

11. Match each heating mechanism (left column) with a corresponding phenomenon (right column). There will be more than one match for some mechanisms.

Mechanism	Phenomenon
a. Conduction	1. You feel cold when wearing wet clothes.
b. Convection	2. A mother rubs a sick child with alcohol.
c. Evaporation	3. A farmer tills the soil around a fruit tree in preparation for a cold winter.
d. Radiation	4. You wear white clothes in the summer.
	5. Ice chests are made of Styrofoam.
	6. An electric heating coil is inserted at the bottom of a water container, not at the top.

12. Your friend says, "Heat rises." Do you agree or disagree? If you disagree, what can you tell your friend to convince him of your opinion?
13. Suggest practical ways for determining the specific heats of different liquids and solids.
14. Suggest practical ways to measure heats of melting and evaporation for ice and water.
15. A solar thermal storage tank holds 2000 kg of water. Approximately what mass of rocks would store the same amount of thermal energy (assuming the same temperature change)?
16. A farmer's fruit storage cellar is unheated. To prevent the fruit from freezing, the farmer places a barrel of water in the cellar. Explain why this helps prevent the fruit from freezing.
17. Why does an egg take the same time interval to cook in water that is just barely boiling as in water that boils vigorously?
18. Why does food cook faster in a pressure cooker than in an open kettle?
19. A potato into which several nails have been pushed bakes faster than a similar potato with no nails. Explain. List the physics ideas and principles that you used to answer this question.
20. Explain why double-paned windows help reduce winter energy losses in a home.
21. The water in a paper cup can be boiled by placing the cup directly over the flame from a candle or from a Bunsen burner. Explain why this is possible without burning a hole in the cup.
22. Provide two reasons why blowing across hot soup or coffee helps lower its temperature. How can you test your explanations?
23. Placing a moistened finger in the wind can help identify the wind's direction. Explain why.
24. Joggers often accumulate large amounts of water on their skin when running with the wind. When running against the wind, their skin may seem almost dry. Give two explanations. How can you test your explanations?
25. Why does covering a keg of beer with wet towels on a warm day help keep the beer cool?
26. Explain why dogs can cool themselves by panting.
27. Some houses are heated by circulating hot oil or water through baseboards. Such a heating system is always positioned close to the floor. Why?
28. If on a hot summer day you place one bare foot on a hot concrete swimming pool deck and the other bare foot on an

adjacent rug at the same temperature as the concrete, the concrete feels hotter. Why?

29. A woman has a cup of hot coffee and a small container of room-temperature milk, which she plans to add to the coffee. If the woman must wait 10 minutes before drinking the coffee, and she wants it to be as hot as possible at that time, should she add the milk and wait 10 minutes or wait 10 minutes and add the milk? Explain.

30. Look carefully at your surroundings for one or more days. Make four recommendations of ways to reduce loss of energy by heat transfer. What physical ideas and principles did you use to make these recommendations?

Problems

Below, BIO indicates a problem with a biological or medical focus. Problems labeled EST ask you to estimate the answer to a quantitative problem rather than develop a specific answer. Problems marked with ✏ require you to make a drawing or graph as part of your solution. Asterisks indicate the level of difficulty of the problem. Problems with no * are considered to be the least difficult. A single * marks moderately difficult problems. Two ** indicate more difficult problems.

12.1 Internal energy and work in gas processes

1. * EST Estimate the thermal energy of the air in your bedroom. List all of the assumptions you make.

2. A helium-filled balloon has a volume of 0.010 m³. The temperature in the room and in the balloon is 20 °C. What are the average speed and the average kinetic energy of a particle of helium inside the balloon? What is the thermal energy of the helium?

3. * Imagine that the helium balloon from the previous problem was placed in an evacuated container of volume 0.020 m³ and that the balloon popped when it touched a sharp edge on the inside of the container. (a) How much work is done on the helium gas? (b) What happens to the temperature of the helium, its density, the gas pressure, the average kinetic energy of each particle, and the thermal energy of the helium gas? Provide quantitative answers.

4. * You accidentally release a helium-filled balloon that rises in the atmosphere. As it rises, the temperature of the helium inside decreases from 20 °C to 10 °C. What happens to the average speed of helium atoms in the balloon and the thermal energy of the helium inside the balloon? Describe the assumptions you made.

5. * Air in a cylinder with a piston and initially at 20 °C expands at constant atmospheric pressure. (a) What is the work that the piston does on the gas if the air expands from 0.030 m³ to 0.043 m³? (b) How many moles of gas are in the container? (c) Suppose that the work leads to a corresponding change in thermal energy (there is no heating). What is the final temperature of the gas?

6. * In an empty rubber raft the pressure is approximately constant. You push on a large air pump that pushes 1.0 L (1.0 × 10⁻³ m³) of air into the raft. You exert a 20-N force while pushing the pump handle 0.02 m. (a) Determine the work done on the gas. (b) If all of the work is converted to thermal energy of the 1.0 L of gas, what is the temperature increase of the gas?

12.2–12.4 Two ways to change the energy of a system; First law of thermodynamics; Specific heat

7. * EST BIO **Body temperature change** A drop in temperature of the human body core from 37 °C to about 31 °C can be fatal. Estimate the thermal energy that must be removed from a human body to cause this temperature change.

8. * BIO **Temperature change of a person** A 50-kg person consumes about 2000 kcal of food in one day. If 10% of this food energy is converted to thermal energy that does not leave the body, what is the person's temperature change? What assumptions did you make?

9. Determine the amount of thermal energy provided by heating to raise the temperature of (a) 0.50 kg of water by 10 °C, (b) 0.50 kg of ethanol by 10 °C, and (c) 0.50 kg of iron by 10 °C.

10. EST Estimate the time interval required for a 600-kg cast iron car engine to warm from 30 °C to 1500 °C (approximately the melting temperature of iron) if burning fuel in the engine as it idles produces thermal energy at a rate of 8000 J/s and none of the energy escapes the car engine.

11. * A lead bullet of mass m traveling at v_i penetrates a wooden block and stops. (a) Represent the process with a bar chart. What system did you choose? (b) Assuming that 50% of the initial kinetic energy of the bullet is converted into thermal energy in the bullet, write an expression that would allow you to determine the block's temperature increase. (c) List all of the physics ideas that you used to solve this problem.

12. * BIO **Exercising warms body** A 50-kg woman repeatedly lifts a 20-kg barbell 0.80 m from her chest to an extended position above her head. (a) If her body retains 10 J of thermal energy for each joule of work done while lifting, how many times must she lift the barbell to warm her body 0.50 °C? (b) State any assumptions you used. (c) List all of the physics ideas that you used to solve this problem.

13. * You add 25 g of milk at 10 °C to 200 g of coffee (essentially water) at 70 °C. The coffee is in a Styrofoam cup. If the specific heat of milk is 3800 J/kg · °C, by how much will the coffee temperature decrease when the milk is added? Indicate any assumptions you made.

14. * You add 20 °C water to 0.20 kg of 40 °C soup. After a little mixing, the water and soup mixture is at 34 °C. The specific heat of the soup is 3800 J/kg · °C. Determine everything you can using this information.

15. BIO **Cooling a hot child** A 30-kg child has a temperature of 39.0 °C (102.2 °F). How much thermal energy must be removed from the child's body by some heating process to lower his temperature to the normal 37.0 °C (98.6 °F) body temperature?

16. * **Impact of extinction-causing meteorite** Scientists have proposed that 65 million years ago in what is now the Yucatan peninsula of Mexico, a 1.2 × 10¹⁶-kg meteorite moving at speed 11 km/s collided with Earth, and the resulting harsh conditions led to the extinction of many species, including the dinosaurs. (a) Calculate the kinetic energy of the meteorite before the collision. (b) If 20% of this energy was converted to

thermal energy in the meteorite, which had a specific heat of 900 J/kg·°C, by how much did its temperature increase?

17. * You pour 250 g of tea into a Styrofoam cup, initially at 80 °C, and stir in a little sugar using a 100-g aluminum 20 °C spoon and leave the spoon in the cup. What is the highest possible temperature of the spoon when you finally take it out of the cup? What is the temperature of the tea at that time? What assumptions did you make to answer the questions?

18. ** A 500-g aluminum container holds 300 g of water. The water and aluminum are initially at 40 °C. A 200-g iron block at 0 °C is added to the water. What can you determine using this information? State any assumptions you used. List all physics ideas that you used to solve this problem.

19. * A 150-g insulated aluminum container holds 250 g of water initially at 20 °C. A 200-g metal block at 60 °C is added to the water, resulting in a final temperature of 22.8 °C. What type of metal is the block? What assumptions did you make to answer the question?

12.5 Applying the first law of thermodynamics to gas processes

In the problems in this section, clearly describe your system and indicate important objects in its environment. Label the work as $W_{\text{Environment on System}}$, heating as $Q_{\text{Environment to System}}$, and the change in internal energy as ΔU_{int}.

20. ** / Gas in a container with a movable piston initially at volume V_1, pressure P_1, and a very high temperature T_1 expands at constant pressure until its temperature and volume became T_2 and V_2. (a) Describe the process using the concepts of work, heating, and internal energy. (b) Draw a bar chart representing the process. (c) Calculate the work that the environment did on the gas. (d) Explain the process from a microscopic point of view. (e) Represent the process using P-versus-V, P-versus-T, and V-versus-T graphs. (f) Repeat steps (a)–(e) for a situation in which the gas started with the same initial state but expanded at constant temperature instead of constant pressure.

21. ** Gas in a closed container undergoes a cyclic process from state 1 to state 2 and then back to state 1 (**Figure P12.21**). Describe the processes 1-2 and 2-1 qualitatively using the concepts of work, heating, and internal energy. (a) What happened to the thermal energy of the gas as it went from 1 to 2 and then from 2 to 1? What is the net change in the internal energy after the gas returned to state 1? (b) On the P-versus-V graph, show the magnitude of the work that was done on the gas by the environment during process 1-2 and during process 2-1. (c) Was the total work done on the gas positive, negative, or zero during the entire process 1-2-1? (d) Discuss the heating of the gas during process 1-2 and then 2-1. Was the total heating of the gas positive, negative, or zero during the whole process 1-2-1?

Figure P12.21

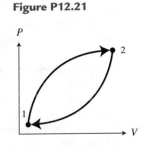

22. * **Jeopardy problem** A gas process is described mathematically as follows: $100 \text{ J} + (-P)(0.001 \text{ m}^3) = 0$. Pose a problem for which this description could be the answer. Describe the process macroscopically and microscopically.

23. * **Jeopardy problem** A gas process is described mathematically as follows: $Q + 120 \text{ J} = 50 \text{ J}$. Pose a problem for which this description could be the answer. Describe the process macroscopically and microscopically.

24. Use the first law of thermodynamics to devise a mathematical description of a process in which gas is being heated (positive heating) but its temperature does not change. Represent the process with a bar chart.

25. * Use the first law of thermodynamics to devise a mathematical description of a process in which gas is being cooled (negative heating) but its temperature increases. Represent the process with a bar chart.

26. * Derive an expression for the amount of energy that must be provided through heating for one mole of gas of molar mass M to have a temperature increase of 1 K. Consider different processes and decide whether the amount of heating is independent of the process.

27. * You are making a table for specific heats of gases. Compared to the specific heats of solid and liquid substances, what additional information do you need to provide when you are listing the values of specific heat for each gas (oxygen, hydrogen, etc.)? Explain your answer.

28. ** / One of the experiments that Joule used to test the idea of energy conservation was measurement of the temperature of a gas during the process of gas expansion into a vessel from which all air was evacuated. (a) Why did he choose this experiment? (b) What outcome would he predict based on the first law of thermodynamics? (c) Draw a picture of the experiment and provide macroscopic and microscopic reasoning for your answer.

29. **EST Temperature change in Carrier Dome** On March 5, 2006, a new college basketball attendance record of 33,633 was set in Syracuse University's Carrier Dome in the last regular-season game against Villanova. The volume of air in the dome is about $1.5 \times 10^6 \text{ m}^3$. Estimate the temperature change of the air in the dome in 2 h, if all the seats in the dome are filled and each person transfers his or her metabolic thermal energy to the air in the dome at a rate of 100 W (100 J/s). Assume that no thermal energy leaves the air through the walls, floor, or ceiling of the dome.

12.6 Changing state

30. Determine the energy needed to change a 0.50-kg block of ice at 0 °C into water at 20 °C.

31. * When $1.4 \times 10^5 \text{ J}$ of energy is removed from 0.60 kg of water initially at 20 °C, will all the water freeze? If not, how much remains unfrozen?

32. An electric heater warms ice at a rate of H. (a) What do you need to do to determine the mass of ice that melts in Δt min? (b) What assumptions did you use? How will your answer change if you use different assumptions?

33. * Determine the number of grams of ice at 0 °C that must be added to a cup with 250 g of tea at 40 °C to cool the tea to 35 °C.

34. An ice-making machine removes thermal energy from ice-cold water at a rate of H. Determine the time interval needed to form m kg of ice at ice melting temperature.

35. **Preventing freezing in canning cellar** A tub containing 50 kg of water is placed in a farmer's canning cellar, initially at 10 °C. On a cold evening the cellar loses thermal energy through the walls at a rate of 1200 J/s. Without the tub of water, the fruit would freeze in 4 h (the fruit freezes at −1 °C

because the sugar in the fruit lowers the freezing temperature). By what time interval does the presence of the water delay the freezing of the fruit?

36. * **Passive solar energy storage material** A Dow Chemical product called TESC-81 (primarily a salt, calcium chloride hexahydrate) is used as an energy-storage material for solar applications. Energy from the Sun raises the temperature of the solid material, causing it to melt at 27 °C (81 °F). At night the energy is released as the salt cools and returns to the solid state. (a) Determine the energy required to raise the temperature of 1.0 kg of solid TESC-81 from 20 °C to the liquid state at 27 °C. (b) How warm would 1.0 kg of water become if it started at 20 °C and absorbed the same energy? (c) Discuss the desirability of TESC-81 as a thermal energy-storage material compared to water. For TESC-81, c (solid) = 1900 J/kg·°C and $L_f = 1.7 \times 10^5$ J/kg.

37. How much energy is required to convert (a) 0.10 kg of water at 100 °C to steam at 100 °C and (b) 0.10 kg of liquid ethanol at 78 °C to ethanol vapor at 78 °C?

38. **Cooling with alcohol rub** During a back rub, 80 g of ethanol (rubbing alcohol) is converted from a liquid to a gas. Determine the thermal energy removed from a person's body by this conversion. Indicate any assumptions you made.

39. **Energy in a lightning flash** A lightning flash releases about 10^{10} J of electrical energy. If all this energy is added to 50 kg of water (the amount of water in a 165-lb person) at 37 °C, what are the final state and temperature of the water?

40. A kettle containing 0.75 kg of boiling water absorbs thermal energy from a gas stove at a rate of 600 J/s. What time interval is required for the water to boil away, leaving a charred kettle?

41. **Cooling nuclear power plant** A nuclear power plant generates waste thermal energy at a rate of 1000 MW = 1000×10^6 W. If this energy is transferred by hot water passing through tubes in the water in an evaporative cooling tower, how much water must evaporate to cool the plant (a) per second and (b) per day?

12.7 Heating mechanisms

In the problems in this section, clearly identify the system and objects in the environment that interact with the system. Label the work as $W_{Environment\ on\ System}$, heating as $Q_{Environment\ to\ System}$, and the change in internal energy as ΔU_{int}.

42. ** **EST** **Energy changes when it rains** Estimate the energy that is released or absorbed as water condenses and falls to Earth. Use the following information. Clouds are formed when moisture in the gaseous state in the air condenses. A rainstorm follows, dropping 2 cm of rain over an area 2 km × 2 km. Note that the mass of 1 m³ of water is 1000 kg.

43. * **Insulating a house** You insulate your house using insulation rated as R-12, which will conduct 1/12 Btu/h of thermal energy through each square foot of surface if there is a 1 °F temperature difference across the material (R-12 insulation is said to have a thermal resistance R of 12 h·ft²·°F/Btu). The conduction rate through a material of area A, across which there is a temperature difference $T_2 - T_1$, is

$$H_{Conduction} = (1/R)A(T_2 - T_1)$$

Use this information to determine the conductive energy flow rate across (a) an 8.0 ft × 16.0 ft R-15 wall; (b) a 3.0 ft × 7.0 ft R-4 door; and (c) a 3.0 ft × 4.0 ft R-1.5 window. Assume that the inside temperature is 68 °F and the

outside temperature is 20 °F. (d) Convert each answer from Btu/h to J/s = W.

44. ** **Igloo thermal energy conduction** A typical snow igloo (thermal conductivity about 1/10 of ice) is shaped like a hemisphere of radius 1.5 m with 0.36-m thick walls. What is the conductive heating rate through the walls if the inside temperature is 10 °C and the outside temperature is −10 °C? [*Hint:* Use the conductive heating equation from the previous problem, but replace $1/R$ by the thermal conductivity K.]

45. After a vigorous workout, you stand in shorts in a 20 °C room in front of a fan that blows air past you. What are the signs (+ or −) of the convective and radiative heating rates? Will you perspire to keep cool? Explain your answer.

46. To cool hot soup, you blow across the top of a bowl of soup. Assuming that the soup is the system, what are the signs (+ or −) of the different heating mechanisms and the effect of this energy transfer on the soup?

47. While blowing across the bowl of soup in the previous problem, you wonder how efficiently the soup can cool by itself through evaporation. You notice that the bowl of hot soup loses 0.40 g of water by evaporation in 1 min. What is the average evaporative heating rate of the soup during that minute? Assume that soup is primarily water.

48. **EST** ** **Solar collector** You wish to install a solar panel that will run at least five lightbulbs, a TV, and a microwave. How large should the panel be if the average sunlight incident on a photoelectric solar collector on a roof for the 8-h time interval is 700 W/m² and the radiant energy is converted to electricity with an efficiency of 20%?

49. **BIO** **Marathon** You are training for a marathon. While training, you lose energy by evaporation at a rate of 380 W. How much water mass do you lose while running for 3.5 h?

50. **Cooling beer keg** A keg of beer is covered with a wet towel. Imagine that the keg gains energy from its surroundings at a rate of 20 W. (a) At what rate in grams per second must water evaporate from a towel placed over the keg to cool the keg at the same rate that energy is being absorbed? (b) How much water in grams is evaporated in 2.0 h?

51. * A canteen is covered with wet canvas. If 15 g of water evaporates from the canvas and if 50% of the thermal energy used to evaporate the water is supplied by the 400 g of water in the canteen, what is the temperature change of the water in the canteen?

52. * **EST** **Evaporative cooling** Each year a layer of water of average depth 0.8 m evaporates from each square meter of Earth's surface. *Estimate* the average energy transfer rate in watts needed to continue this process.

53. * The rate of water evaporation from a fish bowl is 0.050 g/s and the natural thermal energy transfer rate to the bowl by conduction, convection, and radiation is +36 W. What power electric heater must you buy to keep the temperature in the fish bowl constant?

54. **BIO** **Tree leaf** A tree leaf of mass of 0.80 g and specific heat of 3700 J/kg·°C absorbs energy from the sunlight at a rate of 2.8 J/s. If this energy is not removed from the leaf, how much does the temperature of the leaf change in 1 min? [*Note:* Do not be surprised if your answer is large. A leaf clearly needs other heat transfer mechanisms to control its temperature.]

55. **Warming a spaceship** Your friend says that natural convection would not work on a spaceship orbiting the Earth. Do you agree or disagree with her statement? Explain.

56. * **Passive solar energy storage** Solar energy entering the windows of your house is absorbed and stored by a concrete wall of mass m. The wall's temperature increases by $10\,°C$ during the sunlight hours. What mass of water, in terms of m, would have the same temperature increase if it absorbed an equal amount of energy? What assumptions did you make?

12.8 Climate change and controlling body temperature: Putting it all together

57. Which is lighter: dry or wet air? Explain your answer.

58. ** If you drop a burning candle, it stops burning almost instantly. Suggest two explanations for this phenomenon and then propose testing experiments to rule out those explanations.

59. BIO **Losing liquid while running** While running, you need to transfer $320\,J/s$ of thermal energy from your body to the moisture on your skin in order to remain at the same temperature. What mass of perspiration must you evaporate each second? Indicate any assumptions you made.

60. * BIO **Running a marathon** When you run a marathon, the opposing force that air exerts on you does $-150\,J$ of work each second. You convert $1000\,J$ of internal chemical energy to thermal energy each second. (a) How much thermal energy must be removed from your body per second to keep your temperature constant? (b) If 50% of this energy loss is caused by evaporation, how much water do you lose per second? (c) How much water do you lose in 3 h?

61. ** **Global climate change** Assume that because of increasing CO_2 concentration in the atmosphere, the net radiation energy transfer rate for Earth and its atmosphere is $+0.002 \times 1350\,W/m^2$, corresponding to a 0.2% decrease in radiation leaving Earth. (a) Determine the extra thermal energy added to Earth and its atmosphere in 10 years. [*Note:* The radiation falls on an area approximately equal to πr_{Earth}^2 where $r_{Earth} = 6.4 \times 10^6\,m$.] (b) If 30% of this energy is used to melt the polar ice caps, how many kilograms of ice will melt in 10 years? (c) How many cubic meters of ice will melt? (d) By how much will the level of the oceans rise in 10 years?

62. * **Standard house 1** On an average winter day ($3\,°C$ or $38\,°F$) in a typical house, energy already in the house is lost at the following rates: (i) $2.1\,kW$ is lost through partially insulated walls and the roof by conduction; (ii) $0.3\,kW$ is lost through the floor by conduction; and (iii) $1.9\,kW$ is lost by conduction through the windows. Additional heating is also needed at the following rates: (iv) $2.3\,kW$ to heat the air infiltrating the house through cracks, flues, and other openings and (v) $1.1\,kW$ to humidify the incoming air (because warm air must contain more water vapor than cold air for people to be comfortable). What is the total rate at which energy is lost from this house?

63. * **Standard house 2** On the same day in the same house described in the previous problem, some thermal energy is supplied by heating in the following amounts: (i) sunlight through windows, $0.5\,kW$; (ii) thermal energy given off by the inhabitants, $0.2\,kW$; and (iii) thermal energy from appliances, $1.2\,kW$. How many kilowatts must be supplied to this standard house by the heating system to keep its temperature constant?

64. * **Standard house 3** Suppose that the following design changes are made to the house described in the previous two problems: (i) additional insulation of walls, roof, and floors, cutting thermal losses by 60%; (ii) tightly fitting double-glazed windows with selective coatings to reduce the passage of infrared light, cutting conduction losses by 70%; and (iii) elimination of cracks, closing of flues, and so on, cutting infiltration losses by 70%. What is the total rate at which energy is lost from this house?

65. * **Standard house 4** After further improvements (shifting windows from the north to the south sides and replacing outmoded appliances), thermal energy is supplied to the house described above at the following rates: (i) sunlight through windows, $1.0\,kW$; (ii) people's warmth, $0.2\,kW$; and (iii) appliance warmth, $0.8\,kW$. How many kilowatts must be supplied by the heating system of this house to keep it at constant temperature?

General Problems

66. ** EST BIO **Metabolism warms bedroom** Because of its metabolic processes, your body continually emits thermal energy. Suppose that the air in your bedroom absorbs all of this thermal energy during the time you sleep at night. Estimate the temperature change you expect in this air. Indicate any assumptions you make.

67. * EST You have an 850-W electric kettle. Estimate the least amount of time you have to boil water before 10 guests arrive for a tea break. State clearly all numbers that you use in your estimate.

68. * EST **House ventilation** For purposes of ventilation, the inside air in a home should be replaced with outside air once every 2 hours. This air infiltration occurs naturally by leakage through tiny cracks around doors and windows, even in well-caulked and weather-stripped homes. (a) Estimate the mass of air lost every 2 h and each second. (b) Estimate the energy per second needed to warm outside air leaking into the house during a winter night. State your assumptions.

69. BIO **Frostbite** When exposed to very cold temperatures, the human body maintains core body temperature by reducing blood circulation to the skin and extremities, and skin and extremity temperatures drop. This can eventually lead to frostbite. Explain why this helps conserve thermal energy.

70. ** EST **Heating an event center with metabolic energy** Estimate the temperature change in some enclosure on your campus during an athletic event. Assume that there is no thermal energy transfer into or out of the building during the event. Indicate any other assumptions you make.

71. ** EST **Lightning warms body** A lightning flash releases about $10^{10}\,J$ of electrical energy. Quantitatively estimate the effect on your body if you absorbed 10% of the energy. State clearly any assumptions you made.

Reading Passage Problems

Cloud formation Air consists mostly of nitrogen (N_2), with a molecular mass of 28, and oxygen (O_2), with a molecular mass of 32. A water molecule (H_2O) has molecular mass 18. According to the ideal gas law ($N/V = P/kT$), dry air at a particular pressure and temperature has the same particle density (number of particles per unit volume) as humid air at the same pressure and temperature. Consequently, humid air, whose low-mass water molecules replace more massive nitrogen and oxygen molecules, is less dense than dry air—the humid air rises. Atmospheric pressure decreases with elevation since there is less air above that is pushing down. At about 5000 m above the Earth's surface, the pressure is about

0.5 atm. Assuming an ideal gas, the gas volume V increases as pressure P decreases.

What happens to the air temperature as humid air rises? Air is a poor thermal energy conductor, and there is little heating from neighboring air ($Q \approx 0$, an adiabatic expansion process). As the rising gas expands, the neighboring environmental gas pushes in the opposite direction of the increasing volume of this rising gas. The environment does negative work on the rising air system ($W_{\text{by Environment on System}} < 0$). According to the first law of thermodynamics ($W + Q = \Delta U_{\text{System}}$): $-|W| + 0 = \Delta U_{\text{System}}$. The system's internal thermal energy decreases with a corresponding temperature decrease—about $-10\,°C$ for each 1000-m increase in altitude.

When the humid air reaches its dew point temperature, it starts to condense into water droplets (cloud formation). When condensation occurs, energy is released. There is a competition between decreasing thermal energy as the air expands and increasing thermal energy as the water vapor condenses. The air now cools at a lower rate of about $-5\,°C$ for each 1000-m increase in elevation.

72. If no condensation occurred, how high would $40\,°C$ humid air have to rise before its temperature decreased to $10\,°C$?
    (a) 1000 m (b) 2000 m (c) 3000 m
    (d) 6000 m (e) 10,000 m
73. When rising humid air starts to condense, why does its temperature change less rapidly with increasing elevation?
    (a) Its density increases, making it more difficult to change temperature.
    (b) Thermal energy released during condensation causes less thermal energy change.
    (c) The gas expands less, causing less negative work.
    (d) The temperature of the surrounding air is changing less.
74. After crossing a mountain top on a warm sunny day, what should cool dry air do?
    (a) Sink, because it is denser than warmer air below
    (b) Warm, because the surrounding gas does positive work in causing it to contract
    (c) Not change, as the surrounding air transfers little thermal energy by heating ($Q \approx 0$)
    (d) a and b are correct.
75. Why does humid air rise in dry air?
    (a) Water is attracted to the clouds above.
    (b) A water molecule has lower mass than other air molecules.
    (c) $1\,m^3$ of humid air has fewer molecules than $1\,m^3$ of dry air at the same T and P.
    (d) b and c are correct.
    (e) None of the above is correct.
76. If $1\,m^3$ of dry air rises 5000 m and has no temperature change, what would its volume be?
    (a) $0.3\,m^3$ (b) $0.5\,m^3$ (c) $1.5\,m^3$
    (d) $2.0\,m^3$ (e) Too little information to answer
77. EST The magnitude of the thermal energy released from the water molecules to the air if 1.0 g of water vapor condensed to 1.0 g of liquid water is closest to which of the following?
    (a) 300 J (b) 500 J (c) 1000 J
    (d) 2000 J (e) $2 \times 10^6\,J$

Meteorite impact The great Arizona crater was created by the impact of a meteorite of estimated 5×10^8 kg mass. The meteorite's speed before impact was about 10,000 m/s. Large amounts of rock found near the crater appeared to have melted on impact and then solidified as it cooled, indicating that the temperature of the rock during impact reached at least $1700\,°C$, the melting temperature of the rock. Is this possible?

Consider Earth and the meteorite as the system. The initial state of the process is the meteorite moving fast just before hitting Earth's surface. The final state is several minutes after the collision. What types of energy transformation occurred? The meteorite had kinetic energy before impact. In the collision, the meteorite dug a hole in Earth, forming the crater. The displaced soil was raised a distance approximately equal to the diameter of the meteorite. This inelastic collision produced considerable internal energy—thermal energy of the meteorite and of Earth's surface matter at the collision site. If the temperature change was high enough, the meteorite and/or parts of Earth may have undergone one or more phase changes. We summarize the process as follows:

$$K_i = \Delta U_{gf} + \Delta U_{\text{thermal}}$$

In the following questions, estimate different energies and energy changes. In addition to the information already given, you can use the following: $3300\,kg/m^3$ meteorite density; $840\,J/kg \cdot °C$ specific heat for the solid meteorite; $2.7 \times 10^5\,J/kg$ heat of fusion, the same as that of iron; $1000\,J/kg \cdot °C$ specific heat for the liquid meteorite, the same as that of iron; and $6.4 \times 10^6\,J/kg$ heat of vaporization, the same as for iron.

78. EST The initial kinetic energy of the meteorite was closest to
    (a) $2 \times 10^{13}\,J$ (b) $3 \times 10^{15}\,J$
    (c) $3 \times 10^{16}\,J$ (d) $5 \times 10^{16}\,J$
79. EST The radius of the meteorite was closest to
    (a) 10 m (b) 30 m (c) 50 m
    (d) 100 m (e) 1000 m
80. EST The gravitational potential energy change was closest to
    (a) $10^7\,J$ (b) $10^9\,J$ (c) $10^{11}\,J$
    (d) $10^{13}\,J$ (e) $10^{15}\,J$
81. EST The energy needed to warm the solid meteorite to its melting temperature at $1700\,°C$ is closest to
    (a) $2 \times 10^{13}\,J$ (b) $5 \times 10^{14}\,J$
    (c) $7 \times 10^{14}\,J$ (d) $3 \times 10^{15}\,J$
82. EST The energy needed to melt the solid meteorite at its $1700\,°C$ melting temperature is closest to
    (a) $2 \times 10^{13}\,J$ (b) $5 \times 10^{14}\,J$
    (c) $7 \times 10^{14}\,J$ (d) $3 \times 10^{15}\,J$
83. EST The energy needed to vaporize the melted meteorite at its $2600\,°C$ boiling temperature is closest to
    (a) $2 \times 10^{13}\,J$ (b) $5 \times 10^{14}\,J$
    (c) $7 \times 10^{14}\,J$ (d) $3 \times 10^{15}\,J$
84. Is the initial kinetic energy enough to vaporize the meteorite?
    (a) Yes
    (b) No
    (c) Too little information to answer

Second Law of Thermodynamics 13

Why is the statement "we need to conserve energy" incorrect in terms of physics?

How does the inside of a refrigerator stay cold?

You add a pinch of salt to soup and it spreads out evenly in the soup. Why doesn't the salt spontaneously come back together into crystals?

In a gasoline-powered car, the energy of the fuel is converted into kinetic energy when the car moves and into internal thermal energy exhausted to the air. We worry about the increasing cost of a decreasing supply of fossil fuels. Does it mean we will run out of energy? We know from physics that energy does not disappear—it is converted from one form to another. If we consider the whole universe as our system, the amount of energy in the universe is constant no matter what we do. So why can't we just reuse the energy of fuel extracted from the ground? So far we have treated all forms of energy as equal—one form can be converted to another, but the total never changes. However, as we learn in this chapter, this is not entirely true.

Be sure you know how to:

- Identify a system and the initial and final states of a well-defined process (Sections 6.1 and 6.2).
- Determine the energy transferred through heating to, the work done on, and the internal energy change of a system (Sections 12.1 and 12.3).
- Apply the first law of thermodynamics to a process (Section 12.5).

Our discussions of energy until now have been based on the work-heating-energy principle. If a part of the universe chosen as the system gains energy because of work done on it by external forces or because of energy transfer through heating, then the environment surrounding the system loses an equal amount of energy, or vice versa. In short, the total energy of a system plus its environment is constant. However, there is something more subtle occurring. Some processes, despite being consistent with energy conservation, never occur. A ball that you hold above the floor will fall, bounce a few times, and then stop. If we choose the system to be Earth, the ball, and the floor, then we would say that the gravitational potential energy of the system is converted to internal energy after the ball stops bouncing (both the ball and the floor have gotten a little warmer). However, you never observe a ball that has been resting on the floor spontaneously lose thermal energy (cool down) and gain the equal amount of kinetic energy needed to jump up into the air. Such a process does not violate energy conservation, so why doesn't it occur? That question is the subject of this chapter.

Figure 13.1 The warmer water at the bottom of the waterfall never cools and goes back to the top of the waterfall.

13.1 Irreversible processes

As we have already learned, the total energy of a system and its environment is constant. Energy can be transferred between a system and its environment through the mechanisms of mechanical work or by heating. There can also be changes in the forms of energy within a system. For example, the gravitational potential energy of the water-Earth system at the top of a waterfall converts into kinetic energy as the water falls and then to thermal energy when it hits the rocks at the bottom of the falls (**Figure 13.1**). But the warmer water at the bottom never cools and flows back up in the reverse direction. Are our ideas about energy incomplete?

Let's think more deeply about kinetic energy. Imagine a boulder falling off a cliff. All of the particles within the boulder follow more or less the same path. The same is true of the particles in a moving car or in a flying airplane. Thus we can say that the kinetic energy of these objects is organized. What about the motion of water molecules in a glass of water? The molecules also have kinetic energy; however, their motion is random rather than organized—"less organized" than the kinetic energy of a falling boulder. Now that we've made this distinction, consider the experiments in Testing Experiment **Table 13.1**.

TESTING EXPERIMENT TABLE

13.1 Use the first law of thermodynamics to predict what might happen in the following processes and reverse processes.

VIDEO 13.1

Testing experiment	Prediction based on the first law of thermodynamics (energy transfer through work and heating)	Outcome
Experiment 1a. A pendulum bob is raised to the side and released. What happens to the bob?	**1a.** We predict that as the bob moves back and forth, the size of the swing will decrease due to air resistance and friction in the bearing at the top of the string. Eventually, the bob will stop, and the bearing and the air will be slightly warmer. 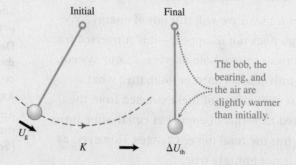	The predicted outcome for Experiment 1a always occurs.

Experiment 1b Reverse process. A pendulum bob hangs straight down at rest. What could happen to the bob that is consistent with the first law of thermodynamics?	**1b.** The bob, air, and bearing at top could cool and the bob would convert the decreasing thermal energy to kinetic energy and start swinging. 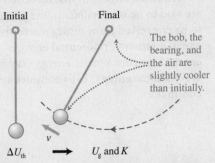 The bob, the bearing, and the air are slightly cooler than initially.	The predicted outcome for Experiment 1b never occurs.
Experiment 2a. A car is coasting on a horizontal road with its motor off. What will happen to the car?	**2a.** We predict that the car will gradually slow down and stop. The initial kinetic energy of the car will convert into internal thermal energy due to friction. The car tires and road will become warmer. 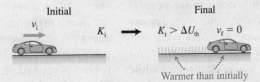 Warmer than initially	The predicted outcome of Experiment 2a occurs.
Experiment 2b Reverse process. A car with its motor off is at rest on a horizontal road. What might happen to the car that is consistent with the first law of thermodynamics?	**2b.** The car tires and road could cool and the decreased thermal energy would be converted into the car's organized kinetic energy as it gradually begins moving. 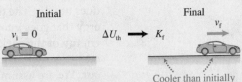 Cooler than initially	The predicted outcome of Experiment 2b never occurs.
Experiment 3a. An ice cube is sitting on a tabletop. What will eventually happen to the ice cube?	**3a.** We predict that the ice cube will gain thermal energy from the tabletop and air, causing the ice to melt and leaving a puddle of water on the tabletop. The tabletop and surrounding air will get a little cooler. Melted ice cube. Tabletop and air slightly cooler than initially. $U_{\text{th tabletop and air}} \rightarrow U_{\text{chem ice cube}}$	The predicted outcome of Experiment 3a occurs.
Experiment 3b Reverse process. A small puddle of cold water sits on a tabletop. What might happen to this puddle over time?	**3b.** The puddle could transfer thermal energy to the tabletop and air, causing the puddle's temperature to decrease below freezing so it turns into ice. Ice. Slightly warmer tabletop and air than initially. $U_{\text{chem ice cube}} \rightarrow U_{\text{th tabletop and air}}$	The predicted outcome of Experiment 3b never occurs.

Conclusion

- In the experiments in which the outcome matches the prediction (consistent with the first law of thermodynamics), the organized energy converts into less organized energy.
- In the experiments in which the outcomes do not match the predictions (also consistent with the first law of thermodynamics), the unorganized energy would have to convert into more organized energy. Such processes never occur.

Active Learning Guide ›

We conclude based on the testing experiments in Table 13.1 that some processes, although allowed by the first law of thermodynamics, do not occur in nature. Processes that occur in one direction but never occur in the opposite direction are said to be *irreversible*. In the processes that never occur, energy would have to be converted from disorganized thermal energy into an equal amount of organized kinetic or potential energy. It seems that an isolated system's thermal energy (random kinetic energy) cannot be converted into organized kinetic or gravitational energy. Let's consider several other situations to see if this pattern continues.

CONCEPTUAL EXERCISE 13.1 Irreversible processes?

Use the ideas of "organized" and "random" energy to decide which of the following processes are reversible. Then compare the results with your everyday experience. (a) A car skids to a stop; (b) a rock is thrown upward and reaches a height *h* above the ground; and (c) two fast-moving cars collide and stick together while stopping.

Sketch and translate The initial and final states of the processes and the choices of systems are sketched below.

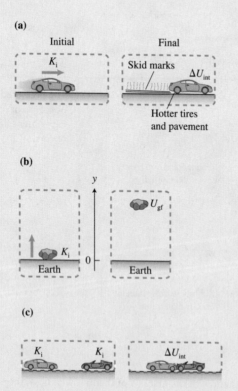

Simplify and diagram According to the pattern we observed in Table 13.1, a process in which a system goes from more organized energy to less organized random energy is not reversible. In (a), the kinetic energy of the car (more organized energy) converts into the thermal energy (random energy) of the tires and road as the car is stopping and into internal potential energy as the rubber is rubbed off the tires while skidding to a stop. Thermal and internal potential energy are less organized forms of energy; thus the reverse process (the hot road and tires and the skid marks coming back together to get the car moving) should not occur, and definitely does not. (b) The thrown rock had organized kinetic energy that goes into the organized gravitational potential energy of the rock-Earth system; thus the reverse process (the rock falling back down) should be able to occur. We are assuming that air resistance is small, since the increase in thermal energy of the air would be a conversion from more organized to less organized energy. (c) Before the collision, the cars have organized kinetic energy. After the collision, the cars are deformed (a change in the internal energy of the molecular structure of the cars) and the temperatures of the cars and the road have increased. Both the deformation and the temperature increase reflect more random, disorganized forms of energy. Thus, the process is not reversible (the warmed road and tires and deformed cars will not spontaneously be converted back into two undamaged moving cars).

Try it yourself: You place a droplet of food coloring in a glass of water. Describe the process and decide whether it is reversible or not based on the idea of organized energy.

Answer: The food coloring spreads throughout the water. We never see it spontaneously reform into its original organized droplet. The process is irreversible.

Reversible and irreversible processes Many processes in nature occur in only one direction—they are called *irreversible processes*. In the allowed direction, energy converts from more organized forms to less organized forms. In the reverse, unallowed processes, energy would have to convert from a less organized form to a more organized form. However, the unallowed reverse process does not contradict energy conservation.

Usefulness of different types of energy in doing work

You may have noticed that in all processes investigated so far in this chapter, no work was done on the system by the environment. Let's investigate a situation in which we wish to do some work—to compress a spring (the system object) a small distance. Imagine that a cart of mass m (not in the system) moves at a speed v. The cart has enough kinetic energy (if considered as a separate system) so that if it hits the spring, it will compress the spring a reasonable amount (**Figure 13.2a**). If the cart has a mass of 10 kg and moves at a speed of 5 m/s before hitting the spring, the kinetic energy of the cart would be 125 J. Some of this energy is transferred to the spring during the collision so that the spring gains elastic potential energy.

   Imagine a second process. A container filled with 1 kg of water is warmed from room temperature (20 °C) to boiling temperature. This requires

$$\Delta U = mc\Delta T = (1\ \text{kg})(4186\ \text{J/kg} \cdot {}^{\circ}\text{C})(100\ {}^{\circ}\text{C} - 20\ {}^{\circ}\text{C}) \approx 330{,}000\ \text{J}$$

of energy transferred through heating—far more than the energy needed to compress the spring by the same amount as the moving cart compressed it. You place the spring in the water. The hot water does not compress the spring at all. The thermal energy of the spring (disorganized energy) increases as it warms, but its elastic potential energy (organized energy) does not change. Apparently, disorganized forms of energy (such as thermal energy) do not have the ability to do work nearly as well as organized forms of energy.

‹ Active Learning Guide

(a)

The organized motion of the cart can do work on the spring and compress it.

(b)

The much greater thermal energy of the hot water cannot convert into the elastic potential energy of the spring.

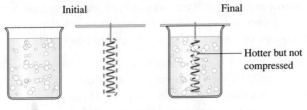

Figure 13.2 The organized energy in (a) can do work, whereas the disorganized energy in (b) cannot.

CONCEPTUAL EXERCISE 13.2 **What type of energy is best for doing work?**

Below we describe three systems that have the same amount of energy. Imagine that each of these systems can interact with another system of interest and transfer energy to that other system by doing work on it. Arrange these systems in the approximate order of how much work they can do on another system, listing the best system for doing work first.

1. 1000 J of chemical potential energy stored in the molecules of gasoline
2. 1000 J of thermal energy in the air particles of the room where you now reside
3. 1000 J of gravitational potential energy between Earth and a 50-kg barbell held 2.0 m above the ground

Sketch and translate The three systems are sketched below.

Simplify and diagram Possible energy conversion processes for each of the above are described at right.

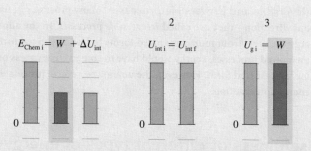

1. The chemical potential energy in the gasoline can do some work (for example, to run a small motor that partially compresses a spring). But the gasoline burning also increases the thermal energy in its surroundings.
2. The thermal energy in the air cannot run a motor or cause a spring to compress if they are at the same temperature as the air in the room. The thermal energy in the air does not change and is not a useful form of energy for doing work.
3. The gravitational potential energy between the barbell and Earth is useful, and nearly all of it is available for doing work on a spring or turning a paddle in water.

The bar charts represent these possible energy conversion processes. In summary, the ranking for the usefulness of the energy is 3 > 1 > 2.

Try it yourself: Which system is more useful for doing work on another system: a block moving upward with 10 J of kinetic energy or a block-Earth system in which the block is lifted so that the system has 10 J of gravitational potential energy relative to the Earth's surface? Ignore air resistance.

Answer: They are equally useful. Both systems have the same amount of organized energy (10 J).

Now we can combine our understanding of irreversible processes, organized energy, and the possibility to do work into one rule.

Active Learning Guide ➤

> **Usefulness of energy** As the energy of a system becomes less organized, the amount of work it can do on other systems decreases.

Carnot's principle

The person who first understood the link between how organized the energy is in a system and how much work it can do was the French engineer Sadi Carnot (1796–1832). Carnot was interested in the maximum work that a steam engine with a certain amount of fuel could do. He proposed two things:

1. It is impossible to use the engine's fuel directly to do work. The fuel needs to first warm a gas. This gas then expands and pushes a piston that moves in an environment that is cooler than the gas. If the piston is considered the system of interest, then the gas does work on the piston.
2. If the fuel transfers a certain amount of energy to the gas, only a fraction of this energy can do work on the piston. There is a theoretical limit

on the work that can be done. The remaining energy is transferred to the environment, where it manifests as thermal energy. Additionally, all real-world engines fail to reach the theoretical limit when they operate.

Carnot devised a rule similar to the one we just did: that the energy in a closed system evolves from more useful to less useful forms for doing work. He also said that energy in a less useful form cannot be converted entirely into the same amount of energy in a more useful form.

Carnot's principle It is impossible to build an engine that uses all of its thermal energy to do mechanical work on its environment.

Another idea about the direction of energy conversion

Observational Experiment **Table 13.2** indicates another way to think about the preferred direction for energy transfer during irreversible processes.

OBSERVATIONAL EXPERIMENT TABLE

13.2   Direction of energy transfer.

 VIDEO 13.2

Observational experiment	Analysis
Experiment 1. You place a room temperature spoon in a cup of hot tea. The spoon gets warmer and the tea gets cooler until they reach the same temperature.	Energy is transferred from the hot tea to the cool spoon until they are at the same temperature.
Experiment 2. The air conditioning in a building fails on a very hot day. The temperature in the building increases until it reaches a similar temperature to the outside temperature.	Energy is transferred through the walls from the hot outside of the building to the cool inside until the inside and outside are at the same temperature.
Pattern	
In both experiments, energy is transferred from a high-temperature region to a low-temperature region.	

In the experiments in Table 13.2, you do not observe the reverse of the processes, in which a cool object becomes cooler and a hot object hotter. Although the first law of thermodynamics allows this, such processes do not happen. This leads us to another formulation concerning the direction of energy conversion in isolated systems.

Direction of thermal energy transfer In an isolated system, energy always transfers from a warmer region to a cooler region.

The second law of thermodynamics

Let's review the boxed statements in this section. We have determined that as the energy of a system becomes less organized, the amount of work it can do on other systems decreases. Carnot's principle tells us that it is impossible to build an engine that uses all of its thermal energy to do mechanical work on its environment. Finally, we have concluded that in an isolated system, energy always transfers from a warmer region to a cooler region. All of these statements are different versions of **the second law of thermodynamics**. The second law describes what processes can and cannot occur in an isolated system. While the first law of thermodynamics is a principle about energy conservation, the second law is a principle of *energy transfer* between two regions at different temperatures and *energy conversion* from one form of energy to another. In order to make a more quantitative version of the second law, we must construct a new physical quantity that describes the degree of organization of a system's energy.

Review Question 13.1 When you burn gasoline in your car, the car can gain organized kinetic energy as it moves faster and faster. Does this contradict the second law of thermodynamics?

13.2 Statistical approach to irreversible processes

Active Learning Guide >

One way to understand the preference of systems for certain forms of energy is to do a statistical analysis of the probability that a system is in one state versus another. Consider a very simple thought experiment with a box holding identical atoms that are labeled so we can tell them apart. We drop a thin divider down a slot in the middle of the box, thus dividing it in half. Let's start with four atoms in the box. At regular time intervals we insert the divider, count the number of atoms on each side, and then remove the divider. The five possible distributions are shown in **Figure 13.3**. Each distribution is characterized using the symbol i, which has integer values 0 through 4 indicating the number of atoms in the left half. For example, the $i = 0$ distribution has zero atoms on the left and four on the right; the $i = 1$ distribution has one on the left and three on the right; and so forth. Each of these five distributions is called a **macrostate** of the system of four atoms. Each macrostate has a particular number of atoms n_l on the left and n_r on the right. In thermodynamics we call the variable quantities n_l and n_r *state variables*; they distinguish one particular macrostate of a system from another.

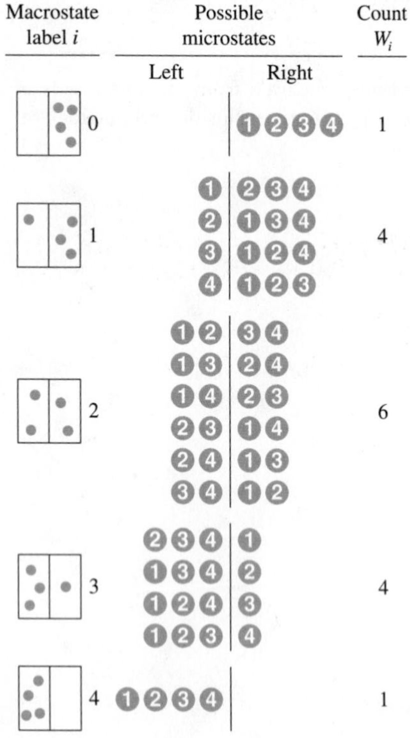

Earlier, we said that each atom is labeled so we can tell them apart. Real atoms don't have labels, but this is a thought experiment so we can construct the situation however we like. Our goal now is to determine the different unique ways for each macrostate to exist. For example, there are four different ways to get one atom on the left and three on the right—each of the four atoms could be alone on the left with the other three on the right. We call this number of ways of getting a particular macrostate the count W_i of that state (note that here we use the letter symbol W for the number of possible microstates, although it is used elsewhere to designate work). Macrostate 0 with zero atoms on the left and four on the right can occur in only one way—if all four atoms are on the right. For macrostate 0, $W_0 = 1$. As we said above, macrostate 1 can occur in four different ways, $W_1 = 4$. Column 3 in Figure 13.3 lists the number of ways of getting the different macrostates. Each of these ways is called a **microstate** of that particular macrostate. All microstates are equally probable. The probability of a macrostate occurring depends on the number of microstates it has. For example, macrostate 2 has six unique ways to arrange four atoms to get two on each side.

In general, for any number of atoms we can determine the number W_i of microstates of the ith macrostate by using the equation

$$W_i = \frac{n!}{(n_l!)(n_r!)} \tag{13.1}$$

where n is the total number of atoms and n_l and n_r are the number of atoms on the left and on the right, respectively.

Figure 13.3 The possible states of four atoms found on each side of a divider after its insertion in the box.

Macrostate label i	Possible microstates		Count W_i
	Left	Right	
0		①②③④	1
1		① ②③④	4
		② ①③④	
		③ ①②④	
		④ ①②③	
2		①② ③④	6
		①③ ②④	
		①④ ②③	
		②③ ①④	
		②④ ①③	
		③④ ①②	
3		②③④ ①	4
		①③④ ②	
		①②④ ③	
		①②③ ④	
4	①②③④		1

We can check that Eq. (13.1) works for the four-atom example, whose results are shown in Figure 13.3. Consider the number of microstates for macrostate 2:

$$W_2 = \frac{4 \times 3 \times 2 \times 1}{(2 \times 1)(2 \times 1)} = 6$$

The equation works. You can try the equation for the other macrostates.

If we add all the numbers in the third column of Figure 13.3, we find that there are 16 microstates, the number of unique ways in which the four atoms can be arranged in the two halves of the box. The *probability* P_i of a particular macrostate occurring is the number of microstates for that macrostate (the count) divided by the total number of microstates. For example, the probability of finding zero atoms in the left box is 1 in 16; the probability of finding one atom in the left half is 4 in 16; and the probability of observing two atoms in each half is 6 in 16.

For the box with four atoms, there is a small chance (1 in 16) that you will observe all atoms on one side of the box. With a larger number of atoms, the situation is quite different. With 10 atoms, for instance, the state with 10 on the left and 0 on the right is 252 times less likely to occur than the macrostate with 5 on each side. That is,

$$\frac{W_0}{W_5} = \frac{1}{252}$$

With 100 atoms, the state with 100 on the left and zero on the right is $1/10^{29}$ as likely to occur as the state with 50 on each side. If we increase the number of atoms to one million, states with equal numbers of atoms on each side are overwhelmingly more probable than states with an unequal number on each side. For example, for one million atoms there is almost a 10^{87} times better chance of observing an equal number on each side than of observing 51% of the atoms on the left and 49% on the right. This 50-50 distribution is the state with the highest probability—the state that is overwhelmingly most likely to occur. This most probable state is the state with the greatest number of microstates and has the greatest randomness especially compared to the state with all particles on the same side.

> **TIP** The exclamation marks (!) in Eq. (13.1) indicate the factorial function. It means that we multiply the number before the exclamation mark by all the positive integers equal to and smaller than the number. Thus, $8! = 8 \times 7 \times 6 \times 5 \times 4 \times 3 \times 2 \times 1$, and $2! = 2 \times 1$. The notation 8! is called eight factorial, and $n!$ is called n factorial. Zero factorial is defined as one ($0! = 1$).

QUANTITATIVE EXERCISE 13.3 Five atoms
Suppose the box with the divider contains five atoms. (a) Identify the six macrostates that can occur. (b) Determine the count of the 0 on left, 5 on right macrostate and that of the 3 on left, 2 on right macrostate.

Represent mathematically Use Eq. (13.1) with $n = 5$ to determine the count for the different macrostates:

$$W_i = \frac{5 \times 4 \times 3 \times 2 \times 1}{(n_l!)(n_r!)}.$$

Solve and evaluate (a) The macrostates (n_l, n_r) are (0,5), (1,4), (2,3), (3,2), (4,1), and (5,0).

(b) The count for the $i = 0$ macrostate [the (0,5) state] is

$$W_0 = \frac{5 \times 4 \times 3 \times 2 \times 1}{(0!)(5 \times 4 \times 3 \times 2 \times 1)} = 1$$

The count for the $i = 3$ macrostate [the (3,2) state] is

$$W_3 = \frac{5 \times 4 \times 3 \times 2 \times 1}{(3 \times 2 \times 1)(2 \times 1)} = 10$$

Try it yourself: Determine the count for a container with seven atoms with 2 on the left side and 5 on the right side.

Answer: 21.

Connecting atoms in a box to physics

In general, an isolated system with a large numbers of particles will evolve toward states that have a higher probability of occurring. The highest probability state is the one in which the particles are evenly distributed—in other words, the state with the highest level of randomness. This state of maximum probability is often called the *equilibrium state* of the system. An isolated system in the most probable state is said to have reached equilibrium.

The same rules apply in the real world. The state of a system with particles moving randomly (random thermal energy) is much more probable than the state with the particles all moving together in the same direction. To characterize this randomness or disorder in the system, physicists use a physical quantity called **entropy,** a measure of the probability of a particular macrostate.

> **Entropy** is a physical quantity that quantifies the probability of a state occurring and of the degree of disorder or disorganization of a system when in that state. In statistical thermodynamics, the entropy S_i of the ith macrostate is defined in terms of the count W_i of that macrostate:
>
> $$\text{Entropy of } i\text{th macrostate} = S_i = k \ln W_i \qquad (13.2)$$
>
> in which $\ln W_i$ is the natural logarithm of the count W_i and k is a proportionality constant (Boltzmann's constant) with a value $k = 1.38 \times 10^{-23}$ J/K (joules/kelvin).*

The entropy of the five macrostates for the distribution of four atoms on the left or right side of a box is $S_0 = 0$, $S_1 = 1.4k$, $S_2 = 1.8k$, $S_3 = 1.4k$, and $S_4 = 0$. Try to use Eq. (13.2) to confirm these results. (The use of logarithms is reviewed briefly in the appendix.)

At the beginning of this section we set out to construct a physical quantity that described the degree of organization of a system's energy. We learned that isolated systems spontaneously evolve toward states whose energy is more disorganized. We now have a way to quantify this statement: isolated systems evolve toward states with higher entropy.

Entropy of an expanding gas—statistical approach

Active Learning Guide >

Ultimately, we would like a quantitative way to understand the concept of entropy in terms of macroscopic quantities such as temperature T and heating Q rather than in terms of counts of microstates. In working toward this, let's first examine from a statistical point of view a familiar process: a gas expanding at constant temperature.

EXAMPLE 13.4 Entropy change of an expanding gas

Suppose you have a gas composed of just six atoms, each moving randomly and with the same amount of energy. The atoms start grouped together in one tiny box (situation 1). The box then doubles in size (situation 2), triples in size (situation 3), and finally reaches six times the original size (situation 4). The gas has expanded its volume to six times its original value. Determine the count and the entropy of the maximum probability state for each of the four situations.

Sketch and translate The situations are pictured at right. For a gas with six atoms and a box that is twice as big, the maximum probability state has three atoms in each half of the box. With a box that is three times

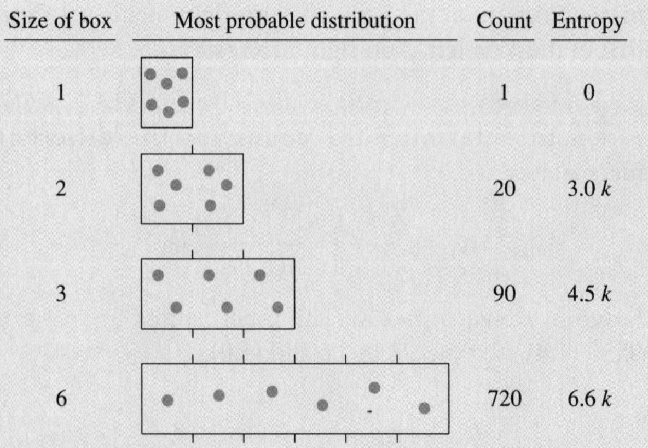

Size of box	Most probable distribution	Count	Entropy
1		1	0
2		20	3.0 k
3		90	4.5 k
6		720	6.6 k

As the box gets bigger, the count of the most probable distribution increases dramatically.

---

*A J/K is sometimes called an entropy unit (eu). Thus, 1 J/K = 1 eu.

bigger, the maximum probability state has two atoms in each third of the box. The maximum probability state with a six times bigger box has one atom in each sixth of the box.

Simplify and diagram We assume that the atoms are labeled. Then for situation 2, the distribution 1, 2, 3 on side 1 and 4, 5, 6 on side 2 differs from 1, 2, 4 on side 1 and 3, 5, 6 on side 2—these are two distinct microstates.

Represent mathematically The count of the number of microstates with n atoms in N distinct parts of a container is

$$W_N = \frac{n!}{n_1! n_2! \cdots n_N!}$$

Solve and evaluate The count for each maximum probability macrostate, as calculated using the above equation, is reported in the figure above. For example, for situation 3 the count is

$$W_3 = \frac{6!}{2! 2! 2!} = 90$$

You can check the numbers for the other maximum probability macrostates. The entropy of each state is determined using

$$S_n = k \ln W_n$$

and is reported in the figure on the previous page. Note that as the number of possible locations for the atoms increases, the maximum probability count and entropy increases dramatically. As the gas expands, its entropy increases. From an energy point of view, we would say that the energy of the gas evolved from being localized in a smaller region and potentially being more useful for doing work to being distributed evenly over a larger region and being much less useful.

Try it yourself: Four atoms start in a small box. They then expand into a box that is twice as large and finally into a box that is four times as large. Determine the count of the maximum probability macrostate for each situation. Label the first box as a single location, the second box as having two places the atoms can reside, and the largest box as having four places the atoms can reside.

Answer: 1; 6; 24.

This last example illustrates quantitatively the second law of thermodynamics expressed using entropy.

> **Second law of thermodynamics** Spontaneous processes in an isolated system tend to proceed in the direction of increasing entropy.

Remember, entropy is a measure of the probability of a particular macrostate. High-entropy states are more disorganized than lower entropy states. These high-probability disorganized states with higher entropy are less able to do work on their environment than less probable organized states with lower entropy.

Review Question 13.2 Explain why turning on an extra lightbulb in your apartment does not prevent energy from being conserved but still affects the entropy.

13.3 Connecting the statistical and macroscopic approaches to irreversible processes

So far, we have related entropy to the number of microstates (the count) for a particular macrostate. The greater the number of microstates, the greater the disorder of the system when in that macrostate and the greater the entropy of the system. In systems with large entropy, the energy is very disorganized (for example, there is more random thermal energy) and the system can do less work on its environment.

These ideas are all based on our observations of real-world phenomena and on the statistical analysis of simple systems. We can actually calculate the entropy of these simple systems (involving 10 or even 100 particles). But real macroscopic systems involve vastly larger numbers of particles (10^{23} or more, for example). It would be difficult to count the number of microstates in a macrostate of such a large system. We need another method to determine the entropy in real-world macroscopic systems with large numbers of particles. Can we express entropy using macroscopic physical quantities such as temperature?

Macroscopic definition of entropy change

Consider a macroscopic process during which a gas (modeled as an ideal gas) is heated by a Bunsen burner flame and expands at constant temperature. When the gas in a system expands, it pushes outward on the environment. Therefore, the environment (in this case, a piston) exerts a force on the system that does negative work (**Figure 13.4a**). The bar chart in Figure 13.4b depicts this process. The system's internal energy does not change since this process is isothermal. Since the environment is doing negative work on the system, the energy provided to the gas by the environment through heating must transfer energy so that the system's thermal energy remains constant.

The volume of the gas increases and the gas particles now have more places to reside than before—the entropy of the system should increase even though the internal energy does not change. We can understand this connection through a limiting case analysis. Imagine that at the beginning, the volume of the gas is almost zero—all particles are crammed together and the number of microstates of that macrostate of the system is 1. Now, if the gas expands, the particles have more room and can occupy this room in many different configurations. Thus, the count of microstates and therefore the entropy of the system increases.

We can use the bar chart in Figure 13.4b to apply the first law of thermodynamics to this isothermal process:

$$+Q_{\text{Env to Sys}} + W_{\text{Env on Sys}} = \Delta U_{\text{int}} = 0$$

or

$$Q_{\text{Env to Sys}} = -W_{\text{Env on Sys}}$$

We know that the work done by the environment on the gas is $W_{\text{Env on Sys}} = -P\,\Delta V$. Thus,

$$Q_{\text{Env to Sys}} = +P\Delta V$$

As the gas expands at constant temperature, its volume increases and its pressure decreases. We use the ideal gas law $P = nRT/V$ to substitute for P in the above:

$$Q_{\text{Env to Sys}} = +P\Delta V = \frac{nRT}{V}\Delta V$$

Solving for $\Delta V/V$, we get

$$\frac{\Delta V}{V} = \left(\frac{1}{nR}\right)\frac{Q_{\text{Env to Sys}}}{T}$$

Remember that this is an isothermal process in which positive energy transfer through heating is required to balance the negative work done by the environment as the gas expands. Because of the increasing volume, more space is available for the same number of gas particles at the same temperature. We can say that the entropy increases because

$$\frac{\Delta V}{V} > 0$$

It appears that either $\Delta V/V$ or Q/T could be used as macroscopic measures of the increase in disorder of the system—measures of the system's entropy change. The units of Q/T are J/K, the same as the units of entropy from the microscopic entropy definition using Eq. (13.2): $S_i = k \ln W_i$. Note that the natural logarithm

Figure 13.4 Isothermal expansion. (a) The gas expands isothermally. (b) The negative work done by the piston on the gas is balanced by the positive heating.

(a)

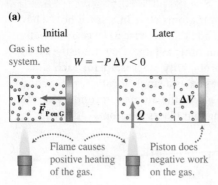

Initial   Later

Gas is the system. $W = -P\,\Delta V < 0$

Flame causes positive heating of the gas.   Piston does negative work on the gas.

(b)

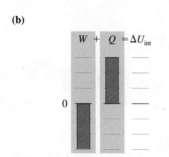

$W + Q = \Delta U_{\text{int}}$

0

of any quantity is unitless, and k (Boltzmann's constant) has units J/K. Thus, the unit of statistical entropy is the same as the unit of Q/T. It seems reasonable to define the entropy change of a macroscopic system as Q/T.

Often, when energy is transferred between an environment and a system through heating, the temperature of the system changes. What temperature should be used to determine the system's entropy change? To deal with this, let's define the entropy change ΔS as a small amount of heating Q when the system is at a particular temperature T. Since this is a small amount of energy, the temperature of the system will not change significantly during the process. Alternatively, we can use the average temperature of the process to find the change in entropy. Here's the summary.

Macroscopic definition of entropy change The entropy change ΔS_{Sys} of a system during a process equals the ratio of the energy transferred to the system from the environment through heating Q divided by the system's temperature T_{Sys} while the process is occurring:

$$\Delta S_{Sys} = \frac{Q_{Env\ to\ Sys}}{T_{Sys}} \qquad (13.3)$$

T_{Sys} is either the constant temperature of the system or the average temperature if the temperature changes.

> **TIP** When you use Eq. (13.3), the temperature must have units of kelvin.

Consider a very simple example. You leave a 1-L water bottle outside on a sunny day. Let's estimate the change in the entropy of the water in the bottle. Assume that the mass of the water in the bottle is about 1 kg, and the temperature inside changes from 20 °C to 30 °C. To change the temperature of water, the environment must have transferred energy to it through heating:

$$\Delta U_{Sys} = Q_{Env\ to\ Sys} = mc\Delta T = \left(4200\frac{J}{kg \cdot °C}\right)(1\ kg)(10\ °C) = 42{,}000\ J$$

The average temperature is 25 °C. Thus the entropy change is

$$\Delta S_{Sys} = \frac{Q_{Env\ to\ Sys}}{T_{Sys}} = \frac{42{,}000\ J}{(25 + 273)\ K} = 140\frac{J}{K}$$

Let's apply this definition to a process in which objects at different temperatures reach thermal equilibrium.

Entropy change of a system and its environment

Consider a process in which a warm object in the environment at temperature T_{Env} transfers a small amount of energy through heating $Q > 0$ to a system that is at a lower temperature T_{Sys}. What was the net entropy change for the system plus the environment, known as the entropy change of the universe: $\Delta S_{Universe} = \Delta S_{Sys} + \Delta S_{Env}$?

$$\Delta S_{Sys} = \frac{Q_{Env\ to\ Sys}}{T_{Sys}} = \frac{Q}{T_{Sys}}$$

and

$$\Delta S_{Env} = \frac{Q_{Sys\ to\ Env}}{T_{Env}} = \frac{-Q}{T_{Env}}$$

or

$$\Delta S_{Universe} = \frac{Q}{T_{Sys}} + \frac{(-Q)}{T_{Env}} = Q\left(\frac{1}{T_{Sys}} - \frac{1}{T_{Env}}\right) = Q\left(\frac{T_{Env} - T_{Sys}}{T_{Sys}T_{Env}}\right) > 0$$

The heating occurred because the environment was at a higher temperature than the system $T_{Env} > T_{Sys}$. But this also means that the magnitude of the system's entropy change was greater than the magnitude of the environment's entropy change. Thus, the entropy change of the system and its environment is positive; the entropy of the system and environment increased. This process occurred when objects of different temperature came in contact and achieved thermal equilibrium. This leads to yet another version of the second law of thermodynamics.

Active Learning Guide >

Second law of thermodynamics During any process that involves the transfer of energy through heating, the net change in entropy of the system and its environment is always greater than zero.

EXAMPLE 13.5 Mixing hot and cool water

We add 0.50 kg of water at 70 °C to 0.50 kg of water at 30 °C. Estimate the net entropy change of all the water once the system has come to equilibrium.

Sketch and translate The process is sketched in the figure below. The warm water cools and the cool water warms.

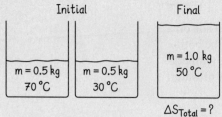

Equal amounts of warm and cool water combine to form intermediate temperature water.

Simplify and diagram Consider this as a two-system process; the warm water (the warm system) transfers energy to the cool water (the cool system) through heating. Assume that the combined system is isolated from its environment. Notice that in this process, the temperature of both systems continually changes. Therefore, we use the average temperature of each system as they undergo the cooling or warming process. Since there are equal amounts of water, the equilibrium temperature will be 50 °C. The average temperature of the warm system is then

$$\frac{70\,°C + 50\,°C}{2} = 60\,°C$$

while the average temperature of the cool system is

$$\frac{30\,°C + 50\,°C}{2} = 40\,°C$$

Represent mathematically The thermal energy transferred from the warm water to the cool water is determined using Eq. (12.5):

$$Q_{Warm\ to\ Cool} = mc_{water}(50\,°C - 30\,°C).$$

Similarly, the negative heating of the warm water is

$$Q_{Cool\ to\ Warm} = mc_{water}(50\,°C - 70\,°C)$$

We now use Eq. (13.3) to add the entropy changes of the warm water and of the cool water:

$$\Delta S_{Total} = \Delta S_{Cool} + \Delta S_{Warm} \approx \frac{Q_{Warm\ to\ Cool}}{T_{Cool}} + \frac{Q_{Cool\ to\ Warm}}{T_{Warm}}$$

Solve and evaluate The specific heat of water is $c = 4180\ J/kg \cdot °C$. The magnitude of the energy transferred through heating is

$$Q_{Warm\ to\ Cool} = mc_{water}(50\,°C - 30\,°C)$$
$$= (0.50\ kg)\left(\frac{4180\ J}{kg \cdot °C}\right)(20\,°C) = 41{,}800\ J$$

The cool to warm water heating is $-41{,}800\ J$. We can now make an estimate of the total entropy change:

$$\Delta S_{Total} = \Delta S_{Cool} + \Delta S_{Warm}$$
$$\approx \frac{+41{,}800\ J}{(273 + 40)K} + \frac{-41{,}800\ J}{(273 + 60)K}$$
$$= +134\ J/K - 126\ J/K = +8\ J/K$$

Since the net entropy of the combined system increases, this process can occur spontaneously (the second law of thermodynamics.)

Suppose you started with all the water at 50 °C and waited for half the water to cool to 30 °C while the other half warmed to 70 °C (the reverse of the above process). This process is consistent with energy conservation (the first law of thermodynamics). The entropy change for such a process would be

$$\Delta S_{Total} = \Delta S_{Cool} + \Delta S_{Warm}$$
$$\approx \frac{-41{,}800\ J}{(273 + 40)K} + \frac{+41{,}800\ J}{(273 + 60)K}$$
$$= -134\ J/K + 126\ J/K = -8\ J/K$$

The entropy of the system would decrease. This process will not occur spontaneously (the second law of thermodynamics.)

Try it yourself: Determine the entropy change of a 0.10-kg chunk of ice at 0 °C that melts and becomes a 0.10-kg puddle of liquid water at 0 °C. The heat of fusion of water is 3.35×10^5 J/kg.

Answer: +120 J/K.

Entropy and complex organic species

Living beings develop from less complex forms to more complex forms. Does this contradict what we have learned about entropy and the second law of thermodynamics? What about when an organism grows? Clearly, its organization and complexity increase.

The growth of an organism does not violate the second law of thermodynamics. The key reason is that the second law applies only to an isolated system; a growing organism is definitely not an isolated system. It interacts with the environment. If we consider the environment as part of the system, then the analysis is quite different.

Plant tissues, for example, absorb energy from the Sun and carbon in the air and water from the soil (**Figure 13.5**), initiating complicated processes that lead to the eventual formation of carbohydrates, some amino acids, fatty acids, and thousands of other compounds by a myriad of other reactions. The formation of these complex useful compounds seems like a decrease in entropy and a violation of the second law of thermodynamics. In each reaction in the complex set of reactions, two or more atoms or molecules combine to form a more complex, lower entropy molecule. However, when these reactions occur, they release energy through heating into the environment. The corresponding entropy increase of the environment is over two times greater than the corresponding decrease due to the formation of the larger molecule. Photosynthesis produces a net increase in the entropy of the Earth system.

What about humans? A person grows and develops into a complex, highly ordered adult by consuming and converting complex molecules with low entropy, such as carbohydrates, fats, and proteins, into simpler molecules with higher entropy. Each year a healthy adult consumes about 500 kg, or a half-ton, of this low-entropy food. During consumption over the course of a year a human transfers roughly two billion joules of energy to the environment, increasing its thermal energy, a form of energy that has little ability to do work. The entropy of the human may decrease, becoming more complex and organized, but the entropy of the environment increases significantly, more than enough to compensate.

In industrialized countries that burn low-entropy fuels such as petroleum and coal to meet their energy needs, the entropy increase of the environment is almost 100 times greater per person than in nonindustrialized countries. The scientific basis of the "energy crisis" rests with the second law of thermodynamics and the increase in entropy of the universe as a result of the conversion of useful energy to much less useful forms. Once we convert the organized energy of petroleum to less organized forms of internal energy, the entropy of the universe has increased and the potential of using that energy for useful purposes is lost forever.

Figure 13.5 The flower is not an isolated system.

Review Question 13.3 The internal energy of a system is a state function, but work and heating are not. Is entropy a state function? Explain.

13.4 Thermodynamic engines and pumps

We have learned that over time an isolated system's energy converts into less useful forms—for example, mechanical energy converts into thermal energy. Is it possible, though, to use the thermal energy of one system to do work on another system?

Suppose that a system has two "reservoirs" at different temperatures. A reservoir could be a pond of cool water, a flame that converts water into steam, or the burning fuel in the cylinder of a car engine. If the hot and

Figure 13.6 Thermodynamic engine. (a) A schematic illustration of the primary components of a thermodynamic engine. (b) A bar chart for the net change during a thermodynamic engine cycle. (c) An electric power plant is a thermodynamic engine.

(a)

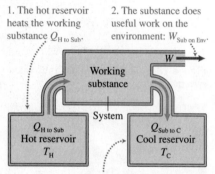

1. The hot reservoir heats the working substance $Q_{H \text{ to Sub}}$.

2. The substance does useful work on the environment: $W_{\text{Sub on Env}}$.

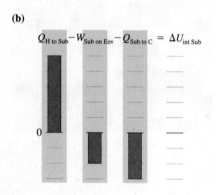

3. The remaining energy is exhausted by heating the cold reservoir $Q_{\text{Sub to C}}$.

(b)

$Q_{H \text{ to Sub}} - W_{\text{Sub on Env}} - Q_{\text{Sub to C}} = \Delta U_{\text{int Sub}}$

(c)

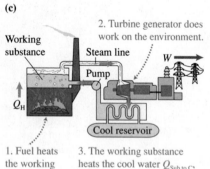

2. Turbine generator does work on the environment.

1. Fuel heats the working substance $Q_{H \text{ to Sub}}$.

3. The working substance heats the cool water $Q_{\text{Sub to C}}$, causing the working substance to condense.

cold reservoirs are in contact, thermal energy flows spontaneously from the warmer reservoir to the cooler one. We can harness the energy of this device in order to work on another system. Such device is called a **thermodynamic engine.** Even more surprising, we can create a device that makes a cool object even colder, such as a refrigerator.

We will begin by examining how a thermodynamic engine works. We will then take a look at how a refrigerator works.

Thermodynamic engine

A thermodynamic engine (**Figure 13.6a**) has three main parts: (1) an input reservoir with hot material at temperature T_H, (2) some kind of working substance that can expand and contract and do useful mechanical work, and (3) an output reservoir with cool material at a lower temperature T_C. The working substance travels through the engine and undergoes thermodynamic processes that cause some desired effect—for example, doing useful work such as turning an electric generator. At the end of each cycle, the working substance returns to its starting state. It then repeats the sequence.

The hot reservoir is maintained at temperature T_H usually by burning some kind of fuel (for example, coal, natural gas, or nuclear fuel in a power plant, or the gasoline in a car engine). The working substance is usually a gas that can expand significantly to do work. The cold reservoir may be a pool of cold water that continually flows to transfer energy away from the working substance.

In the first step of the cycle, the hot reservoir transfers thermal energy to the working substance by heating $Q_{H \text{ to Sub}}$. In the second step, the working substance, which is now hot and has expanded, pushes a piston or turns the turbine of a generator. Thus, the working substance does positive work on the environment $W_{\text{Sub on Env}}$. According to the second law of thermodynamics, it is impossible to use 100% of the energy transferred by heating from the hot reservoir into work done by the working substance. What happens to the rest of the energy?

In the third step, the energy is transferred from the substance to the cold reservoir $Q_{\text{Sub to C}}$. Since the working substance is the system, this would be negative heating of the substance by the cold reservoir $Q_{C \text{ to Sub}}$. The working substance, now cooler, contracts. A compressor also helps in this contraction, doing some work on the substance but somewhat less work than the substance did on the environment in step 2. In other words, during step 3 the environment does positive work on the working substance; however, the net work done by the environment on the working substance during the whole process is negative.

In summary, we have positive heating of the substance by the hot reservoir ($Q_{H \text{ to Sub}}$), the positive net work done by the working substance on the environment ($W_{\text{Sub on Env}}$), and the positive heating of the cold reservoir by the substance ($Q_{\text{Sub to C}}$). The bar chart in Figure 13.6b represents the process. Since the working substance returns to the same state at the end of each loop around the engine, the net internal energy change of the working substance is zero. Thus, we can put this into an equation form using the first law of thermodynamics as applied to the working substance:

$$Q_{H \text{ to Sub}} + W_{\text{Env on Sub}} + Q_{C \text{ to Sub}} = 0$$

The second and third terms in this equation are negative (see the bar chart). We can switch their subscripts and the signs of those two terms:

$$Q_{H \text{ to Sub}} - W_{\text{Sub on Env}} - Q_{\text{Sub to C}} = 0$$

Thus, the net positive work that the substance does on the environment is

$$W_{\text{Sub on Env}} = Q_{\text{H to Sub}} - Q_{\text{Sub to C}} \qquad (13.4)$$

All of the quantities in the above equation have positive values. A schematic model of the thermodynamic engine in an electric power plant is shown in Figure 13.6c.

The efficiency of a thermodynamic engine is the work done by the working substance on the environment divided by the energy transferred from the hot reservoir to the working substance:

$$e = \frac{W_{\text{Sub on Env}}}{Q_{\text{H to Sub}}} \qquad (13.5)$$

EXAMPLE 13.6 Efficiency of a thermodynamic engine

In a car engine, gasoline is ignited by a spark plug to generate an explosion in a cylinder. Each explosion transfers 700 J of energy from the burning fuel to the gas in the cylinder. During the expansion (and later contraction) of the gas, the gas does a net 200 J of mechanical work on the environment (on a piston that pushes outward and helps propel the car). (a) How much energy is transferred through heating to the car's cooling system during this cycle? (b) What is the efficiency of the engine?

Sketch and translate The working substance is the gas (mostly air) inside the cylinder. The hot reservoir is the burning fuel in the cylinder that is ignited by the spark plug. The cooling system is the cold reservoir. One cycle of the process can be represented as shown in the figure below. $Q_{\text{H to Sub}} = +700\,\text{J}$ (the energy transferred from the burning fuel to the gas inside the cylinder) and $W_{\text{Sub to Env}} = +200\,\text{J}$ (the hot air in the cylinder expands and contracts doing positive work against the piston in the cylinder, which ultimately causes the wheels to turn.) We need to find the energy transferred through heating to the cooling system of the car and the efficiency of the engine.

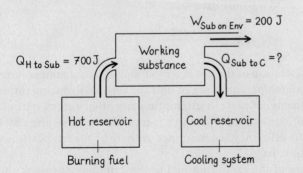

Simplify and diagram We model the car's engine as a thermodynamic engine and represent the process with a bar chart in the figure at right.

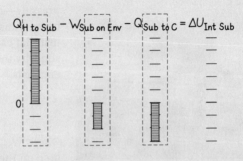

$$Q_{\text{H to Sub}} - W_{\text{Sub on Env}} - Q_{\text{Sub to C}} = \Delta U_{\text{Int Sub}}$$

Represent mathematically Rearrange Eq. (13.4) to determine how much energy is transferred through heating to the car's cooling system:

$$W_{\text{Sub on Env}} = Q_{\text{H to Sub}} - Q_{\text{Sub to C}}$$
$$\Rightarrow W_{\text{Sub}} = Q_{\text{H}} - Q_{\text{C}}$$
$$\Rightarrow Q_{\text{C}} = Q_{\text{H}} - W_{\text{Sub}}$$

Then use Eq. (13.5) to determine the efficiency:

$$e = \frac{W_{\text{Sub}}}{Q_{\text{H}}}$$

Solve and evaluate Using the above equations, and noting that the substance does positive work on the environment (the environment does negative work on the system), we find that

$$Q_{\text{C}} = Q_{\text{H}} - W_{\text{Sub}} = +700\,\text{J} - 200\,\text{J} = +500\,\text{J}$$

Over 70% of the energy that the substance receives from the hot reservoir is moved to the cold reservoir. The engine's efficiency is

$$e = \frac{W_{\text{Sub on Env}}}{Q_{\text{H to Sub}}} = \frac{200\,\text{J}}{700\,\text{J}} = 0.29$$

The result agrees with what you know from your everyday experience when you stand near a parked car when the engine is running. The car's exhaust system pumps hot gas into the atmosphere, carrying away energy that has served

(continued)

no useful purpose. The hood and radiator grill of the car is also warm, releasing even more energy to the environment.

Try it yourself: Another car engine has the same efficiency as the car in this example. If the engine does work on its environment at a rate of 40 hp (1 hp = 746 J/s), how much work does the engine do on its environment each second, and how much energy through heating must be transferred from the burning fuel to the gas in the cylinders each second?

Answer: $W_{\text{Sub on Env}} = (40\text{ hp})(1.0\text{ s})\left(\dfrac{746\text{ J/s}}{1\text{ hp}}\right) = 30{,}000\text{ J}$

and

$$Q_{\text{H to Sub}} = \frac{W_{\text{Sub on Env}}}{e} = \frac{30{,}000\text{ J}}{0.29} = 1.0 \times 10^5\text{ J}.$$

During this second, there might be as many as 100 explosions in the cylinders of the engine.

Maximum efficiency of a thermodynamic engine

Diesel engines are more efficient than standard gasoline engines.

The 29% efficiency of the automobile engine in the last example is a little depressing—most of the chemical potential energy stored in the gasoline ends up being transferred to the environment rather than being used to do work. Is it just that car engines are poorly designed, or is there a more fundamental reason why they are so inefficient? Recall that the efficiency of a thermodynamic engine is

$$e = \frac{W_{\text{Sub on Env}}}{Q_{\text{H to Sub}}}$$

where $W_{\text{Sub on Env}}$ is the useful work done by the working substance in the engine on the environment and $Q_{\text{H to Sub}}$ is the energy transferred to the engine by the hot reservoir.

According to the first law of thermodynamics, the net work done by the engine on its environment is the energy transferred from the hot reservoir to the working substance through heating minus the energy transferred from the working substance to the cold reservoir:

$$W_{\text{Sub on Env}} = Q_{\text{H to Sub}} - Q_{\text{Sub to C}}$$

Thus, the efficiency of a thermodynamic engine can be written as

$$e = \frac{W_{\text{Sub on Env}}}{Q_{\text{H to Sub}}} = \frac{Q_{\text{H to Sub}} - Q_{\text{Sub to C}}}{Q_{\text{H to Sub}}}$$

In England, when thermodynamic engines were being developed to pump water out of mines, scientists and engineers wondered if they could improve the efficiency of the engines. Sadi Carnot, using ideas of entropy and the second law of thermodynamics, showed that the maximum efficiency that a thermodynamic engine could possibly have was

$$e_{\text{max}} = \frac{T_{\text{H}} - T_{\text{C}}}{T_{\text{H}}} \tag{13.6}$$

where T_{H} is the temperature of the hot reservoir and T_{C} is the temperature of the cold reservoir (all temperatures are in units of kelvin). Reducing internal friction between the engine parts or attempting any other sort of optimization may improve efficiency somewhat but never above the value given by Eq. (13.6). The maximum efficiency is determined entirely by the temperatures of the hot and cold reservoirs.

The temperature of the exploding fuel in the cylinders of cars is about 200 °C, while the temperature of the exhaust system of the car is about 60 °C. Thus, the maximum efficiency of such an engine is

$$e_{\text{max}} = \frac{T_{\text{H}} - T_{\text{C}}}{T_{\text{H}}} = \frac{(200 + 273)\text{K} - (60 + 273)\text{K}}{(200 + 273)\text{K}} = \frac{473\text{ K} - 333\text{ K}}{473\text{ K}} \approx 0.30$$

We know that gas mileage in cars can vary significantly in efficiency. The above number is for a relatively low-efficiency car—the number can vary from below 0.3 to above 0.4.

The equation on the previous page indicates ways we might increase the maximum efficiency of an engine.

1. Increase the temperature of the hot reservoir. For example, a pebble bed nuclear power plant reactor under development can operate at a temperature up to 1600 °C with efficiency up to 50% (supposedly more safely than traditional nuclear reactors). The "pebble bed" name comes from the shape of the fuel elements, which are tennis ball-sized pebbles. Diesel fuel is burned at a higher temperature than gasoline, making diesel car engines more efficient than gasoline engines.

2. Reduce the temperature of the cold reservoir. Doing so requires a more advanced cooling system. For example, turbocharged and supercharged car engines have a device known as an intercooler, a special radiator through which the compressed air passes and is cooled before it enters the cylinder.

Thermodynamic pumps (refrigerators or air conditioners)

What happens if we reverse the thermodynamic engine process so that the work done by the environment on the gas is positive? There is then less heating into the working substance than heating out of the working substance. We investigate this process in Testing Experiment **Table 13.3**.

TESTING EXPERIMENT TABLE

13.3   Reversing the operation of a thermodynamic engine.

Testing experiment	Prediction	Outcome
We make a machine that has the same cold and hot reservoirs as the thermodynamic engine but performs all the operations in reverse order. The environment will do positive work on the working substance, and there will be net heating out of the working substance.	If we reverse the energy bar chart used to analyze thermodynamic engines, we conclude that the cold reservoir gets cooler and the hot reservoir gets warmer.	We can build a refrigerator—it makes the cold food inside the refrigerator cooler and the air outside the refrigerator warmer.

$$Q_{\text{C to Sub}} + W_{\text{Comp on Sub}} + Q_{\text{H to Sub}} = \Delta U_{\text{Sub}}$$

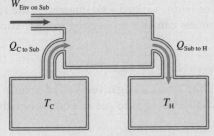

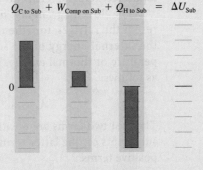

Conclusion

It is possible to create a machine that will transfer thermal energy from a cooler reservoir to a hotter reservoir if the environment does positive work on the working substance. This machine cools the working substance and warms the environment.

Figure 13.7 A schematic drawing of a refrigerator.

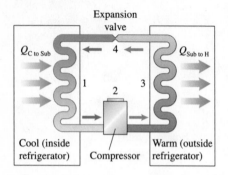

By reversing the thermodynamic engine process, we have invented a machine called a **thermodynamic pump**. A refrigerator is a type of thermodynamic pump. So is an air conditioner. As you can see from the schematic drawing in **Figure 13.7**, in a thermodynamic pump a working substance called a refrigerant takes thermal energy from a "cold reservoir" (position 1 in Figure 13.7). The substance moves through the pipes in the direction of the small blue and red arrows and delivers some of the thermal energy to a hot reservoir outside the pump (position 3 in the figure). The net effect is that thermal energy is transferred from the cool reservoir to the substance ($Q_{C \text{ to Sub}}$), which transfers some of the thermal energy to the warmer air outside the pump ($Q_{\text{Sub to H}}$).

The substance can take energy from the cold reservoir that is at lower temperature and deliver it to the hot reservoir at higher temperature because it is not an isolated system. In addition to the two reservoirs at low and high temperatures, the substance is connected to a motor (a compressor) that does positive work on it ($W_{\text{Comp on Sub}}$).

Let's consider in detail how a refrigerator works. The refrigerant is typically a fluid with a boiling point below 0 °C, such as a hydrofluorocarbon (HFC). The boiling temperature of HFCs is −29.9 °C; below this temperature, most of the substance is in liquid form; above this temperature, it is a gas. The food compartment is the "cold reservoir," which transfers thermal energy to the substance, which runs through coils in the appliance. As the substance's temperature increases, it partially evaporates ($Q_{C \text{ to Ref}}$). As the substance flows through the inside of the refrigerator, more of it evaporates. Energy transfers from the food to the substance, leading to a decrease in the food temperature. As the substance leaves the interior of the refrigerator, it is warmer and much more of it is in the gaseous state than when it entered.

The refrigerant then passes through a compressor (position 2) that does positive work on the substance ($W_{\text{Comp on Sub}}$), adiabatically compressing it to higher internal energy and temperature ($Q = 0$, $\Delta V < 0$, and $P|\Delta V| + 0 = \Delta U_{\text{Sub}}$). The substance is now hotter than the environment outside the refrigerator (considered the "hot reservoir"). The substance passes through a condenser coil (position 3) on the outside of the refrigerator, where it transfers thermal energy to the outside environment ($Q_{\text{Sub to H}}$), which is the hot reservoir. At this point, the substance temperature decreases, and it condenses in part back from a gas to a liquid.

The condensed substance passes through a pressure-lowering device, the expansion valve (position 4). There it expands adiabatically ($Q = 0$, $\Delta V > 0$, and $-P\Delta V + 0 = \Delta U_{\text{Sub}}$), causing its internal energy to decrease. At this point, even more of the substance condenses into a cool liquid, and this liquid reenters the refrigerator for another trip around the cycle.

The flow diagram and bar chart in Testing Experiment Table 13.3 represent the refrigerator (the thermodynamic pump). At the end of one cycle, the internal energy of the substance is unchanged—no change in the temperature or thermal energy. The process can be summarized mathematically as follows:

$$Q_{C \text{ to Sub}} + W_{\text{Comp on Sub}} + Q_{H \text{ to Sub}} = 0$$

The first two terms are positive and the latter is negative. We can reverse the subscripts on the latter term and change its sign to get an equation with all positive terms:

$$Q_{C \text{ to Sub}} + W_{\text{Comp on Sub}} - Q_{\text{Sub to H}} = 0 \tag{13.7}$$

Air conditioners operate in precisely the same way. The efficiencies of refrigerators and air conditioners are rated in terms of a performance coefficient K:

$$K_{\text{Ref}} = \frac{Q_{C \text{ to Sub}}}{W_{\text{Comp on Sub}}} = \frac{Q_{C \text{ to Sub}}}{Q_{\text{Sub to H}} - Q_{C \text{ to Sub}}} \tag{13.8}$$

The second law of thermodynamics limits this performance coefficient:

$$K_{\text{Ref max}} \leq \frac{T_C}{T_H - T_C} \qquad (13.9)$$

The temperatures must be in kelvin (K).

Using a thermodynamic pump to warm a house

A thermodynamic pump can also be used to warm a home in the winter. The only difference is that the outside of the house (cold in the winter) plays the role of the inside of the refrigerator. Energy is transferred through heating from the cooler air outside (the cold reservoir at temperature T_C) to the warmer air inside (the warm reservoir at temperature T_H). A compressor plugged into a wall outlet does work on the working substance (a refrigerant) to make the transfer possible. A sketch of this operation is shown in **Figure 13.8a**. The process is represented by a flow diagram in Figure 13.8b and by a bar chart in Figure 13.8c. The system is the working substance (like the refrigerant). The goal of this device is to warm something (the house) rather than cool something. The compressor does the work that makes possible the transfer of energy through heating to the house. Thus, the performance coefficient $K_{\text{Home pump}}$ is the ratio of the energy provided through heating to the house and the work that must be done for this energy transfer to occur:

$$K_{\text{Home Pump}} = \frac{Q_{\text{Sub to H}}}{W_{\text{Comp on Sub}}} \leq \frac{T_H}{T_H - T_C} \qquad (13.10)$$

Again, the temperatures must be in kelvin (K). From Eq. (13.10), you see that the performance coefficient is always larger than 1.

> **TIP** The term "thermodynamic pump" does not mean that any sort of physical material is being moved around. It is energy that is being "pumped" from a cooler region to a warmer region. This transfer is possible because of the work being done on the working substance by the environment—by the compressor.

Figure 13.8 Thermodynamic pump.
(a) A thermodynamic pump on the outside wall of a house. (b) A schematic for the pump and (c) a bar chart for a cycle of the pump.

(a)

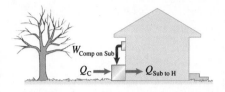

$W_{\text{Comp on Sub}}$

$Q_C \rightarrow$ $\rightarrow Q_{\text{Sub to H}}$

(b)

1. Outside air heats the working substance. $W_{\text{Comp on Sub}}$ 2. The compressor does work W to make heating $Q_{C \text{ on Sub}}$ possible.

Input T_C → Q_C → Working substance → $Q_{\text{Sub to H}}$ → Output T_H

3. The working substance (refrigerant) heats inside of house $Q_{\text{Sub to H}}$.

(c)

$Q_{C \text{ to Sub}} + W_{\text{Comp on Sub}} - Q_{\text{Sub to H}} = \Delta U_{\text{Sub}}$

QUANTITATIVE EXERCISE 13.7 Warming a house in the winter

A thermodynamic pump transfers 14,000 J of energy through heating into a house. The air temperature outside is $-10\,°C$ and inside is $20\,°C$. Determine the maximum performance coefficient of the thermodynamic pump and the minimum work that the pump's compressor must do on the working substance to cause this energy transfer.

Represent mathematically Energy $Q_{\text{Sub to W}} = +14{,}000$ J is transferred from the working substance through heating to the house. Our goal is to determine the maximum performance coefficient of the pump and the work $W_{\text{Env on Sub}}$ that the pump's compressor must do if the pump is operating at maximum efficiency. The maximum performance coefficient is given by Eq. (13.10):

$$K_{\text{Pump max}} = \frac{T_H}{T_H - T_C}$$

We can now determine the minimum work that must be done using the definition of the performance coefficient.

$$K_{\text{Pump}} = \frac{Q_{\text{Sub to H}}}{W_{\text{Comp on Sub}}} \Rightarrow W_{\text{Comp on Sub}} = \frac{Q_{\text{Sub to H}}}{K_{\text{Pump max}}}$$

Solve and evaluate

$$K_{\text{Pump max}} = \frac{T_H}{T_H - T_C} = \frac{(273 + 20)\text{K}}{(273 + 20)\text{K} - (273 - 10)\text{K}} = 9.8$$

$$W_{\text{Comp on Sub}} = \frac{Q_{\text{Sub to H}}}{K_{\text{Pump max}}} = \frac{14{,}000 \text{ J}}{9.8} = 1430 \text{ J}$$

The units are correct for the respective quantities and the numbers are reasonable. Unfortunately, thermodynamic pumps do not usually work at maximum performance, but they are still efficient ways to warm up a house in the winter and cool it in the summer (when behaving as an air conditioner).

Try it yourself: A refrigerator has a performance coefficient of 4. How much work must the compressor do to transfer 4000 J of energy from inside the refrigerator to the outside air?

Answer: 1000 J.

Review Question13.4 Why is there no contradiction between the second law of thermodynamics and the operation of refrigerators and air conditioners that transfer thermal energy from cooler objects to warmer objects?

13.5   Automobile efficiency and power plants: Putting it all together

People often talk about energy conservation in terms of saving energy. As we know now, the problem is not about running out of energy on Earth. Energy cannot be destroyed or used up. From a sustainability perspective, "energy conservation" means making efficient use of useful (low-entropy) forms of energy. In this section we consider two areas in which significant amounts of energy have been converted from useful forms to less useful forms with less then desired efficiency.

Automobile efficiency

Transportation accounts for about 28% of U.S. energy use, and almost all of this demand is satisfied by burning some form of petroleum product (such as unleaded gasoline). Some of the chemical potential energy in the fuel input to cars is eventually transformed into less useful thermal energy in the engine and the drivetrain and through various forms of friction or unproductive use such as idling at a stop light (see **Figure 13.9**). The conversion of useful energy into less useful forms is known as **energy degradation**. Reducing energy degradation improves automobile efficiency.

Active Learning Guide ≻

Energy degradation in the engine (62%) The maximum efficiency of gasoline-powered internal combustion engines depends on the temperatures of the combustion chamber and the exhaust system, but is not very high (30–40% maximum). Thus, 60–70% of the initial chemical energy of the gasoline is converted to thermal energy. Additional energy degradation is caused by engine friction (pistons rubbing against cylinders and friction in all other moving parts of the engine) and energy used to pump air into and out of the engine. Diesel engines operate at higher pressure and temperature of the burning fuel and as a result are more efficient than gasoline engines—up to 50% efficiency.

Drivetrain losses (6%) Energy is lost in the transmission and in the moving parts of the drivetrain. Technologies under development such as automated manual transmissions and continuously variable transmissions can reduce those losses.

Other friction-like losses (13%) Energy is converted to thermal energy due to air drag resistance (a form of friction), rolling resistance (another form of friction that involves the continual flexing of the tires as they rotate), and friction

Figure 13.9 Energy degradation in an automobile.

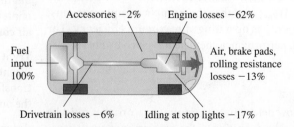

Accessories −2%   Engine losses −62%

Fuel
input
100%

Air, brake pads,
rolling resistance
losses −13%

Drivetrain losses −6%   Idling at stop lights −17%

produced in the disk brakes when stopping the car. Any time you use your brakes to stop the car, the car's kinetic energy is converted to useless thermal energy. Newer braking systems convert a car's kinetic energy while stopping into the electrical potential energy of a charged battery or rotational kinetic energy of a turning flywheel. This stored energy can be reused to get the car moving again.

Energy degradation from idling (17%) Significant energy degradation occurs when a car is idling at stoplights or when it is parked with its engine running. Technologies such as integrated starter/generator systems help reduce these transformations by automatically turning the engine off when the vehicle comes to a stop and restarting it instantaneously when the accelerator is pressed. In some hybrid cars, the gasoline engine turns off completely when the car stops and the only working motor is electric.

Accessories (2%) Additional energy is used to run accessories such as the air conditioner, power steering, and windshield wipers.

A final important efficiency factor for automobiles involves energy used to gain speed. A vehicle's driveline must provide enough energy for the wheels to turn fast enough so the road surface can cause the vehicle to accelerate (the car's kinetic energy increases). The acceleration is directly related to the mass of the car. The less mass a vehicle has, the less work must be done to increase the car's kinetic energy. A smaller car will typically have less mass than a larger one, and manual transmissions have less mass than automatic transmissions.

QUANTITATIVE EXERCISE 13.8 Temperature difference in engine

Note in Figure 13.9 that 62% of useful fuel energy input to that particular car's internal combustion engine is transformed into less useful forms—the engine of that particular car is only about 38% efficient before other degradations that occur, such as braking. If this is the maximum possible efficiency of this engine, what is the temperature difference between the burning fuel in the engine cylinders and the temperature of the air when expelled from the engine?

Represent mathematically The maximum efficiency of a thermodynamic engine is

$$e_{max} = \frac{T_H - T_C}{T_H}$$

Because of the T_H in the denominator, we cannot solve directly for the temperature difference $T_H - T_C$. Instead, we make a rough estimate of T_C and then determine T_H. We'll estimate T_C by estimating the temperature of the exhaust pipe that leaves the car engine. The exhaust pipe is part of the car's cooling system, and we assume it to have a temperature $T_C = 80\,°C$. Multiplying both sides of the equation for the efficiency by T_H, we get

$$T_H e_{max} = T_H - T_C$$
$$\Rightarrow T_C = T_H - T_H e_{max} = T_H(1 - e_{max})$$
$$\Rightarrow T_H = \frac{T_C}{1 - e_{max}}$$

Solve and evaluate The maximum efficiency of this particular engine is 0.38. The temperature of the burning fuel in the cylinder is then

$$T_H = \frac{(273 + 80)K}{1 - 0.38} = 570\ K$$

Remember that kelvin (K) temperatures must be used in efficiency equations. The above temperature in degrees Celsius is $570\,K - 273\,K = 300\,°C$. However, gasoline automatically ignites at 260 °C. With a spark from a spark plug, it would ignite at a lower temperature. A real engine is probably not operating as hot as calculated and consequently is not as efficient. But our outcome seems reasonable as an estimate.

The result is depressing. The efficiencies of typical internal combustion engines are limited by thermodynamics to less than 0.4. To improve automobile efficiency, we need to change to a different type of engine—diesel, hybrid, electric, hydrogen, low mass, or perhaps something even more advanced.

Try it yourself: What temperature changes would result in increased efficiency of a thermodynamic engine?

Answer: Increase the temperature difference between the hot and cold reservoirs. Reducing the temperature of the cold reservoir also helps.

Efficiency of a coal power plant

A power plant is another example of a thermodynamic engine. The Tennessee Valley Authority's (TVA) Kingston Fossil Plant is a typical coal-fired power plant that burns about 14,000 tons (1 ton is 1000 kg) of coal per day in a boiler (the plant's high-temperature reservoir). Here are the steps in the plant's operation:

1. The boiler transfers thermal energy to water (the working substance), converting it to steam at a temperature of about 540 °C.

2. The steam, at very high pressure, flows through a turbine, which spins generators to produce electric energy (the working substance does work on the environment in this step).

3. After leaving the generator, the steam is cooled in a condenser, which is in contact with cold river water (the cold temperature reservoir). The condenser converts the steam back into water at about 32–38 °C and at lower pressure. The low temperature and pressure on the output side of the turbine help move the high-pressure steam at the input to the turbine through the turbine.

4. After the condenser, the cooled water is returned to the boiler and the cycle begins again.

The Kingston plant generates power at an average rate of about 9.1×10^8 watts, enough electric energy to supply 540,000 homes.

EXAMPLE 13.9 Energy from coal

Using the information in the previous discussion of the Kingston coal plant, (a) estimate its maximum possible efficiency, (b) determine the energy transfer rate through heating from the boiler to the working substance, (c) determine the plant's actual efficiency. Note that 1 kg of coal releases roughly 24×10^6 J of energy when burned.

Sketch and translate We already have a sketch of the situation in Figure 13.6c. Information about this particular plant is provided in the text above. We need to find (a) the maximum efficiency e_{max}, (b) the rate

$$\frac{Q_{\text{H to Sub}}}{\Delta t}$$

that energy transfers through heating from the hot reservoir to the working substance, (c) the actual efficiency e, (d) and the rate

$$\frac{Q_{\text{Sub to C}}}{\Delta t}$$

that energy transfers through heating from the working substance to the cool reservoir. The working substance is the water that is boiling and condensing, the hot reservoir is the boiler (at 540 °C), and the cool reservoir is the river water (at 35 °C.)

Simplify and diagram We can model the plant as a thermodynamic engine, such as the one shown in Figure 13.6c. Assume that the rate

$$\frac{W_{\text{Sub on Env}}}{\Delta t}$$

by which work is done by the working substance on the environment equals the electric power generated by the Kingston plant. Also assume that all the chemical potential energy released by the burning coal is transferred to the working substance.

Represent mathematically The maximum efficiency of a thermodynamic engine is

$$e_{max} = \frac{T_H - T_C}{T_H}$$

The first law of thermodynamics written in the form of a rate equation and applied to this process is

$$\left(\frac{Q_{\text{H to Sub}}}{\Delta t} \right) - \left(\frac{W_{\text{Sub on Env}}}{\Delta t} \right) - \left(\frac{Q_{\text{Sub to C}}}{\Delta t} \right) = 0$$

All of the terms in parentheses are positive.

Solve and evaluate (a) The maximum thermodynamic efficiency of the plant depends on the temperatures of the hot and cold reservoirs and is

$$e_{max} = \frac{T_H - T_C}{T_H} = \frac{(540 + 273)K - (35 + 273)K}{(540 + 273)K}$$

$$= 0.62$$

(b) We know that the average rate of electric energy produced by the TVA plant is about

$$\frac{W_{Sub\ on\ Env}}{\Delta t} = 9.1 \times 10^8\ watts = 9.1 \times 10^8\ J/s$$

The mass of coal burned per second is determined from the tons of coal burned each day:

$$\frac{\Delta m}{\Delta t} = \frac{(14,000\ tons/day)(1000\ kg/ton)}{(24\ h/1\ day)(3600\ s/1\ h)} = 162\ kg/s$$

Thus, the rate of thermal energy transfer by burning coal (the hot reservoir) to the working substance is

$$\frac{Q_{H\ to\ Sub}}{\Delta t} = (162\ kg/s)(24 \times 10^6\ J/kg) = 3.9 \times 10^9\ J/s$$

(c) The actual efficiency is the useful output divided by the input:

$$\frac{W_{Sub\ on\ Env}/\Delta t}{Q_{H\ to\ Sub}/\Delta t} = \frac{9.1 \times 10^8\ J/s}{3.9 \times 10^9\ J/s} = 0.23$$

The plant is only operating at half the maximum possible efficiency allowed by the second law of thermodynamics.

Try it yourself: Determine the energy transfer rate from the working substance to the river water in the above power plant.

Answer: $\left(\dfrac{Q_{Sub\ to\ C}}{\Delta t}\right) = \left(\dfrac{Q_{H\ to\ Sub}}{\Delta t}\right) - \left(\dfrac{W_{Sub\ on\ Env}}{\Delta t}\right)$

$$= 3.9 \times 10^9\ J/s - 9.1 \times 10^8\ J/s = 3.0 \times 10^9\ J/s$$

The plant is transferring more energy into the environment as disorganized thermal energy than organized electric energy.

Review Question 13.5 A thermodynamic engine does useful work equal to only 30% or so of the energy supplied to the engine. The rest is transferred to the environment as useless thermal energy. What are the reasons for this inefficiency?

Summary

Words	Pictorial and physical representations	Mathematical representation

Statistical definition of entropy The entropy S of an energy state is a measure of the probability of that state occurring. The greater the entropy, the less useful the energy in the system is for doing work. The entropy depends on the count W_i of a state. (Section 13.2)

Microstate arrangements	Count W_i	Entropy S_i
① ②③④	4	1.4 k
② ①③④		
③ ①②④		
④ ①②③		

$W_i = \dfrac{n!}{n_l! n_r!}$ Eq. (13.1)

$S_i = k \ln W_i$ Eq. (13.2)

where

$k = 1.38 \times 10^{23} \, \text{J/K}$

Thermodynamic definition of entropy A system's entropy changes by ΔS_{Sys} when energy is transferred through heating $Q_{Env\,to\,Sys}$ from the environment while the system is at a temperature T_{Sys}. (Section 13.3)

$\Delta S_{Sys} = \dfrac{Q_{Env\,to\,Sys}}{T_{Sys}}$ Eq. (13.3)

where T_{Sys} is in kelvins.

Second law of thermodynamics The entropy of any system plus its environment increases during any irreversible process. The increase in entropy causes energy to be transferred into more probable forms that have less ability to do useful work. (Sections 13.2; 13.3)

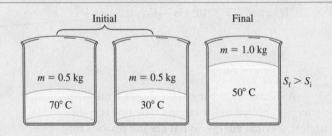

Initial   Final

$m = 0.5$ kg 70° C $m = 0.5$ kg 30° C $m = 1.0$ kg 50° C $S_f > S_i$

During any irreversible process

$\Delta S_{Sys} + \Delta S_{Env} > 0.$

Thermodynamic engines and pumps Energy transferred through heating $Q_{H\,to\,Sub}$ from a hot reservoir at temperature T_H to a working substance allows the substance to do work $W_{Sub\,on\,Env}$ on the environment. Energy is then transferred through heating from the working substance to a cold reservoir $Q_{Sub\,to\,C}$ at temperature T_C. During one cycle of operation, the working substance's internal energy does not change $\Delta U_{Sub} = 0$. The maximum efficiency of this process is e_{max}. A thermodynamic pump is the reverse of this process. (Section 13.6)

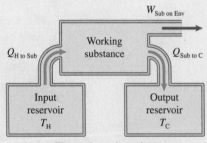

$W_{Sub\,on\,Env}$

$Q_{H\,to\,Sub}$ Working substance $Q_{Sub\,to\,C}$

Input reservoir T_H Output reservoir T_C

Thermodynamic engine

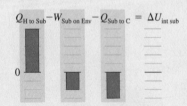

$Q_{H\,to\,Sub} - W_{Sub\,on\,Env} - Q_{Sub\,to\,C} = \Delta U_{int\,sub}$

$Q_{H\,to\,Sub} - W_{Sub\,on\,Env} - Q_{Sub\,to\,C} = 0$ Eq. (13.4)

All quantities are positive.

$e = \dfrac{W_{Sub\,on\,Env}}{Q_{H\,to\,Sub}}$ Eq. (13.5)

$e_{max} = \dfrac{T_H - T_C}{T_H}$ Eq. (13.6)

where T is in kelvins.

$Q_{C\,to\,Sub} + W_{Env\,on\,Sub} - Q_{Sub\,to\,H} = 0$

Refrigeration:

$K = \dfrac{Q_{C\,to\,Sub}}{W_{Env\,on\,Sub}} \leq \dfrac{T_C}{T_H - T_C}$ Eq. (13.8; 13.9)

Home heating:

$K = \dfrac{Q_{Sub\,to\,H}}{W_{Env\,on\,Sub}} \leq \dfrac{T_H}{T_H - T_C}$

Eq. (13.10)

MP° For instructor-assigned homework, go to
MasteringPhysics.

Questions

Multiple Choice Questions

1. Which of the following processes is reversible?
   (a) The room temperature collision of two atoms
   (b) The collision of two cars
   (c) The collision of a pool cue with a pool ball
   (d) The collision of two billiard balls

2. In physics the collision of billiard balls is usually considered to be elastic. What *observable* reason could you have for not choosing the collision of billiard balls as an example of a reversible process?
   (a) You hear sound produced by the collision.
   (b) There is change in the internal energy of the billiard balls.
   (c) All macroscopic processes are irreversible.

3. Why can't all of the thermal energy in a system be converted into mechanical work done on an environment?
   (a) The environment would do equal negative work on the system.
   (b) Energy would not be conserved.
   (c) The disordered energy in the system cannot be completely converted into ordered energy that can do work.

4. Which form of energy listed below is the most useful?
   (a) 1000 J of thermal energy in very hot gas in a cylinder closed by a moveable piston
   (b) 1000 J of thermal energy in cold gas in a cylinder closed by a moveable piston
   (c) 1000 J of thermal energy in water at 15 °C in the rocks at the bottom of a waterfall
   (d) 1000 J of thermal energy in the water in a boiler at 700 °C
   (e) 1000 J of chemical energy stored in the chemical bonds of a piece of wood

5. When driving a car (the system), what object does work on the car to make it move at constant speed in the presence of air resistance and rolling friction?
   (a) The engine
   (b) The surface of the road
   (c) The fuel in the car's tank

6. The law of energy conservation says that energy is always conserved; it can only be transferred from one system to another or converted from one form to another. Why then do environmentalists argue for energy conservation?
   (a) Because they are talking about isolated systems
   (b) Because the environmentalists mean the energy that is useful for doing work
   (c) Because when a car burns fuel, its energy is used up

7. Choose the best slogan(s) from the physics point of view for protecting energy resources. Give the reasons for your choice.
   (a) Conserve energy.
   (b) Reduce entropy increase.
   (c) Conserve useful energy.

8. When the engine in a car overheats, you can help lower the temperature of the engine by doing what?
   (a) Turning the air conditioner on
   (b) Turning off all fanning mechanisms
   (c) Turning on the heater

9. If you leave a refrigerator door open for a long time, the temperature in the room does what?
   (a) Increases   (b) Decreases   (c) Stays the same

10. Entropy can be calculated using which of the following expressions?
    (a) Q/T (b) 10^6 (c) $\Delta Q/T$
    (d) None of these

11. In what way does human life contradict the second law of thermodynamics?
    (a) The law does not apply to living organisms.
    (b) The human body is an isolated system.
    (c) When a human body interacts with the environment, the entropy of the body decreases.
    (d) There is no contradiction.

12. When a drop of ink enters a glass of water and spreads out, the entropy of the ink-water system does what?
    (a) Increases   (b) Decreases   (c) Stays the same

13. Choose the best reason for why the following statement is incorrect: When you clean your room and decrease its disorder, you violate the second law of thermodynamics.
    (a) The room is an isolated system. Entropy does not change.
    (b) Rooms are macroscopic objects; the second law applies only to molecules.
    (c) The room is not an isolated system. Your work leads to considerable entropy increase.

Conceptual Questions

14. Describe five everyday examples of processes that involve increases in entropy. Be sure to state all parts of the system and environment involved in these entropy increases.

15. A cup of hot coffee sits on your desk. Assume that some of the thermal energy in the coffee can be converted to gravitational potential energy of the coffee-Earth system—enough so that the coffee rises up above the rim of the cup. Estimate the drop in temperature needed to accomplish this energy conversion. Why does this conversion not occur?

16. In terms of the statistical definition of entropy, why is there a good chance that the entropy due to the distribution of five atoms in a box can decrease, whereas the chance of an entropy decrease for redistribution of 10^6 atoms in a box is negligible?

17. The entropy of the molecules that form leaves on a tree decreases in the spring of each year. Is this a contradiction of the second law of thermodynamics? Explain.

18. Give three examples of processes, other than those described in this book, in which potentially useful forms of energy are converted to less useful forms of energy.

Problems

Below, BIO indicates a problem with a biological or medical focus. Problems labeled EST ask you to estimate the answer to a quantitative problem rather than develop a specific answer. Problems marked with / require you to make a drawing or graph as part of your solution. Asterisks indicate the level of difficulty of the problem. Problems with no * are considered to be the least difficult. A single * marks moderately difficult problems. Two ** indicate more difficult problems.

13.1 Irreversible processes

1. * BIO **Types of energy and reversibility of a process** Describe the types of energy that change, the work done on the system, and the energy transferred through heating during the following processes. Indicate whether a reverse process can occur. (a) Water at the top of Niagara Falls cascades onto the blades of an electric generator near the bottom of the falls, rotating the blades and generating an electric current that causes a light-bulb to glow. The water, generator, lightbulb, and Earth are the system. (b) Each second, your body converts 100 J of meta-bolic energy (converting complex molecules from food) to thermal energy transferred to the air surrounding your body. The system is your body and the surrounding air. (c) The hot gas in a cylinder pushes a piston, which causes the blades of an electric generator to turn, which in turn causes a lightbulb to glow briefly. The system is the original hot gas (which cools while pushing the piston), the generator, and the lightbulb (which first glows and then stops glowing and cools down).

2. * For the following processes, choose the initial and final states and describe the process using the physical quantities internal energy, work, and heating. Explain why the process is irreversible. (a) A large foam ball is moving vertically up at speed v and reaches a maximum height h' somewhat less than $\sqrt{v^2/2g}$. The ball, Earth, and air are the system. (b) Two cups of water, one cold and the other hot, are mixed in an insulated bowl. The mixture reaches an intermediate temperature. The water in the cups is the system.

3. **Hourglass** An hourglass starts with all of the sand in the top bulb. During the next hour, the sand slowly leaks into the bottom bulb. Describe the energy changes in a system that includes the glass, sand, and Earth. Is this a reversible or ir-reversible process? Explain.

4. **Car hits tree** Your car slides on ice and runs into a tree, caus-ing the front of the car to become slightly hotter and crum-pled. The car and ice are the system. Indicate what object does the work on the system. Indicate whether heating occurs. Identify the types of energy that change. Are these quantities positive or negative? Explain.

5. EST BIO **Human metabolism** A 60-kg person consumes about 2000 kcal of food in one day. If 10% of this food energy is con-verted to thermal energy and cannot leave the body, estimate the temperature change of the person. [*Note:* 1 kcal = 4180 J.] Is this a reversible or irreversible process?

13.2 Statistical approach to irreversible processes

6. * (a) Identify all of the macrostate distributions for five atoms located in a box with two halves. (b) Determine the number of microstates for each macrostate. (c) Determine the entropy of each state.

7. * Repeat the previous problem for a system with six atoms.

8. * Determine the ratio of the number of microstates (count) of a system of eight atoms when in a macrostate with four atoms on the left side of a container and four on the right side and when in a macrostate with seven atoms on the left and one on the right. Which state has the greatest entropy? Explain.

9. * **Person lost on island** The probability that a lost person wandering about an island will be on the north part is one-half and on the south part is also one-half. (a) Determine the prob-ability that three lost people wandering about independently will all be on the south half. (b) Repeat part (a) for the proba-bility of six lost people all being on the south half of the island.

10. * **Parachutists landing on island** Parachutists have an equal chance of landing on the south half of a small island or the north half. If eight parachutists jump at one time, what is the ratio of the probability that six land on the north half and two on the south half to the probability that four land on each half?

11. * Determine the ratio of the counts of a system of 20 atoms when in a macrostate with 10 atoms on the left half of a box and 10 on the right half and when in a macrostate with 18 at-oms on the left and 2 on the right. (b) Do the same for a system with 10 coins for the states with 5 coins on the left and 5 on the right compared to 9 coins on the left and 1 on the right. (c) When you compare your answers to parts (a) and (b), what do you infer about a similar ratio for a system with 10,000 atoms?

12. * Nine numbered balls are dropped randomly into three boxes. The numbers of balls falling into each box are labeled n_1, n_2, and n_3. (a) Identify five of the many possible arrange-ments or macrostates of the balls. (b) Determine the ratio of the count for the equal distribution ($n_1 = 3$, $n_2 = 3$, and $n_3 = 3$) and for the 0, 0, 9 distribution. (c) Determine the ratio of the count for the equal distribution and for the 2, 3, 4 distribution. [*Note:* The count is given by

$$W = \frac{n!}{n_1! \, n_2! \, n_3!},$$

where n is the total number of balls.]

13. ** **Rolling dice** Two dice are rolled. Macrostates of these dice are distinguished by the total number for each roll (that is, 2, 3, 4, . . . , 12). (a) Determine the number of microstates for each macrostate. For example, there are three microstates for macrostate 4: (2, 2), (3, 1), and (1, 3). (b) What is the macro-state with greatest entropy? (c) What is the macrostate with least entropy? Explain.

14. * (a) Apply your knowledge of probability to explain why a drop of food coloring in a glass of clear water spreads out so that all of the water has an even color after some time. (b) Dis-cuss whether after the food coloring spreads evenly in a glass of clear water it could condense back to an original droplet.

15. Explain using your knowledge of probability why a gas always occupies the entire volume of its container.

13.3 Connecting the statistical and macroscopic approaches to irreversible processes

16. * EST Estimate the total change in entropy of two containers of water. One container holds 0.1 kg of water at 70 °C and is warmed to 90 °C by heating from contact with the other con-tainer. The other container, also holding 0.1 kg of water, cools from 30 °C to 10 °C. Is this energy transfer process allowed by the first law of thermodynamics? By the second?

17. *EST (a) You add 0.1 kg of water at 0 °C to 0.3 kg of iced tea at 70 °C. Determine the final temperature of the mixture after it reaches equilibrium. The specific heat of iced tea is the same as water. (b) Estimate the entropy change of this system during this process. Is it allowed by the second law of thermodynamics? Explain.

18. * **Entropy change of a house** A house at 20 °C transfers 1.0×10^5 J of thermal energy to the outside air, which has a temperature of −15 °C. Determine the entropy change of the house-outside air system. Is this process allowed by the second law of thermodynamics?

19. * **Barrel of water in cellar in winter** A barrel containing 200 kg of water sits in a cellar during the winter. On a cold day, the water freezes, releasing thermal energy to the room. This energy passes from the cellar to the outside air, which has a temperature of −20 °C. Determine the entropy change for this process if the cellar remains at 0 °C.

20. *EST (a) Determine the final temperature when 0.1 kg of water at 10 °C is added to 0.3 kg of soup at 50 °C. What assumptions did you make? (b) Estimate the entropy change of this water-soup system during the process. Does the second law of thermodynamics allow this process?

21. * A 5.0-kg block slides on a level surface and stops because of friction. Its initial speed is 10 m/s and the temperature of the surface is 20 °C. Determine the entropy change of the block, which is the system in this process.

22. ** A 5.0-kg block slides from an initial speed of 8.0 m/s to a final speed of zero. It travels 12 m down a plane inclined at 15° with the horizontal. Determine the entropy change of the block-inclined plane-Earth system for this process if originally the block and the inclined plane were at 27 °C. Why did the block stop?

13.4 Thermodynamic engines and pumps

23. **Maximum efficiencies** Determine the maximum efficiencies of the thermodynamic engines described below. (a) Burning coal heats the gas in the turbine of an electric power plant to 700 K. After turning the blades of the generator, the gas is cooled in cooling towers to 350 K. (b) An inventor claims to have a thermodynamic engine that attaches to a car's exhaust system. The temperature of the exhaust gas is 90 °C and the temperature of the output of this proposed heat engine is 20 °C. (c) Near Bermuda, ocean water is about 24 °C at the surface and about 10 °C at a depth of 800 m.

24. * BIO **Efficiency of woman walking** A 60-kg woman walking on level ground at 1 m/s metabolizes energy at a rate of 230 W. When she walks up a 5° incline at the same speed, her metabolic rate increases to 370 W. Determine her efficiency at converting chemical energy into gravitational potential energy.

25. * **Nuclear power plant** A nuclear power plant operates between a high-temperature heat reservoir at 560 °C and a low-temperature stream at 20 °C. (a) Determine the maximum possible efficiency of this thermodynamic engine. (b) Determine the heating rate (J/s) from the high-temperature heat reservoir to the power plant so that it produces 1000 MW of power (work/time).

26. ** A cyclic process involving 1 mole of ideal gas is shown in **Figure P13.26.** (a) Determine the work done on the

Figure P13.26

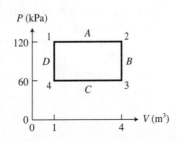

gas by the environment during each step (A, B, C, and D) of the cycle. (b) Determine the net work done on the gas $W_{\text{Env on Gas}}$. (c) Find the work that the gas does on the environment; it is equal to $W_{\text{Gas on Env}} = -W_{\text{Env on Gas}}$. (d) Use the ideal gas law to determine the temperature of the gas at each corner of the process (1, 2, 3, and 4). (e) Use the temperatures found in (d) to determine the thermal energy of the gas at each corner of the process and the change in internal thermal energy during each step of the process. (f) Use the results recorded in the previous parts and the first law of thermodynamics to determine the heating during each step of the process. (g) Determine the efficiency of the process.

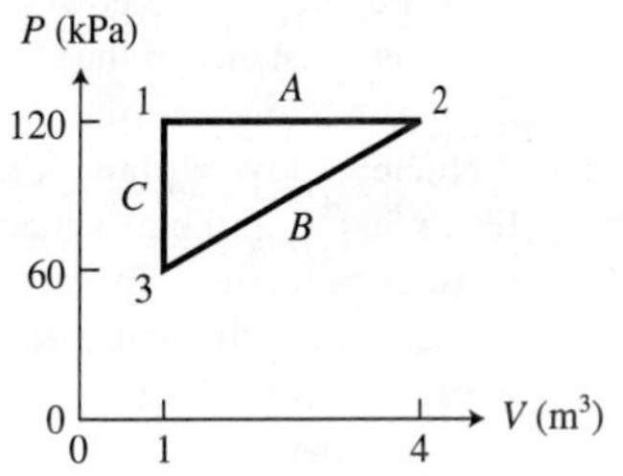

27. ** A cyclic process involving 1 mole of ideal gas is shown in **Figure P13.27** (a) Determine the work done on the gas by the environment during each step (A, B, and C) of the cycle. (b) Determine the net work done on the gas $W_{\text{Env on Gas}}$. (c) Find the work that the gas does on the environment equal to $W_{\text{Gas on Env}} = -W_{\text{Env on Gas}}$. (d) Use the ideal gas law to determine the temperature of the gas at each corner of the process (1, 2, and 3). (e) Use the temperatures found in (d) to determine the thermal energy of the gas at each corner of the process and the change in internal thermal energy during each step of the process. (f) Use the results recorded in the previous parts and the first law of thermodynamics to determine the heating during each step of the process. (g) Determine the efficiency of the process.

Figure P13.27

P (kPa)

120 ⌐ 1 · · · · · A · · · · 2

C

B

60 ⌐ 3

0 ┴────┴────┴→ V (m³)
0   1   4

28. * **Home thermodynamic pump** A heat pump collects thermal energy from outside air at 5 °C and delivers it into a house at 40 °C. (a) Determine the maximum coefficient of performance (the maximum coefficient is determined in the same way as for a thermodynamic engine). (b) If the motor of the heating pump uses 1000 J of electrical energy to do work during a certain time interval, how much thermal energy is delivered into the house through heating, assuming the heating pump works at the maximum coefficient of performance? (c) Repeat (b) for a coefficient of performance of 2.0.

29. ** **Ice-making machine** An ice-making machine needs to convert 0.20 kg of water at 0 °C to 0.20 kg of ice at 0 °C. The room temperature surrounding the ice machine is 20 °C. (a) How much thermal energy must be removed from the water? (b) Determine the minimum work needed to extract this energy by the ice-making machine. (c) How much energy is deposited in the room?

13.5 Automobile efficiency and power plants: Putting it all together

30. * **Automobile engine** An automobile engine has a power output for doing work of 150 kW (about 200 hp). The efficiency of the engine is 0.32. Determine the heating input per second to the engine by burning gasoline and the heating rate of the engine to the environment.

31. **Diesel car engine** A diesel engine in a car does 1000 J of work due to 2800 J of heating caused by the combustion of diesel fuel in its cylinders. Determine the efficiency of the engine and the thermal energy emitted by the engine to the environment.

32. ** **Gas used in car's engine** During each cycle of an automobile's gasoline engine operation, the gasoline burned

provides 12,000 J of energy through heating to the engine and the engine does 3600 J of work. Gasoline provides 4.4×10^7 J of energy for each kilogram of gasoline burned. (a) Determine the efficiency of the engine. (b) Determine the thermal energy exhausted from the engine during each cycle. (c) Determine the mass of the gasoline burned during each cycle. (d) If the engine has 80 cycles/s, how much gasoline does the engine use in 1 h in kilograms and in gallons? The density of gasoline is 737 kg/m³.

33. **Nuclear power plant** A nuclear power plant does useful work generating electric energy at a rate of 500 MW. The energy transfer rate to the electric generator from the high-temperature nuclear fuel (the hot reservoir) is 1200 MW. Determine the efficiency of the power plant and the rate at which the working substance in the plant transfers energy through heating to the cold water (the cool reservoir).

34. * **Nuclear power plant** A nuclear power plant warms water to 500 °C and emits it at 100 °C. You want to get useful work done by the plant at a rate of 1.0×10^9 J/s. Determine the rate at which the nuclear fuel must provide energy to the working substance and the rate of thermal energy exhausted from the plant to the environment.

General Problems

35. ** **EST BIO** **Body efficiency—experiment design** Describe an experiment that can be performed to estimate the average efficiency of a human body.

36. **✲ The following equations represent the four parts (A, B, C, and D) of a cyclic process with a gas. In this case we consider the work done by the system on the environment and consequently $Q - W = \Delta U_{int}$.

$$W = (3.0 \times 10^5 \text{ N/m}^2)(0.020 \text{ m}^3 - 0.010 \text{ m}^3) + 0$$
$$+ (1.0 \times 10^5 \text{ N/m}^2)(0.010 \text{ m}^3 - 0.020 \text{ m}^3) + 0$$

$$\Delta U_{int} = (3/2)(1.0 \text{ mole})(8.3 \text{ J/mole} \cdot \text{K})[(700 \text{ K} - 360 \text{ K})$$
$$+ (480 \text{ K} - 700 \text{ K}) + (240 \text{ K} - 480 \text{ K})$$
$$+ (360 \text{ K} - 240 \text{ K})]$$

$$Q = Q_A + Q_B + Q_C + Q_D$$

(a) Draw a P-versus-V graph for the process with labeled axes (including a scale).
(b) Determine the net change in the internal energy during the entire cycle.
(c) Determine the heating of the system for each of the four parts of the process.

Reading Passage Problem

Fuel used to counter air resistance The resistive drag force that the air exerts on an automobile is

$$F_{\text{Air on Car}} = (1/2)CA\rho v^2$$

where C is a drag coefficient that depends on how streamlined the car is, ρ is the density of the air, A is the cross-sectional area of the car along its line of motion, and v is the car's speed relative to the air. Consider the effect of air resistance on fuel consumption in a typical car during a 160 km (100 mi) trip while traveling at 22 m/s (50 mi/h). The resistive force exerted by the air on the car is

about 200 N. Thermal energy produced due to this resistive force during the trip is

$$\Delta U_{\text{thermal}} = F_{\text{Air on Car}}d = (200 \text{ N})(1.6 \times 10^5 \text{ m})$$
$$= 3.2 \times 10^7 \text{ J}$$

One liter of gas releases about 3.25×10^7 J of energy. So the gasoline energy that is equivalent to the resistive thermal energy produced during this trip is

$$(3.2 \times 10^7 \text{ J})/(3.25 \times 10^7 \text{ J/L}) = 0.98 \text{ L} = 0.26 \text{ gallons}$$

In summary, you use about one-quarter gallon of gasoline to counter air resistance during the 160 km trip. Unfortunately, the car is only about 13% efficient (see Figure 13.9), where

$$\text{Efficiency} = \frac{\text{useful energy output}}{\text{energy input}} = 0.13$$

Consequently, your useful output of 0.98 liters requires an energy input of (0.98 L)/0.13 = 7.5 L of gas to overcome this air resistance. This is about half of the gasoline a typical car consumes when driving at a steady speed for 160 km. The other half is needed to overcome rolling resistance.

37. Assuming a 200-N drag force when traveling at 22 m/s through air of density 1.3 kg/m³, what is the closest value to the product CA in the drag force equation for the vehicle?
    (a) 0.50 m²   (b) 0.62 m²   (c) 0.86 m²
    (d) 1.1 m²   (e) 1.5 m²

38. At 22 m/s the magnitude of the resistive force that air exerts on the car is about 200 N. Which answer below is closest to the magnitude of the drag force when the car is traveling at 31 m/s?
    (a) 100 N   (b) 140 N   (c) 280 N
    (d) 400 N   (e) 520 N

39. The amount of fuel used to counter air resistance should do what?
    (a) Increase in proportion to the speed squared
    (b) Increase in proportion to the speed
    (c) Be the same independent of the speed
    (d) Decrease in proportion to the inverse of the speed
    (e) Decrease in proportion to the inverse of the speed squared

40. The 200-N resistive force of the air in this problem is closest to which answer below?
    (a) 800 lb   (b) 90 lb   (c) 60 lb
    (d) 45 lb   (e) 30 lb

41. Why is the resistive force of the air on a Hummer H3 traveling at 22 m/s about 600 N instead of 200 N?
    (a) The Hummer has a bulky, less streamlined shape.
    (b) The Hummer has a greater cross-sectional area along the line of motion.
    (c) The Hummer has greater mass.
    (d) a and b
    (e) a, b, and c

42. The value of CA for a Ford Escape Hybrid is 1.08 m². Which answer below is closest to the drag force on this car when traveling at 22 m/s?
    (a) 130 N   (b) 180 N   (c) 270 N
    (d) 350 N   (e) 440 N

APPENDIX A

Mathematics Review

A study of physics at the level of this textbook requires some basic math skills. The relevant math topics are summarized in this appendix. We strongly recommend that you review this material and become comfortable with it as quickly as possible so that, during your physics course, you can focus on the physics concepts and procedures that are being introduced, without being distracted by unfamiliarity with the math that is being used.

A.1 Exponents

Exponents are used frequently in physics. When we write 3^4, the superscript 4 is called an **exponent** and the **base number** 3 is said to be raised to the fourth power. The quantity 3^4 is equal to $3 \times 3 \times 3 \times 3 = 81$. Algebraic symbols can also be raised to a power—for example, x^4. There are special names for the operation when the exponent is 2 or 3. When the exponent is 2, we say that the quantity is **squared**; thus, x^2 is x squared. When the exponent is 3, the quantity is **cubed**; hence, x^3 is x cubed.

Note that $x^1 = x$, and the exponent is typically not written. Any quantity raised to the zero power is defined to be unity (that is, 1). Negative exponents are used for reciprocals: $x^{-4} = 1/x^4$. The exponent can also be a fraction, as in $x^{1/4}$. The exponent $\frac{1}{2}$ is called a **square root**, and the exponent $\frac{1}{3}$ is called a **cube root**.

For example, $\sqrt{6}$ can also be written as $6^{1/2}$. Most calculators have special keys for calculating numbers raised to a power—for example, a key labeled y^x or one labeled x^2.

Exponents obey several simple rules, which follow directly from the meaning of raising a quantity to a power:

1. When the product of two powers of the same quantity are multiplied, the exponents are added:

$$(x^n)(x^m) = x^{n+m}.$$

   For example, $(3^2)(3^3) = 3^5 = 243$. To verify this result, note that $3^2 = 9$, $3^3 = 27$, and $(9)(27) = 243$.
   A special case of this rule is $(x^n)(x^{-n}) = x^{n+(-n)} = x^0 = 1$.

2. The product of two different base numbers raised to the same power is the product of the base numbers, raised to that power:

$$(x^n)(y^n) = (xy)^n.$$

   For example, $(2^4)(3^4) = 6^4 = 1296$. To verify this result, note that $2^4 = 16$, $3^4 = 81$, and $(16)(81) = 1296$.

3. When a power is raised to another power, the exponents are multiplied:

$$(x^n)^m = x^{nm}.$$

   For example, $(2^2)^3 = 2^6 = 64$. To verify this result, note that $2^2 = 4$, so $(2^2)^3 = (4)^3 = 64$.

If the base number is negative, it is helpful to know that $(-x)^n = (-1)^n x^n$, and $(-1)^n$ is $+1$ if n is even and -1 if n is odd.

EXAMPLE A.1 Simplifying an exponential expression

Simplify the expression $\dfrac{x^3 y^{-3} x y^{4/3}}{x^{-4} y^{1/3}(x^2)^3}$, and calculate its numerical value when $x = 6$ and $y = 3$.

Represent mathematically, solve, and evaluate
We simplify the expression as follows:

$$\frac{x^3 x}{x^{-4}(x^2)^3} = x^3 x^1 x^4 x^{-6} = x^{3+1+4-6} = x^2;$$

$$\frac{y^{-3} y^{4/3}}{y^{1/3}} = y^{-3+\frac{4}{3}-\frac{1}{3}} = y^{-2}.$$

$$\frac{x^3 y^{-3} x y^{4/3}}{x^{-4} y^{1/3}(x^2)^3} = x^2 y^{-2} = x^2\left(\frac{1}{y}\right)^2 = \left(\frac{x}{y}\right)^2.$$

For $x = 6$ and $y = 3$, $\left(\dfrac{x}{y}\right)^2 = \left(\dfrac{6}{3}\right)^2 = 4.$

If we evaluate the original expression directly, we obtain

$$\frac{x^3 y^{-3} x y^{4/3}}{x^{-4} y^{1/3}(x^2)^3} = \frac{(6^3)(3^{-3})(6)(3^{4/3})}{(6^{-4})(3^{1/3})([6^2]^3)}$$

$$= \frac{(216)(1/27)(6)(4.33)}{(1/1296)(1.44)(46,656)} = 4.00,$$

which checks.

This example demonstrates the usefulness of the rules for manipulating exponents.

EXAMPLE A.2 Solving an exponential expression for the base number

If $x^4 = 81$, what is x?

Represent mathematically, solve, and evaluate
We raise each side of the equation to the $\frac{1}{4}$ power:

$(x^4)^{1/4} = (81)^{1/4}$. $(x^4)^{1/4} = x^1 = x$, so $x = (81)^{1/4}$ and $x = +3$ or $x = -3$. Either of these values of x gives $x^4 = 81$.

Notice that we raised *both sides* of the equation to the $\frac{1}{4}$ power. As explained in Section A.3, an operation performed on both sides of an equation does not affect the equation's validity.

A.2 Scientific Notation and Powers of 10

In physics, we frequently encounter very large and very small numbers, and it is important to use the proper number of significant figures when expressing a quantity. We can deal with both these issues by using **scientific notation**, in which a quantity is expressed as a decimal number with one digit to the left of the decimal point, multiplied by the appropriate power of 10. If the power of 10 is positive, it is the number of places the decimal point is moved to the right to obtain the fully written-out number. For example, $6.3 \times 10^4 = 63,000$. If the power of 10 is negative, it is the number of places the decimal point is moved to the left to obtain the fully written-out number. For example, $6.56 \times 10^{-3} = 0.00656$. In going from 6.56 to 0.00656, the decimal point is moved three places to the left, so 10^{-3} is the correct power of 10 to use when the number is written in scientific notation. Most calculators have keys for expressing a number in either decimal (floating-point) or scientific notation.

When two numbers written in scientific notation are multiplied (or divided), multiply (or divide) the decimal parts to get the decimal part of the result, and multiply (or divide) the powers of 10 to get the power-of-10 portion of the result. You may have to adjust the location of the decimal point in the answer to express it in scientific notation. For example,

$$(8.43 \times 10^8)(2.21 \times 10^{-5}) = (8.43 \times 2.21)(10^8 \times 10^{-5})$$

$$= (18.6) \times (10^{8-5}) = 18.6 \times 10^3$$

$$= 1.86 \times 10^4.$$

Similarly,

$$\frac{5.6 \times 10^{-3}}{2.8 \times 10^{-6}} = \left(\frac{5.6}{2.8}\right) \times \left(\frac{10^{-3}}{10^{-6}}\right) = 2.0 \times 10^{-3-(-6)} = 2.0 \times 10^3.$$

Your calculator can handle these operations for you automatically, but it is important for you to develop good "number sense" for scientific notation manipulations.

A.3 Algebra

Solving Equations

Equations written in terms of symbols that represent quantities are frequently used in physics. An **equation** consists of an equals sign and quantities to its left and to its right. Every equation tells us that the combination of quantities on the left of the equals sign has the same value as (is equal to) the combination on the right of the equals sign. For example, the equation $y + 4 = x^2 + 8$ tells us that $y + 4$ has the same value as $x^2 + 8$. If $x = 3$, then the equation $y + 4 = x^2 + 8$ says that $y = 13$.

Often, one of the symbols in an equation is considered to be the *unknown,* and we wish to solve for the unknown in terms of the other symbols or quantities. For example, we might wish to solve the equation $2x^2 + 4 = 22$ for the value of x. Or we might wish to solve the equation $x = v_0 t + \frac{1}{2} at^2$ for the unknown a in terms of x, t, and v_0.

An equation can be solved by using the following rule:

> **An equation remains true if any operation performed on one side of the equation is also performed on the other side.** The operations could be (a) adding or subtracting a number or symbol, (b) multiplying or dividing by a number or symbol, or (c) raising each side of the equation to the same power.

EXAMPLE A.3 Solving a numerical equation

Solve the equation $2x^2 + 4 = 22$ for x.

Represent mathematically, solve, and evaluate
First we subtract 4 from both sides. This gives $2x^2 = 18$. Then we divide both sides by 2 to get $x^2 = 9$. Finally, we raise both sides of the equation to the $\frac{1}{2}$ power. (In other words, we take the square root of both sides of the equation.) This gives

$x = \pm \sqrt{9} = \pm 3$. That is, $x = +3$ or $x = -3$. We can verify our answers by substituting our result back into the original equation: $2x^2 + 4 = 2(\pm 3)^2 + 4 = 2(9) + 4 = 18 + 4 = 22$, so $x = \pm 3$ does satisfy the equation.

Notice that a square root always has *two* possible values, one positive and one negative. For instance, $\sqrt{4} = \pm 2$, because $(2)(2) = 4$ and $(-2)(-2) = 4$. Your calculator will give you only a positive root; it's up to you to remember that there are actually two. Both roots are correct mathematically, but in a physics problem only one may represent the answer. For instance, if you can get dressed in $\sqrt{4}$ minutes, the only physically meaningful root is 2 minutes!

EXAMPLE A.4 Solving a symbolic equation

Solve the equation $x = v_0 t + \frac{1}{2} at^2$ for a.

Represent mathematically, solve, and evaluate
We subtract $v_0 t$ from both sides. This gives $x - v_0 t = \frac{1}{2} at^2$. Now we multiply both sides by 2 and divide both sides by t^2,

giving $a = \dfrac{2(x - v_0 t)}{t^2}$.

As we've indicated, it makes no difference whether the quantities in an equation are represented by variables (such as x, v, and t) or by numerical values.

The Quadratic Formula

Using the methods of the previous subsection, we can easily solve the equation $ax^2 + c = 0$ for x:

$$x = \pm \sqrt{\frac{-c}{a}}.$$

For example, if $a = 2$ and $c = -8$, the equation is $2x^2 - 8 = 0$ and the solution is

$$x = \pm \sqrt{\frac{-(-8)}{2}} = \pm \sqrt{4} = \pm 2.$$

The equation $ax^2 + bx = 0$ is also easily solved by factoring out an x on the left side of the equation, giving $x(ax + b) = 0$. (To *factor out* a quantity means to isolate it so that the rest of the expression is either multiplied or divided by that quantity.) The equation $x(ax + b) = 0$ is true (that is, the left side equals zero) if either $x = 0$ or $x = -\frac{b}{a}$. Those are the equation's two solutions. For example, if $a = 2$ and $b = 8$, the equation is $2x^2 + 8x = 0$ and the solutions are $x = 0$ and $x = -\frac{8}{2} = -4$.

But if the equation is in the form $ax^2 + bx + c = 0$, with a, b, and c all nonzero, we cannot use our standard methods to solve for x. Such an equation is called a **quadratic equation**, and its solutions are expressed by the **quadratic formula**:

Quadratic formula For a quadratic equation in the form $ax^2 + bx + c = 0$, where a, b, and c are real numbers and $a \neq 0$, the solutions are given by the quadratic formula:

$$x = \frac{-b \pm \sqrt{b^2 - 4ac}}{2a}$$

In general, a quadratic equation has two roots (solutions). But if $b^2 - 4ac = 0$, then the two roots are equal. By contrast, if $b^2 < 4ac$, then $b^2 - 4ac$ is negative, and both roots are complex numbers and cannot represent physical quantities. In such a case, the original quadratic equation has mathematical solutions, but no physical solutions.

EXAMPLE A.5 Solving a quadratic equation

Find the values of x that satisfy the equation $2x^2 - 2x = 24$.

Represent mathematically, solve, and evaluate
First we write the equation in the standard form $ax^2 + bx + c = 0$: $2x^2 - 2x - 24 = 0$. Then $a = 2$, $b = -2$, and $c = -24$. Next, the quadratic formula gives the two roots as

$$x = \frac{-(-2) \pm \sqrt{(-2)^2 - 4(2)(-24)}}{(2)(2)}$$

$$= \frac{+2 \pm \sqrt{4 + 192}}{4} = \frac{2 \pm 14}{4},$$

so $x = 4$ or $x = -3$. If x represents a physical quantity that takes only nonnegative values, then the negative root $x = -3$ is nonphysical and is discarded.

As we've mentioned, when an equation has more than one mathematical solution or root, it's up to *you* to decide whether one or the other or both represent the true physical answer. (If neither solution seems physically plausible, you should review your work.)

Simultaneous Equations

If a problem has two unknowns—for example, x and y—then it takes two independent equations in x and y (that is, two equations for x and y, where one equation is not simply a multiple of the other) to determine their values uniquely. Such equations are called **simultaneous equations** (because you solve them together). A typical procedure is to solve one equation for x in terms of y and then substitute the result into the second equation to obtain an equation in which y is the only unknown. You then solve this equation for y and use the value of y in either of the original equations in order to solve for x. In general, to solve for n unknowns, we must have n independent equations.

EXAMPLE A.6 Solving two equations in two unknowns

Solve the following pair of equations for x and y:

$$x + 4y = 14$$
$$3x - 5y = -9$$

Represent mathematically, solve, and evaluate
The first equation gives $x = 14 - 4y$. Substituting this for x in the second equation yields, successively, $3(14 - 4y) - 5y = -9$, $42 - 12y - 5y = -9$, and $-17y = -51$. Thus, $y = \frac{-51}{-17} = 3$. Then $x = 14 - 4y = 14 - 12 = 2$. We can verify that $x = 2$, $y = 3$ satisfies both equations.

An alternative approach is to multiply the first equation by -3, yielding $-3x - 12y = -42$. Adding this to the second equation gives, successively, $3x - 5y + (-3x) + (-12y) = -9 + (-42)$, $-17y = -51$, and $y = 3$, which agrees with our previous result.

As shown by the alternative approach, simultaneous equations can be solved in more than one way. The basic method we describe is easy to keep straight; other methods may be quicker, but may require more insight or forethought. Use the method you're comfortable with.

A pair of equations in which all quantities are symbols can be combined to eliminate one of the common unknowns.

EXAMPLE A.7 Solving two symbolic equations in two unknowns

Use the equations $v = v_0 + at$ and $x = v_0 t + \frac{1}{2}at^2$ to obtain an equation for x that does not contain a.

Represent mathematically, solve, and evaluate
We solve the first equation for a:

$$a = \frac{v - v_0}{t}.$$

We substitute this expression into the second equation:

$$x = v_0 t + \frac{1}{2}\left(\frac{v - v_0}{t}\right)t^2 = v_0 t + \frac{1}{2}vt - \frac{1}{2}v_0 t$$

$$= \frac{1}{2}v_0 t + \frac{1}{2}vt = \left(\frac{v_0 + v}{2}\right)t.$$

When you solve a physics problem, it's often best to work with symbols for all but the final step of the problem. Once you've arrived at the final equation, you can plug in numerical values and solve for an answer.

A.4 Logarithmic and Exponential Functions

The base-10 logarithm, or **common logarithm** (log), of a number y is the power to which 10 must be raised to obtain y: $y = 10^{\log y}$. For example, $1000 = 10^3$, so $\log(1000) = 3$; you must raise 10 to the power 3 to obtain 1000. Most calculators have a key for calculating the log of a number.

Sometimes we are given the log of a number and are asked to find the number. That is, if $\log y = x$ and x is given, what is y? To solve for y, write an equation in which 10 is raised to the power equal to either side of the original equation: $10^{\log y} = 10^x$. But $10^{\log y} = y$, so $y = 10^x$. In this case, y is called the **antilog** of x. For example, if $\log y = -2.0$, then $y = 10^{-2.0} = 1.0 \times 10^{-2.0} = 0.010$.

The log of a number is positive if the number is greater than 1. The log of a number is negative if the number is less than 1, but greater than zero. The log of zero or of a negative number is not defined, and $\log 1 = 0$.

Another base that occurs frequently in physics is the quantity $e = 2.718\ldots$. The **natural logarithm** (ln) of a number y is the power to which e must be raised to obtain y: $y = e^{\ln y}$. If $x = \ln y$, then $y = e^x$. Most calculators have keys for $\ln x$ and for e^x. For example, $\ln 10.0 = 2.30$, and if $\ln x = 3.00$, then $x = 10^{3.00} = 20.1$. Note that $\ln 1 = 0$.

Logarithms with any choice of base, including base 10 or base e, obey several simple and useful rules:

1. $\log(ab) = \log a + \log b$.

2. $\log\left(\dfrac{a}{b}\right) = \log a - \log b$.

3. $\log(a^n) = n \log a$.

A particular example of the second rule is

$$\log\left(\frac{1}{a}\right) = \log 1 - \log a = -\log a,$$

since $\log 1 = 0$.

EXAMPLE A.8 Solving a logarithmic equation

If $\frac{1}{2} = e^{-\alpha T}$, solve for T in terms of α.

Represent mathematically, solve, and evaluate
We take the natural logarithm of both sides of the equation: $\ln\left(\frac{1}{2}\right) = -\ln 2$ and $\ln\left(e^{-\alpha T}\right) = -\alpha T$. The equation thus becomes $-\alpha T = -\ln 2$, and it follows that $T = \frac{\ln 2}{\alpha}$.

The equation $y = e^{\alpha x}$ expresses y in terms of the exponential function $e^{\alpha x}$. The general rules for exponents in Appendix A.1 apply when the base is e, so $e^x e^y = e^{x+y}$, $e^x e^{-x} = e^{x+(-x)} = e^0 = 1$, and $(e^x)^2 = e^{2x}$.

Figure A.1

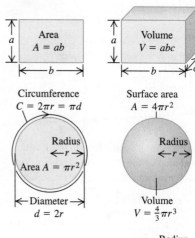

A.5 Areas and Volumes

Figure A.1 illustrates the formulas for the areas and volumes of common geometric shapes:

- A rectangle with length a and width b has area $A = ab$.
- A rectangular solid (a box) with length a, width b, and height c has volume $V = abc$.
- A circle with radius r has diameter $d = 2r$, circumference $C = 2\pi r = \pi d$, and area $A = \pi r^2 = \pi d^2/4$.
- A sphere with radius r has surface area $A = 4\pi r^2$ and volume $V = \frac{4}{3}\pi r^3$.
- A cylinder with radius r and height h has volume $V = \pi r^2 h$.

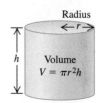

A.6 Plane Geometry and Trigonometry

Following are some useful results about angles:

1. Interior angles formed when two straight lines intersect are equal. For example, in **Figure A.2**, the two angles θ and ϕ are equal.
2. When two parallel lines are intersected by a diagonal straight line, the alternate interior angles are equal. For example, in **Figure A.3**, the two angles θ and ϕ are equal.
3. When the sides of one angle are each perpendicular to the corresponding sides of a second angle, then the two angles are equal. For example, in **Figure A.4**, the two angles θ and ϕ are equal.
4. The sum of the angles on one side of a straight line is 180°. In **Figure A.5**, $\theta + \phi = 180°$.
5. The sum of the angles in any triangle is 180°.

Similar Triangles

Triangles are **similar** if they have the same shape, but different sizes or orientations. Similar triangles have equal angles and equal ratios of corresponding sides. If the two triangles in **Figure A.6** are similar, then $\theta_1 = \theta_2$, $\phi_1 = \phi_2$, $\gamma_1 = \gamma_2$, and $\dfrac{a_1}{a_2} = \dfrac{b_1}{b_2} = \dfrac{c_1}{c_2}$.

Figure A.2

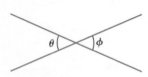

Interior angles formed when two straight lines intersect are equal:
$\theta = \phi$

Figure A.3

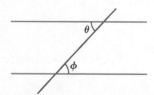

When two parallel lines are intersected by a diagonal straight line, the alternate interior angles are equal:
$\theta = \phi$

If two similar triangles have the same size, they are said to be **congruent**. If triangles are congruent, one can be rotated to where it can be placed precisely on top of the other.

Right Triangles and Trig Functions

In a **right triangle**, one angle is 90°. Therefore, the other two acute angles (*acute* means less than 90°) have a sum of 90°. In **Figure A.9**, $\theta + \phi = 90°$. The side opposite the right angle is called the **hypotenuse** (side c in the figure). In a right triangle, the square of the length of the hypotenuse equals the sum of the squares of the lengths of the other two sides. For the triangle in Figure A.9, $c^2 = a^2 + b^2$. This formula is called the **Pythagorean theorem**.

If two right triangles have the same value for one acute angle, then the two triangles are similar and have the same ratio of corresponding sides. This true statement allows us to define the functions **sine**, **cosine**, and **tangent** that are ratios of a pair of sides. These functions, called **trigonometric** functions or **trig functions**, depend only on one of the angles in the right triangle. For an angle θ, these functions are written $\sin\theta$, $\cos\theta$, and $\tan\theta$.

In terms of the triangle in Figure A.9, the sine, cosine, and tangent of the angle θ are as follows:

$$\sin\theta = \frac{\text{opposite side}}{\text{hypotenuse}} = \frac{a}{c}.$$

$$\cos\theta = \frac{\text{adjacent side}}{\text{hypotenuse}} = \frac{b}{c}.$$

$$\tan\theta = \frac{\text{opposite side}}{\text{adjacent side}} = \frac{a}{b}.$$

Note that $\tan\theta = \dfrac{\sin\theta}{\cos\theta}$. For angle ϕ, $\sin\phi = \dfrac{b}{c}$, $\cos\phi = \dfrac{a}{c}$, and $\tan\phi = \dfrac{b}{a}$.

In physics, angles are expressed in either degrees or radians, where π radians = 180°. Most calculators have a key for switching between degrees and radians. Always be sure that your calculator is set to the appropriate angular measure.

Inverse trig functions, denoted, for example, by $\sin^{-1}x$ (or $\arcsin x$), have a value equal to the angle that has the value x for the trig function. For example, $\sin 30° = 0.500$, so $\sin^{-1}(0.500) = \arcsin(0.500) = 30°$. Note that $\sin^{-1}x$ does *not* mean $\dfrac{1}{\sin x}$.

Figure A.4

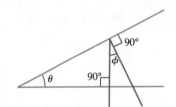

When the sides of one angle are each perpendicular to the corresponding sides of a second angle, then the two angles are equal:
$$\theta = \phi$$

Figure A.5

The sum of the angles on one side of a straight line is 180°:
$$\theta + \phi = 180°$$

Figure A.6

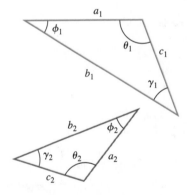

Two similar triangles: Same shape but not necessarily the same size

---

EXAMPLE A.9 Using trigonometry I

A right triangle has one angle of 30° and one side with length 8.0 cm, as shown in **Figure A.7**. What is the angle ϕ, and what are the lengths x and y of the other two sides of the triangle?

Figure A.7

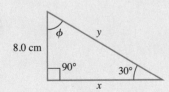

Represent mathematically, solve, and evaluate
$\phi + 30° = 90°$, so $\phi = 60°$.

$$\tan 30° = \frac{8.0\text{ cm}}{x}, \text{ so } x = \frac{8.0\text{ cm}}{\tan 30°} = 13.9\text{ cm}.$$

To find y, we use the Pythagorean theorem: $y^2 = (8.0\text{ cm})^2 + (13.9\text{ cm})^2$, so $y = 16.0$ cm.

Or we can say $\sin 30° = 8.0\text{ cm}/y$, so $y = 8.0\text{ cm}/\sin 30° = 16.0$ cm, which agrees with the previous result.

Notice how we used the Pythagorean theorem in combination with a trig function. You will use these tools constantly in physics, so make sure that you can employ them with confidence.

EXAMPLE A.10 Using trigonometry II

A right triangle has two sides with lengths as specified in **Figure A.8**. What is the length x of the third side of the triangle, and what is the angle θ, in degrees?

Figure A.8

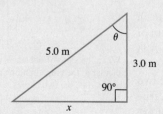

Represent mathematically, solve, and evaluate

The Pythagorean theorem applied to this right triangle gives $(3.0\,\text{m})^2 + x^2 = (5.0\,\text{m})^2$, so

$x = \sqrt{(5.0\,\text{m})^2 - (3.0\,\text{m})^2} = 4.0\,\text{m}$. (Since x is a length, we take the positive root of the equation.) We also have

$$\cos\theta = \frac{3.0\,\text{m}}{5.0\,\text{m}} = 0.600,\ so\ \theta = \cos^{-1}(0.600) = 53.1°.$$

In this case, we knew the lengths of two sides, but none of the acute angles, so we used the Pythagorean theorem first and then an appropriate trig function.

Figure A.9

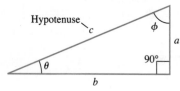

For a right triangle:
$\theta + \phi = 90°$
$c^2 = a^2 + b^2$ (Pythagorean theorem)

Figure A.10

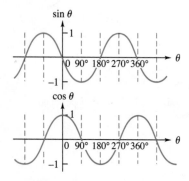

Figure A.11

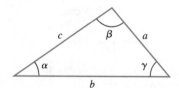

In a right triangle, all angles are in the range from 0° to 90°, and the sine, cosine, and tangent of the angles are all positive. This must be the case, since the trig functions are ratios of lengths. But for other applications, such as finding the components of vectors, calculating the oscillatory motion of a mass on a spring, or describing wave motion, it is useful to define the sine, cosine, and tangent for angles outside that range. Graphs of $\sin\theta$ and $\cos\theta$ are given in **Figure A.10**. The values of $\sin\theta$ and $\cos\theta$ vary between $+1$ and -1. Each function is periodic, with a period of 360°. Note the range of angles between 0° and 360° for which each function is positive and negative. The two functions $\sin\theta$ and $\cos\theta$ are 90° out of phase (that is, out of step). When one is zero, the other has its maximum magnitude.

For any triangle (see **Figure A.11**)—in other words, not necessarily a right triangle—the following two relations apply:

1. $\dfrac{\sin\alpha}{a} = \dfrac{\sin\beta}{b} = \dfrac{\sin\gamma}{c}$ (law of sines).

2. $c^2 = a^2 + b^2 - 2ab\cos\gamma$ (law of cosines).

Some of the relations among trig functions are called trig identities. The following table lists only a few, those most useful in introductory physics:

Useful trigonometric identities
$\sin(-\theta) = -\sin(\theta)$ ($\sin\theta$ is an odd function)
$\cos(-\theta) = \cos(\theta)$ ($\cos\theta$ is an even function)
$\sin 2\theta = 2\sin\theta\cos\theta$
$\cos 2\theta = \cos^2\theta - \sin^2\theta = 2\cos^2\theta - 1 = 1 - 2\sin^2\theta$
$\sin(\theta \pm \phi) = \sin\theta\cos\phi \pm \cos\theta\sin\phi$
$\cos(\theta \pm \phi) = \cos\theta\cos\phi \mp \sin\theta\sin\phi$
$\sin(180° - \theta) = \sin\theta$
$\cos(180° - \theta) = -\cos\theta$
$\sin(90° - \theta) = \cos\theta$
$\cos(90° - \theta) = \sin\theta$

APPENDIX B
Working with Vectors

Graphical Representation of Vectors

We use arrows to graphically represent vector quantities. The arrow's direction indicates the direction of the vector, and the arrow's length indicates the vector's magnitude. We need to use an appropriate scale when drawing the length of the arrow. Imagine that you need to walk from point M to point N (**Figure B.1a**). The distance between the points is 200 m and the direction is 25° North of East. We represent your walk with an arrow or a vector called a displacement vector $\vec{D}$. If 1.0 cm represents a displacement of 50 m, then the length of the vector will be 4.0 cm and its direction is 25° North of East (Figure B.1b). This is the first very important observation about vectors—it is not important where the vector starts or ends, as long as it reflects the same direction and magnitude as the quantity it represents. Notice the difference between the vectors $\vec{A}$ and $\vec{B}$ and $\vec{A}$ and $\vec{C}$ in **Figures B.2a** and b. $\vec{A}$ and $\vec{B}$ in Figure B.2a have the same magnitude but different directions and $\vec{A}$ and $\vec{C}$ in Figure B.2b have the same direction but different magnitudes.

Graphical Addition and Subtraction of Vectors

Addition Because vector quantities have both magnitude and direction, we cannot use the normal rules of algebraic addition and subtraction to add or subtract them. Suppose, for example, that you take a two-day trip. The first day you travel 500 km toward the east along a straight road. The second day involves another 500-km displacement, but not necessarily in the same direction. In what direction and how far from your starting position is your final location? The answer, of course, depends on the direction of the second day's trip. You could be 1000 km east of your starting position if you continued traveling east during the second day (**Figure B.3a**). However, if you traveled west on the second day, your net displacement would be zero as you would be back where you started (Figure B.3b). If you traveled north during the second day, your net displacement would be a little over 700 km to the northeast of your starting position (Figure B.3c). When adding displacement vectors or vectors of any type, we are concerned only with their net result.

To add two vectors we can use the following graphical technique. Suppose we want to add the two vectors $\vec{A}$ and $\vec{C}$ shown in **Figure B.4a**. To add

Figure B.1 Representing a displacement with a vector.

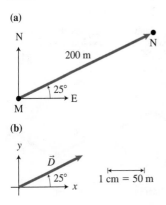

Figure B.2 (a) Equal magnitude but different direction vectors. (b) Equal direction but different magnitude vectors.

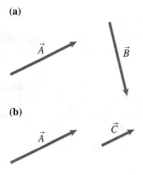

Figure B.3 The net displacement for two 500-km successive trips depends on the relative direction of each day's trip.

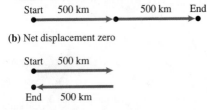

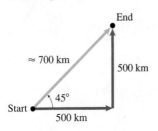

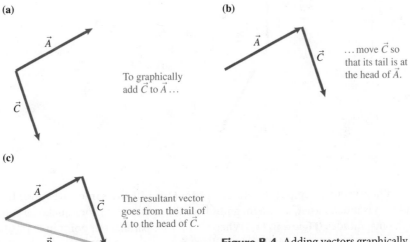

Figure B.4 Adding vectors graphically.

Figure B.5 Graphical vector addition.

(a)

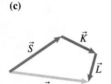

To add $\vec{S}$, $\vec{K}$, and $\vec{L}$ graphically...

(b)

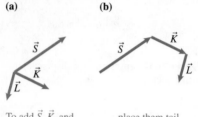

...place them tail to head one after the other.

(c)

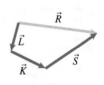

The resultant goes from the tail of the first vector to the head of the last vector.

(d)

The order in which you add the vectors makes no difference in the result.

them graphically, we place the tail of $\vec{C}$ at the head of $\vec{A}$, as in Figure B.4b. We can move a vector from one location to another as long as we do not change its magnitude or direction; therefore moving vector $\vec{C}$ as shown did not change it. Having moved vector $\vec{C}$, we draw another vector $\vec{R}$ from the tail of $\vec{A}$ to the head of $\vec{C}$, as in Figure B.4c. This vector $\vec{R}$ represents the result of the addition of two vectors $\vec{A}$ and $\vec{C}$. We can write the resultant vector as a mathematical equation: $\vec{R} = \vec{A} + \vec{C}$. As you see in Figure B.4c, the magnitude of the resultant vector is not equal to the sum of the magnitudes of $\vec{A}$ and $\vec{C}$.

Graphical addition of vectors: To graphically add two or more vectors (**Figure B.5a**), place the vectors tail to head one at a time, as illustrated in Figure B.5b. The magnitudes and directions of the vectors must not be changed as they are moved about. The resultant vector $\vec{R}$ goes from the tail of the first vector to the head of the last vector, as illustrated in Figure B.5c, and equals the sum of the vectors $\vec{R} = \vec{S} + \vec{K} + \vec{L}$. The order in which you add the vectors makes no difference (for example, $\vec{S} + \vec{K} + \vec{L} = \vec{L} + \vec{K} + \vec{S}$, as seen by comparing Figures B.5c and d).

Usually, the graphical technique for adding vectors is used as a rough check on another vector addition technique introduced later. However, if done with care using a ruler and protractor, the graphical vector addition technique is a fairly accurate method for determining the resultant of several vectors.

EXAMPLE B.1

A car travels 200 km west, 100 km south, and finally 150 km at an angle 60° south of east. Determine the net displacement of the car.

Reasoning Use the graphical addition technique with the three displacements drawn tail to head, as shown in **Figure B.6a**. To find the resultant displacement $\vec{R}$, draw an arrow

from the tail of the first displacement to the head of the last (Figure B.6b). We measure the magnitude of the resultant with a ruler and find that its length is 5.2 cm. Since each centimeter represents 50 km, the magnitude of the resultant displacement is (5.2 cm)(50 km/cm) = 260 km. Using a protractor, we confirm that the direction of the resultant is 60° south of west.

Figure B.6 (a) Three displacement vectors placed tail to head. (b) The resultant displacement.

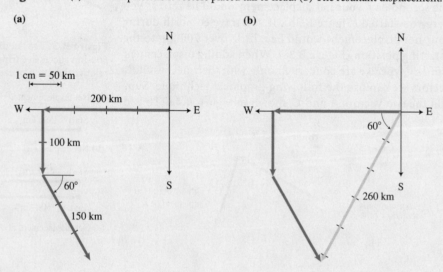

When vectors are parallel to each other and in the same direction, the addition is much easier, as the magnitude of the resultant vector equals the sum of the magnitudes (**Figure B.7a**). When vectors are parallel but point in opposite directions, the magnitude of the resultant vector equals the difference in the magnitudes of the vectors being added (Figure B.7b).

Subtraction Sometimes we need to subtract two vectors $(\vec{A} - \vec{B})$. We can view the procedure of subtraction as the familiar procedure of addition where instead of adding a vector $\vec{B}$ to the vector $\vec{A}$, we add $(-\vec{B})$ to $\vec{A}$: $\vec{R} = \vec{A} - \vec{B} = \vec{A} + (-\vec{B})$. To draw $(-\vec{B})$, we simply reverse the direction of vector $\vec{B}$. Vector subtraction is illustrated in **Figure B.8**, where vector $\vec{B}$ is subtracted from $\vec{A}$.

> Graphical subtraction of vectors: To graphically subtract vector $\vec{B}$ from vector $\vec{A}$ (see Figure B.8a), first reverse the direction of $\vec{B}$, thus producing $-\vec{B}$ (Figure B.8b). Then add $\vec{A}$ and $-\vec{B}$ (Figure B.8c). The resultant vector goes from the tail of $\vec{A}$ to the head of $-\vec{B}$ (Figure B.8d) and is the difference of $\vec{A}$ and $\vec{B}$:

$$\vec{R} = \vec{A} + (-\vec{B}) = \vec{A} - \vec{B}$$

Components of a Vector

The graphical method of adding and subtracting vectors takes considerable time if done accurately. There is a different vector addition technique, which we describe in this and the next sub-sections. It is more productive for calculating the length and direction of the resultant vector and is also faster and more convenient for solving a variety of interesting problems involving vectors. The component method, as it is called, uses a principle that any vector can be represented as the sum of two vectors, called *vector components*, which are perpendicular to each other (**Figure B.9**).

A graphical method for finding these two vectors whose sum equals vector $\vec{A}$ is illustrated in **Figure B.10a**. Draw a coordinate system so its x-axis points east and y-axis points north. The x component of $\vec{A}$ is a vector $\vec{A}_x$ that points along the x-axis and whose length equals the projection of $\vec{A}$ on that axis, as shown in Figure B.10b. The y component of the vector $\vec{A}_y$ is the projection of $\vec{A}$ on the y-axis in Figure B.10b. It is important to notice that the vector sum of $\vec{A}_x$ and $\vec{A}_y$ equals $\vec{A}$, as shown in Figure B.10c.

We often work with vectors whose tails are located at the origin of a coordinate system—see **Figure B.11**. This is especially true for force vectors representing the forces that other objects exert on an object of interest. The vector components of these vectors can be converted to scalar components, which represent vectors using numbers for the magnitude and the signs for direction. For example, in Figure B.11 the head of the x component $\vec{F}_x$ is located 5 units in the negative x direction. The scalar x component $F_x = -5$ units. The head of the y component $\vec{F}_y$ is located 5 units in the positive y direction. The scalar y component $F_y = +5$ units. Note that if a vector component points in the positive x or y direction, its scalar component is positive; if the vector component points in the negative x or y direction, the scalar component is negative. We can calculate the scalar

Figure B.7 (a) The magnitudes of parallel vectors can be added. (b) The magnitudes of anti-parallel vectors can be subtracted.

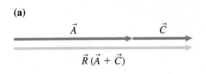

(a)

(b)

Figure B.8 Graphical subtraction of vectors.

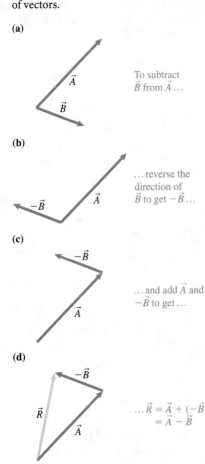

(a)

To subtract $\vec{B}$ from $\vec{A}$ ...

(b)

... reverse the direction of $\vec{B}$ to get $-\vec{B}$ ...

(c)

... and add $\vec{A}$ and $-\vec{B}$ to get ...

(d)

... $\vec{R} = \vec{A} + (-\vec{B})$ $= \vec{A} - \vec{B}$

Figure B.9 A vector $\vec{F}$ equals the sum of its x and y-components $\vec{F}_x$ and $\vec{F}_y$.

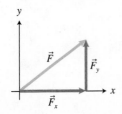

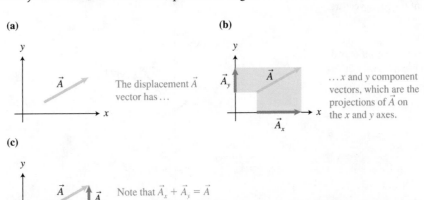

(a) The displacement $\vec{A}$ vector has ...

(b) ... x and y component vectors, which are the projections of $\vec{A}$ on the x and y axes.

(c) Note that $\vec{A}_x + \vec{A}_y = \vec{A}$

Figure B.10 The component vectors $\vec{A}_x$ and $\vec{A}_y$ of a displacement $\vec{A}$.

Figure B.11 The scalar components of a vector.

The scalar y-component of $\vec{F}$ is $F_y = +5$ units.

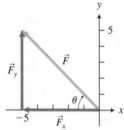

The scalar x-component of $\vec{F}$ is $F_x = -5$ units.

components using the triangle shown in **Figure B.11** and the trigonometric sin and cos functions.

x and y scalar components of a vector: The x and y scalar components of a vector are:

$$F_x = \pm F \cos\theta$$
$$F_y = \pm F \sin\theta$$

where F is the magnitude of the vector and θ is the angle ($90°$ or less) that $\vec{F}$ makes with the x axis. F_x is positive if $\vec{F}_x$ points in the positive x direction and negative if $\vec{F}_x$ points in the negative x direction. F_y is either positive or negative, depending on the direction of $\vec{F}_y$ relative to the y axis.

EXAMPLE B.2

Suppose we have a vector $\vec{N}$ of the magnitude of 10 units that is directed at $20°$ above the horizontal direction as shown in **Figure B.12a**. The hypotenuse and the x and y scalar components of $\vec{N}$ form a triangle for which:

$$\cos\theta = \frac{\text{adjacent side}}{\text{hypotenuse}} = \frac{N_x}{N} \quad \text{or} \quad N_x = N\cos\theta$$

$$\sin\theta = \frac{\text{opposite side}}{\text{hypotenuse}} = \frac{N_y}{N} \quad \text{or} \quad N_y = N\sin\theta$$

where N is the magnitude of the vector $\vec{N}$. Using the magnitude N and the angle, we can calculate the values of the scalar components of $\vec{N}$:

$$N_x = N\cos\theta = (10 \text{ units}) \cos 20° = 9.4 \text{ units}$$
$$N_y = N\sin\theta = (10 \text{ units}) \sin 20° = 3.4 \text{ units}$$

If the vector pointed in the opposite direction, as in Figure B.12b, the magnitude of the scalar stays the same but the signs are negative:

$$N_x = -N\cos\theta = -9.4 \text{ units}$$
$$N_y = -N\sin\theta = -3.4 \text{ units}$$

The signs depend on the orientation of the scalar components relative to the x and y axes.

Figure B.12 N_x and N_y can be calculated from known N and θ.

(a)

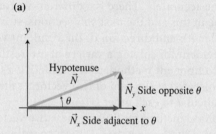

(b)

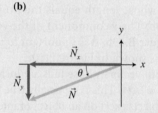

EXAMPLE B.3

Determine the x and y components of each of the force vectors shown in **Figure B.13**.

Reasoning $F_{1x} = +24$ N and $F_{1y} = +32$ N; $F_{2x} = -35$ N and $F_{2y} = +35$ N. Note that the angle between $\vec{F}_3$ and the negative x axis is $30°$ and not $60°$. Thus, $F_{3x} = -26$ N and $F_{3y} = -15$ N.

Figure B.13

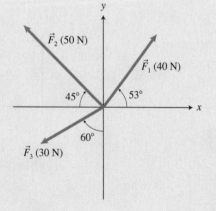

Vector addition by components

Let us see how the vector and scalar components allow us to add vectors. In **Figure B.14a** we add two vectors $\vec{K}$ and $\vec{L}$ to form the resultant vector $\vec{R}$. We first determine the scalar components of the vectors being added (the scalar components have been given in Figure B.14b—normally we would have to calculate the scalar components from the known directions and magnitudes of $\vec{K}$ and $\vec{L}$). Suppose that for vector $\vec{K}, K_x = +3$ units and $K_y = (-1)$ unit. For vector $\vec{L}, L_x = +2$ units and $L_y = (-3)$ units.

Next, we add the scalar components of $\vec{K}$ and $\vec{L}$ to find the scalar components of vector $\vec{R}$ (Figure B.14c):

$$R_x = K_x + L_x = 3 + 2 = 5 \text{ units}$$
$$R_y = K_y + L_y = (-1) + (-3) = (-4) \text{ units.}$$

The negative sign for the y component of the resultant vector means it points in the negative direction relative to the y-axis, as seen in Figure B.14d.

To find the length of the resultant vector we can use our knowledge of geometry. Examine the right triangle (Figure B.14d) that is formed by the scalar components of $\vec{R}$: R_x and R_y. In this triangle, R_x and R_y are the sides and the magnitude R of $\vec{R}$ is the hypotenuse. Using the Pythagorean theorem we find the length of the hypotenuse if we know the length of the sides: $R^2 = R_x^2 + R_y^2$ or $R = \sqrt{R_x^2 + R_y^2}$. Apply this reasoning to our situation:

$$R = \sqrt{R_x^2 + R_y^2} = \sqrt{5^2 + (-4)^2} = \sqrt{25 + 16} = 6.4 \text{ units.}$$

Notice that all numbers here have two significant digits.

Figure B.14 Scalar component vector addition.

(a)

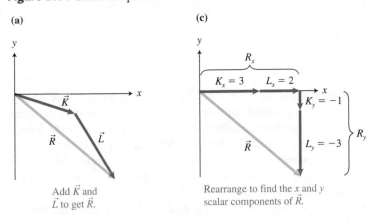

Add $\vec{K}$ and $\vec{L}$ to get $\vec{R}$.

Rearrange to find the x and y scalar components of $\vec{R}$.

(b)

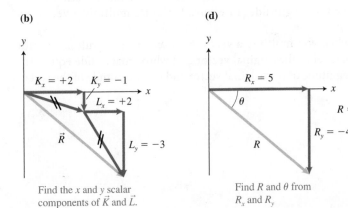

Find the x and y scalar components of $\vec{K}$ and $\vec{L}$.

Find R and θ from R_x and R_y

The tan function allows us to determine the angle between vector $\vec{R}$ and the positive or negative x-axis: $\tan\theta = R_y/R_x = (-4)/5 = (-0.80)$. Using the arctan function on the calculator, we find that the angle whose tan is (-0.80) equals $-39°$. Therefore the vector $\vec{R}$ points $39°$ *below* the direction of the $+x$ axis. Recall that in the right triangle (Figure B.13d) formed by vector $\vec{R}$ and its components $\vec{R}_x$ and $\vec{R}_y$, R_x is the adjacent side and R_y is the opposite side.

To determine a vector if its scalar components are known: If the scalar components B_x and B_y are known, the magnitude of the vector $\vec{B}$ is

$$B = \sqrt{B_x^{\,2} + B_y^{\,2}}$$

and the angle θ that $\vec{B}$ makes with the horizontal direction is

$$\theta = \arctan\frac{B_y}{B_x}.$$

We can now summarize this operation.

EXAMPLE B.4

Determine the sum of the three forces shown in Figure B.13. You can use the results of the component calculation in Example B.3.

Reasoning The x and y components of the resultant force are:

$$R_x = F_{1x} + F_{2x} + F_{3x} = +24\,\text{N} + (-35\,\text{N}) + (-26\,\text{N})$$
$$= -37\,\text{N}$$

$$R_y = F_{1y} + F_{2y} + F_{3y} = +32\,\text{N} + 35\,\text{N} + (-15\,\text{N})$$
$$= +52\,\text{N}$$

The magnitude of the resultant force is:

$$R = \sqrt{(-37\,\text{N})^2 + (52\,\text{N})^2} = 64\,\text{N}.$$

The resultant vector is in the second quadrant (it has a negative x component and a positive y component) and makes the following angle above the negative x axis:

$$\theta = \arctan\frac{52\,\text{N}}{37\,\text{N}} = \arctan 1.41 = 75°.$$

Thus, the sum of the three forces has a magnitude of 64 N and points $75°$ above the negative x-axis.

Figure B.15 Multiplying a vector by a scalar (a) changes its length (a positive number) or (b) changes its length and reverses its direction (a negative number).

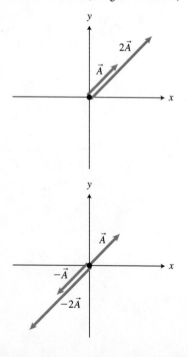

Multiplying a vector by a number

You may need to multiply a vector by a positive or negative number (a scalar). To multiply a vector by a positive number means to draw a vector in the same direction as the original one but of a different magnitude. The magnitude is determined by multiplying the magnitude of the vector by the magnitude of the number. For example in **Figure B.15a** we multiply vector $\vec{A}$ by 2: vector $2\vec{A}$ is in the same direction as $\vec{A}$ but is two times longer. If we need to draw a vector $\vec{A}/3$ the operation is similar to multiplying $\vec{A}$ by 1/3: $\vec{A}/3 = 1/3\vec{A}$. Thus all we need to do is to draw a vector of one-third $\vec{A}$'s magnitude and in the same direction.

To multiply a vector by a negative number, reverse the direction of the vector and change the magnitude. In Figure B.15b we multiply a vector by (-1) and by (-2).

In general, when you multiply a vector by a scalar, the result is a vector parallel or anti-parallel to the original vector and whose magnitude equals the product of the magnitude of the original vector and the magnitude of the scalar.

APPENDIX C

Base Units of SI System

The SI system of units is built on the seven base units listed below. All other derived units are combinations of two or more of these base units.

Table C.1 Base Units of the SI System

Base Quantity	Name	Abbreviation
Length	meter	m
Mass	kilogram	kg
Time	second	s
Electric current	ampere	A
Temperature	kelvin	K
Amount of substance	mole	mol
Luminous intensity	candela	cd

An example of a derived unit is the unit for kinetic energy, the joule (J), which is expressed in terms of the kg, m, and s:

$$1\,J = 1\,kg\left(\frac{1\,m}{1\,s}\right)^2$$

Definitions of the base units

meter (m) The distance that light travels in a vacuum in a time interval of $\left(\dfrac{1}{299{,}792{,}458}\right)$s.

kilogram (kg) The mass of an international standard cylindrical platinum-iridium alloy stored in a vault in Sèvres, France by the International Bureau of Weights and Measures.

second (s) The duration of 9,192,631,770 vibrations of radiation emitted by a particular transition of a cesium-133 atom.

ampere (A) The constant electric current that, when flowing in two very long parallel straight wires placed 1 m apart in a vacuum, causes them to exert a force on each other of 2×10^{-7} newton per meter length of wire.

kelvin (K) 1/273.16 of the thermodynamic temperature of the triple point of water.

mole (mol) The amount of a substance that contains as many elementary entities as there are carbon atoms in 0.012 kg of the carbon isotope carbon 12. The elementary entities could be atoms, molecules, ions, electrons, other particles, or specified groups of such particles.

candela (cd) The luminous intensity in a given direction of a light source that emits monochromatic radiation of frequency 540×10^{12} Hz and that has a radiant intensity in that direction of 1/683 W per steradian.

Atomic and Nuclear Data

Atomic number (Z)	Element	Symbol	Mass number (A)	Atomic mass (u)	Percent abundance	Decay mode	Half-life $t_{1/2}$
0	(Neutron)	n	1	1.008 665		β^-	10.4 min
1	Hydrogen	H	1	1.007 825	99.985	stable	
	Deuterium	D	2	2.014 102	0.015	stable	
	Tritium	T	3	3.016 049		β^-	12.33 yr
2	Helium	He	3	3.016 029	0.000 1	stable	
			4	4.002 602	99.999 9	stable	
			6	6.018 886		β^-	0.81 s
3	Lithium	Li	6	6.015 121	7.50	stable	
			7	7.016 003	92.50	stable	
			8	8.022 486		β^-	0.84 s
4	Beryllium	Be	7	7.016 928		EC	53.3 days
			9	9.012 174	100	stable	
			10	10.013 534		β^-	1.5×10^6 yr
5	Boron	B	10	10.012 936	19.90	stable	
			11	11.009 305	80.10	stable	
			12	12.014 352		β^-	0.020 2 s
6	Carbon	C	10	10.016 854		β^+	19.3 s
			11	11.011 433		β^+	20.4 min
			12	12.000 000	98.90	stable	
			13	13.003 355	1.10	stable	
			14	14.003 242		β^-	5 730 yr
			15	15.010 599		β^-	2.45 s
7	Nitrogen	N	12	12.018 613		β^+	0.011 0 s
			13	13.005 738		β^+	9.96 min
			14	14.003 074	99.63	stable	
			15	15.000 108	0.37	stable	
			16	16.006 100		β^-	7.13 s
			17	17.008 450		β^-	4.17 s
8	Oxygen	O	14	14.008 595		EC	70.6 s
			15	15.003 065		β^+	122 s
			16	15.994 915	99.76	stable	
			17	16.999 132	0.04	stable	
			18	17.999 160	0.20	stable	
			19	19.003 577		β^-	26.9 s
9	Fluorine	F	17	17.002 094		EC	64.5 s
			18	18.000 937		β^+	109.8 min
			19	18.998 404	100	stable	
			20	19.999 982		β^-	11.0 s
10	Neon	Ne	19	19.001 880		β^+	17.2 s
			20	19.992 435	90.48	stable	
			21	20.993 841	0.27	stable	
			22	21.991 383	9.25	stable	

Atomic number (Z)	Element	Symbol	Mass number (A)	Atomic mass (u)	Percent abundance	Decay mode	Half-life $t_{1/2}$
11	Sodium	Na	22	21.994 434		β^+	2.61 yr
			23	22.989 770	100	stable	
			24	23.990 961		β^-	14.96 hr
12	Magnesium	Mg	24	23.985 042	78.99	stable	
			25	24.985 838	10.00	stable	
			26	25.982 594	11.01	stable	
13	Aluminum	Al	27	26.981 538	100	stable	
			28	27.981 910		β^-	2.24 min
14	Silicon	Si	28	27.976 927	92.23	stable	
			29	28.976 495	4.67	stable	
			30	29.973 770	3.10	stable	
			31	30.975 362		β^-	2.62 hr
15	Phosphorus	P	30	29.978 307		β^+	2.50 min
			31	30.973 762	100	stable	
			32	31.973 908		β^-	14.26 days
16	Sulfur	S	32	31.972 071	95.02	stable	
			33	32.971 459	0.75	stable	
			34	33.967 867	4.21	stable	
			35	34.969 033		β^-	87.5 days
			36	35.967 081	0.02	stable	
17	Chlorine	Cl	35	34.968 853	75.77	stable	
			36	35.968 307		β^-	3.0×10^5 yr
			37	36.965 903	24.23	stable	
18	Argon	Ar	36	35.967 547	0.34	stable	
			38	37.962 732	0.06	stable	
			39	38.964 314		β^-	269 yr
			40	39.962 384	99.60	stable	
			42	41.963 049		β^-	33 yr
19	Potassium	K	39	38.963 708	93.26	stable	
			40	39.964 000	0.01	β^+	1.28×10^9 yr
			41	40.961 827	6.73	stable	
20	Calcium	Ca	40	39.962 591	96.94	stable	
			42	41.958 618	0.64	stable	
			43	42.958 767	0.13	stable	
			44	43.955 481	2.08	stable	
			47	46.954 547		β^-	4.5 days
			48	47.952 534	0.18	stable	
24	Chromium	Cr	50	49.946 047	4.34	stable	
			52	51.940 511	83.79	stable	
			53	52.940 652	9.50	stable	
			54	53.938 883	2.36	stable	
26	Iron	Fe	54	53.939 613	5.9	stable	
			55	54.938 297		EC	2.7 yr
			56	55.934 940	91.72	stable	
			57	56.935 396	2.1	stable	
			58	57.933 278	0.28	stable	

Atomic number (Z)	Element	Symbol	Mass number (A)	Atomic mass (u)	Percent abundance	Decay mode	Half-life $t_{1/2}$
27	Cobalt	Co	59	58.933 198	100	stable	
			60	59.933 820		β^-	5.27 yr
28	Nickel	Ni	58	57.935 346	68.08	stable	
			60	59.930 789	26.22	stable	
			61	60.931 058	1.14	stable	
			62	61.928 346	3.63	stable	
			64	63.927 967	0.92	stable	
29	Copper	Cu	63	62.929 599	69.17	stable	
			65	64.927 791	30.83	stable	
37	Rubidium	Rb	96	95.93427			33 min
38	Strontium	Sr	90	89.907 320		α	28.9 y
47	Silver	Ag	107	106.905 091	51.84	stable	
			109	108.904 754	48.16	stable	
48	Cadmium	Cd	106	105.906 457	1.25	stable	
			109	108.904 984		EC	462 days
			110	109.903 004	12.49	stable	
			111	110.904 182	12.80	stable	
			112	111.902 760	24.13	stable	
			113	112.904 401	12.22	stable	
			114	113.903 359	28.73	stable	
			116	115.904 755	7.49	stable	
50	Tin	Sn	120	119.902 197	32.4		
53	Iodine	I	127	126.904 474	100	stable	
			129	128.904 984		β^-	1.6×10^7 yr
			131	130.906 124		β^-	8.03 days
54	Xenon	Xe	128	127.903 531	1.9	stable	
			129	128.904 779	26.4	stable	
			130	129.903 509	4.1	stable	
			131	130.905 069	21.2	stable	
			132	131.904 141	26.9	stable	
			133	132.905 906		β^-	5.4 days
			134	133.905 394	10.4	stable	
			136	135.907 215	8.9	stable	
55	Cesium	Cs	133	132.905 436	100	stable	
			137	136.907 078		β^-	30 yr
			138	137.911 017		β^-	32.2 min
56	Barium	Ba	131	130.906 931		EC	12 days
			133	132.905 990		EC	10.5 yr
			134	133.904 492	2.42	stable	
			135	134.905 671	6.59	stable	
			136	135.904 559	7.85	stable	
			137	136.905 816	11.23	stable	
			138	137.905 236	71.70	stable	
79	Gold	Au	197	196.966 543	100	stable	
			198	197.968 242		β^-	2.7 d
81	Thallium	Tl	203	202.972 320	29.524	stable	
			205	204.974 400	70.476	stable	
			207	206.977 403		β^-	4.77 min

Atomic number (Z)	Element	Symbol	Mass number (A)	Atomic mass (u)	Percent abundance	Decay mode	Half-life $t_{1/2}$
82	Lead	Pb	204	203.973 020	1.4	stable	
			205	204.974 457		EC	1.5×10^7 yr
			206	205.974 440	24.1	stable	
			207	206.975 871	22.1	stable	
			208	207.976 627	52.4	stable	
			210	209.984 163		α, β^-	22.3 yr
			211	210.988 734		β^-	36.1 min
83	Bismuth	Bi	208	207.979 717		EC	3.7×10^5 yr
			209	208.980 374	100	stable	
			211	210.987 254		α	2.14 min
			215	215.001 836		β^-	7.4 min
84	Polonium	Po	209	208.982 405		α	102 yr
			210	209.982 848		α	138.38 days
			215	214.999 418		α	0.001 8 s
			218	218.008 965		α, β^-	3.10 min
85	Astatine	At	218	218.008 685		α, β^-	1.6 s
			219	219.011 294		α, β^-	0.9 min
86	Radon	Rn	219	219.009 477		α	3.96 s
			220	220.011 369		α	55.6 s
			222	222.017 571		α, β^-	3.823 days
87	Francium	Fr	223	223.019 733		α, β^-	22 min
88	Radium	Ra	223	223.018 499		α	11.43 days
			224	224.020 187		α	3.66 days
			226	226.025 402		α	1 600 yr
			228	228.031 064		β^-	5.75 yr
89	Actinium	Ac	227	227.027 749		α, β^-	21.77 yr
			228	228.031 015		β^-	6.15 hr
90	Thorium	Th	227	227.027 701		α	18.72 days
			228	228.028 716		α	1.913 yr
			229	229.031 757		α	7 300 yr
			230	230.033 127		α	75.000 yr
			231	231.036 299		α, β^-	25.52 hr
			232	232.038 051	100	α	1.40×10^{10} yr
			234	234.043 593		β^-	24.1 days
91	Protactinium	Pa	231	231.035 880		α	32.760 yr
			234	234.043 300		β^-	6.7 hr
92	Uranium	U	232	232.03713		α	72 y
			233	233.039 630		α	1.59×10^5 yr
			234	234.040 946		α	2.45×10^5 yr
			235	235.043 924	0.72	α	7.04×10^8 yr
			236	236.045 562		α	2.34×10^7 yr
			238	238.050 784	99.28	α	4.47×10^9 yr
93	Neptunium	Np	236	236.046 560		EC	1.15×10^5 yr
			237	237.048 168		α	2.14×10^6 yr
94	Plutonium	Pu	238	238.049 555		α	87.7 yr
			239	239.052 157		α	2.412×10^4 yr
			240	240.053 808		α	6 560 yr
			242	242.058 737		α	3.73×10^6 yr

Answers to Review Questions

Chapter 1

1.1 It is true because an observer A can see an object moving and observer B can see the same object not moving. When you are sitting on a train, you are not moving with respect to the train but are moving with respect to the trees on the ground.

1.2 We can decide if the object is moving at a constant rate, moving faster and faster, or slowing down. We can also decide if the direction of the motion is changing.

1.3 Ten kilometers is the distance, and 16 km is the path length.

1.4 To find the position we need to find the value of x corresponding to the value of t. We find that the object's position is $x = 60$ m at $t = 2.0$ s and $x = 20$ m at $t = 5.0$ s.

1.5 Because the area of a rectangle is equal to the product of length and width. In this case it is $v_x \cdot \Delta t$, which is equal to the displacement.

1.6 (a) A person (the object of interest) standing in a parking lot as seen by a person in a car that is leaving the lot at increasing speed (the forward direction is positive). (b) While standing in a parking lot, you observe a car moving past you in the negative direction and slowing down. The direction of the car's acceleration is opposite its velocity in the positive direction.

1.7 Mike is right about the initial position, $x_0 = -48$ m, but is not right about the acceleration; it is $a_x = -4$ m/s^2.

1.8 You can choose the positive direction for your coordinate axis as either up or as down at your convenience. If the axis points down, the component of acceleration is positive ($g_y = +9.8$ m/s^2); if the axis points up, the component of acceleration is negative ($g_y = -9.8$ m/s^2).

1.9 The first car starts to slow down at the same time that the tailgating car sees the first car's brake lights. Only after the reaction time does the tailgating car's speed start to decrease. The front car is now moving slower and continues to move slower than the tailgating car until they collide.

Chapter 2

2.1 Earth, the floor, and the air surrounding you, although the interaction with the air as you are sliding might be very weak.

2.2 About 64 ± 1 units.

2.3 We conducted experiments with the bowling ball and the person on rollerblades and analyzed them using force diagrams and motion diagrams. Our analysis showed that when the sum of the forces exerted on the object was zero, the object moved at constant velocity.

2.4 The magnitude of the upward force exerted by the cable on the elevator $F_{C\,on\,El}$ is exactly equal to the magnitude of the downward force exerted by Earth on the elevator $F_{E\,on\,El}$. The cable and Earth are the only two objects interacting with the elevator. If the forces they exert on the elevator have the same magnitudes and opposite directions, the sum of the force is zero, and the elevator should move at constant velocity.

2.5 Observers in inertial reference frames find that if an object does not interact with any other objects or all interactions add to zero, the object's velocity does not change. However, an observer in a noninertial reference frame can find an object whose velocity changes even when there are no other objects interacting with it or when all interactions are balanced (add to zero). For example, a passenger on a bus observes her purse slide off her lap (accelerating away from her) without any extra objects pushing the purse. A person standing beside the bus sees the purse moving forward at constant speed.

2.6 $\vec{a}$ is a measure of how fast velocity changes and is a consequence of the sum of all forces exerted on the object. The product of mass and acceleration is *not* an additional force and thus should not be on the force diagram.

2.7 The force that Earth exerts on any object is proportional to the object's mass. The acceleration is inversely proportional to the object's mass. Thus the effect of the mass cancels.

2.8 Mike is correct; the scale reads the force that it exerts on the person. We know this because the reading of the scale changes as the person rides the elevator while the gravitational force that Earth exerts on the person (the weight of the person) does not change. It does not read the sum of the forces. If it did, then the reading should be zero when the elevator is at rest.

2.9 The rollerblader pushes the floor, and the floor pushes the rollerblader ($\vec{F}_{R\,on\,F} = -\vec{F}_{F\,on\,R}$). You push the refrigerator and the refrigerator pushes you ($\vec{F}_{Y\,on\,R} = -\vec{F}_{R\,on\,Y}$). The truck pulls the car and the car pulls the truck ($\vec{F}_{T\,on\,C} = -\vec{F}_{C\,on\,T}$).

2.10 By extending the stopping distance and consequently reducing the acceleration during the stopping process.

Chapter 3

3.1 A scalar component is positive if the vector projects in the positive direction on the axis and negative if the vector projects in the negative direction on the axis.

3.2 P—person, S—surface, E—Earth

$x\colon$ $ma_x = +F_{P\,on\,C}\cos 30° + N_{S\,on\,C}\cos 90° + F_{E\,on\,C}\cos 90°$
or $ma_x = +F_{P\,on\,C}\,0.87 + 0 + 0$
$y\colon$ $ma_y = -F_{P\,on\,C}(\sin 30°) + N_{S\,on\,C}\sin 90° + (-F_{E\,on\,C}\sin 90°)$
or $m(0) = (-F_{P\,on\,C}\,0.50) + N_{S\,on\,C} - F_{E\,on\,C}$

3.3 For objects moving along inclined surfaces, the acceleration is parallel to the inclined surface. Thus, if we choose the x-axis parallel to the surface, the acceleration is entirely along that axis, and is zero in the direction of the y-axis, which is perpendicular to the surface. This makes problem solving much easier than if we choose horizontal and vertical axes, in which case there is a nonzero component of the acceleration along each axis.

3.4 The friction force is zero if no one is pushing or pulling the refrigerator. However if you do need to move it, you will have to push rather hard since the force needed to get it moving will have to be greater than $f_{s\,max} = \mu_s N = \mu_s (mg) = 350$ N (or about 80 lb).

3.5 We analyze the constant velocity horizontal motion and the constant acceleration vertical motion independently. We need to use the appropriate initial velocities (the velocity components) for each motion.

3.6 The engine is a part of the system; thus it cannot exert an external force that will accelerate the car system. The ground is the main external object interacting with the car that exerts a force on it.

Chapter 4

4.1 The velocity vector is tangent to the circle. Using the velocity change technique, we find that the velocity change vector $\Delta \vec{v}$ points toward the center. The acceleration vector equals the velocity change vector divided by the time interval during which the change occurred: $\vec{a} = \Delta \vec{v}/\Delta t$. So the acceleration vector points in the same direction as the velocity change vector, toward the center of the circle.

4.2 At all times during the ball's motion, the downward force exerted by Earth on the ball and the upward normal force exerted by the surface on the ball balance so that there is no net force in the vertical direction. But in the horizontal plane of the table while the ball is in contact with the semicircular ring, the ring exerts an inward force on the ball toward the center of the circle. This force causes the ball to accelerate toward the center of the circle while it contacts the semicircle. Before and after the semicircle, the ball is not accelerating, and the net force that other objects exert on the ball is zero. After the ball exits the barrier it will move in direction B.

4.3 For radial acceleration in terms of speed and radius, we have $a = v^2/r$. The dimensions on the right are $(L/T)^2/L = L/T^2$, which are the correct dimensions for acceleration: length L over time squared T^2. Centripetal acceleration in terms of radius and period is $a = 4\pi^2 r/T^2$. The dimensions are L/T^2, which again are correct for acceleration.

4.4 The woman's velocity is tangent to the drum. She would normally fly forward and fall (like a projectile). But the drum's surface intercepts her forward path and pushes in on her, causing her to move in a circular path. The friction force exerted by the drum prevents her from falling.

4.5 The friend can imagine that we take two point-like objects each with a mass of 1 kg and place them apart at a distance of 1 m. They will attract each other, exerting a force of magnitude 6.67×10^{-11} N—a very tiny force. It looks like such a force should not affect the interactions of objects of masses comparable to masses of people, houses, trees, etc. However, when the mass of one of the objects is on the order of 10^{26} kg, as the mass of Earth is, the force becomes significant even though the separation between the center of Earth and any object is huge—more than 6000 km.

4.6 The Moon actually does fall toward Earth all the time, if the word "fall" implies the motion of an object when the only force exerted on it is the force exerted by Earth. But it also flies forward all the time. Thus, the Moon combines two motions, flying forward and at the same time falling toward Earth. The net result is that it continually "lands" on its circular path around Earth.

Chapter 5

5.1 We can define the system as the log and all of the air that participated in burning.

5.2 Momentum is a vector quantity. Before the collision, the carts of the same mass were moving in the opposite directions at the same speeds. The vector sum of their momenta was zero. After the collision, it was zero again because both carts had zero velocities.

5.3 The momentum of the apple increases because of the impulse of the force exerted on it by Earth. If you consider the apple and Earth as a system, then the momentum is constant: the apple gains a downward momentum and Earth gains an upward momentum. We do not observe Earth moving upward because its mass is huge.

5.4 *Sad ball:* $m_b v_i + 0 = m_b \cdot 0 + m_{board} v_{board\,fx}$

5.4 *Happy ball:* $m_b v_i + 0 = m_b(-v_i) + m_{board}(v_{board\,fx})$

5.5 We considered the block and the bullet as a system. The force exerted on the bullet by the block is an internal force.

5.6 A 2.0-kg skateboard was rolling in the negative direction at a speed of 8 m/s when Marsha (58 kg) jumped on it. What is the new speed of the skateboard and Marsha together?

5.7 When we consider the momentum of a system, mass is as important as velocity. The mass of Earth is huge compared to the mass of the meteorite. Consider the meteorite and Earth to be the system. When the meteorite hits Earth, they both continue moving in the direction of motion of the meteorite, so that the initial momentum of the meteorite-Earth system is constant before and after the collision; however, the velocity of the system is tiny due to the huge mass of Earth.

Chapter 6

6.1 (1) You push a heavy crate sitting on the floor and it does not move—the work is zero because the displacement is zero. (2) Earth does zero work on the orbiting Moon because the force that Earth exerts on the Moon is perpendicular to the Moon's displacement.

6.2 Work involves *a process* that occurs when a force is exerted by some external object on the object in the system as the system object moves. Thus the system does not possess work. On the other hand, the state of a system can be characterized by the amount of each type of energy in the system.

6.3 We can do it in one of two ways. (1) We include Earth in the system and consider the change in gravitational potential energy of the system that changes as the elevation of the object relative to Earth changes. (2) We exclude Earth from the system and include the work done on an object by the external force that Earth exerts on the object as it changes elevation. The first method is usually easier to use in problem solving.

6.4 The force that you would exert on the spring while stretching it is not constant in magnitude. It increases linearly from 0 to kx, and the average force is $(1/2)kx$.

6.5 We could exclude a stationary surface from the system and determine the negative work done by friction—a common practice. However, this method has been shown not to account for the thermal energy change of the touching surface of an object in the system. It is easier, we think, to include the surfaces in the system and account for the energy change as an internal energy change—a change that we can see and feel.

6.6 The system would have gravitational potential energy but Earth would do work on the system—the work-energy bar chart would have changed and instead of $U_{gi} = U_{sf}$ we would write $W_{Earth\,on\,system} = U_{sf}$. However, as the work done by Earth is equal to mgy_i, the final answer will not change.

6.7 Calculate the total kinetic energy of the system (the two colliding objects) before the collision and after the collision. If the numbers are the same, then no kinetic energy went into internal energy—the collision was elastic. If the kinetic energy of the system changed, then the collision was partially or totally inelastic.

6.8 If we neglect rolling friction and air friction (which are pretty small in this case), there is no work done on Jim by any forces and there are no energy changes (no matter what system we choose). So the power is zero.

6.9 The potential energy of two objects can be negative if the objects exert attractive forces on each other, and we choose their zero energy reference separation to be very far apart—for example, zero potential energy when infinitely far apart. An example is the gravitational potential energy of two objects that attract each other. Positive work is required to pull them far apart where their energy is approximately zero. Thus, they must have started with negative energy.

Chapter 7

7.1 The nail should be along a vertical line that passes through the center of mass of the painting when it is in the correct orientation. The force that Earth exerts on the painting will then not cause the painting to rotate.

7.2 (a) The torque of the force that Earth exerts on an any object about its center of mass is zero, but it is not zero about any other point. Thus when we support an object at the center of mass, it is in equilibrium, but if we put the support at any other location, it tips. (b) Suspend a meter stick and pull it along the direction of the meter stick. (c) Two people sitting on a seesaw so that their torques with respect to the fulcrum are the same in magnitude but opposite in signs.

7.3 We can choose any point as a possible axis of rotation to write the torque condition of equilibrium (because when the object is in equilibrium, it does not rotate about any axis at all, so we can choose the one we want to use). However, it is useful to choose a location for which the largest number of forces has zero torque.

7.4 One way to do it is to put the person face down on a big exercise ball and ask him to keep his muscles tense so that his body remains straight. Then slowly roll the ball under him until it is in a position where the person can balance horizontally on the ball. The ball will be under his center of mass.

7.5 The torque exerted by the muscle needs to balance the torque exerted by the backpack. The straps of the backpack are longer than the muscle; thus the distance from the force exerted by the backpack to the axis of rotation is larger than the distance between the force exerted by the muscle and the axis of rotation. In addition, the trapezius exerts a force at an angle, thus diminishing its rotational effect. Both factors lead to the increase of the magnitude of the force exerted by the muscle to balance the effect of the force exerted by the backpack.

7.6 To answer this question, consider the forces exerted on the ball and on the pencil when in equilibrium (Figures 7.43a and c) and when the equilibrium is disturbed (Figures 7.43b and d). Earth exerts a gravitational force on the string-ball system, which returns the ball toward its stable equilibrium position. The gravitational force exerts a torque on the pencil, which moves the pencil farther away from its unstable equilibrium position.

7.7 The distance between the muscle force exerted on the bone and the axis of rotation is somewhat less than the distance between the force exerted by the load being lifted or supported and the axis of rotation.

Chapter 8

8.1 When an ice skater is rotating faster and faster in a clockwise direction, her rotational velocity ω is negative and rotational acceleration α is negative. When she starts slowing down, the rotational velocity is still negative, but the rotational acceleration becomes positive.

8.2 We tested both ideas in the experiments with the rotating cylinder when we exerted the same magnitude force at different locations for the axis of rotation. If the force determines the magnitude of the rotational acceleration, the outcomes of both experiments would be the same, but this is not what we found. When the same force was exerted farther from the axis of rotation, the acceleration was larger. From these experiments we concluded that it was the torque that affects angular acceleration.

8.3 The wooden ball has more mass farther from the axis of rotation and thus a higher rotational inertia. Thus the same torque will produce a smaller change in motion in the wooden ball.

8.4 Both laws represent a cause-effect relationship between an interaction of an object with external objects, the object's inertial properties, and the change in its motion due to this external interaction. The change in motion (rotational or translational acceleration) is directly proportional to the measure of the interaction (external torque or force) and inversely proportional to the inertial properties of the object (rotational inertia or mass).

8.5 A person jumping on the merry-go-round right before landing had zero rotational velocity. Her landing increases the rotational inertia of the system; thus the rotational speed decreases. A person jumping off is originally moving with the carousel and has rotational momentum. The total rotational momentum of the merry-go-round and the person was the sum of both. When the person steps off, she takes her rotational momentum with her, and the rotational momentum of the carousel alone remains the same.

8.6 The chicken soup can is similar to the bottle filled with water, in which the mass of the water does not rotate as the bottle rolls down the ramp. The clam chowder can is similar to a bottle filled with ice, which rotates as it rolls down. The chicken soup rolls faster!

8.7 Choose the solid Earth and the ocean water as the system but not the tidal bulges that rise due to the Moon's greater gravitational force on water nearer the Moon. The tidal bulges rubbing against water below cause an external friction force that slows Earth's rotation.

Chapter 9

9.1 Moist objects dry when randomly moving liquid particles leave the object's surface.

9.2 The distances by themselves do not tell us much. We need to compare them to the sizes of the particles. As the size of the particles is about 10^{-8} cm, the average 3×10^{-7}-cm distance between particles is about 30 times larger than their sizes.

9.3 We can think of the $m_p v^2$ as the product of two terms—as $m_p v$ times v. If the momentum $m_p v$ is greater, then the particles exert a greater force when hitting the wall. A greater force per collision causes greater pressure. Finally, if the particles move back and forth faster (larger v), they have more frequent collisions, leading to greater pressure. So the equation seems reasonable.

9.4 Assuming the same particle temperature (the same average kinetic energy $1/2 m_p v^2$ of the molecules), the less massive molecules (smaller m_p) will have higher average speed (greater v^2).

9.5 The right side of the first equation involves microscopic quantities that are not measured directly, while the right side of the second equation involves macroscopic quantities that can be measured directly.

9.6 Stern's experiment tested the quantitative relations of the theory. Stern had the clear goal of designing an experiment to see if the particle speeds matched the predictions based on the ideal gas model and Newton's laws.

9.7 When the mass of the gas is constant and one of the following parameters (P, V, or T) stays constant, then one can use gas laws to describe the behavior of the gas.

9.8 We calculated that the time during which the Sun has emitted thermal energy is about 40 million years. The age of Earth is about 4.5 billion years. Thus, the Sun must have some other source of energy that would allow it to have emitted light for at least 4.5 billion years.

Chapter 10

10.1 Determine the mass of the object in kilograms using a scale. Then submerge the object in water in a graduated cylinder and note the change in the water level. Use the volume of the displaced water to determine the volume of the object in cubic meters. Divide mass by volume to find density.

10.2 When you squeeze the closed end of the tube, you exert an additional pressure that is transferred uniformly in all directions.

10.3 The pressure differs at different heights because liquid layers at lower elevations support the liquid above it. An increased pressure in one part of the liquid (for example, by a plunger) causes a uniform increase throughout the liquid. At the same height, the pressure is the same in all directions—up, down, right, and left.

10.4 It means that a column of mercury of density $\rho = 13{,}600 \text{ kg/m}^3$ and height $h = 760$ mm exerts the same pressure $P = \rho g h$ as Earth's atmosphere (about $1.0 \times 10^5 \text{ N/m}^2$).

10.5 The pressure that the fluid exerts on the walls of a container increases as the depth of the fluid increases. When an object is submerged in a fluid, the upward pressure of the fluid on the bottom surface of the object is greater than the downward fluid pressure on the top of the object.

10.6 The forces are the same and are independent of the mass of the object. The forces are determined by the volume of the displaced fluid.

10.7 You can measure how much water the ship displaces if it submerges to the waterline. The mass of this water will be exactly equal to the mass of cargo that you can put on the ship for it to submerge to the waterline level.

Chapter 11

11.1 The lightbulbs came together because the pressure of the air in between them is lower than the pressure outside. Thus the outside pressure pushes them closer together.

11.2 The cross-sectional area of the river before the outlet is much greater than the area of the cross section at the outlet. Water flows at higher speed through the outlet with the smaller cross-sectional area.

11.3 In most cases we would prefer streamline flow. There is less friction-like resistance to this flow and less pressure needed to cause the flow—the heart does not have to work as hard.

11.4 They are both based on the work-energy equation. However, the work-energy charts apply mainly to processes involving solid objects and the Bernoulli charts to fluid processes. The latter involve energy densities in the fluids and pressures that cause the fluid energy density to change. They also apply to the pressure and energy densities at particular positions in the fluid, whereas the work-energy bar charts apply to initial and final situations in a process.

11.5 The water is open to the air pressure at both levels. We neglect the change in the atmospheric pressure with height because the bottle is very small compared to the height of the atmosphere.

11.6 The blood flows more easily; thus your blood pressure can be lower as your heart needs to do less work in pumping blood. You may also have fewer heartbeats per minute.

11.7 The air pressure above is reduced because the air's kinetic energy density is greater and the pressure is lower. Thus, the normal pressure from inside the house pushes the roof up and off the house.

11.8 For an object to move at constant velocity with respect to an inertial reference frame (in this case, Earth), the magnitudes of the forces exerted on it in opposite directions should be equal. In the case of skydiving, the net force is zero and the diver moves down at constant terminal speed. Note that we are neglecting the buoyant force exerted by the air on the diver. If we take the buoyant force into account, the sum of the upward drag and buoyant forces should exactly equal the magnitude of the gravitational force that Earth exerts downward.

Chapter 12

12.1 $W = -P_{\text{atm}} \Delta V$

12.2 About 1674 Joules.

12.3 The work-energy equation allows us to find the final energy of a system from knowing its initial energy and the work done on the system. The first law of thermodynamics also tells us how to include the energy transferred to the system through heating.

12.4 When a hot aluminum or iron block is added to cool water of the same mass, the water and block reach the same final temperature, which is much closer to the initial temperature of the water. Because of its much greater specific heat, the water temperature changes much less.

12.5 Imagine that the same amount of energy is provided by heating a gas at constant volume and then at constant pressure. At constant volume, the gas reaches a higher temperature than at constant pressure, because the environment does negative work on the gas while the gas expands. Because of this negative work, the change in the internal energy of the gas in the latter case is less than in the former; hence the change in the temperature is less.

12.6 Specific heat characterizes how much energy should be supplied to a unit mass of a substance to change its temperature by $1 \, ^\circ\text{C}$ while heat of fusion or vaporization characterizes energy supplied at constant temperatures to a unit mass of a substance to change its state. The unit of specific heat is $\text{J/kg} \cdot {}^\circ\text{C}$. The unit of heat of fusion or vaporization is J/kg.

12.7 Conduction is efficient in transferring energy from atom to atom or molecule to molecule in solids with close contact between neighboring particles. Convection is very efficient in transferring energy through liquids and gases if the warm parts of the liquid or gas move toward cooler parts. Radiation is the best way to transfer energy through a vacuum, in which there are no particles to transfer energy by interactions or by moving from one place to another.

12.8 Carbon dioxide absorbs long-wavelength infrared radiation but does not absorb much short-wavelength infrared radiation and visible light. This means that carbon dioxide does not reduce energy Earth receives from the Sun, but it reduces energy emitted by Earth. This reduces the cooling rate of Earth and its atmosphere and thus contributes to the increase of Earth's temperature.

Chapter 13

13.1 The gasoline molecules are relatively large and contain considerable chemical potential energy in their bonds. When they burn with oxygen, chemical products are formed that have somewhat less chemical energy. Part of the energy released is converted by the car's engine into kinetic energy, and about 80% is converted into thermal energy. Thus, the final energy is much less organized, even though a relatively small part is converted to organized kinetic energy.

13.2 The lightbulb converts electrical energy into some light and considerable thermal energy (energy is conserved). The final state of the energy is more disordered with greater entropy. Energy was conserved, but entropy increased.

13.3 Entropy is a state function; the equation Q/T characterizes the change in entropy, not the entropy itself.

13.4 The second law is formulated for isolated systems. In a refrigerator or air conditioner, there is an external object (a pump) that does work on the system.

13.5 The engine efficiency is limited by a second law of thermodynamics expression that depends on the difference in temperature between the hot and cold heating reservoirs. In addition, there are other factors that limit the efficiency, such as friction in moving parts (generators and compressors) and burning of the fuel used for the hot reservoir.

Answers to Select Odd-Numbered Problems

Chapter 1

Multiple-Choice Questions

1. (c)   3. (b)   5. (b)   7. (d)   9. (b)   11. (c)   13. (b)   15. (c)

Problems

1.

11. (a) 36 mph $=$ 57.9 km/h $=$ 16.1 m/s;
(b) 349 km/h $=$ 217 mph $=$ 96.9 m/s;
(c) 980 m/s $=$ 3528 km/h $=$ 2192 mph
13. (a) 60 mph
17. (1.4 ± 0.1) km
19. 3.99×10^{16} m; uncertainty: 0.009×10^{16} m
21. (a) Gabriele: 160 m, Xena: 120 m; (b) 30 s;
(c) 60 m $-$ $(2.0 \text{ m/s})t$
25. (a) Gabriele: 3.0×10^3 m $-$ $(8.0 \text{ m/s})t$;
Xena: 1.0×10^3 m $+$ $(6.0 \text{ m/s})t$; (b) 1.9 km;
(c) -2.0×10^3 m $+$ $(14 \text{ m/s})t$
27. 3.6 mph
29. impossible to reach this average speed
31. (a) $x_1(t) = 30$ m $+ (-8.33 \text{ m/s})t$,
$x_2(t) = -10$ m, $x_3(t) = -10$ m $+ (5.0 \text{ m/s})t$,
$x_4(t) = -10$ m $+ (-3.33 \text{ m/s})t$
37. (a)

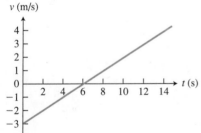

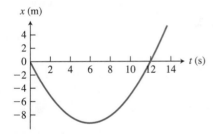

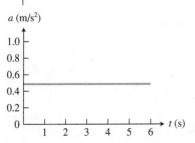

39. -154 m/s^2
41. $57.2 \text{ m/s}^2; -9.2 \text{ m/s}^2$
43. The distance between the runners will increase.

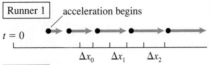

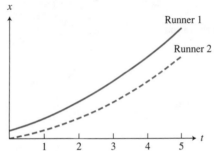

45. $4.0 \times 10^3 \text{ m/s}^2$, 2.0 mm
49. (a) 5.6 m/s^2; (b) 11.2 m; (d) 7.93 s;
(e) 9.9 s, the uncertainty is 0.1 s
51. 42 m
53. yes, $a = 2.0 \text{ m/s}^2$
55. $x_A(t) = 200$ m $- (20 \text{ m/s})t$, $x_B(t) = -200$ m $+ (10 \text{ m/s})t$
57. (a) constant acceleration;
(b) $x_0 = -100$ m, $v_0 = +30$ m/s, $a = 6.0 \text{ m/s}^2$;
(d)

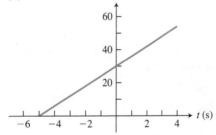

(e) $v(t) = 30$ m/s $+ (6.0 \text{ m/s}^2)t$, $v = 0$ at $t = -5.0$ s
59. 2.2 s, 1.2 m

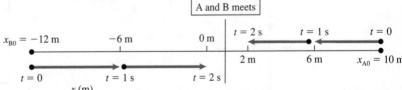

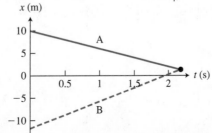

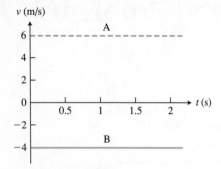

61. (b) 83.3 m; (c) 14 m/s
63. (a) 3.5 s; (b) 1.0 s
65. (a) 1.75 s; (b) 17.1 m/s
69. 9.9 m/s, 2.02 s
71. 0.156 s
73. 5.33 m/s
75. 22.2 m/s
79. 5.0 m (two floors) above you
81. 24.7 m/s^2
85. (c)
87. (a)
89. (d)
91. (a)
93. (c)

Chapter 2

Multiple-Choice Questions

1. (c)   3. (a)   5. (c)   7. (d)   9. (a)   11. (a)
13. (b)   15. (c)   17. (d)

Problems

1. (1) (d); (2) (a); (3) (d); (4) (b); (5) (a)
3. (a)

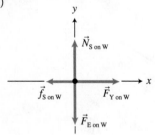

(b)

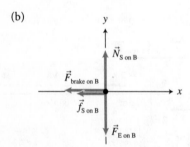

(c)

5. (a)

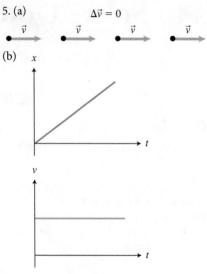

(b)

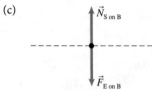

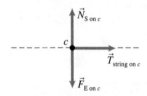

(c)

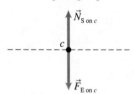

7. (a) After the pulling stops, the cart continues to move with constant velocity, in accordance with Newton's first law.
(b)

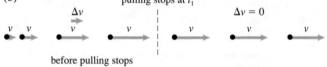

before pulling stops

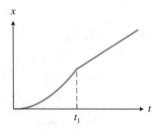

after pulling stops

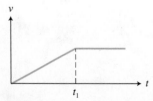

9. (a) After the applied force has been reduced by half, the cart will continue to accelerate. If the initial acceleration is a_0, then the subsequent acceleration would be $a_0/2$.

(b)

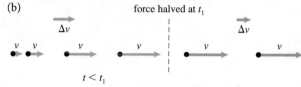
force halved at t_1

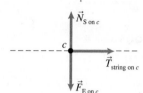
$t < t_1$

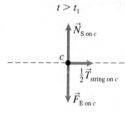
$t > t_1$

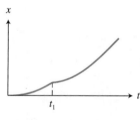

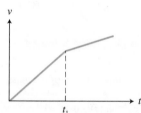

11. Both force diagrams are the same in each case.

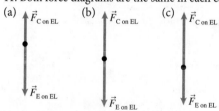

15. All force diagrams are the same.

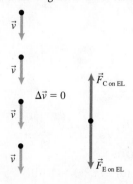

21. 10 N
29. (a) 490 N; (b) 590 N; (c) 390 N; (d) 0
31. 168 N
37. 33 N
39. (a) 1.0 N, upward; (b) Earth acceleration: 1.67×10^{-25} m/s^2
45. (b) 0.67 m
47. (a) 0.9 N; (b) 30%
49. -1.55×10^3 N
55. (d)
57. (b)
59. (c)
61. (b)
63. (a)
65. (d)

Chapter 3

Multiple-Choice Questions

1. (c)   3. (b)   5. (b)   7. (a)   9. (a)   11. (b)

Problems

1. $N_{\text{S on C}x} = 0$, $N_{\text{S on C}y} = 250$ N, $F_{\text{E on C}x} = 0$, $F_{\text{E on C}y} = -150$ N, $F_{\text{P on C}x} = 173$ N, $F_{\text{P on C}y} = -100$ N
3. $A_x = -7.07$ m, $A_y = 7.07$ m, $B_x = 5.0$ m, $B_y = 0$
5. (a) 223.6 N, 153°

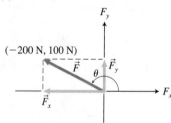
$(-200$ N, 100 N$)$

(b) 500 N, 53°

$(300$ N, 400 N$)$

(c) 500 N, 37° below $+ x$ axis

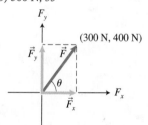
$(400$ N, -300 N$)$

7. $T_{2 \text{ on K}} = 500$ N, $T_{3 \text{ on K}} = 300$ N
9. (a) $\Sigma F_{\text{on W}x} = T_{\text{R on W}} \cos \theta - f_{\text{S on W}} = m_{\text{W}} a_x$,
$\Sigma F_{\text{on W}y} = T_{\text{R on W}} \sin \theta + N_{\text{S on W}} - F_{\text{E on W}} = 0$;
(b) $\Sigma F_{\text{on B}x} = -f_{\text{S on B}} = m_{\text{B}} a_x$,
$\Sigma F_{\text{on B}y} = N_{\text{S on B}} - F_{\text{E on B}} = m_{\text{B}} a_y = 0$;
(c) $\Sigma F_{\text{on Y}x} = F_{\text{E on Y}} \sin \theta - f_{\text{S on Y}} = m a_x$,
$\Sigma F_{\text{on Y}y} = N_{\text{S on Y}} - F_{\text{E on Y}} \cos \theta = m_{\text{W}} a_y = 0$;

(d) $\sum F_{\text{on B}\,y} = F_{\text{Y on B}} - F_{\text{E on B}} = m_{\text{B}}a_y = 0$;

(e) m_1: $\sum F_{\text{on m1}\,x} = T_{\text{R on m1}} - T_{\text{m2 on m1}} - f_{\text{S on m1}} = m_1 a_x$,

$\sum F_{\text{on m1}\,y} = N_{\text{S on m1}} - F_{\text{E on m1}} = m_1 a_y = 0$,

m_2: $\sum F_{\text{on m2}\,x} = T_{\text{m1 on m2}} - f_{\text{S on m2}} = m_2 a_x$,

$\sum F_{\text{on m2}\,y} = N_{\text{S on m2}} - F_{\text{E on m2}} = m_2 a_y = 0$

11. (a) $\sum F_{\text{on B}\,y} = F_{\text{Y on B}} - F_{\text{E on B}} = 0$;

(b) $\sum F_{\text{on sled}\,x} = T_{\text{R on sled}} \cos\theta - f_{\text{S on sled}} = m_{\text{sled}}a_x$,

$\sum F_{\text{on sled}\,y} = T_{\text{R on W}} \sin\theta + N_{\text{S on sled}} - F_{\text{E on sled}} = m_{\text{sled}}a_y = 0$;

(c) $\sum F_{\text{on sled}\,x} = F_{\text{R on sled}} - F_{\text{E on sled}} \sin\theta - f_{\text{S on sled}} = m_{\text{sled}}a_x$,

$\sum F_{\text{on sled}\,y} = N_{\text{S on sled}} - F_{\text{E on sled}} \cos\theta = m_{\text{sled}}a_y = 0$;

(d) $\sum F_{\text{on SD}\,y} = F_{\text{air on SD}} - F_{\text{E on SD}} = m_{\text{SD}}a_y = 0$

13. $\sum F_{\text{on O}\,x} = N_{\text{R2 on O}\,x} + F_{\text{E on O}\,x} + F_{\text{R1 on O}\,x} = 0 + 0 + (30\ \text{N}) \cos 245° = -12.7\ \text{N}$,

$\sum F_{\text{on O}\,y} = N_{\text{R2 on O}\,y} + F_{\text{E on O}\,y} + F_{\text{R1 on O}\,y}$
$= 40\ \text{N} + (-10\ \text{N}) + (30\ \text{N}) \sin 245° = 2.8\ \text{N}$

19. 5.94 m/s

23. 101.5 N

25. 1350 N, 1690 N

27. 150.3 N, 60 N

31. 1.73 m/s^2; 242 N; 3.4 s

33. 0.67 m/s^2; 104.7 N

39. 0.41

41. 0.76, static

45. g/μ_s

47. 98.8 N, $\mu_k = 0.68$

49. 0.84

51. 2.43 m/s^2

55. $m_2 = \mu_k m_1$

57. On table

In air

59.

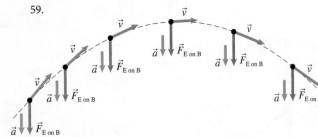

61. 44.6 m/s

63. 57.1°, 23.0 m/s

65. (a) from a horizontal distance of 723 m; (b) later

67. Projectile fired at $\theta_2 = 60°$ has longer flight time, and hence, greater air resistance.

69. 4.9 m

71. (a) Frictional force by road surface on tires propels minivan to move forward; (b) 3420 N

75. 6.13 m/s^2

77. $a = 0$, no motion

81. 5.35 s

89. (c)

91. (a)

93. (b)

95. (e)

97. (a)

Chapter 4

Multiple-Choice Questions

1. (d)   3. (b)   5. (a)   7. (a)   9. (a)   11. (b)   13. (a)

Problems

1. bottom: $mg + mv^2/r$; top: $mg - mv^2/r$

5. 5.9×10^{-3} m/s^2

7. 92.2 m/s

9. 2.20 m/s^2, or 0.22 g

13. person at outermost radius

15.

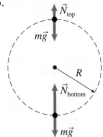

19. 138 N

21. 2.70 m/s

25. 214 N

27. 9.3 m/s

29. 90 N, radially inward

31. 9000 N

35. (a) $v = \sqrt{rg\tan\theta}$; (b) Speed is independent of the mass of the object.

39. (a) 4.3×10^{20} N; (b) 2.0×10^{20} N; (c) 2.0×10^{20} N

41. 200 N

43. 27.4 days

45. 37.6% of what it is on Earth

47. 1.8×10^{27} kg

49. 1.67 h

51. 2.0×10^7 m above Earth's surface

53. 45 s

55. 1764 N

65. 1.4 h

67. 1.7 km/s

69. (c)

71. (b)

73. (c)

75. (a)

77. (a)

79. (b)

Chapter 5

Multiple-Choice Questions

1. (b)   3. (e)   5. (a)   7. (c)   9. (d)   11. (b)   13. (d)

Problems

1. (a) 1.71 kg·m/s; (b) 5.34 m/s; (c) 298.3 kg·m/s
3. the ball that rebounds
7. $v_2 = -\dfrac{10v}{7}$
11. 0.313 N·s, 0.13 N·s
15. (a) 4.0×10^5 N; (b) 100 s
19. 160 N
21. 0.22 N·s, 22 N
23. (a) 25%; (b) −10.6%
31. $p_{Bix} + J_{W \text{ on } Bx} = p_{Bfx}$
Earth is the object of reference.

$p_{Bix} + J_{W \text{ on } Bx} = p_{Bfx}$

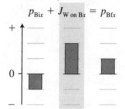

33.

$p_{Hiy} + J_{G \text{ on } Hy} + J_{E \text{ on } Hy} = p_{Hfy}$

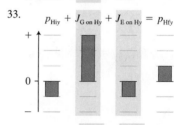

$p_{Siy} + J_{G \text{ on } Sy} + J_{E \text{ on } Sy} = p_{Sfy}$

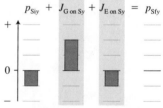

37. (a) 6.12 kg·m/s upward; (b) 56.9 N
39. 43.8 m/s
41. −0.19 m/s
43. 1.33 m/s
47. 165 km/h
49. 2×10^4 N
53. (a) 7500 N; (b) yes; (c) 0.167 s
57. -1.4×10^7 m/s
59. 8.27 m/s, 14.7° north of east
61. 56.3° north of east, 6.18 m/s
63. (a) 0.75 m/s; (b) 13.3 s
65. (a)

$p_{Rix} + p_{Fix} + J_x = p_{Rfx} + p_{Ffx}$

(b)

$p_{Rix} + p_{Fix} + J_x = p_{Rfx} + p_{Ffx}$

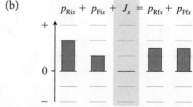

(c)

$p_{Rix} + J_x = p_{Rfx}$

$p_{Rix} + J_x = p_{Rfx}$

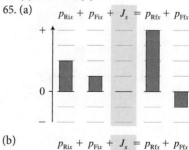

67. 1×10^{-14} m/s
77. $\tan \theta = \dfrac{m_1 v_1}{m_2 v_2}, \ d = \dfrac{1}{2\mu_k g} \dfrac{m_1^2 v_1^2 + m_2^2 v_2^2}{(m_1 + m_2)^2}$
81. (d)
83. (e)
85. (b)
87. (c)
89. (e)
91. (c)

Chapter 6

Multiple-Choice Questions

1. (b)   3. (c)   5. (b)   7. (b)   9. (e)   11. (c)

Problems

1. 540 J
3. (a) 7500 J; (b) −7500 J; c) 7500 J
5. lifting 196 J, carrying 0, setting down −196 J, total 0
7. 6.2 m/s
9. (a) $\dfrac{K_P}{K_C} = \dfrac{m_C}{m_P} = \dfrac{1100 \text{ kg}}{2268 \text{ kg}} = 0.485$; (b) $K_P/K_C = 1$
11. yes
15. (a) 3×10^9 J; (b) 5×10^6 m
17. 6×10^{-4} m
19. (a) 720 N; (b) 21.6 J; (c) −16.2 J; (d) −5.4 J
21. $k = 8.9 \times 10^5$ N/m
23. $k = 330$ N/m
25. 4.4 m/s
27. $v_f = \sqrt{v_i^2 + 2gh - \dfrac{2fl}{m}}$
29. 6.4 m/s, 46 m, increase if the friction force remains the same
33. 8.4 m/s
43. $v_f = 22$ m/s
45. (b) decrease of 33.3%
47. (a) 0.635 m/s; (b) 0.28 m
49. 222 m/s
51. 36 kg, −55 J
53. (a) 90 m/s; (b) 540 J; (c) Falcons have strong feet that can deliver a greater force (impulse) to strike their prey.

55. $v_{1f} = \dfrac{m_1 - m_2}{m_1 + m_2}v, \quad v_{2f} = \dfrac{2m_1}{m_1 + m_2}v$

57. (a) 9.8 W; (b) 196 W; (c) 196 W; (d) 196 W

59. (a) 1.15×10^7 J; (b) 47 s

61. (a) 4.34×10^5 J; (b) 60.3 W

65. 1 hp

67.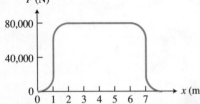

69. (a) 1; (b) 4; (c) more hydrogen molecules can attain speeds higher than v_{esc}

71. 4.21×10^4 m/s

77. (a) 11 m/s; (b) 1.8 mg

79. 690 N

81. (a) 3.3×10^{-6} m/s; (b) 1.1×10^{20} N;
(c) 1.1×10^{23} J $= 1.7 \times 10^9$ atomic bombs

85. (c) and (d)

87. (c)

89. (b)

91. (a)

93. (d)

95. (b)

Chapter 7

Multiple-Choice Questions

1. (b)   3. (a)   5. (a)   7. (d)   9. (c)

Problems

1. $\tau_1 = 0, \tau_2 = -183.85$ N $\cdot$ m, $\tau_3 = +240$ N $\cdot$ m, $\tau_4 = -240$ N $\cdot$ m

3. ± 10.2 N $\cdot$ m

7. 218 N, 113°

9. $T_1 = 11.8$ N, $T_2 = 19.5$ N, $T_3 = 15.6$ N

11. $m = 7.82$ kg, $T_1 = 77$ N, $T_3 = 64$ N

13. 3250 N, 15° above horizontal

15. (a)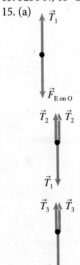

(b) $T_1 = 9800$ N, $T_2 = 4900$ N, $T_3 = 2450$ N, $T_4 = 4900$ N

17. 10.2°

19. (a) 823 N $\cdot$ m; (b) 1.56 m

21. 0.56 m from the center, towards Tahreen

23. 250 N

27. left 617 N, right 274 N

29. 0.77 m

31. (a) 2.1 m; (b) It did not move.

33. $-\dfrac{r^2 a}{R^2 - r^2}$

37. $x_{cm} = \dfrac{m_b b/2}{m_a + m_b}, y_{cm} = \dfrac{m_a a/2}{m_a + m_b}$

39. 550 N

41. 1250 N, 1050 N

43. 220 lb, 200 lb

45. 171 N

47. $T = 37.3$ N, $F_{\text{H on B}\,x} = 37.3$ N, $F_{\text{H on B}\,y} = 39.2$ N, $F_{\text{H on B}} = 54.1$ N, $\phi = 46.4°$

49. 980 N

53. 35 cm from the clay

57. (a) $T = 1570$ N, $F = 1620$ N

59. 2240 N

61. 1650 N, 1630 N

67. 5.0 m up the ladder

71. (a)

73. (c)

75. (a)

77. (b) or (c)

79. (e)

Chapter 8

Multiple-Choice Questions

1. (a)   3. (b)   5. (d)   7. (b)   9. (d)

Problems

1. (a) 0.105 rad/s; (b) 0.021 m/s; (c) zero

3. (a) -0.058 rad/s^2; (b) 0.231 rad/s^2;
(c) 0.625 rad/s^2; (d) 0.094 m/s^2

5. 8.38 s

7. 300 rad/s

9. (a) 200 m/s^2; (b) 133 rad/s^2

13. 8300

15. (a) $\alpha = \dfrac{\omega^2}{4\pi}$; (b) $\Delta t_1 = \dfrac{4\pi}{\omega}$; (c) $\Delta t_2 - \Delta t_1 = \dfrac{4\pi}{\omega}(\sqrt{2} - 1)$;
(d) $\Delta s = 4\pi l$

17. -1.2×10^{-14} rad/s^2

19. 0.040 N

21. 205 N $\cdot$ m

23. (a) $\dfrac{T_2}{T_1} = \dfrac{4}{3}$; (b) $T_2 > \dfrac{4}{3}T_1$; (c) $T_2 < \dfrac{4}{3}T_1$

25. (a) $\alpha = 4a$; (b) $\omega = 16a$

29. 8m

31. 5.9×10^4 N $\cdot$ m

33. 1.9 s

35. $1.2 \text{ kg} \cdot \text{m}^2$

37. -780 N

39. (a)

(b) $mg - T = ma$; (c) $rT = I\alpha$;
(d) $a = 3.27 \text{ m/s}^2$, $T = 196$ N

41. (a)

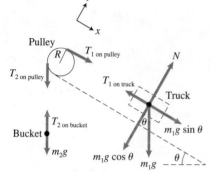

(b) $T_{1 \text{ on truck}} = m_1 g \sin \theta$,
$T_{2 \text{ on bucket}} = m_2 g$; (c) the same

43. $3.2 \text{ kg} \cdot \text{m}^2$

47. $7.8 \times 10^6 \text{ s} \approx 90$ days

49. $\omega = \sqrt{\dfrac{2K}{I}}$

51. 96%

53. (a) 12.3 rad/s; (b) -1.74×10^{-3} J; (c) Negative work is done as the beetle moves from the center to the edge.

55. 16.8 J

59. (a) 1.78 rad/s; (b) -35.6 J; (c) -67.2 J; (d) $+31.6$ J;
(e) $\Delta K = \Delta K_b + \Delta K_m$

61. (a) 1.67 rad/s; (b) -80 J; (c) -97.8 J; (d) $+17.8$ J;
(e) $\Delta K = \Delta K_b + \Delta K_m$

63. (a) 1.28 J; (b) 9.23 rad/s; (c) 1.38 m/s; (d) -0.43 J; (e) $+0.19$ J;
(f) -0.24 J

65. 142 N

67. 7.7×10^3 N

71. (a) it does not move; (b) to the left; (c) to the right

73. 0.035 rad/s

75. (a)

77. (c)

79. (e)

81. (c)

83. (b)

Chapter 9

Multiple-Choice Questions

1. (c)   3. (c)   5. (d)   7. (c)   9. (c)   11. (b)   13. (b)   15. (d)

Problems

3. 8.4×10^{56}

5. (a) 2.68×10^{25} molecules/m^3; (b) $D/d \approx 30$; (c) yes

7. 3.0×10^{-26} kg, 4.8×10^{-26} kg

9. 6.2×10^{-3} kg

11. (a) 4.8 N; (b) 1.6 N/m^2

15. 440 m/s

17. $P_1 = 2P_2$

21. 198 °F, atmospheric pressure is lower

23. 254 °C

25. 5.65×10^{-21} J

27. 0.041 mol

29. $v_{1f} = v_{2i}$, $v_{2f} = v_{1i}$

31. 7.2 cm^3

33. $1.4 \times 10^{-16} \text{ N/m}^2$

43. 1.1 m

45. about 3 times more frequently

49. decrease the absolute temperature by a factor of 1.5

51. $6.6 \times 10^6 \text{ N/m}^2$

57. (a) 2.1×10^{-18} J; (b) $6.2 \times 10^{12} \text{ s} = 2.0 \times 10^5$ years

59. $U = \dfrac{3mkT}{2M}$, $P_2 = \frac{1}{2}P_1$, temperature and thermal energy are unchanged

67. $2.6 \times 10^{21} \text{ O}_2$ molecules per breath

71. 68.5 °C

73. (d)

75. (a)

77. (e)

79. (b)

81. (b)

Chapter 10

Multiple-Choice Questions

1. (c)   3. (c)   5. (d)   7. (b)   9. (a)   11. (c)   13. (a)
15. (b)   17. (c)   19. (b)

Problems

1. 0.047 m^3

3. 7850 m

5. 6.9% decrease

7. Density of oil is 900 kg/m^3.

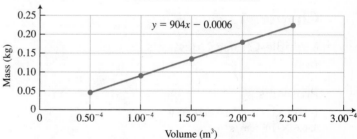

Oil mass as a function of volume

9. (a) $\rho_A = \rho_B = \rho_C$ (b) $V_A = a^3$, $V_B = (2a)^3 = 8a^3$, and
$V_C = (3a)^3 = 27a^3$. Thus, $V_A < V_B < V_C$.
(c) $S_A = 6a^2$, $S_B = 6(2a)^2 = 24a^2$, and $S_C = 6(3a)^2 = 54a^2$.
Thus, $S_A < S_B < S_C$.
(d) $A_A = a^2$, $A_B = (2a)^2 = 4a^2$, and $A_C = (3a)^2 = 9a^2$.
Thus, $A_A < A_B < A_C$.
(e) $m_A = \rho V_A = \rho a^3$, $m_B = \rho V_B = 8\rho a^3$, and
$m_C = \rho V_C = 27\rho a^3$. Thus, $m_A < m_B < m_C$.

11. 200 kg/m^3

13. 2300 kg/m^3

17. both are wrong; 6×10^3 N

19. 8.0×10^{-3} m^2

21. $\dfrac{A_1}{A_2} = 0.034$; 8.82 m

25. 2.32×10^3 N

29. $P_A = P_B = P_C; F_A > F_C > F_B$
31. $5.1 \times 10^5 \, \text{N/m}^2$
35. $-2.2 \times 10^4 \, \text{N/m}^2$
39. 0.336 m
43. 14.4 cm
47. 10.74 m
51. Liquid B has higher density than liquid A.
57. The upward buoyant force exerted by the water on the brick makes it feel lighter when you hold it.

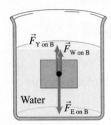

61. 0.774 N
63. The kerosene side will tilt down, and the water side will tilt up.
65. $400 \, \text{kg/m}^3$
69. 582.6 N; The person will sink.
71. The pressure will not change.
75. 200 kg
77. $\frac{1}{4}(\rho_{\text{water}} - \rho_{\text{log}})\pi d^2 L$
79. 92%
85. $8.6 \times 10^5 \, \text{N}$
91. $1.05 \times 10^4 \text{m}^2$
93. $V_2 = \frac{5}{8} V_1$
95. (d)
97. (c)
99. (d)
101. (a)
103. (c)
105. (d)

Chapter 11

Multiple-Choice Questions

1. (b)   3. (a)   5. (c)   7. (b)   9. (c)   11. (a)

Problems

1. (a) $1.45 \times 10^{-4} \, \text{m}^3/\text{s}$; (b) 0.46 m/s
5. $6.6 \times 10^{-3} \text{cm/s}$
7.

$$K_1 + P_1 = P_2 + K_2 + U_{g2}$$

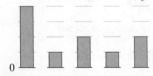

$$P_1 + K_1 + U_{g1} = P_2 + K_2 + U_{g2} \text{ or}$$

$$P_1 + \frac{1}{2}\rho v_1^2 + 0 = P_2 + \frac{1}{2}\rho v_2^2 + \rho g y_2$$

17.

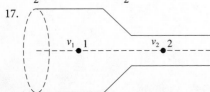

$$K_1 + U_{g1} + (P_1 - P_2) = K_2 + U_{g2}$$

$$P_1 + \frac{1}{2}\rho v_1^2 = P_2 + \frac{1}{2}\rho v_2^2$$

19. (a) 5.24 m/s; (b) 1.41 cm; (c) The speed remains unchanged.
21. (a) 4.77 m/s; (b) 5.3 m/s; (c) $4.95 \times 10^5 \text{N/m}^2$
25. 0.0041 m
27. $9.22 \times 10^3 \, \text{N/m}^2$
31. 0.50 cm
33. 46 m/s
35. $\frac{1}{32} = 0.03125$
39. (a) 630 N; (b) $1.0 \times 10^3 \, \text{N}$
41. $1.9 \times 10^{-12} \text{N}$
43. $2.0 \times 10^{-5} \, \text{m}$
45. $v_{\text{terminal}} = \sqrt{\dfrac{mg}{0.03}} = 18.1\sqrt{m}$
53. $33 \, \text{m}^2$
59. (c)
61. (d)
63. (c)
65. (c)
67. (c)
69. (a)

Chapter 12

Multiple-Choice Questions

1. (b), (d)   3. (e)   5. (b)   7. (c)   9. (c)

Problems

3. (a) zero; (b) The temperature remains the same, the density and pressure each decrease by half, the average kinetic energy and the thermal energy remain the same.
5. (a) -1300 J; (b) 1.23 mol; (c) 208 K
9. (a) 2.1×10^4 J; (b) 1.2×10^4 J; (c) 2.3×10^3 J
11. (b) $\dfrac{v_i^2}{4(130 \, \text{J/kg} \cdot {}^\circ\text{C})}$
13. $6.1 \, {}^\circ\text{C}$
15. $2.1 \times 10^5 \, \text{J}$
17. $75.2 \, {}^\circ\text{C}$
19. The material is likely to be iron.
29. $13 \, {}^\circ\text{C}$
31. no; 0.33 kg
33. 0.011 kg
35. $1.57 \times 10^4 \text{s} = 4.36 \, \text{h}$
37. (a) $2.26 \times 10^5 \, \text{J}$; (b) $8.54 \times 10^4 \, \text{J}$
39. steam at $1 \times 10^5 \, {}^\circ\text{C}$
41. (a) 443 kg/s; (b) $3.94 \times 10^7 \, \text{kg/day}$
43. (a) $-410 \, \text{Btu/h} = -120 \, \text{W}$; (b) $-252 \, \text{Btu/h} = -73.8 \, \text{W}$; (c) $-384 \, \text{Btu/h} = -113 \, \text{W}$
45. $-$ (negative); You will perspire until your body temperature becomes normal.
47. 16 J/s
49. 2.0 kg
51. $-11 \, {}^\circ\text{C}$
53. 84 W
59. $1.33 \times 10^{-4} \, \text{kg/s}$

61. (a) 1.1×10^{23} J; (b) 9.8×10^{16} kg; (c) 1.1×10^{14} m^3;
(d) up to 0.27 m
63. 5.8 kW
65. 0.55 kW
73. (b)
75. (b)
77. (d)
79. (b)
81. (c)
83. (d)

Chapter 13

Multiple-Choice Questions

1. (a)   3. (c)   5. (b)   7. (c)   9. (a)   11. (d)   13. (c)

Problems

5. 4.0 °C; irreversible
7. (a) (6, 0), (5, 1), (4, 2), (3, 3), (2, 4), (1, 5), (0, 6)
(b) and (c)

Macrostate (n_L, n_R)	$W_{n_L} = W(n_L, n_R) = \dfrac{n!}{n_L! n_R!}$	$S_i = k \ln W_i$
(0, 6)	$W_0 = W(0, 6) = \dfrac{6!}{0!6!} = 1$	0
(1, 5)	$W_1 = W(1, 5) = \dfrac{6!}{1!5!} = 6$	$k \ln 6 = 1.79k$
(2, 4)	$W_2 = W(2, 4) = \dfrac{6!}{2!4!} = 15$	$k \ln 15 = 2.71k$
(3, 3)	$W_3 = W(3, 3) = \dfrac{6!}{3!3!} = 20$	$k \ln 20 = 3.00k$
(4, 2)	$W_4 = W(4, 2) = \dfrac{6!}{4!2!} = 15$	$k \ln 15 = 2.71k$
(5, 1)	$W_5 = W(5, 1) = \dfrac{6!}{5!1!} = 6$	$k \ln 6 = 1.79k$
(6, 0)	$W_6 = W(6, 0) = \dfrac{6!}{6!0!} = 1$	0

9. (a) $\dfrac{1}{8} = 0.125$; (b) $\dfrac{1}{64} = 0.016$
11. (a) 972.4; (b) 25.2; (c) The ratio will be very large.
13. (a)

n	2	3	4	5	6	7	8	9	10	11	12
W_n	1	2	3	4	5	6	5	4	3	2	1

(b) $n = 7$; (c) $n = 2$ and $n = 12$
17. (a) 52.5 °C; (b) +7.7 J/K
19. $+1.94 \times 10^4$ J/K
21. $+0.853$ J/K
23. (a) 0.50; (b) 0.19; (c) 0.047
25. (a) 0.648; (b) 1.54×10^9 W
27. (a) $+2.7 \times 10^5$ J; (b) -9.0×10^4 J; (c) $+9.0 \times 10^4$ J;
(d) $T_1 = 1.4 \times 10^4$ K, $T_2 = 5.8 \times 10^4$ K, $T_3 = 7.2 \times 10^3$ K;
(e) $U_{\text{thermal 1}} = 1.8 \times 10^5$ J, $U_{\text{thermal 2}} = 7.2 \times 10^5$ J,
$U_{\text{thermal 3}} = 9.0 \times 10^4$ J, $\Delta U_A = 5.4 \times 10^5$ J,
$\Delta U_B = -6.3 \times 10^5$ J, $\Delta U_C = 9.0 \times 10^4$ J;
(f) $Q_A = +9.0 \times 10^5$ J, $Q_B = -9.0 \times 10^5$ J, $Q_C = +9.0 \times 10^4$ J;
(g) 0.091
29. (a) 6.71×10^4 J; (b) 4.9×10^3 J; (c) 7.2×10^4 J
31. 0.36, 1800 J
33. 0.42, 700 MW
37. (b)
39. (a)
41. (d)

Credits

Introducing Physics

Opener: Reuters/Sergei Ilnitsky/Pool; p. xxxiv: John Reader/Photo Researchers, Inc.; p. xxxv: Gary Yim/Shutterstock; p. xxxviii (top); Goddard Space Flight Center/NASA; p. xxxviii (middle): Reuters/Frank Polich; p. xxxviii (bottom): Reuters/Toby Melvile; p. xlii: Sadequl Hussain/Shutterstock; p. xliii: Fotolia.

Chapter 1

Opener: Transtock Inc./Alamy.

Chapter 2

Opener: Reuters/ADAC; Fig. 2.6: Reuters/Anwar Mirza; p. 71: Cheryl Power/Photo Researchers, Inc.; Fig. 2.8: fStop/Alamy; Fig. P2.33: Seaman Justin E. Yarborough/U.S. Navy; Fig. P2.52: Berc/iStockphoto.

Chapter 3

Opener: AP Photo/The News Tribune, Janet Jensen; p. 93: PhotoStock-Israel/Alamy; Fig. 3.17: HP Canada/Alamy; Fig. 3.18: Reuters/Stefan Wermuth.

Chapter 4

Opener: vesilvio/Shutterstock.

Chapter 5

Opener: NASA; Fig. 5.5: Ted Kinsman/Photo Researchers, Inc.; Fig. 5.8: NASA; Fig. 5.9: Walter G Arce/Shutterstock.

Chapter 6

Opener: J-L Charmet/Photo Researchers, Inc.; p. 194: Vladimir Wrangel/Shutterstock; Fig 6.10: Ted Foxx/Alamy; Fig. P6.23: EPA/Horacio Villalobos/Newscom; Fig. P6.78: Exactostock/SuperStock; Fog. P6.79: BrandonR; Fig. P6.82: billdayone/Alamy.

Chapter 7

Opener: Leo Mason sports photos/Alamy; Fig 7.1: Ayakovlev/Shutterstock; Fig. 7.13: Reuters/Dylan Martinez; Fig 7.15: GIPhotoStock/Photo Researchers, Inc.; Leo Mason sports photos/Alamy; Fig. 7.21: Hogar/Shutterstock; Fig. 7.23: Reuters/Stringer; p. 257: Pearson Science/Eric Schrader; Fig. 7.24: 4×6/iStockphoto; p. 265: Hogar/Shutterstock.

Chapter 8

Opener: Matt Tilghman/Shutterstock; p. 280: NASA; Fig. 8.8: Ivonne Wierink; Fig. 8.19: Associated Press/Aman Sharma; Fig 8.21: JP5/ZOB/WENN/Newscom.

Chapter 9

Opener: F1online digitale Bildagentur GmbH/Alamy; Fig. 9.9: Richard Megna/Fundamental Photographs; Fig 9.19: Frances M. Roberts/Alamy.

Chapter 10

Opener: Steve Bower/Shutterstock; Fig. 10.15: SuperStock/SuperStock; Fig. 10.16: Richard Megna/Fundamental Photographs; Fig. P10.44: John Kershner/Shutterstock.

Chapter 11

Opener: Biophoto Associates/Photo Researchers, Inc.; Fig. 11.1: Pearson Science/Eric Schrader; Fig. 11.7: Thomas Otto/Fotolia; Fig. 11.8: Dmitry Naumov/Fotolia; Fig. 11.9: Ustyujanin/Shutterstock.

Chapter 12

Opener: Bruce Mitchell/Getty; Fig. 12.1 (Top): USGS; Fig. 12.1 (Bottom): USGS; p. 445: Dorling Kindersley Media Library; p. 446: Marc Mueller/dpa/picture-alliance/Newscom; Fig. 12.14: Ted Kinsman/Photo Researchers, Inc.

Chapter 13

Opener: Richard Megna/Fundamental Photographs; Fig. 13.1: iPics/Fotolia; Fig. 13.5: misu/Fotolia.

Index

A

Absolute (Kelvin) temperature scale, 333–334
Absorbed dose, ionizing radiation, 1072
Absorption process, 1006
Absorption spectrum, 1016
Accelerating universe, 1101, 1102
Acceleration, 19–25
    acceleration-versus-force graphs, 56
    acceleration-versus-time graphs, 31
    average acceleration, 20, 34, 64
    in circular motion, 122, 124–128
    defined, 20, 34
    dependence of on radius of curved
        path, 126–127
    free fall, 31
    free-fall acceleration, 62
    invariance, 926, 927
    mass and, 57–58, 152
    motion at constant acceleration,
        19–25, 34
    motion at constant nonzero
        acceleration, 27–28
    Newton's second law, 55–57, 58–59,
        63–67, 75
    operational definition, 59
    radial acceleration, 122, 124–128,
        135, 141, 144, 1004
    rotational acceleration, 278–279,
        309
    simple harmonic motion (SMH),
        704–705
    speed and, 124–125
    of train, 91
    translational acceleration, 233
    velocity change determined
        from, 21
    vibrational motion, 703–704
    See also Velocity
Acceleration-versus-force graphs, 56
Acceleration-versus-time graphs, 31
Adiabatic processes, 435
Age of the universe, 947–948
Air bags, 43, 72, 73
Air conditioners, 480–481
Alpha decay, 1045, 1060–1061,
    1063, 1075
Alpha particles, 1042, 1045, 1046–1047,
    1060
Alpha rays, 1045
Alpher, Ralph, 1097
Alternating current (AC), 609, 682
Altitude sickness, 380
AM radio, 903
Ammeters, 583–584, 634–635
Ampere (unit), 579, 630

Ampére, André-Marie, 630
Amplitude
    vibrational motion, 698, 725, 738
    waves, 738, 744–746
Anderson, Carl, 1083, 1084
Aneroid barometer, 324–325
Angle of incidence, 781, 801
Angle of reflection, 781, 801
Angle of refraction, 786
Angular acceleration, 278–279
Angular magnification, 837–838, 841, 844
Angular position, 275–276
Angular size, 837
Angular velocity, 277–278
Anode, 635, 968, 982
"Anomalous" Zeeman effect, 1020
Antennas, 897–899
Antielectrons, 1082–1083
Antimatter, 1062, 1082
Antimuons, 1091
Antineutrinos, 1062, 1082
Antineutrons, 1086, 1094
Antinodes, 759
Antiparticles, 1082–1086
    beta-plus decay, 1081
    defined, 1103
    pair annihilation, 1085
    pair production, 1084
    positron emission tomography (PET),
        1081–1082, 1086
Antitau particles, 1091
Arc length, 276, 277
Archimedes, 359
Archimedes' principle, 374
Area of support, 253
Arrow, shooting, 200
Astronauts
    Body Mass Measurement Device
        (BMMD), 714–716
    heartbeat rate, 932
    weightlessness, 142–143
    See also Space flight
Astrophysics
    accelerating universe, 1101, 1102
    age of the universe, 947–948
    black holes, 217–218, 280, 950
    cosmic rays, 519, 638–639, 938
    cosmological constant, 1101
    dark energy, 1101–1102, 1104
    dark matter, 1099–1101, 1102, 1104
    "dark stars", 218
    escape speed, 216–217
    expansion of the universe, 946–947,
        948, 1096, 1101–1102
    gravitational potential energy,
        214–218

    Hubble's law, 947
    MACHOs (massive compact halo
        objects), 1100
    precession, 949
    pulsars, 275, 298–299
    red shift, 864, 945, 1097
    solar constant, 908
    spectrometers, 864
    starlight deflection, 949
    stars: color of, 962
    stars: continuous spectra, 1014–1016
    stars: formation of, 1098–1099
    stars: fusion in, 1057
    stars: gravitational potential energy,
        215–218
    Sun: as black hole, 218
    Sun: atmospheric magnetic field,
        1020
    Sun: composition, 1014
    Sun: mass of, 981
    Sun: power of surface radiation from,
        962–963
    Sun: surface temperature of, 962
    Sun: thermal energy of, 347–349
    supernovas, 1057, 1099
    WIMPs (weakly interacting massive
        particles), 1100
    See also Cosmology; Universe
Atmospheric pressure, 369–372
    diving bell, 371–372
    measuring, 369–371
    normal atmospheric pressure, 753
Atomic mass, measurement of, 1048
Atomic mass unit, 1048, 1052
Atomic nucleus, 1042
    atomic structure and, 1000
    binding energy, 1050, 1052–1053,
        1075
    describing, 1047–1048
    early model, 1045
    notation, 1047–1048, 1075
    nuclear force, 1049–1050
    size of, 1045–1046, 1049
    structure of, 1047
Atomic number, 1047
Atomic physics, 997–1035
    "anomalous" Zeeman effect, 1020
    atomic wave functions, 1027
    de Broglie waves, 1024–1026
    lasers, 1016–1018
    multi-electron atoms, 1027–1028
    periodic table, 1029, 1030, 1049
    quantum numbers, 1005, 1018–1023,
        1027, 1035
    spectral analysis, 1009–1016
    tunneling, 1033–1034